IB	IIB	IIIA	IVA	VA	VIA	VIIA	VIIIA
								2 **He** 4.00260
			5 **B** 10.81	6 **C** 12.011	7 **N** 14.0067	8 **O** 15.9994	9 **F** 18.998403	10 **Ne** 20.179
			13 **Al** 26.98154	14 **Si** 28.0855	15 **P** 30.97376	16 **S** 32.06	17 **Cl** 35.453	18 **Ar** 39.948
28 **Ni** 58.69	29 **Cu** 63.546	30 **Zn** 65.38	31 **Ga** 69.72	32 **Ge** 72.59	33 **As** 74.9216	34 **Se** 78.96	35 **Br** 79.904	36 **Kr** 83.80
46 **Pd** 106.42	47 **Ag** 107.868	48 **Cd** 112.41	49 **In** 114.82	50 **Sn** 118.69	51 **Sb** 121.75	52 **Te** 127.60	53 **I** 126.9045	54 **Xe** 131.29
78 **Pt** 195.08	79 **Au** 196.9665	80 **Hg** 200.59	81 **Tl** 204.383	82 **Pb** 207.2	83 **Bi** 208.9804	84 **Po** (209)	85 **At** (210)	86 **Rn** (222)

63 **Eu** 151.96	64 **Gd** 157.25	65 **Tb** 158.9254	66 **Dy** 162.50	67 **Ho** 164.9304	68 **Er** 167.26	69 **Tm** 168.9342	70 **Yb** 173.04	71 **Lu** 174.967
95 **Am** (243)	96 **Cm** (247)	97 **Bk** (247)	98 **Cf** (251)	99 **Es** (252)	100 **Fm** (257)	101 **Md** (258)	102 **No** (259)	103 **Lr** (260)

ORGANIC CHEMISTRY

A Brief Introduction

ORGANIC CHEMISTRY

A BRIEF INTRODUCTION

Robert J. Ouellette

The Ohio State University

With Contributions by

J. David Rawn

Towson State University

PRENTICE HALL
Englewood Cliffs, NJ 07632

Library of Congress Cataloging-in-Publication Data
Ouellette, Robert J.,
Organic chemistry: a brief introduction / Robert J. Ouellette.
p. cm.
Includes index.
ISBN 0-02-389591-8
1. Chemistry, Organic. I. Title.
QD253.088 1994
547—dc20

Editor: Paul F. Corey
Production Supervisor: Elisabeth Belfer
Production Manager: Nicholas Sklitsis
Text Designer: Eileen Burke

This book was set in 10/12 Palatino by York Graphic Services, Inc.

Printed in the United States of America

10 9 8 7 6 5 4

ISBN 0-02-389591-8

Prentice-Hall International (UK) Limited, *London*
Prentice-Hall of Australia Pty. Limited, *Sydney*
Prentice-Hall Canada Inc., *Toronto*
Prentice-Hall Hispanoamericana, S.A., *Mexico*
Prentice-Hall of India Private Limited, *New Delhi*
Prentice-Hall of Japan, Inc., *Tokyo*
Simon & Schuster Asia Pte. Ltd., *Singapore*
Editora Prentice-Hall do Brasil, Ltda., *Rio de Janeiro*

PREFACE

This Brief Introduction to Organic Chemistry has evolved from the author's experience in teaching a one term "sophomore" organic chemistry course as well as a two term "freshman" general, organic, and biological chemistry course. The students in these courses are interested in chemistry only as it applies to their major. These majors include agriculture, biology, health sciences, home economics, medical technology, nutrition, physical therapy, and zoology. Thus, these students appreciate efforts made to illustrate the relationship of organic chemistry to their areas of interest.

One of the dominant characteristics of all organic chemistry texts, regardless of level or the intended audience, is their length. Teachers generally feel that books are too big and that the subject matter deemed to be vital has increased each year. Thus, this author has made decisions on content that leave out a certain number of "favorites" and decrease the detail in some areas. The goal is not to provide another "short" classical organic chemistry text but to write a text that is useful to the student.

ORGANIZATION

This text is organized to integrate classical, synthetic organic chemistry with biochemistry, medicine, pharmacy, and the chemistry of life. Even in the first few chapters, in which essential chemical concepts are reviewed, there are numerous examples of the physical and chemical properties of bioorganic molecules.

The author's experience of dealing with "sophomore" students, who have either forgotten general chemistry or claim never to have been exposed to certain fundamental chemical concepts, is not unique. However, unlike many organic chemists, the author also has considerable experience in teaching general chemistry and has developed some ideas about how to bridge the gap between the two levels of chemistry. This text incorporates those ideas. Foremost among them is the belief that the solid review given in Chapters 1 and 2, which incorporates or integrates the two levels of chemistry, is worth the time and effort. To teach organic chemistry without a review of the chemical fundamentals is a mistake because students then fail to develop an appreciation for organic chemistry. For example, Chapter 2 contains material that provides a foundation for the introduction of mechanisms in later chapters. In addition, there is an overview of acid-base chemistry and oxidation-reduction reactions as encountered in organic chemistry.

The presentation of complex organic structures early and continuously through the various chapters is a distinctive feature of the book. The author has found that students can focus on relevant functional groups quite early before they are overwhelmed with the details of chemical reactions. In addition they appreciate organic chemistry more if they are accustomed to viewing relevant structures rather than the "irrelevant" simple structures commonly used in introductory texts. Finally, the author has found that students taught with a classical organic chemistry approach often are resentful when they find out only at the end of the course that many organic molecules have interesting biological properties.

The author's belief that "biochemistry" is not a separate subject and should be incorporated even in the one term "organic" chemistry course is reflected both within each chapter and in the order of chapters. Relevant applications of organic chemistry to pharmaceutical and medicinal chemistry are incorporated within each chapter and emphasized further in separate essays that stand out in special type. Each essay appears close to where the related chemistry is developed within the chapter. In addition, certain classes of compounds usually relegated to later chapters are placed in early chapters. For example, steroids appear in Chapter 3 with cycloalkanes, and terpenes are described in Chapter 4 with alkenes.

The most prominent indication of the integration of the classical functional group approach to organic chemistry and its relationship to "biochemistry" is the order of chapters. Rather than place the carbohydrate chapter after all of the functional group chapters, the subject is located immediately after the chapters on alcohols and carbonyl compounds. Similarly, the lipid chapter follows the chapter on carboxylic acids and esters. Only the protein and nucleic acid chapters, which follow after the amine chapter, remain in their traditional last place.

SPECTROSCOPY

The role of spectroscopic methods in organic chemistry, as well as the location of this subject in the two term organic chemistry text, is still an open question with as many solutions as there are texts. The problems are compounded in the one term course. Some faculty prefer to use spectroscopy early and often, and others feel that the subject is best learned after the study of the chemistry of a substantial number of functional groups. Rather than interrupt the flow of describing the relationship between structure and properties of organic compounds, this author has placed spectroscopy in the last chapter. Ultraviolet, infrared, and nuclear magnetic resonance are covered in this chapter. The kind of information that can be derived from each technique is illustrated using selected examples rather than presenting a complete spectral discussion of all functional groups. The goal is to show students how we know that molecules have the structures that we have assigned them.

EXERCISES

All faculty teaching organic chemistry agree that students need to be drilled in applications of the subject matter. Accordingly, 127 solved Examples are placed within sections containing relevant subject matter. In addition, the 870 Exercises at the ends of chapters are arranged by categories and in pairs of questions on related material. Although the difficulty of the questions varies, the primary emphasis is on drill using the fundamental concepts discussed in the chapter. These Exercises account for approximately 20% of the text pages. Answers to the Exercises are given in a separate Study Guide, which also has Learning Objectives and additional solved Examples arranged by chapter.

ESSAYS

Essays are set off from the main sections of the chapters. These essays vary in content and style, but are related to the subject of the chapter. Many of them form part of a continuing story which can be told in greater detail based on the new material presented within that chapter.

ILLUSTRATIONS

Although there are abundant Figures and Tables, a substantial number of structures, with accompanying legends and arrows pointing to special features, are located within the textual material. Their placement within the text rather than in separate figures lessens the problem of trying to read material on one page while referring to a figure that may not even be on the same page. The individual reactions in multi-step processes are presented and discussed within the textual material. The proximity of a reaction and the explanation of what is occurring in that step should facilitate learning the material.

ACKNOWLEDGMENTS

The helpful comments of J. David Rawn and his contributions in rewriting the second draft of the manuscript are gratefully acknowledged, as are the comments and suggestions of Macmillan's reviewers:

Henry Abrash
California State University, Northridge

Keith Chenault
University of Delaware

David Gani
University of St. Andrews (U.K.)

R. A. Hill
University of Glasgow (U.K.)

John Huffman
Clemson University

A. M. Jones
University of Northumbria at Newcastle (U.K.)

Richard Larock
Iowa State University

Russell Linderman
North Carolina State University

Derek Nonhebel
University of Strathclyde (U.K.)

Daniel O'Brien
Texas A & M University

Kevin Smith
University of California, Davis

Frank Weaver
South Carolina State University

The author is particularly indebted to Mary Bailey for her continuing contributions in checking all Exercises in this text and preparing the Study Guide. In addition she read the manuscript and all page proofs with the eye of a non-organic chemist. Her skill in knowing what I meant to say but did not is a unique quality that has improved the final product. It has been a pleasure to have Elisabeth Belfer as the Production Supervisor for the fifth time. She not only provides control on consistency of style and terminology but her carefully phrased questions about the actual chemistry content continue to be immensely helpful.

R. J. O.

CONTENTS

1 BONDING AND STRUCTURE OF ORGANIC COMPOUNDS 1

2 CHEMICAL REACTIONS 50

3 ALKANES AND CYCLOALKANES 92

4 ALKENES AND ALKYNES 131

5 AROMATIC COMPOUNDS 182

6 STEREOCHEMISTRY 221

7 HALOALKANES 250

8 ALCOHOLS AND PHENOLS 278

9 ETHERS AND EPOXIDES 315

10 ALDEHYDES AND KETONES 334

11 CARBOHYDRATES 373

12 CARBOXYLIC ACIDS AND DERIVATIVES 408

13 LIPIDS 449

14 AMINES AND AMIDES 468

15 AMINO ACIDS AND PROTEINS 503

16 NUCLEIC ACIDS 533

17 STRUCTURE DETERMINATION AND SPECTROSCOPY 560

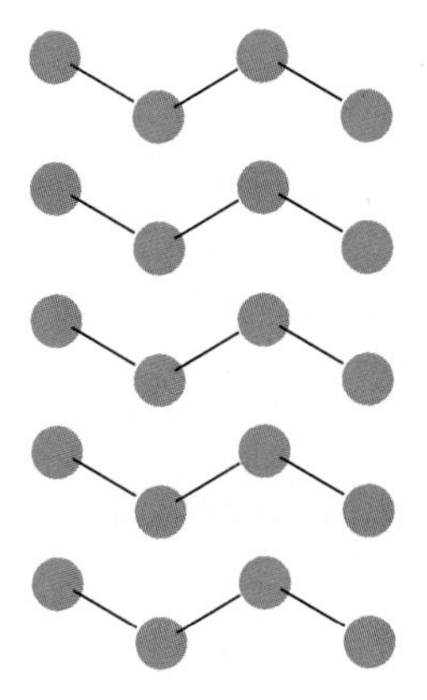

1

BONDING AND STRUCTURE OF ORGANIC COMPOUNDS

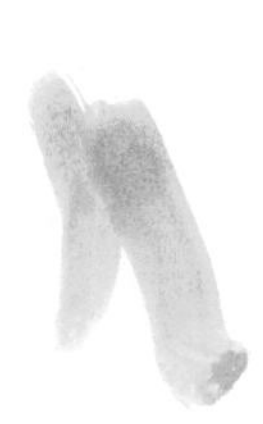

1.1 INORGANIC AND ORGANIC COMPOUNDS

In the late eighteenth century, substances were divided into two classes called inorganic and organic compounds. Inorganic chemistry dealt with compounds derived from mineral sources, whereas organic compounds were obtained only from plants or animals. Organic compounds were more difficult to work with in the laboratory, and decomposed more easily, than inorganic compounds. The differences between inorganic and organic compounds were attributed to a "vital force" associated with organic compounds. This unusual attribute was thought to exist only in living matter. It was believed that without the vital force, organic compounds could not be synthesized in the laboratory. However, by the mid-nineteenth century, chemists had learned not only how to work with organic compounds in the laboratory but how to synthesize them as well.

The distinction between inorganic and organic compounds is now based on chemical composition. The compositions of inorganic compounds are quite varied and include most of the elements. Organic compounds always contain carbon and a limited number of other elements such as hydrogen, oxygen, and nitrogen. Most compounds of carbon contain many more atoms than inorganic compounds and have more complex structures. Common examples of organic compounds include the sugar sucrose ($C_{12}H_{22}O_{11}$), vitamin B_2 ($C_{17}H_{20}N_4O_6$), cholesterol ($C_{27}H_{46}O$), and the fat glycerol tripalmitate ($C_{51}H_{98}O_6$). Some organic substances are gigantic. For example, researchers have prepared the self-replicating poliovirus in a test tube. Its molecular formula is $C_{332652}H_{492388}N_{98245}O_{131196}P_{7501}S_{2340}$!

The physical properties of inorganic and organic compounds differ substantially. Inorganic compounds generally have high melting points, but many organic compounds are gases, liquids, or low-melting-point solids. The solubilities of inorganic and organic compounds also differ. Many inorganic compounds are soluble in water, but most organic compounds are insoluble in water. On the other hand, inorganic compounds are not soluble in organic liquids such as ether, alcohol, or carbon tetrachloride, but most organic compounds easily dissolve in these solvents.

Inorganic and organic compounds also have different electrical conductivities. Most inorganic compounds exist as ions in water, and their solutions conduct an electric current. In contrast, the organic compounds that dissolve in water, such as ethanol (alcohol) and sucrose (table sugar), do not produce ions when they dissolve, and their solutions do not conduct electricity. Other organic compounds, such as acetic acid, dissolve in water to give some ions, and their solutions weakly conduct electricity.

Based on the physical characteristics of compounds, chemists have proposed that the atoms of the elements are bonded in compounds in two principal ways—ionic bonds and covalent bonds. Both types of bonds result from a change in the electronic structure of atoms as they associate with each other. Thus, the number and type of bonds formed and the resultant shape of the molecule depend on the electron configuration of the atoms. Therefore, we will review some of the electronic features of atoms and the periodic properties of the elements before describing the structures of organic compounds.

1.2 ATOMIC STRUCTURE

Atoms have a central, small, dense nucleus that contains protons and neutrons; electrons are located in space about the nucleus. Protons have a +1 charge; electrons have a −1 charge. Atoms have an equal number of protons and electrons and are electrically neutral.

The atoms of different elements contain unique numbers of subatomic particles. The number of protons determines the identity of an atom. For example, a hydrogen atom has one proton; the carbon, nitrogen, and oxygen atoms have six, seven, and eight protons, respectively. The number of protons in an atom define its **atomic number.** Because the atomic number equals the number of protons in the nucleus, it also equals the number of electrons in the atom. The number of electrons in the hydrogen, carbon, nitrogen, and oxygen atoms are one, six, seven, and eight, respectively.

The periodic table of the elements is arranged by atomic number. The elements are arrayed in horizontal rows called **periods** and vertical columns called **groups.** In this text we will emphasize hydrogen in the first period and the elements carbon, nitrogen, and oxygen in the second period. The electronic configurations of these atoms are the basis for their chemical reactivity.

Atomic Orbitals

Electrons about the nucleus of an atom are found in **atomic orbitals.** Each orbital can contain a maximum of two electrons. The orbitals, designated by the letters *s*, *p*, *d*, and *f*, differ in energy, shape, and orientation. We need to consider only the *s* and *p* orbitals for elements such as carbon, oxygen, and nitrogen.

Orbitals are grouped in shells of increasing energy designated by the integers $n = 1, 2, 3, 4, \ldots n$. These integers are called **principal quantum numbers.** With few exceptions, we need consider only the orbitals of the first three shells for the common elements found in organic compounds.

Each shell contains unique numbers and types of orbitals. The first shell contains only one orbital—the *s* orbital. It is designated 1*s*. The second shell contains two types of orbitals—one *s* orbital and three *p* orbitals. Each orbital regardless of type can have no more than two electrons.

An *s* orbital is a spherical region of space centered around the nucleus (Figure 1.1). The electrons in a 2*s* orbital are higher in energy than those in a 1*s* orbital. The 2*s* orbital is larger than the 1*s* orbital, and its electrons are farther from the nucleus. The three *p* orbitals in a shell are shaped like "dumbbells". However, they have different orientations with respect to the nucleus (Figure 1.1). The orbitals are often designated p_x, p_y, and p_z to emphasize that they are mutually perpendicular to one another. Although the

FIGURE 1.1 Shapes of *s* and *p* Orbitals

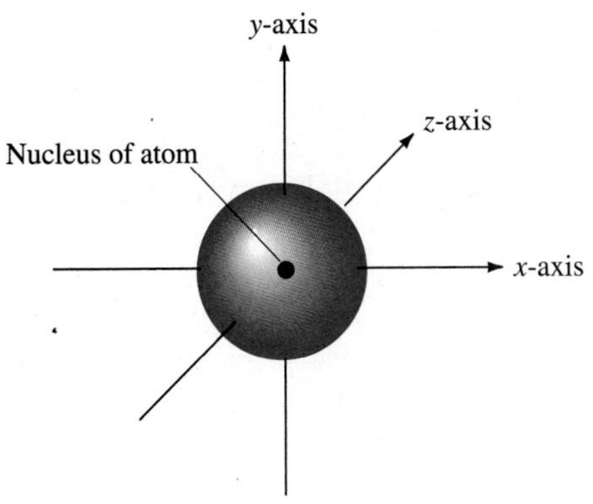

Boundary surface enclosing a volume where electrons in an *s* orbital may be located.

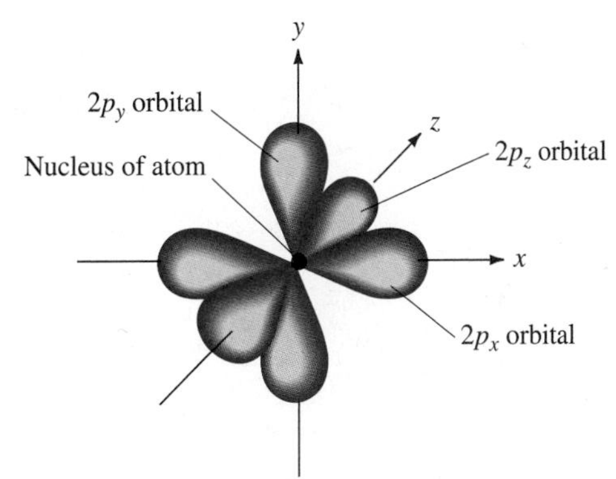

Boundary surfaces of the three mutually perpendicular 2*p* orbitals. Each orbital may be occupied by a maximum of two electrons.

orientations of the *p* orbitals are different, the electrons in each *p* orbital have equal energies.

Orbitals of the same type within a shell are often considered as a group called a **subshell.** There is only one orbital in an *s* subshell. An *s* subshell can contain only two electrons but a *p* subshell can contain $3 \times 2 =$ six electrons.

Electrons are located in subshells of successively higher energies so that the total energy of all electrons are as low as possible. The order of increasing energy of subshells is $1s < 2s < 2p < 3s < 3p$ for elements of low atomic number. The number and location of electrons for the first 18 elements is given in Table 1.1. The location of electrons in atomic orbitals is the **electron configuration** of an atom.

Valence-Shell Electrons

Electrons in filled, lower energy shells of atoms have no role in determining the structure of molecules, nor do they participate in chemical reactions. Only the higher energy electrons located in the outermost shell, the **valence shell,** participate in chemical reactions. Electrons in the valence shell are **valence electrons.** For example, the single electron of the hydrogen atom is a valence electron. The number of valence electrons for the common atoms contained in organic molecules are given by their group number in the periodic table. Thus carbon, nitrogen, and oxygen have four, five, and six

TABLE 1.1 Electron Configuration

		Electrons in subshells				
Atomic number	Element	1*s*	2*s*	2*p*	3*s*	3*p*
1	H	1				
2	He	2				
3	Li	2	1			
4	Be	2	2			
5	B	2	2	1		
6	C	2	2	2		
7	N	2	2	3		
8	O	2	2	4		
9	F	2	2	5		
10	Ne	2	2	6		
11	Na	2	2	6	1	
12	Mg	2	2	6	2	
13	Al	2	2	6	2	1
14	Si	2	2	6	2	2
15	P	2	2	6	2	3
16	S	2	2	6	2	4
17	Cl	2	2	6	2	5
18	Ar	2	2	6	2	6

valence electrons, respectively. With this information we can understand how these elements combine to form the structure of organic compounds.

1.3 ATOMIC PROPERTIES

The physical and chemical properties of an element may be estimated from its position in the periodic table. Two properties that help us explain the properties of organic compounds are atomic radius and electronegativity.

The overall shape of an atom is spherical, and the volume of the atom depends on the number of electrons and the energies of the electrons in occupied orbitals. The size of an atom is expressed as the **atomic radius,** which is the distance from the nucleus to the region of space containing the highest energy electrons. Figure 1.2 lists the radii of some atoms given in angstroms. The atomic radius for an atom does not vary significantly from one compound to another.

Atomic radii increase from top to bottom in a group of the periodic table. Each successive member of a group has one additional energy level containing electrons located at larger distances from the nucleus. Thus, the atomic radius of sulfur is greater than that of oxygen, and the radii of the halogens increase in the order F < Cl < Br < I.

The atomic radius decreases from left to right across a period. Although electrons are located in the same energy level within the *s* and *p* orbitals of the elements, the nuclear charge increases from left to right within a period. As a consequence, the nucleus draws the electrons inward and the radius decreases. The radii of the common elements in organic compounds are in the order C > N > O.

Electronegativity

Electronegativity is a measure of the attraction of an atom for bonding electrons in molecules compared to that of other atoms. The electronegativity values devised by Linus Pauling, an American chemist, are dimensionless quantities that range from slightly less than one for the alkali metals to a

FIGURE 1.2
Atomic Radii (in angstrom units)

H 0.37						
Li 1.52	Be 1.11	B 0.88	C 0.77	N 0.70	O 0.66	F 0.64
Na 1.86	Mg 1.60	Al 1.43	Si 1.17	P 1.10	S 1.04	Cl 0.99
						Br 1.14
						I 1.33

FIGURE 1.3
Electronegativity

H 2.1						
Li 1.0	Be 1.5	B 2.0	C 2.5	N 3.0	O 3.5	F 4.0
Na 0.9	Mg 1.2	Al 1.5	Si 1.8	P 2.1	S 2.5	Cl 3.0
						Br 2.8
						I 2.5

maximum of four for fluorine. Large electronegativity values indicate a stronger attraction for electrons than small electronegativity values.

Electronegativities increase from left to right across the periodic table (Figure 1.3). Elements on the left of the periodic table have low electronegativities and are often called **electropositive** elements. The order of electronegativities F > O > N > C is an important property that we will use to explain the chemical properties of organic compounds.

Electronegativities decrease from top to bottom within a group of elements. The order of decreasing electronegativities F > Cl > Br > I is another sequence that we will use to interpret the chemical and physical properties of organic compounds.

1.4 TYPES OF BONDS

In 1916, the American chemist G. N. Lewis proposed that second-period elements tend to react to obtain an electron configuration of eight electrons so that they electronically resemble the inert gases. This hypothesis is summarized in the **Lewis octet rule:** Second-period atoms tend to combine and form bonds by transferring or sharing electrons until each atom is surrounded by eight electrons in its valence shell. Note that hydrogen requires only two electrons to complete its valence shell.

Ionic Bonds

Ionic bonds are formed between two or more atoms by the transfer of one or more electrons between atoms. Electron transfer produces negative ions called **anions** and positive ions called **cations.** These ions attract each other.

Let us examine the ionic bond in sodium chloride. A sodium atom, which has 11 protons and 11 electrons, has a single valence electron in its 3*s* subshell. A chlorine atom, which has 17 protons and 17 electrons, has seven valence electrons in its third shell, represented by $3s^23p^5$. In forming an ionic bond, the sodium atom, which is electropositive, loses its valence electron to chlorine. As a result, the sodium atom obtains the same electron configuration as neon ($1s^2 2s^2 2p^6$) and develops a +1 charge because there are 11 protons in the nucleus, but only 10 electrons about the nucleus of the ion.

The chlorine atom, which has a high electronegativity, gains an electron and is converted into a chloride ion that has the same electron configuration as argon ($1s^2 2s^2 2p^6 3s^2 3p^6$). The chloride ion has a −1 charge because there are 17 protons in the nucleus, but there are 18 electrons about the nucleus of the ion.

The formation of sodium chloride from the sodium and chlorine atoms can be shown by Lewis structures. **Lewis structures** represent only the valence electrons; electron pairs are shown as pairs of dots.

$$\mathrm{Na}\cdot + :\ddot{\underset{\cdot\cdot}{\mathrm{Cl}}}\cdot \longrightarrow \mathrm{Na}^+ + :\ddot{\underset{\cdot\cdot}{\mathrm{Cl}}}:^-$$

Note that, by convention, the complete octet is shown for anions formed from electronegative elements. However, the filled outer shell of cations that results from loss of electrons from electropositive elements is not shown.

Metals are electropositive and tend to lose electrons, whereas nonmetals are electronegative and tend to gain electrons. A metal atom loses one or more electrons to form a cation with an octet of electrons. The same number of electrons are accepted by the appropriate number of atoms of a nonmetal to form an octet in the anion to give an ionic compound. In general, ionic compounds result from combinations of metallic elements, located on the left side of the periodic table, with nonmetals, located on the upper right side of the periodic table.

Polyatomic ions consist of several bonded atoms and may be positive or negative. Some of these ions, such as cyanide (CN^-), hydroxide (OH^-), nitrate (NO_3^-), sulfate (SO_4^{2-}), phosphate (PO_4^{3-}), hydronium (H_3O^+), and ammonium (NH_4^+), will be encountered in the study of organic chemistry.

EXAMPLE 1.1 Magnesium forms ionic compounds. What is the electronic configuration of the magnesium ion? What are the chemical formulas for magnesium hydroxide (Milk of Magnesia) and magnesium sulfate (Epsom Salts)?

Solution The electron configuration of the valence-shell electrons of magnesium is $3s^2$. Therefore, magnesium loses two electrons to produce the Mg^{2+} cation and achieve the stable electron configuration of neon: $1s^2 2s^2 2p^6$.

$$\mathrm{Mg}(1s^2 2s^2 2p^6 3s^2) \longrightarrow \mathrm{Mg}^{2+}(1s^2 2s^2 2p^6) + 2\,e^-$$

The charges of the hydroxide ion and sulfate ion are −1 and −2, respectively. Thus, two hydroxide ions are required to balance the +2 charge of the magnesium ion in magnesium hydroxide, which is written as $Mg(OH)_2$. The −2 charge of the sulfate ion is balanced by the +2 charge of the magnesium ion when the ions are present in a 1:1 ratio. The formula of magnesium sulfate is $MgSO_4$.

Covalent Bonds

A **covalent bond** consists of the mutual sharing of one or more pairs of electrons between two atoms. These electrons are simultaneously attracted by the two atomic nuclei. A covalent bond forms when the difference between the electronegativities of two atoms is too small for an electron transfer to occur to form ions. Shared electrons located in the space between the two nuclei are called **bonding electrons.** The bonded pair is the "glue" that holds the atoms together in molecular units.

Let's consider the covalent bond in the hydrogen molecule. It forms from two hydrogen atoms, each with one electron in a 1*s* orbital. Both hydrogen atoms share the two electrons in the covalent bond, and each acquires a helium-like electron configuration.

$$\mathrm{H\cdot + \cdot H \longrightarrow H\cdot\cdot H}$$

A similar bond forms in F_2. The two fluorine atoms in the fluorine molecule are joined by a shared pair of electrons. Each fluorine atom has seven valence electrons in the second energy level and requires one more electron to form a neonlike electron configuration. Each fluorine atom contributes one electron to the bonded pair shared by the two atoms. The remaining six valence electrons of each fluorine atom are not involved in bonding and are concentrated around their respective atoms. These valence electrons, customarily shown as pairs of electrons, are variously called **nonbonded electrons, lone-pair electrons,** or **unshared electron pairs.**

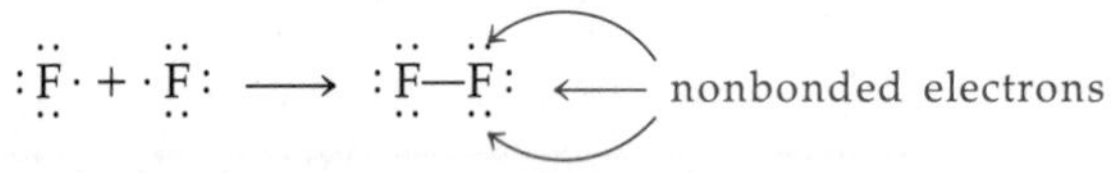

$$\mathrm{:\ddot{F}\cdot + \cdot\ddot{F}: \longrightarrow :\ddot{F}{-}\ddot{F}:} \longleftarrow \text{nonbonded electrons}$$

A covalent bond is drawn as a dash in a **Lewis structure** to distinguish the bonded pair from the lone-pair electrons. Lewis structures show the nonbonded electrons as pairs of dots located about the atomic symbols for the atoms. The Lewis structures of four simple organic compounds, methane, methylamine, methanol, and chloromethane, are drawn below to show both bonded and nonbonded electrons. In these compounds carbon, nitrogen, oxygen, and chlorine have four, three, two, and one bonds, respectively.

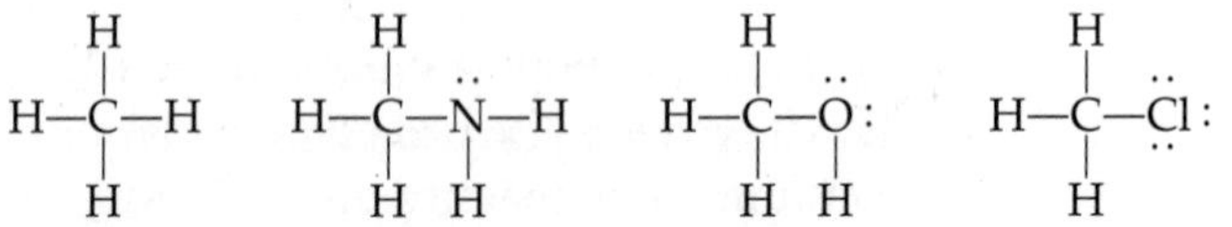

```
    H              H                  H               H
    |              |   ..             |   ..          |   ..
 H—C—H         H—C—N—H          H—C—O:         H—C—Cl:
    |              |   |              |   |           |   ..
    H              H   H              H   H           H
 methane       methylamine         methanol      chloromethane
```

The hydrogen atom and the halogen atoms form only one covalent bond to other atoms in most stable neutral compounds. However, the carbon, oxygen, and nitrogen atoms can simultaneously bond to more than one atom. The number of such bonds is the **valence** of the atom. The valences of carbon, nitrogen, and oxygen are four, three, and two, respectively.

Multiple Covalent Bonds

In some molecules more than one pair of electrons is shared between pairs of atoms. If four electrons (two pairs) or six electrons (three pairs) are shared, the bonds are called **double** and **triple** bonds, respectively. A carbon atom can form single, double, or triple bonds with other carbon atoms as well as with some other elements. Single, double, and triple covalent bonds link two carbon atoms in ethane, ethylene, and acetylene, respectively. Each carbon atom in these compounds shares one, two, and three electrons, respectively, with each other. The remaining valence electrons of the carbon atoms are contained in the single bonds with hydrogen atoms.

single bond sharing 2 e^- — double bond sharing 4 e^- — triple bond sharing 6 e^-

$$H_3C{-}CH_3 \qquad H_2C{=}CH_2 \qquad H{-}C{\equiv}C{-}H$$

ethane — ethylene — acetylene

Polar Covalent Bonds

A **polar covalent bond** exists when atoms with different electronegativities share electrons in a covalent bond. Consider the hydrogen chloride molecule (HCl). Each atom in HCl requires one more electron to form an inert gas electron configuration. Chlorine has a larger electronegativity than hydrogen, but the chlorine atom's attraction for electrons is not sufficient to remove an electron from hydrogen. Consequently, the bonding electrons in hydrogen chloride are shared unequally in a polar covalent bond. The molecule is represented by the conventional Lewis structure, even though the shared electron pair is associated to a larger extent with chlorine than with hydrogen. The unequal sharing of the bonded pair results in a partial negative charge on the chlorine atom and a partial positive charge on the hydrogen atom. The symbol δ (Greek lowercase delta) denotes these fractional charges.

$$^{\delta+}H{-}\ddot{\underset{\cdot\cdot}{Cl}}\!:^{\delta-}$$

The hydrogen chloride molecule has a **dipole** (two poles), which consists of a pair of opposite charges separated from each other. The dipole is shown by an arrow with a cross at one end. The cross is near the end of the molecule that is partially positive and the arrowhead is near the partially negative end of the molecule.

$$\overset{+\!\!\longrightarrow}{H{-}Cl}$$

Single or multiple bonds between carbon atoms are nonpolar. Hydrogen and carbon have similar electronegativity values and the C—H bond is not normally considered a polar covalent bond. Thus, ethane, ethylene, and acetylene have nonpolar covalent bonds and the compounds are nonpolar. Organic compounds with bonds between carbon and other elements such as oxygen and nitrogen are polar.

The polarity of a bond depends upon the electronegativities of the bonded atoms. Increasing the difference between the electronegativities of the bonded atoms increases the polarity of a bond. Thus, the direction of the polarity of common bonds found in organic molecules is easily predicted. The common nonmetals are more electronegative than carbon. Therefore, when a carbon atom is bonded to common nonmetal atoms, it has a partial positive charge.

$$\overset{+\!\!\longrightarrow}{\text{C—N}} \quad \overset{+\!\!\longrightarrow}{\text{C—O}} \quad \overset{+\!\!\longrightarrow}{\text{C—F}} \quad \overset{+\!\!\longrightarrow}{\text{C—S}} \quad \overset{+\!\!\longrightarrow}{\text{C—Cl}}$$

Hydrogen is less electronegative than the common nonmetals. Therefore, when a hydrogen atom is bonded to common nonmetals, the resulting polar bond has a partial positive charge on the hydrogen atom.

$$\overset{+\!\!\longrightarrow}{\text{H—N}} \quad \overset{+\!\!\longrightarrow}{\text{H—O}} \quad \overset{+\!\!\longrightarrow}{\text{H—S}}$$

The magnitude of the polarity of a bond is the **bond moment.** This value is reported in Debye units (D). The bond moments of several bond types are given in Table 1.2. The bond moment of a specific bond is relatively constant from compound to compound. Note that multiple bonds

TABLE 1.2 Bond Moments (Debye)

Hydrogen bonds to common elements			
H—C	H—N	H—O	
0.4	1.3	1.5	
Single bonds of carbon to common elements			
C—C	C—N	C—O	C—F
0.0	0.22	0.74	1.41
Multiple bonds of carbon to common elements			
C=C	C≡N	C=O	
0.0	3.5	2.3	

between carbon and oxygen and between carbon and nitrogen are highly polar. There is a polar double bond to oxygen in formaldehyde and a polar triple bond to nitrogen in acetonitrile.

```
H                       H
 \     ..               |
  C=O:           H—C—C≡N:
 /                      |
H                       H
formaldehyde       acetonitrile
```

Note that the number of covalent bonds that an atom forms in a molecule (not an ion) is a unique feature for each atom. The nitrogen atom has three single covalent bonds in methylamine and a triple bond in acetonitrile. The oxygen atom has two single covalent bonds in methanol and a double bond in formaldehyde.

EXAMPLE 1.2 Chloroethane (CH_3CH_2Cl) is a topical anesthetic that boils at 12 °C. It is used to numb the skin when the liquid under pressure is released from a spray can. Describe the bonding in this compound. (Refer to the structure of ethane.)

Solution In ethane each carbon atom is bonded to three hydrogen atoms.

```
   H H
   | |
H—C—C—H
   | |
   H H
```

Chlorine has seven valence electrons and requires only one additional electron to form a Lewis octet. Thus, a bond can form between a carbon atom and a chlorine atom by sharing one electron from each.

```
   H H
   | |  ..
H—C—C—Cl:
   | |  ..
   H H
chloroethane
```

Chlorine is more electronegative than carbon, and as a consequence the covalent carbon–chlorine bond is polar. Chlorine shares one electron in the carbon–chlorine bond; the remaining six valence electrons of chlorine are shown as nonbonded electrons.

The individual hydrogen atoms provide one electron to form the indicated covalent bonds with the carbon atoms in chloroethane.

Coordinate Covalent Bonds

A **coordinate covalent bond** is formed when both electrons of a bonded pair are provided by one atom. For example, a coordinate covalent bond forms in the reaction of ammonia with boron trifluoride.

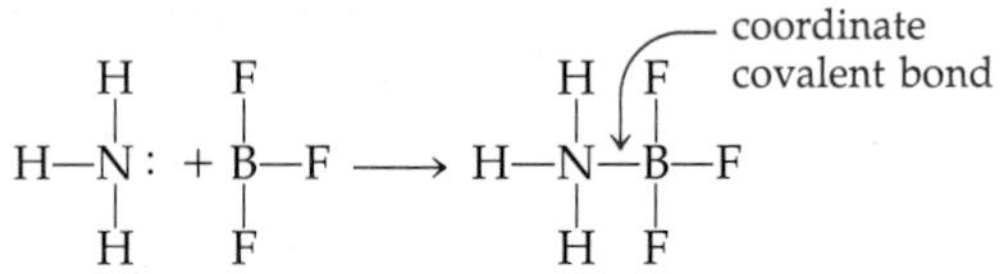

$$\begin{array}{ccccccccccc} & & \mathrm{H} & & & \mathrm{F} & & & & \mathrm{H} & \mathrm{F} & \\ & & | & & & | & & & & | & | & \\ \mathrm{H} & - & \mathrm{N}: & + & & \mathrm{B} & -\mathrm{F} & \longrightarrow & \mathrm{H}- & \mathrm{N} \rightarrow & \mathrm{B} & -\mathrm{F} \\ & & | & & & | & & & & | & | & \\ & & \mathrm{H} & & & \mathrm{F} & & & & \mathrm{H} & \mathrm{F} & \end{array}$$

coordinate covalent bond

In ammonia, the nitrogen atom shares three of its five valence electrons with three hydrogen atoms. The remaining two valence-shell electrons are an unshared pair that forms a coordinate covalent bond with the electron-deficient boron atom of BF_3. Boron in BF_3 has only six electrons in its valence shell and needs the two electrons provided by the nitrogen atom to form a covalent bond and achieve a Lewis octet.

EXAMPLE 1.3 Dimethyl sulfoxide is a liquid that is readily absorbed through the skin. It was once considered as a solvent to deliver drugs by direct application to the skin, but it is too toxic for this use. Compare its structure to dimethyl sulfide, and describe the sulfur–oxygen bond.

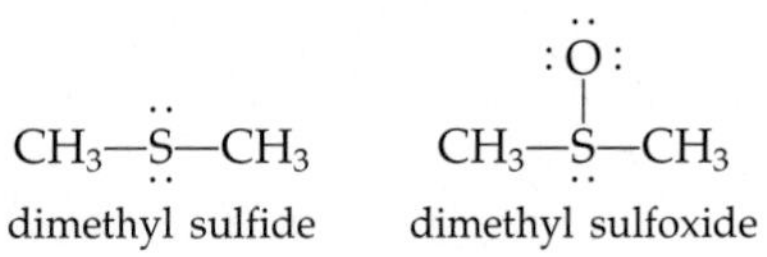

$$CH_3{-}\ddot{\underset{\cdot\cdot}{S}}{-}CH_3 \qquad \begin{array}{c} :\ddot{O}: \\ | \\ CH_3{-}\underset{\cdot\cdot}{S}{-}CH_3 \end{array}$$

dimethyl sulfide   dimethyl sulfoxide

Solution The sulfur atom in dimethyl sulfide has eight valence electrons—two bonded pairs and two nonbonded pairs. The sulfur atom in dimethyl sulfoxide has three bonded pairs and one nonbonded pair of electrons. The oxygen atom in dimethyl sulfoxide has three nonbonded pairs of electrons and one bonded pair. The six valence electrons of oxygen occur as nonbonded pairs in dimethyl sulfoxide; therefore, the electrons in the sulfur–oxygen bond are derived from the sulfur atom. A coordinate covalent bond results when the sulfur atom shares one pair of electrons with the oxygen atom.

1.5 STRATEGY FOR WRITING LEWIS STRUCTURES

When we write Lewis structures, all the valence electrons of the atoms are shown. Hydrogen is limited to two electrons in a bonded pair. Usually each second-row element has an octet of electrons in the form of nonbonded pairs or as bonded pairs that may be in single, double, or triple bonds. We can use the following strategy to write such structures systematically.

1. Determine the total number of valence electrons by adding the valence electrons in the constituent atoms.
2. Write a skeleton structure linking the necessary atoms with single covalent bonds. This structure has the minimum number of bonding electrons.
3. For each bond, subtract two electrons from the total number of valence electrons to give the number of electrons available as nonbonded electrons or to form multiple bonds.
4. Determine the number of electrons necessary to complete the octet about each atom (except for hydrogen, which requires only two electrons.) If this number equals the number calculated in step 3, place the electrons around the appropriate atoms to complete the structure.
5. If the number of electrons determined in step 3 does not provide all atoms with octets, multiple bonds must be used. If the deficiency is two, a double bond must be used. If the deficiency is four, either two double bonds or a triple bond must be used.
6. Modify the structure with the appropriate number of multiple bonds. The remaining electrons are nonbonded electrons that satisfy the individual electronic requirements of each atom.

Let's apply these rules to nitrosomethane, CH_3NO, which has the following arrangement of atoms.

```
     H
     |
   H—C—N—O
     |
     H
```

The total number of valence electrons is $3(1) + 4 + 5 + 6 = 18$ for the hydrogen, carbon, nitrogen, and oxygen atoms, respectively. A total of 10 electrons are depicted in the skeletal structure. The number of "unused" electrons is $18 - 10 = 8$. Now determine the number of electrons needed by each atom to complete its octet.

Atom	Electrons present	Electrons needed
hydrogen	2 in each case	0
carbon	$4 \times 2 = 8$	0
nitrogen	$2 \times 2 = 4$	4
oxygen	2	6

Because the 10 electrons required to form octets exceed the 8 electrons available after forming the indicated single bonds, it is necessary to use a double bond in the structure. The carbon atom has its required octet in the structure, so the double bond can be placed only between nitrogen and oxygen.

```
    H
    |
H—C—N=O
    |
    H
```

Based on this structure, the number of electrons needed by each atom is again calculated.

Atom	Electrons present	Electrons needed
hydrogen	2 in each case	0
carbon	$4 \times 2 = 8$	0
nitrogen	$2 + 4 = 6$	2
oxygen	4	4

The number of electrons present in the structure is now 12, and 6 additional electrons are required to complete the necessary octets. The total number 12 + 6 corresponds to the number of valence electrons originally available. Thus, the structure is

```
    H
    |   ..
H—C—N=O:
    |     ..
    H
```

EXAMPLE 1.4 Write the Lewis structure for carbon disulfide (CS_2). Each sulfur atom is bonded to the central carbon atom.

Solution The total number of valence electrons is 2(6) + 4 = 16 electrons for the two sulfur atoms and one carbon atom. The arrangement of atoms with the minimum number of bonding electrons is

S—C—S

Four electrons are used to form the two bonds. The remaining 16 − 4 = 12 electrons are available as nonbonded pairs or in multiple bonds. The carbon atom has four electrons in two bonds and requires four additional electrons or two unshared pairs of electrons. Each sulfur atom has two electrons in its bonded pair. Thus, each sulfur atom requires six electrons as three unshared pairs to form a Lewis octet. The total number of electrons required is 4 + 2(6) = 16, which exceeds by 4 the number available after the two single bonds are formed. Now consider using double bonds to bond the carbon atom to each sulfur atom. The partial working structure is

S=C=S

Now there are eight bonding electrons. Thus, 16 − 8 = 8 electrons are available as nonbonded pairs of electrons to provide an octet for each atom. The carbon atom has its Lewis octet; each sulfur atom needs four more electrons to form an octet. The eight available electrons form two nonbonded pairs of electrons on each sulfur atom.

:S̈=C=S̈:

1.6 FORMAL CHARGE

Although most organic molecules are represented by Lewis structures containing the "normal" number of bonds, some organic ions and even molecules contain less than or more than the customary number of bonds. First let's remember the structures of some "inorganic" ions. The valence of the oxygen atom is two—it normally forms two bonds. However, there are three bonds in the hydronium ion and one in the hydroxide ion.

H—Ö⁺—H (with H bonded below O)

This cation has one more bond than "normal".

H—Ö:⁻ (with a lone pair below O)

This anion has one less bond than "normal".

How do we predict the charge of the ions? Second, what atoms bear the charge? There is a useful formalism for answering both of these questions. Each atom is assigned a **formal charge** by a bookkeeping method counting electrons. The method is also used for neutral molecules that have unusual numbers of bonds. In such cases, centers of both positive and negative charge are located at specific atoms.

The **formal charge** of an atom is equal to the number of its valence electrons as a free atom minus the number of electrons that it "owns" in the Lewis structure.

$$\text{formal charge} = \left[\begin{array}{l}\text{number of valence}\\ \text{electrons in free}\\ \text{atom}\end{array}\right] - \left[\begin{array}{l}\text{number of valence}\\ \text{electrons in bonded}\\ \text{atom}\end{array}\right]$$

The question of ownership is decided by two simple rules. Unshared electrons belong exclusively to the parent atom. One-half of the bonded electrons between a pair of atoms is assigned to each atom. Thus, the total number of electrons "owned" by an atom in the Lewis structure equals the number of nonbonded electrons plus half the number of bonded electrons. Therefore, we write

$$\text{formal charge} = \begin{bmatrix}\text{number of}\\ \text{valence}\\ \text{electrons}\end{bmatrix} - \begin{bmatrix}\text{number of}\\ \text{nonbonded}\\ \text{electrons}\end{bmatrix} - \frac{1}{2}\begin{bmatrix}\text{number of}\\ \text{bonded}\\ \text{electrons}\end{bmatrix}$$

The formal charge of each atom is zero in most organic molecules. However, the formal charge may also be negative or positive. The sum of the formal charges of each atom in a molecule equals zero; the sum of the formal charges of each atom in an ion equals the charge of the ion. Let's consider the molecule HNC.

$$H{-}N{\equiv}C{:}$$

Lone-pair electrons assign both to C

2 bonding electrons assign 1 to H assign 1 to N

6 bonding electrons assign 3 to N assign 3 to C

The formal charge of each atom may be calculated by substitution into the formula.

$$\text{formal charge of hydrogen} = 1 - 0 - \tfrac{1}{2}(2) = 0$$
$$\text{formal charge of carbon} = 4 - 2 - \tfrac{1}{2}(6) = -1$$
$$\text{formal charge of nitrogen} = 5 - 0 - \tfrac{1}{2}(8) = +1$$

The formal charges of the carbon and nitrogen atoms in this compound are not zero. However, note that the sum of the formal charges of the atoms equals the net charge of the species, which in this case is zero.

There are often important chemical consequences when a neutral molecule contains centers whose formal charges are not zero. It is important that you be able to recognize these situations in order to understand the chemical reactivity of such molecules.

EXAMPLE 1.5 Consider the structure of dimethyl sulfoxide given in Example 1.3, and calculate the formal charges of the sulfur and oxygen atoms.

Solution There are six valence electrons in both sulfur and oxygen. The sulfur atom in the compound has three bonded pairs of electrons and one nonbonded pair of electrons. The formal charge of the sulfur atom is

$$6 - [2 + \tfrac{1}{2}(6)] = +1$$

The oxygen atom has one bonded pair and three nonbonded pairs of electrons. Its formal charge is

$$6 - [6 + \tfrac{1}{2}(2)] = -1$$

The sum of the formal charges of the sulfur atom and the oxygen atom is (+1) + (−1) = 0, as required for a neutral compound.

1.7 RESONANCE STRUCTURES

In the Lewis structures for the molecules shown to this point, the electrons have been pictured as either between two nuclei or about a specific atom. These electrons are **localized.** The electronic structures of molecules are written to be consistent with their physical properties. However, the electronic structures of some molecules cannot be represented adequately by a single Lewis structure. For example, the Lewis structure of ozone, O_3, has one double bond and one single bond.

$$:\ddot{O}=\ddot{O}-\ddot{\underset{\cdot\cdot}{O}}:$$

However, single and double bonds have different bond lengths—a double bond between two atoms is shorter than a single bond. The above Lewis structure implies that there is one "long" O—O bond and a "short" O=O bond in ozone. But both oxygen–oxygen bond lengths in the ozone molecule are actually 1.28 angstroms! This fact indicates that the bonds are equal and that the terminal oxygen atoms are structurally equivalent. Therefore, the above Lewis structure with single and double bonds does not accurately describe the ozone molecule. Under these circumstances, the concept of **resonance** is used. We say that a molecule is **resonance-stabilized** if two or more Lewis structures can be written that have the identical arrangement of atoms but different arrangements of electrons. Ozone is such a molecule. The real structure of ozone can be represented better as a **hybrid** of two Lewis structures, neither of which is completely correct.

$$:\ddot{O}=\ddot{O}-\ddot{\underset{\cdot\cdot}{O}}: \longleftrightarrow :\ddot{\underset{\cdot\cdot}{O}}-\ddot{O}=\ddot{O}:$$

A double-headed arrow between two Lewis structures indicates that the actual structure is similar in part to the two simple structures, but lies somewhere between them. The individual Lewis structures are called contributing structures or **resonance structures.** Each resonance structure for ozone has one O=O bond and one O—O bond. The arrangements of the atoms are the same, but the structures differ in the arrangement of electrons.

Curved arrows can be used to keep track of the electrons when writing resonance structures. The tail of the arrow is located near the bonded or nonbonded pair of electrons to be "moved" or "pushed", and the arrowhead shows the final destination of the electron pair.

:Ö=Ö–Ö: pushing electrons gives :Ö–Ö=Ö:

structure 1 structure 2

In resonance structure 1, the nonbonded pair of electrons on the right oxygen atom is moved to form a double bond with the central oxygen atom. A bonded pair of electrons between the central oxygen atom and the oxygen atom on the left is also moved to form a nonbonded pair of electrons on the left oxygen atom. The result is resonance structure 2. This procedure of "pushing" electrons from one position to another is only a bookkeeping formalism. Electrons do not really move this way! The actual molecule has **delocalized** electrons distributed over three atoms—a phenomenon that cannot be shown by a single Lewis structure.

Electrons can be delocalized over many atoms. For example, benzene, C_6H_6, consists of six equivalent carbon atoms contained in a ring in which all carbon–carbon bonds are identical. Each carbon atom is bonded to a hydrogen atom. A single Lewis structure containing alternating single and double bonds can be written to satisfy the Lewis octet requirements.

However, single and double bonds have different bond lengths. In the actual benzene molecule all carbon–carbon bonds are the same length. Like ozone, benzene is represented by two contributing resonance structures separated by a double-headed arrow. The positions of the alternating single and double bonds are interchanged in the two resonance structures.

The electrons in benzene are delocalized over the six carbon atoms in the ring to result in a unique structure. There are no single or double bonds in benzene; its bonds are of an intermediate type that cannot be represented with a single structure.

EXAMPLE 1.6 Consider the structure of nitromethane, a compound used to increase the power in specialized race car engines. A nitrogen–oxygen single-bond length is 1.36 angstroms; a nitrogen–oxygen double-bond length is 1.14 angstroms. The nitrogen–oxygen bonds in nitromethane are equal and are 1.22 angstroms. Explain.

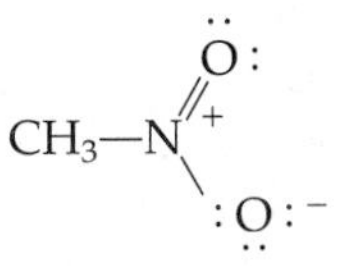

$$CH_3{-}N^{+}(=O){-}O^{-}$$

Solution The actual nitrogen–oxygen bonds are neither single nor double bonds. Two resonance forms can be written to represent nitromethane. They result from "moving" a nonbonded pair of electrons from the single-bonded oxygen atom to form a double bond with the nitrogen atom. One of the bonded pairs of electrons to the double-bonded oxygen atom is moved to the other oxygen atom. The structures differ only in the location of the single and double bonds.

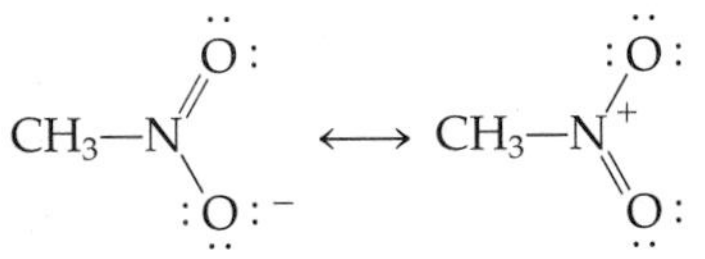

$$CH_3{-}N^{+}(=O){-}O^{-} \longleftrightarrow CH_3{-}N^{+}(-O^{-}){=}O$$

The two structures are contributing resonance forms. The actual molecule has delocalized electrons distributed over the nitrogen atom and the two oxygen atoms.

1.8 PREDICTING THE SHAPES OF SIMPLE MOLECULES

Up to this point, we have considered the distribution of bonded electrons and nonbonded electrons within molecules without regard to their location in three-dimensional space. But, molecules have characteristic shapes that reflect the spatial arrangement of electrons in bonds.

We can describe the geometry of simple molecules using **valence-shell electron-pair repulsion** (abbreviated VSEPR) theory. This theory is based on the idea that bonded and nonbonded electron pairs about a central atom repel each other. VSEPR theory predicts that electron pairs in molecules should be arranged as far apart as possible. Thus, two electron pairs should be arranged at 180° to each other; three pairs should be at 120° in a common plane; four electron pairs should have a tetrahedral arrangement with angles of 109.5°.

two electron pairs   three electron pairs   four electron pairs

VSEPR theory explains why carbon dioxide is a linear molecule, formaldehyde, CH_2O, is a trigonal planar molecule, and methane, CH_4, is a tetrahedral molecule. Organic molecules may have single, double, or triple bonds. Each type of bond may be regarded as a region that contains electrons that should be arranged as far apart as possible.

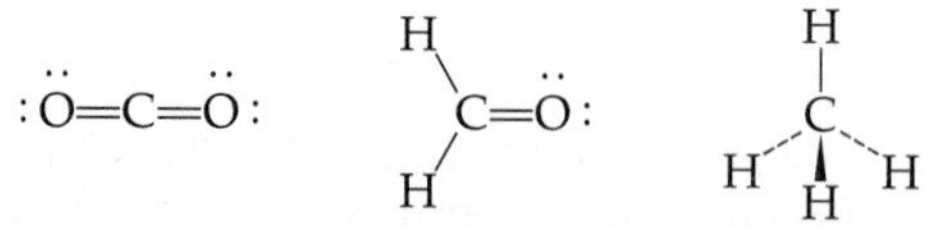

carbon dioxide   formaldehyde   methane

All of the valence electrons about the central carbon atom in carbon dioxide, formaldehyde, and methane are in bonds. Carbon dioxide has two double bonds; the double bonds are separated by the maximum distance, and the resulting angle between the bonds is 180°. Formaldehyde has a double bond and two single bonds to the central carbon atom; these bonds correspond to three regions containing electrons and are separated by the maximum distance in a trigonal planar arrangement with bond angles of 120°. Methane has four bonded electron pairs, and they are best located in a tetrahedral arrangement. Each H—C—H bond angle is predicted to be 109.5°, in agreement with the experimental value.

Now let us consider molecules that have both bonded and nonbonded pairs of electrons in the valence shell of the central atom. Both water and ammonia have four electron pairs about the central atom, as does methane. However water and ammonia have shapes described as angular and trigonal pyramidal, respectively.

angular molecule   trigonal pyramidal molecule

Some of the electron pairs in water and ammonia are bonded to hydrogen atoms, but there are also unshared electron pairs. Methane also has four pairs of electrons about the central atom, but they are all bonded pairs.

VSEPR theory describes the distribution of electron pairs including the nonbonded pairs. However, molecular structure is defined by the positions of the nuclei. Although the four pairs of electrons in both water and ammo-

FIGURE 1.4 VSEPR Model to Predict Geometry About a Central Atom
All electron pairs in methane, ammonia, and water are directed to the corners of a tetrahedron. However, the ammonia molecule is described as trigonal pyramidal; the water molecule is angular.

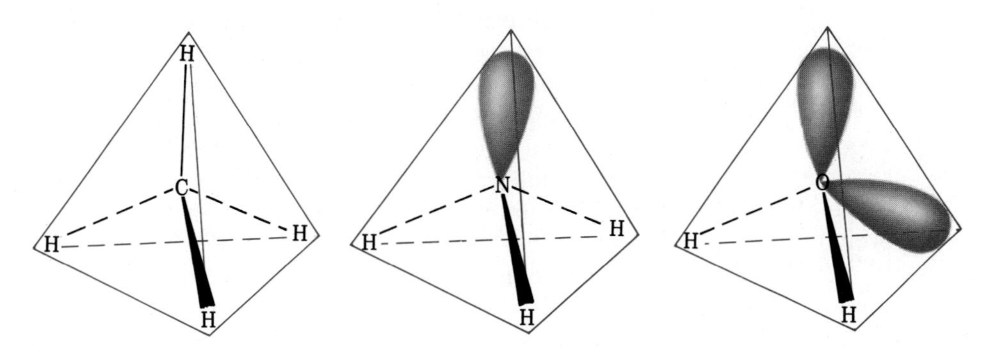

nia are tetrahedrally arranged, water and ammonia are angular and pyramidal molecules, respectively (Figure 1.4).

The arrangements of bonds to the oxygen atom of alcohols and the nitrogen atom of amines are similar to those in water and ammonia, respectively. The groups bonded to the oxygen atom of an alcohol are arranged to form angular molecules. The groups bonded to the nitrogen atom of an amine are arranged to form a pyramid.

EXAMPLE 1.7 The electronic structure of allyl isothiocyanate ($CH_2{=}CH{-}CH_2{-}N{=}C{=}S$), a flavor ingredient in horseradish, is shown below. What are the C—N=C and N=C=S bond angles?

$$CH_2{=}CH{-}CH_2{-}\ddot{N}{=}C{=}\ddot{S}:$$

Solution The C—N=C bond angle is determined by the electrons associated with the nitrogen atom. This atom has a single bond, a double bond, and a nonbonded pair of electrons. These three regions containing electrons have a trigonal planar arrangement. Only two of the electron-containing regions are bonding, but the C—N=C bond angle must still be 120°.

```
C
 \
 .N=C=S
 .
```

The N=C=S bond angle is determined by the electrons associated with the carbon atom. This atom has two double bonds. These electron-containing regions are arranged in a straight line; the N=C=S bond angle is 180°.

1.9 ORBITALS AND MOLECULAR SHAPES

Because electrons form bonds between atoms, the shapes of molecules depend on the location of the electrons in the orbitals of the various atoms. Two electrons in a covalent bond are shared in a region of space common to the bonding atoms. This region of space is pictured as an overlap or merging of two atomic orbitals. For example, the covalent bond in H_2 results from the overlap of two *s* orbitals to give a sigma (σ) bond (Figure 1.5). This bond is symmetrical around an axis joining the two nuclei—hence, the name sigma. If the hydrogen molecule is rotated about its axis, the appearance of the orbital remains unchanged.

The simple picture of bonding described for H_2 has to be modified somewhat for carbon-containing compounds. Carbon has the electron configuration $1s^2 2s^2 2p^2$, which suggests that only the two electrons in the $2p$ orbitals would be available to form two covalent bonds. If this were so, the molecular formula for a compound of carbon and hydrogen would be CH_2. This substance would have a 90° bond angle, and the carbon atom would not have a Lewis octet.

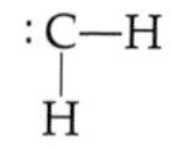

:C—H
 |
 H

However, there are four equivalent C—H bonds in methane, CH_4. All carbon compounds presented in this chapter have a Lewis octet about the carbon atoms, and each carbon atom has four bonds. How can the difference between these structural facts and predictions based on the atomic orbitals of carbon be resolved? In this section we will examine this question by considering the bonding in methane. We will also discuss the bonding in ethylene and acetylene. The bonds in organic compounds are formed from **hybrid orbitals** that result from the "mixing" of two or more orbitals in the bonded atoms. This mixing process—called **orbital hybridization**—was proposed by Pauling to account for the formation of bonds by using orbitals with the geometry appropriate for the actual molecule. As a result of hybridization, two or more hybrid orbitals can be formed from atomic orbitals that

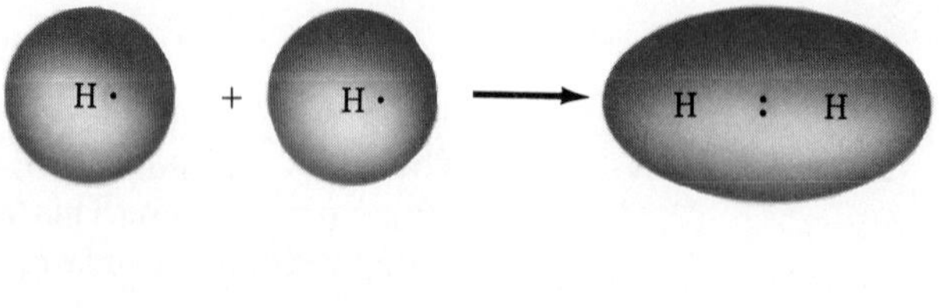

H· + H· ⟶ H : H

FIGURE 1.5
Overlap of *s* Orbitals in the Hydrogen Molecule
The region of space occupied by the electron pair is symmetrical about both hydrogen nuclei. Although the two electrons may be located anywhere within the volume, it is most probable that they are between the two nuclei.

are not identical. The number of hybrid orbitals created equals the number of atomic orbitals used in hybridization.

sp^3 Hybridization of Carbon

Let's consider the bonding in methane, CH_4. Methane has a tetrahedral structure, and all four C—H bonds are equivalent. Pauling suggested that the tetrahedral geometry of methane results from hybridization of the 2*s* and the three 2*p* orbitals of carbon, which combine to form four equivalent hybrid orbitals. The four single bonds and the tetrahedral shape of CH_4 can be explained by these hybrid orbitals. To form four tetrahedral bonds in CH_4, the four electrons of the second shell, ($2s^22p^2$), are located in four hybrid orbitals. Each hybrid orbital contains one electron. These orbitals extend toward the corners of a tetrahedron so there is maximum separation of the electrons. Each hybrid atomic orbital then overlaps with a hydrogen 1*s* orbital to form a σ bond.

The formation of the hybrid orbitals is illustrated in Figure 1.6. The four new orbitals are called **sp^3 hybrid orbitals** because they result from the combination of one *s* and three *p* orbitals. Each sp^3 orbital has the same shape, and the electrons in each orbital have the same energy. The orbitals differ only in their position in space.

sp^2 Hybridization of Carbon

Now let us consider the bonding electrons in the double bond of ethylene in which each carbon atom is bonded to three atoms. All six nuclei lie in a plane, and all the bond angles are close to 120° (Figure 1.7).

Each carbon atom in ethylene is pictured with three **sp^2 hybrid orbitals** and one remaining 2*p* orbital. The three sp^2 hybrid orbitals result from "mixing" a single *s* orbital and two 2*p* orbitals. Each sp^2 orbital has the same shape, and the electrons in each orbital have the same energy. The orbitals differ only in their position in space. They are separated by 120° and are directed to the corners of a triangle to have maximum separation of the electrons. The four valence electrons are distributed as indicated in Figure

FIGURE 1.6 **sp^3-Hybridized Carbon Atom**

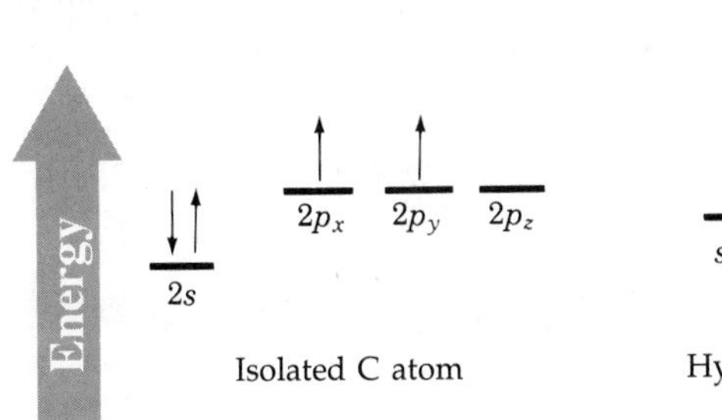

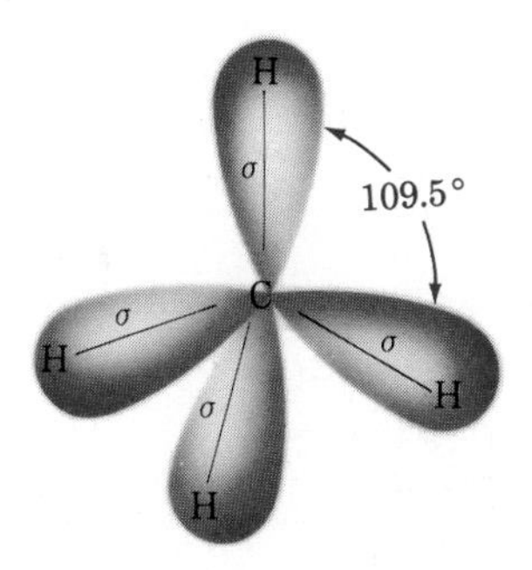

The σ bond between the carbon atom and each hydrogen atom is formed by overlap of the sp^3 orbitals of carbon with the individual *s* orbitals of hydrogen.

FIGURE 1.7
sp^2-Hybridized Carbon Atom

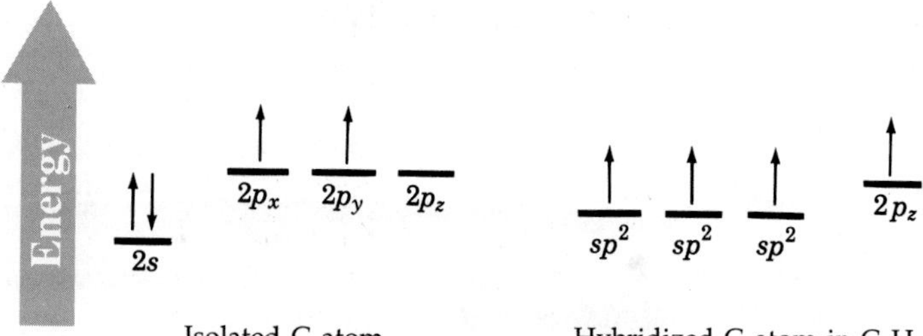

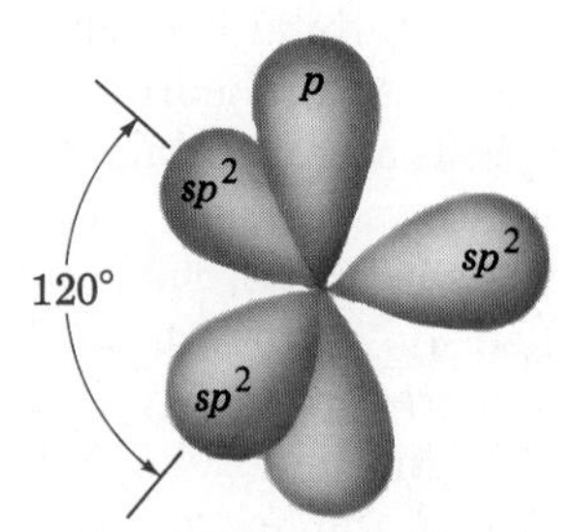

The three sp^2 hybrid orbitals lie in a plane with 120° angles between them. The remaining p orbital is perpendicular to the plane of the sp^2 orbitals.

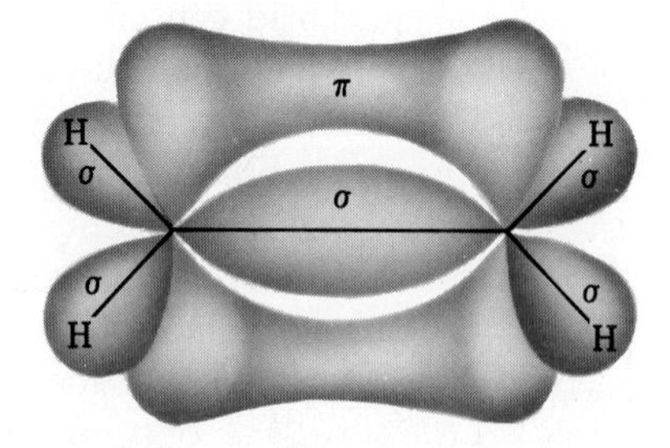

The σ bond between the carbon atoms is formed by overlap of sp^2 orbitals. The σ bonds to hydrogen are formed by overlap of the sp^2 orbitals of carbon with the s orbital of individual hydrogen atoms.

The π bond is formed by sideways overlap of parallel p orbitals.

1.7. The three sp^2-hybridized orbitals are used to make σ bonds. Two of the sp^2 orbitals, containing one electron each, form σ bonds with hydrogen. The third sp^2 orbital, which also contains one electron, forms a σ bond with the other carbon atom in ethylene.

The second bond of the double bond in ethylene results from a lateral (side-by-side) overlap of the p orbitals of each carbon atom. Each p orbital is perpendicular to the plane containing the sp^2 orbitals. The $2p$ orbital of each atom provides one electron to the electron pair for the second bond. A bond formed by sideways overlap of p orbitals is a **π (pi)** bond. Note that the electrons in the π bond are not concentrated along an axis between the two atoms but are shared in regions of space both above and below the plane defined by the sp^2 orbitals. It is, nevertheless, only one bond.

sp Hybridization of Carbon

Now let us consider the triple bond of acetylene in which each carbon atom is bonded to two atoms. All four nuclei are arranged in a line, and all the bond angles are 180° (Figure 1.8).

In acetylene, we mix a $2s$ orbital with a $2p$ orbital to give two ***sp* hybrid orbitals** of equal energy. The remaining two $2p$ orbitals are unchanged (Fig-

FIGURE 1.8 ***sp*-Hybridized Carbon Atom**

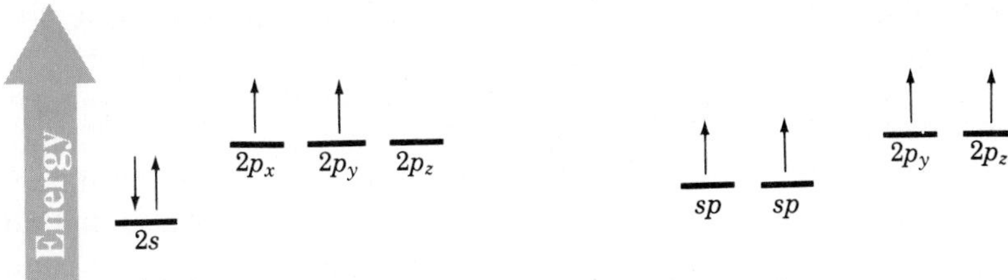

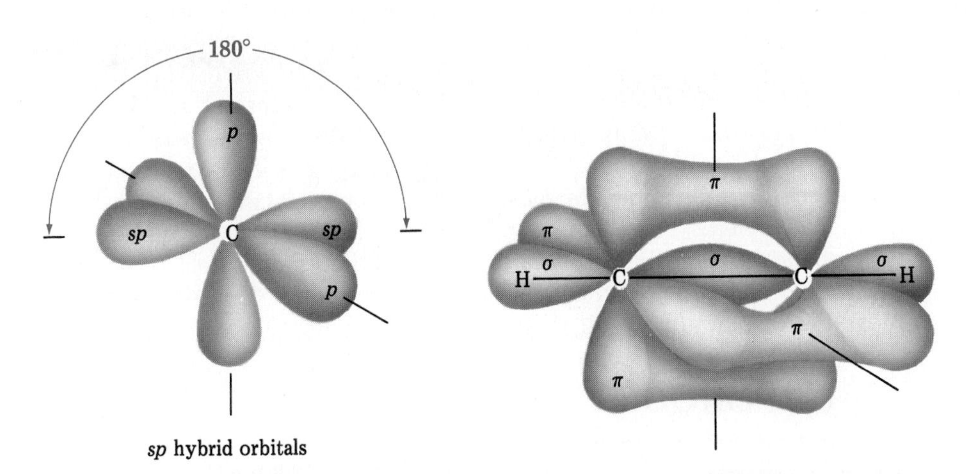

***sp* hybrid orbitals**

The two *sp* hybrid orbitals are at 180° to each other. The two *p* orbitals are mutually perpendicular to each other and to the axis of the *sp* orbitals.

A σ bond is formed by end-to-end overlap of one *sp* orbital from each carbon atom. Two sets of parallel-oriented *p* orbitals form two mutually perpendicular π bonds. The remaining *sp* orbital of each carbon atom forms a σ bond to a hydrogen atom.

ure 1.8). The *sp* orbitals have the same shape, and the electrons in each orbital have the same energy. The orbitals differ only in their position in space; they are at 180° angles to each other—again to provide for maximum separation of the electrons.

Each carbon atom in acetylene has four valence electrons. The two *sp* hybrid orbitals of each carbon atom each contain one electron, and the two 2*p* orbitals of each carbon atom each contain one electron. The carbon atoms in acetylene are linked by one σ bond and two π bonds to give a triple bond. One *sp* orbital and its electron form a bond with hydrogen; the other *sp* orbital forms a σ bond with the second carbon atom. The second and third bonds between carbon atoms result from sideways overlap of 2*p* orbitals. One set of 2*p* orbitals overlaps in front and back of the molecule to form one π bond. The second set of 2*p* orbitals overlaps above and below the molecule to form the second π bond.

Effect of Hybridization on Bond Length

The hybridization of carbon in methane, ethylene, and acetylene affects the C—H and C—C bond lengths. Note in Table 1.3 that the length of the C—H bond decreases in the order $sp^3 > sp^2 > sp$. What causes this order? The energy of the 2*s* orbital is lower than that of the 2*p* orbital, and on average, the 2*s* orbital is closer to the nucleus than the 2*p* orbital. The average distance of hybrid orbitals from the nucleus depends on the percent contribution of the *s* and *p* orbitals. The contribution of the *s* orbital is 25% in an sp^3 hybrid orbital because one *s* and three *p* orbitals are replaced by the four hybrid orbitals. Similarly, the contribution of the *s* orbital is 33% and 50% for the sp^2 and *sp* orbitals, respectively. Because an sp^3 hybrid orbital has a smaller *s* character than an sp^2 or *sp* hybrid orbital, the electrons in an sp^3 orbital are generally farther from the nucleus. As a consequence, a bond formed with an sp^3 orbital is longer than bonds involving sp^2 or *sp* hybrid orbitals.

The length of the carbon–carbon bond also decreases in the order $sp^3 > sp^2 > sp$. This trend partly reflects the effect of the closer approach to the nucleus of the σ-bonding electrons as the percent *s* character increases. However, the substantial decrease in the carbon–carbon bond length of ethane > ethylene > acetylene also is a consequence of the increased number of bonds joining the carbon atoms. Two shared pairs of electrons draw the carbon atoms closer together than a single bond. Three shared pairs move the carbon atoms still closer.

1.10 FUNCTIONAL GROUPS

Atoms or groups of bonded atoms responsible for similar physical and chemical properties are **functional groups.** It would be difficult to study the millions of organic compounds and their reactions without using a classification system based on the functional groups contained in molecules.

Some functional groups are a part of the molecular framework. These include the carbon–carbon double bond in compounds such as ethylene and the carbon–carbon triple bond in compounds such as acetylene. Double or triple bonds are also present in various molecules that have more complex structures. Although benzene is represented as a series of alternating carbon–carbon single and double bonds, it reacts differently from ethylene

TABLE 1.3 Effect of Hybridization on Bond Lengths

	$H_3C{-}CH_3$	$H_2C{=}CH_2$	H—C≡C—H
C to H bond length	1.10 Å	1.08 Å	1.06 Å
C to C bond length	1.54 Å	1.33 Å	1.20 Å

(Chapter 5). The benzene ring is often incorporated in more complex structures, but it is easy to recognize.

$$\underset{\text{ethylene}}{H_2C{=}CH_2} \qquad \underset{\text{acetylene}}{H{-}C{\equiv}C{-}H} \qquad \underset{\text{benzene}}{C_6H_6}$$

Chemical reactions in organic compounds occur at functional groups, while the rest of the structure remains unchanged. Once you learn the properties and reactions of one functional group, you will know the properties and reactions of thousands of compounds in that class. For example, the carbon–carbon double bond in ethylene reacts with hydrogen in the presence of a platinum catalyst to give ethane.

$$\underset{\text{ethylene}}{H_2C{=}CH_2} + H_2 \xrightarrow{Pt} \underset{\text{ethane}}{H_3C{-}CH_3}$$

Any compound containing a carbon–carbon double bond undergoes a similar reaction.

Functional groups can contain a variety of elements. The most common elements in functional groups are oxygen and nitrogen, although sulfur or the halogens may also be present. Alcohols and ethers are two classes of compounds that contain carbon–oxygen single bonds. The —OH unit in alcohols is the **hydroxyl group.**

$$-\overset{|}{\underset{|}{C}}-\ddot{\underset{..}{O}}-H \quad \text{as in methyl alcohol} \quad \underset{\text{an alcohol}}{H_3C{-}\ddot{\underset{..}{O}}{-}H}$$

$$-\overset{|}{\underset{|}{C}}-\ddot{\underset{..}{O}}-\overset{|}{\underset{|}{C}}- \quad \text{as in dimethyl ether} \quad \underset{\text{an ether}}{H_3C{-}\ddot{\underset{..}{O}}{-}CH_3}$$

Aldehydes and ketones contain double bonds to oxygen. The unit C=O is called the **carbonyl group.** The carbon atom of the carbonyl group is called the **carbonyl carbon atom,** and the oxygen atom is called the **carbonyl oxygen atom.** Note that an **aldehyde** has at least one hydrogen atom bonded to the carbonyl carbon atom. In **ketones,** the carbonyl carbon atom is bonded to two other carbon atoms.

```
  :Ö                          H :Ö
   ‖                          |  ‖
 —C—H   as in acetaldehyde  H—C—C—H
                              |
                              H
                          an aldehyde

    :Ö                     H :Ö  H
  |  ‖  |                  |  ‖  |
 —C—C—C—   as in acetone H—C—C—C—H
  |     |                  |     |
                           H     H
                            a ketone
```

Carboxylic acids and esters contain both single and double bonds from a carbon atom to oxygen atoms. In a carboxylic acid, the carbonyl group is bonded to a hydroxyl group and either hydrogen or carbon atoms. In an ester, the carbonyl group is bonded to an "OR" group, where "R" contains one or more carbon atoms, and to either a hydrogen or carbon atom.

```
  :Ö                          H :Ö
   ‖                          |  ‖
 —C—Ö—H  as in acetic acid  H—C—C—Ö—H
                              |
                              H
                        a carboxylic acid

  :Ö                              H :Ö     H
   ‖     |                        |  ‖     |
 —C—Ö—C—   as in methyl acetate H—C—C—Ö—C—H
         |                        |        |
                                  H        H
                                   an ester
```

A nitrogen atom can form single, double, or triple bonds to a carbon atom. Compounds with carbon–nitrogen single bonds are amines. The remaining two bonds to nitrogen can be to hydrogen or carbon atoms.

```
                                  H
  |                               |
 —C—N̈—H   as in methylamine  H—C—N̈—H
  |  |                            |  |
                                  H  H
                                an amine
```

Compounds with carbon–nitrogen double and triple bonds are imines and nitriles, respectively.

```
     :N—H                       H :N—H
  |   ‖                         |  ‖
 —C—C—H   as in ethylimine   H—C—C—H
  |                             |
                                H
                             an imine
```

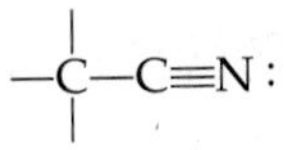

$-\mathrm{C}-\mathrm{C}\equiv\mathrm{N}:$ as in acetonitrile $\mathrm{H-CH_2-C\equiv N:}$

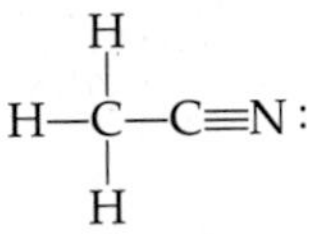

a nitrile

Amides are functional groups in which a carbonyl carbon atom is linked by a single bond to a nitrogen atom and either a hydrogen or carbon atom. The remaining two bonds to the nitrogen atom may be to either hydrogen or carbon atoms.

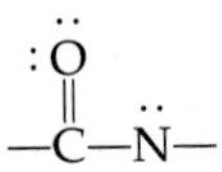

$-\mathrm{C(=O)}-\mathrm{N}-$ as in acetamide $\mathrm{H-CH_2-C(=O)-NH-H}$

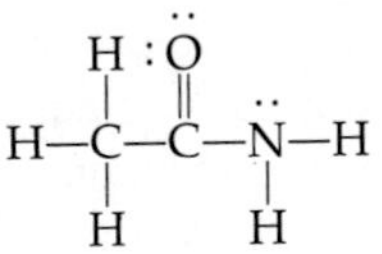

an amide

Sulfur forms single bonds to carbon in two classes of compounds. Thiols (also called mercaptans) and thioethers (also called sulfides) structurally resemble alcohols and ethers, which contain oxygen, another element in the same group as sulfur.

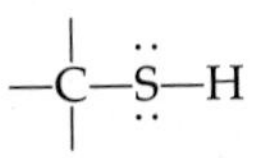

$-\mathrm{C}-\mathrm{S}-\mathrm{H}$ as in methanethiol $\mathrm{H-CH_2-S-H}$

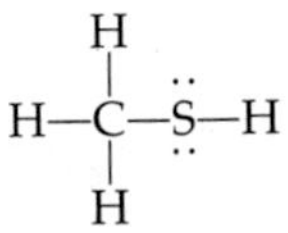

a thiol

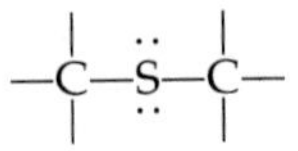

$-\mathrm{C}-\mathrm{S}-\mathrm{C}-$ as in dimethyl sulfide $\mathrm{H-CH_2-S-CH_2-H}$

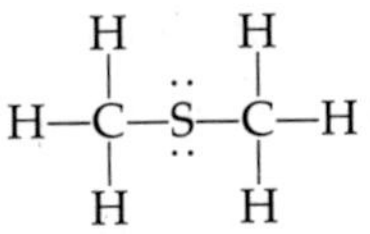

a thioether

The halogens form single bonds to carbon. Chlorine and bromine are the more common halogens in organic compounds.

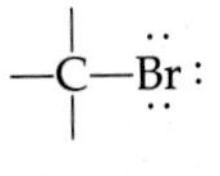

$-\mathrm{C}-\mathrm{Br}:$ as in bromoethane $\mathrm{H-CH_2-CH_2-Br:}$

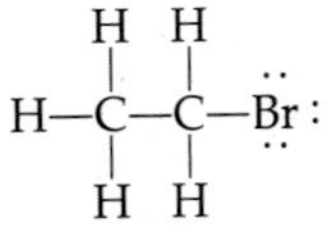

a haloalkane

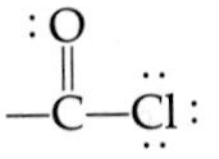

$-\mathrm{C(=O)}-\mathrm{Cl}:$ as in acetyl chloride $\mathrm{H-CH_2-C(=O)-Cl:}$

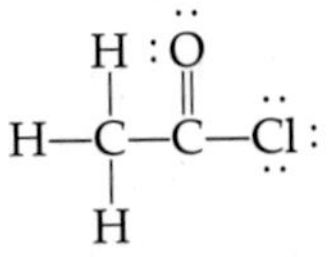

an acyl halide

Pheromones—Chemical Communications in the Insect World

The scope of organic chemistry is rapidly changing and contributes to many fields. For example, we cannot understand modern biology without a foundation in organic chemistry and indirectly without an understanding of functional groups. Organic chemistry underlies all life forms. As an example, we will consider the structure and functional groups of some pheromones. Pheromones (Gk., *pherein*, to transfer + *hormon*, to excite) are compounds (occasionally mixtures of compounds) that insects and other animals use to communicate. The major identified species that use pheromones are insects, but it is thought that even higher animals such as mammals may emit pheromones.

Pheromones are used to mark trails, warn of dangers, cause aggregation of species, for defensive purposes, and to attract members of the opposite sex. The whip scorpion ejects a defensive spray that it uses to ward off predators. Some species of ants warn other ants of danger by an alarm pheromone. Bark beetles responsible for Dutch Elm disease emit an aggregation pheromone that results in the gathering of a large number of beetles and causes infestation by a species transmitting a fungus that kills the tree. The sex attractants, usually emitted by the female of the species, attract members of the opposite sex. Pheromones are signals that the female is ready to mate. They also aid the male in locating the female, often from great distances.

All moths that have been studied have sex attractants that are species specific. The compounds usually are derived from long chains of carbon atoms. However, the functional groups in pheromones vary considerably. Two examples are the sex attractants of the gypsy moth and the grape berry moth. Their structures are shown below. The ether oxygen atom in the three-membered ring of the sex attractant of the gypsy moth is the only functional group in the molecule. The oxygen atoms in the sex attractant of the grape berry moth are part of an ester functional group. Note that this compound also contains a second functional group, a double bond.

When the structures of sex attractants were determined, some scientists predicted that it might be possible to bait traps with the compound and, by removal of one sex, break the reproductive cycle. Unfortunately, this "ideal" way to control insects and eliminate the use of pesticides has not proved effective for most species. However, there is some evidence that pheromone traps cause some confusion and may inhibit mating. Also pheromone traps serve as an early warning of possible infestation of that species. The ultimate goal of replacing pesticides with pheromones may yet be possible.

$$CH_3CH_2CH_2CH_2CH_2CH_2CH_2CH_2CH_2CH_2CH_2\underset{\diagdown O \diagup}{CH{-}CH}CH_2CH_2CH_2CH_2CH(CH_3)_2$$

sex attractant of gypsy moth

$$CH_3CH_2{-}CH{=}CH{-}CH_2CH_2CH_2CH_2CH_2CH_2CH_2CH_2{-}O{-}\overset{\overset{\displaystyle O}{\|}}{C}{-}CH_3$$

sex attractant of grape berry moth

1.11 STRUCTURAL FORMULAS

The **molecular formula** of a compound indicates its atomic composition. For example, the molecular formula of butane is C_4H_{10}. However, to understand the chemistry of organic compounds, it is necessary to represent the structure of a molecule by a **structural formula** that shows the arrangement of atoms and bonds.

To save time and space, chemists draw abbreviated or condensed versions of structural formulas. **Condensed structural formulas** show only specific bonds; other bonds are "left out" but implied. The degree of condensation depends on which bonds are shown and which are implied. For example, because hydrogen forms only a single bond to carbon, the C—H bond need not be shown in the condensed structure of a molecule such as butane.

$$CH_3—CH_2—CH_2—CH_3$$

butane

One carbon–carbon bond is shown between a terminal carbon atom and an internal carbon atom. The terminal carbon atoms are understood to have single bonds to three hydrogen atoms. Each carbon atom in the interior of the molecule has the two carbon–carbon bonds shown; the two carbon–hydrogen bonds are implied but not shown. Note that by convention the symbol for the hydrogen atom is written to the right of the symbol for the carbon atom.

In a further condensation of a structural formula, the C—C bonds are "left out".

This carbon atom is bonded to one carbon atom and three hydrogen atoms.

$$CH_3CH_2CH_2CH_3$$

This carbon atom is bonded to the carbon atoms on either side and to two hydrogen atoms.

In this representation, the carbon atom on the left is understood to be bonded to the three hydrogen atoms and the carbon atom to its right. The second carbon atom from the left is bonded to the two hydrogen atoms to the right. That carbon atom is also bonded to a carbon atom to its immediate right and a carbon atom to its left.

Large structures may have repeated structural subunits that are represented by grouping the subunits within parentheses. The number of times the unit is repeated is given by a subscript after the closing parenthesis. For example, butane is represented by

$$CH_3(CH_2)_2CH_3$$

The —CH_2— unit is a **methylene** group. It occurs twice in butane. Because the methylene units are linked in a repeating chain, they are placed within the parentheses.

Two or more identical groups of atoms bonded to a common central atom may also be represented within parentheses with an appropriate subscript in a condensed formula. The parentheses may be placed to the right or left, depending on the way in which the molecule is drawn.

$$\underset{}{CH_3-\overset{\overset{\displaystyle CH_3}{|}}{C}H-CH_2-CH_2-CH_2-CH_3} \quad \text{is} \quad (CH_3)_2CHCH_2CH_2CH_2CH_3$$

$$CH_3-CH_2-CH_2-\underset{\underset{\displaystyle CH_3}{|}}{\overset{\overset{\displaystyle CH_3}{|}}{C}}-CH_3 \quad \text{is} \quad CH_3CH_2CH_2C(CH_3)_3$$

EXAMPLE 1.8 A species of cockroach secretes the following substance as a signal for other cockroaches to congregate. Write three condensed structural formulas for the substance.

$$H-\underset{\underset{\displaystyle H}{|}}{\overset{\overset{\displaystyle H}{|}}{C}}-\underset{\underset{\displaystyle H}{|}}{\overset{\overset{\displaystyle H}{|}}{C}}-\underset{\underset{\displaystyle H}{|}}{\overset{\overset{\displaystyle H}{|}}{C}}-\underset{\underset{\displaystyle H}{|}}{\overset{\overset{\displaystyle H}{|}}{C}}-\underset{\underset{\displaystyle H}{|}}{\overset{\overset{\displaystyle H}{|}}{C}}-\underset{\underset{\displaystyle H}{|}}{\overset{\overset{\displaystyle H}{|}}{C}}-\underset{\underset{\displaystyle H}{|}}{\overset{\overset{\displaystyle H}{|}}{C}}-\underset{\underset{\displaystyle H}{|}}{\overset{\overset{\displaystyle H}{|}}{C}}-\underset{\underset{\displaystyle H}{|}}{\overset{\overset{\displaystyle H}{|}}{C}}-\underset{\underset{\displaystyle H}{|}}{\overset{\overset{\displaystyle H}{|}}{C}}-\underset{\underset{\displaystyle H}{|}}{\overset{\overset{\displaystyle H}{|}}{C}}-H$$

Solution With the C—H bonds understood, we write

$$CH_3-CH_2-CH_2-CH_2-CH_2-CH_2-CH_2-CH_2-CH_2-CH_2-CH_3$$

With both the C—H and C—C bonds understood, we write

$$CH_3CH_2CH_2CH_2CH_2CH_2CH_2CH_2CH_2CH_2CH_3$$

In the most condensed version, the nine methylene units are represented within parentheses.

$$CH_3(CH_2)_9CH_3$$

Bond-Line Structures

Condensed structural formulas are convenient but still require considerable time to draw compared to yet another shorthand method using **bond-line structures.** The bond-line structure method also results in a less cluttered drawing. However, the reader has to mentally add many more features to understand the structure. The rules for drawing bond-line structures are

1. Carbon atoms are not shown unless needed for special emphasis or clarity.
2. All atoms other than carbon and hydrogen are explicitly shown.
3. A carbon atom is assumed to be located at the end of each line segment or at the intersection of two or more lines, which are used to depict bonds.
4. Multiple bonds are shown with multiple lines.

To draw a bond-line structure, it is best to arrange the carbon atoms in a zigzag manner and then mentally remove the carbon atoms.

$CH_3—CH_2—CHBr—CH_3$ is CH_3 CH_2 $CH(Br)$ CH_3 is Br

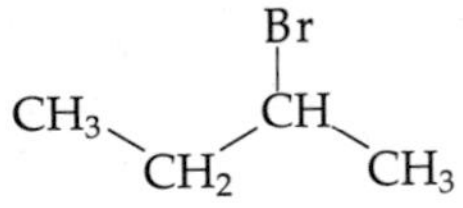

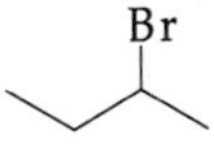

Bond-line formulas are also used to show cyclic structures. Rings of carbon atoms are shown by polygons such as an equilateral triangle, square, pentagon, and hexagon.

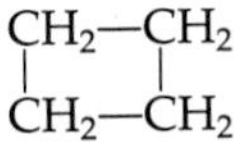

$CH_2—CH_2$ / $CH_2—CH_2$ is

Note the difference in the representation of two different multiple bonds. Atoms such as oxygen must be shown in carbonyl groups, but a terminal double-bonded carbon atom is implied by the structure and is not written.

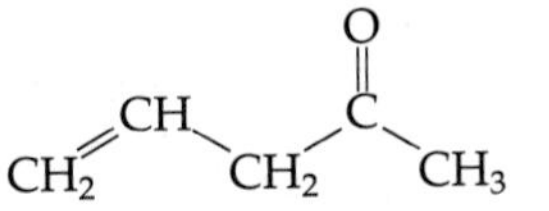

CH_2 CH CH_2 $C(=O)$ CH_3 is O

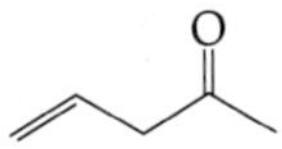

It is important to remember the normal number of bonds formed by each common atom in an organic compound. Carbon, nitrogen, and oxygen form four, three, and two bonds, respectively.

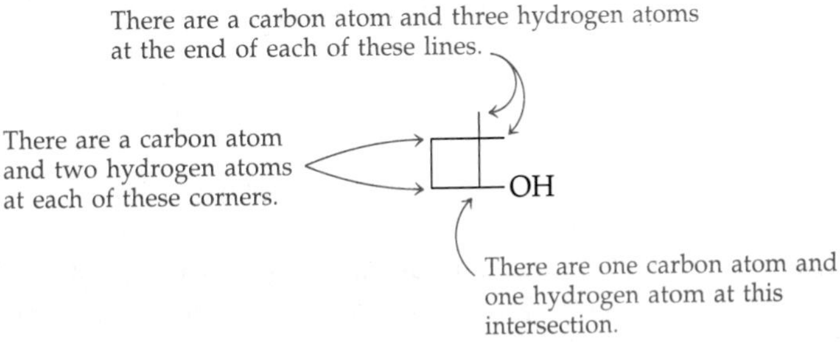

There are a carbon atom and three hydrogen atoms at the end of each of these lines.

There are a carbon atom and two hydrogen atoms at each of these corners.

OH

There are one carbon atom and one hydrogen atom at this intersection.

EXAMPLE 1.9 What is the molecular formula of carvone, which is found in oil of caraway?

O

Solution There are 10 carbon atoms in the structure, located at the ends or intersections of line segments. An oxygen atom is located at the end of a segment representing the double bond of a carbonyl group. Hydrogen atoms are counted by determining the number of bonds from each carbon atom to other atoms. Note that three carbon atoms have no hydrogen atoms. The molecular formula is $C_{10}H_{14}O$.

3 H
O
1 H →
2 H → ← 2 H
1 H
2 H → ← 3 H

Three-Dimensional Structures and Models

Because structure is so important to understanding chemical reactions, chemists construct models of molecules that can be viewed from a variety of angles. You may find it useful to purchase a molecular model kit to help you understand the structures of organic molecules.

Ball-and-stick models and space-filling models are two types of molecular models; each has certain advantages and disadvantages. Ball-and-stick models show the molecular framework and bond angles: the balls represent the nuclei of the atoms, and the sticks represent the bonds (Figure 1.9). The actual volume occupied by the molecule is not shown realistically.

Space-filling models show the entire volume occupied by the electrons

FIGURE 1.9 Perspective Structural Formula and Molecular Models

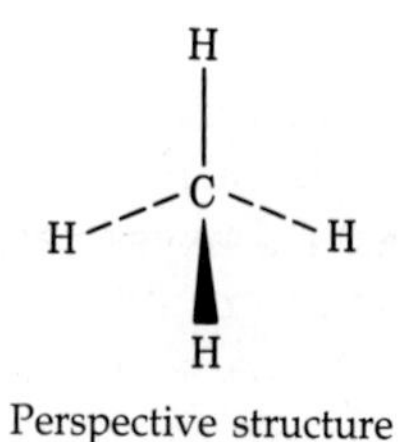

Perspective structure

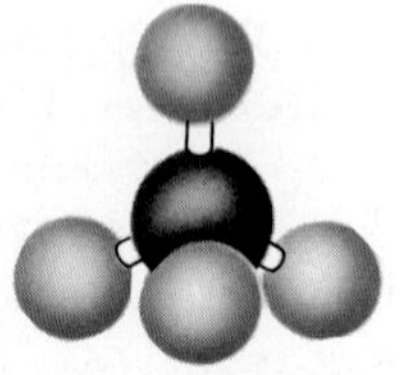

Ball-and-stick model

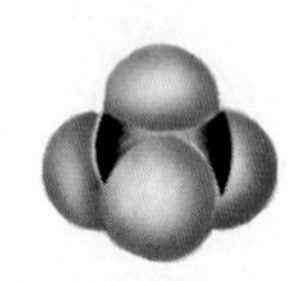

Space-filling model

surrounding each atom, but as a consequence the carbon skeleton and its bond angles are obscured.

On paper, the three-dimensional shape of molecules is shown with wedges and dashed lines (Figure 1.9). A wedge is viewed as a bond extending out of the plane of the page toward the reader. A dashed line represents a bond directed behind the plane of the page. A solid line is a bond in the plane of the page. Three-dimensional representations of molecules using wedge and dashed lines are **perspective structural formulas.**

Recognizing Structural Features

The structural features that allow chemists to predict the physical and chemical properties of naturally occurring molecules are often only a small part of a larger structure. These large structures are written in condensed forms that are meaningful because certain conventions are used. Regardless of the size and complexity of a molecule, you should glance over the molecule, ignore the many lines indicating the carbon–carbon bonds, and learn to focus on the important parts. Are there multiple bonds? Are there atoms, such as oxygen and nitrogen, that are part of functional groups? How are these atoms bonded, and what other atoms are nearby? For example, if a C=O is present, it may be part of an aldehyde, ketone, acid, ester, or amide. The distinction between these functional groups can be decided by looking at the atoms bonded to the carbonyl carbon atom.

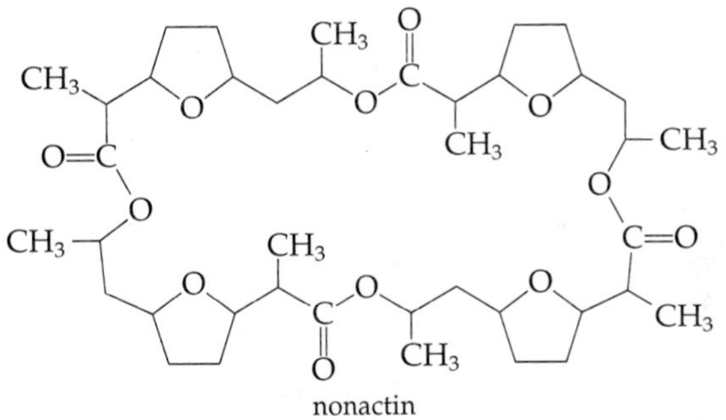

nonactin

Consider the structure for nonactin, an ionophorous antibiotic. It binds potassium ions through the many oxygen atoms in the large ring of atoms. It transports potassium ions across bacterial cell membranes, and the cells die.

What are the oxygen-containing functional groups in this complex structure? Concentrate on one oxygen atom at a time. Some oxygen atoms are part of a C=O group, a carbonyl group; in fact there are four such locations. Now look at the atoms bonded to the carbonyl carbon atom of the C=O groups. One bond is to carbon and the other to oxygen. Both carboxylic acids and esters have such features. The oxygen atom of carboxylic acids is in an —OH group, whereas the oxygen atom of esters is bonded to another carbon atom. Convince yourself that four ester groups are in this molecule.

Biodiversity and Pharmaceutical Chemicals

Compounds obtained from plant sources dominated early chemical research in organic chemistry. Some of these crude plant extracts were mixtures of organic compounds that were used as medicines. How effective were these herbal medicines of primitive societies? Folklore indicates that some medicines were useful, but they were discovered by a "hit or miss" approach. Often they had no effect, and sometimes herbal medicines caused great harm.

Eventually, organic chemists in the last century isolated and purified the active ingredients of some herbal remedies. Among these are digitalis—a cardiac steroid, morphine—a potent analgesic, and quinine—an antimalarial agent. A major part of organic chemistry is the determination of the structure of useful compounds obtained from nature. Once a given structure has been determined, it may then be possible to synthesize it in the laboratory—often at a lower cost and in higher purity.

Chemists continue to search for compounds in nature that might have pharmaceutical activity. The biodiversity of plants in unexplored areas such as tropical rain forests offers opportunities for the discovery of new drugs. The destruction of rain forests may forever eliminate plant forms that contain undiscovered compounds with diverse structures that could be used to combat disease.

Chemists continually modify pharmaceutical products to improve their therapeutic properties. Some minor modifications involve changing the structure either to increase the water solubility of the compound or to allow more effective transport across biological membranes to the desired site. Other changes may be more substantial and involve alterations of the essential structural unit responsible for the therapeutic action of a drug.

Penicillins and cephalosporins (see figure), two classes of antibiotics, are excellent examples of how the chemical modification of structural units provides similar but varied pharmacological character. In each class of compounds, significant portions of the molecule, the essential structural unit, are identical. In penicillins, the four-membered ring that contains an amide functional group is essential for antibiotic activity. This ring is fused to a five-membered ring that contains a sulfur atom. Cephalosporins also contain a four-membered ring with an amide functional group. However, this ring is fused to a six-membered ring that contains a sulfur atom.

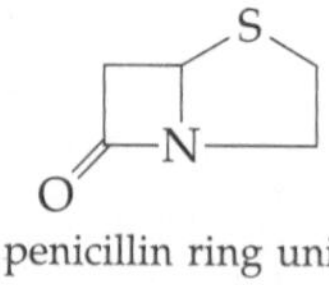

S N O
penicillin ring unit

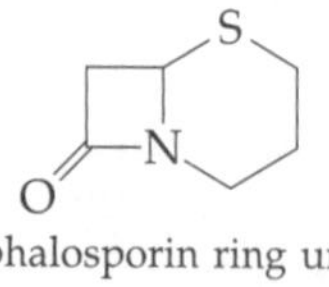

S N O
cephalosporin ring unit

Now concentrate on the second type of oxygen-containing functional group in the molecule. There are oxygen atoms contained as part of a five-membered ring—four times in fact. Each oxygen atom is bonded by single bonds to two carbon atoms. The functional groups are ethers.

1.12 ISOMERS

Compounds that have the same molecular formula but different structures are **isomers.** Structure refers to the linkage of the atoms. As we examine the structure of organic compounds in increasing detail, you will learn how

Structures of Penicillins and Cephalosporins

Penicillins

O
—O—CH$_2$—C—NH—
S
CH$_3$
CH$_3$
N
O
C—OH
O

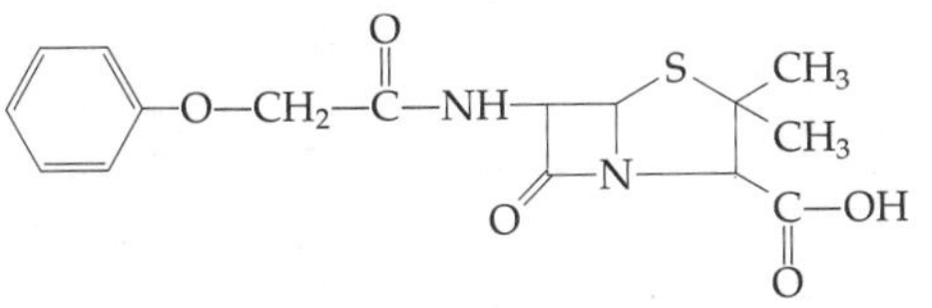

penicillin V

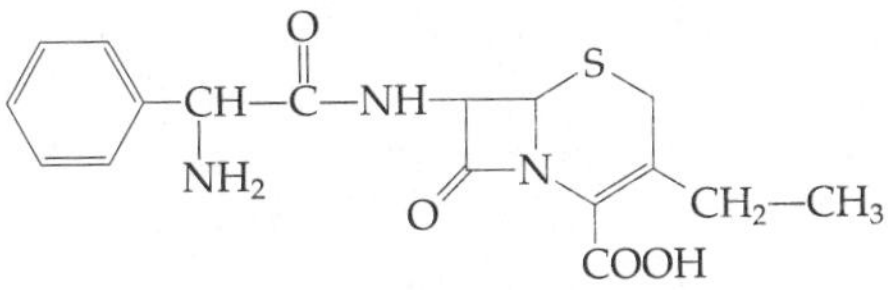

O
C—NH—
N
O
CH$_3$
O
S
CH$_3$
CH$_3$
N
C—OH
O

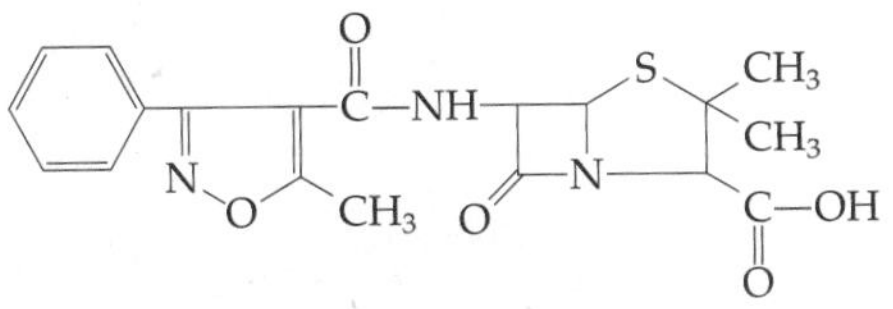

oxacillin

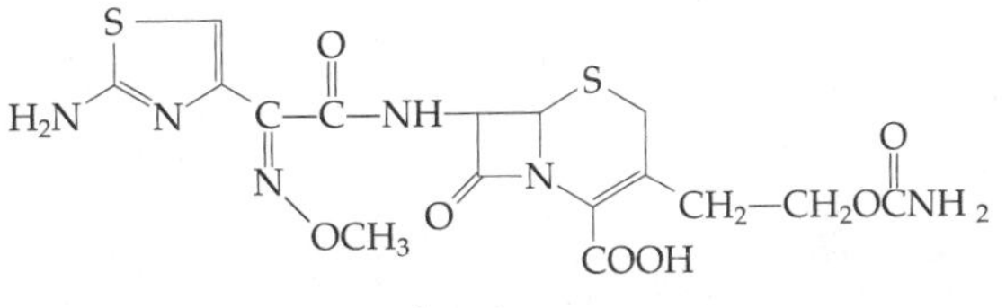

S
O
CH—C—NH—
CO$_2$H
S
CH$_3$
CH$_3$
N
O
C—OH
O

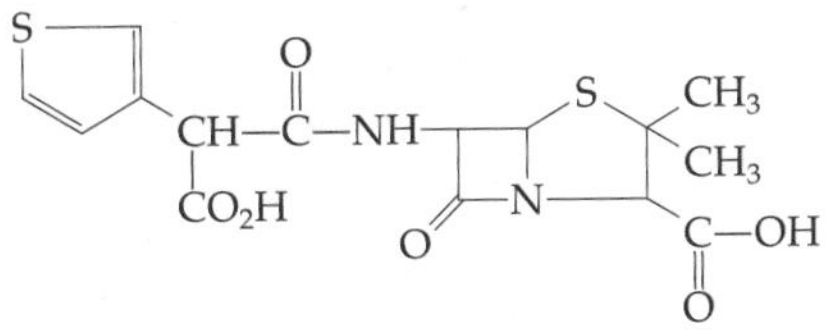

ticarcillin

Cephalosporins

O
—CH—C—NH—
NH$_2$
S
N
O
CH$_2$—CH$_3$
COOH

cephalexin

S
H$_2$N
N
O
C—C—NH—
N
OCH$_3$
S
N
O
O
CH$_2$—CH$_2$OCNH$_2$
COOH

cefotaxime

N
N
N—CH$_2$—C—NH—
O
S
N
O
N—N
CH$_2$—CH$_2$S
S
CH$_3$
COOH

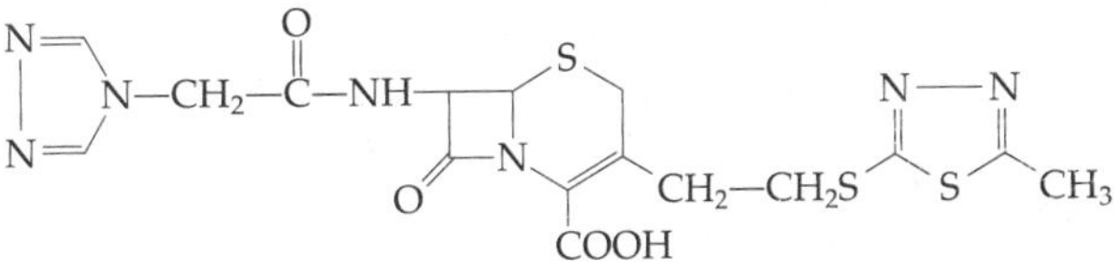

cefazolin

There are substantial differences in the structural units bonded to the basic rings of penicillins and cephalosporins. These structural variations were selected to increase the effectiveness of the antibiotic and to overcome the resistance developed by bacteria when one antibiotic is used too frequently.

subtle structural differences in isomers affect the physical and chemical properties of compounds.

We can divide isomers into several groups. The number of groups will increase in later chapters. Isomers that differ in their carbon skeleton are **skeletal isomers.** Consider the structural differences in the two isomers of C_4H_{10}, butane and isobutane. Butane has an uninterrupted chain of carbon atoms (Figure 1.10), but isobutane has only three carbon atoms connected in sequence and a fourth carbon atom appended to the chain. The boiling points of butane and isobutane are −1 and −12 °C, respectively; the chemical properties of the two compounds are similar but differ somewhat.

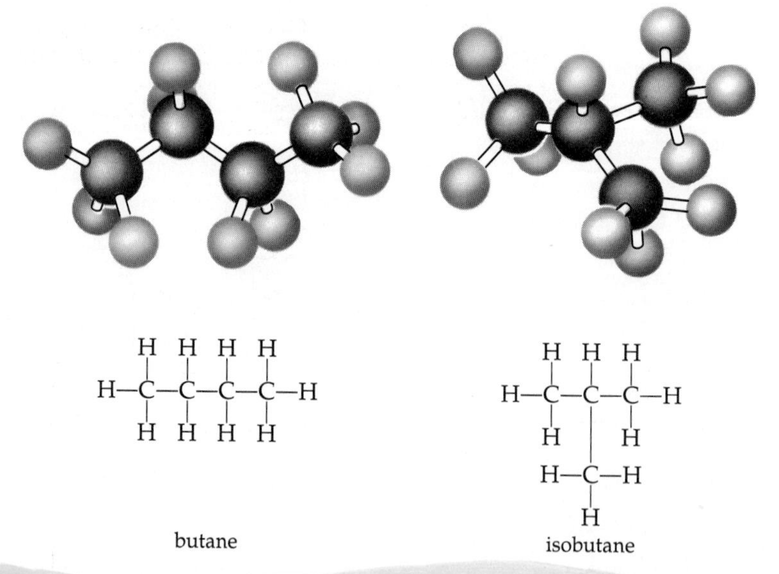

FIGURE 1.10
Structural Formulas and Molecular Models of Isomers

Isomers that have different functional groups are **functional group isomers.** The molecular formula for ethyl alcohol and dimethyl ether is C_2H_6O. Although the compositions of the two compounds are identical, their functional groups differ. The atomic sequence is C—C—O in ethyl alcohol, and the oxygen atom is present as an alcohol. The C—O—C sequence in the isomer corresponds to an ether.

$CH_3{-}CH_2{-}OH$	$CH_3{-}O{-}CH_3$
ethyl alcohol	dimethyl ether
(bp 78.5 °C)	(bp −24 °C)

The physical properties of these two functional group isomers, as exemplified by their boiling points, are very different. These substances also have different chemical properties because their functional groups differ. For example, ethyl alcohol reacts with sodium to produce hydrogen gas, whereas dimethyl ether does not react with sodium.

$$2\,CH_3{-}CH_2{-}OH + 2\,Na \longrightarrow 2\,CH_3{-}CH_2{-}O^-\,Na^+ + H_2$$
$$CH_3{-}O{-}CH_3 + Na \longrightarrow \text{no reaction}$$

Positional isomers are compounds that have the same functional groups in different positions in the carbon skeleton. For example, the isomeric alcohols 1-propanol and 2-propanol differ in the location of the hydroxyl group.

1-propanol

$CH_3{-}CH_2{-}CH_2{-}OH$

—OH group on the end of chain

2-propanol

$CH_3{-}\underset{}{\overset{OH}{\overset{|}{CH}}}{-}CH_3$

—OH group in the middle of the chain

Isomerism is not always immediately obvious. Sometimes two structures appear to be isomers when in fact the structures are the same compound written in slightly different ways. It is important to be able to recognize isomers and distinguish them from equivalent representations of the same compound. For example, 1,2-dichloroethane can be written in several ways. In each formula, the bonding sequence is Cl—C—C—Cl.

$$\begin{array}{ccccccc} & & H & & H & & \\ & & | & & | & & \\ Cl & - & C & - & C & - & Cl \\ & & | & & | & & \\ & & H & & H & & \end{array} \quad \text{or} \quad \begin{array}{ccccccc} & & H & & Cl & & \\ & & | & & | & & \\ Cl & - & C & - & C & - & H \\ & & | & & | & & \\ & & H & & H & & \end{array} \quad \text{or} \quad \begin{array}{ccccccc} & & H & & Cl & & \\ & & | & & | & & \\ H & - & C & - & C & - & H \\ & & | & & | & & \\ & & Cl & & H & & \end{array}$$

1,2-dichloroethane (CH_2ClCH_2Cl)

The isomer of 1,2-dichloroethane is 1,1-dichloroethane.

$$\begin{array}{ccccccc} & & Cl & & H & & \\ & & | & & | & & \\ Cl & - & C & - & C & - & H \\ & & | & & | & & \\ & & H & & H & & \end{array}$$

1,1-dichloroethane ($CHCl_2CH_3$)

In 1,1-dichloroethane, the two chlorine atoms are bonded to the same carbon atom. In 1,2-dichloroethane, the two chlorine atoms are bonded to different carbon atoms. The different condensed structural formulas, $CHCl_2CH_3$ and CH_2ClCH_2Cl, also convey information about the different structures.

EXAMPLE 1.10 Consider the following structural formulas for two compounds used as general anesthetics. Do they represent isomers? How do they differ?

$$\begin{array}{ccccccccc} & & F & & F & & & & F & & \\ & & | & & | & & & & | & & \\ Cl & - & C & - & C & - & O & - & C & - & F \\ & & | & & | & & & & | & & \\ & & H & & F & & & & H & & \end{array} \qquad \begin{array}{ccccccccccc} & & F & & Cl & & & & H & & \\ & & | & & | & & & & | & & \\ F & - & C & - & C & - & O & - & C & - & F \\ & & | & & | & & & & | & & \\ & & F & & H & & & & F & & \end{array}$$

Solution The atomic compositions given in these structural formulas are identical; the molecular formula is $C_3H_2F_5Cl$. Therefore, the compounds are isomers. The carbon skeletons are identical, and the compounds are both ethers.

Both isomers have a CHF_2 unit on the right side of the ether oxygen atom in spite of the different ways in which the fluorine and hydrogen are written—this is not the basis for isomerism. The two-carbon unit on the left of the oxygen atom has the halogen atoms distributed in two different ways. That is, they are positional isomers. The structure on the left has two fluorine atoms bonded to the carbon atom appended to the oxygen atom. The carbon atom on the left has a fluorine and a chlorine atom bonded to it. The structure on the right has one chlorine atom bonded to the carbon appended to the oxygen atom. The carbon atom on the left has three fluorine atoms bonded to it.

1.13 NOMENCLATURE

Nomenclature refers to a systematic method of naming materials. In chemistry, the nomenclature of compounds is exceedingly important. The phenomenon of isomerism easily illustrates this point. The common names butane and isobutane of the two isomeric C_4H_{10} compounds are easy to learn. However, there are 75 isomers of $C_{10}H_{22}$ and 62,491,178,805,831 isomers of $C_{40}H_{82}$. Without a system of naming compounds, organic chemistry would be difficult, if not impossible, to comprehend.

At a meeting in Geneva, Switzerland, in 1892 chemists devised a systematic nomenclature for all compounds, including organic compounds. Compounds are now named by rules developed by the International Union of Pure and Applied Chemistry (IUPAC). The rules result in a clear and definitive name for each compound. A universal system for naming organic compounds was needed because different names had often been given to the same compound. For example, CH_3CH_2OH had been called not only alcohol but also spirits, grain alcohol, ethyl alcohol, methyl carbinol, and ethanol. Furthermore, a variety of names developed in each language.

A chemical name consists of three parts: prefix, parent, and suffix. The parent indicates how many carbon atoms are in the main carbon skeleton. The suffix identifies most of the functional groups present in the molecule. Examples of suffixes are *-ol* for alcohols, *-al* for aldehydes, and *-one* for ketones. The prefix specifies the location of the functional group designated in the suffix, as well as some other types of substituents on the main parent chain.

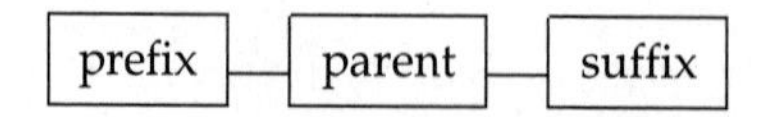

Once the rules are applied, there is only one name for each structure, and one structure for each name. For example, a compound that is partly responsible for the odor of a skunk is 3-methyl-1-butanethiol.

$$\begin{array}{c} \quad CH_3 \\ \quad | \\ CH_3{-}CH{-}CH_2{-}CH_2{-}SH \end{array}$$

Butane is the parent name of the four-carbon unit that is written horizontally. The prefix *3-methyl* refers to the $-CH_3$ written above the chain of carbon atoms. The prefix *1-* and the suffix *-thiol* refer to the position and identity of the —SH group. This method of assigning numbers to the carbon chain and other features of the IUPAC system will be discussed further in subsequent chapters.

In spite of the IUPAC system, many common names are so well-established that both common and IUPAC names must be recognized. The IUPAC name for CH_3CH_2OH is ethanol, but the common name ethyl alcohol is still used.

As we introduce the nomenclature of each class of organic compounds, we will see that the subject is natural and ordered when the rules are followed.

EXERCISES

Atomic Properties

1.1 How many valence-shell electrons are in each of the following elements?
(a) N (b) F (c) C (d) O (e) Cl (f) Br (g) S (h) P

1.2 Which of the following atoms has the larger electronegativity? Which has the larger atomic radius?
(a) Cl or Br (b) O or S (c) C or N (d) N or O (e) I or Br (f) C or F (g) C or O (h) O or I

Ions and Ionic Compounds

1.3 The formula of the dihydrogen phosphate ion, an ion eliminated in urine to control the pH of cellular fluids, is $H_2PO_4^-$. What is the formula of calcium dihydrogen phosphate?

1.4 Ionic arsenic compounds contain ions such as arsenite, AsO_3^{3-}, which react with thiol groups of proteins. What is the formula of magnesium arsenite?

1.5 Write a Lewis structure for each of the following ions.
(a) OH^- (b) CN^- (c) H_3O^+ (d) NH_4^+ (e) NO_3^-

1.6 Write a Lewis structure for each of the following ions.
(a) NO_2^- (b) SO_3^{2-} (c) SO_4^{2-} (d) NH_2^- (e) CO_3^{2-}

Lewis Structures of Covalent Compounds

1.7 Write a Lewis structure for each of the following compounds.
(a) NH_2OH (b) CH_3CH_3 (c) CH_3OH (d) CH_3NH_2 (e) CH_3Cl (f) CH_3SH

1.8 Write a Lewis structure for each of the following compounds.
(a) HCN (b) HNNH (c) CH_2NH (d) CH_3NO (e) CH_2NOH (f) CH_2NNH_2

1.9 Place any required unshared pairs of electrons that are missing from the following formulas.

(a) $CH_3—C(=O)—OH$ (b) $CH_3—C(=O)—O—CH_3$ (c) $H—C(=O)—NH—CH_3$

(d) $CH_3—S—CH{=}CH_2$ (e) $CH_3—C(=N—H)—CH_3$ (f) $N{\equiv}C—CH_2—C{\equiv}N$

1.10 Place any required unshared pairs of electrons that are missing from the following formulas.

(a) $CH_3—C(=O)—Cl$ (b) $CH_3—O—CH{=}CH_2$ (c) $CH_3—C(=O)—SH$

```
              O—CH3                         O
              |                             ||
(d) CH3—CH—O—CH3     (e) NH2—C—O—CH3
```

(f) CH_3—O—CH_2—O—CH_3

1.11 Using the number of valence electrons in the constituent atoms and the given arrangement of atoms in the compound, write the Lewis structure for each of the following molecules.

```
                  H                  O
     H            |                  |
(a)    >C—N—C—H   (b) Cl—C—Cl
     H            |
                  H

           O                    H  S
           |                    |  |
(c) H—N—C—N—H      (d) H—C—C—O—H
      |     |                   |
      H     H                   H
```

1.12 Using the number of valence electrons in the constituent atoms and the given arrangement of atoms in the compound, write the Lewis structure for each of the following molecules.

```
       H          H                H  O
       |          |                |  |
(a) H—C—S—S—C—H    (b) H—C—C—S—H
       |          |                |
       H          H                H

       H     H                H       O
       |     |                |       |
(c) H—C—O—C—Cl     (d) H—C—O—C—N—H
       |     |                |         |
       H     H                H         H
```

1.13 Two compounds used as dry cleaning agents have the molecular formulas C_2Cl_4 and C_2HCl_3. Write the Lewis structure for each compound.

1.14 Acrylonitrile, a compound used to produce polymeric fibers for rugs, is represented by the condensed formula CH_2CHCN. Write the Lewis structure for the compound.

Formal Charge

1.15 Assign the formal charges for the atoms other than hydrogen in each of the following species.

```
       ..                    ..
(a) H—O—C≡N:     (b) H—O—N≡C:
       ..                    ..

          CH3
          |   ..                    ..     ..
(c) CH3—N—O:     (d) CH3—N=N=N:
          |   ..
          CH3
```

1.16 Assign the formal charges for the atoms other than carbon and hydrogen in each of the following species.

```
                                  CH3
          ..                      |
(a) CH3—O—BF3     (b) CH3—N—AlCl3
          |                       |
          CH3                     CH3
```

(c) 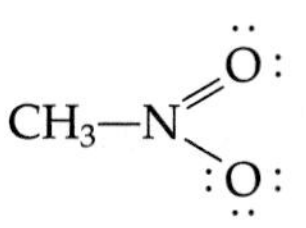 $CH_3—N(=\ddot{O}:)—\ddot{O}:$ (d) 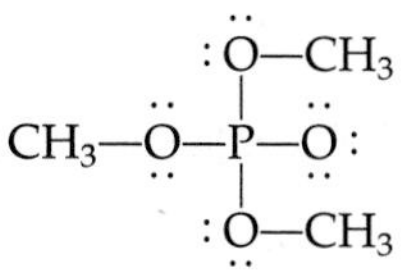 $CH_3—\ddot{O}—P(—\ddot{O}—CH_3)(—\ddot{O}—CH_3)—\ddot{O}:$

1.17 The following species are isoelectronic; that is, they have the same number of electrons bonding the same number of atoms. Determine which atoms have a formal charge. Calculate the net charge for each species.

(a) $:C{\equiv}O:$ (b) $:N{\equiv}O:$ (c) $:C{\equiv}N:$ (d) $:C{\equiv}C:$

1.18 The following species are isoelectronic; that is, they have the same number of electrons bonding the same number of atoms. Determine which atoms have a formal charge. Calculate the net charge for each species.

(a) $:\ddot{N}{=}N{=}\ddot{N}:$ (b) $:\ddot{O}{=}N{=}\ddot{O}:$

1.19 Acetylcholine, a compound involved in the transfer of nerve impulses, has the following structure. What is the formal charge of the nitrogen atom? What is the net charge of the species?

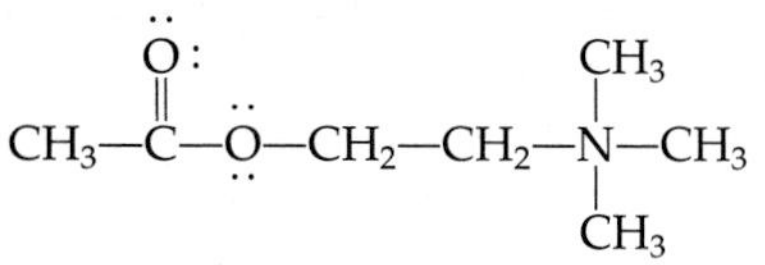

$CH_3—C(=\ddot{O}:)—\ddot{O}—CH_2—CH_2—N(CH_3)_2—CH_3$

1.20 Sarin, a nerve gas, has the following structure. What is the formal charge of the phosphorus atom?

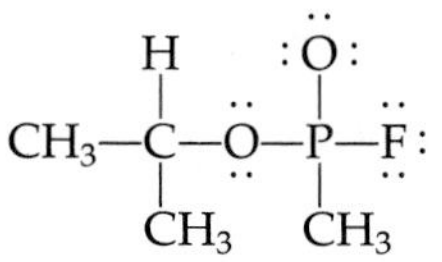

$CH_3—C(H)(CH_3)—\ddot{O}—P(—\ddot{O}:)(CH_3)—\ddot{F}:$

Resonance

1.21 The small amounts of cyanide ion contained in the seeds of some fruits are eliminated from the body as SCN^-. Draw two possible resonance forms for the ion. Which atom has the formal negative charge in each form?

1.22 Are the following pairs contributing resonance forms of a single species or not? Explain.

(a) $:\ddot{N}—N{\equiv}N:$ and $:\ddot{N}{=}N{=}\ddot{N}:$

(b) $H—C{\equiv}N—\ddot{O}:$ and $H—\ddot{C}{=}N{=}\ddot{O}:$

1.23 Write the resonance structure that results when electrons are moved in the direction indicated by the curved arrows for the following amide. Calculate any formal charges that may result.

$CH_3—C(=\ddot{O}:)—\ddot{N}H_2$

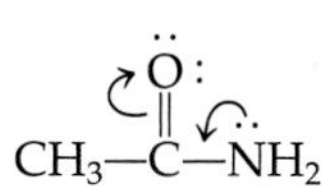

1.24 Write the resonance structure that results when electrons are moved in the direction indicated by the curved arrows for the acetate ion. Calculate any formal charges that may result.

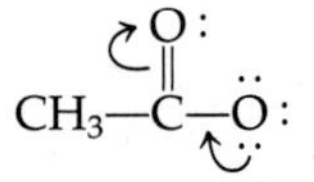

$$CH_3-\overset{\overset{\Large\ddot{O}:}{\|}}{C}-\ddot{O}:$$

Molecular Shapes

1.25 Based on VSEPR theory, what is the expected value of the indicated bond angle in each of the following compounds?
(a) C—C—N in $CH_3-C\equiv N$ (b) C—O—C in CH_3-O-CH_3
(c) C—N—C in $CH_3-NH-CH_3$ (d) C—C—C in $CH_2=C=CH_2$
(e) C—C—C in $CH_3-C\equiv C-H$

1.26 Based on VSEPR theory, what is the expected value of the indicated bond angle in each of the following compounds?
(a) C—O—H in $CH_3-OH_2^+$ (b) C—N—H in $CH_3-NH_3^+$
(c) C—N—C in $(CH_3)_2NH_2^+$ (d) C—O—C in $(CH_3)_2OH^+$

1.27 Based on VSEPR theory, what is the expected value of the indicated bond angle in each of the following compounds?

(a) C—C—O in $CH_3-\overset{\overset{O}{\|}}{C}-H$ (b) C—C—C in $CH_3-CH=CH_2$

(c) O—C—O in $CH_3-\overset{\overset{O}{\|}}{C}-OH$

1.28 Based on VSEPR theory, what is the expected value of the indicated bond angle in each of the following compounds?

(a) C—O—C in $CH_3-\overset{\overset{O}{\|}}{C}-O-CH_3$

(b) O—C—N in $H-\overset{\overset{O}{\|}}{C}-NH_2$

(c) O—C—O in $CH_3-\overset{\overset{O}{\|}}{C}-O-CH_3$

1.29 What is the C—N=N bond angle in Prontosil, an antibiotic?

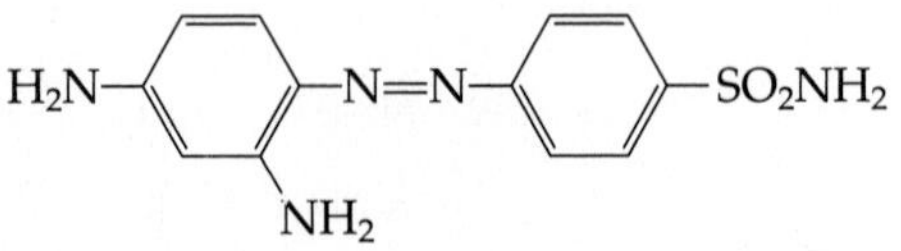

H_2N—(benzene ring with NH_2)—N=N—(benzene ring)—SO_2NH_2

1.30 What is the S—C—S bond angle in dibenthiozole disulfide, a catalyst used in the vulcanization of rubber?

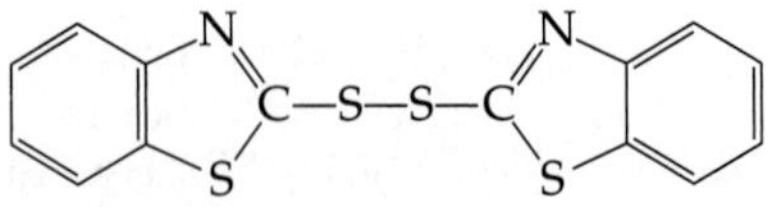

(benzothiazole: N, S)C—S—S—C(benzothiazole: N, S)

Hybridization

1.31 What is the hybridization of each carbon atom in each of the following compounds?

(a) $CH_3{-}\overset{\overset{O}{\|}}{C}{-}H$ (b) $CH_3{-}CH{=}CH_2$ (c) $CH_3{-}\overset{\overset{O}{\|}}{C}{-}OH$

(d) $CH_3{-}\overset{\overset{O}{\|}}{C}{-}O{-}CH_3$

1.32 What is the hybridization of each carbon atom in each of the following compounds?

(a) $H{-}\overset{\overset{O}{\|}}{C}{-}NH{-}CH_3$ (b) $CH_3{-}S{-}CH{=}CH_2$ (c) $CH_3{-}\overset{\overset{N-H}{\|}}{C}{-}CH_3$

(d) $N{\equiv}C{-}CH_2{-}C{\equiv}N$

1.33 Carbocations and carbanions are classes of unstable organic species with a positive and negative charge on carbon, respectively. What is the hybridization of the carbon atom in each ion?

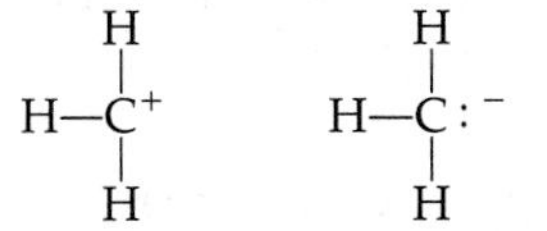

$$H{-}\overset{\overset{H}{|}}{\underset{\underset{H}{|}}{C^+}} \qquad H{-}\overset{\overset{H}{|}}{\underset{\underset{H}{|}}{C}}{:}^-$$

a carbocation   a carbanion

1.34 The structural formula of allene is $CH_2{=}C{=}CH_2$. What is the hybridization of the central carbon atom?

1.35 What is the hybridization of each of the carbon atoms bonded to two oxygen atoms in aspirin?

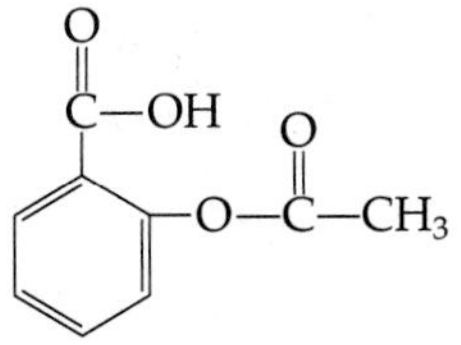

C—OH, O—C(=O)—CH₃ (structure of aspirin)

1.36 What is the hybridization of the carbon atom bonded to the nitrogen atom in L-dopa, a drug that is used in the treatment of Parkinson's disease?

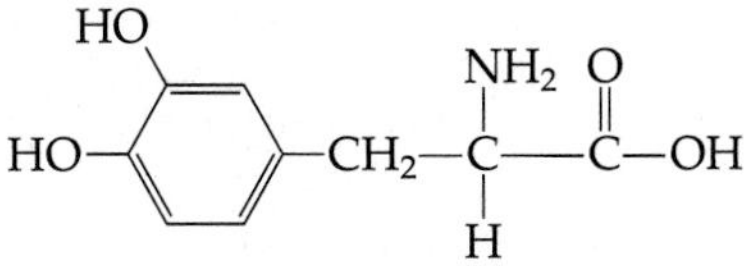

HO, HO, CH₂—C(NH₂)(H)—C(=O)—OH (structure of L-dopa)

Molecular Formulas

1.37 Write the molecular formula for each of the following.

(a) $CH_3{-}CH_2{-}CH_2{-}CH_2{-}CH_3$ (b) $CH_3{-}CH_2{-}CH_2{-}CH_3$

(c) $CH_2{=}CH{-}CH_2{-}CH_3$ (d) $CH_3{-}CH_2{-}C{\equiv}C{-}H$

(e) $CH_3{-}CH_2{-}CH_2{-}CH{=}CH_2$ (f) $CH_3{-}CH_2{-}C{\equiv}C{-}CH_3$

1.38 Write the molecular formula for each of the following.

(a) $CH_3CH_2CH_2CH_2CH_2CH_2CH_2CH_2CH_3$

(b) $CH_3CH_2CH_2CH_2CH_2CH_2CH_3$ (c) $CH_3CH_2C{\equiv}CH$

(d) $CH_3CH_2C{\equiv}CCH_3$ (e) $CH_3CH_2CH_2CH{=}CHCH_3$

(f) $CH_2{=}CHCH_2CH_2CH_3$

1.39 Write the molecular formula for each of the following.
(a) $CH_3—CH_2—CHCl_2$ (b) $CH_3—CCl_2—CH_3$
(c) $Br—CH_2—CH_2—Br$ (d) $CH_3—CHBr—CHBr_2$
(e) $CH_3—CF_2—CH_2F$ (f) $F—CH_2—CHF—CH_2—F$

1.40 Write the molecular formula for each of the following.
(a) $CH_3—CH_2—CH_2—OH$ (b) $CH_3—CH_2—O—CH_2—CH_3$
(c) $CH_3—CH_2—SH$ (d) $CH_3—CH_2—S—CH_3$
(e) $CH_3—CH_2—CH_2—NH_2$ (f) $CH_3—CH_2—NH—CH_3$

Condensed Structural Formulas

1.41 Write condensed structural formulas in which only the bonds to hydrogen are not shown.

```
          H H                 H H H H H
          | |                 | | | | |
(a) Br—C—C—Br     (b) H—C—C—C—C—C—H
          | |                 | | | | |
          H H                 H H H H H

        H H H                    H H H H
        | | |                    | | | |
(c) H—C—C—C—S—H     (d) H—C—C—C—N—H
        | | |                    | | |
        H H H                    H H H
```

1.42 Write condensed structural formulas in which only the bonds to hydrogen are not shown.

```
        H H H H H                    H H H   H
        | | | | |                    | | |   |
(a) H—C—C—C—N—C—H     (b) H—C—C—C—O—C—H
        | | |   |                    | | |   |
        H H H   H                    H H H   H

        H H H H Cl                   H H H H H
        | | | | |                    | | | | |
(c) H—C—C—C—C—C—Cl    (d) H—C—C—N—C—C—H
        | | | | |                    | |   | |
        H H H H Cl                   H H   H H
```

1.43 Write a condensed structural formula in which no bonds are shown for each substance in 1.41.

1.44 Write a condensed structural formula in which no bonds are shown for each substance in 1.42.

1.45 Write a complete structural formula showing all bonds for each of the following condensed formulas.
(a) $CH_3CH_2CH_2CH_3$ (b) $CH_3CH_2CH_2Cl$ (c) $CH_3CHClCH_2CH_3$
(d) $CH_3CH_2CHBrCH_3$ (e) $CH_3CH_2CHBr_2$ (f) $CH_3CBr_2CH_2CH_2CH_3$

1.46 Write a complete structural formula showing all bonds for each of the following condensed formulas.
(a) $CH_3CH_2CH_3$ (b) $CH_3CH_2CHCl_2$ (c) $CH_3CH_2CH_2CH_2SH$
(d) $CH_3CH_2C{\equiv}CCH_3$ (e) $CH_3CH_2OCH_2CH_2CH_3$
(f) $CH_3CH_2CH_2C{\equiv}CH$

Bond-Line Structures

1.47 What is the molecular formula for each of the following bond-line representations?

(a)

(b)

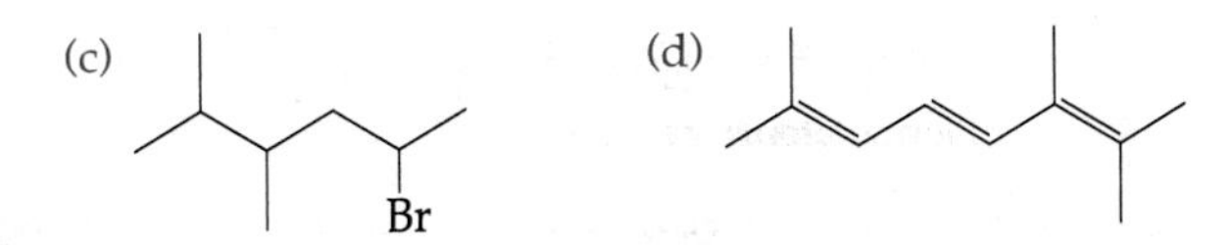

1.48 What is the molecular formula for each of the following bond-line representations?

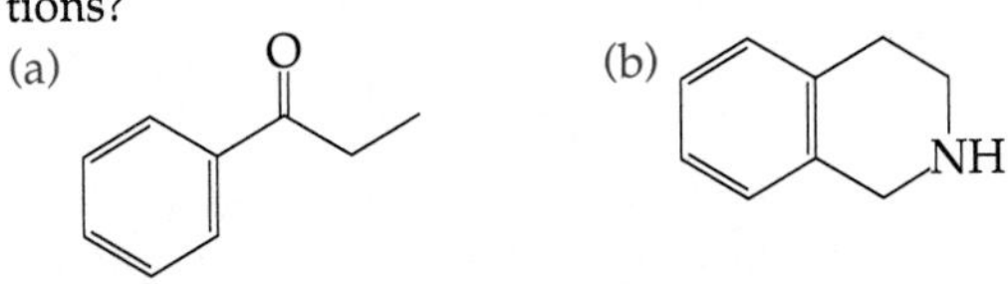

1.49 What is the molecular formula for each of the following bond-line representations?

(a) a scent marker of the red fox

(b) a compound responsible for the odor of the iris

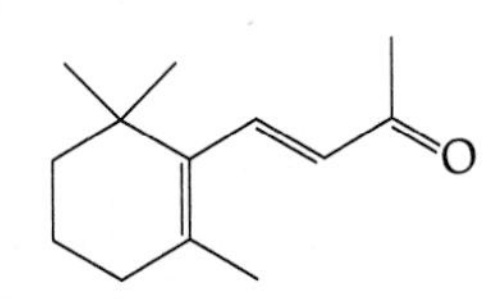

(c) a defense pheromone of some ants

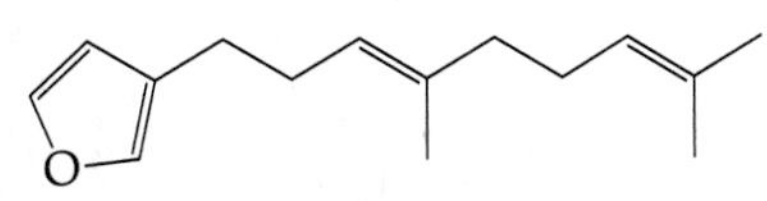

1.50 What is the molecular formula for each of the following bond-line representations?

(a) a compound found in clover and grasses

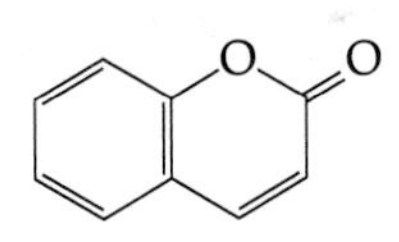

(b) an oil found in citrus fruits

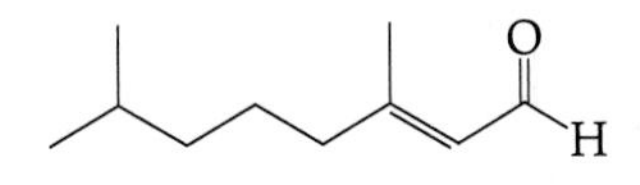

(c) a male sex hormone

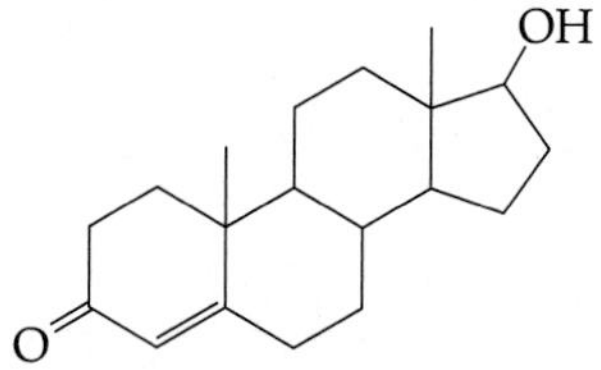

OH

O

Functional Groups

1.51 Identify the functional groups contained in each of the following structures.
(a) caprolactam, a compound used to produce a type of nylon

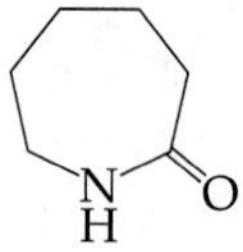

N
H
O

(b) civetone, a compound in the scent gland of the civet cat

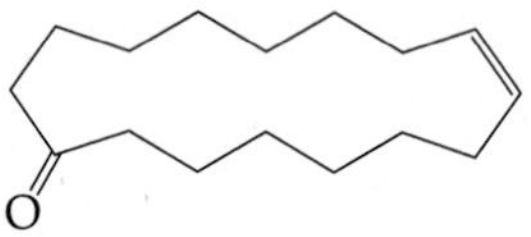

O

(c) DEET, the active ingredient in some insect repellents

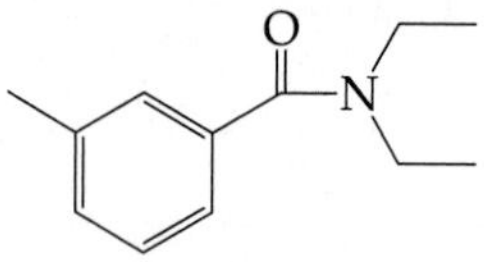

O
N

1.52 Identify the oxygen-containing functional groups in each of the following compounds.
(a) isoimpinellin, a carcinogen found in diseased celery

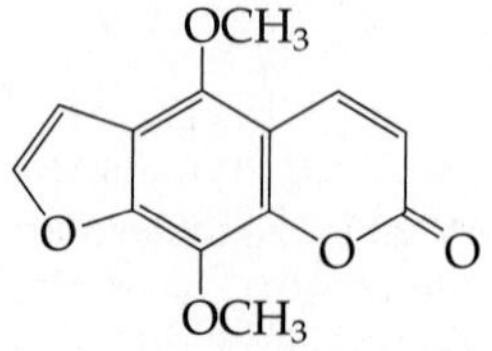

OCH_3

O O O

OCH_3

(b) aflatoxin B_1, a carcinogen found in moldy foods

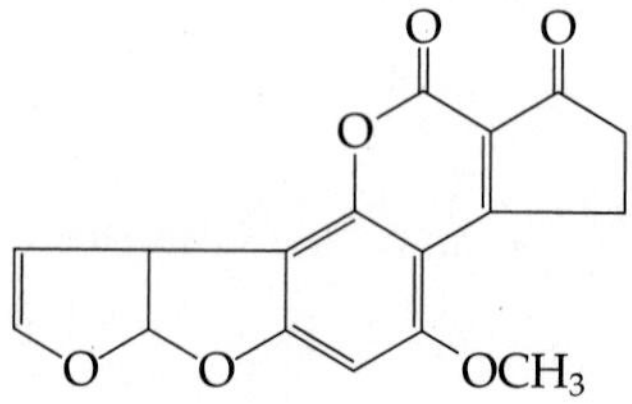

O O

O

O O OCH_3

(c) penicillin G, an antibiotic first isolated from a mold

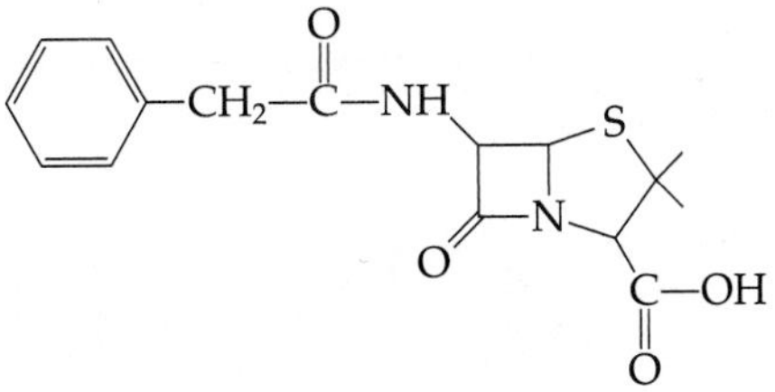

O
—CH_2—C—NH
S
N
O
C—OH
O

Isomerism

1.53 Indicate whether the following pairs of structures are isomers or different representations of the same compound.

(a) 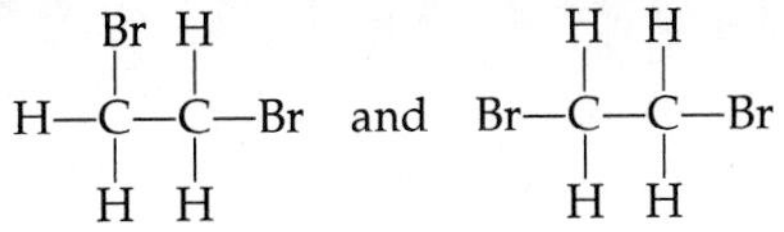 $\begin{array}{c}\text{Br}\;\;\text{H}\\ \text{H—C—C—Br}\\ \text{H}\;\;\text{H}\end{array}$ and $\begin{array}{c}\text{H}\;\;\text{H}\\ \text{Br—C—C—Br}\\ \text{H}\;\;\text{H}\end{array}$

(b) $\begin{array}{l}CH_3\text{—}CH_2\\ \qquad\; |\\ \qquad CH_2\text{—}Cl\end{array}$ and CH_3—CH_2—CH_2—Cl

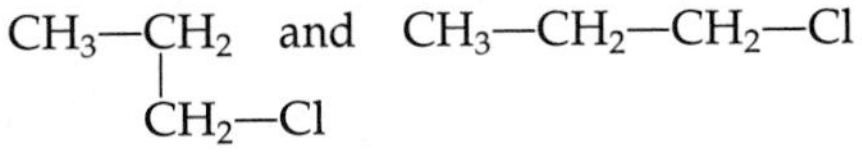

(c) $\begin{array}{l}CH_3\text{—}CH\text{—}Cl\\ \qquad\; |\\ \qquad CH_3\end{array}$ and CH_3—CH_2—CH_2—Cl

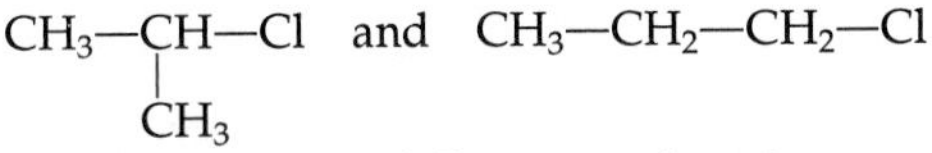

1.54 Indicate whether the following pairs of structures are isomers or different representations of the same compound.

(a) 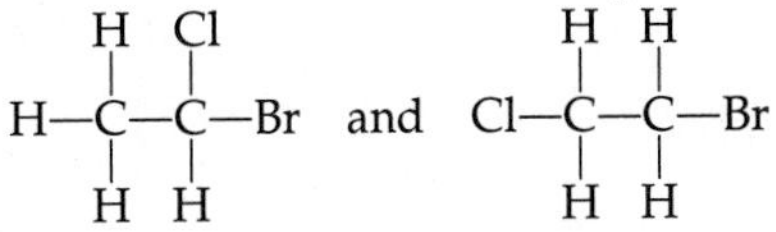 $\begin{array}{c}\text{H}\;\;\text{Cl}\\ \text{H—C—C—Br}\\ \text{H}\;\;\text{H}\end{array}$ and $\begin{array}{c}\text{H}\;\;\text{H}\\ \text{Cl—C—C—Br}\\ \text{H}\;\;\text{H}\end{array}$

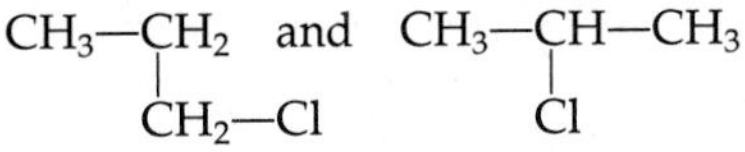

(b) $\begin{array}{l}CH_3\text{—}CH_2\\ \qquad\; |\\ \qquad CH_2\text{—}Cl\end{array}$ and $\begin{array}{l}CH_3\text{—}CH\text{—}CH_3\\ \qquad\;\; |\\ \qquad\;\; Cl\end{array}$

(c) $\begin{array}{l}CH_3\text{—}CH\text{—}CH_2\text{—}Cl\\ \qquad\; |\\ \qquad CH_3\end{array}$ and CH_3—CH_2—CH_2—CH_2—Cl

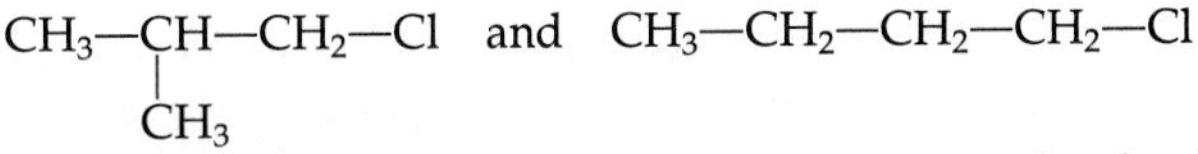

1.55 There are two isomers for each of the following molecular formulas. Draw their structural formulas.
(a) $C_2H_4Br_2$ (b) C_2H_6O (c) C_2H_4BrCl (d) C_3H_7Cl (e) C_2H_7N
(f) $C_2H_3Br_3$

1.56 There are three isomers for each of the following molecular formulas. Draw their structural formulas.
(a) $C_2H_3Br_2Cl$ (b) C_3H_8O (c) C_3H_8S

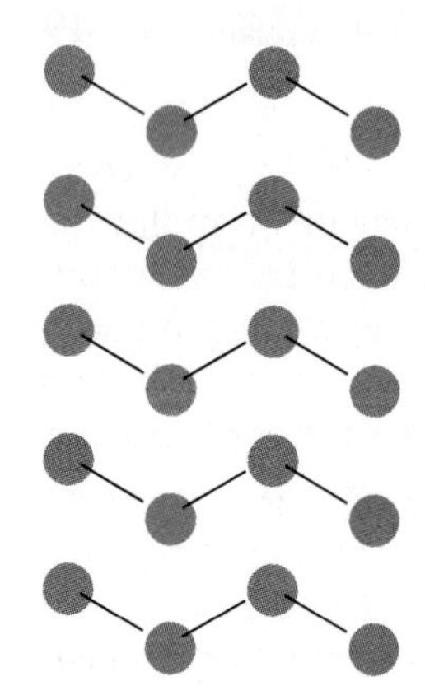

2

PROPERTIES OF ORGANIC COMPOUNDS

2.1 STRUCTURE AND PHYSICAL PROPERTIES

Approximately nine million organic compounds are currently known. Each has unique physical and chemical properties. Thus, we might expect that understanding the relationships between the structure of compounds and their physical properties such as melting point, boiling point, and solubility could be handled only by computers—certainly not by human beings. Yet, we can make reasonable guesses about the physical properties of a compound based on its structure because all organic compounds belong to a relatively few classes of substances characterized by their functional groups. These structural units within a molecule are largely responsible for its properties. These properties reflect the attractive **intermolecular** (between molecules) forces characteristic of the bonding in the functional groups.

Weak intermolecular forces between individual molecules hold them close together in the liquid and solid states. These forces, which are usually smaller than 10 kcal/mole, are smaller than the typical bond energies (80–100 kcal/mole) that hold the atoms together in a molecule. Intermolecular forces are of three types: **dipole–dipole forces, London forces,** and **hydrogen–bonding** forces. To understand these forces, we must understand chemical bonding and know how to predict molecular geometry.

Dipole–Dipole Forces

The bonding electrons in polar covalent bonds are not shared equally, and a bond moment results. However, a molecule may be polar or nonpolar depending on its geometry. For example, tetrachloromethane (carbon tetrachloride, CCl_4) has polar C—Cl bonds, but the tetrahedral arrangement of the four bonds about the central carbon atom causes the individual bond moments to cancel. In contrast, dichloromethane (methylene chloride, CH_2Cl_2) is a polar molecule with a net polarity away from the partially positive carbon atom toward the partially negative chlorine atoms.

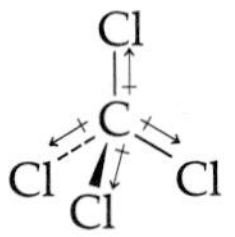

The bond moments cancel and no net polarity results.

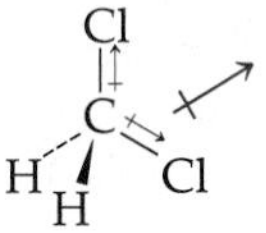

The bond moments do not cancel and a net polarity results.

Polar molecules have a negative "end" and a positive "end" and tend to associate closely. The positive end of one molecule attracts the negative end of another molecule (Figure 2.1). The physical properties of polar molecules reflect this association. An increased association between molecules decreases their vapor pressure, which in turn results in a higher boiling point because more energy is required to vaporize the molecules.

The molecular weights of ethanal (acetaldehyde, CH_3CHO) and propane ($CH_3CH_2CH_3$) are similar, but ethanal boils at a higher temperature than propane. Ethanal contains a polar carbonyl group, whereas propane is a nonpolar molecule. The higher boiling point of ethanal results from the dipole–dipole interaction between ethanal molecules.

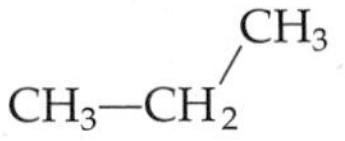

propane (nonpolar) (bp −42 °C)

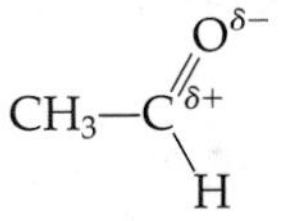

ethanal (polar) (bp 20 °C)

London Forces

Intermolecular forces exist even between nonpolar molecules. In a nonpolar molecule, the electrons, on average, are distributed uniformly about the molecule. However, the electrons at some instant may be distributed closer to one atom in a molecule or toward one side of a molecule. At that instant, a **temporary dipole** is present (Figure 2.2). A temporary dipole exerts an influ-

FIGURE 2.1
Intermolecular Forces in Polar Molecules
The carbon–chlorine bond in chloromethane is polar. As a consequence, the positively charged carbon atom in one molecule is attracted to the negatively charged chlorine atom in a neighboring molecule. The resulting intermolecular forces cause an increase in the boiling point compared to that of a nonpolar molecule of similar molecular weight.

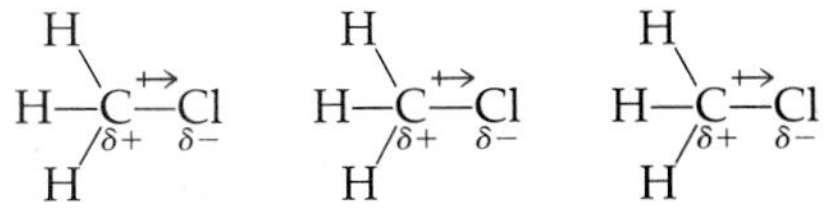

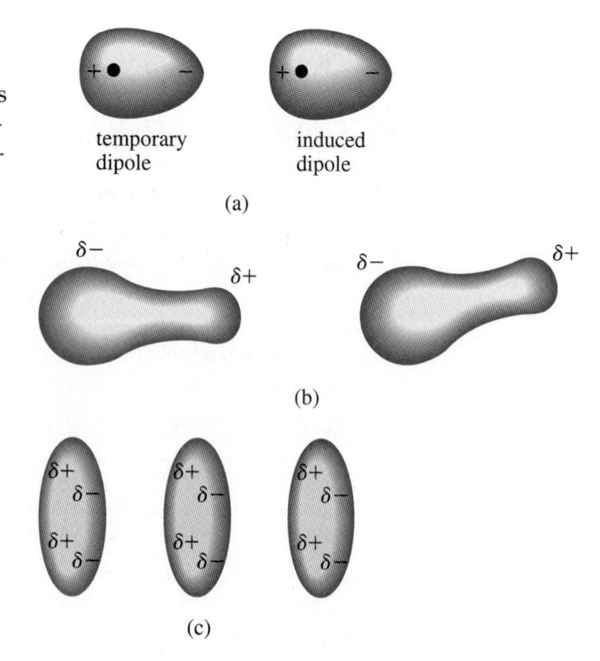

FIGURE 2.2
London Forces in Nonpolar Molecules
In (a) the electron distribution in an atom becomes distorted and a temporary dipole results. An adjacent atom then has a dipole induced by the movement of electrons, resulting in a net attraction between atoms.
In a molecule represented by (b), the electrons are distorted toward one end of the molecule.
In a molecule represented by (c), the electrons are distorted toward one side of the molecule.

ence on nearby molecules; it polarizes neighboring molecules and results in an **induced dipole.** The resultant attractive forces between a temporary dipole and an induced dipole are called **London forces.** The ease with which an electron cloud is distorted by nearby charges or dipoles is called **polarizability.**

The attractive forces between the temporary dipoles in otherwise nonpolar molecules are small and have a short lifetime at any given site in the sample. However, the cumulative effect of these attractive forces holds a collection of molecules together in the condensed state. The attractive forces can be overcome when sufficient heat is added, and a solid melts or a liquid boils.

The strength of London forces depends on the number of electrons in a molecule and on the types of atoms containing those electrons. Electrons that are far from atomic nuclei are more easily distorted or polarizable than electrons that are closer to atomic nuclei. For example, the polarizability of the halogens increases in the order F < Cl < Br < I.

The boiling point of bromoethane is higher than the boiling point of chloroethane. Because a C—Cl bond is more polar than a C—Br bond we might have expected the more polar chloroethane to have a higher boiling point than bromoethane. But polarity isn't everything. First, the molecular weights of the two compounds are substantially different. Second, the electrons of the bromine atom are more polarizable than the electrons of the chlorine atom. Thus, the order of boiling points reflects the relative polarizability of the molecules and the larger London attractive forces of bromoethane.

Br is more polarizable than Cl.

	$CH_3{-}CH_2{-}Br$	$CH_3{-}CH_2{-}Cl$
bp	38.4 °C	12.3 °C
mol wt	109 amu	64.5 amu

Even when the types of atoms in molecules are the same, London forces differ because either the molecular weights or the shapes of the compounds are different. For example, the boiling points of pentane and hexane are 36 and 69 °C, respectively. These two nonpolar molecules contain the same types of atoms, but the numbers of atoms differ. Hexane is a larger molecule whose chain has more surface area to interact with neighboring molecules. As a result, the London forces are stronger in hexane than in pentane. This increased attraction between molecules decreases the vapor pressure of hexane, and its boiling point is higher than the boiling point of pentane.

$CH_3CH_2CH_2CH_2CH_3$ pentane (bp 36 °C)

$CH_3CH_2CH_2CH_2CH_2CH_3$ hexane (bp 69 °C)

London forces also depend on molecular shape. For example, the boiling point of isobutane is lower than that of butane. Isobutane is more spherical, and it therefore has less surface area than the more cylindrical butane molecule (Figure 2.3). As a consequence, there is less effective contact between isobutane molecules, and the London forces are smaller and less effective.

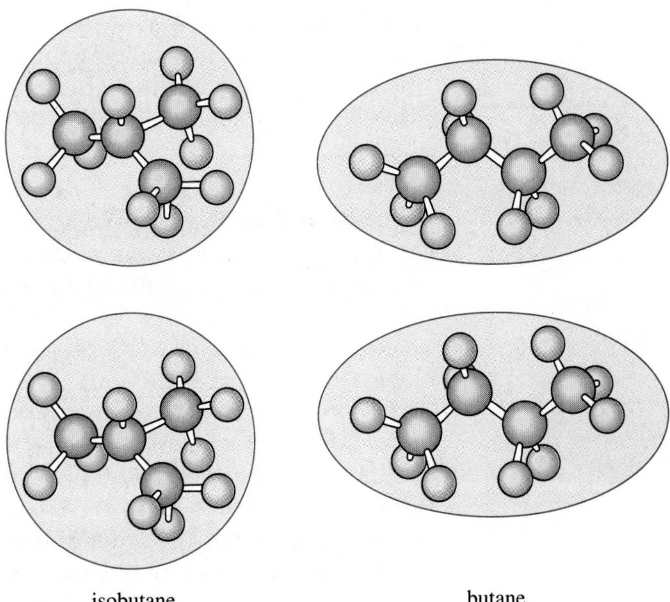

FIGURE 2.3
London Forces and Molecular Shape
The shape of isobutane is nearly spherical, and the points of contact between neighboring molecules are limited. Butane is an extended molecule with an ellipsoidal shape, and neighboring molecules are in closer contact. The attractive forces are larger in butane than in isobutane.

$$\underset{\text{isobutane (bp } -12\ ^\circ\text{C)}}{CH_3{-}\underset{\displaystyle CH_3}{\underset{|}{CH}}{-}CH_3} \qquad \underset{\text{butane (bp } -1\ ^\circ\text{C)}}{CH_3{-}CH_2{-}CH_2{-}CH_3}$$

EXAMPLE 2.1 Based on the boiling points of pentane and hexane, predict the boiling point of heptane $CH_3CH_2CH_2CH_2CH_2CH_2CH_3$.

Solution The boiling points of pentane and hexane are 36 and 69 °C, respectively, a difference of 33 °C. The boiling point of heptane should be higher than that of hexane. The effect of the extra $-CH_2-$ group on the boiling point could be predicted to be an additional 33 °C by assuming a linear relationship between molecular weight and boiling point. The predicted boiling point would be 102 °C. The actual boiling point is 98 °C.

EXAMPLE 2.2 The boiling points of CCl_4 and $CHCl_3$ are 77 and 62 °C, respectively. Which compound is the more polar? Is the polarity consistent with the boiling points?

Solution The CCl_4 molecule is nonpolar, and the intermolecular forces would be expected to be smaller than those of a polar molecule of comparable molecule weight. The $CHCl_3$ molecule is polar and should have higher intermolecular forces than a nonpolar molecule of comparable molecule weight. However, the two molecules do not have comparable molecular weights, and thus the boiling points cannot be compared easily. Apparently, the London forces of the higher molecular weight nonpolar CCl_4 molecule exceed the dipole–dipole forces of the polar $CHCl_3$ molecule.

Hydrogen-Bonding Forces

Compounds containing hydrogen bonded to fluorine, oxygen, and nitrogen, such as HF, H_2O, and NH_3, interact by very strong intermolecular forces. This interaction is called a **hydrogen bond.**

hydrogen bonds

$$H{-}\ddot{\underset{..}{F}}{:}\cdots H{-}\ddot{\underset{..}{F}}{:} \qquad H{-}\underset{..}{\overset{H}{O}}{:}\cdots H{-}\underset{..}{\overset{H}{O}}{:} \qquad H{-}\underset{H}{\overset{H}{N}}{:}\cdots H{-}\underset{H}{\overset{H}{N}}{:}$$

The hydrogen atom in a polar covalent bond to an electronegative element has a partial positive charge. As a result there is an attraction between the hydrogen atom and the lone-pair electrons of another molecule. The strengths of hydrogen bonds are about 5 kcal/mole, whereas the energy of a covalent bond such as C—H or O—H is about 100 kcal/mole.

The O—H or N—H groups in organic compounds can form hydrogen bonds. For example, the physical properties of alcohols and amines are strongly affected by hydrogen bonds. The boiling point of 1-butanol, an alcohol, is substantially higher than the boiling point of diethyl ether, which has the same molecular weight.

	$CH_3—CH_2—CH_2—CH_2—OH$	$CH_3—CH_2—O—CH_2—CH_3$
bp	117.7 °C	34.6 °C
mol wt	74 amu	74 amu

Because the numbers of atoms are the same and the shapes of the molecules are similar, the boiling point difference cannot be due to differences in London forces. Both molecules have polar bonds, and the dipole–dipole forces should be similar. The high boiling point of the 1-butanol results from the hydrogen bonding of the hydroxyl groups of neighboring molecules.

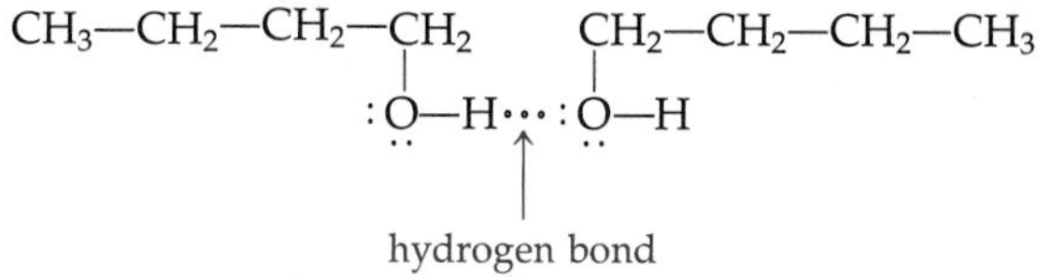

$CH_3—CH_2—CH_2—CH_2$ (bonded to) :Ö—H···:Ö—H (bonded to) $CH_2—CH_2—CH_2—CH_3$

hydrogen bond

EXAMPLE 2.3 The boiling point of 1,2-ethanediol (ethylene glycol), which is used as antifreeze, is 190 °C. Why is the boiling point higher than that of 1-propanol (97 °C)?

$HO—CH_2—CH_2—OH$	$CH_3—CH_2—CH_2—OH$
1,2-ethanediol	1-propanol

Solution Both molecules have similar molecular weights and should have comparable London forces. However, 1,2-ethanediol has two hydroxyl groups per molecule, compared to only one per molecule in 1-propanol. As a consequence, liquid 1,2-ethanediol has twice as many hydrogen bonds. The increased number of hydrogen bonds decreases the vapor pressure of 1,2-ethanediol and leads to a higher boiling point.

Solubility

A maxim of the chemistry laboratory is that "like dissolves like." This generalization is reasonable because molecules of solute that are similar to molecules of solvent should interact by similar intermolecular attractive forces. Carbon tetrachloride, CCl_4, a nonpolar substance, does not dissolve ionic

Water-Soluble and Fat-Soluble Vitamins

The different solubilities of two vitamins illustrate the maxim that "like dissolves like." Vitamin C is a water-soluble vitamin, whereas vitamin A is a fat-soluble vitamin.

The vitamin C molecule is small and has many —OH groups that can form hydrogen bonds to water. The vitamin A molecule, with the exception of one —OH group, is nonpolar. The vitamin A molecule is not "like" water and has a very low solubility in water. On the other hand, vitamin A, which structurally resembles the carbon compounds in fats, dissolves in fatty tissue.

Because it is water-soluble, vitamin C is not stored in the body and should be taken in as part of one's daily diet. Unneeded vitamin C is eliminated from the body. Fat-soluble vitamins are stored by the body for future use. If excessive quantities of fat-soluble vitamins are consumed in vitamin supplements, illness can result. The condition is known as hypervitaminosis.

vitamin C

vitamin A

compounds such as sodium chloride. However, this nonpolar solvent is a good solvent for nonpolar compounds such as fats and waxes. Water, which is quite polar, is a good solvent for ionic compounds and substances that can produce ions in water. Water dissolves a limited number of low molecular weight organic compounds if they are sufficiently polar or can form hydrogen bonds with water.

Liquids that dissolve in each other in all proportions are said to be **miscible.** Liquids that do not dissolve in each other are **immiscible.** Immiscible liquids form separate layers in a container. For example, ethyl alcohol is miscible with water, but carbon tetrachloride is immiscible with water. The miscibility of ethyl alcohol with water is explained by its structure.

```
   H  H
   |  |  ..
H—C—C—O:
   |  |  |
   H  H  H
```

Ethyl alcohol, like water, has an —OH group. Ethyl alcohol is polar, as is water. The nonbonded electron pairs on the oxygen atom in ethyl alcohol and the hydroxyl hydrogen atom form hydrogen bonds with water and thus make it soluble.

2.2 STRUCTURE AND CHEMICAL CHANGE

Predicting the chemical properties of compounds, that is, their reactivity with other compounds, is potentially an even more overwhelming problem than that of predicting physical properties. Each chemical reaction is a unique event. Bonds in the reactants are broken, and new bonds are formed in the products. The number of known and potential reactions among the dozens of functional groups in the millions of organic compounds is astronomically large. However, we can understand these myriad reactions by learning a few fundamental concepts that underlie all organic chemical reactions. In other words, certain patterns of chemical behavior unify many facts into a few classes of chemical reactions.

In Sections 2.3 and 2.4 we will review acid–base and oxidation–reduction reactions and illustrate how these concepts apply to organic chemical reactions. In Section 2.5, several other classes of organic reactions will be briefly illustrated.

Describing Chemical Reactions

All chemical reactions are reversible to some degree, and some reactions result in an equilibrium mixture containing substantial amounts of reactants as well as products. If the amount of reactant remaining is less than about 0.1%, the reaction is said to have gone to completion.

In industry, experimental conditions are sought that convert as much reactant to product as possible. Not only is the inefficient conversion of chemicals costly, but the unwanted material must be removed to purify the product. Impure materials cannot be tolerated for many products, especially those for human consumption.

The study of the speed or rate of a chemical reaction is called **kinetics.** The kinetics of chemical reactions is of concern in industry, where it is important to understand how to form a desired chemical product rapidly and in preference to other products. A reaction may be too slow to be economically practical, or a reaction may be so fast that it is dangerous and difficult to control. In biological systems, virtually all reactions produce only a single product, and the reactions are very fast.

The study of the kinetics of a reaction helps us determine the **mechanism** of the process. A mechanism details the order of bond cleavage and bond formation that occurs during the reaction. This information establishes general guidelines, which allow chemists to extrapolate a few observations on selected reactions to many other reactions. The prediction of reaction mechanisms for new reactions, based on other reactions with well-established mechanisms, is a powerful tool in understanding organic chemistry.

2.3 ACID–BASE REACTIONS

According to the Brønsted–Lowry theory an **acid** is a substance that can donate a proton (H^+); a **base** is a substance that can accept a proton. For example, when gaseous hydrogen chloride dissolves in water, virtually all of the HCl molecules transfer a proton to water, and a solution of hydronium ions and chloride ions results.

$$H_2\ddot{O}: + H{-}\ddot{Cl}: \longrightarrow H_3O^+ + :\ddot{Cl}:^-$$

This bond is cleaved. This bond is formed.

This reaction is illustrated by curved arrows. This convention was also used in the previous chapter to write resonance forms of a single molecule. A similar notation that shows the direction of electron flow is used to depict the sequence of events in chemical reactions. Electrons are pictured as flowing from the start of the arrow toward the arrowhead. The nonbonded pair of electrons of the oxygen atom forms a bond to the hydrogen atom, and the bonded pair of electrons in HCl is fully transferred to the chlorine atom.

The most common base is the hydroxide ion, which exists as an ion in compounds such as NaOH, KOH, and $Ca(OH)_2$. Hydroxide ion is a base because it can accept a proton from an acid such as a hydronium ion. Ammonia is also a base because it can accept a proton from an acid. Ammonia has a nonbonded pair of electrons on its nitrogen atom that forms a bond to the hydrogen atom of H_3O^+. One of the bonded hydrogen atoms in H_3O^+ is transferred to the nitrogen atom in this process. A curved arrow shows the movement of the pair of electrons.

$$H_3N: + H{-}\ddot{O}^+{-}H\,(H) \dashrightarrow H_3N^+{-}H + :\ddot{O}{-}H\,(H)$$

This bond is cleaved. This bond is formed.

Organic acids and bases behave similarly. Carboxylic acids contain a carboxyl group, which can donate a proton to a base such as water. Using the curved-arrow formalism to depict electron movement shows how the nonbonded electron pair of the water molecule forms a bond with the hydrogen atom of the carboxyl group.

$$CH_3{-}C(=\ddot{O}:){-}\ddot{O}{-}H + :\ddot{O}(H){-}H \longrightarrow CH_3{-}C(=\ddot{O}:){-}\ddot{O}:^- + H{-}\ddot{O}^+(H){-}H$$

acetic acid (an acid)	water (a base)	acetate ion (a conjugate base)	hydronium ion (a conjugate acid)

Amines are a class of organic bases whose acid–base chemistry is like that of ammonia. Thus, methylamine behaves as a base because the nonbonded electron pair of the nitrogen atom can accept a proton from an acid such as the hydronium ion.

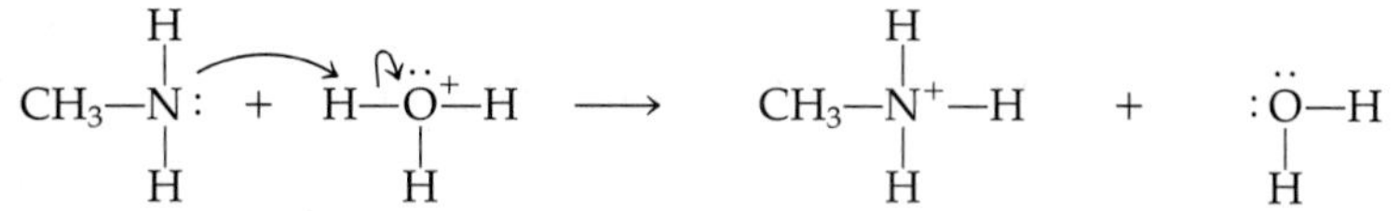

$$CH_3\text{—}\ddot{N}H_2 + H\text{—}\ddot{O}^+H\text{—}H \longrightarrow CH_3\text{—}NH_2^+\text{—}H + :\ddot{O}H\text{—}H$$

methylamine (a base) | hydronium ion (an acid) | methylammonium ion (a conjugate acid) | water (a conjugate base)

When an acid transfers a proton to a base, another base and acid are produced. The acid loses a proton and becomes a **conjugate base.** For example, the conjugate base of acetic acid is the acetate ion. When a base accepts a proton, the substance formed is a **conjugate acid.** Thus, the conjugate acid of methylamine is the methylammonium ion.

EXAMPLE 2.4 Assume that methanol behaves as an acid. What is the conjugate base of methanol?

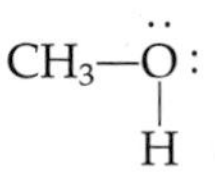

$$CH_3\text{—}\ddot{O}\text{:—}H$$

Solution Loss of a proton (H^+) from an electrically neutral acid results in a conjugate base with a negative charge. The electron pair of the O—H bond remains with the oxygen atom.

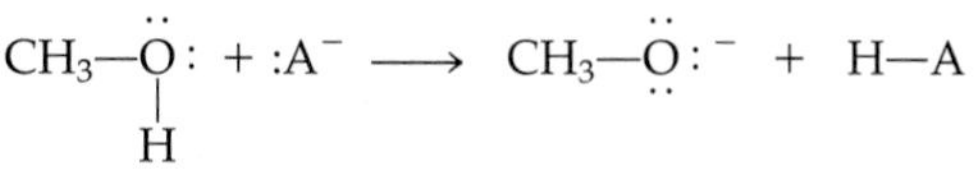

$$CH_3\text{—}\ddot{O}\text{:—}H + :A^- \longrightarrow CH_3\text{—}\ddot{O}:^- + H\text{—}A$$

acid | base | conjugate base | conjugate acid

Lewis Acids and Bases

Some chemical reactions do not occur with proton transfer, but are also regarded as acid–base reactions. These reactions can be explained in terms of Lewis acids and Lewis bases. The Lewis concept of acids and bases focuses on the behavior of an electron pair. A **Lewis acid** is a substance that accepts an electron pair; a **Lewis base** is a substance that donates an electron pair. Thus HCl, which is an acid in the Brønsted–Lowry sense, is also a Lewis acid because it contains a proton that can accept an electron pair. Similarly, ammonia is a Lewis base as well as a Brønsted–Lowry base because it can donate an electron pair. However, the Lewis classification of acids and bases is more extensive because it is not restricted to protons.

Boron trifluoride (BF_3) and aluminum trichloride ($AlCl_3$) are two common Lewis acids encountered in organic chemical reactions. The central atom in each has only six electrons in its valence shell, and each can thus accept an electron pair from a Lewis base.

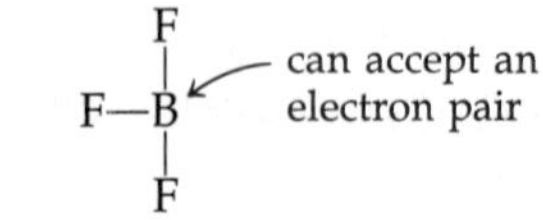

$$\mathrm{BF_3}\ \text{(can accept an electron pair)} \qquad \mathrm{AlCl_3}\ \text{(can accept an electron pair)}$$

boron trifluoride 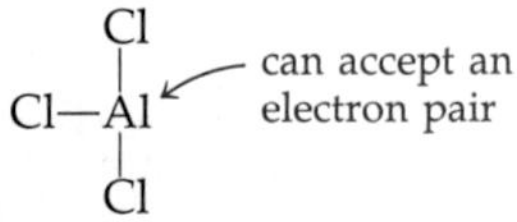 — aluminum trichloride

Other Lewis acids include transition metal compounds such as $FeBr_3$ that react by accepting a pair of electrons. For example, $FeBr_3$ reacts with molecular bromine to accept a bromide ion via a pair of electrons. In this reaction, $FeBr_3$ behaves as a Lewis acid, and bromine behaves as a Lewis base.

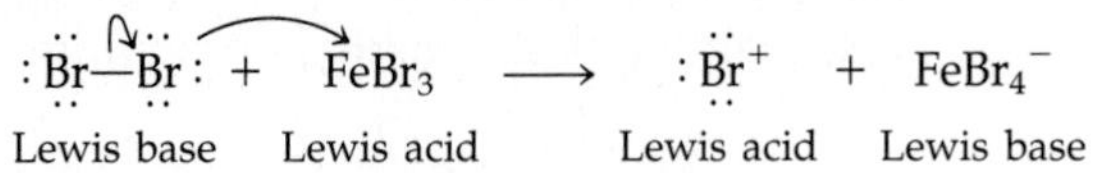

$$:\ddot{\mathrm{Br}}-\ddot{\mathrm{Br}}: + \mathrm{FeBr_3} \longrightarrow :\ddot{\mathrm{Br}}^+ + \mathrm{FeBr_4}^-$$

Lewis base — Lewis acid — Lewis acid — Lewis base

Many organic compounds that contain oxygen and nitrogen atoms can act as Lewis bases because these atoms have nonbonded electrons that can react with Lewis acids. For example, ethers react with boron trifluoride to give a product with a coordinate covalent bond between boron and oxygen.

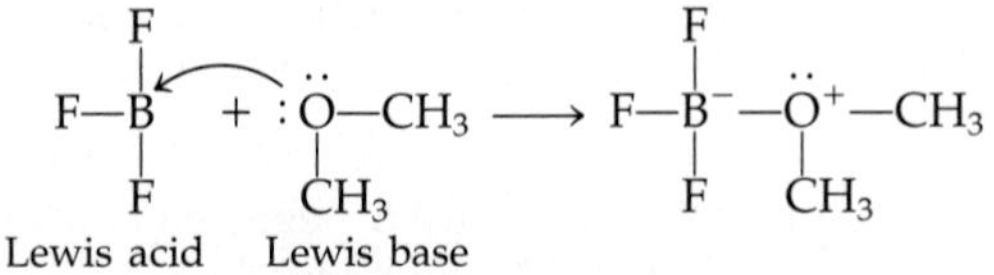

$$\mathrm{BF_3} + :\ddot{\mathrm{O}}(\mathrm{CH_3})-\mathrm{CH_3} \longrightarrow \mathrm{F_3B^-}-\ddot{\mathrm{O}}^+(\mathrm{CH_3})-\mathrm{CH_3}$$

Lewis acid — Lewis base

Another somewhat less obvious example of a Lewis acid–base reaction is the reaction of the cyanide ion with a carbonyl group of ethanal (acetaldehyde). The cyanide ion is a Lewis base, and the carbonyl carbon atom of ethanal is a Lewis acid.

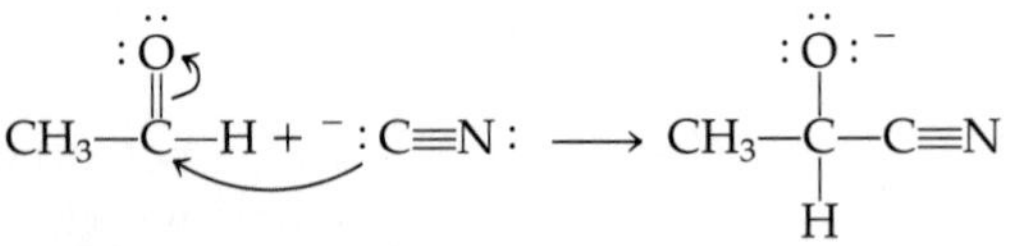

$$\mathrm{CH_3}-\mathrm{C}(=\ddot{\mathrm{O}}:)-\mathrm{H} + {}^-:\mathrm{C}\equiv\mathrm{N}: \longrightarrow \mathrm{CH_3}-\mathrm{C}(:\ddot{\mathrm{O}}:^-)(\mathrm{H})-\mathrm{C}\equiv\mathrm{N}$$

EXAMPLE 2.5 Consider the reaction of a hydrogen ion with ethylene to give a charged intermediate called a carbocation. Classify the reactants according to Lewis acid–base nomenclature.

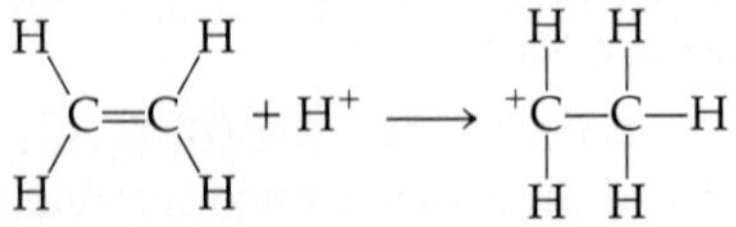

$$\mathrm{H_2C}=\mathrm{CH_2} + \mathrm{H^+} \longrightarrow {}^+\mathrm{CH_2}-\mathrm{CH_2}-\mathrm{H}$$

Solution When a substance acts as a Lewis acid, it accepts an electron pair from a Lewis base. The hydrogen ion does not have any electrons, and it accepts an electron pair from the π bond of ethylene to form a carbon–hydrogen bond. Thus, the hydrogen ion acts as a Lewis acid, and ethylene acts as the Lewis base.

2.4 OXIDATION–REDUCTION REACTIONS

Oxidation is the loss of electrons by a substance or an increase in its oxidation number. **Reduction** is the gain of electrons by a substance or a decrease in its oxidation number. From a slightly different point of view, it follows that when a substance is reduced, it gains the electrons from a substance that becomes oxidized.

The close and necessary relationship between oxidation and reduction is emphasized further in the terms *oxidizing agent* and *reducing agent*. In an oxidation–reduction reaction, the substance that is reduced is the **oxidizing agent** because, by gaining electrons, it causes oxidation of another substance. The substance that is oxidized is called the **reducing agent** because, by losing its electrons, it causes the reduction of another substance.

In organic chemistry, oxidation numbers are not as easily assigned as in inorganic chemistry. Fortunately, we can decide on the change in the oxidation state of a compound in many chemical reactions by accounting for the number of hydrogen atoms or oxygen atoms gained or lost. The oxidation state of a molecule increases (oxidation) if its hydrogen content decreases or its oxygen content increases. Conversely, the oxidation state of a molecule decreases (reduction) if its hydrogen content increases or its oxygen content decreases. For example, the reaction of methanol (CH_3OH) to produce methanal (formaldehyde, CH_2O) is an oxidation because methanol loses two hydrogen atoms. Further reaction of methanal to produce methanoic acid (formic acid) occurs with an increase in the oxygen content and is also an oxidation process.

$$H-\underset{H}{\overset{H}{\underset{|}{\overset{|}{C}}}}-\ddot{\underset{..}{O}}-H \xrightarrow{[O]} H-\overset{\ddot{O}:}{\overset{\|}{C}}-H \xrightarrow{[O]} H-\overset{\ddot{O}:}{\overset{\|}{C}}-\ddot{\underset{..}{O}}-H$$

methanol — methanal (formaldehyde) — methanoic acid (formic acid)

In these reactions, the symbol [O] represents an unspecified oxidizing agent. Note that the equations are not balanced. The focus in organic chemistry is on the conversion of the organic compound. Oxidizing agents such as potassium dichromate, which might be used in the above oxidation reaction, are seldom balanced and are placed above the reaction arrow.

$$CH_3-\underset{CH_3}{\overset{H}{\underset{|}{\overset{|}{C}}}}-CH_2-OH \xrightarrow{K_2Cr_2O_7} CH_3-\underset{CH_3}{\overset{H}{\underset{|}{\overset{|}{C}}}}-\overset{O}{\overset{\|}{C}}-OH$$

P-450—The Liver and Oxidative Transformations

Foreign compounds (xenobiotics) and many common drugs are eliminated from the body by metabolic reactions. Water-soluble substances are easily excreted, but most organic compounds are nonpolar and are **lipid-soluble,** that is, they dissolve in the fatty components of cells. If **lipophilic** (lipid-loving) xenobiotics or drugs were not eliminated, they would accumulate and an organism would eventually become a living (and soon dead) "toxic dump."

Organisms ordinarily transform lipophilic substances into more polar water-soluble products that can be excreted. The metabolic reactions usually are oxidation reactions. Although the process is designed by nature to detoxify material and eliminate it, some pharmaceutical products are converted into active drugs by the same process. These drugs, known as **prodrugs,** are often oxidized into pharmacologically active products. Nevertheless, even these products must be further metabolized to rid the body eventually of substances that are not part of any ordinary life processes.

The liver is the most important organ for the oxidation of xenobiotics and drugs. Other tissues have some metabolizing capabilities, but they are limited to reactions of a small number of substrates. The oxidation of compounds, represented as R—H, in the liver requires molecular oxygen and nicotinamide adenosine dinucleotide phosphate, NAPDH (see figure). One of the oxygen atoms is incorporated in the substrate and the other oxygen atom in water.

$$\text{R—H} + \text{NADPH} + O_2 + H^+ \longrightarrow \text{R—OH} + \text{NADP}^+ + H_2O$$

The enzyme responsible for catalyzing the reaction is cytochrome P-450, an iron-containing heme-protein (see figure). Heme contains an iron atom bound to four nitrogen atoms. The iron is coordinatively bonded to oxygen and to the surrounding protein by a nitrogen atom of the amino acid lysine. Substrates are accommodated at a site within the protein that allows them to be oxidized by the oxygen bonded to iron.

Most enzymes are quite specific in their actions, but cytochrome P-450 is not a single species and has greater versatility. It exists in different forms with different three-dimensional structures—tertiary structures (Chapter 15)—as a result of different amino acids in the protein chain. These structural differences account for the efficiency of oxidation of a variety of different compounds.

In this reaction, the conversion of an alcohol into a carboxylic acid, an oxidation has occurred, and the substance above the arrow must be an oxidizing agent.

The conversion of a carbon–carbon triple bond into a double bond and finally into a single bond involves reduction because the hydrogen content increases in each step. The symbol [H] represents an unspecified reducing agent, such as hydrogen gas in the presence of a platinum catalyst.

$$\text{H—C≡C—H} \xrightarrow{[H]} \text{H}_2\text{C=CH}_2 \xrightarrow{[H]} \text{H}_3\text{C—CH}_3$$

ethyne (acetylene) — ethene (ethylene) — ethane

P-450—The Site of Oxidative Reactions in the Liver
(a) The structure of the site for substrate binding varies in the various P-450 forms. Oxygen is bonded to the iron atom of the heme. The protein bonds to the heme by a coordinate covalent bond.
(b) Nicotinamide adenine dinucleotide phosphate contains two units of ribose (a carbohydrate), one adenine unit (a nitrogen base found in DNA), and nicotinamide (a vitamin).

(a)

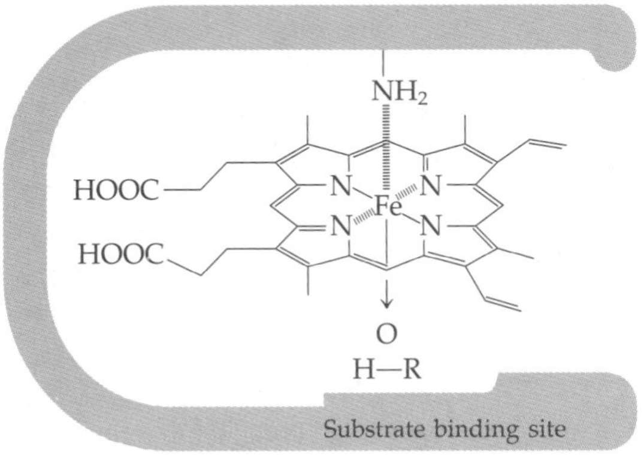

HOOC
HOOC
NH_2
N
N
Fe
N
N
O
H—R
Substrate binding site

(b)

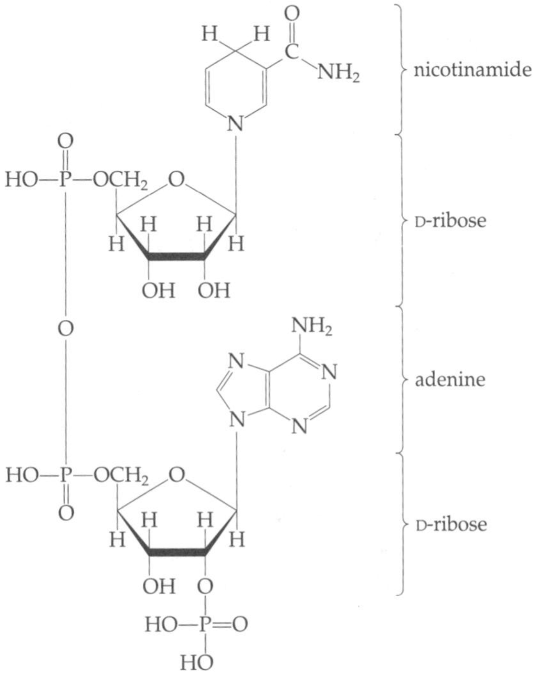

NH_2
nicotinamide
HO—P—OCH_2
D-ribose
OH OH
O
NH_2
adenine
HO—P—OCH_2
D-ribose
OH O
HO—P=O
HO

The simultaneous increase or decrease of two hydrogen atoms and one oxygen atom in a reactant is neither reduction nor oxidation. Thus, the conversion of ethene to ethanol is not an oxidation–reduction reaction.

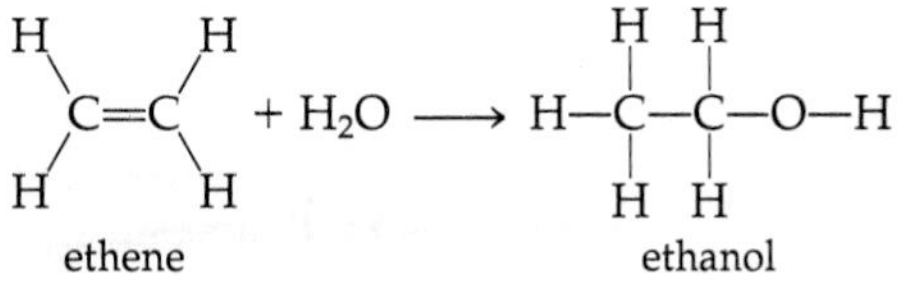

$$H_2C{=}CH_2 + H_2O \longrightarrow H{-}CH_2{-}CH_2{-}O{-}H$$

ethene ethanol

The concept of oxidation in organic chemistry based on change in molecular composition is extended to elements other than oxygen. An increase in the number of electronegative elements such as nitrogen or the halogens is also defined as an oxidation process. Conversely, a decrease in the number of these atoms in a molecule is reduction.

EXAMPLE 2.6 Ethylene oxide is used to sterilize medical equipment that is temperature-sensitive and cannot be heated in an autoclave. It is produced from ethylene by the following process. Classify the type of reaction. Is an oxidizing or reducing agent required?

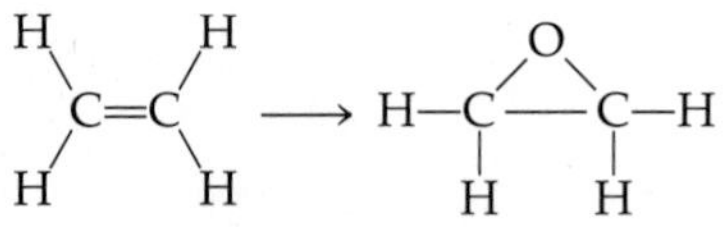

$$H_2C{=}CH_2 \longrightarrow H{-}\overset{}{CH}{-}O{-}CH{-}H \text{ (ring: C—C bridged by O)}$$

Solution The oxygen content of the ethylene is increased, and thus the reaction involves oxidation. An oxidizing agent is required for the reaction.

2.5 CLASSIFICATION OF ORGANIC REACTIONS

In the preceding two sections we have reviewed two classes of reactions that you learned in your first course in chemistry. Now we will look at several more common examples of organic reactions. These reactions will be discussed in greater detail in subsequent chapters.

Addition reactions occur when two reactants combine to give a single product. An example of an addition reaction is the reaction of ethylene with HBr to form bromoethane. The hydrogen and bromine atoms are added to adjacent atoms, a common characteristic of addition reactions.

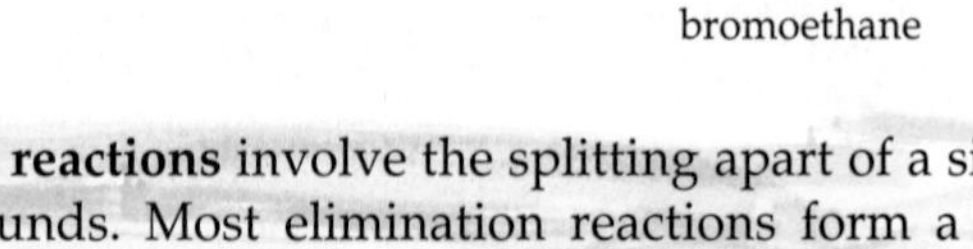

These two atoms are added to adjacent atoms.

$$H_2C{=}CH_2 + H{-}Br \longrightarrow H{-}CH_2{-}CH_2Br$$

bromoethane

Elimination reactions involve the splitting apart of a single compound into two compounds. Most elimination reactions form a product with a double bond containing the majority of the atoms in the reactant, and a second smaller molecule such as H_2O or HCl. The atoms eliminated to form the smaller molecule are usually located on adjacent carbon atoms in the reactant. For example, 2-propanol reacts with concentrated sulfuric acid to produce propene; water is eliminated in this reaction.

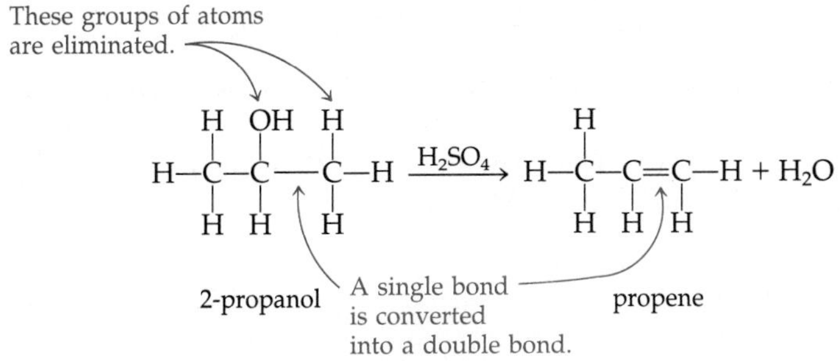

These groups of atoms are eliminated.

$$CH_3{-}CH(OH){-}CH_3 \xrightarrow{H_2SO_4} CH_3{-}CH{=}CH_2 + H_2O$$

2-propanol propene

A single bond is converted into a double bond.

In **substitution reactions,** one atom or group of atoms displaces a second atom or group of atoms. A generalized reaction is as follows.

$$A{-}X + Y \longrightarrow A{-}Y + X$$

In the above reaction, Y substitutes for X. An example of a substitution reaction is the conversion of bromomethane into methanol.

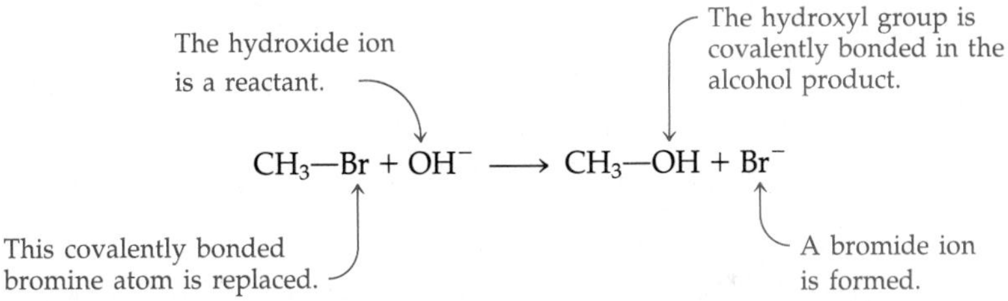

The hydroxide ion is a reactant.

The hydroxyl group is covalently bonded in the alcohol product.

$$CH_3{-}Br + OH^- \longrightarrow CH_3{-}OH + Br^-$$

This covalently bonded bromine atom is replaced.

A bromide ion is formed.

In **hydrolysis reactions** (Gk. *hydro,* water + *lysis,* splitting) water splits a large reactant molecule into two smaller product molecules. The generalized reaction is

$$A{-}B + H_2O \longrightarrow A{-}H + HO{-}B$$

One product molecule contains a hydrogen atom derived from water. The other product contains an —OH group derived from water. The hydrolysis of an amide to produce a carboxylic acid and an amine is an example of this process.

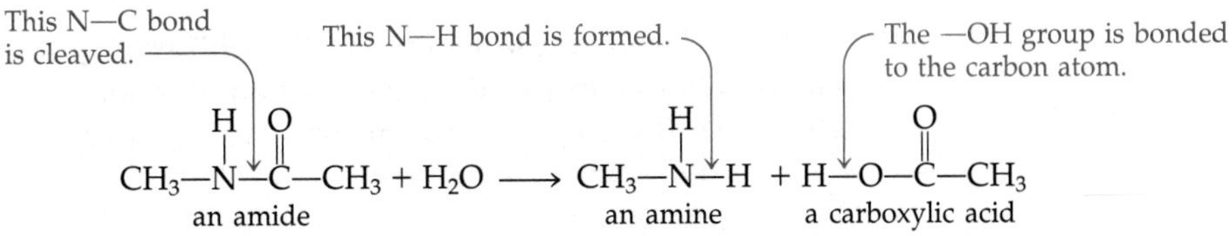

This N—C bond is cleaved.

This N—H bond is formed.

The —OH group is bonded to the carbon atom.

$$CH_3{-}NH{-}C(=O){-}CH_3 + H_2O \longrightarrow CH_3{-}NH{-}H + H{-}O{-}C(=O){-}CH_3$$

an amide an amine a carboxylic acid

In **condensation reactions,** two reactants combine to form one larger product with the simultaneous formation of a second, smaller product such as water. When the second product is water, the reaction is the reverse of a hydrolysis reaction. The general reaction is

$$\text{A—H} + \text{H—O—B} \longrightarrow \text{A—B} + \text{H}_2\text{O}$$

The formation of an ester from an alcohol and a carboxylic acid is an example of this process.

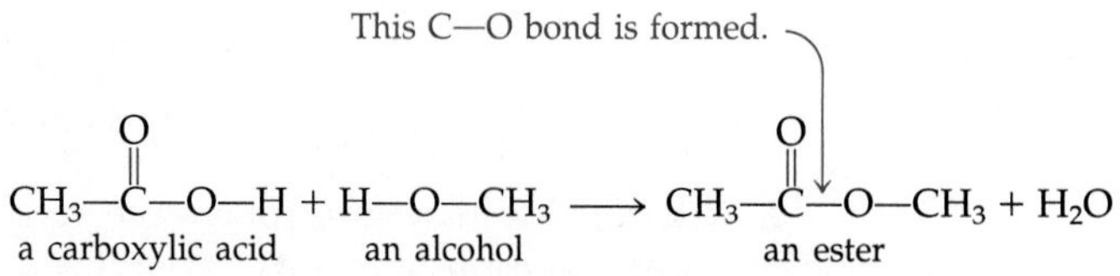

This C—O bond is formed.

$$\underset{\text{a carboxylic acid}}{\text{CH}_3\text{—}\overset{\overset{\text{O}}{\|}}{\text{C}}\text{—O—H}} + \underset{\text{an alcohol}}{\text{H—O—CH}_3} \longrightarrow \underset{\text{an ester}}{\text{CH}_3\text{—}\overset{\overset{\text{O}}{\|}}{\text{C}}\text{—O—CH}_3} + \text{H}_2\text{O}$$

Rearrangement reactions result from the reorganization of bonds within a single reactant to give an isomeric product. This type of reaction will not be encountered as frequently in this text as the other reactions described in this section. One example is a rearrangement in which the location of a double bond changes to give an isomer.

$$\text{CH}_3\text{—}\overset{\overset{\text{Br}}{|}}{\text{CH}}\text{—CH=CH}_2 \longrightarrow \text{CH}_3\text{—CH=CH—CH}_2\text{—Br}$$

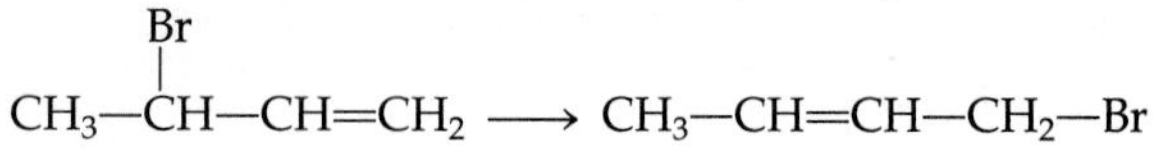

Note that it is also necessary to relocate the bromine atom to form the product.

EXAMPLE 2.7 Classify each of the following reactions.

(a) 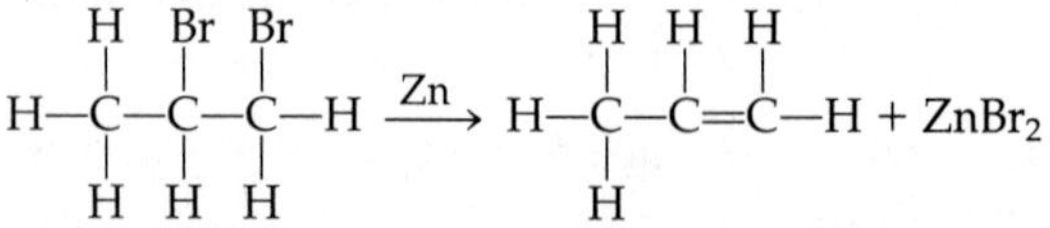 $$\text{H—}\underset{\underset{\text{H}}{|}}{\overset{\overset{\text{H}}{|}}{\text{C}}}\text{—}\underset{\underset{\text{H}}{|}}{\overset{\overset{\text{Br}}{|}}{\text{C}}}\text{—}\underset{\underset{\text{H}}{|}}{\overset{\overset{\text{Br}}{|}}{\text{C}}}\text{—H} \xrightarrow{\text{Zn}} \text{H—}\underset{\underset{\text{H}}{|}}{\overset{\overset{\text{H}}{|}}{\text{C}}}\text{—}\overset{\overset{\text{H}}{|}}{\text{C}}\text{=}\overset{\overset{\text{H}}{|}}{\text{C}}\text{—H} + \text{ZnBr}_2$$

(b) 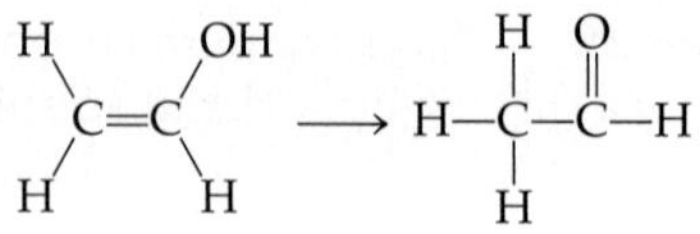 $$\text{(H)(H)C=C(OH)(H)} \longrightarrow \text{H—}\underset{\underset{\text{H}}{|}}{\overset{\overset{\text{H}}{|}}{\text{C}}}\text{—}\overset{\overset{\text{O}}{\|}}{\text{C}}\text{—H}$$

Solution Reaction (a) involves the elimination reaction of atoms located on adjacent carbon atoms in this reactant. The organic product contains the majority of the atoms of the reactant.

Reaction (b) is a rearrangement reaction in which the location of a double bond changes to give an isomer. A hydrogen atom also changes location.

EXAMPLE 2.8 A drug called clofibrate is used to lower plasma triacylglycerol and cholesterol levels. The following reaction occurs in the body to yield a related compound that is more effective in its action. What is the reaction type? What reactant is required?

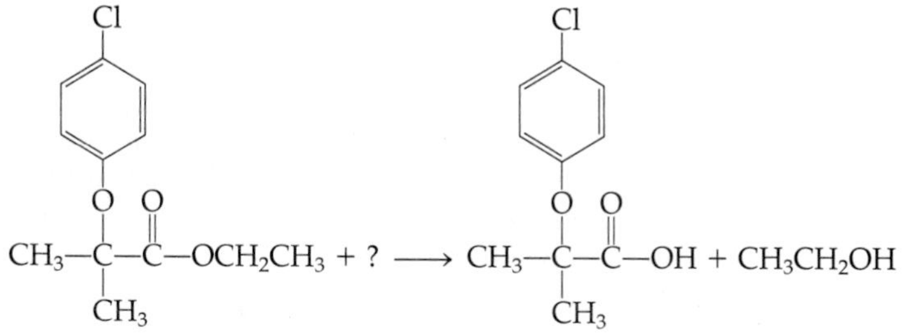

$$CH_3{-}C(CH_3)(O{-}C_6H_4{-}Cl){-}C({=}O){-}OCH_2CH_3 + ? \longrightarrow CH_3{-}C(CH_3)(O{-}C_6H_4{-}Cl){-}C({=}O){-}OH + CH_3CH_2OH$$

Solution First find the part of the molecule that changes during the reaction. The functional group undergoing reaction is an ester. The compound is cleaved into two smaller products. A carboxylic acid and an alcohol are produced. The carboxylic acid contains an —OH group derived from water and the alcohol contains a hydrogen atom derived from water. The net difference in the number of atoms shown is H_2O, which is the reactant. Thus, the reaction is hydrolysis.

2.6 CHEMICAL EQUILIBRIUM

Chemical reactions do not proceed in one direction only. As a reaction occurs, product molecules can revert to reactant molecules. Thus, two opposing reactions occur. When the rate of product formation is equal to the rate of reactant formation, an equilibrium is established.

Equilibrium Constant

Consider a general chemical reaction at equilibrium, where A and B are reactants, X and Y are products, and m, n, p, and q are coefficients.

$$m\,\mathrm{A} + n\,\mathrm{B} \underset{\text{reverse reaction}}{\overset{\text{forward reaction}}{\rightleftharpoons}} p\,\mathrm{X} + q\,\mathrm{Y}$$

The following expression for the reaction is a constant at a specific temperature. The brackets indicate molar concentrations, and the exponents are the coefficients in the balanced chemical equation.

$$\frac{[\mathrm{X}]^p[\mathrm{Y}]^q}{[\mathrm{A}]^m[\mathrm{B}]^n} = K$$

If $K > 1$, little reactant is present at equilibrium. If $K < 1$, little product is present at equilibrium. Consider the equilibrium constant for the addition reaction of gaseous ethylene with gaseous hydrogen bromide.

$$CH_2{=}CH_2 + HBr \rightleftharpoons CH_3CH_2Br$$

$$\frac{[CH_3CH_2Br]}{[CH_2{=}CH_2][HBr]} = K = 10^8$$

Because the equilibrium constant is very large, the reaction "goes to completion." That is, for all practical purposes, no reactant remains at equilibrium. Reactions with equilibrium constants greater than 10^3 can be considered to be "complete" because the amount of reactant is about 0.1%, or less.

Now let's consider the condensation reaction of acetic acid and ethyl alcohol to produce ethyl acetate, and the related equilibrium constant expression.

$$CH_3{-}\overset{\overset{\large :\ddot{O}}{\|}}{C}{-}\ddot{\underset{..}{O}}{-}H + CH_3CH_2{-}\ddot{\underset{..}{O}}H \rightleftharpoons CH_3{-}\overset{\overset{\large :\ddot{O}}{\|}}{C}{-}\ddot{\underset{..}{O}}{-}CH_2CH_3 + H_2O$$

$$\frac{[CH_3CO_2CH_2CH_3][H_2O]}{[CH_3CO_2H][CH_3CH_2OH]} = K = 4.0$$

In this reaction, significant concentrations of reactants are present at equilibrium. Thus, the product yield is less than 100% based on the balanced equation.

Catalysts do not change the position of a chemical equilibrium or the value of the equilibrium constant. A **catalyst** increases the rates of the forward and reverse reactions equally, and K does not change. In the reaction of ethanol and acetic acid, the reaction is acid-catalyzed. Thus, the equilibrium is established in a shorter time period at the same temperature in the presence of an acid such as HCl.

EXAMPLE 2.9 Chloromethane reacts with sodium hydroxide in aqueous solution to produce methanol and sodium chloride. Write the equilibrium constant expression for this substitution reaction. The equilibrium constant is 10^{16}. What can be concluded about the degree of completion of the reaction?

$$OH^- + CH_3Cl \rightleftharpoons CH_3OH + Cl^-$$

Solution The equilibrium constant expression contains the products in the numerator and the reactants in the denominator.

$$\frac{[CH_3OH][Cl^-]}{[CH_3Cl][OH^-]} = K$$

The equilibrium constant, 10^{16}, is greater than the equilibrium constant given for the addition reaction of HBr with ethylene. Thus, the reaction is even more "complete".

Le Châtelier's Principle

Le Châtelier's principle states that a change in the conditions of a chemical equilibrium causes a shift in the concentrations of reactants and products to result in a new equilibrium system. If additional reactant is added to a chemical system at equilibrium, the concentrations of both reactants and products change to establish a new equilibrium system, but the equilibrium constant is unchanged. After adding reactant, the total concentration of reactant is initially increased, but then decreases to establish a new equilibrium. As a result, the concentration of the products increases. In short, the change imposed on the system by adding reactants is offset when more reactants are converted to products. If a product is removed from a chemical system at equilibrium, the forward reaction occurs to give more product. Regardless of the condition imposed on the system at equilibrium, the concentrations change to maintain the same value of the equilibrium constant. Consider the equilibrium in the formation of ethyl acetate.

$$CH_3{-}\overset{\overset{\large :\ddot{O}}{\|}}{C}{-}\ddot{\underset{..}{O}}{-}H + CH_3CH_2{-}\ddot{\underset{..}{O}}H \rightleftharpoons CH_3{-}\overset{\overset{\large :\ddot{O}}{\|}}{C}{-}\ddot{\underset{..}{O}}{-}CH_2CH_3 + H_2O$$

Adding alcohol "pushes" the reaction to the right.

Removing water "pulls" the reaction to the right.

As noted before, significant concentrations of reactants are present at equilibrium for this reaction. If water is removed from the system by some means, the equilibrium is disturbed and the equilibrium position of the reaction would shift to the right to produce more water and ethyl acetate. If the amount of alcohol is increased, a larger amount of the carboxylic acid will be converted into products.

2.7 EQUILIBRIA IN ACID–BASE REACTIONS

Throughout this text we will see how the acid–base properties of members of a class of organic compounds are affected by changes in their structure. In addition, we will find that the mechanisms of chemical reactions often depend on intermediates that function as acids or bases.

Water is the reference solvent commonly used to compare the strengths of acids or bases quantitatively. The strengths of acids are measured by their tendencies to transfer protons to water.

$$HA + H_2O \rightleftharpoons H_3O^+ + A^-$$

K_a and pK_a

A quantitative measure of the acidity of an acid with the general formula HA is given by the equilibrium constant for ionization, which is obtained from the equation for ionization.

$$K = \frac{[H_3O^+][A^-]}{[HA][H_2O]}$$

The concentration of water, about 55 M, is so large compared to that of the other components of the equilibrium that its value changes very little when the acid HA is added. Therefore, the concentration of water is included in the acid ionization constant K_a.

$$K_a = K[H_2O] = \frac{[H_3O^+][A^-]}{[HA]}$$

Acids with $K_a > 10$ are strong acids. Most organic acids have $K_a < 10^{-4}$ and are weak acids. Acid dissociation constants are conveniently expressed as pK_a values. The pK_a expressed as the logarithm of base ten is

$$pK_a = -\log K_a$$

Note that the pK_a values listed in Table 2.1 decrease as the acidity increases.

Weak acids do not transfer their protons completely to water, and few ions are produced. An example of a weak acid is acetic acid, which ionizes in water to give acetate ions and hydronium ions.

TABLE 2.1
pK_a Values

Compound	pK_a
CH_4	49
$CH_2{=}CH_2$	44
NH_3	36
$HC{\equiv}CH$	25
CCl_3H	25
H_2O	15.7
CH_3OH	15.5
CH_3CH_2OH	15.9
$(CH_3)_3COH$	18.0
CF_3CH_2OH	12.4
CH_3CH_2SH	10.6
CH_3CO_2H	4.7

weaker acid than H_3O^+ — stronger base than H_2O (conjugate pair)

$$CH_3CO_2H + H_2O \rightleftharpoons CH_3CO_2^- + H_3O^+$$

weaker base than $CH_3CO_2^-$ — stronger acid than CH_3CO_2H (conjugate pair)

Acetic acid is a weaker acid than H_3O^+, and $CH_3CO_2^-$ is a stronger base than H_2O. The equilibrium position favors the side containing the weaker acid and weaker base. The equilibria between acids and bases and their conjugate bases and acids can be viewed as a "contest" for protons. A strong acid, with its great tendency to lose protons, is paired with a weak conjugate base that has a low affinity for protons. Thus, as the tendency of an acid to lose a proton increases, the tendency of its conjugate base to accept a proton decreases.

K_b and pK_b

There is a close relationship between the acidity of acids and the basicity of bases. When an acid dissociates, a base is formed that can react in the reverse direction by accepting a proton. Thus, we can discuss the acidity of the

TABLE 2.2
pK_b Values

Compound	pK_b
C_6H_5–NH_2	9.4
$CH_3CO_2^-$	9.3
CN^-	4.8
NH_3	4.8
CH_3NH_2	3.4
CH_3O^-	−1.5

acid HA or the basicity of the base A^-. As in the case of acids, the basicity of bases is both qualitatively and quantitatively compared to the properties of water. A base, A^-, removes a proton from water to form hydroxide ion and the conjugate acid HA. The base dissociation constant, K_b, for the reaction is

$$A^- + H_2O \rightleftharpoons HA + OH^-$$

$$K_b = \frac{[HA][OH^-]}{[A^-]}$$

The K_b values of bases are conveniently expressed as pK_b values. The pK_b is defined as

$$pK_b = -\log K_b$$

pK_b values increase with decreasing basicity. The values for some organic bases are listed in Table 2.2.

A strong base has a large K_b (small pK_b) and completely removes the proton of an acid. The most common strong base is hydroxide ion, which will remove and accept protons from weak acids such as acetic acid.

$$CH_3CO_2H + OH^- \longrightarrow CH_3CO_2^- + H_2O$$

stronger acid than H_2O — conjugate pair — weaker base than OH^-

stronger base than $CH_3CO_2^-$ — conjugate pair — weaker acid than CH_3CO_2H

Weak bases do not have a large attraction for the protons of an acid. Only a small fraction of the molecules of a weak base in a sample will accept protons at equilibrium. For example, methylamine is a weak base. When it dissolves in water, a low concentration of methylammonium ions forms.

$$CH_3NH_2 + H_2O \rightleftharpoons CH_3NH_3^+ + OH^-$$

weaker base than OH^- — conjugate pair — stronger acid than H_2O

weaker acid than $CH_3NH_3^+$ — conjugate pair — stronger base than CH_3NH_2

Structure and Acidity

Removing a proton from an electrically neutral acid in a solvent requires breaking a bond to hydrogen and placing a negative charge on the resulting conjugate base. Thus, K_a values depend on both the strength of the H—A bond and the stability of A^- in the solvent.

Let's recall trends in the periodic table. First, the acidities of acids, HA, increase as we move down a column of the periodic table. The strength of the H—A bond decreases because of less effective overlap of the more diffuse atomic orbitals of A with the hydrogen 1*s* orbital. For example, the acidities of the halogen acids increase in the order HF < HCl < HBr < HI. For the same reasons, H_2O is also a weaker acid than H_2S.

Second, in a given row of the periodic table, acidity increases from left to right. The order of increasing acidity is $CH_4 < NH_3 < H_2O < HF$. This trend reflects the stability of the negative charge on the electronegative element of the conjugate base. That is, the order of increasing strength of conjugate bases is $F^- < OH^- < NH_2^- < CH_3^-$.

Many organic compounds are structurally related to inorganic acids and bases. As a consequence, we can predict the acid–base properties by making an appropriate comparison. For example, methanesulfonic acid has an O—H bond that is structurally similar to the O—H bond in sulfuric acid. Because sulfuric acid is a strong acid, it is reasonable to expect methanesulfonic acid to be a strong acid. Using the concept of functional groups discussed in Chapter 1, we should expect all sulfonic acids to be strong acids.

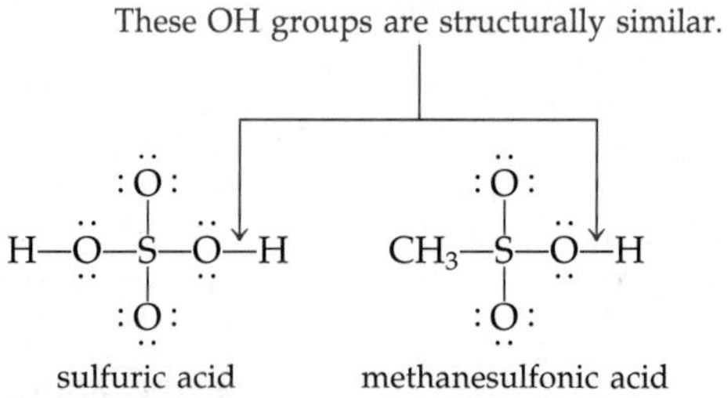

These OH groups are structurally similar.

sulfuric acid   methanesulfonic acid

Ethylamine, $CH_3CH_2NH_2$, is structurally related to ammonia, which is a weak base. Thus, ethylamine and other amines are expected to be weak bases. As we examine various functional groups in detail, we will find that the acid–base properties do vary somewhat with structure.

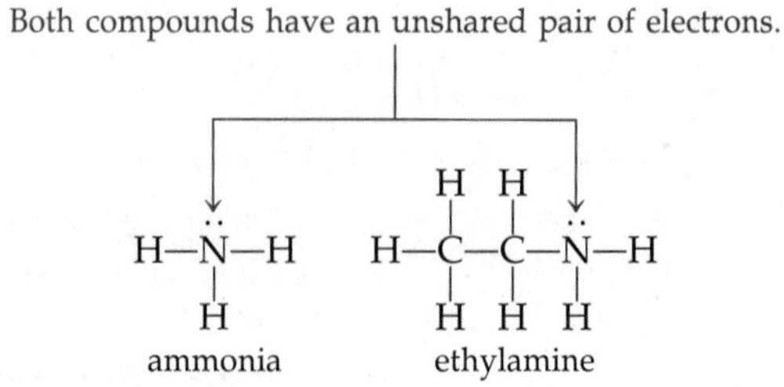

Both compounds have an unshared pair of electrons.

ammonia   ethylamine

EXAMPLE 2.10 Which is the stronger acid, CH_3OH or CH_3SH? Why?

Solution The acidity of OH and SH bonds in the two compounds would be expected to be similar to those in H_2O and H_2S, respectively. Sulfur is below oxygen in the same family of the periodic table. Within families, the stronger acid is the compound with the weaker bond to hydrogen. The HS bond energy is less than the bond energy of the OH bond. The stronger acid is CH_3SH.

A reaction in which relatively unstable reactants are converted to more stable products has a large equilibrium constant. Thus, stabilizing the negative charge in the conjugate base formed from an acid increases K_a. When an anion produced by ionization of an acid has resonance stabilization, acid strength increases by a substantial amount. For example, both methanol and acetic acid ionize to form conjugate bases that have a negative charge on oxygen. However, acetic acid is about 10^{10} times more acidic than methanol.

$$CH_3O{-}H + H_2O \rightleftharpoons CH_3O^- + H_3O^+ \qquad K_a = 10^{-16}$$
$$CH_3CO_2{-}H + H_2O \rightleftharpoons CH_3CO_2^- + H_3O^+ \qquad K_a = 1.8 \times 10^{-5}$$

The greater acidity of acetic acid is the result of resonance stabilization of the negative charge in the conjugate base, acetate ion. Each oxygen atom in the acetate ion bears one-half of the negative charge, whereas in the methoxide ion (CH_3O^-) the negative charge is concentrated on a single oxygen atom. Thus, resonance stabilization spreads out or delocalizes the negative charge of the conjugate base.

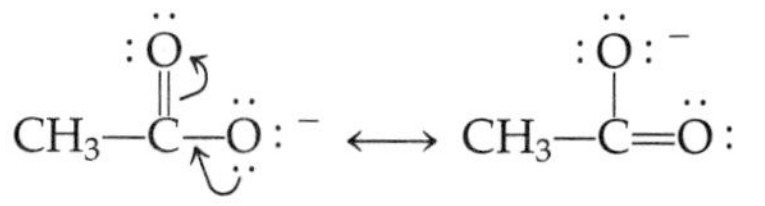

$$CH_3{-}C(=\ddot{O}{:}){-}\ddot{O}{:}^- \longleftrightarrow CH_3{-}C(-\ddot{O}{:}^-){=}\ddot{O}{:}$$

Acidity also reflects the ability of an atom to polarize neighboring bonds by an **inductive effect.** For example, sulfuric acid is a strong acid, whereas sulfurous acid is a weak acid.

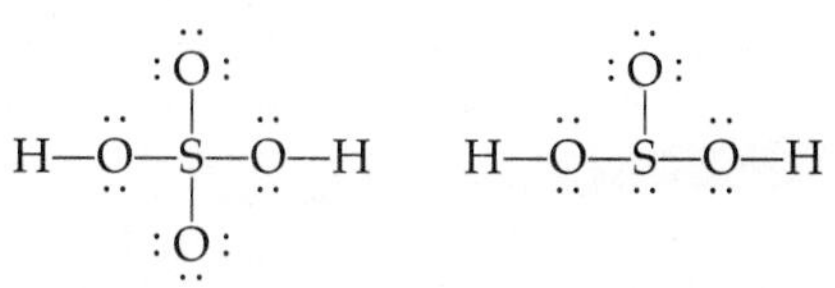

$$H{-}\ddot{O}{-}S(=\ddot{O}{:})(=\ddot{O}{:}){-}\ddot{O}{-}H \qquad H{-}\ddot{O}{-}\ddot{S}(-\ddot{O}{:}){-}\ddot{O}{-}H$$

sulfuric acid   sulfurous acid

In sulfuric acid and sulfurous acid, the formal charges of sulfur are +2 and +1, respectively. As a consequence, the sulfur atom of sulfuric acid attracts the bonding electrons of attached oxygen atoms more strongly than does the sulfur atom in sulfurous acid. The shifting of electrons from oxygen toward sulfur is called **inductive electron withdrawal** of electrons. The electrons of the oxygen atom are drawn away from the O—H bond, and the proton of sulfuric acid can therefore ionize more easily.

Any structural feature that withdraws electron density from the bond between hydrogen and another atom causes an increase in its acidity. Inductive effects as well as resonance effects play major roles in chemical reactions. We will employ these two concepts many times throughout this text.

EXAMPLE 2.11 The pK_a of ethanol and 2,2,2-trifluoroethanol are 15.9 and 12.4, respectively. What is responsible for this difference?

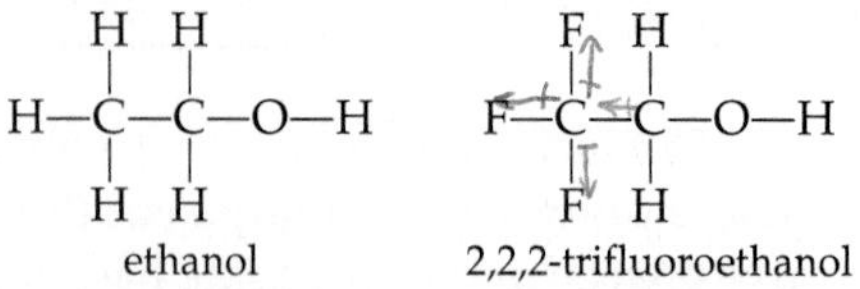

$$\mathrm{CH_3CH_2OH}\ \text{(ethanol)} \qquad \mathrm{CF_3CH_2OH}\ \text{(2,2,2-trifluoroethanol)}$$

Solution 2,2,2-Trifluoroethanol is the stronger acid. The carbon atom with three fluorine atoms has a partial positive charge as a consequence of inductive electron withdrawal by the three polar covalent C—F bonds. This carbon atom, in turn, inductively attracts electrons from the other carbon atom and, indirectly, from the oxygen atom. Thus, the oxygen–hydrogen bond is more strongly polarized, and the compound is more acidic.

EXAMPLE 2.12 The pK_a of nitromethane is 10.2, whereas the pK_a of methane is approximately 49. Explain why nitromethane is more acidic.

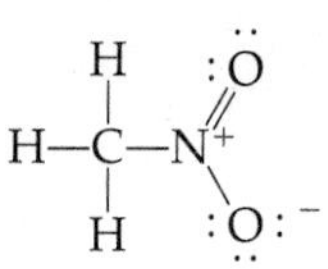

$$\mathrm{H_3C{-}N^{+}(=O){-}O^{-}}$$

Solution When methane ionizes, its conjugate base, CH_3^-, has a negative charge on the carbon atom. The formal positive charge on the nitrogen atom in CH_3NO_2 withdraws electrons from the carbon–hydrogen bond and increases its acidity. In addition, when nitromethane ionizes, its conjugate base is resonance-stabilized. A resonance form can be written that has negative charges on both oxygen atoms of the nitro group. The conjugate base of nitromethane is stabilized by resonance delocalization of electrons.

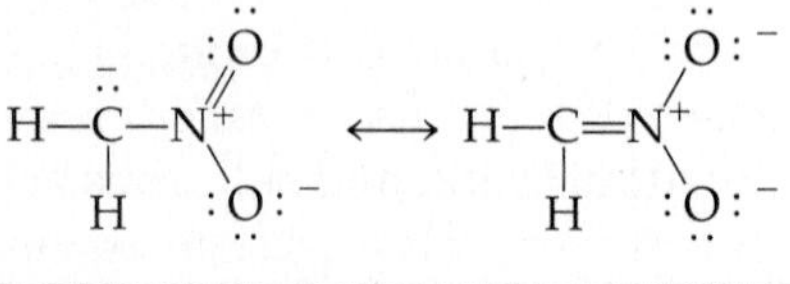

$$\mathrm{H_2\ddot{C}^{-}{-}N^{+}(=O){-}O^{-}} \longleftrightarrow \mathrm{H_2C{=}N^{+}(O^{-}){-}O^{-}}$$

2.8 REACTION MECHANISMS

The description of the individual steps of a reaction is called the **reaction mechanism.** It describes the various stages in a reaction and shows the order in which bonds are broken in the reactant and made in the product.

Concerted and Multistep Reactions

Some reactions occur in a single step and bonds form and break simultaneously. Such processes are **concerted reactions.** The reaction mechanism thus resembles that of an ordinary chemical equation.

$$\text{A (reactant)} \longrightarrow \text{B (product)}$$

Many reactions occur in a series of steps. For example, the conversion of reactant A into product B may involve two steps in which an intermediate M is formed and then reacts.

$$\text{A (reactant)} \xrightarrow{\text{step 1}} \text{M (intermediate)}$$
$$\text{M (intermediate)} \xrightarrow{\text{step 2}} \text{B (product)}$$

In a multistep reaction, the individual steps usually have different rates. The overall rate of conversion of reactant into product can occur no faster than the slowest individual step. The slowest step is called the **rate-determining step.** This step controls the overall rate of conversion of reactants to products.

Types of Bond Cleavage and Formation

When a bond is broken so that one electron remains with each of the two fragments, the process is **homolytic** cleavage. A fragment with an unpaired electron is a **radical.**

$$\text{X—Y} \longrightarrow \text{X}\cdot + \cdot\text{Y} \qquad \text{homolytic cleavage}$$

Homolytic cleavage of a bond to carbon produces a **carbon radical** that is highly reactive and tends to react to obtain another electron. This electron is usually acquired along with another atom or group of atoms that become bonded to the carbon atom.

$$CH_3\text{—Y} \xrightarrow{\text{homolysis}} \underset{\text{methyl radical}}{CH_3\cdot} + \cdot\text{Y}$$

When a bond is broken so that one fragment gains both bonding electrons, the process is called **heterolytic** cleavage. The fragment that gains electrons has a negative charge. The second fragment is electron deficient and has a positive charge. Because a heterolytic process yields ions, it is also called a polar mechanism.

$$\text{X—Y} \longrightarrow X^+ + :Y^- \qquad \text{heterolytic cleavage}$$

Heterolytic cleavage of a bond to carbon can produce either of two different carbon species. If the bond breaks so that its electrons remain with the carbon atom, a negatively charged **carbanion** results. The carbanion has an octet of electrons. If the bond breaks so that its electrons are lost by the carbon atom, a positively charged **carbocation** results. The carbocation has a sextet of electrons and is an electron-deficient species.

$$CH_3{-}Y \xrightarrow{\text{heterolysis}} \begin{cases} CH_3{:}^- + Y^+ & \text{methyl carbanion} \\ CH_3^+ + {:}Y^- & \text{methyl carbocation} \end{cases}$$

The mode of heterolytic cleavage of a C—Y bond depends on the electronegativity of Y. If Y is a more electronegative element than carbon—a halogen atom, for example—the carbon bears a partial positive charge, and the carbon bond tends to break heterolytically to form a carbocation. Conversely, if Y is a less electronegative element, such as a metal, the bond has the opposite polarity and tends to break heterolytically to form a carbanion.

$$\overset{\delta+}{CH_3}{-}\overset{\delta-}{Br} \qquad \overset{\delta-}{CH_3}{-}\overset{\delta+}{Li}$$

There are two ways to form two-electron covalent bonds from fragments. These processes are the reverse of the two cleavage reactions. Bond formation from fragments that each contain one electron is a **homogenic** process. Formation of a two-electron bond from oppositely charged fragments is a **heterogenic** process.

$$X\cdot + Y\cdot \longrightarrow X{-}Y \qquad \text{homogenic bond formation}$$
$$X^+ + Y{:}^- \longrightarrow X{-}Y \qquad \text{heterogenic bond formation}$$

Heterogenic reactions are more common than homogenic reactions in organic chemistry. In organic reactions, a carbocation behaves as an **electrophile** (electron-loving species). It seeks a negatively charged center to neutralize its positive charge and to obtain a stable octet of electrons. On the other hand, a carbanion has an electron pair that causes it to react as a **nucleophile** (nucleus-loving species). It seeks a positively charged center to neutralize its negative charge.

Many organic reactions can be depicted by the following equation, in which E^+ and $:Nu^-$ represent an electrophile and nucleophile, respectively.

$$\underset{\text{electrophile}}{E^+} + \underset{\text{nucleophile}}{{:}Nu^-} \xrightarrow{\text{heterogenic process}} E{-}Nu$$

The curved-arrow notation shows the movement of a pair of electrons from the nucleophile to the electrophile. This notation is exactly like that used to show the reaction between a Lewis base and a Lewis acid. In fact, Lewis bases can act as nucleophiles, and Lewis acids can behave as electrophiles.

The mechanism of a chemical reaction is not revealed by a balanced chemical equation. For example, consider the following two substitution reactions.

$$CH_3{-}Cl + NaOH \longrightarrow CH_3{-}OH + NaCl$$
$$CH_3{-}H + Cl_2 \longrightarrow CH_3{-}Cl + H{-}Cl$$

In the first equation a chlorine atom is replaced by a hydroxyl group; in the second equation, a hydrogen atom is replaced by a chlorine atom. These processes occur by very different mechanisms. The first reaction occurs in one step in which OH^- replaces Cl^-. The second reaction occurs in several steps and involves homolytic bond cleavage and homogenic bond formation.

Nucleophilic Substitution

Reactions in which a nucleophile "attacks" a carbon atom and replaces another group are **nucleophilic substitution reactions.** The "leaving group" displaced from the carbon center is symbolized by L. A leaving group is often an electronegative atom or a group that can exist as a stable anion. R represents the remainder or the rest of the molecule.

$$Nu{:}^- + R{-}L \longrightarrow Nu{-}R + {:}L^-$$

Note that the nucleophile has an unshared pair of electrons that bonds to the carbon residue. Thus, bond formation is a heterogenic process. The leaving group departs with an electron pair, and cleavage of the bond between the leaving group and carbon is a heterolytic process.

An example of this type of nucleophilic substitution process is the reaction of chloromethane with hydroxide ion.

$$OH^- + \overset{\delta+}{C}H_3{-}\overset{\delta-}{Cl} \longrightarrow CH_3{-}OH + Cl^-$$

In this reaction, the nucleophile approaches the carbon atom, which is made somewhat positive by the electronegative chlorine atom. The nucleophile has a nonbonded pair of electrons that begins to bond to the carbon atom. As the nucleophile approaches the carbon atom, the bond between carbon and the chloride ion, a leaving group, weakens. The entire process is **concerted**—that is, both bond breaking and bond formation occur simultaneously. The mechanism of nucleophilic substitution depends on many factors to be discussed in Chapter 7.

Chlorination of an Alkane

Methane reacts with chlorine gas at elevated temperatures or in the presence of ultraviolet light as an energy source. In this reaction, a chlorine atom replaces a hydrogen atom.

$$CH_3{-}H + Cl_2 \longrightarrow CH_3{-}Cl + H{-}Cl$$

The mechanism of this reaction involves homolytic bond cleavage and homogenic bond formation. In the first step, a chlorine molecule absorbs either heat or light energy and the Cl—Cl bond is broken to give two chlorine atoms. They are electron-deficient radicals and highly reactive. This step starts the reaction and is called the **initiation step.**

step 1 (initiation)

$$:\ddot{Cl}-\ddot{Cl}: \longrightarrow :\ddot{Cl}\cdot + \cdot\ddot{Cl}:$$

chlorine atoms (radicals)

After the initial homolytic bond cleavage, the other steps occur. In step 2, a C—H bond is broken and an H—Cl bond is formed; in step 3, a Cl—Cl bond is broken and a C—Cl bond is formed. In each step, a radical reacts and a radical is produced.

step 2 (propagation)

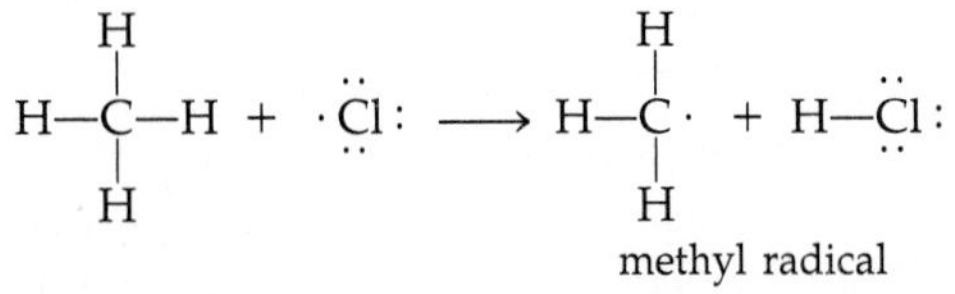

$$CH_4 + \cdot\ddot{Cl}: \longrightarrow CH_3\cdot + H-\ddot{Cl}:$$

methyl radical

step 3 (propagation)

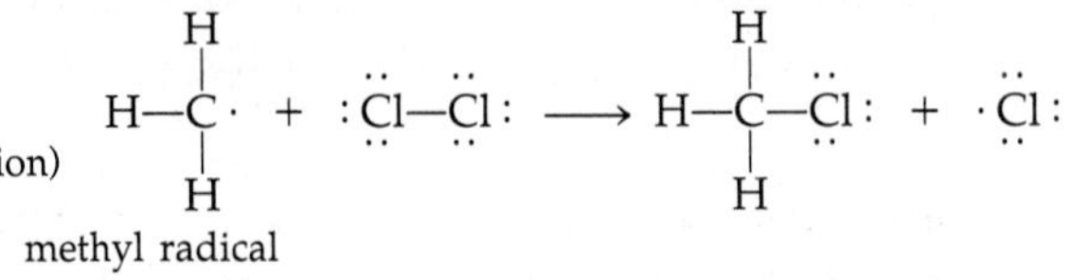

$$CH_3\cdot + :\ddot{Cl}-\ddot{Cl}: \longrightarrow CH_3-\ddot{Cl}: + \cdot\ddot{Cl}:$$

methyl radical

Steps 2 and 3 continue the reaction and are known as **propagation steps.** Thus, one radical generates another in this **chain propagation** sequence. The process continues as long as radicals and a supply of both reactants are available.

2.9 WHAT AFFECTS REACTION RATES?

In a reaction, the reactant molecules collide with each other, some bonds rupture, and others form. The factors that increase the rate of a reaction are (1) nature of the reactants, (2) temperature, (3) concentration of reactants, and (4) presence of substances called catalysts.

The nature of the reactants is the most important feature controlling a chemical reaction. For example, in the addition reaction of ethylene (C_2H_4) with HBr, a bond must be broken between hydrogen and bromine atoms. Bonds must be formed between carbon and the hydrogen atom and between carbon and the bromine atom.

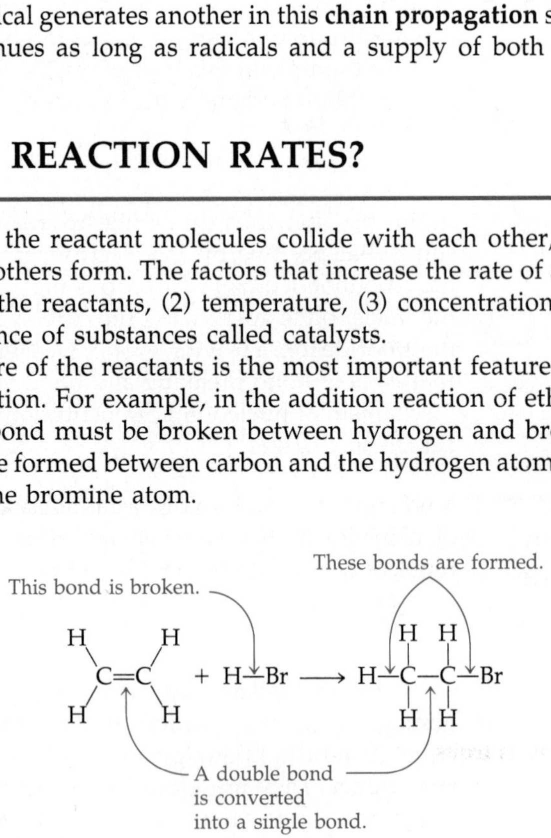

$$H_2C{=}CH_2 + H-Br \longrightarrow H-CH_2-CH_2-Br$$

This bond is broken.

These bonds are formed.

A double bond is converted into a single bond.

Although you may not be able to predict the order of reactivity, you should expect that the addition reaction of HCl to ethylene would occur at a rate different from that of the analogous reaction with H—Br. The bonding that occurs now involves carbon and chlorine atoms and the bond broken is between hydrogen and chlorine atoms. The energy requirements associated with the reorganization of these atoms and bonds must be different.

The rates of chemical reactions increase with a rise in temperature because the reactant molecules collide more frequently and with a greater energy. As a "rule of thumb," the rate approximately doubles for a 10 °C rise in temperature. As the concentration of reactants is increased, the reaction velocity increases because reactant molecules are more likely to collide.

A **catalyst** is a substance that increases a reaction rate. A catalyst is said to catalyze the reaction, and its effect is known as catalysis. Catalysts are usually required only in small amounts. The catalyst is present in the same amount before and after the reaction takes place, even though it must interact with the reactant during the reaction. Although a catalyst increases the rate of a reaction, it does not change the equilibrium constant for the reaction.

2.10 REACTION RATE THEORY

Not every collision between reactant molecules results in the formation of product. In most collisions the molecules simply bounce off each other. Collisions between molecules that cause a chemical reaction are called **effective collisions,** and the minimum energy required for an effective collision is the **activation energy.** The activation energy for a given reaction depends on the types of bonds broken and made in the reaction.

During a reaction, the arrangements of the atoms change as bonds are distorted and eventually broken while new bonds form. During this process, some repulsion occurs when reactant atoms move close together. This repulsion results from the proximity of the electrons surrounding each atom. During a reaction, each specific intermediate arrangement of atoms has an associated energy. All such intermediate arrangements of atoms have energies higher than the initial energy of the reactant. The atomic arrangement whose structure has the maximum energy is the **transition state.**

The transition state in the nucleophilic substitution of chloromethane by hydroxide ion has both the hydroxide and chloride ions bonded to some degree to carbon.

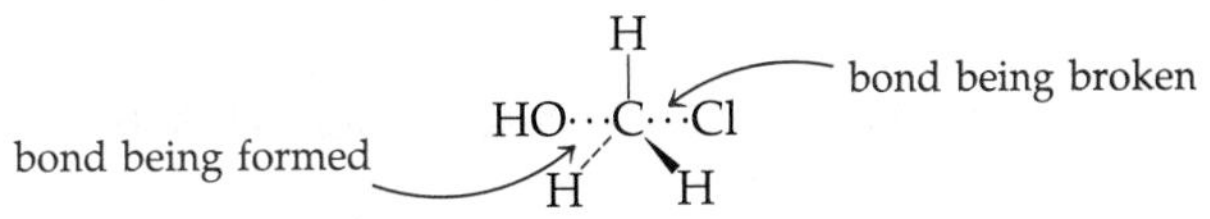

H
HO···C···Cl
H H
bond being broken
bond being formed

The carbon–chlorine bond breaks on one side of the transition state structure, while the carbon–oxygen bond forms on the other side. Transition states cannot be isolated because they exist for only short periods of time during a reaction. However, the structure of the transition state for a reaction can be inferred from various kinds of experimental data.

Reaction Coordinate Diagrams

Reaction coordinate diagrams are used to represent the progress of a reaction. The vertical axis gives the total energy of the reacting system; the horizontal axis qualitatively represents the progress of an exothermic reaction ($\Delta H^\circ < 0$) from the reactants (left) to the products (right) (Figure 2.4).

The difference between the energy of the reactants and the transition state is the activation energy ($E_a > 0$). In the transition state for the nucleophilic substitution of chloromethane by hydroxide ion, both hydroxide and chloride are partially bonded to the carbon atom.

A large activation energy results in a slow reaction because only a small fraction of the molecules collide with sufficient energy to reach the transition state. Once the molecules reach this point, energy is released as the reaction proceeds to form products. The energy released is equal to the activation energy originally added plus an amount equal to that characteristic for the exothermic reaction.

The kinetic energy of molecules increases with increasing temperature. As the kinetic energy increases, the chances increase for molecular collisions with energy equal to the activation energy, and the rate of reaction increases.

Some reactions occur in two or more steps, as in the case of the two-step addition reaction of HBr with ethylene.

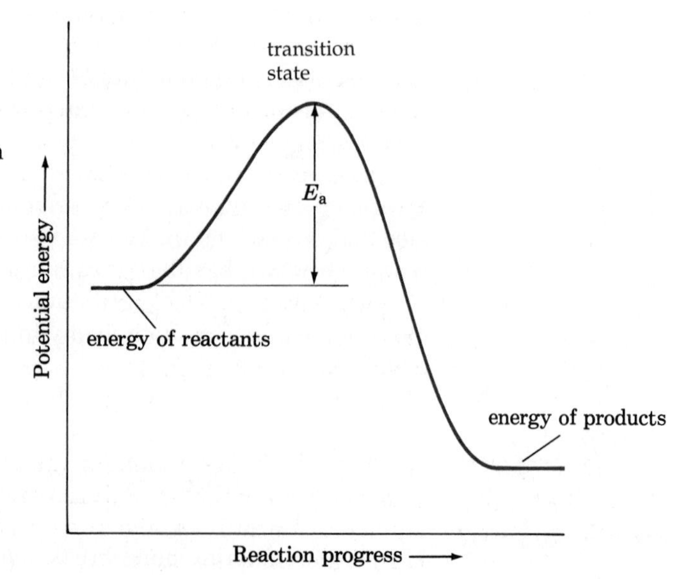

FIGURE 2.4
Reaction Coordinate Diagram for a Single-Step Reaction
This reaction coordinate diagram shows that the potential energy of the transition state is greater than the energy of either the reactants or the products.

FIGURE 2.5
Reaction Coordinate Diagram for a Two-Step Reaction

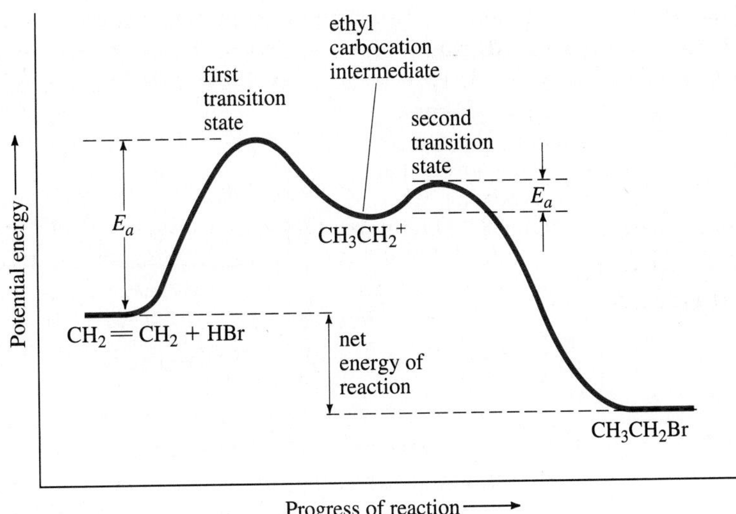

$$H^+ + \underset{H}{\overset{H}{}}C{=}C\underset{H}{\overset{H}{}} \longrightarrow H{-}\underset{H}{\overset{H}{C}}{-}\underset{H}{\overset{H}{C^+}}$$

$$H{-}\underset{H}{\overset{H}{C}}{-}\underset{H}{\overset{H}{C^+}} + :\ddot{\underset{..}{Br}}:^- \longrightarrow H{-}\underset{H}{\overset{H}{C}}{-}\underset{H}{\overset{H}{C}}{-}\ddot{\underset{..}{Br}}:$$

In the first step, a proton acts as an electrophile. It forms a bond to carbon using the π electrons of the double bond. As a consequence, an intermediate carbocation is formed. It then reacts in a second step with the bromide ion, which is a nucleophile. Each step is shown in the reaction coordinate diagram in Figure 2.5.

In the first transition state, a hydrogen ion begins to bond to carbon as π electrons are removed from the double bond. The energy then decreases until an intermediate carbocation is formed. In the second step, which has its own activation energy, the carbocation starts to bond to the nucleophilic bromide ion. Note that the energy of the carbocation is lower than the energy of the two transition states. Finally, as the carbon–bromine bond becomes fully formed, the reaction coordinate diagram shows that the energy of the products is lower than the energy of the reactants.

How Do Catalysts Function?

A catalyst provides a path for the progress of the reaction that is different from the path of the uncatalyzed reaction. The path starts at the same reactants and concludes at the same products. However, the path for the catalyzed reaction has a different, lower activation energy (Figure 2.6).

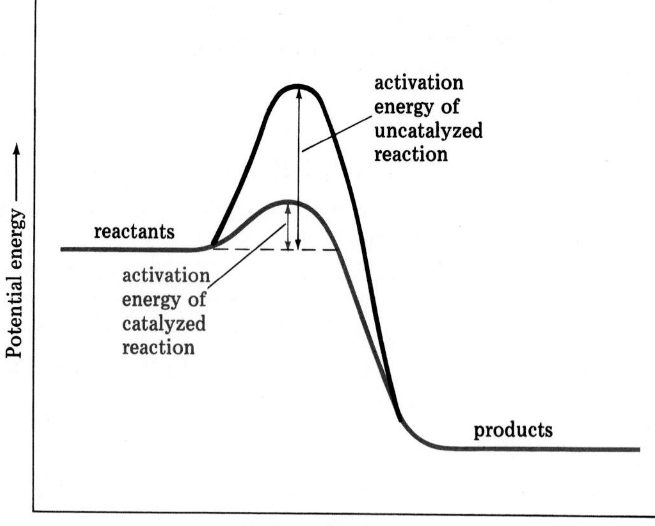

FIGURE 2.6
Effect of a Catalyst on the Reaction Pathway
The catalyst provides an alternate pathway for a reaction. The activation energy for the process is lower than for the uncatalyzed reaction. This lower energy requirement results in a faster reaction.

To illustrate the effect of a catalyst on the path of a reaction, consider the hypothetical reaction of A and B, a reaction with a high activation energy.

$$A + B \longrightarrow X$$

The transition state can be reached by only a few high-energy molecules and as a consequence the reaction is slow. However, in the presence of a catalyst, represented by C, the following reactions may occur.

step 1 $\quad A + C \longrightarrow A{-}C$
step 2 $\quad A{-}C + B \longrightarrow X + C$

The catalyst may combine with A in a reaction with a low activation energy. Similarly, the reaction of A—C with B may require little energy. If the activation energy of each step is low, a larger fraction of molecules will be able to react faster via this catalyzed pathway than could react without the catalyst at the same temperature.

EXERCISES

Physical Properties

2.1 Suggest a reason for the difference in boiling points between the following isomeric pairs of compounds. (Several structural features may be responsible.)

(a) $CH_3{-}CH_2{-}CH_2{-}O{-}CH_2{-}CH_2{-}CH_3$ (bp 90.5 °C) and $CH_3{-}CH(CH_3){-}O{-}CH(CH_3){-}CH_3$ (bp 68 °C)

(b) 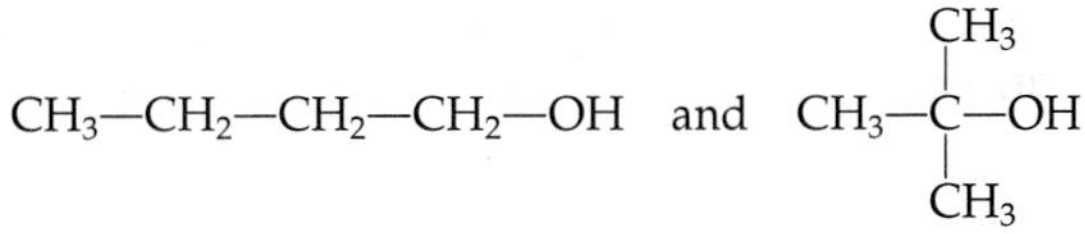 $CH_3-CH_2-CH_2-CH_2-OH$ and $CH_3-C(CH_3)_2-OH$

bp 117.7 °C; bp 82.5 °C

(c) 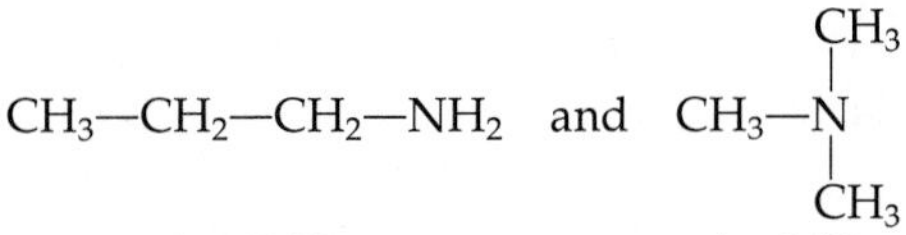 $CH_3-CH_2-CH_2-NH_2$ and $CH_3-N(CH_3)-CH_3$

bp 49 °C; bp 3 °C

(d) 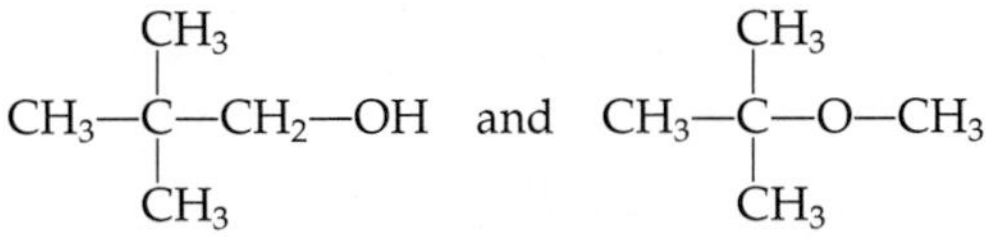 $CH_3-C(CH_3)_2-CH_2-OH$ and $CH_3-C(CH_3)_2-O-CH_3$

bp 113 °C; bp 55 °C

2.2 The boiling points of the following pairs of isomeric compounds are very similar. Explain why.

(a) CH_3-CH_2-SH and CH_3-S-CH_3

bp 35 °C; bp 37 °C

(b) $CH_3-CH_2-CH_2-S-CH_3$ and $CH_3-CH_2-S-CH_2-CH_3$

bp 95.5 °C; bp 92.1 °C

(c) 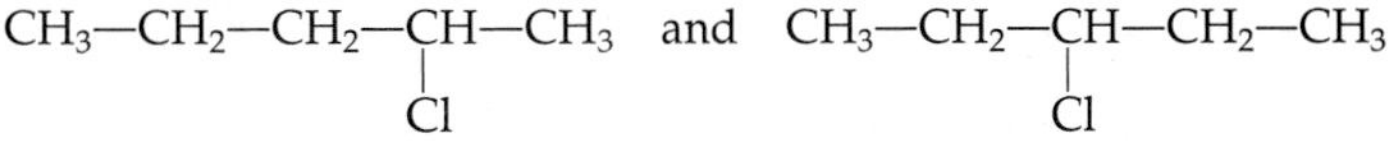 $CH_3-CH_2-CH_2-CH(Cl)-CH_3$ and $CH_3-CH_2-CH(Cl)-CH_2-CH_3$

bp 96.9 °C; bp 97.8 °C

(d) 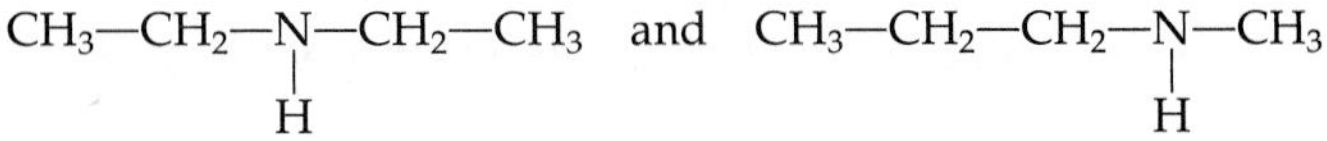 $CH_3-CH_2-N(H)-CH_2-CH_3$ and $CH_3-CH_2-CH_2-N(H)-CH_3$

bp 56 °C; bp 61 °C

2.3 Suggest reasons for the order of boiling points for the following compounds.

$CH_3-CH_2-CH_2-CH_2-F$ (bp 2.5 °C) $\quad$ $CH_3-CH_2-O-CH_2-CH_3$ (bp 34.6 °C) $\quad$ $CH_3-CH_2-S-CH_3$ (bp 66.6 °C)

2.4 Suggest reasons for the order of boiling points for the following compounds.

$F-CH_2-CH_2-F$ (bp 30.7 °C) $\quad$ $NH_2-CH_2-CH_2-NH_2$ (bp 116 °C) $\quad$ $HO-CH_2-CH_2-OH$ (bp 190 °C)

2.5 The ethene molecule is planar and all bond angles are close to 120°. There are three isomeric dichloroethenes. Two isomers have dipole moments and the third does not. Which of the following three is nonpolar? Explain why.

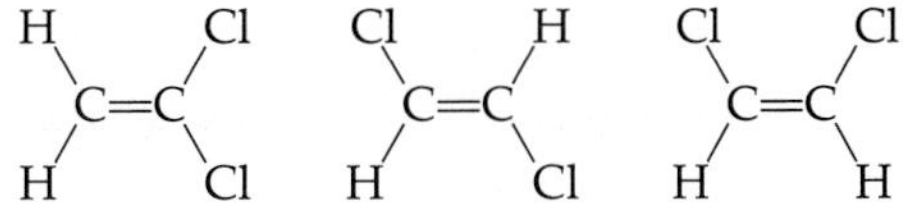

$(H)(H)C=C(Cl)(Cl)$ $\quad$ $(Cl)(H)C=C(H)(Cl)$ $\quad$ $(Cl)(H)C=C(Cl)(H)$

2.6 One of the following compounds has a dipole moment and the other does not. Select the polar compound and explain why the other compound has no dipole moment.

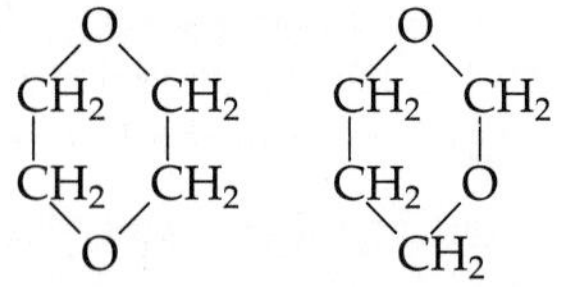

```
        O              O
      /   \          /   \
   CH2     CH2    CH2     CH2
    |       |      |       |
   CH2     CH2    CH2      O
      \   /          \   /
        O             CH2
```

2.7 Propylene glycol is miscible with water but the solubility of 1-butanol is only 7.9 g/100 mL of water. Explain why.

```
CH3—CH—CH2—OH        CH3—CH2—CH2—CH2—OH
     |
     OH
 propylene glycol          1-butanol
```

2.8 Butanoic acid is miscible with water, but ethyl ethanoate is not. Explain why.

```
              O                 O
              ‖                 ‖
CH3—CH2—CH2—C—O—H        CH3—C—O—CH2—CH3
   butanoic acid           ethyl ethanoate
```

Conjugate Acids and Bases

2.9 Write the structure of the conjugate acid of each of the following species.
(a) H—O—O—H (b) NH_2—NH_2 (c) CH_3—S—CH_3
(d) CH_3—O—CH_3 (e) CH_3—NH_2 (f) CH_3—OH

2.10 Write the structure of the conjugate base of each of the following species.
(a) CH_3—SH (b) CH_3—NH_2 (c) CH_3—O—SO_3H (d) CH_2═CH_2
(e) CH≡CH (f) CH_3—CN

2.11 Write the structure of the two conjugate acids of hydroxylamine, NH_2—OH. Which is the more acidic?

2.12 Write the structure of the two conjugate bases of hydroxylamine, NH_2—OH. Which is the more basic?

2.13 Write the structure of the conjugate acid of each of the following species.

```
                                                        O
                                                        ‖
(a) H2C═O  (b) CH3—NH—CH3  (c) CH2═NH  (d) CH3—C—OH
```

2.14 Write the structure of the conjugate acid of each of the following species.

```
      O              CH2—CH2            CH2              O
     / \              |    |            /  \              ‖
(a) CH2—CH2   (b)  CH2    CH2   (c) CH═N    (d) CH3—C—CH3
                      \  /
                       S
```

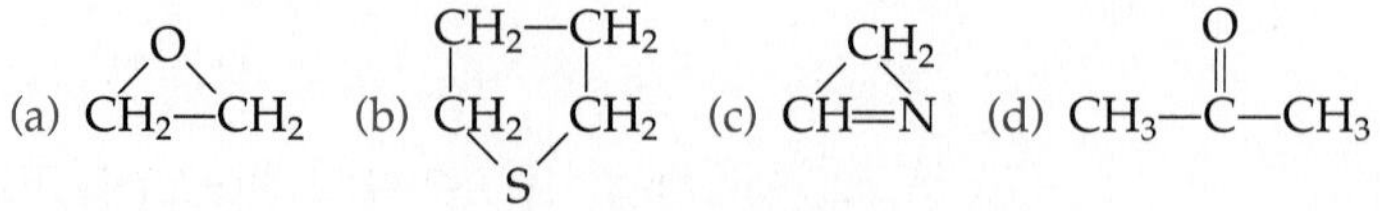

Lewis Acids and Bases

2.15 Identify the Lewis acid and Lewis base in each of the following reactions.
(a) CH_3—CH_2—Cl + $AlCl_3$ ⟶ CH_3—CH_2^+ + $AlCl_4^-$
(b) CH_3—CH_2—SH + CH_3—O^- ⟶ CH_3—CH_2—S^- + CH_3—OH
(c) CH_3—CH_2—OH + NH_2^- ⟶ CH_3—CH_2—O^- + NH_3
(d) $(CH_3)_2N^-$ + CH_3—OH ⟶ $(CH_3)_2NH$ + CH_3—O^-

2.16 Identify the Lewis acid and Lewis base in each of the following reactions.
(a) $(CH_3)_2O$ + HI ⟶ $(CH_3)_2OH^+$ + I^-
(b) CH_3—CH_2^+ + H_2O ⟶ CH_3—CH_2—OH_2^+

(c) $CH_3{-}CH{=}CH_2 + HBr \longrightarrow (CH_3)_2CH^+ + Br^-$
(d) $CH_3{-}C{\equiv}CH + CH_3{-}NH^- \longrightarrow CH_3{-}C{\equiv}C^- + CH_3{-}NH_2$

pK_a and Acid Strength

2.17 The approximate pK_a values of CH_4 and CH_3OH are 49 and 16, respectively. Which is the stronger acid? Will the equilibrium position of the following reaction lie to the left or right?

$$CH_4 + CH_3O^- \rightleftharpoons CH_3^- + CH_3OH$$

2.18 The approximate pK_a values of NH_3 and CH_3OH are 36 and 16, respectively. Which is the stronger acid? Will the equilibrium position of the following reaction lie to the left or right?

$$CH_3OH + NH_2^- \rightleftharpoons NH_3 + CH_3O^-$$

2.19 The pK_a of acetic acid ($CH_3{-}CO_2H$) is 4.7. Explain why the carboxylic acid group of amoxicillin (pK_a = 2.4), a synthetic penicillin, is more acidic than acetic acid, whereas the carboxylic acid group of indomethacin (pK_a = 4.5), an anti-inflammatory analgesic used to treat rheumatoid arthritis, is of comparable acidity.

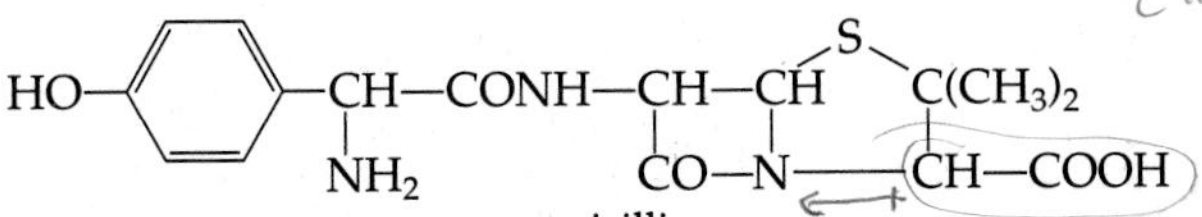

HO—C6H4—CH(NH2)—CONH—CH—CH—S—C(CH3)2; CO—N—CH—COOH

amoxicillin

CH3O, CH2COOH, N, CH3, C=O, Cl

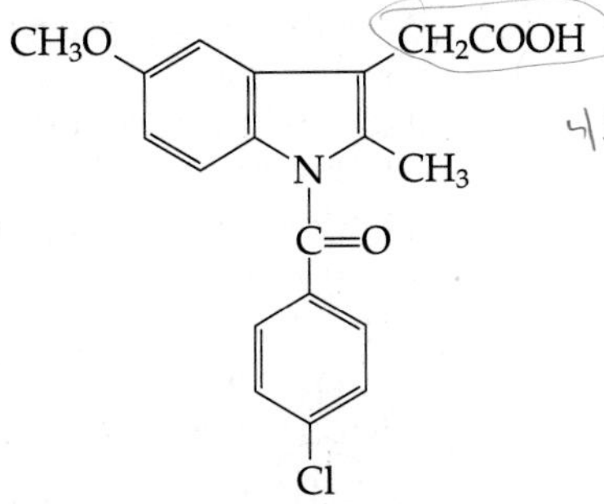

indomethacin

2.20 The pK_a of the —OH group of phenobarbital is 7.5, whereas the pK_a of CH_3OH is 16. Explain why phenobarbital is significantly more acidic.

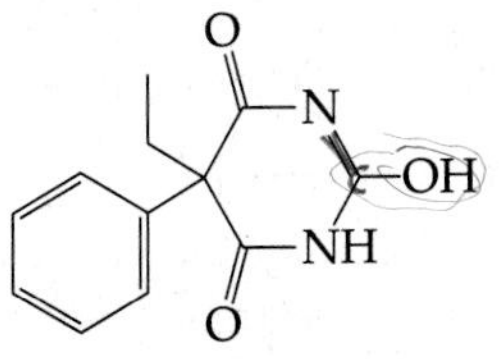

O, N, OH, NH, O

phenobarbital

2.21 The N—H bond is not very acidic; the pK_a of ammonia is 36. However the pK_a of sulfanilamide, a sulfa drug, is 10.4. Suggest a reason for the higher acidity of sulfanilamide.

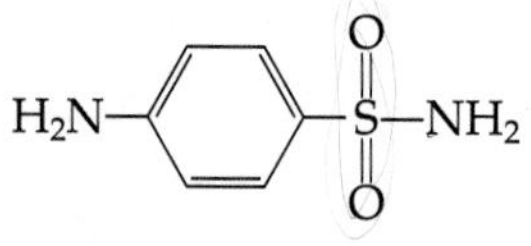

$$H_2N\text{—}C_6H_4\text{—}S(=O)_2\text{—}NH_2$$

2.22 The pK_a of sulfadiazine, a sulfa drug, is 6.5. Why is this compound more acidic than sulfanilamide?

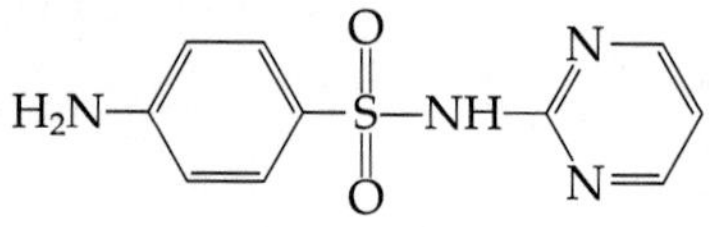

$$H_2N\text{—}C_6H_4\text{—}S(=O)_2\text{—}NH\text{—}C_4H_3N_2$$

Oxidation–Reduction Reactions

2.23 Determine if each of the following transformations given by unbalanced equations involves oxidation, reduction, or neither.

(a) $CH_3\text{—}C\equiv N \longrightarrow CH_3\text{—}CH_2\text{—}NH_2$

(b) $2\ CH_3\text{—}SH \longrightarrow CH_3\text{—}S\text{—}S\text{—}CH_3$

(c) $CH_3\text{—}S\text{—}CH_3 \longrightarrow CH_3\text{—}S(\text{—}O)\text{—}CH_3$

(d) $CH_3\text{—}S(\text{—}O)_2\text{—}CH_3 \longrightarrow CH_3\text{—}S(\text{—}O)\text{—}CH_3$

2.24 None of the following reactions involves oxidation or reduction, although they may appear to be redox reactions. Explain why.

(a) $CH_3\text{—}CH\text{=}CH_2 \longrightarrow CH_3\text{—}CH(OH)\text{—}CH_3$

(b) $CH_3\text{—}C\equiv CH \longrightarrow CH_3\text{—}C(=O)\text{—}CH_3$

(c) $CH_3\text{—}C(=NH)\text{—}CH_3 \longrightarrow CH_3\text{—}C(=O)\text{—}CH_3$

(d) $CH_3\text{—}C(O\text{—}CH_3)_2\text{—}CH_3 \longrightarrow CH_3\text{—}C(=O)\text{—}CH_3$

2.25 Consider each of the following reactions for the metabolism of drugs. What type of reaction occurs?

(a) carbamazepine, an anticonvulsant

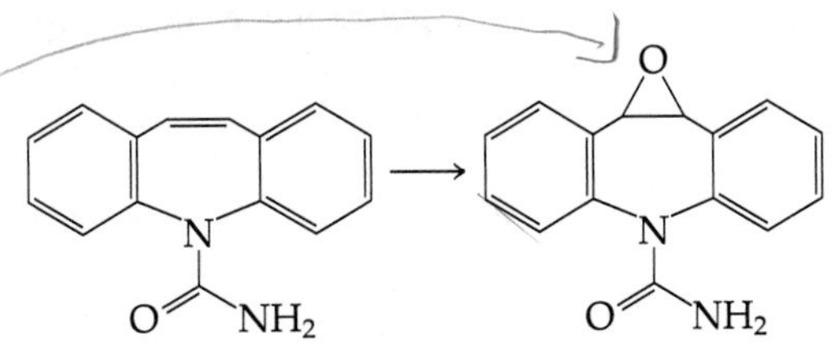

oxidation

(b) tolmetin, an anti-inflammatory drug

H_3C–C$_6$H$_4$–C(=O)–(N-methylpyrrol-2-yl)–CH_2COOH ⟶ HOOC–C$_6$H$_4$–C(=O)–(N-methylpyrrol-2-yl)–CH_2COOH

oxidation

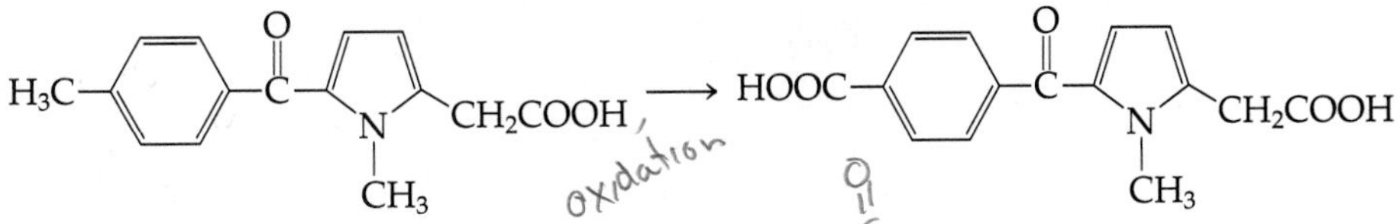

(c) dantrolene, a muscle relaxant

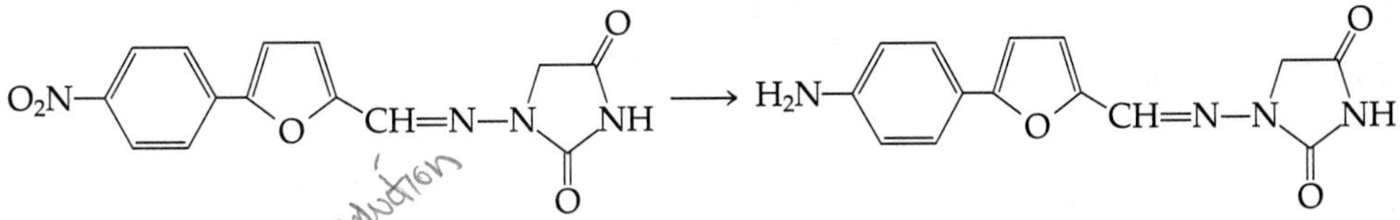

O_2N–C$_6$H$_4$–(furyl)–CH=N–N(hydantoin) ⟶ H_2N–C$_6$H$_4$–(furyl)–CH=N–N(hydantoin)

reduction

2.26 Consider each of the following reactions for the metabolism of drugs. What type of reaction occurs?

(a) sulindac, an anti-inflammatory drug

(F, CH_2COOH, CH_3, H, $CH_3S(=O)$) ⟶ (F, CH_2COOH, CH_3, H, CH_3S)

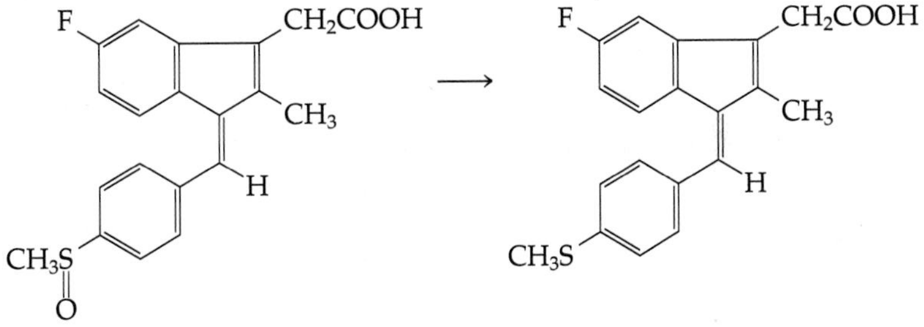

(b) ibuprofen, an analgesic

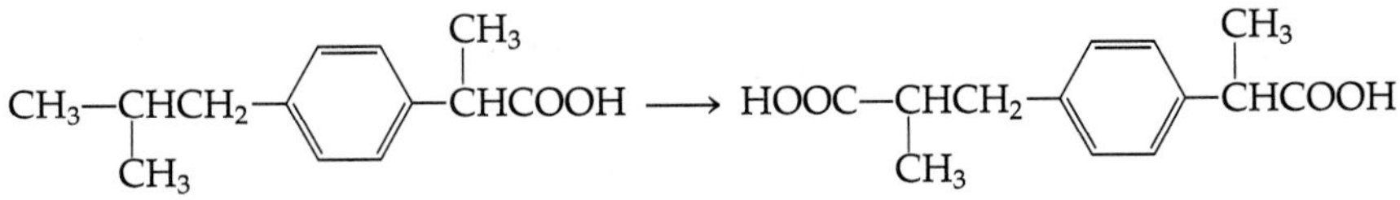

CH_3–CH(CH_3)CH_2–C$_6$H$_4$–CH(CH_3)COOH ⟶ HOOC–CH(CH_3)CH_2–C$_6$H$_4$–CH(CH_3)COOH

(c) disulfiram, a drug used in treating alcoholism

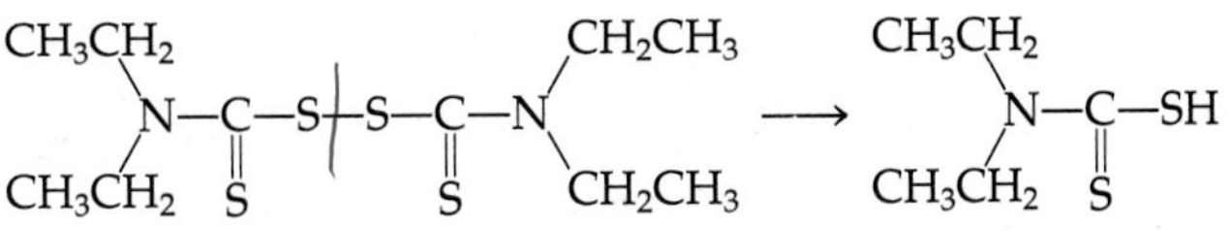

$(CH_3CH_2)_2N$–C(=S)–S–S–C(=S)–N$(CH_2CH_3)_2$ ⟶ $(CH_3CH_2)_2N$–C(=S)–SH

Types of Organic Reactions

2.27 Classify the type of reaction represented by each of the following unbalanced equations. Identify any additional reagents required for the reaction or any additional products that are formed.

(a) $$CH_3{-}\overset{\overset{\displaystyle O}{\|}}{C}{-}CH_3 \longrightarrow CH_3{-}\overset{\overset{\displaystyle OH}{|}}{\underset{\underset{\displaystyle OCH_3}{|}}{C}}{-}CH_3$$

(b) $$CH_3{-}CH_2{-}Br \longrightarrow CH_3{-}CH_2{-}CN$$

(c) $$\begin{matrix} CH_2{-}CH{-}OH \\ | \quad\;\; | \\ CH_2{-}CH_2 \end{matrix} \longrightarrow \begin{matrix} CH_2{-}CH \\ | \quad\;\; \| \\ CH_2{-}CH \end{matrix}$$

(d) $$CH_3{-}C{\equiv}CH \longrightarrow CH_2{=}C{=}CH_2$$

2.28 Classify the type of reaction represented by each of the following unbalanced equations. Identify any additional reagents required for the reaction or any additional products formed.

(a) $$2\ CH_3{-}CH_2{-}O{-}H \longrightarrow CH_3{-}CH_2{-}O{-}CH_2{-}CH_3$$

(b) $$CH_3{-}\overset{\overset{\displaystyle O}{\|}}{C}{-}S{-}CH_3 \longrightarrow CH_3{-}\overset{\overset{\displaystyle O}{\|}}{C}{-}O{-}H + CH_3SH$$

(c) $$CH_3{-}\overset{\overset{\displaystyle OH}{|}}{\underset{\underset{\displaystyle OCH_3}{|}}{C}}{-}CH_3 \longrightarrow CH_3{-}\overset{\overset{\displaystyle OCH_3}{|}}{\underset{\underset{\displaystyle OCH_3}{|}}{C}}{-}CH_3$$

(d) $$CH_3{-}CH{=}CH_2 \longrightarrow CH_3{-}\overset{\overset{\displaystyle OH}{|}}{CH}{-}CH_3$$

2.29 The metabolism of fatty acids (long-chain carboxylic acids) involves several steps. Indicate the type of reaction involved in each step. (The R represents a chain of carbon atoms. The CoA represents coenzyme A.)

(a) $$R{-}CH_2{-}CH_2{-}\overset{\overset{\displaystyle O}{\|}}{C}{-}SCoA \longrightarrow R{-}CH{=}CH{-}\overset{\overset{\displaystyle O}{\|}}{C}{-}SCoA$$

(b) $$R{-}CH{=}CH{-}\overset{\overset{\displaystyle O}{\|}}{C}{-}SCoA \longrightarrow R{-}\overset{\overset{\displaystyle OH}{|}}{CH}{-}CH_2{-}\overset{\overset{\displaystyle O}{\|}}{C}{-}SCoA$$

(c) $$R{-}\overset{\overset{\displaystyle OH}{|}}{CH}{-}CH_2{-}\overset{\overset{\displaystyle O}{\|}}{C}{-}SCoA \longrightarrow R{-}\overset{\overset{\displaystyle O}{\|}}{C}{-}CH_2{-}\overset{\overset{\displaystyle O}{\|}}{C}{-}SCoA$$

2.30 A series of ten steps involved in glycolysis (metabolism of glucose) includes the following three steps. Indicate the type of reaction involved in each step.

(a) $$\begin{matrix} CH_2OH \\ | \\ C{=}O \\ | \\ CH_2OPO_3^{2-} \end{matrix} \longrightarrow \begin{matrix} O{=}C{-}H \\ | \\ H{-}C{-}OH \\ | \\ CH_2OPO_3^{2-} \end{matrix}$$

(b) H—C(—COO⁻)—OH / H—C—OPO_3^{2-} / H ⟶ H—C(—COO⁻)—OPO_3^{2-} / H—C—OH / H

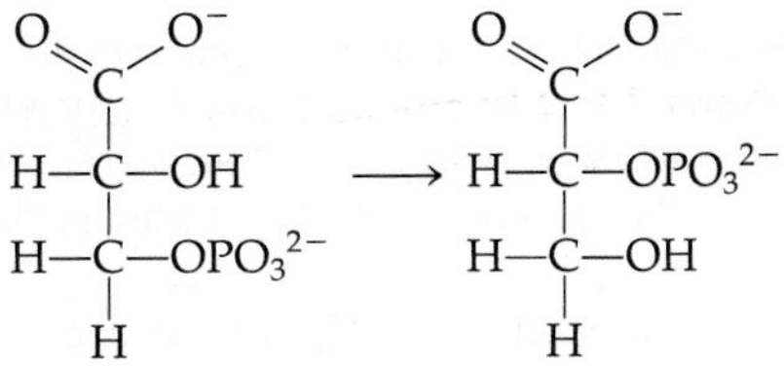

(c) H—C(—COO⁻)—OPO_3^{2-} / H—C—OH / H ⟶ C(—COO⁻)—OPO_3^{2-} ‖ H—C / H

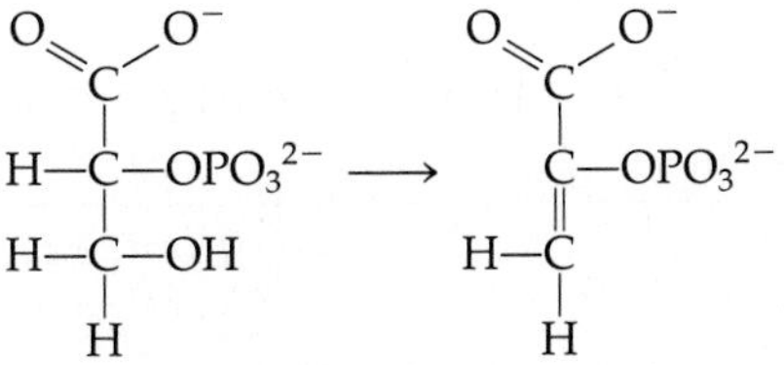

2.31 The antibacterial prontosil, a prodrug, is metabolized to produce sulfanilamide, the actual antibacterial compound. What type of reaction occurs to produce sulfanilamide?

H_2NO_2S—C₆H₄—N=N—C₆H₃(NH_2)—NH_2 ⟶ H_2NO_2S—C₆H₄—NH_2 + H_2N—C₆H₃(NH_2)—NH_2

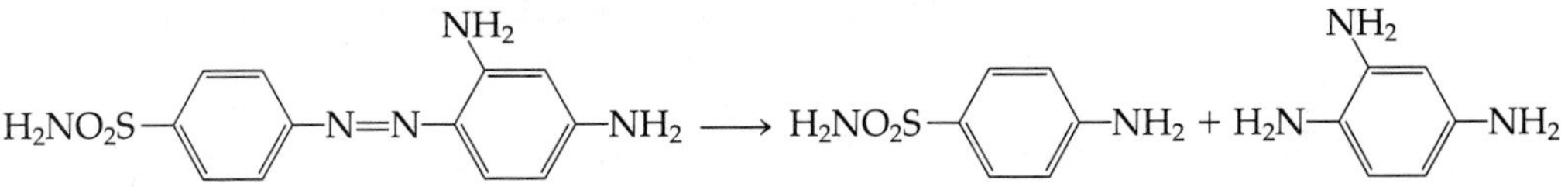

sulfanilamide

2.32 Tolbutamide, a hypoglycemic agent used to lower blood sugar in diabetics, is metabolized in a series of steps to the indicated product. How is the net overall reaction classified?

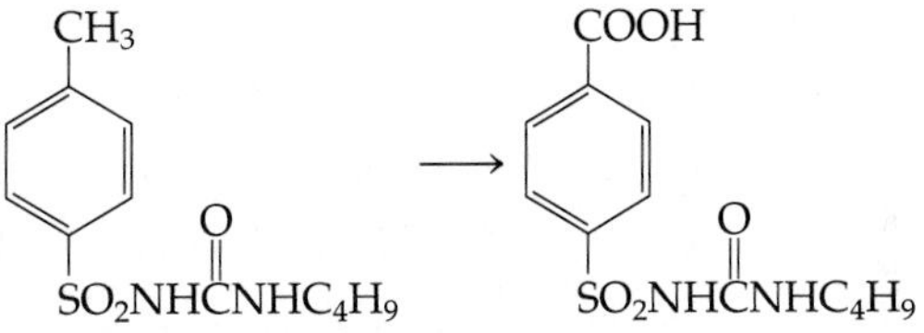

CH_3—C₆H₄—$SO_2NHC(=O)NHC_4H_9$ ⟶ COOH—C₆H₄—$SO_2NHC(=O)NHC_4H_9$

2.33 Chloroform is metabolized via an intermediate to phosgene, a compound that causes liver damage. What type of reactions are involved in the formation and decomposition of the intermediate?

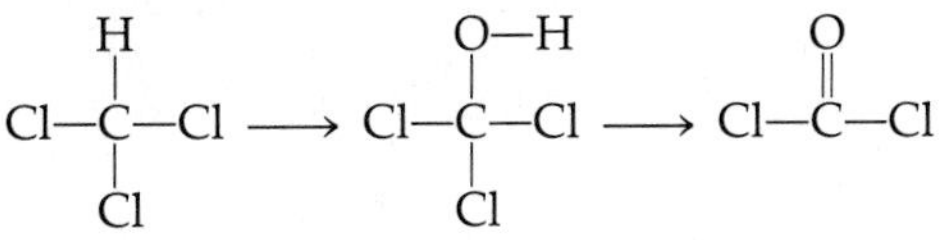

$CHCl_3$ ⟶ Cl—C(O—H)(Cl)—Cl ⟶ Cl—C(=O)—Cl

chloroform   phosgene

2.34 The sedative–hypnotic chloral hydrate is metabolized as follows. What type of reaction occurs in each step?

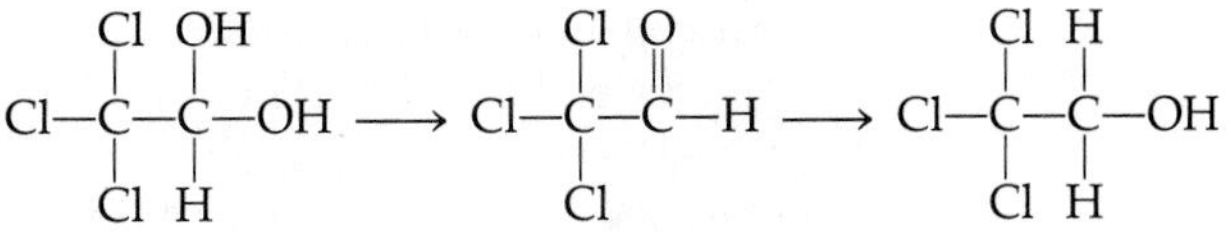

Cl_3C—CH(OH)—OH ⟶ Cl_3C—C(=O)—H ⟶ Cl_3C—CH_2—OH

Equilibria and Rates of Reaction

2.35 A reaction has $K = 1 \times 10^{-5}$. Are the products more or less stable than the reactants? Is the reaction exothermic or endothermic?

2.36 Could a reaction have $K = 1$? What relationship would exist between the energies of the reactants and products as shown in the reaction progress diagram?

2.37 Consider the following information about two reactions. Which reaction will occur at the faster rate at a common temperature?

Reaction	ΔH°	E_a
A → X	−30 kcal/mole	+25 kcal/mole
B → Y	−25 kcal/mole	+30 kcal/mole

2.38 Consider the information given in exercise 2.37. Which reaction is more exothermic?

2.39 Distinguish between an intermediate and a transition state.

2.40 A reaction occurs in three steps. How many transiton states are there? How many intermediates form?

Reaction Rates and Mechanisms

2.41 Identify the processes of bond cleavage and bond formation for each of the following reactions.

(a) $CH_4 + \cdot\ddot{Br}: \longrightarrow H_3C\cdot + H{-}\ddot{Br}:$

(b) $H_3C\cdot + :\ddot{Br}{-}\ddot{Br}: \longrightarrow H_3C{-}\ddot{Br}: + \cdot\ddot{Br}:$

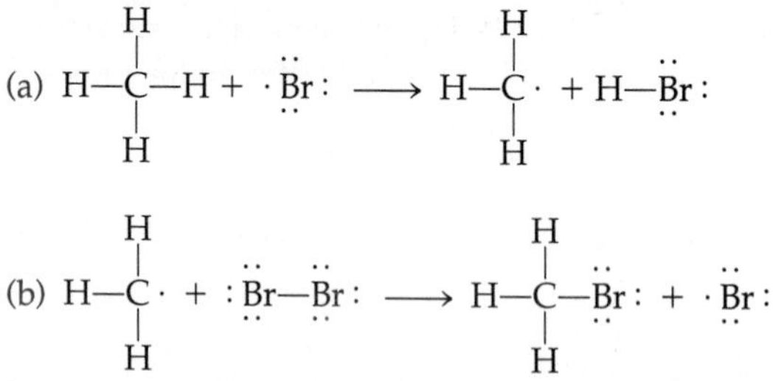

2.42 Identify the processes of bond cleavage and bond formation for each of the following reactions.

(a) $(CH_3)_3C^+ + HO^- \longrightarrow (CH_3)_3C{-}OH$

(b) $(CH_3)_3C{-}Cl \longrightarrow (CH_3)_3C^+ + Cl^-$

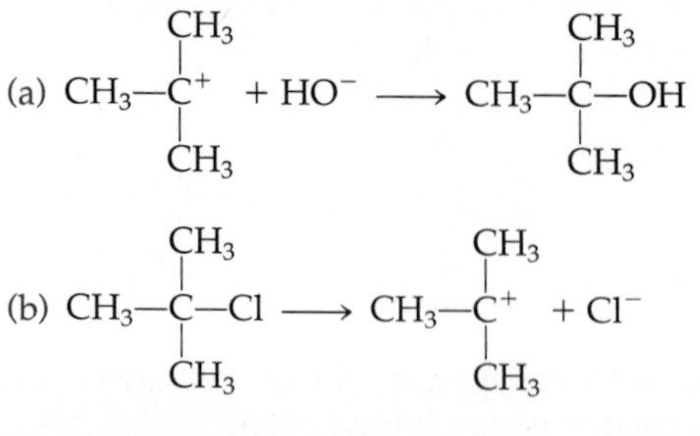

2.43 Benzoyl peroxide is used in creams to control acne. It is an irritant that causes proliferation of epithelial cells. It undergoes a homolytic cleavage of the oxygen–oxygen bond. Write the structure of the product, indicating all of the electrons present on the oxygen atom. What type of reactions might the product initiate?

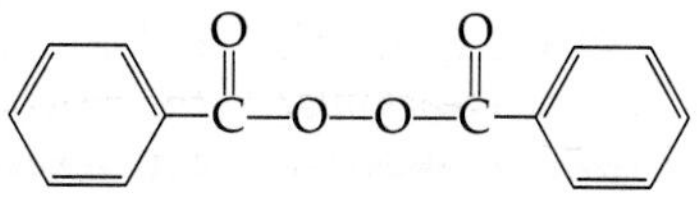

O‖—C—O—O—C—‖O (with phenyl rings on each end)

2.44 The oxygen–chlorine bond of hypochlorites, such as CH_3—O—Cl, can cleave heterolytically. Based on the electronegativity values of chlorine and oxygen, predict the charges on the cleavage products.

2.45 Hydrogen peroxide (H—O—O—H) reacts with a proton to give a conjugate acid that undergoes heterolytic oxygen–oxygen bond cleavage to yield water. What is the second product?

2.46 Chloromethane (CH_3—Cl) reacts with the Lewis acid $AlCl_3$ to give $AlCl_4^-$ and a carbon intermediate. What is the intermediate? What type of bond cleavage occurred?

2.47 The following alcohol acts as a base with a strong acid. The resulting conjugate acid produces water and an intermediate. Write the structure of the intermediate. What type of bond cleavage occurred?

CH_3
|
CH_3—C—OH
|
CH_3

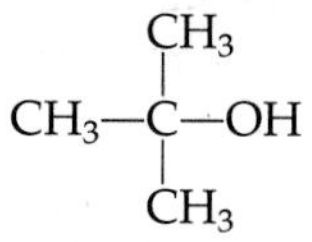

2.48 Bromine adds to the double bond of ethylene (CH_2=CH_2). The reaction occurs via an intermediate that results from the heterolytic cleavage of Br_2. Write a plausible two-step mechanism for the reaction.

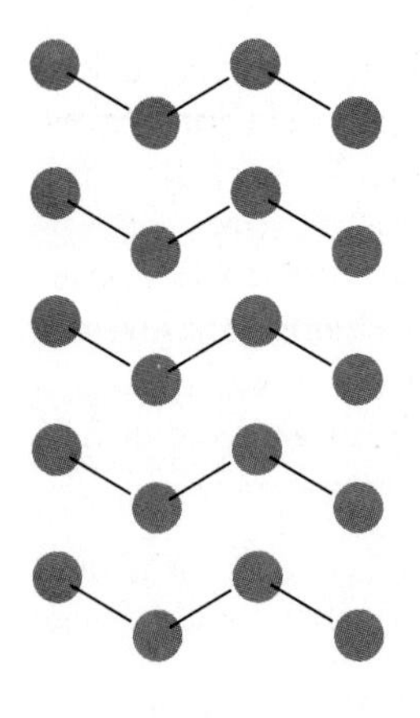

3

ALKANES AND CYCLOALKANES

3.1 CLASSES OF HYDROCARBONS

Now that we have reviewed some principles of chemistry and have seen how these principles can be extended to the study of organic chemistry, it is time to begin a systematic approach to the various classes of organic compounds. **Hydrocarbons** are compounds that contain only hydrogen and carbon. They occur as mixtures in natural gas, petroleum, and coal, which are collectively known as fossil fuels.

Hydrocarbons fall into several classes based on the types of bonds between the carbon atoms. A hydrocarbon that has only carbon–carbon single bonds and the maximum number of hydrogen atoms bonded to the carbon atoms is **saturated.** Hydrocarbons that contain carbon–carbon multiple bonds are **unsaturated.**

Saturated hydrocarbons are of two types: alkanes and cycloalkanes. **Alkanes** have carbon atoms bonded in chains; **cycloalkanes** have carbon atoms bonded to form a ring.

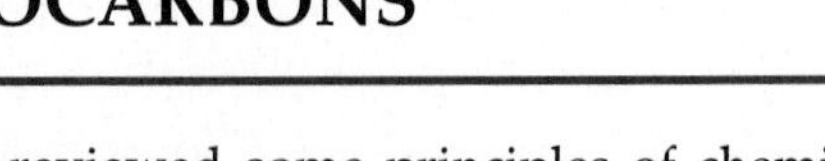

$CH_3—CH_2—CH_2—CH_3$

butane
(an alkane)

$$\begin{array}{c} CH_2—CH_2 \\ | \quad\quad | \\ CH_2—CH_2 \end{array}$$

cyclobutane
(a cycloalkane)

These two classes of hydrocarbons provide the molecular framework of most organic compounds. Compounds that have a chain of carbon atoms, some of which are attached to functional groups, are called **acyclic** compounds, meaning *not cyclic*. Compounds that contain rings of carbon atoms and may also contain functional groups are **carbocyclic** compounds, commonly called cyclic compounds. However, some cyclic compounds contain at least one atom in the ring that is not a carbon atom; those atoms are called **heteroatoms.** Cyclic compounds containing one or more heteroatoms are called **heterocyclic** compounds (Chapters 5 and 14).

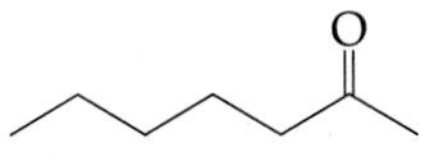

O N N CH_3

2-heptanone (in oil of cloves) | carvone (in spearmint oil) | nicotine (in tobacco)

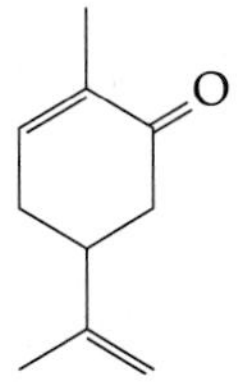

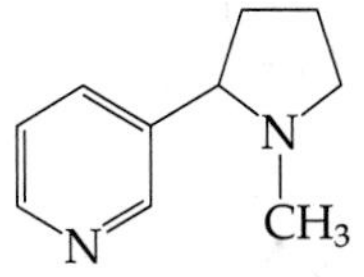

3.2 ALKANES

Alkanes contain carbon atoms bonded either to other carbon atoms or to hydrogen atoms by single covalent bonds. Each carbon atom in an alkane has four sp^3 orbitals arranged in a tetrahedron to form four σ bonds to four atoms.

Normal and Branched Alkanes

Hydrocarbons with a continuous chain of carbon atoms are **normal alkanes.** An example of a normal alkane is octane. Normal alkanes are usually drawn with the carbon chain in a horizontal straight line.

$$CH_3—CH_2—CH_2—CH_2—CH_2—CH_2—CH_2—CH_3$$

octane (a normal alkane)

The names and condensed structural formulas of 20 normal alkanes are given in Table 3.1. The first four compounds have common names. The names of the higher molecular weight compounds are derived from Greek numbers that indicate the number of carbon atoms. Each name has the suffix *-ane,* which identifies the compound as an alkane.

Hydrocarbons that have carbon atoms bonded to more than two other carbon atoms are called **branched alkanes.** The carbon atom bonded to three or four other carbon atoms is the branching point. The carbon atom attached to the main chain of carbon atoms at the branching point is part of

TABLE 3.1 Names of Normal Alkanes

Number of carbon atoms	Name	Molecular formula
1	methane	CH_4
2	ethane	C_2H_6
3	propane	C_3H_8
4	butane	C_4H_{10}
5	pentane	C_5H_{12}
6	hexane	C_6H_{14}
7	heptane	C_7H_{16}
8	octane	C_8H_{18}
9	nonane	C_9H_{20}
10	decane	$C_{10}H_{22}$
11	undecane	$C_{11}H_{24}$
12	dodecane	$C_{12}H_{26}$
13	tridecane	$C_{13}H_{28}$
14	tetradecane	$C_{14}H_{30}$
15	pentadecane	$C_{15}H_{32}$
16	hexadecane	$C_{16}H_{34}$
17	heptadecane	$C_{17}H_{36}$
18	octadecane	$C_{18}H_{38}$
19	nonadecane	$C_{19}H_{40}$
20	eicosane	$C_{20}H_{42}$

an **alkyl group.** An example of a branched alkane is isobutane. It has three carbon atoms in the main chain and one branch, a $—CH_3$ group.

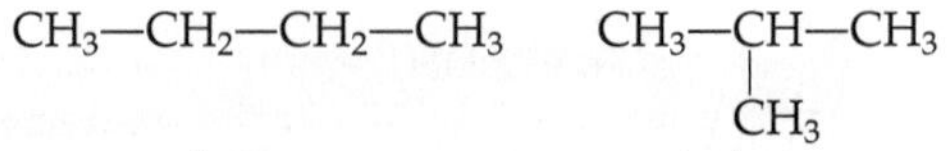

$CH_3—CH_2—CH_2—CH_3$
butane

$CH_3—CH(CH_3)—CH_3$
isobutane

Isopentane is a branched alkane that is an isomer of pentane. It has four carbon atoms in the main chain and a $—CH_3$ alkyl group. Pentane has another isomer called neopentane. It has three carbon atoms in the chain with two $—CH_3$ groups bonded to the central carbon atom. Pentane, isopentane, and neopentane have different physical properties; their boiling points are 36, 28, and 10 °C, respectively.

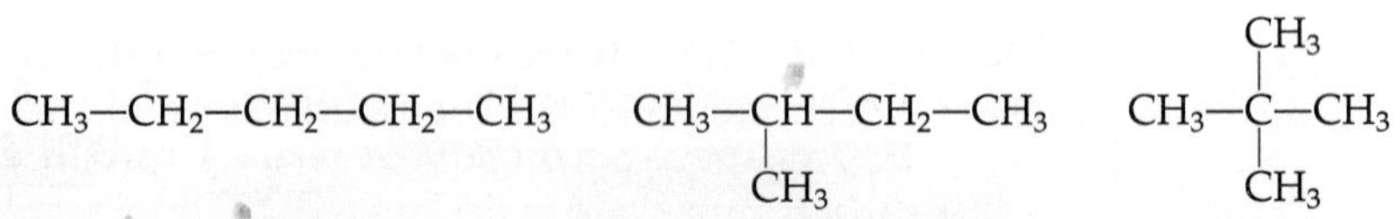

$CH_3—CH_2—CH_2—CH_2—CH_3$
pentane
(bp 36 °C)

$CH_3—CH(CH_3)—CH_2—CH_3$
isopentane
(bp 28 °C)

$CH_3—C(CH_3)_2—CH_3$
neopentane
(bp 10 °C)

Both normal and branched alkanes have the general molecular formula C_nH_{2n+2}. For example, the molecular formula of hexane is C_6H_{14}.

$$
\begin{array}{c}
\boxed{\begin{array}{cccccc} H & H & H & H & H & H \\ | & | & | & | & | & | \\ C- & C- & C- & C- & C- & C \\ | & | & | & | & | & | \\ H & H & H & H & H & H \end{array}} \\
H - \cdots - H \\
C_nH_{2n} \\
H_1 + C_6H_{12} + H_1 = C_6H_{14}
\end{array}
$$

Each carbon atom in this normal alkane, where $n = 6$, has at least two hydrogen atoms bonded to it, which accounts for the $2n$ in the general formula. Each of the two terminal carbon atoms has another hydrogen atom bonded to it, which accounts for the +2 in the subscript on hydrogen.

The molecular formulas of the series of alkanes differ from one another in the number of $—CH_2—$ units. A series of compounds whose members differ from adjacent members by a repeating unit is called a **homologous series.**

EXAMPLE 3.1 One of the components of the wax of a cabbage leaf is a normal alkane containing 29 carbon atoms. What is the molecular formula of the compound?

Solution The value of n is 29. There must be $(2 \times 29) + 2$ hydrogen atoms. The molecular formula is $C_{29}H_{60}$. Any alkane, normal or branched, with 29 carbon atoms would have 60 hydrogen atoms.

Classification of Carbon Atoms

We can classify parts of a hydrocarbon structure according to the number of carbon atoms directly bonded to a specific carbon atom. This classification is used to describe the reactivity of functional groups attached at the various carbon atoms in a structure. As we will see many times in later chapters, reactivity is related to structure.

A carbon atom bonded to only one other carbon atom is called a **primary carbon atom.** A primary carbon atom is designated by the symbol 1°. The carbon atom at each end of a carbon chain is primary. For example, ethane and propane each have two primary carbon atoms. In contrast, the middle carbon atom in propane is not primary because it is bonded to two other carbon atoms (Figure 3.1).

A carbon atom that is bonded to two other carbon atoms is a **secondary carbon atom,** designated by the symbol 2°. For example, the middle carbon atom of propane is secondary. A **tertiary carbon atom** is bonded to three other carbon atoms and is designated 3°. For example, when we examine the structure of isobutane, we see that one of the four carbon atoms is tertiary;

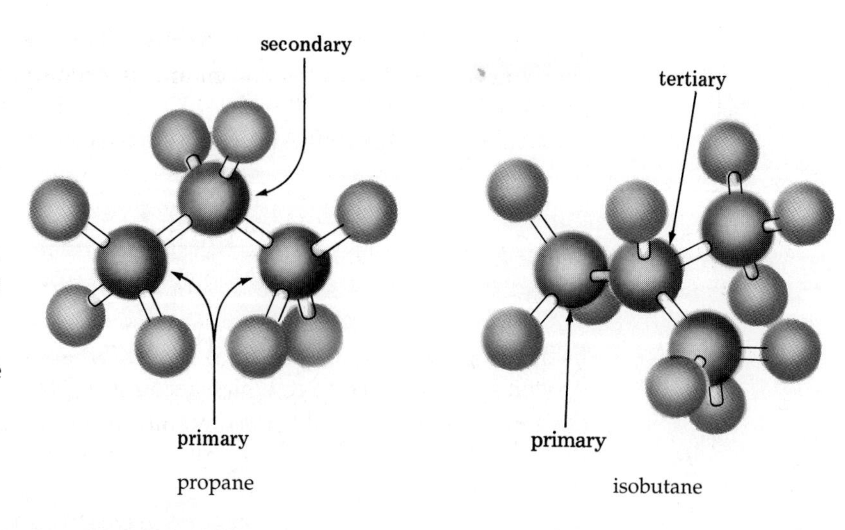

FIGURE 3.1 Classification of Carbon Atoms The terminal carbon atoms of propane are primary; they are directly bonded to only one other carbon atom. The internal carbon atom is secondary; it is bonded to two carbon atoms. The terminal carbon atoms of isobutane are primary; they are directly bonded to only one other carbon atom. The internal carbon atom is tertiary; it is bonded to three carbon atoms.

the other three are primary (Figure 3.1). A **quaternary carbon atom** is bonded to four other carbon atoms.

EXAMPLE 3.2 The following compound is a sex attractant released by the female tiger moth. Classify the carbon atoms in this compound as 1°, 2°, or 3°.

$$\begin{array}{l} \quad\;\; CH_3 \\ \quad\;\; | \\ CH_3CHCH_2CH_2CH_2CH_2CH_2CH_2CH_2CH_2CH_2CH_2CH_2CH_2CH_2CH_2CH_3 \end{array}$$

Solution The two terminal carbon atoms and the branching CH_3— group each are primary carbon atoms because they are each bonded to only one other carbon atom. The second carbon atom from the left is bonded to two atoms in the chain as well as the branching CH_3— group and is tertiary. All remaining carbon atoms are bonded to two carbon atoms. These 14 carbon atoms are all secondary.

EXAMPLE 3.3 Pentaerythritol tetranitrate is used to reduce the frequency and severity of angina attacks. Classify the carbon atoms in this compound.

$$\begin{array}{c} CH_2\text{—}O\text{—}NO_2 \\ | \\ O_2N\text{—}O\text{—}CH_2\text{—}C\text{—}CH_2\text{—}O\text{—}NO_2 \\ | \\ CH_2\text{—}O\text{—}NO_2 \end{array}$$

Solution Each of the four carbon atoms bearing an $—O—NO_2$ group has one bond to another carbon atom and is primary. The center carbon atom has all four bonds to $—CH_2—O—NO_2$ units. The four carbon–carbon bonds define the central carbon atom as quaternary.

3.3 NOMENCLATURE OF ALKANES

There are two isomeric C_4H_{10} alkanes and three isomeric C_5H_{12} alkanes. As the number of carbon atoms in an alkane increases, the number of isomers also increases (Table 3.2). Although many of these possible isomers have never been found in petroleum, any one of them could conceivably be made in the laboratory. A system for naming the many isomeric alkanes is clearly needed.

IUPAC Rules

Alkanes are named by the rules set forth by the International Union of Pure and Applied Chemistry (IUPAC). When these rules are followed, a unique name describes each compound. The IUPAC name is constructed of three parts: prefix, parent, and suffix.

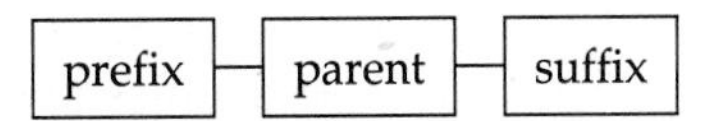

The **parent** is the longest continuous carbon chain in a molecule. A parent alkane has the ending *-ane*. Other suffixes identify functional groups such as a hydroxyl group. Some functional groups such as the halogens are identi-

TABLE 3.2 Number of Isomers of the Alkanes

Molecular formula	Number of isomers
C_4H_{10}	2
C_5H_{12}	3
C_6H_{14}	5
C_7H_{16}	9
C_8H_{18}	18
C_9H_{20}	35
$C_{10}H_{22}$	75
$C_{20}H_{42}$	336,319
$C_{30}H_{62}$	4,111,846,763
$C_{40}H_{82}$	62,491,178,805,831

fied in the prefix. For example, the prefixes *chloro-* and *bromo-* identify chlorine and bromine, respectively.

The prefix also indicates the identity and location of any branching alkyl groups. An alkane that has "lost" one hydrogen atom is called an **alkyl group**. Alkyl groups are named by replacing the *-ane* ending of an alkane by *-yl*. The parent name of CH_4 is methane. Thus, CH_3— is a methyl group. The parent name of C_2H_6 is ethane and therefore, CH_3CH_2— is an ethyl group.

$$\begin{array}{c} H \\ | \\ H{-}C{-}H \\ | \\ H \end{array} \quad \text{removing H gives} \quad \begin{array}{c} H \\ | \\ H{-}C{-} \\ | \\ H \end{array} \quad \text{or} \quad CH_3{-}$$

methyl group

$$\begin{array}{c} H\ \ H \\ |\ \ \ | \\ H{-}C{-}C{-}H \\ |\ \ \ | \\ H\ \ H \end{array} \quad \text{removing H gives} \quad \begin{array}{c} H\ \ H \\ |\ \ \ | \\ H{-}C{-}C{-} \\ |\ \ \ | \\ H\ \ H \end{array} \quad \text{or} \quad CH_3CH_2{-}$$

ethyl group

The general shorthand representation of an alkyl group is R—, which stands for the "rest" or "remainder" of the molecule.

The names of alkanes specify the length of the carbon chain and the location and identity of alkyl groups attached to it. The IUPAC rules for naming alkanes are as follows.

1. The longest continuous chain of carbon atoms is the parent. This chain is not always immediately apparent.

$$\begin{array}{l} \qquad\qquad\qquad\ \ CH_2{-}CH_3 \\ \qquad\qquad\qquad\ \ | \\ CH_3{-}CH_2{-}CH{-}CH_3 \end{array}$$

There are five carbon atoms in the longest carbon chain—not four carbon atoms.

If two possible chains have the same number of carbon atoms, the parent is the one with the larger number of branch points.

$$\begin{array}{l} \qquad\qquad\quad CH_2{-}CH_3 \\ \qquad\qquad\quad | \\ CH_3{-}CH{-}CH{-}CH_2{-}CH_2{-}CH_3 \\ \qquad\quad | \\ \qquad\quad CH_3 \end{array}$$

This compound should be considered a six-carbon parent chain with two branches, a methyl group and an ethyl group.

$$\begin{array}{l} \qquad\qquad\quad CH_2{-}CH_3 \\ \qquad\qquad\quad | \\ CH_3{-}CH{-}CH{-}CH_2{-}CH_2{-}CH_3 \\ \qquad\quad | \\ \qquad\quad CH_3 \end{array}$$

A six-carbon parent chain with only a single three-carbon alkyl group is not a correct choice.

2. Number the carbon atoms in the longest continuous chain starting from the end of the chain nearer the first branch.

end nearer first branch

$$\overset{1}{CH_3}-\underset{\underset{CH_3}{|}}{\overset{2}{CH}}-\underset{3}{CH_2}-\underset{4}{\overset{CH_2-CH_3}{\overset{|}{CH}}}-\underset{5}{CH_2}-\underset{6}{CH_3}$$

This substituted hexane has a methyl group at C-2 and an ethyl group at C-4, not an ethyl group at C-3 and a methyl group at C-5.

If branching occurs at an equal distance from each end of the chain, number from the end that is nearer the second branch.

$$\overset{8}{CH_3}-\underset{\underset{CH_3}{|}}{\overset{7}{CH}}-\overset{6}{CH_2}-\underset{5}{CH_2}-\underset{4}{\overset{CH_3-CH_2}{\overset{|}{CH}}}-\underset{3}{CH_2}-\underset{2}{\overset{CH_3}{\overset{|}{CH}}}-\underset{1}{CH_3}$$

The ethyl group is closer to the right side of the molecule.

3. Each branch or substituent has a number that indicates its location on the parent chain. When two substituents are located on the same carbon atom, each must be assigned the same number. There must be as many numbers as there are branches or substituents.

$$\overset{8}{CH_3}-\overset{7}{CH_2}-\underset{\underset{CH_3}{|}}{\overset{6}{CH}}-\underset{5}{CH_2}-\underset{\underset{CH_3}{|}}{\overset{CH_3-CH_2}{\overset{|}{\underset{4}{C}}}}-\underset{3}{CH_2}-\underset{2}{\overset{CH_3}{\overset{|}{CH}}}-\underset{1}{CH_3}$$

This octane has methyl groups on the C-2, C-4, and C-6 atoms and an ethyl group on the C-4 atom.

4. The number for the position of each alkyl group is placed immediately before the name of the group and is joined to the name by a hyphen. Alkyl groups are listed in alphabetical order.

$$\overset{1}{CH_3}-\underset{\underset{CH_3}{|}}{\overset{2}{CH}}-\underset{3}{CH_2}-\underset{4}{\overset{CH_2-CH_3}{\overset{|}{CH}}}-\underset{5}{CH_2}-\underset{6}{CH_3}$$

This is 4-ethyl-2-methylhexane, not 2-methyl-4-ethylhexane.

If two or more groups of the same type are present, this is indicated by the prefixes *di-*, *tri-*, etc. The numbers that indicate the locations of the branches are separated by commas.

$$\underset{1}{CH_3}-\underset{2}{\overset{CH_3}{\overset{|}{CH}}}-\underset{3}{CH_2}-\underset{\underset{CH_3}{|}}{\overset{4}{CH}}-\overset{5}{CH_2}-\overset{6}{CH_3}$$

This is 2,4-dimethylhexane.

5. The prefixes di-, tri-, tetra-, etc., do not alter the alphabetical ordering of the alkyl groups.

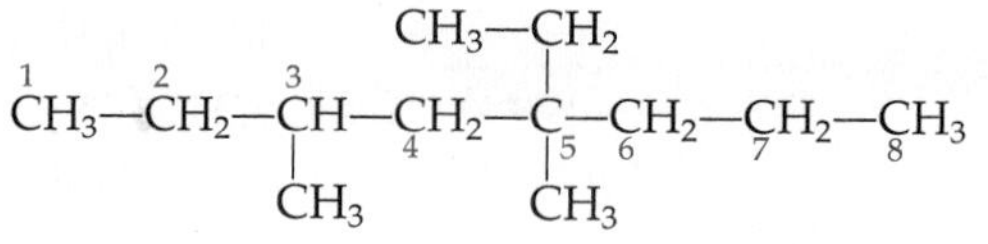

$$
\begin{array}{c}
CH_3-CH_2 \\
| \\
\overset{1}{C}H_3-\overset{2}{C}H_2-\overset{3}{C}H-\underset{4}{C}H_2-\underset{5}{C}-\underset{6}{C}H_2-\underset{7}{C}H_2-\underset{8}{C}H_3 \\
|| \\
CH_3CH_3
\end{array}
$$

This is 5-ethyl-3,5-dimethyloctane, not 3,5-dimethyl-5-ethyloctane.

EXAMPLE 3.4 Name the following compound, which is produced by the alga *Spirogyra*.

$$
\begin{array}{c}
CH_3CH_3CH_3CH_3 \\
|||| \\
CH_3CHCH_2CH_2CH_2CHCH_2CH_2CH_2CHCH_2CH_2CH_2CHCH_2CH_3
\end{array}
$$

Solution The longest continuous chain has sixteen carbon atoms and is named as a substituted hexadecane. The chain is numbered from left to right to locate the four methyl groups at positions 2, 6, 10, and 14. The compound is 2,6,10,14-tetramethylhexadecane.

EXAMPLE 3.5 Name the following compound.

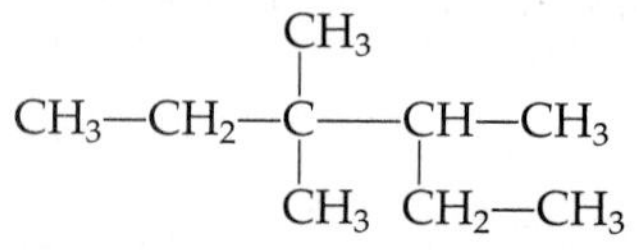

$$
\begin{array}{c}
CH_3 \\
| \\
CH_3-CH_2-C-CH-CH_3 \\
|| \\
CH_3\;CH_2-CH_3
\end{array}
$$

Solution The longest continuous chain has six carbon atoms. The chain is numbered to locate the methyl groups at positions 3, 3, and 4. The compound is 3,3,4-trimethylhexane.

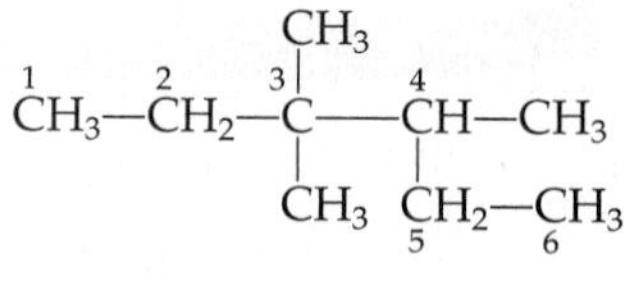

$$
\begin{array}{c}
CH_3 \\
| \\
\overset{1}{C}H_3-\overset{2}{C}H_2-\overset{3}{C}-\overset{4}{C}H-CH_3 \\
|| \\
CH_3\;\underset{5}{C}H_2-\underset{6}{C}H_3
\end{array}
$$

Selecting the five carbon atoms written in a straight line would give 2-ethyl-3,3-dimethylpentane, which is incorrect.

Names of Alkyl Groups

There is only one alkyl group derived from methane or ethane. However, for a longer chain of carbon atoms, there are usually several isomeric alkyl groups depending on which carbon atom "loses" a hydrogen atom. For example, propane has two 1° carbon atoms and a 2° carbon atom. If a primary carbon atom loses a hydrogen atom, a primary alkyl group, *propyl*, is produced. Propyl and other primary alkyl groups derived from normal alkanes are **normal alkyl groups.** If the 2° carbon atom of propane loses a hydrogen atom, a secondary alkyl group known as the *isopropyl* group is formed.

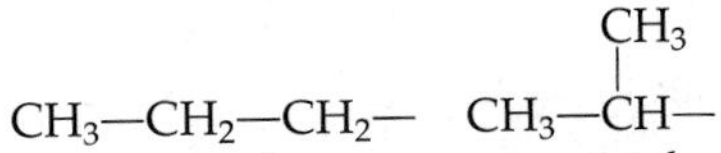

$$\underset{\text{propyl}}{CH_3{-}CH_2{-}CH_2{-}} \qquad \underset{\text{isopropyl}}{CH_3{-}\overset{\overset{\displaystyle CH_3}{|}}{C}H{-}}$$

Next, let's look at the alkyl groups that can be derived from the two isomers of butane, C_4H_{10}. These alkyl groups all have the formula C_4H_9. Two alkyl groups are derived from butane and two from isobutane. If a primary carbon atom of butane loses a hydrogen atom, a *butyl* group results; if a secondary carbon atom of butane loses a hydrogen atom, a secondary alkyl group called the *sec-butyl* group forms.

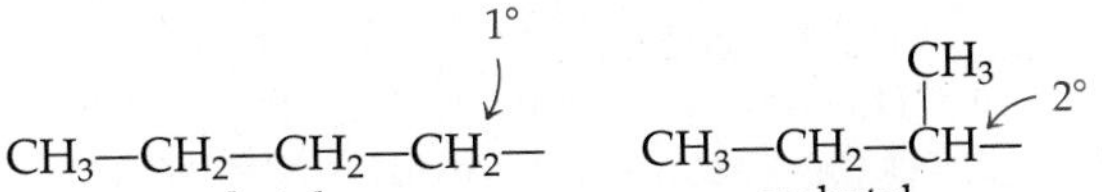

$$\underset{\text{butyl}}{CH_3{-}CH_2{-}CH_2{-}\overset{1^\circ}{CH_2}{-}} \qquad \underset{sec\text{-butyl}}{CH_3{-}CH_2{-}\overset{\overset{\displaystyle CH_3}{|}}{C}H{-}\;(2^\circ)}$$

Removal of a hydrogen atom from a primary carbon atom of isobutane gives a primary alkyl group called the *isobutyl* group. Removal of a hydrogen atom from the tertiary carbon atom of isobutane gives a tertiary alkyl group called the *tert-butyl* (*t*-butyl) group. Thus, there are four isomeric C_4H_9 alkyl groups.

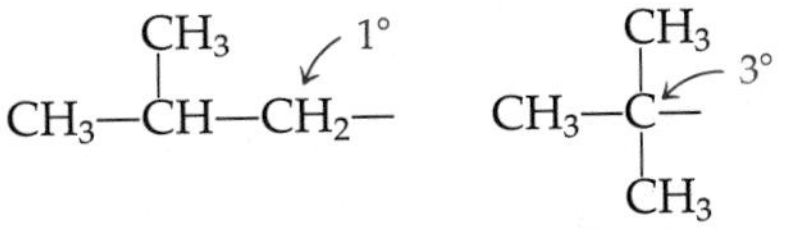

$$\underset{\text{isobutyl}}{CH_3{-}\overset{\overset{\displaystyle CH_3}{|}}{C}H{-}\overset{1^\circ}{CH_2}{-}} \qquad \underset{tert\text{-butyl}}{CH_3{-}\overset{\overset{\displaystyle CH_3}{|}}{\underset{\underset{\displaystyle CH_3}{|}}{C}}{-}\;(3^\circ)}$$

EXAMPLE 3.6 Identify the alkyl group on the left of the benzene ring in ibuprofen, an analgesic present in Nuprin, Advil, and Motrin.

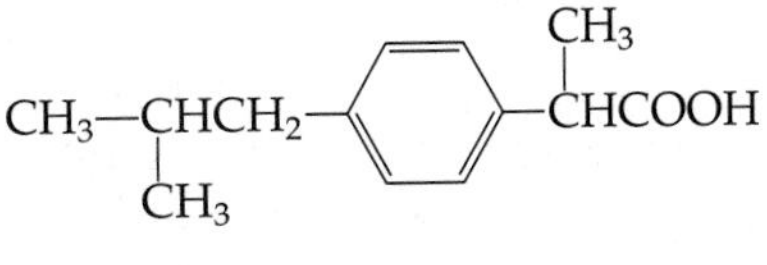

$$CH_3{-}\underset{\underset{\displaystyle CH_3}{|}}{C}HCH_2{-}C_6H_4{-}\overset{\overset{\displaystyle CH_3}{|}}{C}HCOOH$$

Solution There are four carbon atoms in the alkyl group, which is derived from isobutane—not butane. The benzene ring is bonded to the terminal carbon atom—not the internal carbon atom. This group is the isobutyl group.

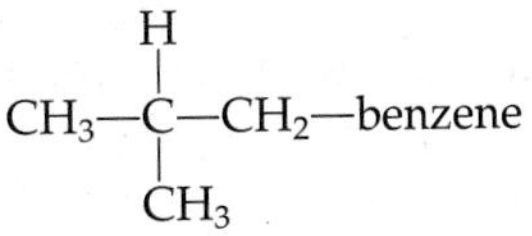

$$CH_3{-}\overset{\overset{\displaystyle H}{|}}{\underset{\underset{\displaystyle CH_3}{|}}{C}}{-}CH_2{-}\text{benzene}$$

Complex alkyl groups are named by an IUPAC procedure similar to that used to name alkanes. Complex alkyl groups are named by the longest continuous chain beginning at the branch point. Thus, the IUPAC name for an isopropyl group is 1-methylethyl, and the IUPAC name for an isobutyl group is 2-methylpropyl.

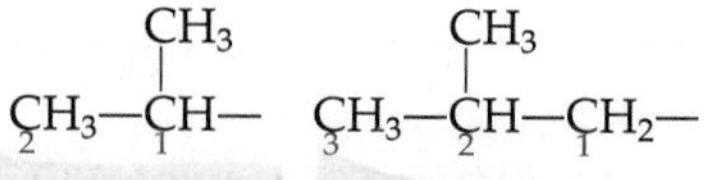

$$\underset{2}{CH_3}-\underset{1}{\overset{\overset{\displaystyle CH_3}{|}}{CH}}- \qquad \underset{3}{CH_3}-\underset{2}{\overset{\overset{\displaystyle CH_3}{|}}{CH}}-\underset{1}{CH_2}-$$

1-methylethyl　　2-methylpropyl

Complex alkyl groups are enclosed within parentheses when used to name hydrocarbons. Thus 4-isopropylheptane is also 4-(1-methylethyl)heptane. The nonsystematic names for the alkyl groups containing three and four carbon atoms are commonly used, and the IUPAC rules allow for their continued use.

EXAMPLE 3.7 The common name for the five-carbon alkyl group with a quaternary carbon atom is neopentyl. What is its systematic name?

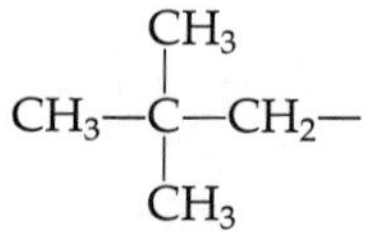

$$CH_3-\underset{\underset{\displaystyle CH_3}{|}}{\overset{\overset{\displaystyle CH_3}{|}}{C}}-CH_2-$$

neopentyl

Solution The "branch point" for this alkyl group, that is, the point at which it would be attached to a longer parent alkane chain, is on the right of the structure. Numbering the chain from right to left places both methyl groups at the 2 position. The name is 2,2-dimethylpropyl.

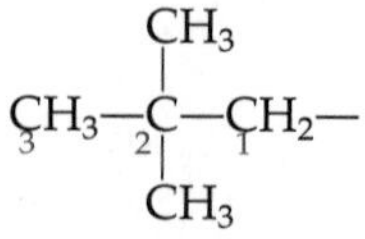

$$\underset{3}{CH_3}-\underset{\underset{\displaystyle CH_3}{|}}{\overset{\overset{\displaystyle CH_3}{|}}{\underset{2}{C}}}-\underset{1}{CH_2}-$$

EXAMPLE 3.8 Name the alkyl group bonded to the five-membered ring of cholesterol.

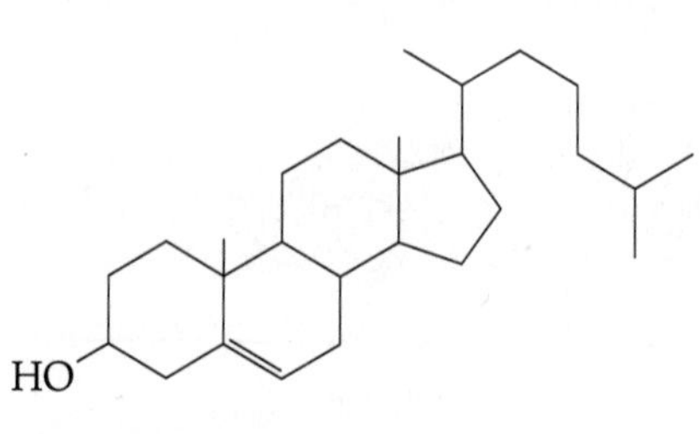

HO

Solution There are eight carbon atoms in the alkyl group. Six of these carbon atoms can be arranged in a continuous chain starting from the five-membered ring. The remaining two carbon atoms are methyl branches located at the 1 and 5 positions. The name is 1,5-dimethylhexyl.

$$\text{ring}-\underset{1}{\overset{\overset{\large CH_3}{|}}{C}H}-\underset{2}{CH_2}-\underset{3}{CH_2}-\underset{4}{CH_2}-\underset{5}{\overset{\overset{\large CH_3}{|}}{C}H}-\underset{6}{CH_3}$$

3.4 CONFORMATIONS OF ALKANES

Sometimes we can construct two molecular models that appear to be different but are not, in fact, structural isomers. Consider the models of ethane shown in Figure 3.2. In both examples, the two carbon atoms are bonded to each other, and each carbon atom is bonded to three hydrogen atoms. The two representations differ in the positions of the hydrogen atoms of one

FIGURE 3.2 Conformations of Ethane
Rotation of the methyl group on the right by 60° converts a staggered conformation into an eclipsed conformation. Viewing the carbon–carbon bond end-on in the eclipsed conformation, the observer would see only the carbon atom and the three hydrogen atoms on the right. The left carbon atom and its three hydrogen atoms would be hidden.

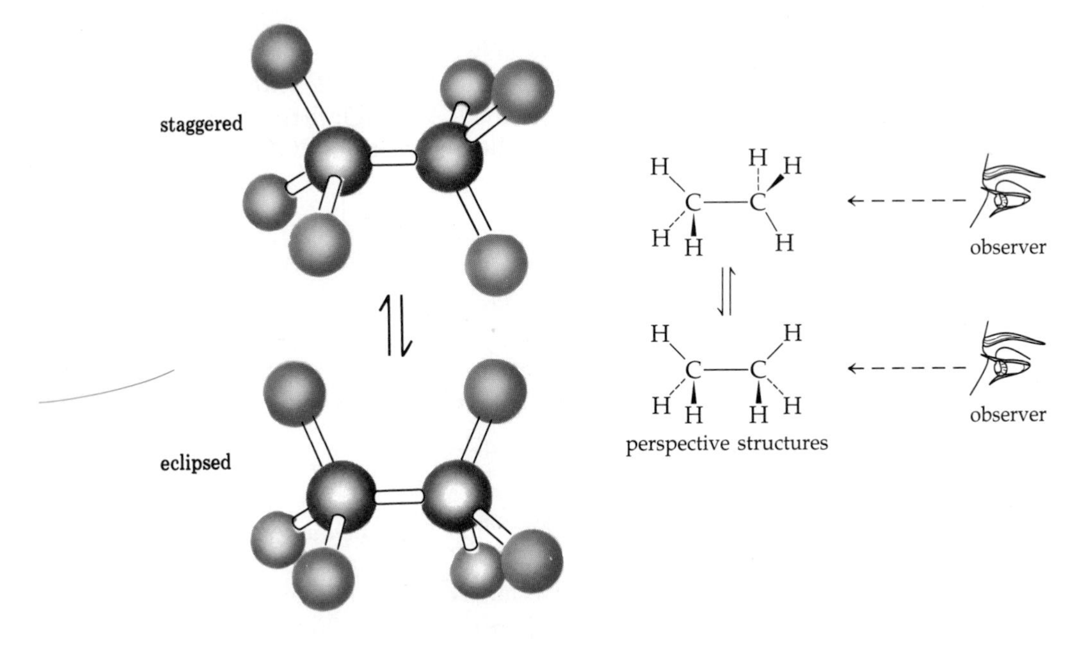

carbon atom relative to those of the other carbon atom. Which form represents ethane? The answer is that both do to some extent. Ethane and other molecules can exist in different orientations or **conformations** by rotation about σ bonds.

A single bond between two carbon atoms forms when an sp^3 hybrid orbital of one carbon atom overlaps an sp^3 hybrid orbital of another carbon atom along the axis between the two nuclei. Each carbon atom with its three bonded hydrogen atoms can rotate about the carbon–carbon σ bond. Rotation alters the spatial positions of the hydrogen atoms.

Rotation around the carbon–carbon σ bond occurs constantly in alkanes such as ethane. However, this rotation does not alter the connectivity of the carbon–carbon or carbon–hydrogen bonds. The motion is like the twisting and turning of your body while dancing. You may look different, but the parts of your body are still connected to the normal places. Only the orientation of your limbs is changing.

Ethane can exist in many conformations. The conformation in which the hydrogen atoms and the bonding electrons are the farthest away from one another has the lowest energy. This conformation is said to be **staggered.** The conformation in which the hydrogen atoms are closest to each other has the highest energy. This conformation is said to be **eclipsed.** In the eclipsed conformation each C—H bond on one carbon atom eclipses a C—H bond on another carbon atom, as the moon sometimes eclipses the sun. Any other intermediate conformation is a **skew** conformation.

Newman Projection Formulas

Conformations of alkanes are often depicted by a representation called a Newman projection formula. This structure concentrates on the two carbon atoms about which rotation may occur. The two atoms are viewed end-on. The front atom—that nearest the viewer—is represented by a point with three bonds. The back atom is represented by a circle with three bonds that reach only to the perimeter of the circle. Although there is a bond between the two carbon atoms, it is hidden because it is located along the viewing axis.

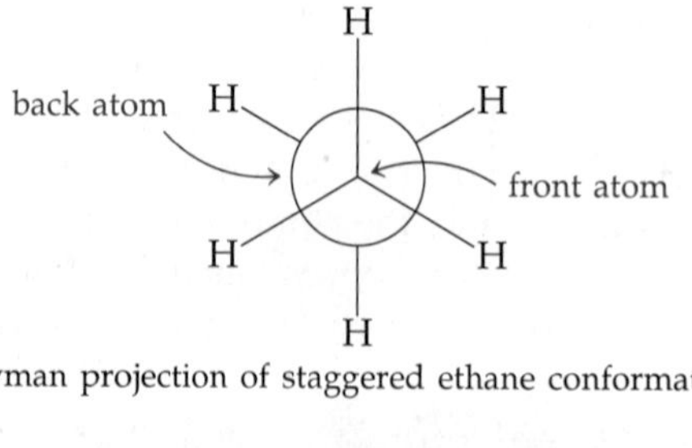

Newman projection of staggered ethane conformation

A Newman projection of the eclipsed conformation of ethane shows only the three C—H bonds of the front carbon atom. The rear bonds and hydrogen atoms are not visible because they are hidden by the front eclipsing bonds and hydrogen atoms. However, the bonded hydrogen atoms of the back carbon atom can be shown by viewing the conformation slightly off the bond axis. Viewing the eclipsed ethane model slightly to the right results in a convenient representation in which all bonds can be seen.

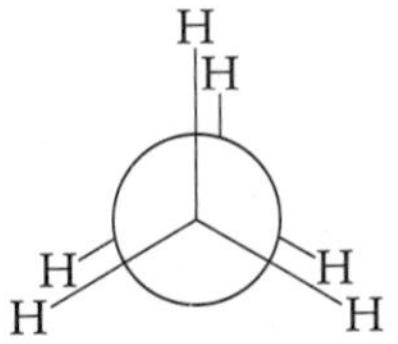

Barrier to Rotation

Conformations interconvert by rotation around σ bonds. Thus, when the eclipsed conformation of ethane is rotated by 60° about the C—C axis, the staggered conformation is produced. Continued rotation by another 60° gives a new eclipsed conformation, which is identical with the first eclipsed conformation.

rotate ⇌ rotate ⇌

eclipsed staggered eclipsed

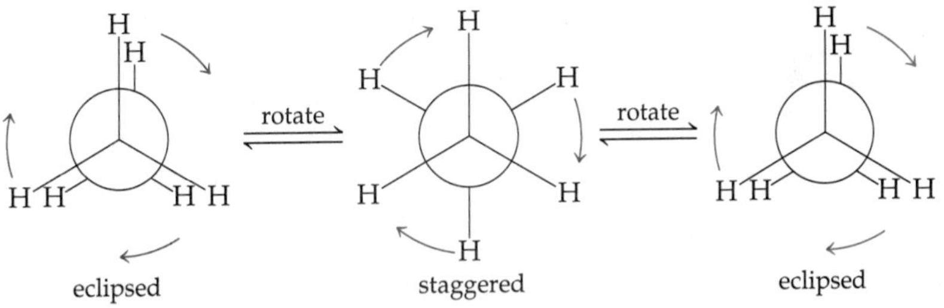

Continued rotation results in a series of staggered and eclipsed conformations. A plot of potential energy versus the angle of rotation for a complete 360° rotation about the C—C bond is shown in Figure 3.3.

We recall that the VSEPR model predicts that electron pairs and bonded atoms should be separated by the maximum distance. In fact, the staggered conformation of ethane is more stable than the eclipsed conformation. The eclipsed conformation has a higher energy because of **torsional strain.** This strain is due to the small repulsion between the bonded electrons in the C—H bonds as they approach and pass each other in the eclipsed conformation. Each hydrogen–hydrogen eclipsing interaction amounts to 1 kcal/mole. Thus, the rotational barrier is 3 kcal/mole.

The difference in energy between eclipsed and staggered conformations is small, and there is enough thermal energy at room temperature to allow rapid interconversion between these conformations. Thus, we say that rotation about the C—C bond is virtually free or unrestricted.

Conformations of ethane are not different compounds but different forms of a single molecule. Ethane is a mixture of conformations, but they cannot be separated from each other. However, at room temperature the ratio of staggered to eclipsed conformations is about 99:1.

FIGURE 3.3 Rotational Barrier for Conformations of Ethane
Rotation about the carbon–carbon bond of ethane in 60° increments starting from an eclipsed conformation gives a series of alternating eclipsed and staggered conformations. The eclipsed conformation is 3 kcal/mole higher in energy than the staggered conformation.

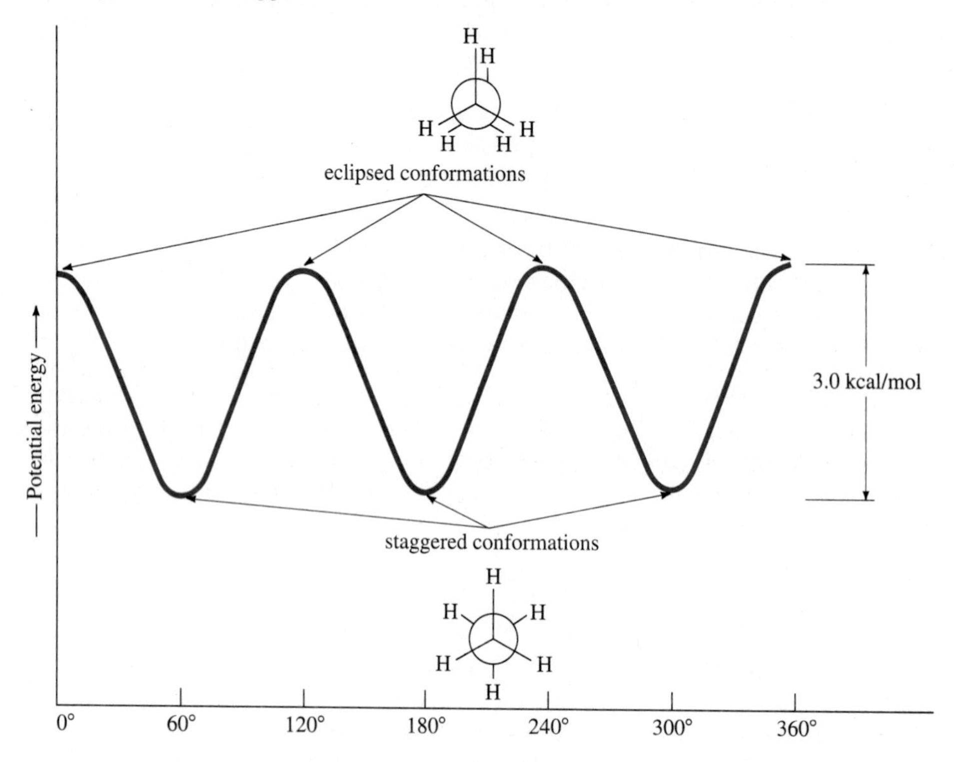

Propane and Butane

All acyclic alkanes exist as mixtures of conformations that result from rotation about every single bond in their structures. Consider propane, which can exist in both eclipsed and staggered conformations.

staggered propane eclipsed propane

The eclipsed conformation of propane has two hydrogen–hydrogen eclipsing interactions and one hydrogen–methyl group eclipsing interaction. Because methyl is larger than hydrogen, the energy difference between the eclipsed and staggered conformation of propane is greater than the 3.0 kcal/mole difference in conformations of ethane. The energy difference is 3.4 kcal/mole. Because each hydrogen–hydrogen eclipsing interaction is 1.0 kcal/mole, we can conclude that the hydrogen–methyl interaction is 1.4 kcal/mole.

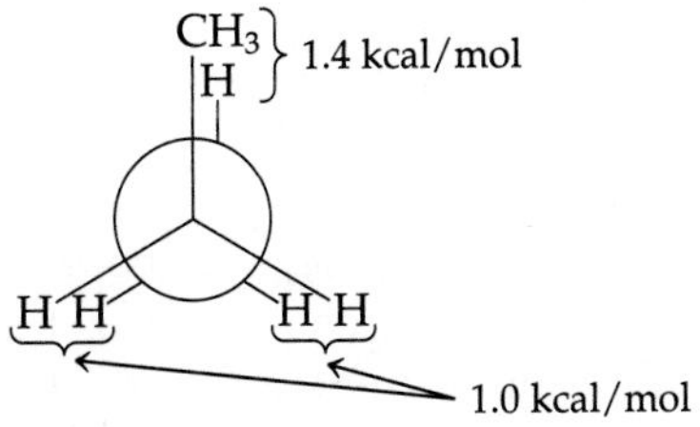

CH_3 / H } 1.4 kcal/mol

H H   H H

1.0 kcal/mol

The most stable conformation of butane has a zigzag arrangement of the carbon–carbon bonds in which all bonds are staggered. A Newman projection viewed along the C-2 to C-3 bond shows that the two methyl groups in this staggered conformation are the maximum distance apart in an **anti conformation.** A second staggered conformation called the **gauche conformation** is also possible.

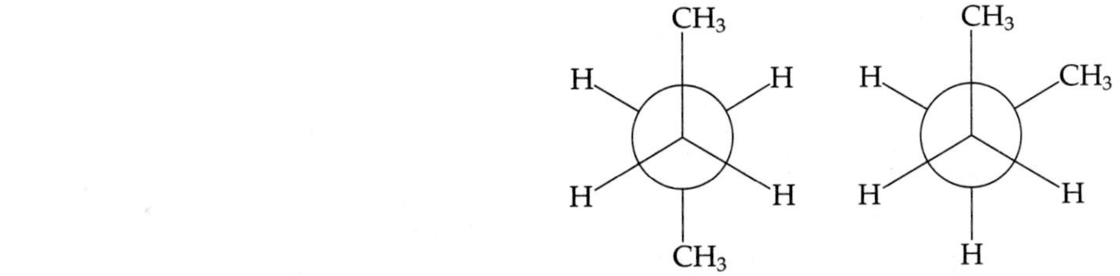

CH_3 H H H H CH_3 — anti conformation

CH_3 H CH_3 H H H — gauche conformation

There is no torsional strain in these two conformations because the angle between the bonded pairs is 60° in both cases. What accounts for the difference in the energy of the two conformations? The methyl groups are closer to each other in the gauche than in the anti conformation. The interference and repulsion of the electron clouds of the atoms of the two methyl groups with each other is called **steric strain.**

The anti conformation of butane is more stable than the gauche conformation by 0.9 kcal/mole. The anti and gauche conformations interconvert rapidly, but the ratio of anti to gauche conformations is about 2:1. Of course, both conformations are more stable than any eclipsed conformations, which are present to only a limited extent. In the eclipsed conformation of butane, the two methyl groups experience a combination of torsional strain and steric strain amounting to about 6 kcal/mole.

steric hindrance

eclipsed conformation of butane

The principles discussed for propane and butane also apply for higher alkanes. The energy of eclipsing conformations increases in the order hydrogen–hydrogen < hydrogen–alkyl < alkyl–alkyl. The most favored conformation is the one in which all carbon–carbon bonds have a staggered arrangement. For the most part we will consider only staggered conformations of acyclic compounds in this text.

all-anti conformation of hexane

3.5 CYCLOALKANES

Cycloalkanes are compounds that contain carbon atoms in a ring. We recall that the general formula for an acyclic alkane is C_nH_{2n+2}. Compounds with one ring have the general formula C_nH_{2n}. Cycloalkanes have fewer hydrogen atoms because another carbon–carbon bond is needed to form the ring. The cycloalkanes contain only carbon–carbon single bonds and are saturated hydrocarbons.

Cycloalkanes are usually drawn as simple polygons. The sides of the polygon represent the carbon–carbon bonds, and it is understood that each corner of the polygon is a carbon atom attached to two hydrogen atoms.

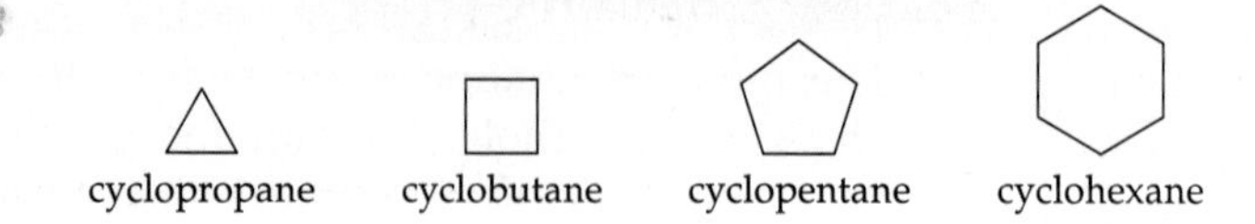

cyclopropane cyclobutane cyclopentane cyclohexane

Two or more rings can be part of the same molecule and share no common atoms. However, rings in a molecule can also share one or more common atoms. **Spirocyclic compounds** share one carbon atom between two rings. These compounds are relatively rare in nature. **Fused-ring com-**

pounds share two common atoms and the bond between them. These compounds are prevalent in nature. For example, steroids (Section 3.7) contain four fused rings. **Bridged-ring compounds** share two nonadjacent carbon atoms, which are called the **bridgehead carbon** atoms. These compounds are less prevalent in nature than fused-ring compounds. Examples of compounds containing two rings sharing common atoms are shown below.

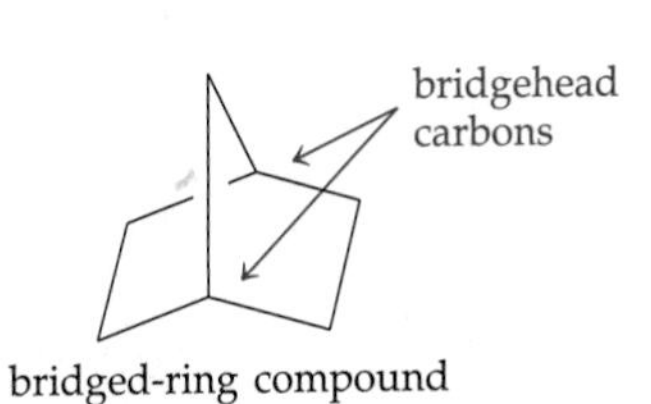

spirocyclic compound fused-ring compound bridged-ring compound

---

EXAMPLE 3.9 Adamantane has a carbon skeleton that is also found as part of the structure of diamond. Amantadine, which contains an amino group bonded to the adamantane structure, is useful in the prevention of infection by influenza A viruses. What are the molecular formulas of adamantane and amantadine?

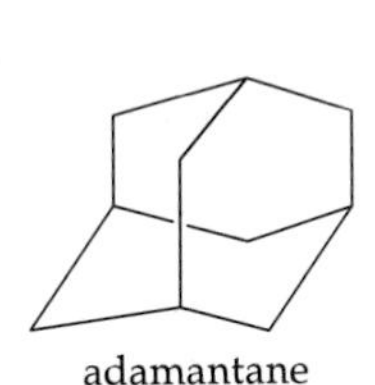

adamantane

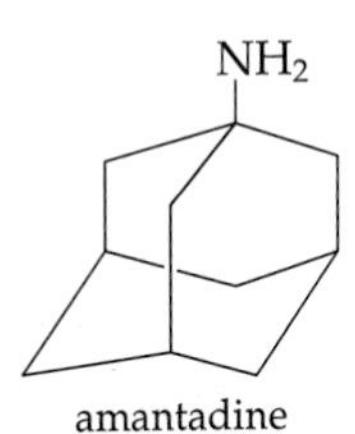

amantadine

Solution Adamantane has 10 carbon atoms. Four of these carbon atoms are tertiary; they have bonds to three other carbon atoms and one bond to a hydrogen atom. The remaining six carbon atoms are secondary; they have bonds to two other carbon atoms and two bonds to hydrogen atoms. The total number of hydrogen atoms is 4(1) + 6(2) = 16. The molecular formula of adamantane is $C_{10}H_{16}$.

Amantadine has an amino group, —NH_2, in place of a hydrogen atom at one of the tertiary carbon atoms. Thus, the molecular formula of amantadine differs from adamantane by one nitrogen and one hydrogen atom. The molecular formula is $C_{10}H_{17}N$.

Geometric Isomerism

In Chapter 1 we saw that compounds can exist as isomers with different carbon skeletons, functional groups, and functional group locations. These isomers have different sequential arrangements of atoms. Now let us consider a different type of isomerism. Compounds that have the same sequential arrangement of atoms but different spatial arrangements are **geometric isomers.**

We have seen that free rotation around the carbon–carbon bonds of alkanes gives different spatial arrangements called conformations. In cycloalkanes, some rotation about the carbon–carbon bond is possible, but there is less conformational freedom. The carbon–carbon bonds cannot rotate fully because they are restricted by the ring. Hence, cycloalkanes have two sides: a "top" and a "bottom".

Consider the "top" and "bottom" of cyclopropane, whose three carbon atoms are in a single plane. Any group attached to the ring may be held "above" or "below" the plane of the ring. If we attach two bromine atoms on adjacent carbon atoms on the same side of the plane of the ring, the substance is called a **cis isomer;** it is *cis*-1,2-dibromocyclopropane. If the two bromine atoms are attached on the opposite sides of the plane of the ring, the compound is the **trans isomer.** Thus, 1,2-dibromocyclopropane exists as both cis and trans isomers; these isomers are geometric isomers. In the structures shown below, the cyclopropane ring is viewed as perpendicular to the plane of the page and the CH_2 is pointed toward the viewer.

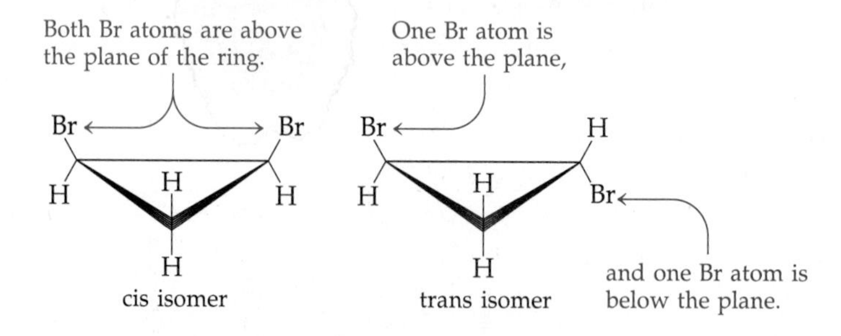

cis isomer
trans isomer

In the structures shown below, the cyclopropane ring is viewed in the plane of the page. Wedge-shaped lines denote bonds above the plane of the ring and dashed lines show bonds below the plane of the ring.

H H Br Br — cis isomer

H Br Br H — trans isomer

Note that cis and trans compounds are not two conformations of the same molecule but isomeric substances that have different physical properties. It is impossible to convert one isomer into the other without breaking a bond.

EXAMPLE 3.10 Disparlure, the sex-attractant pheromone of the female gypsy moth, has the following general structure. Are geometric isomers possible for this structure?

$$(CH_3)_2CHCH_2CH_2CH_2CH_2\underset{\diagdown\ O\ \diagup}{CH-CH}CH_2CH_2CH_2CH_2CH_2CH_2CH_2CH_2CH_2CH_3$$

Solution The three-membered heterocyclic ring contains two carbon atoms and one oxygen atom. Each carbon atom has a hydrogen atom and a large alkyl group bonded to it. These alkyl groups could be located cis or trans with respect to the plane of the ring. The cis isomer is the biologically active compound.

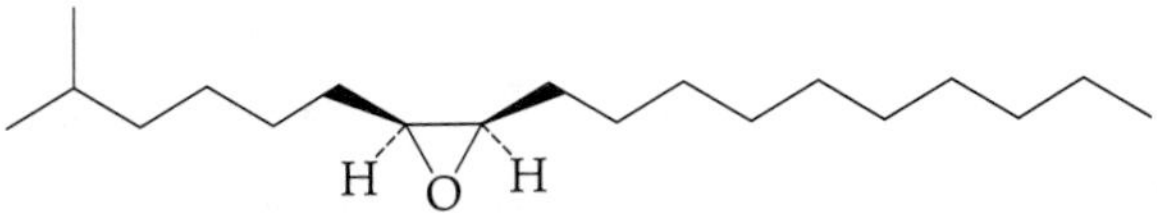

H O H

Cycloalkane Nomenclature

Cycloalkanes are named according to the IUPAC system by using the prefix cyclo-. When only one position contains a functional group or alkyl group, only one compound is possible, and therefore no number is necessary. Thus, ethylcyclopentane and isopropylcyclobutane are the names of the following molecules.

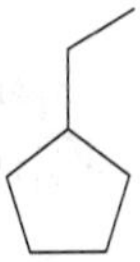

ethylcyclopentane 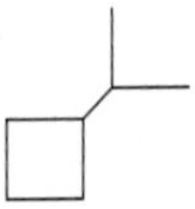 isopropylcyclobutane

When more than one group is attached to the ring, the ring is numbered. One substituent is always at position 1, and the ring is numbered in a clockwise or counterclockwise direction to give the lower number to the position with the next substituent attached to the ring, as in 1,1,4-trichlorocyclodecane. Geometric isomers have the prefix *cis-* or *trans-*. (Note that when used as prefixes in names of compounds, *cis-* and *trans-* are italicized.)

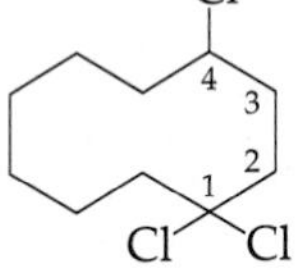

Cl 4 3 2 1 Cl Cl
1,1,4-trichlorocyclodecane

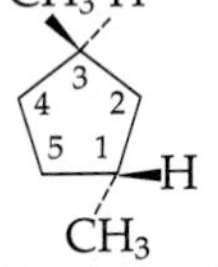

CH_3 H 3 4 2 5 1 H CH_3
trans-1,3-dimethylcyclopentane

EXAMPLE 3.11 What is the name of the following compound?

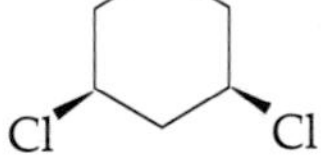

Cl Cl

Solution The ring must be numbered starting from one carbon atom with a chlorine atom and counting toward the other carbon atom with a chlorine atom by the shortest direction. Starting with the carbon atom at the "4 o'clock" position and numbering clockwise gives the number 3 to the atom at the "8 o'clock" position.

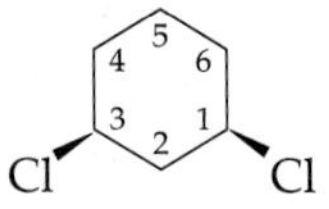

The two chlorine atoms are on the same side of the plane of the ring, and the correct name is *cis*-1,3-dichlorocyclohexane.

3.6 CONFORMATIONS OF CYCLOALKANES

Cyclopropane, with only three carbon atoms, is a planar molecule. The hydrogen atoms of cyclopropane lie above and below the plane of the ring of carbon atoms, and they eclipse hydrogen atoms on adjacent carbon atoms. The C—C—C bond angle is only 60° because the carbon atoms form an equilateral triangle. Cyclobutane and cyclopentane are not planar and exist in slightly "puckered" conformations, which reduces some of the eclipsing of hydrogen atoms on adjacent carbon atoms (Figure 3.4). The conformations of these compounds will not be further considered. For the purposes of this text, the two molecules will be depicted as planar rings of carbon atoms.

The six-membered ring of cyclohexane is not planar; it exists in a puckered conformation in which all C—H bonds on neighboring carbon atoms are staggered. Figure 3.5 shows a bond-line representation of the **chair conformation** of cyclohexane. Note that the hydrogen atoms in this conformation fall into two sets. Six of the hydrogen atoms are **axial;** they point up or down with respect to the average plane of the ring of carbon atoms. Three of the axial hydrogen atoms point up and the other three point down. The up–down relationship alternates from one carbon atom to the next. The

FIGURE 3.4 Conformations of Cyclobutane and Cyclopentane
The conformation of cyclobutane is a slightly bent ring so that the hydrogen atoms on adjacent carbon atoms are not quite eclipsed. The conformation of cyclopentane is described as an "envelope". One of the carbon atoms is out of the plane. This feature decreases the number of eclipsing interactions of hydrogen atoms on adjacent carbon atoms.

cyclobutane

cyclopentane

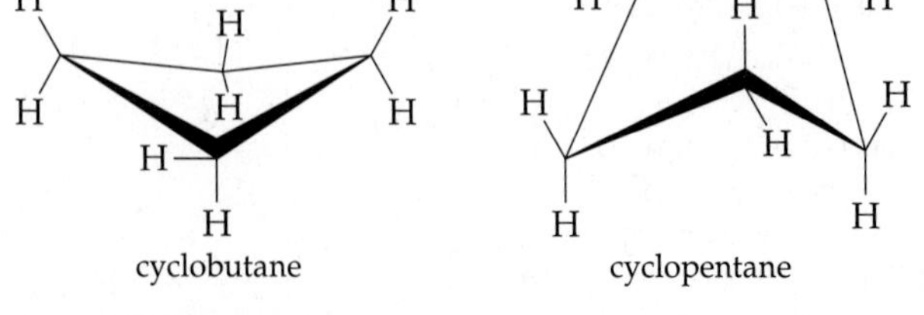

FIGURE 3.5 Axial and Equatorial Hydrogen Atoms in Cyclohexane The equatorial C—H bonds shown in (a) are located in a band around the "equator" of the ring. Each carbon atom has one equatorial hydrogen atom. The six axial C—H bonds are parallel to the axis shown perpendicular to the plane in (b). Three hydrogen atoms are pointed up from the average plane of the ring; three hydrogen atoms are pointed down. The axial hydrogen atoms are located in an alternating up–down relationship. All hydrogen atoms are shown in (c). The axial hydrogen atoms are circled.

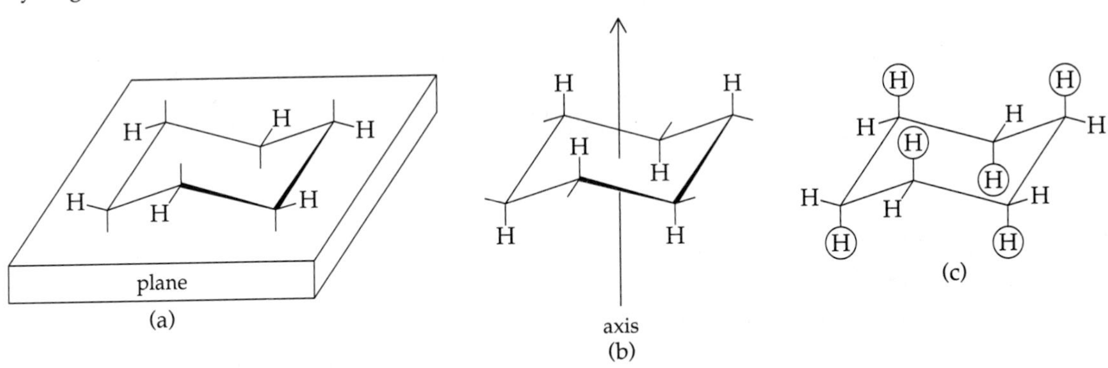

remaining six hydrogen atoms—called **equatorial** atoms—lie approximately in the average plane of the ring. Each carbon atom has one equatorial and one axial C—H bond.

Cyclohexane is conformationally mobile. Different chair conformations can interconvert and, as a result, the axial and equatorial positions become interchanged. This process, known as a **ring flip,** is shown in Figure 3.6. You can see this process more clearly by practicing with molecular models. To flip a cyclohexane ring, hold the four "middle" atoms in place while pushing one "end" carbon downward and the other upward. At each "end" atom, an equatorial position becomes an axial position and vice versa. In addition, although it is not as easy to see, the hydrogen atoms on every other carbon atom undergo the same transformation.

Now consider methylcyclohexane in a chair conformation with an equatorial methyl group. When the ring flips, the equatorial methyl group

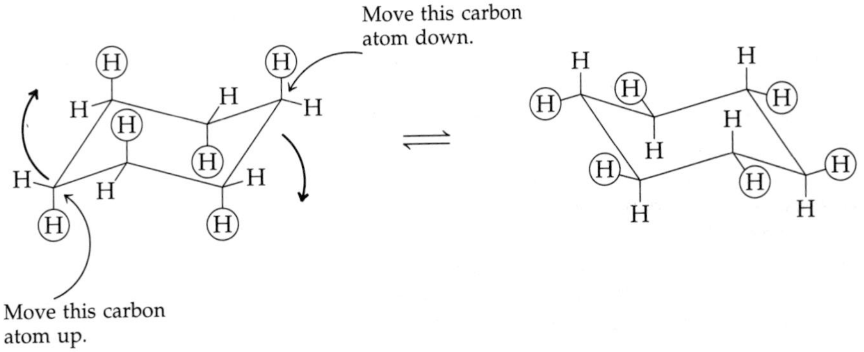

FIGURE 3.6 Conformational Mobility of Cyclohexane When the atoms are moved in the indicated direction, an axial position in one chair conformation becomes an equatorial position in the other conformation and vice versa.

Move this carbon atom down.

Move this carbon atom up.

FIGURE 3.7 Conformations of Methylcyclohexane
Methylcyclohexane interconverts rapidly between two conformations of unequal energy. At any time there is 95% of the equatorial conformation and 5% of the axial conformation. The axial conformation has unfavorable interactions between the methyl group hydrogen atoms and the axial hydrogen atoms on the C-3 and C-5 carbon atoms.

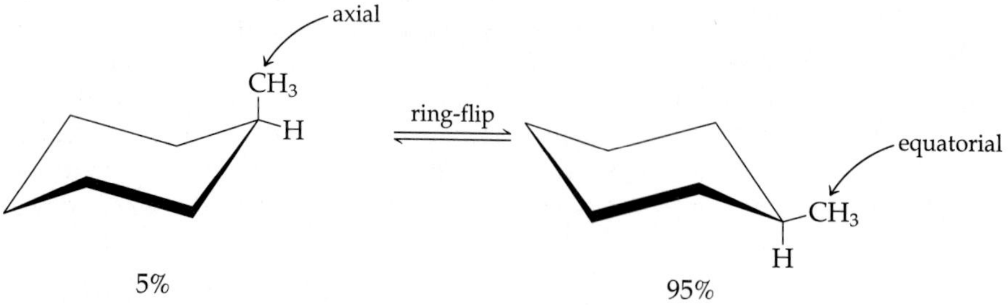

axial
CH_3
H
ring-flip
equatorial
CH_3
H
5%
95%

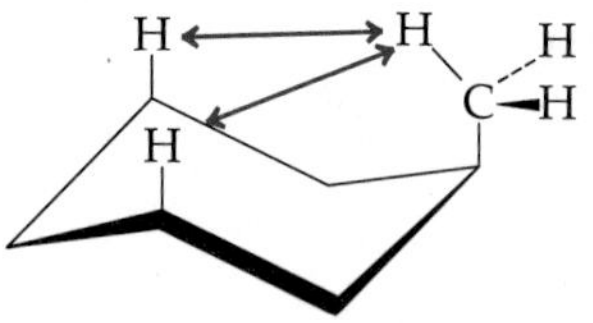

Arrows show the steric repulsion between the methyl group and the axial hydrogen atoms.

moves into an axial position (Figure 3.7). These two structures are different conformations, not isomers. The chair–chair interconversion occurs very rapidly. An equatorial methyl group is more stable than an axial methyl group, and 95% of the mixture at equilibrium has an equatorial methyl group. The conformation with an axial methyl group is less stable because there is steric strain between the methyl group and the axial hydrogen atoms at the C-3 and C-5 atoms. All substituted cyclohexanes behave similarly. Thus, the equatorial conformation is always more stable than the axial one, although the relative amounts of the two conformations vary.

3.7 STEROIDS

Steroids are tetracyclic compounds containing three six-membered rings and a five-membered ring. These compounds contain a variety of functional groups such as hydroxyl, carbonyl, and carbon–carbon double bonds. Each ring is assigned a letter, and the carbon atoms are numbered by a standard system.

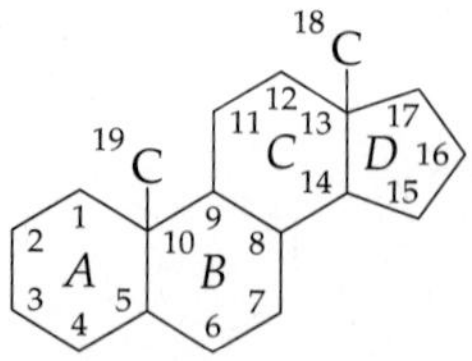

steroid ring system

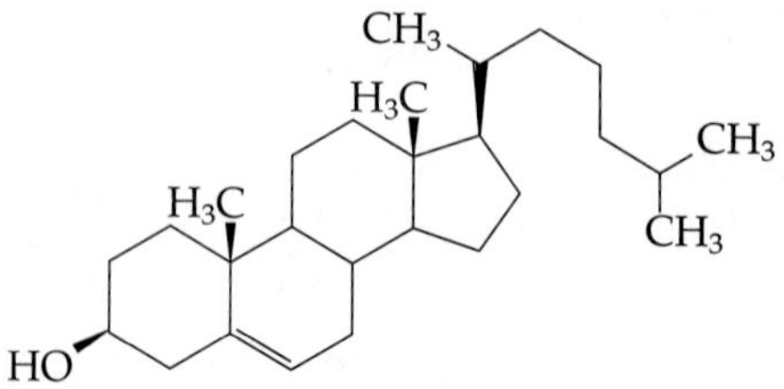

cholesterol

Cholesterol plays a vital biological role in the formation of other steroid hormones; it is converted to progesterone by shortening the chain attached at the C-17 position (ring *D*). Progesterone is converted to corticosteroids and sex hormones (Figure 3.8).

Corticosteroids are produced in the adrenal cortex. They are of two types: *glucocorticoids* and *mineralocorticoids.* Glucocorticoids help to control the glucose level in blood. Cortisol (Figure 3.8) promotes the formation of the storage carbohydrate glycogen in the liver. Mineralocorticoids affect the electrolytic balance of body fluids and, hence, the water balance. Aldosterone, secreted by the adrenal cortex, is the most active mineralocorticoid.

The testes of the male and the ovaries of the female produce steroidal sex hormones that control the growth and development of reproductive

FIGURE 3.8 Progesterone and Derived Steroid Hormones

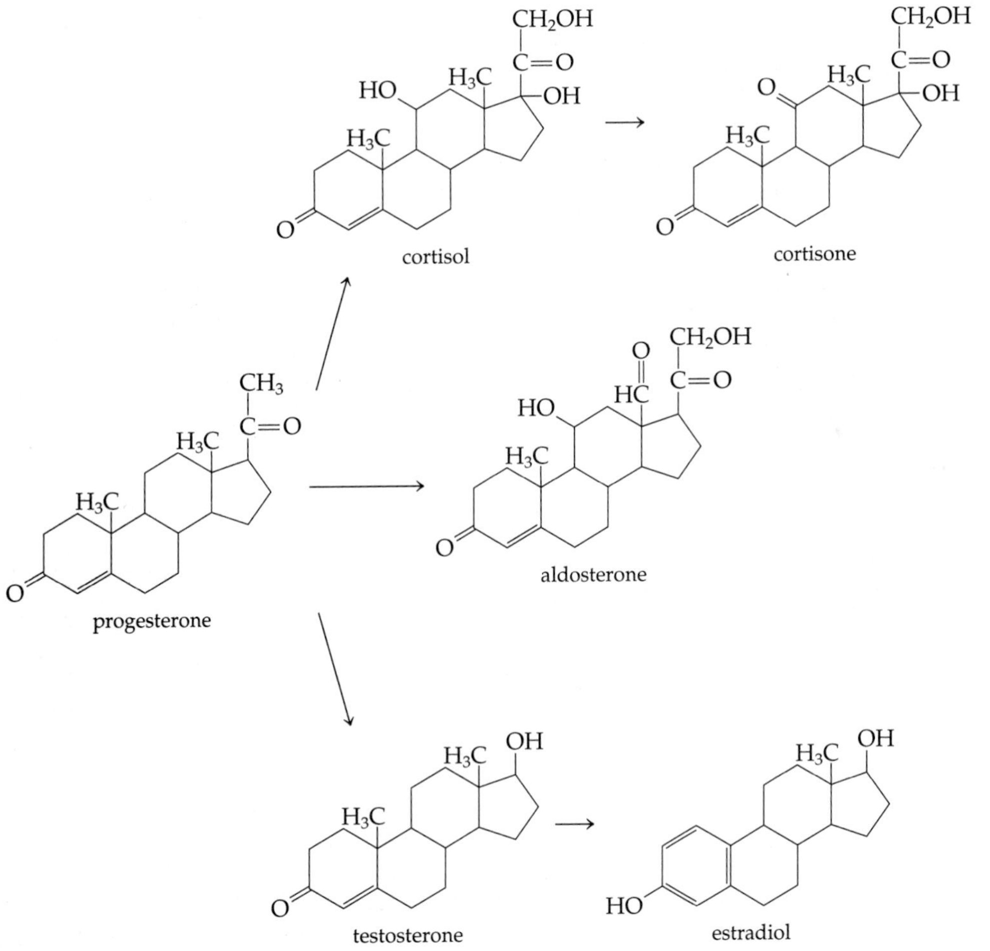

Anabolic Steroids—Use and Abuse

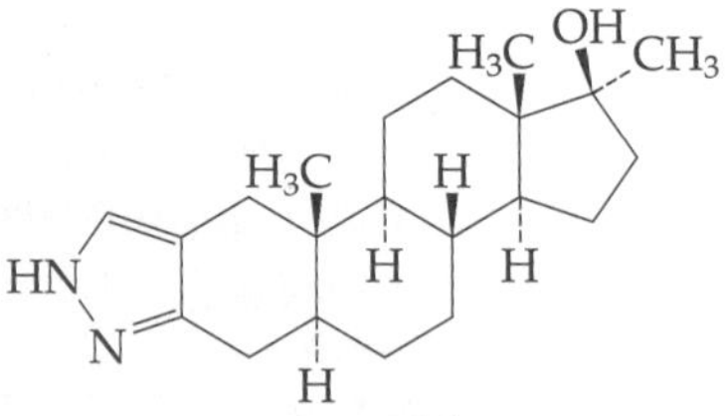

Testosterone has two biological roles: androgenic activity (sex-characteristic building) and anabolic activity (muscle building). It would be medically useful if drugs could be developed that had anabolic activity without androgenic activity. Such drugs would be useful to repair the muscles of severely debilitated individuals. Efforts in this research area have improved the ratio of anabolic to androgenic activity, but no compound synthesized is completely free of androgenic activity.

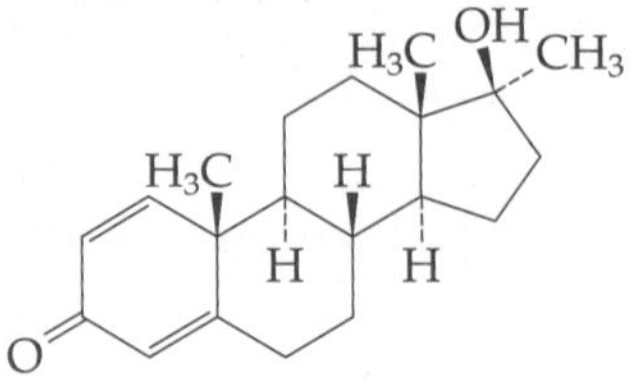

Athletes began to use anabolic steroids about 1950, and the abuse of these compounds rapidly became widespread. As recently as 1985 it was estimated that 90% of competitive weight lifters and body builders were using anabolic steroids. In the 1972 summer Olympics, about 65% of the track and field athletes had used anabolic steroids. The most publicized abuse of these drugs occurred when Ben Johnson of Canada was disqualified as the winner of the 100-meter dash in the 1988 Olympics after his urine was found to contain the anabolic steroid stanozolol.

The anabolic steroids stanozolol and dianabol are legally available only by prescription. Illegal and uncontrolled use of these compounds has severe consequences. In men the side effects include overly aggressive behavior, testicular atrophy, impotence, enlarged prostate, and cancer of the liver. In women significant virilization occurs, resulting in clitoral enlargement, breast diminution, baldness, beard growth, deepened voice, and increased risk of cancer.

stanozolol

dianabol

In spite of all the adverse publicity, these synthetic substances continue to be used by many amateur as well as some professional athletes, such as football players, to promote muscle development. Steroid use is banned by most athletic unions, and it is now standard practice during athletic competitions to test urine samples for the presence of synthetic anabolic steroids.

organs, the development of secondary sex characteristics, and the reproductive cycle.

Male sex hormones are **androgens.** Testosterone (Figure 3.8), the most important androgen, stimulates production of sperm by the testes and promotes the growth of the male sex organs. Testosterone is also responsible for muscle development.

Estrogens are female sex hormones. They are also produced in the adrenal cortex. However, the major source of male and female sex hormones is the gonads. Progesterone and two estrogens, estrone and estradiol, which are produced in the ovaries, control the menstrual cycle. After menstrual flow, estrogen secretion causes growth in the lining of the uterus and the ripening of the ovum. Progesterone secretion prevents other ova from ripening after ovulation and maintains a fertilized egg after implantation.

The estrogens cause the growth of tissues in female sexual organs. Although estrogens are secreted in childhood, the rate increases by 20-fold after puberty. The fallopian tubes, uterus, and vagina all increase in size. The estrogens also initiate growth of the breasts and the breasts' milk-producing ductile system.

3.8 PHYSICAL PROPERTIES OF SATURATED HYDROCARBONS

Alkanes have densities between 0.6 and 0.8 g/cm^3, so they are less dense than water (Table 3.3). Thus gasoline, which is largely a mixture of alkanes, is less dense than water and will float on water. Pure alkanes are colorless, tasteless, and nearly odorless. You may have noted, however, that gasoline does have an odor and some color. Dyes are added to gasoline by refiners to indicate its source and composition. Gasoline also contains aromatic compounds (Chapter 5), which have characteristic odors.

Alkanes contain only carbon–carbon and carbon–hydrogen bonds. Because carbon and hydrogen have similar electronegativity values, the C—H bonds are essentially nonpolar. Thus, alkanes are nonpolar, and they interact only by weak London forces. These forces govern the physical properties of alkanes such as solubility and boiling point.

Solubility of Alkanes

Alkanes are not soluble in water, a polar substance. The two substances do not meet the usual criterion of solubility: "like dissolves like". Water molecules are too strongly attracted to each other by hydrogen bonds to allow nonpolar alkanes to slip in between them and dissolve.

TABLE 3.3 Physical Properties of Some Alkanes

Alkane	Boiling point (°C)	Density (g/mL)
methane	−164.0	
ethane	−88.6	
propane	−42.1	
butane	−0.5	
pentane	36.1	0.6262
hexane	68.9	0.6603
heptane	98.4	0.6837
octane	125.7	0.7025
nonane	150.8	0.7176
decane	174.1	0.7300

Alkanes dissolve nonpolar organic materials such as fats and oils. Vapors of alkanes in gasoline cause severe damage to lung tissue because the fatty material in cell membranes is dissolved.

Body oils maintain the "moisture" of human skin. Long-term contact between skin and low molecular weight alkanes removes skin oils and can cause soreness and blisters. For this reason, you should minimize contact with hydrocarbon solvents such as paint thinner or paint remover.

Boiling Points of Alkanes

The boiling points of the normal alkanes increase with increasing molecular weight (Table 3.3). As the molecular weight increases, London forces increase because more atoms are present to increase the surface area of the molecules. Simply put, there are more points of contact between neighboring molecules, and the London forces are larger.

Normal alkanes can have efficient contact between chains, and the molecules can "stack" together. Branching in alkanes increases the distance between molecules, and the chains of carbon atoms are less able to come close to one another. Also, branched alkanes are more compact and have a smaller surface area than normal alkanes (Figure 3.9). The order of boiling points of the isomeric C_5H_{12} compounds illustrates this phenomenon.

$CH_3—CH_2—CH_2—CH_2—CH_3$
pentane (bp 36 °C)

$CH_3—CH(CH_3)—CH_2—CH_3$
isopentane (bp 28 °C)

$CH_3—C(CH_3)_2—CH_3$
neopentane (bp 10 °C)

Properties of Cycloalkanes

The physical properties of a series of cycloalkanes of increasing molecular weight are similar to those of a series of alkanes. The densities increase as do the boiling points (Table 3.4). The boiling points of the cycloalkanes are higher than those of the alkanes containing the same number of carbon atoms, partly because of the large London forces between the relatively rigid cyclic systems.

TABLE 3.4 Physical Properties of Some Cycloalkanes

Alkane	Boiling point (°C)	Density (g/mL)
cyclopropane	−32.7	
cyclobutane	12.0	
cyclopentane	49.3	0.7547
cyclohexane	80.7	0.7786
cycloheptane	118.5	0.8098
cyclooctane	148.5	0.8349

FIGURE 3.9 Shapes of Alkanes and Surface Area
The shape of neopentane is nearly spherical. Neighboring molecules can not approach each other closely and the points of contact are limited. Pentane is an extended molecule with an ellipsoidal shape. Neighboring molecules can approach each other side-by-side and as a consequence the attractive forces are larger.

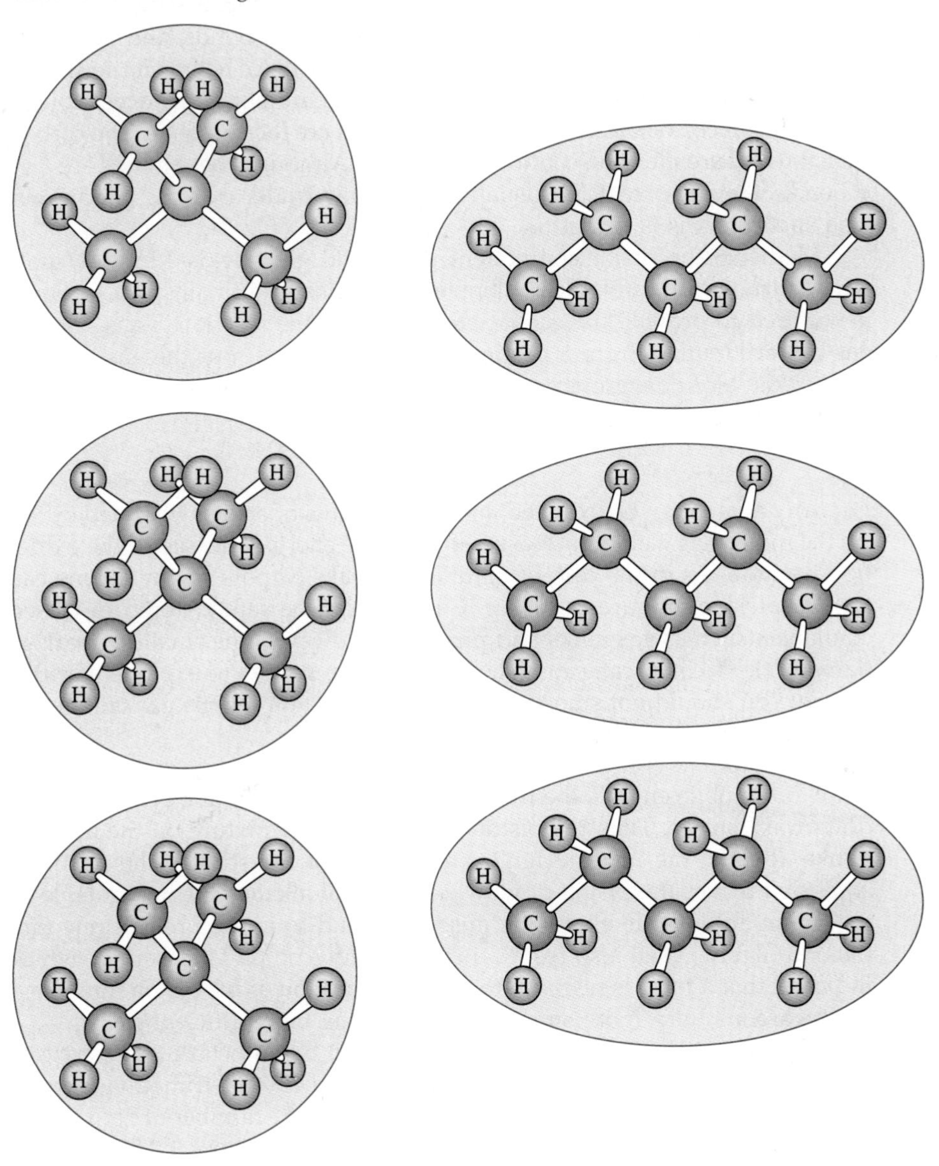

3.9 OXIDATION OF ALKANES AND CYCLOALKANES

The carbon–carbon and carbon–hydrogen bonds of alkanes and cycloalkanes are not very reactive. Alkanes are also called **paraffins** (L. *parum affinis,* little activity). The carbon–carbon bonds are σ bonds, and the bonding electrons are tightly held in a small region of space between the carbon atoms. Thus, electrons in carbon–carbon bonds are not easily accessible to other substances. The carbon–hydrogen bonds are located about the carbon skeleton and are more susceptible to reaction. Although these bonds are also σ bonds, they can react but usually do so only under extreme conditions. One such process is oxidation.

Methane, the major component of natural gas, yields 192 kcal/mole when burned. Although the reaction is spontaneous, a small spark or flame is required to provide the activation energy for the reaction. Thus, natural gas can accumulate from a gas leak and not explode. But the gas–oxygen mixture is very dangerous.

$$CH_4 + 2\,O_2 \longrightarrow CO_2 + 2\,H_2O + 192 \text{ kcal}$$

Any alkane can be oxidized, but the amounts of energy liberated differ. As the molecular weight increases, so does the energy released in the oxidation reaction. As in the case of natural gas leaks, vapors of any alkane can form explosive mixtures with air. Gasoline can be safely stored in closed containers such as gas tanks and pipelines. However, the smallest spark or flame in the vicinity can cause the hydrocarbon vapors to explode. For this reason, you should not smoke while filling your automobile gas tank, lawn mower, or outboard motor with gasoline.

Octane Numbers

In an automobile engine, the fuel and air are drawn into the cylinder on its downward stroke. Then the piston compresses the mixture on the upward stroke. Ideally, the mixture ignites at the top of the stroke. The resulting explosion drives the piston downward. Normal alkanes are not suitable as fuel in an automobile engine because they tend to ignite prematurely and uncontrollably. Their use results in a knocking or pinging sound, which indicates that a force is resisting the upward motion of the piston. Branched hydrocarbons burn more smoothly and are the more efficient fuels.

The burning efficiency of gasoline is rated by an octane number scale (Table 3.5). An octane number of 100 is assigned to 2,2,4-trimethylpentane, an excellent fuel. Heptane, a poor fuel, has an octane number of zero. Gasoline with the same burning characteristics as a 90% mixture of 2,2,4-trimethylpentane and 10% heptane is rated at 90 octane. Hydrocarbons that burn more efficiently than 2,2,4-trimethylpentane have octane numbers greater than 100. Hydrocarbons that burn less efficiently than heptane have negative octane numbers. Octane numbers decrease with increasing molecular

TABLE 3.5 Octane Numbers of Some Alkanes

Formula	Compound	Octane number
C_4H_{10}	butane	94
C_5H_{12}	pentane	62
	2-methylbutane	94
C_6H_{14}	hexane	25
	2-methylpentane	73
	2,2-dimethylbutane	92
C_7H_{16}	heptane	0
	2-methylhexane	42
	2,3-dimethylpentane	90
C_8H_{18}	octane	−19
	2-methylheptane	22
	2,3-dimethylhexane	71
	2,2,4-trimethylpentane	100

weight. In isomeric compounds, increased branching increases the octane number (Table 3.5).

3.10 HALOGENATION OF SATURATED HYDROCARBONS

Alkanes react with halogens at high temperature or in the presence of light. In this reaction, a halogen atom replaces a hydrogen atom. Thus, the process is a substitution reaction. A radical chain mechanism for this reaction was given in Chapter 2. The substitution reaction of hydrogen by chlorine (chlorination) is exothermic, but requires a large activation energy to break the strong carbon–hydrogen bond. Methane reacts with chlorine when heated to a high temperature or when exposed to ultraviolet light.

$$CH_4 + Cl_2 \longrightarrow CH_3Cl + HCl$$

The reaction is difficult to control because the product also has carbon–hydrogen bonds that can continue to react with additional chlorine to produce several substitution products.

$$CH_3Cl + Cl_2 \longrightarrow \underset{\substack{\text{dichloromethane} \\ \text{(methylene chloride)}}}{CH_2Cl_2} + HCl$$

$$CH_2Cl_2 + Cl_2 \longrightarrow \underset{\substack{\text{trichloromethane} \\ \text{(chloroform)}}}{CHCl_3} + HCl$$

$$CHCl_3 + Cl_2 \longrightarrow \underset{\substack{\text{tetrachloromethane} \\ \text{(carbon tetrachloride)}}}{CCl_4} + HCl$$

Freon, Radicals, and the Ozone Layer

Halogenated alkanes are less flammable than alkanes. Extensively halogenated compounds such as carbon tetrachloride will not burn at all. At one time carbon tetrachloride was used in fire extinguishers to provide an inert atmosphere to prevent oxygen from reaching the flames. Reduced combustibility makes haloalkanes useful for many purposes. For example, hydrocarbons have been used as refrigerants and as aerosol propellants, but the danger of combustion is a serious drawback. Thus, certain halogenated alkanes are now used. Freon-12 will not burn and is produced in large quantities for air conditioners. It was also used until recently as an aerosol propellant.

$$\begin{array}{c} F \\ | \\ F-C-Cl \\ | \\ Cl \end{array}$$

freon-12

Unfortunately, although inert on Earth, freon-12 decomposes in the stratosphere to produce radicals. Ultraviolet radiation causes rupture of the carbon–chlorine bond.

$$CF_2Cl_2 \longrightarrow Cl\cdot + \cdot CF_2Cl$$

This process is partially responsible for the destruction of the ozone layer in the stratosphere. Ozone in the stratosphere protects us from solar ultraviolet radiation, which splits ozone molecules into molecular oxygen and atomic oxygen. These products then recombine and release heat energy.

$$O_3 \xrightarrow{\text{UV light}} O_2 + O$$
$$O_2 + O \longrightarrow O_3 + \text{heat energy}$$

As a result of the combination of the two reactions, the Earth is protected from extensive doses of ultraviolet radiation that would be harmful to life and increase the incidence of skin cancer.

The chlorine radical from freon-12 reacts with ozone in the stratosphere. Then the resultant ClO product reacts with atomic oxygen.

$$Cl\cdot + O_3 \longrightarrow ClO\cdot + O_2$$
$$ClO\cdot + O \longrightarrow Cl\cdot + O_2$$

Note that a chlorine radical reacts in the first equation and is a product in the second equation. Thus, the chlorine is a catalyst for the destruction of ozone. The net reaction of these two steps is the destruction of an ozone molecule for each cycle initiated by the chlorine atom.

$$O_3 + O \longrightarrow 2\,O_2$$

These reactions effectively remove ozone and atomic oxygen from the atmosphere, and the protection provided by the ozone layer is diminished. The destruction of the ozone layer is most pronounced at the South Pole; recently a similar effect has been detected at the North Pole. Continued destruction of the ozone layer now appears to be happening at mid-latitudes as well, where more of the Earth's inhabitants live.

Chlorination of higher molecular weight alkanes yields many monosubstituted products. The chlorine atom is so reactive that it is not selective in its substitution of hydrogen atoms. Thus, reaction with pentane yields 1-chloropentane, 2-chloropentane, and 3-chloropentane. A large number of polysubstituted products are also possible.

Bromination of alkanes occurs in the same manner as chlorination, but the reaction is less exothermic. Iodination is not favorable; fluorination is very exothermic and the reaction is difficult to control.

Halogenated hydrocarbons are used for many industrial purposes. Unfortunately, many of these compounds can cause liver damage and cancer. In the past, chloroform was used as an anesthetic, and carbon tetrachloride was used as a dry-cleaning solvent. They are no longer used for these purposes.

EXAMPLE 3.12 How many mono-, di-, and trichlorinated compounds result from the chlorination of ethane?

Solution The two carbon atoms in ethane are equivalent, and only one monochlorinated compound can result.

$$CH_3CH_2Cl$$

The two carbon atoms in this product are not equivalent. Substitution by a second chlorine atom results in two isomers.

CH_3CHCl_2	$ClCH_2CH_2Cl$
1,1-dichloroethane	1,2-dichloroethane

In subsequent reactions of these products, three chlorine atoms may be located on a single carbon atom or two chlorine atoms may be located on the same carbon atom and one on the other carbon atom.

CH_3CCl_3	$ClCH_2CHCl_2$
1,1,1-trichloroethane	1,1,2-trichloroethane

EXERCISES

Molecular Formulas

3.1 Does each of the following molecular formulas for an acyclic hydrocarbon represent a saturated compound?
(a) C_6H_{12} (b) C_5H_{12} (c) C_8H_{16} (d) $C_{10}H_{22}$

3.2 Can each of the following formulas correspond to an actual acyclic or cyclic molecule?
(a) C_6H_{14} (b) $C_{10}H_{23}$ (c) C_7H_{14} (d) C_5H_{14}

3.3 Beeswax contains approximately 10% hentriacontane, which is a normal alkane with 31 carbon atoms. What is the molecular formula of hentriacontane? Write a completely condensed formula of hentriacontane.

3.4 Hectane is a normal alkane with 100 carbon atoms. What is the molecular formula of hectane? Write a completely condensed formula of hectane.

Structural Formulas

3.5 Redraw each of the following so that the longest continuous chain is written horizontally.

(a) 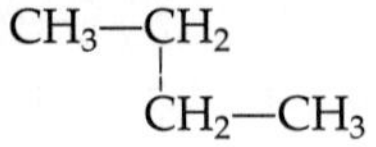 $CH_3—CH_2$
 |
$CH_2—CH_3$

(b) 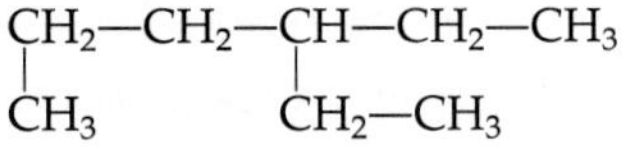 $CH_2—CH_2—CH—CH_2—CH_3$
 | |
CH_3 $CH_2—CH_3$

(c) 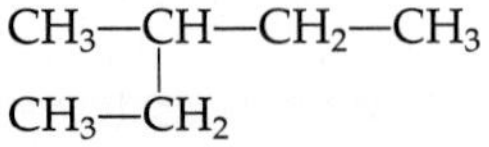 $CH_3—CH—CH_2—CH_3$
 |
$CH_3—CH_2$

(d) 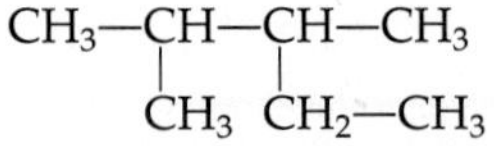 $CH_3—CH—CH—CH_3$
 | |
CH_3 $CH_2—CH_3$

3.6 Redraw each of the following so that the longest continuous chain is written horizontally.

(a) 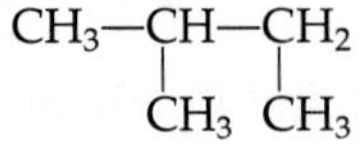 $CH_3—CH—CH_2$
 | |
CH_3 CH_3

(b) 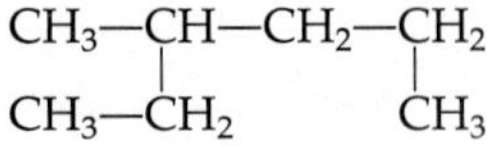 $CH_3—CH—CH_2—CH_2$
 | |
$CH_3—CH_2$ CH_3

(c) 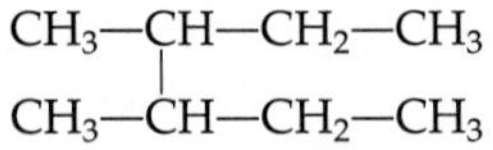 $CH_3—CH—CH_2—CH_3$
 |
$CH_3—CH—CH_2—CH_3$

(d) 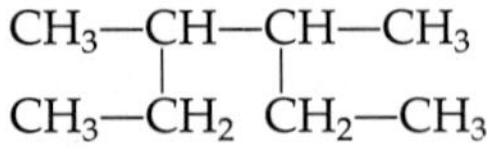 $CH_3—CH—CH—CH_3$
 | |
$CH_3—CH_2$ $CH_2—CH_3$

3.7 Which of the following represent the same compound?

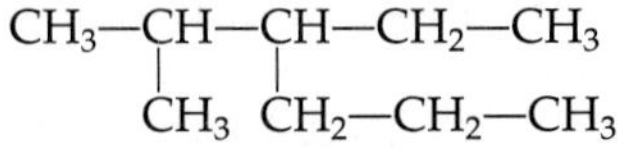

$CH_3—CH—CH—CH_2—CH_3$
 | |
CH_3 $CH_2—CH_2—CH_3$
(I)

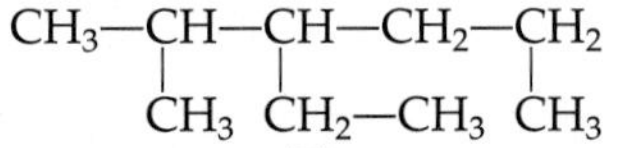

$CH_3—CH—CH—CH_2—CH_2$
 | | |
CH_3 $CH_2—CH_3$ CH_3
(II)

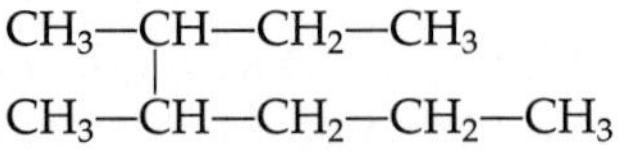

$CH_3—CH—CH_2—CH_3$
 |
$CH_3—CH—CH_2—CH_2—CH_3$
(III)

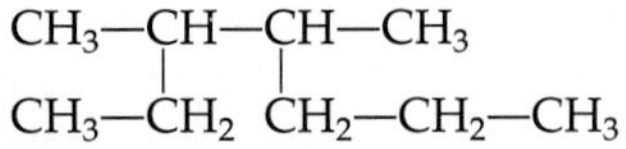

$CH_3—CH—CH—CH_3$
 | |
$CH_3—CH_2$ $CH_2—CH_2—CH_3$
(IV)

3.8 Which of the following represent the same compound?

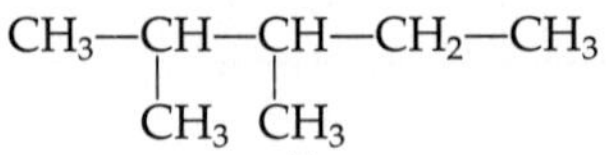

$CH_3—CH—CH—CH_2—CH_3$
 | |
CH_3 CH_3
(I)

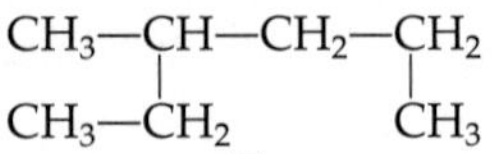

$CH_3—CH—CH_2—CH_2$
 | |
$CH_3—CH_2$ CH_3
(II)

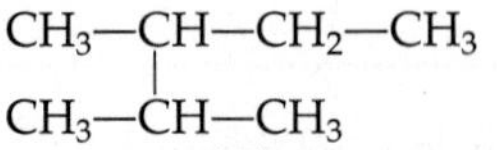

$CH_3—CH—CH_2—CH_3$
 |
$CH_3—CH—CH_3$
(III)

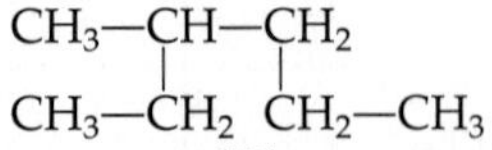

$CH_3—CH—CH_2$
 | |
$CH_3—CH_2$ $CH_2—CH_3$
(IV)

Alkyl Groups

3.9 What is the common name for each of the following alkyl groups?

(a) $CH_3—$

(b) 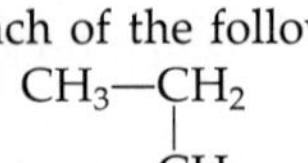 $CH_3—CH_2$
 |
$CH_2—$

(c) 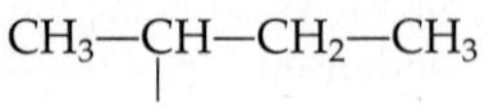 $CH_3—CH—CH_2—CH_3$
 |

(d) 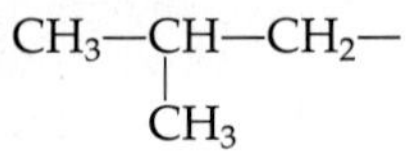 $CH_3—CH—CH_2—$
 |
CH_3

3.10 What is the common name for each of the following alkyl groups?

(a) $CH_3{-}CH_2{-}$

(b) $CH_3{-}\underset{\displaystyle CH_3}{\underset{|}{CH}}{-}$

(c) $CH_3{-}CH_2{-}CH_2{-}CH_2{-}$

(d) $CH_3{-}\overset{|}{\underset{\displaystyle CH_3}{\underset{|}{C}}}{-}CH_3$

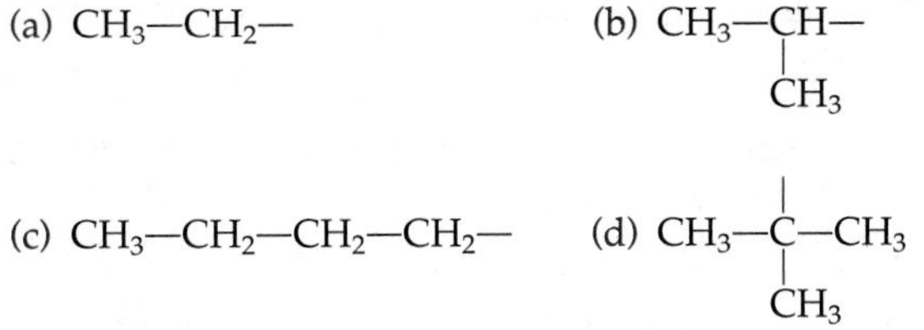

3.11 What is the IUPAC name for each of the following alkyl groups?

(a) $CH_3{-}\underset{\displaystyle CH_2{-}CH_2{-}}{\underset{|}{CH_2}}$

(b) $CH_3{-}\underset{\displaystyle CH_2{-}}{\underset{|}{CH}}{-}CH_3$

(c) $CH_3{-}\underset{\displaystyle CH_2{-}}{\underset{|}{CH}}{-}CH_2{-}CH_3$

(d) $CH_3{-}\underset{\displaystyle CH_3}{\underset{|}{CH}}{-}CH_2{-}CH_2{-}$

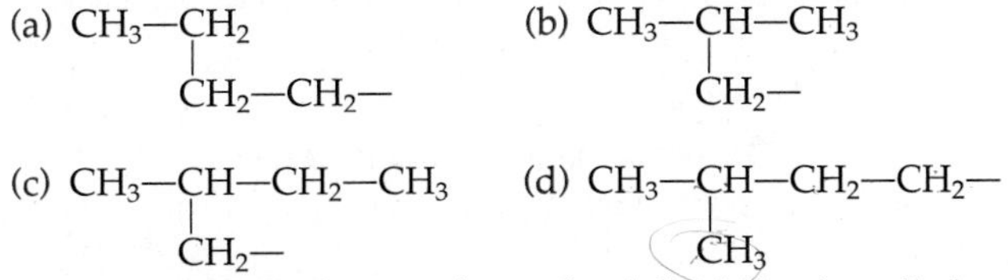

3.12 What is the IUPAC name for each of the following alkyl groups?

(a) $CH_3{-}\underset{\displaystyle CH_3}{\underset{|}{CH}}{-}CH_2{-}$

(b) $CH_3{-}\underset{\displaystyle CH_2{-}CH_3}{\underset{|}{CH}}{-}CH_2{-}$

(c) $CH_3{-}\underset{\displaystyle CH_2{-}CH_3}{\underset{|}{CH}}{-}CH_2{-}CH_2{-}$

(d) $CH_3{-}\overset{|}{\underset{\displaystyle CH_3}{\underset{|}{C}}}{-}CH_2{-}CH_3$

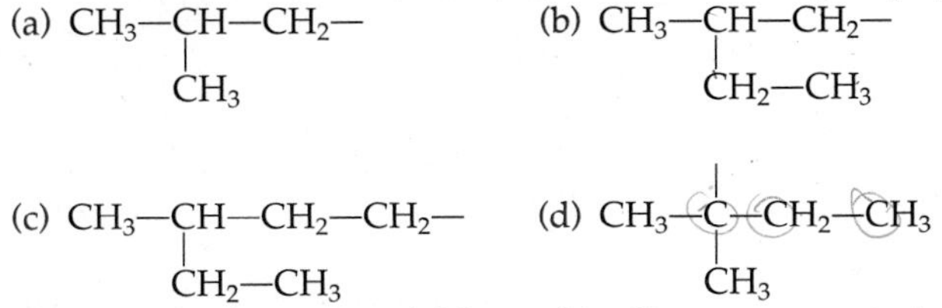

3.13 The spermicide octoxynol-9 is used in diverse contraceptive products. Name the alkyl group to the left of the benzene ring.

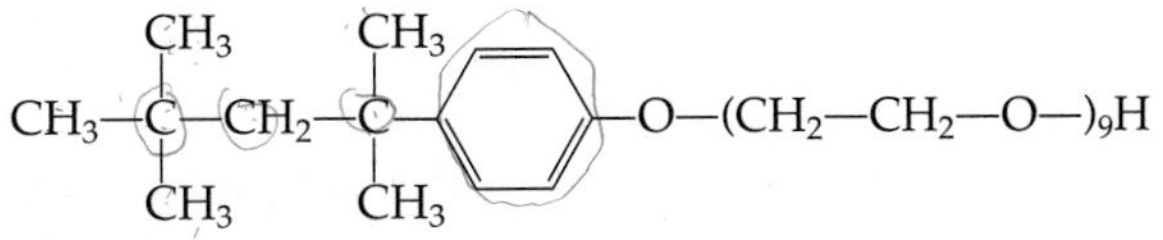

$$CH_3{-}\overset{\displaystyle CH_3}{\overset{|}{\underset{\displaystyle CH_3}{\underset{|}{C}}}}{-}CH_2{-}\overset{\displaystyle CH_3}{\overset{|}{\underset{\displaystyle CH_3}{\underset{|}{C}}}}{-}C_6H_4{-}O{-}(CH_2{-}CH_2{-}O{-})_9H$$

3.14 Vitamin E actually represents a series of closely related compounds called tocopherols. Name the complex alkyl group present in α-tocopherol.

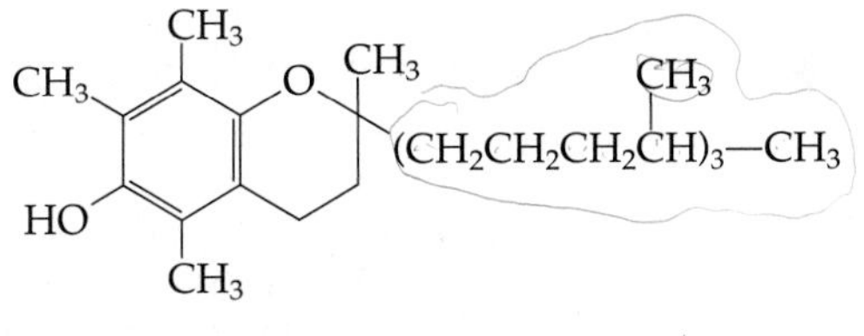

CH₃ groups on the aromatic ring, HO, O, and side chain: $\overset{\displaystyle CH_3}{}$ $(CH_2CH_2CH_2\overset{\displaystyle CH_3}{\overset{|}{C}H})_3{-}CH_3$

Nomenclature of Alkanes

3.15 Give the IUPAC name for each of the following compounds.

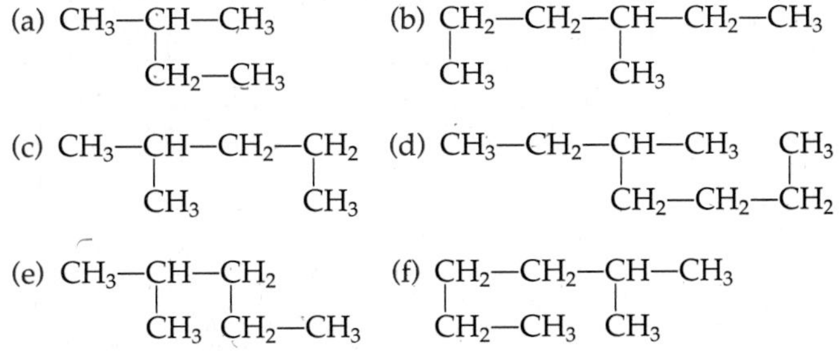

(a) $CH_3{-}\underset{\displaystyle CH_2{-}CH_3}{\underset{|}{CH}}{-}CH_3$

(b) $\underset{\displaystyle CH_3}{\underset{|}{CH_2}}{-}CH_2{-}\underset{\displaystyle CH_3}{\underset{|}{CH}}{-}CH_2{-}CH_3$

(c) $CH_3{-}\underset{\displaystyle CH_3}{\underset{|}{CH}}{-}CH_2{-}\underset{\displaystyle CH_3}{\underset{|}{CH_2}}$

(d) $CH_3{-}CH_2{-}\underset{\displaystyle CH_2{-}CH_2{-}\overset{\displaystyle CH_3}{\overset{|}{CH_2}}}{\underset{|}{CH}}{-}CH_3$

(e) $CH_3{-}\underset{\displaystyle CH_3}{\underset{|}{CH}}{-}\underset{\displaystyle CH_2{-}CH_3}{\underset{|}{CH_2}}$

(f) $\underset{\displaystyle CH_2{-}CH_3}{\underset{|}{CH_2}}{-}CH_2{-}\underset{\displaystyle CH_3}{\underset{|}{CH}}{-}CH_3$

3.16 Give the IUPAC name for each of the following compounds.

(a) $CH_3—CH—CH—CH_3$
with CH_3 on C-2 and $CH_2—CH_2—CH_3$ on C-3

(b) $CH_3—CH—CH_2\ \ CH_3$
with CH_3 on C-2; $CH_2—CH—CH_3$ below (the CH_3 at the right is attached to that CH)

(c) $CH_3—CH—CH_2—CH_3$
$CH_3—CH—CH_2—CH_2—CH_3$ (the two CH groups are bonded)

(d) $CH_3—CH—CH_2—CH—CH_3$
with $CH_3—CH_2$ on C-2 and $CH_2—CH_3$ on C-4

(e) $CH_3—CH—CH_2—CH_2—CH_3$
with $CH_2—CH_3$ on C-2

(f) $CH_3—CH_2—CH—CH_2$
with $CH_3—CH_2$ on C-3 and CH_3 on the terminal CH_2

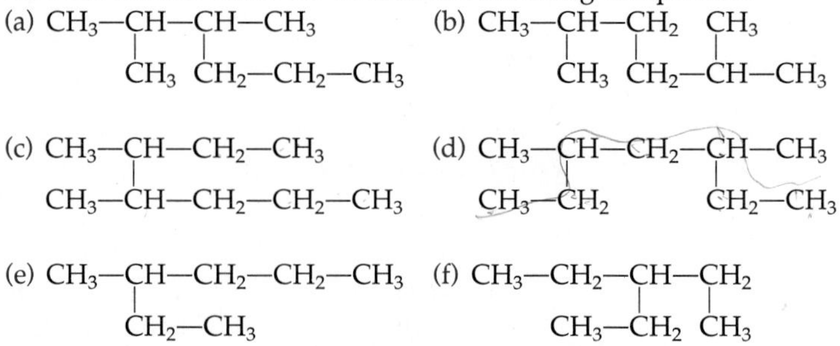

3.17 Give the IUPAC name for the following compound.

$CH_3—CH_2—CH_2—CH_2—CH—CH_2—CH_2—CH_2—CH_2—CH_3$
$CH_3—CH_2—CH—CH_2—CH_3$ (the middle CH bonded to C-5)

3.18 Give the IUPAC name for the following compound.

$CH_3—CH_2—CH_2—CH_2—CH—CH_2—CH_2—CH_2—CH_3$
$CH_3—C—CH_3$ (the C bonded to C-5)
$CH_2—CH_3$ (bonded to that C)

3.19 Write the structural formula for each of the following compounds.
(a) 3-methylpentane (b) 3,4-dimethylhexane
(c) 2,2,3-trimethylpentane (d) 4-ethylheptane
(e) 2,3,4,5-tetramethylhexane

3.20 Write the structural formula for each of the following compounds.
(a) 2-methylpentane (b) 3-ethylhexane
(c) 2,2,4-trimethylhexane (d) 2,4-dimethylheptane
(e) 2,2,3,3-tetramethylpentane

3.21 Write the structural formula for each of the following compounds.
(a) 4-(1-methylethyl)heptane
(b) 4-(1,1-dimethylethyl)octane
(c) 5-(1-methylpropyl)nonane

3.22 Write the structural formula for each of the following compounds.
(a) 5-(2-methylpropyl)nonane
(b) 5-butylnonane
(c) 5-(2,2-dimethylpropyl)nonane

Isomers

3.23 There are nine isomeric C_7H_{16} compounds. Which isomers have a single methyl group as a branch?

3.24 There are nine isomeric C_7H_{16} compounds. Which isomers have two methyl groups as branches and are named as dimethyl-substituted pentanes?

Classification of Carbon Atoms

32.5 Classify each carbon atom in the following compounds.

(a) $CH_3—CH_2—CH_2—CH_2—CH_3$

(b) $CH_3—CH_2—CH—CH_2—CH_3$
with CH_3 on C-3

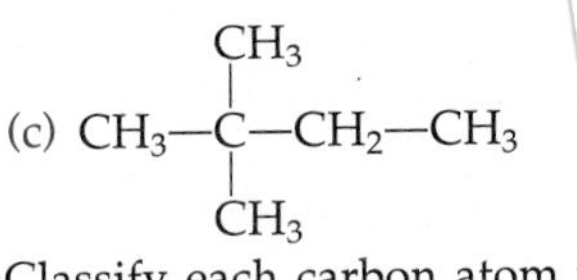

3.26 Classify each carbon atom in
(a) $CH_3-CH(CH_3)-CH_2-CH(CH_3)-C$

(c) $CH_3-CH(CH_3)-CH(CH_3)-CH_3$ (printed as $CH_3-CH-CH_3$ over $CH_3-CH-CH_3$)

3.36 W

3.27 Draw the structure of a compoun
quaternary and four primary car

3.28 Draw the structure of a compounc ...14 that has two
tertiary and four primary carbon

3.29 Determine the number of primary, secondary, tertiary, and quaternary carbon atoms in each of the following compounds.

(a) $CH_3-C(CH_3)_2-CH_3$ (b) $CH_3-CH(CH_3)-CH_2-CH_3$

(c) $CH_3-CH(CH_3)-CH_2-CH_2-CH_3$ (d) $CH_3-CH(CH_3)-CH(CH_3)-CH_3$

3.30 Determine the number of primary, secondary, tertiary, and quaternary carbon atoms in each of the following compounds.

(a) $CH_3-C(CH_3)_2-CH_2-C(CH_3)_2-CH_3$ (b) $CH_3-CH(CH_3)-CH_2-CH(CH_3)-CH_3$

(c) $CH_3-CH_2-CH(CH_3)-CH_2-CH_3$ (d) $CH_3-CH(CH_3)-CH(CH_3)-CH(CH_3)-CH_3$

Conformations of Alkanes

3.31 Draw the Newman projection of the staggered conformation of 2,2-dimethylpropane about the C-1 to C-2 bond.

3.32 Draw the Newman projections of the two possible staggered conformations of 2,3-dimethylbutane about the C-2 to C-3 bond.

3.33 Draw the Newman projections of the two possible staggered conformations of 2-methylbutane about the C-2 to C-3 bond. Which is the more stable?

3.34 Draw the Newman projections of the two possible staggered conformations of 2,2-dimethylpentane about the C-3 to C-4 bond. Which is the more stable?

Cycloalkanes

3.35 Write fully condensed planar formulas for each of the following compounds.
(a) chlorocyclopropane (b) 1,1-dimethylcyclobutane
(c) cyclooctane

rite fully condensed planar formulas for each of the following compounds.
(a) bromocyclopentane (b) 1,1-dichlorocyclopropane
(c) cyclopentane

3.37 Name each of the following compounds.

(a) 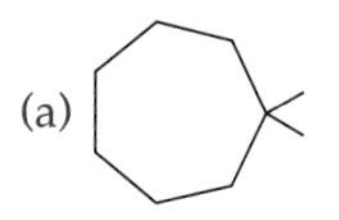
(b) 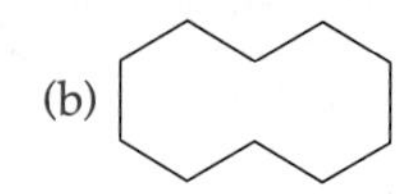
(c) 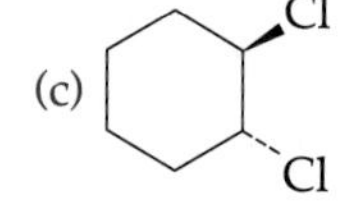

(d) 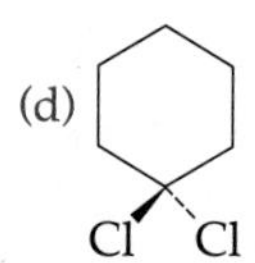

3.38 Name each of the following compounds.

(a) 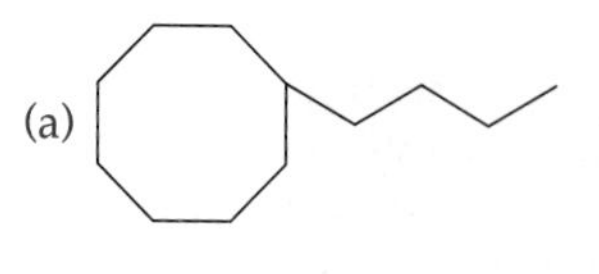
(b)

CH—CH_3
CH_3

(c) 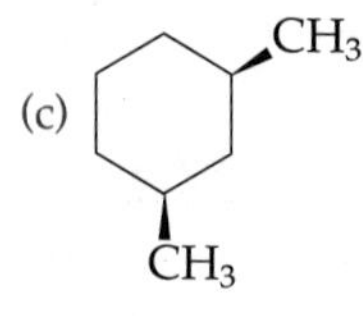

(d)

Cl
Cl Cl

3.39 What is the molecular formula of each of the following compounds?

(a) 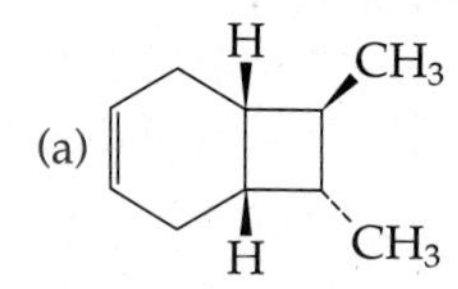

(b) 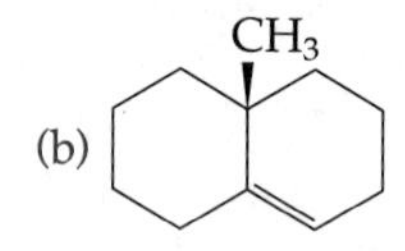

(c) 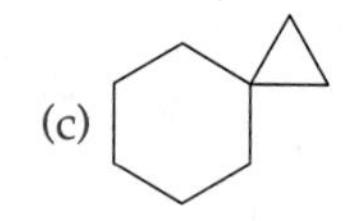
(d) 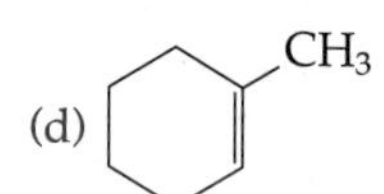

3.40 What is the molecular formula of each of the following compounds?

(a) 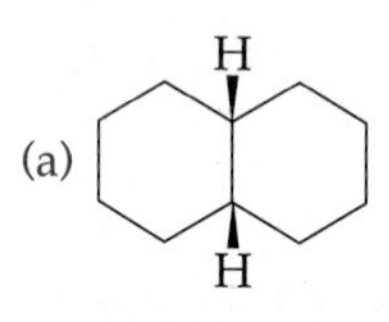

(b) 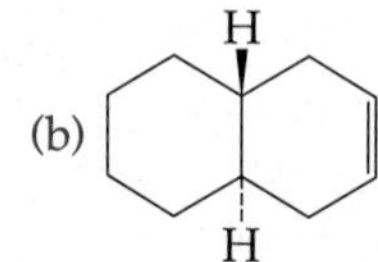

(c) 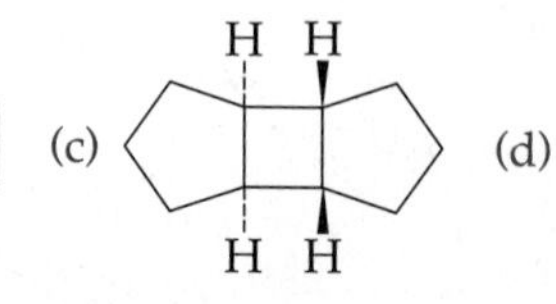

(d) 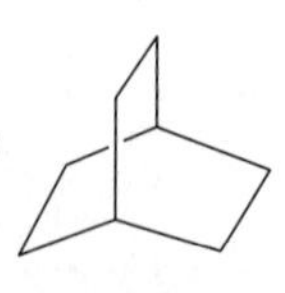

3.41 Are the two structures in each pair isomers or not?

(a) 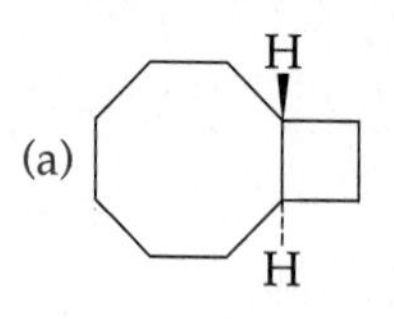

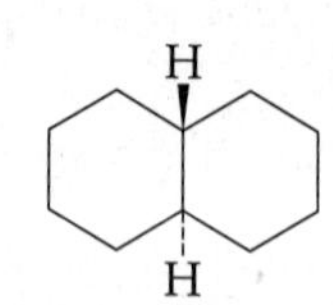

(b) 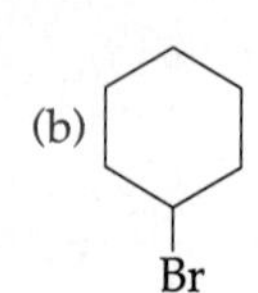

CH_2Br

(c)

CH_3
CH_3

CH_2CH_3

3.42 Are the two structures in each pair isomers or not?

(a) CH_2CH_3 CH_2CH_3 CH_2CH_3 CH_2CH_3

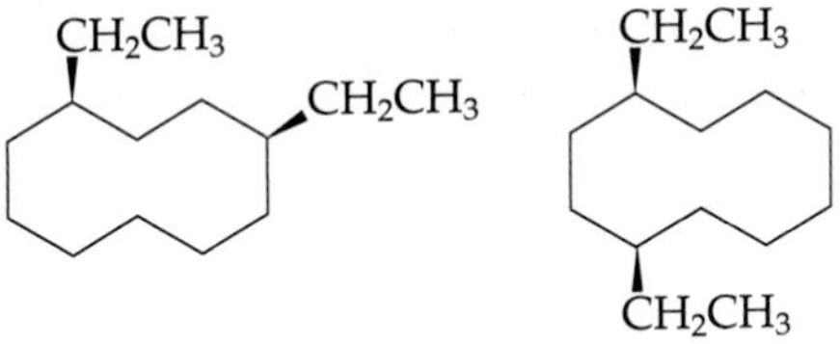

(b) CH_2Br Cl CH—CH_2—Br Cl

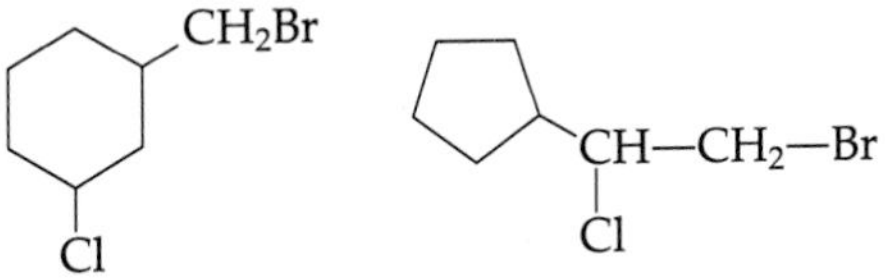

(c) CH_2CH_3 CH_3 CH_3

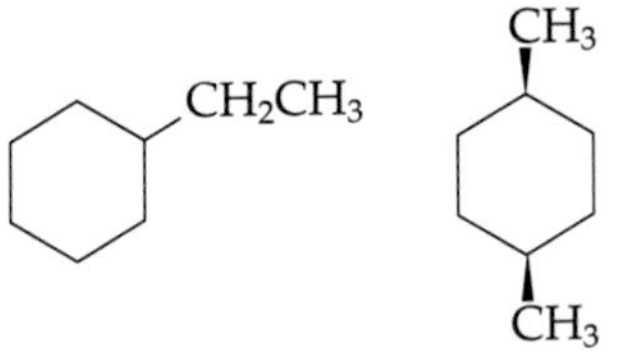

Conformations of Cyclohexanes

3.43 Draw the two chair conformations of fluorocyclohexane. Would you expect the energy difference between these two conformations to be greater or less than the energy difference between the two conformations of methylcyclohexane? Why?

3.44 Draw the two chair conformations of *t*-butylcyclohexane. Would you expect the energy difference between these two conformations to be greater or less than the energy difference between the two conformations of methylcyclohexane? Why?

3.45 Draw the most stable conformation of each of the following compounds.
(a) *trans*-1,4-dimethylcyclohexane
(b) *cis*-1,3-dimethylcyclohexane
(c) *trans*-1,2-dimethylcyclohexane

3.46 Draw the most stable conformation of each of the following compounds.
(a) 1,1,4-trimethylcyclohexane
(b) 1,1,3-trimethylcyclohexane
(c) *cis*-1-fluoro-4-iodocyclohexane

Properties of Hydrocarbons

3.47 Cyclopropane is an anesthetic, but it cannot be used in operations in which electrocauterization of tissue is done. Why?

3.48 Which compound should have the higher octane number, cyclohexane or methylcyclopentane?

3.49 Which of the isomeric C_8H_{18} compounds has the highest boiling point? Which has the lowest boiling point?

3.50 The boiling point of methylcyclopentane is lower than the boiling point of cyclohexane. Suggest a reason why.

Halogenation of Hydrocarbons

3.51 How many products can result from the substitution of a chlorine atom for one hydrogen atom in each of the following compounds?
(a) propane (b) butane (c) methylpropane (d) pentane
(e) cyclohexane

3.52 How many products can result from the substitution of a chlorine atom for one hydrogen atom in each of the following compounds?
(a) 2-methylbutane (b) 2,2-dimethylbutane
(c) 2,3-dimethylbutane (d) 3-methylpentane
(e) cyclopentane

3.53 Halothane, an anesthetic, has the formula C_2HF_3ClBr. Draw structural formulas for the four possible isomers of this molecular formula.

3.54 A saturated refrigerant has the molecular formula C_4F_8. Draw structural formulas for two possible isomers of this compound.

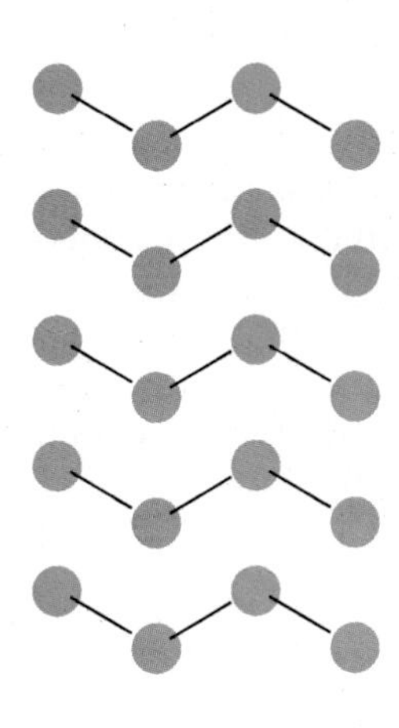

4

ALKENES AND ALKYNES

4.1 UNSATURATED HYDROCARBONS

Organic compounds with carbon–carbon multiple bonds contain fewer hydrogen atoms than structurally related alkanes or cycloalkanes. For this reason, these compounds are said to be **unsaturated.** In this chapter we will focus on two classes of unsaturated compounds: alkenes and alkynes. **Alkenes** contain a carbon–carbon double bond; **alkynes** contain a carbon–carbon triple bond. **Aromatic** hydrocarbons, compounds that contain a benzene ring or structural units that resemble a benzene ring, will be discussed in the next chapter.

Carbon–carbon double and triple bonds contain π electrons, which are the sites of specific reactions. We recall that π bonds involve a side-by-side overlap of $2p$ orbitals on adjacent carbon atoms. The hybridization of carbon and the model of σ and π bonds in ethylene (an alkene) and acetylene (an alkyne) were introduced in Chapter 1.

Alkenes and their chemical cousins, the cycloalkenes, are very common in nature. In fact, one is important to the common housefly *(Musca domestica)*. Muscalure, an unbranched alkene containing 23 carbon atoms, is released by the female to attract males. The compound has been synthesized in the laboratory and can be used to lure male flies to traps.

$$\begin{array}{ccc} CH_3(CH_2)_{11}CH_2 & & CH_2(CH_2)_6CH_3 \\ & C{=}C & \\ H & & H \end{array}$$

muscalure

Alkynes are not as prevalent in nature, but quite a few of them are physiologically active. For example, the triyne ichthyothereol is secreted from the skin of a frog in the Lower Amazon Basin. This compound is apparently a defensive venom and mucosal-tissue irritant that wards off mammals and reptiles. The Indians of the area use the secretion to coat arrowheads. When the arrow pierces the skin of the enemy, the compound causes convulsions.

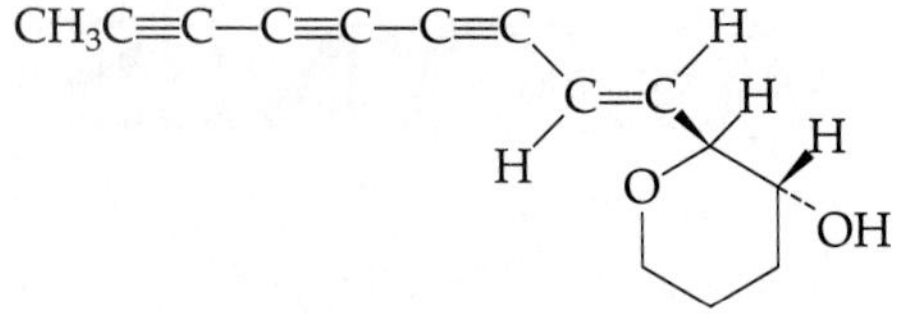

$CH_3C{\equiv}C{-}C{\equiv}C{-}C{\equiv}C$ H C=C H H H O OH

ichthyothereol

In this chapter the chemistry of some alkadienes, informally called dienes, will be examined. We will focus on **conjugated dienes,** compounds in which two double bond units are joined by one single bond. Conjugated dienes undergo reactions that differ from those of individual double bonds. Natural rubber is a polymer of isoprene, a conjugated diene. Some synthetic rubbers called neoprenes are produced from chloroprene and are used in products such as industrial hoses.

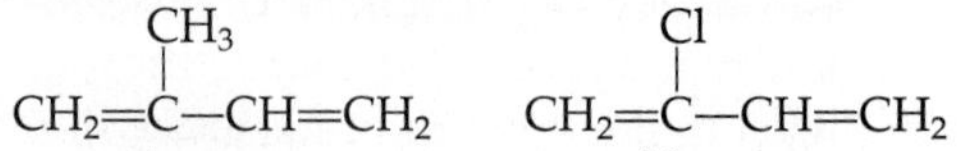

$CH_2{=}C(CH_3){-}CH{=}CH_2$
isoprene

$CH_2{=}C(Cl){-}CH{=}CH_2$
chloroprene

You may have encountered the term polyunsaturated fat (or oil) in various advertisements. A molecule is **polyunsaturated** if it contains several multiple bonds. For example, polyunsaturated oils contain several double bonds; these oils will be discussed in Chapter 13.

Polyunsaturated compounds are common in nature. For example, both β-carotene, which is found in carrots, and vitamin A are polyunsaturated. Note the resemblance of these two compounds. A biochemical reaction in mammals splits and oxidizes β-carotene into two molecules of vitamin A.

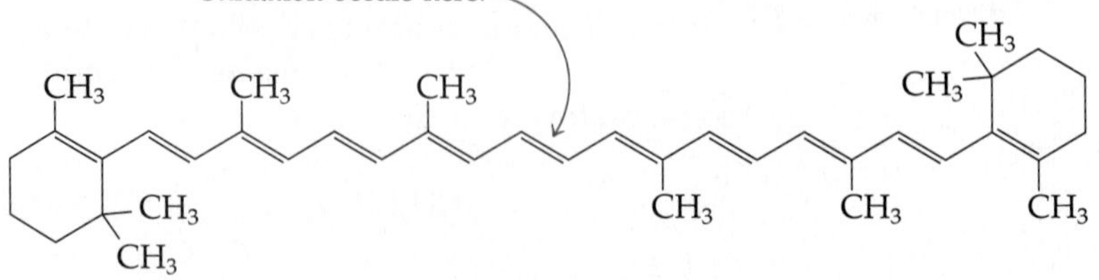

Oxidation occurs here.

CH_3 CH_3 CH_3 CH_3 CH_3 CH_3 CH_3 CH_3 CH_3 CH_3

β-carotene
(in vegetables such as carrots)

vitamin A

Alkenes

The simplest alkene, C_2H_4, is commonly called ethylene. Its IUPAC name is ethene. The IUPAC names of alkenes use the suffix *-ene*. In the structure of ethene shown in Figure 4.1, all six atoms, two carbon atoms and four hydrogen atoms, are located in the same plane. The plane may be written either in the page or perpendicular to it. If the plane is perpendicular to the printed page, the carbon–hydrogen bonds project in front of and in back of the page. As before, wedge-shaped lines represent bonds in front of the page, and dashed lines those behind the page.

A double bond decreases the number of hydrogen atoms in a molecule by two compared to the number in alkanes, so the general formula for an alkene is C_nH_{2n}. Each additional double bond reduces the number of hydrogen atoms by another two.

EXAMPLE 4.1 Caryophyllene, which is responsible for the odor of oil of cloves, contains 15 carbon atoms. The compound has two rings and two double bonds. What is the molecular formula for caryophyllene?

Solution Each ring decreases the number of hydrogen atoms by two compared to an alkane. A compound with two rings would have the formula C_nH_{2n-2}. (Remember that for n carbon atoms in an alkane, there are $2n + 2$ hydrogen atoms.) For each double bond, another two hydrogen atoms must

FIGURE 4.1 Structure of Ethylene (Ethene)
(a) The π bond is formed by sideways overlap of the parallel $2p$ orbitals of adjacent carbon atoms.
(b) Perspective formula with plane of a molecule perpendicular to the plane of the page.
(c) Perspective formula with plane of molecule in the plane of the page.
(d) Space-filling model of ethylene in the plane of the page.

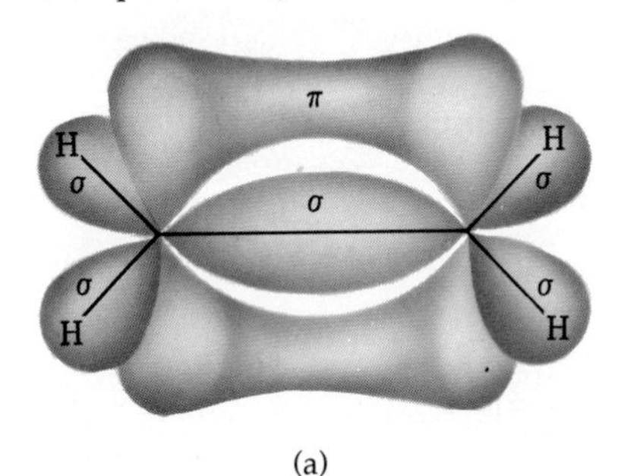

(a)

(b)

(c)

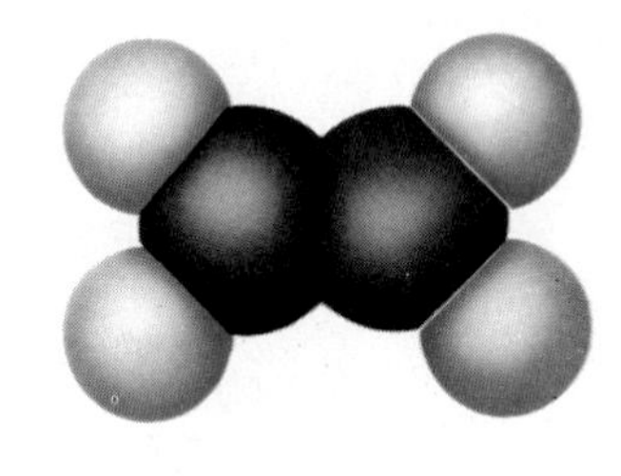

(d)

be subtracted to give C_nH_{2n-6} for two double bonds as well as the two rings. For 15 carbon atoms, the molecular formula must be $C_{15}H_{24}$.

Alkynes

The simplest alkyne, C_2H_2, is commonly called acetylene. Unfortunately, the common name ends in *-ene,* which seems to suggest that the compound contains a double bond. Such confusion is one reason IUPAC names are so important for clear communication in chemistry. The IUPAC name for C_2H_2 is ethyne.

The structure of ethyne is shown in Figure 4.2. All four atoms lie in a straight line. Each H—C≡C bond angle is 180°. In other alkynes also, the two triple-bonded carbon atoms and the two atoms directly attached to them all lie in a straight line.

The triple bond in an alkyne decreases the number of hydrogen atoms in the molecule by four compared to alkanes. As a result, the general molecular formula for alkynes is C_nH_{2n-2}.

Classification of Alkenes and Alkynes

It is often useful to describe alkenes and alkynes based on the number of alkyl groups attached to the double or triple bond. In the case of double bonds, we speak of the degree of substitution at the site of the double bond. A **monosubstituted** alkene has a single alkyl group attached to the sp^2-hybridized carbon atom of the double bond. An alkene whose double bond is at the end of a chain of carbon atoms is also sometimes called a terminal alkene. Alkenes with two, three, and four alkyl groups are **disubstituted, trisubstituted,** and **tetrasubstituted,** respectively.

monosubstituted	$RCH{=}CH_2$
disubstituted	$RCH{=}CHR$
disubstituted	$R_2C{=}CH_2$

FIGURE 4.2 Structure of Acetylene (Ethyne)
(a) Two sets of parallel oriented 2*p* orbitals form two mutually perpendicular π bonds.
(b) Structural formula with all atoms in the plane of the page.
(c) Space-filling model of acetylene.

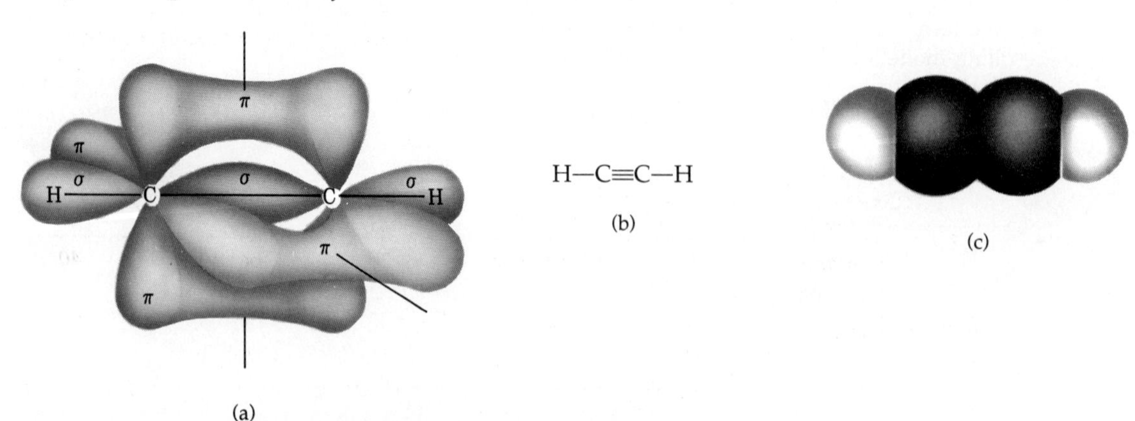

trisubstituted	$R_2C{=}CHR$
tetrasubstituted	$R_2C{=}CR_2$

In general, alkyl groups increase the stability of a double bond in an alkene. Thus, a disubstituted alkene is more stable than a monosubstituted alkene. As a consequence, if a chemical reaction can lead to either of these two products, the disubstituted alkene predominates.

The classes of alkynes are more limited. Only one alkyl group can be bonded to each of the two possible carbon atoms of the triple bond. If one alkyl group is bonded to the *sp*-hybridized carbon atom of the triple bond, the compound is a **monosubstituted** alkyne (R—C≡C—H). It is also called a **terminal** alkyne because the triple bond is at the end of the carbon chain. When alkyl groups are bonded to each carbon atom of the triple bond, the compound is **disubstituted** or an **internal** alkyne (R—C≡C—R).

EXAMPLE 4.2 The urine of the red fox contains a scent marker that is an unsaturated thioether. Classify the double bond of the scent marker.

Solution The left carbon atom of the double bond is a —CH_2 unit. The right carbon atom has two bond lines emanating from it. One of these represents a methyl group; the other is a methylene unit that is part of the main chain. Therefore, two alkyl groups are bonded to the atoms of the double bond, and the compound is a disubstituted alkene of the general class $CH_2{=}CR_2$.

Physical Properties

The physical properties of the homologous series of alkenes (C_nH_{2n}) and alkynes (C_nH_{2n-2}) are similar to those of the homologous series of alkanes (C_nH_{2n+2}). The compounds in both classes of unsaturated hydrocarbons are nonpolar. The members of each series containing fewer than five carbon atoms are gases at room temperature. As in the case of alkanes, the boiling points of the alkenes and alkynes increase with an increase in the number of carbon atoms in the molecule because the London forces increase (Table 4.1).

TABLE 4.1 Boiling Points of Alkanes, Alkenes, and Alkynes

Alkanes	Boiling point (°C)	Alkenes	Boiling point (°C)	Alkynes	Boiling point (°C)
pentane	36	1-pentene	30	1-pentyne	40
hexane	69	1-hexene	63	1-hexyne	71
heptane	98	1-heptene	94	1-heptyne	100
octane	126	1-octene	121	1-octyne	125

4.2 GEOMETRIC ISOMERISM

We know that there is free rotation about carbon–carbon single bonds—about 3 kcal/mole is sufficient to rotate about the σ bond of ethane. Therefore, alkanes can exist in many conformations. Free rotation cannot occur around the carbon–carbon double bond of an alkene because of its electronic structure. To maintain the π bond, the two $2p$ orbitals of the sp^2-hybridized carbon atoms must remain side by side (Figure 4.3). If sufficient energy were added to the molecule to force the two carbon atoms to rotate about the bond axis, the σ bond would remain, but the π bond would be broken. About 60 kcal/mole is required to break a π bond, so it does not break at ordinary reaction temperatures.

The groups bonded to the carbon atoms of the double bond can exist in different spatial or geometric arrangements. These isomers have the same connectivity of atoms, but differ from each other in the geometry about the double bond. Hence, these compounds are called **geometric isomers** or cis-trans isomers.

Consider a general alkene whose formula is CXY=CXY. When we draw a more detailed structural formula, we find that two representations are possible.

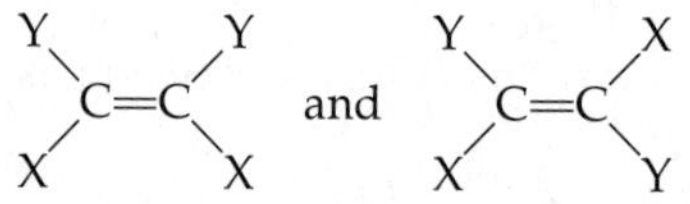

$$\underset{X}{\overset{Y}{}}\!\!\searrow C{=}C \swarrow\!\!\underset{X}{\overset{Y}{}} \quad \text{and} \quad \underset{X}{\overset{Y}{}}\!\!\searrow C{=}C \swarrow\!\!\underset{Y}{\overset{X}{}}$$

These two structures represent different molecules (Figure 4.4). In the structure on the left, two X groups are on the same "side" of the molecule; this is the cis isomer. The X's are on opposite "sides" of the molecule in the structure on the right, which is called the trans isomer.

Cis and trans isomers are possible only if an alkene has two different atoms or groups of atoms attached to each double-bonded carbon atom. Each unsaturated carbon atom in 1,2-dichloroethene has a chlorine atom and a hydrogen atom attached to it. These groups are different.

FIGURE 4.3
Rotation of the π Bond
In order for rotation to occur about a carbon–carbon double bond, the π bond must break. Loss of overlap between parallel p orbitals requires about 60 kcal/mole.

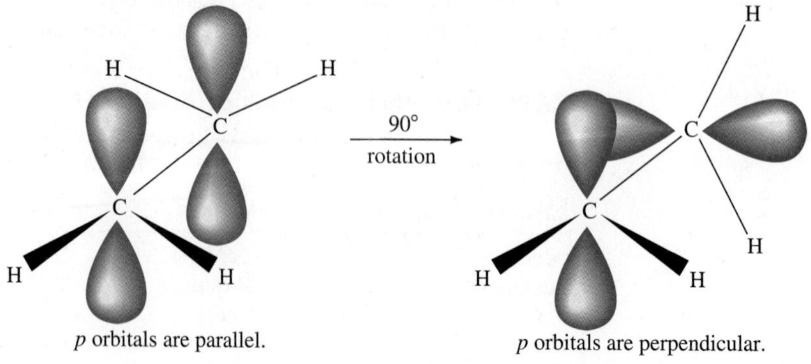

90° rotation

p orbitals are parallel.

p orbitals are perpendicular.

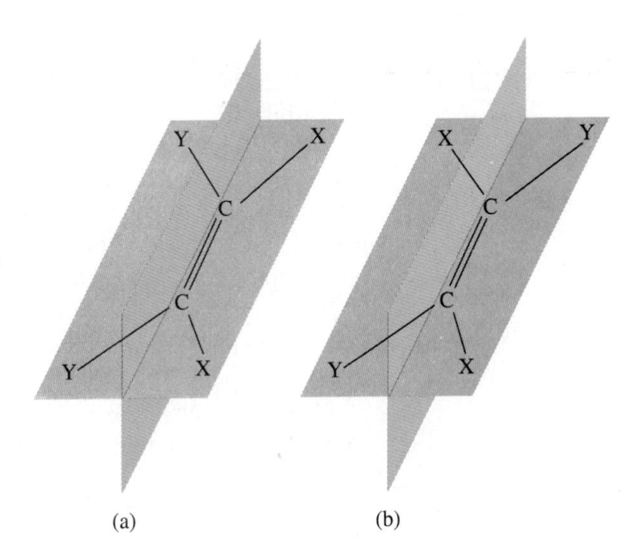

FIGURE 4.4
Geometrical Isomerism of Alkenes
(a) The two X groups are on the same side of the plane that is placed perpendicular to the plane containing the molecule. This is the cis isomer.
(b) The two X groups are on the opposite sides of the plane that is placed perpendicular to the plane containing the molecule. This is the trans isomer.

different groups → H, Cl $\;$ C=C $\;$ H, Cl ← different groups

If one of the unsaturated carbon atoms is attached to two identical groups, cis-trans isomerism is not possible. Neither chloroethene nor 1,1-dichloroethene can exist as a cis-trans pair of geometric isomers. There is only one chloroethene and one 1,1-dichloroethene.

different groups → H, Cl $\;$ C=C $\;$ H, H ← identical groups

chloroethene

identical groups → Cl, Cl $\;$ C=C $\;$ H, H ← identical groups

1,1-dichloroethene

Geometric isomers, like all isomers, have different physical properties. For example, the boiling points of *cis*- and *trans*-1,2-dichloroethene are 60 and 47 °C, respectively. The two C—Cl bond moments of the trans isomer cancel each other and the compound has no dipole moment. The cis compound has a dipole moment because the two C—Cl bond moments reinforce each other. As a consequence the cis isomer is polar and has the higher boiling point.

H, Cl $\;$ C=C $\;$ H, Cl

cis-1,2-dichloroethene

H, Cl $\;$ C=C $\;$ Cl, H

trans-1,2-dichloroethene

EXAMPLE 4.3 Is cis-trans isomerism possible about either of the double bonds of geraniol, a natural oil?

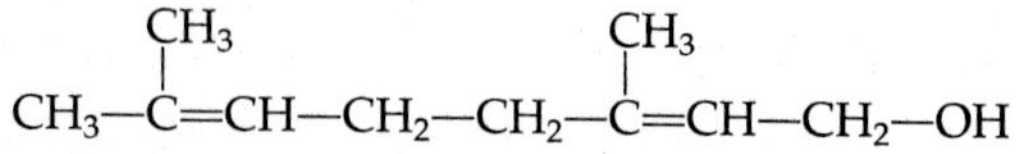

$$CH_3-\overset{\displaystyle CH_3}{\overset{|}{C}}=CH-CH_2-CH_2-\overset{\displaystyle CH_3}{\overset{|}{C}}=CH-CH_2-OH$$

Solution First, consider the double bond near the left end of the molecule. The leftmost carbon atom of that double bond has two $-CH_3$ units bonded to it. On this basis alone geometrical isomerism is not possible, regardless of the groups bonded to the right atom of that double bond.

Next, consider the double bond toward the right end of the molecule. The leftmost carbon of that double bond is bonded to a $-CH_3$ group and a $-CH_2$ unit that is part of the parent chain. The groups are different. The rightmost carbon of that double bond is bonded to two different groups: a hydrogen atom and a $-CH_2$ unit. Thus, geometrical isomerism is possible. The naturally occurring isomer of geraniol has a trans arrangement in the parent chain.

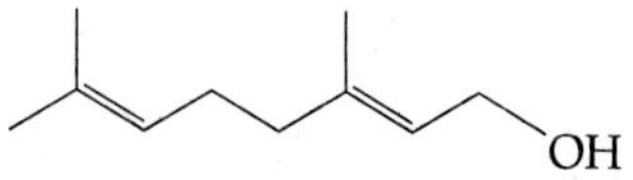

OH

Geometric Isomers and the Sex Life of Moths

We recall that pheromones are released by some animals to transmit information within the species, but not to other species. Some pheromones are sex attractants used to communicate between the male and female. The female usually emits the pheromones, and the male responds by seeking the female. The amount of pheromone required is as little as 1 picogram (1 pg = 10^{-12} g).

The female silkworm moth secretes bombykol, a diene alcohol that has a trans-cis arrangement of its two double bonds. By numbering the carbon chain from the carbon atom bearing the hydroxyl group, we see that there is a trans double bond at the C-10 atom and a cis double bond at the C-12 atom. Only this isome is biologically active.

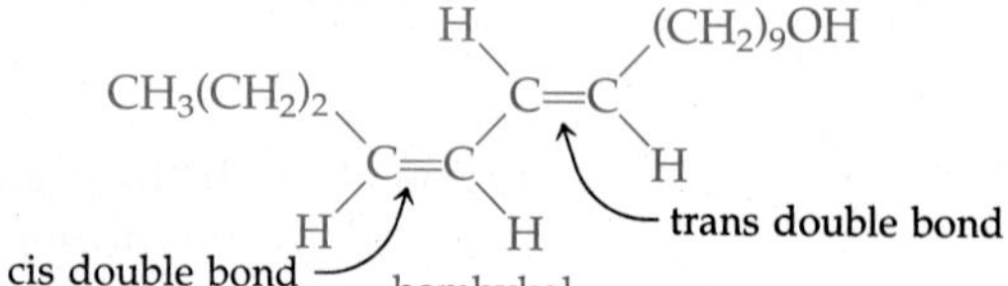

H, $(CH_2)_9OH$, $CH_3(CH_2)_2$, C=C, C=C, H, H, H

cis double bond

trans double bond

bombykol

The other three geometric isomers (cis-cis, cis-trans, and trans-trans), which have been synthesized in the laboratory, do not attract the male silkworm moth.

The structures of pheromones of many pests that damage agricultural crops have been determined. These compounds have been prepared in large quantities in the laboratory. They are used in traps to lure the males of the species and eliminate them from the breeding cycle. This procedure effectively controls the silkworm moth.

4.3 THE *E,Z* DESIGNATION OF GEOMETRIC ISOMERS

In the previous section, the terms cis and trans were applied to denote the relationship of two substituents in 1,2-dichloroethene—a disubstituted alkene. This type of nomenclature can easily be used for any disubstituted alkenes, as in the case of *cis*- and *trans*-2-butene.

CH_3 CH_3 / C=C / H H — *cis*-2-butene

CH_3 H / C=C / H CH_3 — *trans*-2-butene

However, the cis and trans notation fails to describe isomeric trisubstituted and tetrasubstituted alkenes because there is no longer a simple reference giving the relationship of groups one to another. For example, even the following relatively simple compounds cannot be designated as cis or trans isomers.

Br H / C=C / Cl F

Cl H / C=C / Br F

We can distinguish the above isomers and all other tri- and tetrasubstituted alkenes by the *E,Z* system of nomenclature. The *E,Z* system uses **sequence rules** to assign priorities to the groups bonded to the atoms of the double bond of any alkene. The two groups bonded to one carbon atom are arranged by their respective priorities and designated low and high priority. The same consideration is given to the two groups bonded to the other carbon atom. If the higher priority groups on each carbon atom are on the same side of the double bond, the alkene is the *Z* isomer (Ger. *zusammen*, together). If the higher priority groups on each carbon atom are on opposite sides of the double bond, the alkene is the *E* isomer (Ger. *entgegen*, opposite).

high high / C=C / low low — *Z* isomer

high low / C=C / low high — *E* isomer

Sequence Rules

The first rule of assigning priorities to groups is based on the atomic numbers of the atoms directly attached to the carbon atom of the double bond. A high atomic number atom receives a higher priority than a low atomic number atom. Thus, for some common atoms typically attached to a double bond, the priority order is Br > Cl > F > O > N > C > H. Applying these priorities to the following alkene, which contains several halogen atoms, allows us to make the *E,Z* assignment.

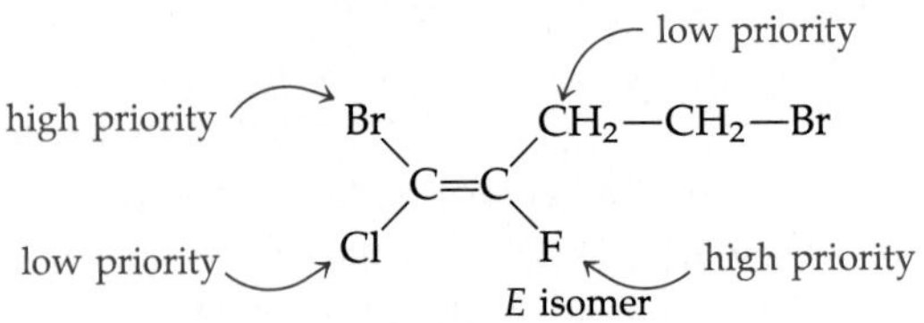

low priority
high priority Br CH₂—CH₂—Br
C=C
low priority Cl F high priority
E isomer

The higher priority of bromine compared to chlorine follows directly from their atomic numbers. The reason for the lower priority of the $—CH_2CH_2Br$ group compared to fluorine may not be as evident. However, the first atom bonded to the double-bonded carbon atom is the most important. In this case we have to compare carbon to fluorine. Because fluorine has a higher atomic number than carbon, it has the higher priority.

If the first atoms have the same priority, then the second, third, and farther atoms are considered until a difference is found. Thus, a methyl and an ethyl group are equivalent by the first rule of the directly bonded atom. However, the ethyl group has a higher priority because the second atom is a carbon atom (and two hydrogen atoms), whereas the methyl group has only hydrogen atoms. Based on this consideration, the order of alkyl groups is *t*-butyl > isopropyl > ethyl > methyl.

In assigning priorities, a double bond is counted as two single bonds for both of the atoms involved. The same principle is used for a triple bond.

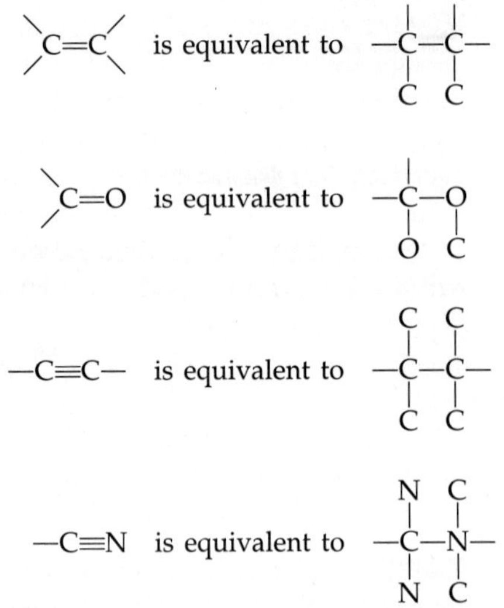

C=C is equivalent to —C—C— (each carbon bearing an additional C)

C=O is equivalent to —C—O (C bearing an additional O; O bearing an additional C)

—C≡C— is equivalent to —C—C— (each carbon bearing two additional C)

—C≡N is equivalent to —C—N— (C bearing two additional N; N bearing two additional C)

EXAMPLE 4.4 Tiglic acid, found in some natural oils, has the following structure. Designate the compound as an *E* or *Z* isomer.

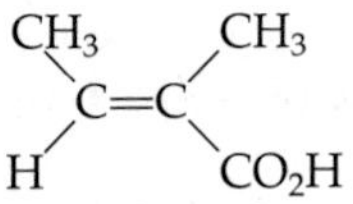

$$\begin{array}{ccc} CH_3 & & CH_3 \\ & C{=}C & \\ H & & CO_2H \end{array}$$

tiglic acid

Solution The double-bonded carbon atom on the left is bonded to a hydrogen atom and a methyl group, whose priorities are low and high, respectively. The right double-bonded carbon atom is bonded to a methyl group and a carboxylic acid (CO_2H) group. The oxygen atoms of the carboxylic acid group give it a higher priority than the methyl group, which has only hydrogen atoms bonded to the carbon atom. Because the higher priority groups are on opposite sides of the double bond, the compound is *E*.

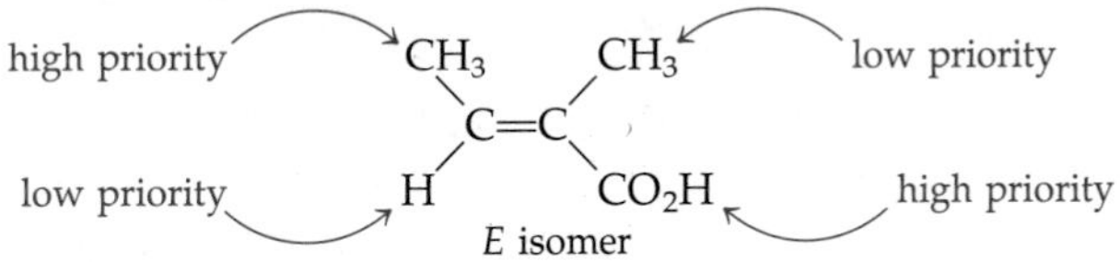

$$\begin{array}{ccccc} \text{high priority} & CH_3 & & CH_3 & \text{low priority} \\ & & C{=}C & & \\ \text{low priority} & H & & CO_2H & \text{high priority} \end{array}$$

E isomer

Note, that if we consider only the positions of the methyl groups, the compound looks like a cis isomer. Thus, one could erroneously classify the compound as the *Z* isomer. This example illustrates the value of the unambiguous *E,Z* system.

4.4 NOMENCLATURE OF ALKENES AND ALKYNES

The IUPAC rules for naming alkenes and alkynes are similar to those for alkanes, but the position of the double or triple bond in the chain and the geometric arrangement of substituents about the double bond must be indicated.

Along with the IUPAC nomenclature, a few older common names are still used to name groups appended to parent chains. Vinyl and allyl are examples that contain a double bond. The propargyl group contains a triple bond.

$CH_2{=}CH{-}$	$CH_2{=}CHCH_2{-}$	$HC{\equiv}C{-}CH_2{-}$
vinyl	allyl	propargyl

Alkenes

1. The longest continuous chain containing the double bond is the parent.

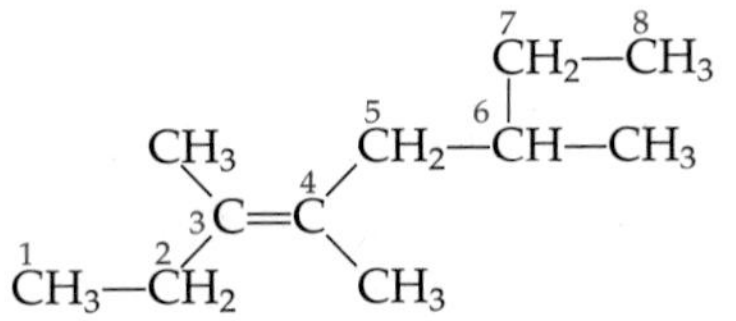

$$\begin{array}{ccccc} & & & \overset{7}{C}H_2{-}\overset{8}{C}H_3 & \\ & & & | & \\ & CH_3 & & \overset{5}{C}H_2{-}\overset{6}{C}H{-}CH_3 & \\ & & {}_3C{=}\overset{4}{C} & & \\ \overset{1}{C}H_3{-}\overset{2}{C}H_2 & & & CH_3 & \end{array}$$

There are eight carbon atoms in this chain.

2. The longest chain is given the same stem name as an alkane, but *-ene* replaces *-ane*. The parent name of the structure given in rule 1 is octene.
3. The carbon atoms are numbered consecutively from the end nearer the double bond. The number of the first carbon atom of the double bond is used as a prefix to the parent name and is separated from it by a hyphen.

$$\overset{1}{C}H_3-\overset{2}{C}H_2(CH_3)\overset{3}{C}=\overset{4}{C}(CH_3)-\overset{5}{C}H_2-\overset{6}{C}H_2-\overset{7}{C}H_3$$

This is a substituted 3-heptene, not a substituted 4-heptene.

4. Alkyl groups and other substituents are named and their positions on the chain are identified according to the numbering established by rule 3. Names and numbers are prefixed to the parent name.

$$\overset{5}{C}H_3-\overset{4}{C}H_2(CH_3)\overset{3}{C}=\overset{2}{C}(CH_3)-\overset{1}{C}H_3$$

This is 2,3-dimethyl-2-pentene, not 3,4-dimethyl-3-pentene.

5. If the compound can exist as an *E* or *Z* isomer, the appropriate prefix followed by a hyphen is placed in parentheses in front of the name.

$$\overset{1}{C}H_3-\overset{2}{C}H_2(CH_3)\overset{3}{C}=\overset{4}{C}(H)-\overset{5}{C}H_2-\overset{6}{C}H_3$$

This is (*E*)-3-methyl-3-hexene.

6. If there is more than one double bond, indicate the location of each double bond by a number. A prefix to *-ene* indicates the number of double bonds.

$$\overset{1}{C}H_2=\overset{2}{C}H-\overset{3}{C}H=\overset{4}{C}H-\overset{5}{C}H=\overset{6}{C}H-\overset{7}{C}H_3$$

1,3,5-heptatriene

7. Name cycloalkenes by numbering the ring to give the double-bonded carbon atoms the numbers 1 and 2. Choose the direction of numbering so that the first substituent on the ring receives the lower number. The position of the double bond is not given because it is known to be between the C-1 and C-2 atoms.

3-methylcyclopentene

1-methylcyclohexene

EXAMPLE 4.5 Name the following compound.

$$\underset{CH_3CH_2CH_2}{\overset{Br}{}}\!\!>C=C<\!\!\underset{CH_3}{\overset{Br}{}}$$

Solution There are six carbon atoms in the longest chain. It is numbered from right to left so that the double bond is at the carbon atom in position 2. The parent is then 2-hexene. The bromine atoms are at the 2- and 3-positions.

higher priority → Br (on C-3); higher priority → Br (on C-2)

$$\overset{6}{C}H_3\overset{5}{C}H_2\overset{4}{C}H_2-\overset{3}{C}(Br)=\overset{2}{C}(Br)-\overset{1}{C}H_3$$

The two different groups of atoms on each unsaturated carbon atom make geometric isomers possible. Because bromine has a higher priority than carbon, this is (Z)-2,3-dibromo-2-hexene.

Alkynes

1. The longest continuous chain containing the triple bond is used as the parent.

$$CH_3-CH_2-C\equiv C-\underset{}{\overset{CH_2-CH_3}{\overset{|}{C}H}}-CH_3$$

There are seven carbon atoms in the longest chain.

2. The longest chain is given the same stem name as an alkane, but *-yne* replaces *-ane*.

$$CH_3-C\equiv C-\overset{Br}{\overset{|}{C}H}-CH_3$$

This is a substituted pentyne.

3. The carbon atoms are numbered consecutively from the end of the chain nearer the triple bond. The number of the first carbon atom with the triple bond is used as a prefix separated by a hyphen from the parent name.

$$\overset{6}{C}H_3-\overset{5}{C}H_2-\overset{4}{C}H_2-\overset{3}{C}\equiv\overset{2}{C}-\overset{1}{C}H_3$$

This compound is numbered from right to left. It is 2-hexyne.

4. Alkyl groups are named and their positions on the chain are determined by the numbering established by rule 3.

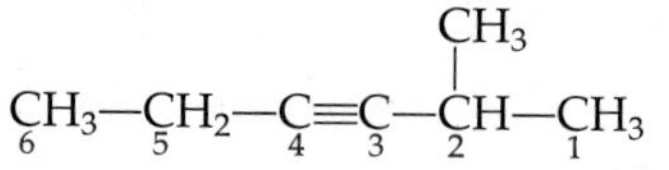

$$\underset{6}{CH_3}-\underset{5}{CH_2}-\underset{4}{C}\equiv\underset{3}{C}-\underset{2}{\overset{\overset{\displaystyle CH_3}{|}}{CH}}-\underset{1}{CH_3}$$

The triple bond is in the middle of the chain. Thus, the methyl group controls the numbering. The compound is 2-methyl-3-hexyne.

5. Compounds with multiple triple bonds are diynes, triynes, and so on. Compounds with both double and triple bonds are enynes—not ynenes. Numbering of compounds with both double and triple bonds starts from the end nearest the first multiple bond regardless of type. When a choice is possible, a double bond is assigned a lower number than a triple bond.

$$CH_2{=}CH-CH_2-C\equiv C-H$$

This is 1-penten-4-yne, not 4-penten-1-yne.

EXAMPLE 4.6 Why is 2-bromo-4-hexyne an incorrect name for the following compound?

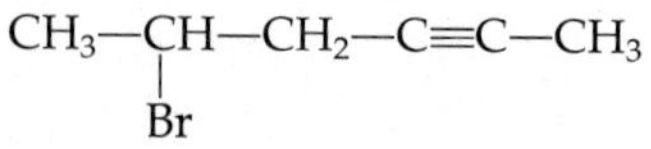

$$CH_3-\underset{\underset{\displaystyle Br}{|}}{CH}-CH_2-C\equiv C-CH_3$$

Solution The chain must be numbered from the end nearer the triple bond. Thus, the chain is numbered from right to left. The correct name is 5-bromo-2-hexyne.

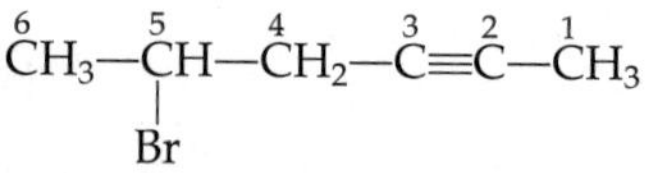

$$\overset{6}{C}H_3-\underset{\underset{\displaystyle Br}{|}}{\overset{5}{C}H}-\overset{4}{C}H_2-\overset{3}{C}\equiv\overset{2}{C}-\overset{1}{C}H_3$$

4.5 ACIDITY OF ALKENES AND ALKYNES

Most of this chapter will be devoted to the chemistry of the π bonds of alkenes and alkynes rather than the carbon–hydrogen bonds in these compounds. However, there is some acidic character in C—H σ bonds, and a hydrogen ion may be removed by a very strong base.

$$R-H + B^- \rightleftharpoons R:^- + BH$$

Removing a proton from an alkane, alkene, or alkyne produces a **carbanion,** an anion with a negative charge on the carbon atom. Formation of a carbanion, the conjugate base of a hydrocarbon, is generally less favorable than the ionization of many acids that you encountered in your first chemistry course. In those acids, the hydrogen atom is bonded to an electronegative atom. Carbon is not electronegative, and the K_a values of hydrocarbons are very small.

$$\underset{K_a = 10^{-49}}{CH_3{-}CH_3} \qquad \underset{K_a = 10^{-44}}{CH_2{=}CH_2} \qquad \underset{K_a = 10^{-25}}{HC{\equiv}CH}$$

The acidity of hydrocarbons is related to the hybridization of the carbon atom. The K_a increases for carbon atoms in the order $sp^3 < sp^2 < sp$. The order of acidities parallels the percent contribution of the 2*s* orbital to the hybrid orbitals in the σ bond. The energy of a 2*s* orbital is lower than that of a 2*p* orbital, and on average a 2*s* orbital is closer to the nucleus than a 2*p* orbital. The average distance of hybrid orbitals from the nucleus depends on the percent contribution of the *s* and *p* orbitals. For an sp^3 hybrid orbital, the contribution of the *s* orbital is 25% because one *s* and three *p* orbitals contribute to the four hybrid orbitals. Similarly, the contribution of the *s* orbital is 33% and 50% for the sp^2 and *sp* hybrid orbitals, respectively. Because an *sp* hybrid orbital has more *s* character than an sp^2 or sp^3 orbital, its electrons are located closer to the nucleus. As a consequence a proton is more easily removed and the electron pair is left on the carbon atom.

Acetylene and terminal alkynes are much stronger acids than other hydrocarbons. However, they are still very weak acids. Hydroxide ion is not a strong enough base to convert a terminal alkyne to its conjugate base in any significant amounts. In fact, the conjugate base of an alkyne is rapidly and quantitatively converted to the alkyne whenever it reacts with compounds containing hydroxyl groups (such as water, alcohols, and carboxylic acids).

$$R{-}C{\equiv}C{:}^- + H_2O \longrightarrow R{-}C{\equiv}C{-}H + HO^-$$

Recall the periodic trends of acidity discussed in Chapter 2. An N—H bond is a weaker acid than an O—H bond. Therefore NH_2^-, the conjugate base of ammonia—a very weak acid—is stronger than OH^-, the conjugate base of water—a weak acid. The K_a of ammonia is 10^{-36}. The K_a of a terminal alkyne is about 10^{-26}. Thus, amide ion quantitatively removes a proton from any terminal alkyne.

$$R{-}C{\equiv}C{-}H + NH_2^- \longrightarrow R{-}C{\equiv}C{:}^- + NH_3$$

4.6 HYDROGENATION OF ALKENES AND ALKYNES

We expect alkenes and alkynes to have chemical similarities because both have π bonds. In general, that expectation is correct. Because alkynes have two π bonds, they often react with twice the amount of reagent that reacts with alkenes, which have only one π bond. The compounds in both classes are unsaturated and react with hydrogen gas to give more saturated compounds.

Alkenes Are Reduced to Alkanes

The reaction of hydrogen gas with an alkene (or cycloalkene) yields a saturated compound. The process is a reduction, but the reaction is also called **hydrogenation.** The hydrogenation of 1-octene yields octane. Hydrogenation requires a catalyst. The catalyst is usually finely divided platinum on finely divided carbon, but nickel and palladium can also be used. The hydrogenation process is heterogeneous, that is, it occurs on the surface of the solid catalyst. The catalyst is symbolized Pt/C, where Pt is the metal and C is carbon.

The double bond is changed into a single bond.

$$\underset{\text{1-octene}}{CH_3(CH_2)_5CH{=}CH_2} + H{-}H \xrightarrow{Pt/C} \underset{\text{octane}}{CH_3(CH_2)_5CH_2{-}CH_3}$$

$$\text{cyclohexene} + H{-}H \xrightarrow{Pt/C} \text{cyclohexane}$$

cyclohexene cyclohexane

Although functional groups such as ketones or esters also have multiple bonds, they are not reduced under the conditions used to hydrogenate a carbon–carbon double bond.

$$CH_3{-}\overset{O}{\overset{\|}{C}}{-}CH_2{-}CH_2{-}CH{=}CH_2 + H_2 \xrightarrow{Pt/C} CH_3{-}\overset{O}{\overset{\|}{C}}{-}CH_2{-}CH_2{-}CH_2{-}CH_3$$

EXAMPLE 4.7 How many moles of hydrogen gas will react with cembrene, which is contained in pine oil? What is the molecular formula of the product?

cembrene

Solution There are four double bonds in the compound. One molar equivalent of the compound will react with four molar equivalents of hydrogen gas.

The product will be a cycloalkane. There are 14 carbon atoms in the ring, three methyl groups, and an isopropyl group for a total of 20 carbon

Hydrogenation of Oils

Fats are glycerol esters of long-chain saturated carboxylic acids. Oils are structurally related esters, but the acids have one or more double bonds. Fats have long been used in cooking and in producing butter substitutes. Initially fats were obtained from animals and tended to have some residual tastes of other animal substances. Oils generally have little flavor. Thus, a way was sought to convert oils into fats. Hydrogenation of oils is now carried out on a large industrial scale. The difference between Crisco oil and its companion solid Crisco is just the degree of saturation of the carboxylic acids in the esters: the solid is more saturated.

Over the years, the extensive use of fats has been discouraged because their consumption is linked to heart disease. Cooking oils are now widely used rather than fats. Soft margarines, which contain more unsaturated compounds, are often used in place of butter.

$$3\,H_2 + \begin{array}{l} H_2C{-}O{-}C({=}O){-}(CH_2)_7{-}CH{=}CH{-}(CH_2)_7{-}CH_3 \\ HC{-}O{-}C({=}O){-}(CH_2)_7{-}CH{=}CH{-}(CH_2)_7{-}CH_3 \\ H_2C{-}O{-}C({=}O){-}(CH_2)_7{-}CH{=}CH{-}(CH_2)_7{-}CH_3 \end{array} \xrightarrow[200\,^\circ C]{Ni} \begin{array}{l} H_2C{-}O{-}C({=}O){-}(CH_2)_7{-}CH_2{-}CH_2{-}(CH_2)_7{-}CH_3 \\ HC{-}O{-}C({=}O){-}(CH_2)_7{-}CH_2{-}CH_2{-}(CH_2)_7{-}CH_3 \\ H_2C{-}O{-}C({=}O){-}(CH_2)_7{-}CH_2{-}CH_2{-}(CH_2)_7{-}CH_3 \end{array}$$

a liquid oil — a solid fat

atoms. Because the general molecular formula for a cycloalkane is C_nH_{2n}, the molecular formula of the product is $C_{20}H_{40}$.

Reduction of Alkynes

Alkynes can be completely reduced to alkanes by reaction with two molar equivalents of hydrogen gas in the presence of a palladium catalyst.

$$CH_3(CH_2)_7C{\equiv}CH + 2\,H_2 \xrightarrow{Pd/C} CH_3(CH_2)_8CH_3$$

The reaction can be stopped after adding one molar equivalent of hydrogen gas to form an alkene if the palladium is specially prepared. In the Lindlar catalyst, palladium is coated on calcium carbonate that contains a small amount of lead acetate. Hydrogenation of an alkyne using the Lindlar catalyst gives cis alkenes.

$$CH_3(CH_2)_3C{\equiv}C(CH_2)_3CH_3 \xrightarrow[\text{Lindlar catalyst}]{H_2} \underset{H}{\overset{CH_3(CH_2)_2CH_2}{}}C{=}C\underset{H}{\overset{CH_2(CH_2)_2CH_3}{}}$$

5-decyne — *cis*-5-decene

In contrast, reduction of an alkyne with lithium metal as the reducing agent in liquid ammonia as the solvent gives the trans isomer. Water is added to the reaction mixture in a second step.

$$CH_3(CH_2)_3C{\equiv}C(CH_2)_3CH_3 \xrightarrow[2.\ H_2O]{1.\ Li/NH_3} \underset{H}{\overset{CH_3(CH_2)_2CH_2}{}}\!\!>C{=}C<\!\!\underset{CH_2(CH_2)_2CH_3}{\overset{H}{}}$$

5-decyne → *trans*-5-decene

4.7 OXIDATION OF ALKENES AND ALKYNES

Oxidizing agents attack the π electrons of the double bonds of alkenes and the triple bonds of alkynes more easily than the σ electrons of the carbon–carbon single bonds and carbon–hydrogen bonds. Thus, in contrast to alkanes, both alkenes and alkynes can easily be oxidized without destroying the carbon chain.

Hydroxylation of Alkenes

Reaction of an alkene with potassium permanganate ($KMnO_4$) in basic solution yields a product that contains a hydroxyl group on each carbon atom of the original double bond. Thus, the alkene is oxidized. Permanganate is reduced to MnO_2.

$$CH_3(CH_2)_3CH{=}CH_2 + KMnO_4 \xrightarrow{NaOH/H_2O} CH_3(CH_2)_3\underset{OH}{\underset{|}{C}H}{-}\underset{OH}{\underset{|}{C}H_2}$$

1-hexene → 1,2-hexanediol

Potassium permanganate is purple in aqueous solution. Manganese dioxide (MnO_2), the product of the reaction, is a brown solid that precipitates from solution. The color change in oxidation with potassium permanganate allows it to be used to test visually for the presence of a double bond. Alkanes and cycloalkanes are not oxidized by $KMnO_4$, so the purple color remains. Alkenes are oxidized by $KMnO_4$, so the purple color fades as a brown precipitate appears.

Ozonolysis of Alkenes and Alkynes

Alkenes and alkynes react rapidly with ozone, O_3. The reaction is carried out in an inert solvent such as dichloromethane. The oxidation occurs in two steps. In the first, an unstable intermediate called an ozonide forms. The intermediate is not isolated but treated in solution by reacting it with zinc and acid. The net result of the two reactions is the cleavage of the carbon–carbon double bond to produce two carbonyl compounds. The overall process is called **ozonolysis.**

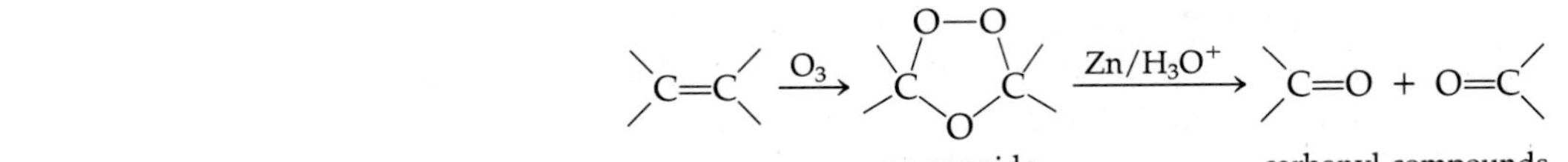

$$>C=C< \xrightarrow{O_3} \text{(ozonide: } >C \text{—O—O—} C< \text{ bridged by O)} \xrightarrow{Zn/H_3O^+} >C=O + O=C<$$

an ozonide carbonyl compounds

Ozonolysis can be used to determine the position of a double bond in an alkene. If one of the double-bonded carbon atoms has two hydrogen atoms bonded to it, the ozonolysis product is methanal (formaldehyde). If an alkyl group and a hydrogen atom are bonded to the double-bonded carbon atom, the product is an aldehyde; if two alkyl groups are bonded to the double-bonded carbon atom, the product is a ketone.

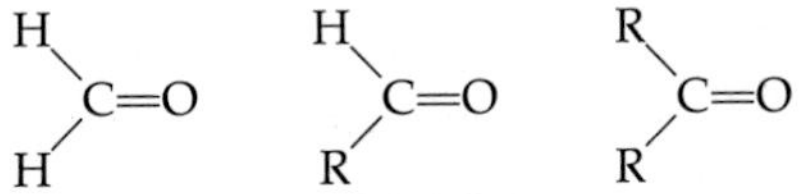

$$(H)(H)C=O \qquad (H)(R)C=O \qquad (R)(R)C=O$$

methanal an aldehyde a ketone

Although less commonly used, the ozonolysis of alkynes also results in cleavage products. Carboxylic acids are obtained from internal alkynes. A terminal alkyne forms one molar equivalent of CO_2.

$$R—C\equiv C—R' \xrightarrow[2.\ Zn/H_3O^+]{1.\ O_3} RCO_2H + R'CO_2H$$

$$R—C\equiv C—H \xrightarrow[2.\ Zn/H_3O^+]{1.\ O_3} RCO_2H + CO_2$$

EXAMPLE 4.8 An alkene of molecular formula C_7H_{14} reacts with ozone to yield the following two carbonyl compounds. What is the structure of the alkene?

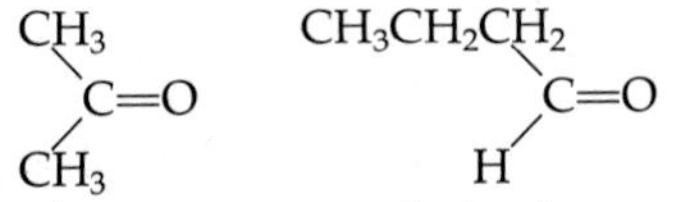

$$(CH_3)(CH_3)C=O \qquad (CH_3CH_2CH_2)(H)C=O$$

propanone (acetone) butanal

Solution Oxygen atoms mark the carbon atoms that were originally part of the double bond. Place the carbon atoms from the two carbonyl compounds near each other. Mentally remove the oxygen atoms and join the carbon atoms with a double bond.

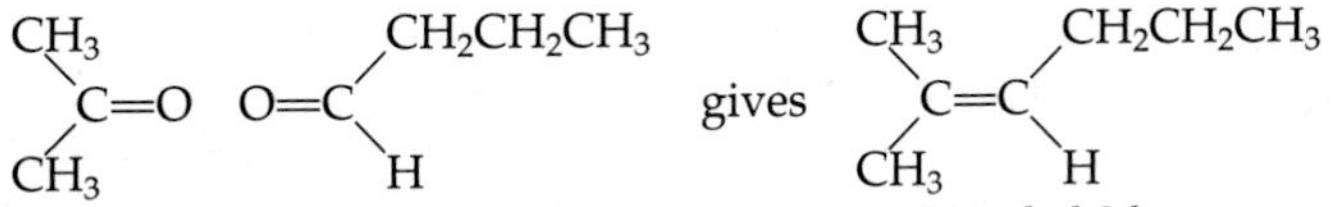

$$(CH_3)(CH_3)C=O \quad O=C(CH_2CH_2CH_3)(H) \quad \text{gives} \quad (CH_3)(CH_3)C=C(CH_2CH_2CH_3)(H)$$

2-methyl-2-hexene

Toxicity of Alkenes in Metabolic Oxidation

Oxidation of a variety of compounds containing double bonds occurs in the liver and requires cytochrome P-450, an iron-containing heme-protein (Section 2.4). The metabolic oxidation of carbon–carbon double bonds produces epoxides. The chemistry of these three-membered heterocyclic compounds will be discussed in Chapter 9. They react with water to produce 1,2-diols, also known as glycols.

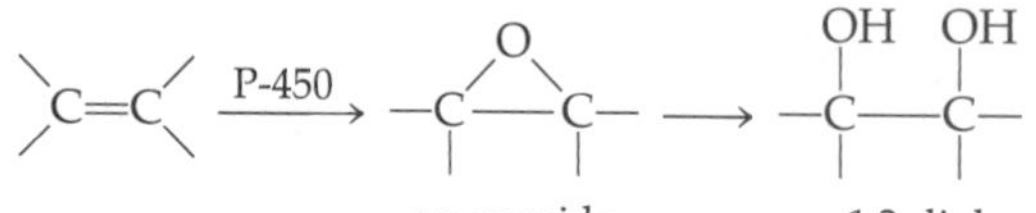

$$\mathrm{>C{=}C<} \xrightarrow{\text{P-450}} \underset{\text{an epoxide}}{-\overset{\overset{O}{\frown}}{C-C}-} \longrightarrow \underset{\text{a 1,2-diol}}{-\overset{OH}{\overset{|}{C}}-\overset{OH}{\overset{|}{C}}-}$$

The diol is quite polar and is often water-soluble. Thus, the oxidation products can be excreted from the body. For example, when the anti-inflammatory agent alcofenac is epoxidized and hydrated, the resulting product is very polar.

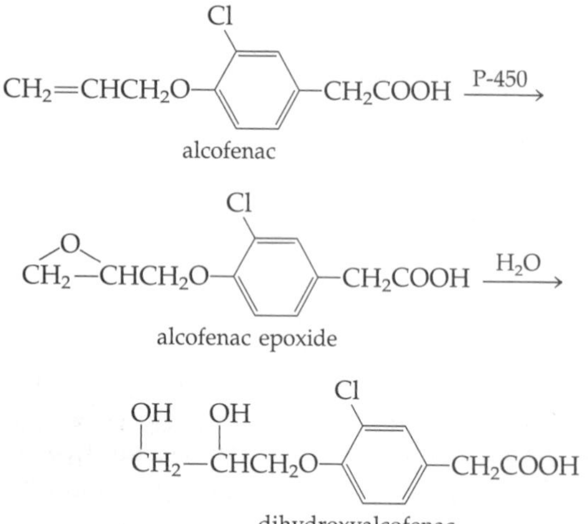

$$CH_2{=}CHCH_2O{-}C_6H_3(Cl){-}CH_2COOH \xrightarrow{\text{P-450}}$$

alcofenac

$$\overset{O}{CH_2{-}CH}CH_2O{-}C_6H_3(Cl){-}CH_2COOH \xrightarrow{H_2O}$$

alcofenac epoxide

$$\overset{OH}{\overset{|}{CH_2}}{-}\overset{OH}{\overset{|}{CH}}CH_2O{-}C_6H_3(Cl){-}CH_2COOH$$

dihydroxyalcofenac

Reactions other than the hydrolysis of epoxides occur. The toxicity of some unsaturated compounds is a consequence of reactions of the chemically reactive epoxides with biomolecules. For example, aflatoxin B_1, a carcinogen found in moldy food, contains a double bond in a five-membered ring that is adjacent to an oxygen atom in a cyclic ether. During metabolic oxidation, this double bond is converted to an epoxide that reacts rapidly with proteins, RNA, and DNA and causes serious cellular damage. The chemistry of these epoxides with nucleophilic sites in biological molecules will be discussed in Chapter 9.

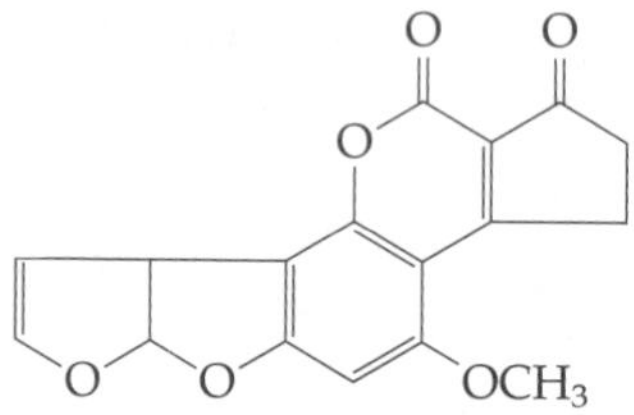

aflatoxin B_1

Some alkenes form epoxides that are so reactive that they covalently bind to the cytochrome P-450 itself! One of the simplest such compounds is fluroxene, an anesthetic agent. Obviously, drugs that disrupt the very enzyme responsible for detoxifying drugs must be carefully administered.

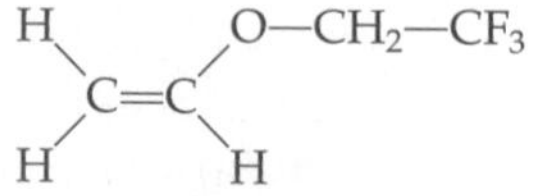

$$H_2C{=}CH{-}O{-}CH_2{-}CF_3$$

fluroxene

4.8 ADDITION REACTIONS OF ALKENES AND ALKYNES

As first described in Chapter 2, an **addition reaction** occurs when two reactants combine to form a single product. No atoms are "left over". The composition of the product is the sum of all of the atoms in the two reactants.

Symmetrical and Unsymmetrical Reagents

Examples of addition reactions of ethene (C_2H_4) with some common reagents are:

$$CH_2{=}CH_2 + Br_2 \longrightarrow BrCH_2CH_2Br$$
$$CH_2{=}CH_2 + HCl \longrightarrow CH_3CH_2Cl$$
$$CH_2{=}CH_2 + H_2O \xrightarrow{H^+} CH_3CH_2OH$$

Symmetrical reagents consist of two identical groups, one of which adds to each carbon atom of the double bond. Bromine is a symmetrical reagent. **Unsymmetrical reagents** consist of different groups, each of which bonds to a different carbon atom of the double bond. Examples of unsymmetrical reagents are HCl and H_2O.

Addition of Halogens

The reaction of ethene with Br_2 to form 1,2-dibromoethane is an addition reaction. The atoms that add to the double bond are located on adjacent carbon atoms, a common characteristic of addition reactions of alkenes.

These two atoms are added to adjacent atoms.

$$H_2C{=}CH_2 + Br{-}Br \longrightarrow H{-}\underset{H}{\overset{Br}{C}}{-}\underset{H}{\overset{Br}{C}}{-}H$$

1,2-dibromoethane

Evidence for the addition of bromine to an alkene is easily seen. Bromine is reddish; it reacts with alkenes to give a colorless product. Hence, the disappearance of the reddish color of bromine can be used to determine if a compound is unsaturated. Drops of Br_2 dissolved in CCl_4 are added to a compound. If the bromine color disappears, the compound is unsaturated.

$$\underset{\text{colorless}}{CH_3(CH_2)_3CH{=}CH_2} + \underset{\text{red}}{Br_2} \longrightarrow \underset{\text{colorless}}{CH_3(CH_2)_3CHBr{-}CH_2Br}$$

Chlorine also adds to a carbon–carbon double bond, but iodine is not sufficiently reactive to give a good yield of addition product. Fluorine is too reactive to control, and several competing reactions also occur.

Alkynes react with chlorine or bromine to produce tetrahaloalkanes, which contain two halogen atoms on each of the original carbon atoms of the triple bond. Thus, two molar equivalents of the halogen are consumed in the reaction.

$$CH_3{-}CH_2{-}C{\equiv}C{-}H \xrightarrow{2\,Cl_2} CH_3{-}CH_2{-}CCl_2{-}CCl_2{-}H$$

If only one molar equivalent of the halogen is used, the reaction product has the halogen atoms on the opposite sides of the double bond.

$$CH_3{-}CH_2{-}C{\equiv}C{-}H \xrightarrow{Br_2} \underset{(E)\text{-1,2-dibromo-1-butene}}{(CH_3{-}CH_2)(Br)C{=}C(Br)(H)}$$

Addition of Hydrogen Halides

The addition of a symmetrical reagent to an alkene yields only one possible product. For example, bromine is a symmetrical molecule, and it reacts with propene to yield a single compound. It makes no difference which bromine atom bonds to which carbon atom.

$$\underset{\text{propene}}{CH_3{-}CH{=}CH_2} + Br_2 \longrightarrow \underset{\text{1,2-dibromopropane}}{CH_3{-}CHBr{-}CH_2Br}$$

Now let's consider the addition reaction of an unsymmetrical reagent, such as a hydrogen halide HX. (Usually only HCl or HBr are used.) With a symmetrical alkene, such as ethylene, only one product is possible because the two carbon atoms are identical.

$$CH_2{=}CH_2 + HCl \longrightarrow H{-}CH_2{-}CH_2{-}Cl \text{ or } Cl{-}CH_2{-}CH_2{-}H$$

identical carbon atoms

identical structures

Two products could conceivably result from the addition of HBr to an unsymmetrical alkene, but only one is actually formed. Addition of HBr to propene could yield either 1-bromopropane or 2-bromopropane. However, only the latter is formed. The × written through a reaction arrow is used to indicate that the reaction does not occur. Thus, the addition of an unsymmetrical reagent to an alkene is a selective process; that is, only one product is formed.

$$\underset{\text{propene}}{CH_3{-}CH{=}CH_2} + HBr \longrightarrow \underset{\text{2-bromopropane}}{CH_3{-}CHBr{-}CH_2H}$$

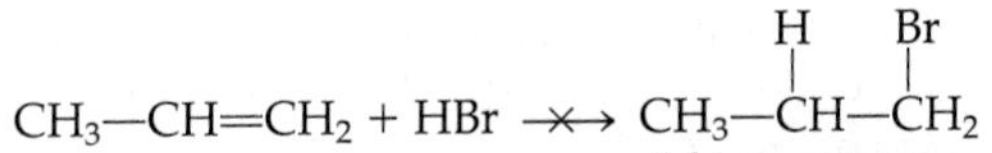

$$CH_3{-}CH{=}CH_2 + HBr \not\longrightarrow CH_3{-}\underset{}{\overset{H}{\overset{|}{C}}}H{-}\overset{Br}{\overset{|}{C}}H_2$$

1-bromopropane
(not observed)

A similar selectivity is observed for the reaction of alkynes with hydrogen halides.

Markovnikov's Rule

The Russian chemist Vladimir Markovnikov observed that unsymmetrical reagents add to unsymmetrical double bonds in a specific way. **Markovnikov's rule** states that a molecule of the general formula HX adds to a double bond so that the hydrogen atom bonds to the unsaturated carbon atom that already has the greater number of directly bonded hydrogen atoms. This is the least substituted carbon atom.

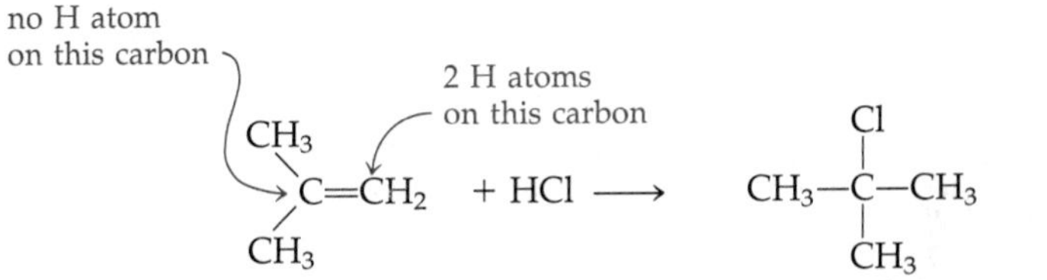

no H atom on this carbon

2 H atoms on this carbon

$$(CH_3)_2C{=}CH_2 + HCl \longrightarrow CH_3{-}\overset{Cl}{\overset{|}{\underset{\underset{CH_3}{|}}{C}}}{-}CH_3$$

2-methylpropene — 2-chloro-2-methylpropane

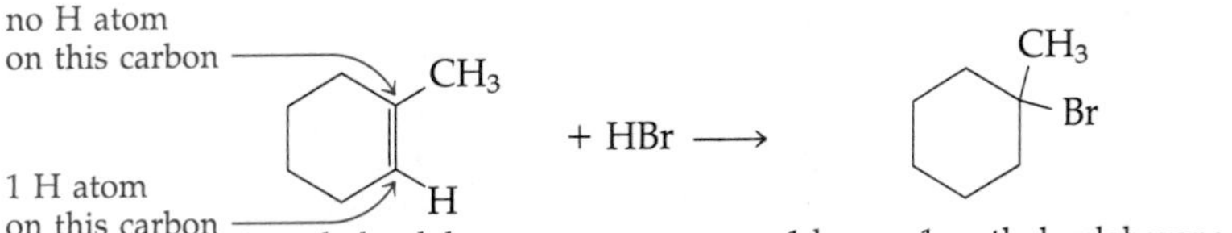

no H atom on this carbon

1 H atom on this carbon

1-methylcyclohexene + HBr ⟶ 1-bromo-1-methylcyclohexane

EXAMPLE 4.9 Predict what product will be formed when HCl is added to 2-methyl-2-butene.

Solution One of the unsaturated carbon atoms has one attached hydrogen atom, the other no attached hydrogen atoms. The predicted product is 2-chloro-2-methylbutane.

Chlorine becomes attached to this carbon.

Hydrogen becomes attached to this carbon.

$$(H_3C)_2C{=}C(H)CH_3 \text{ to give } CH_3{-}\overset{CH_3}{\overset{|}{\underset{\underset{Cl}{|}}{C}}}{-}CH_2CH_3$$

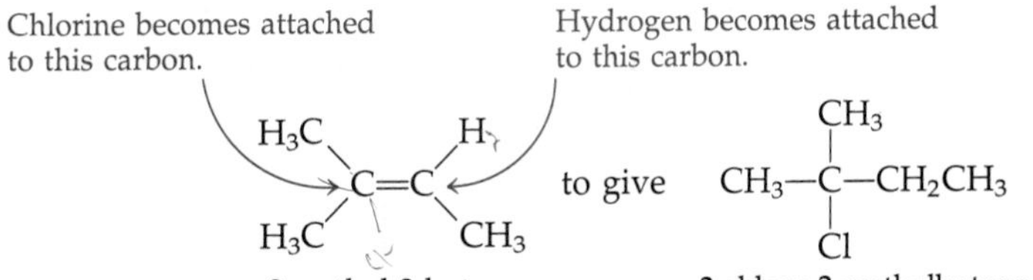

2-methyl-2-butene — 2-chloro-2-methylbutane

4.9 MECHANISTIC BASIS OF MARKOVNIKOV'S RULE

Markovnikov's rule is the result of experimental observations and has predictive value. However, it doesn't tell us why the reagents react the way they do. An understanding of the reaction is provided by the mechanism that has now been established. The π electrons in an alkene allow it to serve as a Lewis base (Chapter 2) and accept an electrophile (electron-loving species). Consider the reaction of propene with HBr. The first step is written by using a curved arrow to show the movement of two electrons in the π bond to form a σ bond to hydrogen. As a result, a positively charged species called an isopropyl carbocation forms.

$$(H)(CH_3)C{=}CH_2 + H{-}\ddot{Br}: \longrightarrow (H)(CH_3)C^{+}{-}CH_3 + :\ddot{Br}:^{-}$$

isopropyl carbocation

These two electrons in the C—H bond were in the π bond.

The movement of the π electrons in this step is like the movement of a swinging gate. One end of the gate stays attached, and the other end "swings" free to bond to the electrophile. The electron pair stays as part of a bond on one of the two carbon atoms.

In the second step of the addition reaction, the isopropyl carbocation acts as a Lewis acid and accepts an electron pair from the bromide ion, which acts as a nucleophile (nucleus-loving species).

These two electrons in the C—Br bond were with the bromide ion.

$$(H)(CH_3)C^{+}{-}CH_3 + :\ddot{Br}:^{-} \longrightarrow CH_3{-}CHBr{-}CH_3$$

isopropyl carbocation — 2-bromopropane

In the first step of this reaction, the hydrogen atom is attached to one of the two possible carbon atoms of the original π bond; this placement accounts for the product predicted by Markovnikov's rule. Why is this particular carbocation, the isopropyl carbocation, formed? If the hydrogen atom had bonded to the interior carbon atom, the electrons would have had to "swing" in the other direction, and a propyl carbocation would have formed.

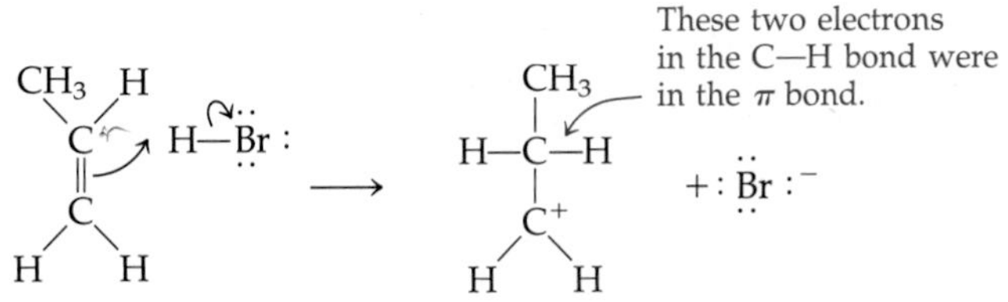

$$CH_3(H)C{=}CH_2 + H{-}\ddot{Br}: \longrightarrow H{-}C(CH_3)(H){-}C^{+}H_2 + :\ddot{Br}:^{-}$$

propyl carbocation

These two electrons in the C—H bond were in the π bond.

Both the isopropyl carbocation and the propyl carbocation are unstable intermediates. However, the isopropyl carbocation is more stable and is formed preferentially. Alkyl groups attached to a positively charged carbon atom help stabilize the charge because the electrons in the carbon–carbon bonds are polarized toward the positive center. The isopropyl carbocation has the charge on a secondary carbon atom, one that has two alkyl groups attached. The propyl carbocation has the charge on a primary carbon atom, one that has only one alkyl group attached.

By the same reasoning, it follows that a tertiary carbocation is more stable than a secondary carbocation because it has three alkyl groups attached to the positive carbon atom. The order of stability of carbocations is

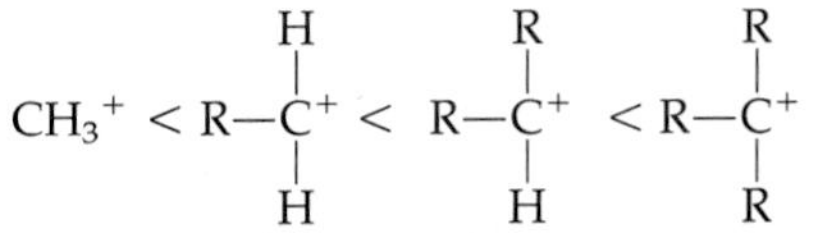

$$\underset{\text{methyl}}{CH_3^{+}} < \underset{\text{primary}}{R{-}CH_2^{+}} < \underset{\text{secondary}}{R{-}CHR^{+}} < \underset{\text{tertiary}}{R{-}CR_2^{+}}$$

——— increasing stability ———→

This order of stability "explains" Markovnikov's rule. Addition of the electrophile always occurs to give the most stable carbocation, which controls the product formed.

4.10 HYDRATION OF ALKENES AND ALKYNES

Water is one of the unsymmetrical reagents listed in Section 4.8 that can add to a π bond. In fact, H_2O fits the general class of HX compounds (H—OH). Thus, water adds to the double bond of an alkene in accordance with Markovnikov's rule.

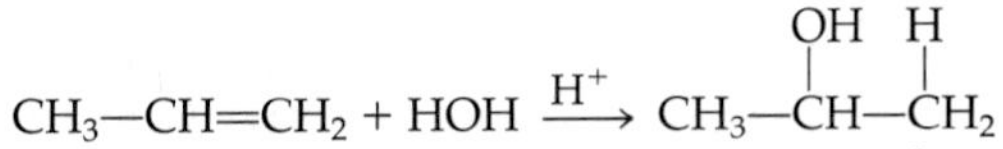

$$\underset{\text{propene}}{CH_3{-}CH{=}CH_2} + HOH \xrightarrow{H^{+}} \underset{\text{2-propanol}}{CH_3{-}CH(OH){-}CH_2(H)}$$

$$CH_3-CH=CH_2 + HOH \not\rightarrow CH_3-\overset{H}{\overset{|}{C}}H-\overset{OH}{\overset{|}{C}}H_2$$

1-propanol
(not observed)

The reverse of the hydration reaction is dehydration, which is an example of an elimination reaction discussed in Section 2.5. Alcohols can be dehydrated (lose water) to produce alkenes.

$$H-\overset{|}{\underset{|}{C}}-\overset{|}{\underset{|}{C}}-OH \xrightarrow{H^+} \gt C=C \lt + H_2O$$

The direction of the reaction—hydration or dehydration—is controlled by conditions predictable by LeChâtelier's principle. Conversion of an alkene to an alcohol requires an excess of water. Dehydration occurs if the water concentration is very low, as in concentrated sulfuric acid, which is about 98% H_2SO_4.

Hydration of Alkynes Produces Carbonyl Compounds

Water adds to one of the π bonds of a triple bond in aqueous sulfuric acid in the presence of mercuric sulfate catalyst. However, the alcohol that forms has its —OH group bonded to the double-bonded carbon atom of an alkene. This type of compound is called an **enol,** a name that includes both the *-ene* suffix of a double bond and the alcohol suffix *-ol.*

$$R-C\equiv C-H \xrightarrow[HgSO_4]{H_2O,\ H_2SO_4} R-\overset{OH}{\overset{|}{C}}=CH_2$$

an enol

Enols are unstable compounds and are rapidly converted to carbonyl compounds in a rearrangement reaction (Section 2.5). We will discuss this reaction further in Chapter 10.

$$R-\overset{OH}{\overset{|}{C}}=CH_2 \longrightarrow R-\overset{O}{\overset{\|}{C}}-CH_3$$

a ketone

The final product of hydration of an alkyne is a ketone. The more substituted carbon atom of the alkyne is converted into a carbonyl carbon atom.

$$C_5H_9-C\equiv C-H \xrightarrow[HgSO_4]{H_2O,\ H_2SO_4} C_5H_9-\overset{O}{\overset{\|}{C}}-CH_3$$

EXAMPLE 4.10 What product(s) will result from the hydration of 2-decyne?

Solution Note that the initial hydration of an alkyne places the hydroxyl group on the more substituted carbon atom. This group is then converted into a carbonyl group of a ketone. In 2-decyne both the C-2 and C-3 atoms are substituted to the same degree.

C-3 is bonded to one alkyl group. C-2 is bonded to one alkyl group.

$$CH_3(CH_2)_6{-}C{\equiv}C{-}CH_3$$

2-decyne

As a consequence hydration can occur either of two ways, and two products, 3-decanone and 2-decanone, are produced.

$$CH_3(CH_2)_6{-}\overset{\overset{\large O}{\|}}{C}{-}CH_2{-}CH_3 \qquad CH_3(CH_2)_6{-}CH_2{-}\overset{\overset{\large O}{\|}}{C}{-}CH_3$$

3-decanone 2-decanone

4.11 SYN–ANTI ADDITION REACTIONS

Consider the general addition of two groups X and Y (or identical groups such as X and X) to a double bond. Do the groups add to the same "face" of the double bond or to opposite "faces" (Figure 4.5)? Addition to the same face is **syn addition;** addition to the opposite face is **anti addition.** This mechanistic question concerning the mode of addition cannot be answered by studying the reactions of acyclic alkenes. For example, the reaction of bromine with 1-butene gives 1,2-dibromobutane, a conformationally flexible molecule. The bromine atoms have added to adjacent carbon atoms, but we cannot tell how the individual atoms approached the plane of the alkene molecule. However, the locations of the groups that add to the double bonds of cycloalkenes can be established.

Consider the reaction of a cyclohexene with the general reagent X—Y. Addition of the two groups to the same face—syn addition—gives a cis compound (Figure 4.5). Addition of one group to a carbon atom from the "top" of the double bond and the other group to an adjacent carbon atom at the "bottom" of the double bond—anti addition—produces the trans isomer. We will now consider the mode of addition in three reactions: hydrogenation, oxidation by potassium permanganate, and addition of bromine.

Hydrogenation of an alkene occurs by the syn addition of hydrogen. This fact is established by the hydrogenation of 1,2-dimethylcyclopentene to produce *cis*-1,2-dimethylcyclopentane.

FIGURE 4.5
Syn–Anti Addition to Alkenes
The two possible modes of addition of a reagent X—Y to an alkene are shown in (a). In syn addition, both groups add to the same "face" of the molecule. In anti addition, the groups add to the opposite "faces" of the molecule. The consequences of syn and anti addition are shown in (b). Geometric isomers can result from the addition of a reagent X—Y to the double bond on a cycloalkene. Syn addition produces a cis product, whereas anti addition produces a trans product.

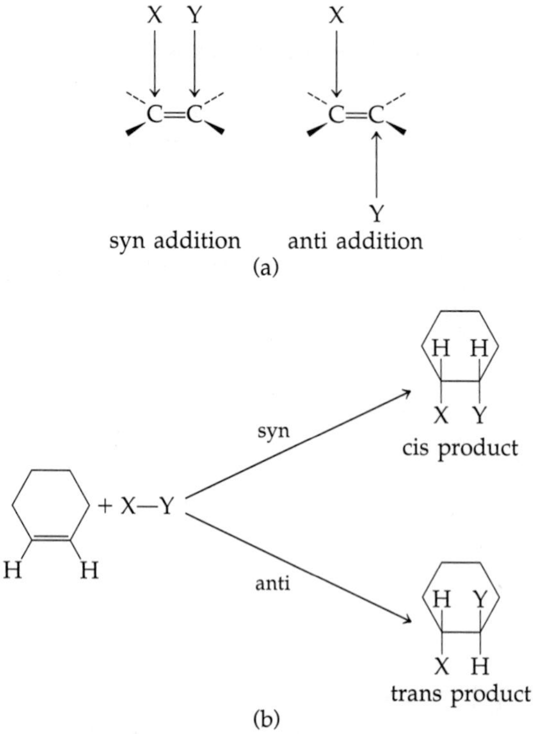

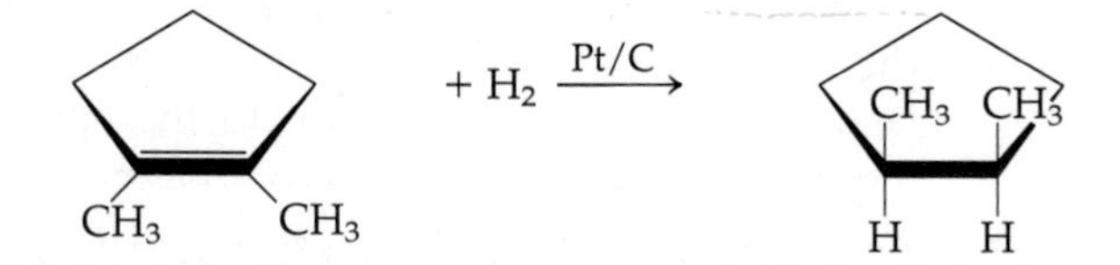

X Y X
C=C C=C
Y
syn addition anti addition
(a)

H H
+ X—Y
syn
anti
H H
X Y
cis product
H Y
X H
trans product
(b)

$+ H_2 \xrightarrow{Pt/C}$
CH_3 CH_3
1,2-dimethylcyclopentene
CH_3 CH_3
H H
cis-1,2-dimethylcyclopentane

The finely divided metal catalyst adsorbs hydrogen gas on the surface, and the hydrogen–hydrogen bond is broken. When the alkene approaches the surface, the hydrogen atoms must add to the same face of the double bond (Figure 4.6).

Oxidation of cyclohexene by potassium permanganate in basic solution yields *cis*-1,2-cyclohexanediol. The cis arrangement of the oxygen atoms indicates that syn addition also occurs in this reaction. The intermediate is a cyclic manganate compound. The five-membered ring can form only if the oxygen atoms become bonded from the same face. Hydrolysis of the intermediate gives the cis-diol product.

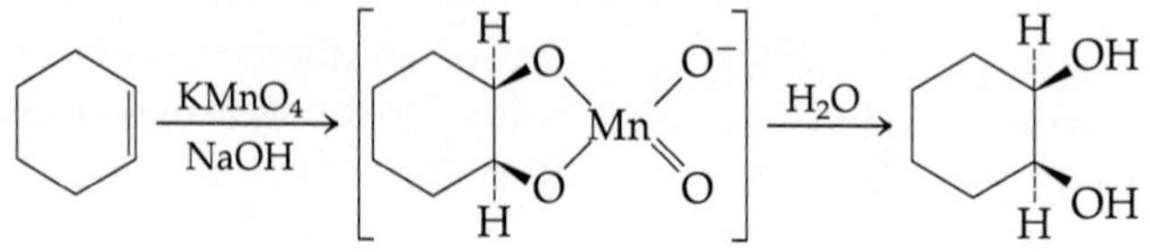

$\xrightarrow[NaOH]{KMnO_4}$
H O O^-
Mn
O O
H
$\xrightarrow{H_2O}$
H OH
H OH

FIGURE 4.6
Syn Addition of Hydrogen to an Alkene
Absorbed hydrogen gas on the surface of the metal catalyst resembles atomic hydrogen. The cycloalkene approaches the catalytic surface and hydrogen atoms add to the unsaturated carbon atoms on the same "face". The result of syn addition is shown. The two methyl groups of the 1,2-dimethylcyclopentene are located cis in the saturated product.

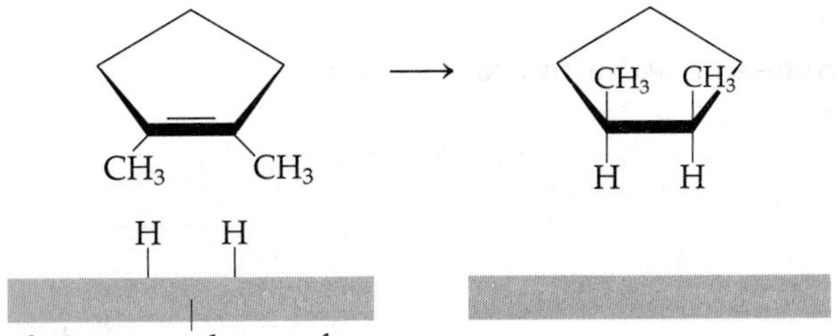

CH_3 CH_3
H H
platinum catalyst surface
CH_3 CH_3
H H

The addition of bromine to cyclopentene yields *trans*-1,2-dibromocyclopentane. The cis isomer is not formed. Thus, this reaction occurs by anti addition of the bromine atoms.

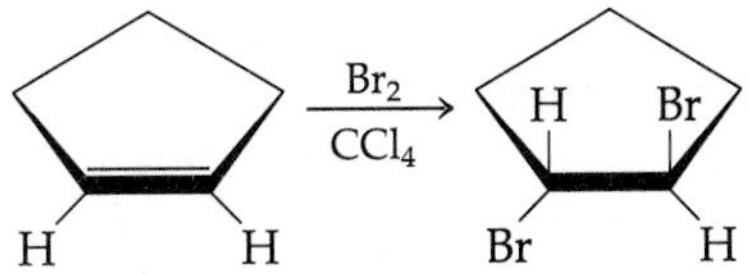

$\xrightarrow[CCl_4]{Br_2}$

The reaction occurs via a bromonium ion intermediate that results from nucleophilic attack of the π electrons of the alkene on one bromine atom (Figure 4.7). A bromide ion is displaced. (This process can also be viewed as electrophilic attack on the double bond by Br^+.) The intermediate has a three-membered ring that contains a bromine atom. Subsequent nucleophilic attack by bromide ion at a carbon atom of the three-membered ring is a displacement reaction in which one of the carbon–bromine bonds of the cyclic intermediate is broken. Because the bromine atom in the three-membered ring of the intermediate "shields" one face of the alkene, the

FIGURE 4.7
Anti Addition of Bromine to Alkenes
The π electrons of the alkene act as a nucleophile to displace bromide ion from bromine. The resulting cyclic bromonium ion can be viewed as the addition of Br^+ to the double bond. Bromine has two covalent bonds and has a formal 1+ charge in this intermediate. Attack of the nucleophilic bromide ion occurs from the opposite face because the bromine atom that is already there blocks approach from the same face.

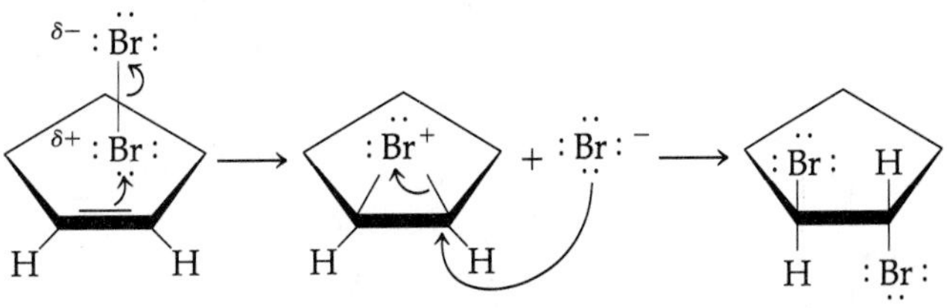

bromide ion can attack only at the opposite face. The net effect is anti addition to produce the trans isomer.

4.12 POLYMERIZATION OF ALKENES

A **polymer** is a high molecular weight molecule created by the repetitive reaction of many thousands of low molecular weight molecules called **monomers.** The process in which monomers join to produce polymers is called **polymerization.** Naturally occurring polymers—**biopolymers**—include carbohydrates, proteins, and nucleic acids. Cellulose, a polymer of glucose, is the major structural material of plants. Lignin, another polymer, occurs within the spaces between the long fibers of cellulose. Proteins, polymers of amino acids, are major constituents of living matter. Some serve a structural role; other are catalysts for virtually all metabolic reactions. Nucleic acids, which are polymers of nucleotides, are carriers of genetic information. Each compound class will be studied later.

Synthetic polymers made in the laboratory are also familiar because we are literally surrounded with these substances. Examples include some of the clothes we wear, containers for foods such as milk, and the computers that we use. Football players depend on polymers in helmets for protection; police use bullet-proof vests made of a polymer; our cars have many polymers in the interior as well as in the bumpers. In this section, we will discuss the remarkably simple reactions that join together monomers to provide these diverse products.

There are two classes of polymers: addition polymers and condensation polymers. These names indicate the type of reaction used to join the monomers. Only addition polymerization will be illustrated in this section. Condensation polymerization to give polyesters and polyamides is described in Chapters 12 and 14, respectively.

Alkenes can be polymerized by a multiple addition reaction. For example, when $CH_2{=}CHCl$ (vinyl chloride) is polymerized, the product is polyvinyl chloride, commonly known as PVC.

$$(H)_2C{=}C(H)(Cl) \longrightarrow \cdots-\overset{H}{\underset{H}{C}}-\overset{Cl}{\underset{H}{C}}-\overset{H}{\underset{H}{C}}-\overset{Cl}{\underset{H}{C}}-\overset{H}{\underset{H}{C}}-\overset{Cl}{\underset{H}{C}}-\cdots$$

An exact formula for a polymer cannot be written because the size of the molecule depends on how it forms. There is no single "polyvinyl chloride" molecule; it is really a mixture of compounds with a range of molecular weights. For this reason, polymers are represented by placing the repeating unit derived from the monomer within a set of parentheses. For polyvinyl chloride, the unit is $(-CH_2CHCl-)$. To show that a large number of units are present, the subscript n is used. For the polymer of vinyl chloride we write

$$\left[CH_2-\underset{\displaystyle Cl}{\underset{|}{CH}} \right]_n$$

The properties of a polymer depend on the monomer used and the molecular weight of the product. A list of some useful addition polymers is given in Table 4.2.

Polymerization of an alkene occurs when a small amount of a suitable initiator is used. The initiator may be a radical, a cation, or an anion, depending on the properties of the monomer. In each case, the initiator reacts with the double bond to form an intermediate that reacts with another double bond to form another reactive intermediate. This process continues and generates the polymeric chain of monomers. Eventually the process is termi-

TABLE 4.2 Uses and Structures of Polymers

Monomer	Polymer	Use
propylene $CH_2{=}CHCH_3$	polypropylene $-CH_2CH(CH_3)CH_2CH(CH_3)-$	carpet fibers, heart valves, bottles
vinyl chloride $CH_2{=}CHCl$	polyvinyl chloride (PVC) $-CH_2CH(Cl)CH_2CH(Cl)CH_2CH(Cl)-$	floor covering, records, garden hoses
dichloroethylene $CH_2{=}CCl_2$	polydichloroethylene $-CH_2CCl_2CH_2CCl_2CH_2CCl_2-$	plastic food wrap
tetrafluoroethylene $CF_2{=}CF_2$	polytetrafluoroethylene $-CF_2CF_2CF_2CF_2CF_2CF_2-$	Teflon, bearings
acrylonitrile $CH_2{=}CHCN$	polyacrylonitrile $-CH_2CH(CN)CH_2CH(CN)CH_2CH(CN)-$	Orlon, Acrilan
styrene $CH_2{=}CHC_6H_5$	polystyrene $-CH_2CH(C_6H_5)-CH_2CH(C_6H_5)-CH_2CH(C_6H_5)-$	toys, styrofoam
methyl methacrylate $CH_2{=}C(CH_3)-C({=}O)OCH_3$	polymethyl methacrylate $-CH_2C(CH_3)(COOCH_3)-CH_2-C(CH_3)(COOCH_3)-CH_2-C(CH_3)(COOCH_3)-$	Plexiglas, Lucite

nated by some reaction that destroys the reactive site. Details of termination reactions will not be discussed.

Polyethylene is produced by **free-radical polymerization** that occurs at pressures of 1000–3000 atm and temperatures of 100–200 °C. A radical initiator is used to generate the radical intermediates responsible for the growth of the polymer chain. The initiator is a peroxide represented as R—O—O—R. Homolytic cleavage of the oxygen–oxygen bond occurs when the peroxide is heated. Reaction of the radical with ethylene in a homogenic reaction forms a carbon radical.

$$\text{R—O—O—R} \longrightarrow 2\,\text{R—O}\cdot$$

$$\text{R—O}\cdot + \text{H}_2\text{C=CH}_2 \longrightarrow \text{R—O—CH}_2\text{—CH}_2\cdot$$

The carbon radical "reactant" continues in a chain reaction with more ethylene units. Each new radical "product" contains one more ethylene unit.

$$\text{ROCH}_2\text{CH}_2\cdot \xrightarrow{\text{CH}_2\text{CH}_2} \text{ROCH}_2\text{CH}_2\text{CH}_2\text{CH}_2\cdot \xrightarrow{\text{CH}_2\text{CH}_2} \text{ROCH}_2\text{CH}_2\text{CH}_2\text{CH}_2\text{CH}_2\text{CH}_2\cdot$$

Although the polymer consists of a chain of —CH_2— units, the repeating unit, based on the structure of the monomer, is —CH_2CH_2—. The polymer is represented by

$$\left[\text{CH}_2\text{—CH}_2\right]_n$$

Polyethylene is the world's most common polymer. Production in the United States alone is over 8 million tons a year. The properties of the polymer depend on the method of production and its molecular weight. Examples include very flexible sandwich bags (20,000 units/molecule), less flexible milk and soft-drink bottles (30,000 units/molecule), and very stiff plastic bottle caps (40,000 units/molecule).

Cationic polymerization involves carbocations rather than radicals. A Lewis acid such as BF_3, $Al(CH_2CH_3)_3$, $TiCl_4$, or $SnCl_4$ is used to react with the alkene to form a carbocation, which in turn reacts with another alkene molecule to form another cation. Consider the reaction with 2-methylpropene (isobutylene) as the monomer. The Lewis acid that acts as an electrophile is represented by E^+.

$$\text{E}^+ + \text{H}_2\text{C=C(CH}_3)_2 \longrightarrow \text{E—CH}_2\text{—C}^+(\text{CH}_3)_2$$

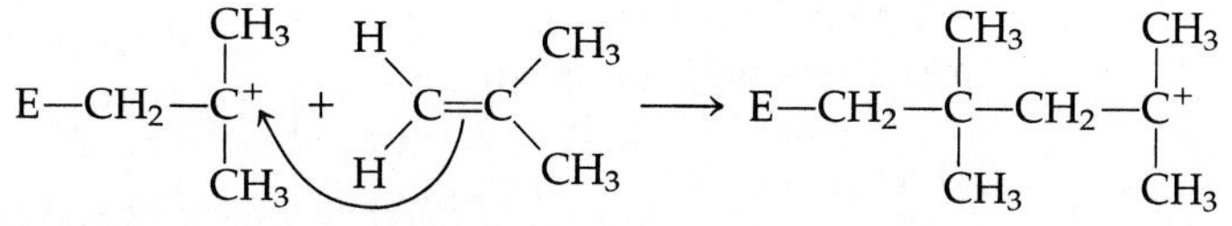

$$E{-}CH_2{-}\overset{CH_3}{\underset{CH_3}{C^+}} + \overset{H}{\underset{H}{}}C{=}C\overset{CH_3}{\underset{CH_3}{}} \longrightarrow E{-}CH_2{-}\overset{CH_3}{\underset{CH_3}{C}}{-}CH_2{-}\overset{CH_3}{\underset{CH_3}{C^+}}$$

Note that addition occurs in the Markovnikov manner. Subsequent reactions continue with the carbocation adding to the less substituted carbon atom. As a consequence, the more stable tertiary carbocation is formed each time. The structure of the polymer is represented as

$$\left[{-}CH_2{-}\overset{CH_3}{\underset{CH_3}{C}}{-}\right]_n$$

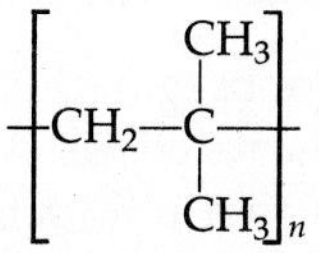

Low molecular weight polyisobutylene is used in lubricating oil and in adhesives for removable paper labels. Higher molecular weight polyisobutylene is used to produce inner tubes for bicycle and truck tires.

Anionic polymerization is initiated by a carbanion that behaves as a nucleophile. One example is the butyl anion, which is provided by butyllithium. The lithium compound has a very polar bond, and the carbon atom has a partially negative charge. In the following reactions, the butyl group is represented as Bu—. The monomer is acrylonitrile.

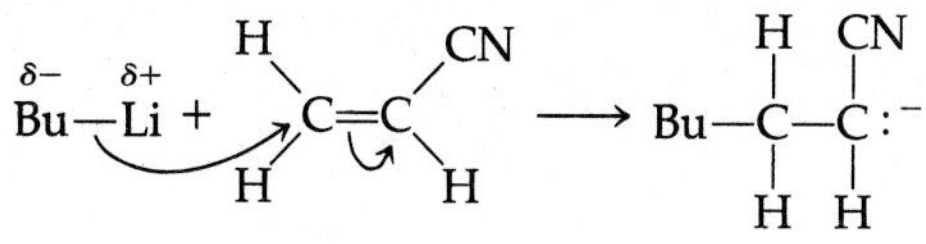

$$\overset{\delta-}{Bu}{-}\overset{\delta+}{Li} + \overset{H}{\underset{H}{}}C{=}C\overset{CN}{\underset{H}{}} \longrightarrow Bu{-}\overset{H}{\underset{H}{C}}{-}\overset{CN}{\underset{H}{C}}{:}^-$$

Addition occurs at the less substituted carbon atom because the resulting carbanion is resonance-stabilized.

$$Bu{-}CH_2{-}\ddot{C}H^-{-}C{\equiv}N{:} \longleftrightarrow Bu{-}CH_2{-}CH{=}C{=}\ddot{N}{:}^-$$

Continued reaction of the nucleophilic carbanion "reactant" gives a carbanion "product", and the length of the polymer chain increases.

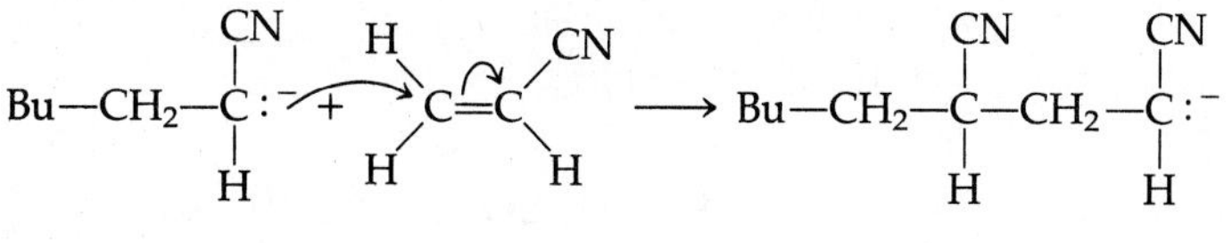

$$Bu{-}CH_2{-}\overset{CN}{\underset{H}{C}}{:}^- + \overset{H}{\underset{H}{}}C{=}C\overset{CN}{\underset{H}{}} \longrightarrow Bu{-}CH_2{-}\overset{CN}{\underset{H}{C}}{-}CH_2{-}\overset{CN}{\underset{H}{C}}{:}^-$$

Subsequent reactions continue to form stabilized carbanions. The structure of the polymer is represented as

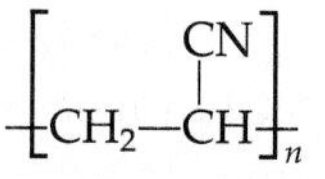

$$\left[CH_2-\underset{}{\overset{CN}{\overset{|}{C}}H} \right]_n$$

Polyacrylonitrile is used in fibers that can be spun to give the textiles Orlon or Acrilan. Some rugs are also produced by using this polymer.

4.13 PREPARATION OF ALKENES AND ALKYNES

Alkenes are prepared from either alcohols or haloalkanes (alkyl halides) by elimination reactions. We recall from Chapter 2 that a single compound splits into two products in an **elimination reaction.** One product usually contains most of the atoms in the reactant, and the remaining atoms are found in a second smaller molecule. The atoms eliminated to form the smaller molecule are usually located on adjacent carbon atoms in the reactant. The mechanism of elimination reactions will be discussed in Chapter 7.

Dehydration of Alcohols

The dehydration of 2-propanol produces propene. The reaction requires concentrated acids such as sulfuric acid, H_2SO_4, or phosphoric acid, H_3PO_4. The reaction is pulled to completion because the water formed in the reaction is solvated with the concentrated acid.

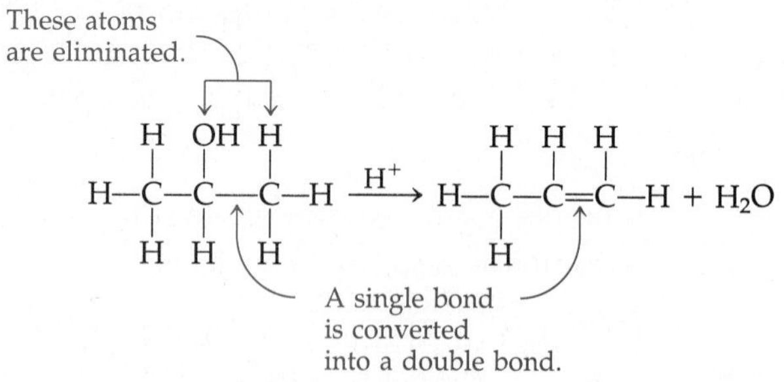

These atoms are eliminated.

$$CH_3-CH(OH)-CH_3 \xrightarrow{H^+} CH_3-CH=CH_2 + H_2O$$

A single bond is converted into a double bond.

The elimination reaction requires breaking the carbon–oxygen bond and a carbon–hydrogen bond on an adjacent carbon atom. For alcohols such as 2-butanol, two different carbon atoms are adjacent to the OH-bearing carbon atom. Each could potentially release a hydrogen atom to form water.

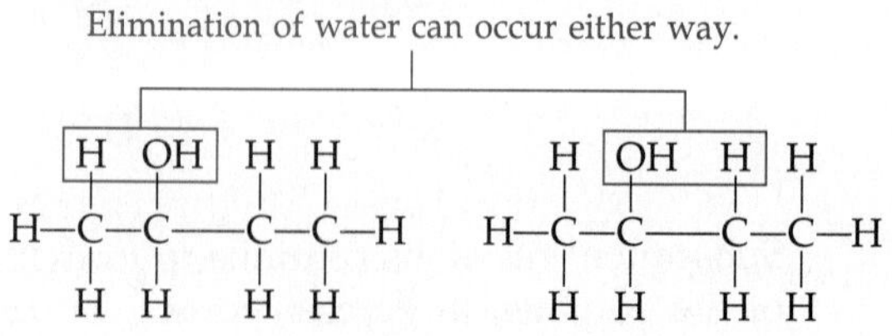

Elimination of water can occur either way.

$$CH_3-CH(OH)-CH_2-CH_3$$

Thus, dehydration produces a mixture of products. The isomer that contains the greater number of alkyl groups attached to the double bond (the more substituted alkene) predominates in the mixture.

Elimination reactions that give the more substituted double bond are said to obey **Zaitsev's rule.** This generalization was discovered by Alexander Zaitsev, a nineteenth century Russian chemist. We recall from Section 4.1 that increasing the number of alkyl groups bonded to unsaturated carbon atoms of an alkene increases its stability. Thus, Zaitsev simply observed that the major product of an elimination reaction is the more stable isomer. Zaitsev's rule also applies to mixtures of geometric isomers. For example, 3-pentanol yields a mixture of *cis-* and *trans*-2-pentene. The trans isomer is the major product.

$$CH_3CH_2CH(OH)CH_2CH_3 \xrightarrow{H_2SO_4} \underset{\textit{cis}\text{-2-pentene (25\%)}}{(CH_3CH_2)(H)C{=}C(CH_3)(H)} + \underset{\textit{trans}\text{-2-pentene (75\%)}}{(CH_3CH_2)(H)C{=}C(H)(CH_3)}$$

In the trans isomer the alkyl groups are well-separated, whereas in the cis isomer the alkyl groups are near each other. The proximity of the alkyl groups causes steric hindrance; that is, the groups repel each other. The trans isomer is more stable, so it predominates.

Dehydrohalogenation of Alkyl Halides

The elimination of the elements H and X as in HCl or HBr from adjacent carbon atoms in an alkyl halide is called **dehydrohalogenation.** The product of the reaction is an alkene.

$$H{-}\overset{|}{\underset{|}{C}}{-}\overset{|}{\underset{|}{C}}{-}X + B^- \longrightarrow \;\rangle C{=}C\langle\; + HB + X^-$$

A base, represented by B^-, is required for the reaction. Although hydroxide ion is sufficiently basic for this reaction, it is usual to use alkoxide, the conjugate base of an alcohol, as the base. Sodium ethoxide is commonly used in combination with ethanol as the solvent. The mechanism of this reaction will be presented in Chapter 7.

$$\underset{\text{chlorocyclohexane}}{C_6H_{11}Cl} \xrightarrow[CH_3CH_2OH]{CH_3CH_2O^-Na^+} \underset{\text{cyclohexene}}{C_6H_{10}}$$

As in the case of dehydration, the more highly substituted alkene predominates when two or more products are possible. When geometric isomers are possible, trans isomers are favored over cis isomers.

$$CH_3C(Cl)(CH_3)CH_2CH_3 \xrightarrow[CH_3CH_2OH]{CH_3CH_2O^-} (CH_3)_2C{=}CHCH_3 + H_2C{=}C(CH_3)CH_2CH_3$$

2-methyl-2-butene (70%) 2-methyl-1-butene (30%)

Elimination Reactions of Dihalides

Alkynes can be prepared by elimination reactions under conditions similar to those used to form alkenes. Because an alkyne has two π bonds, two molar equivalents of HX must be eliminated. The reactant needed for the reaction is a **vicinal** dihalide, a compound with halogen atoms on adjacent carbon atoms. A stronger base than an alkoxide ion is required. The most commonly used base is sodium amide in liquid ammonia as the solvent.

$$CH_3(CH_2)_3{-}CH(Cl){-}CH(Cl){-}H \xrightarrow[NH_3]{NH_2^-} CH_3(CH_2)_3{-}C{\equiv}C{-}H$$

1,2-dichlorohexane 1-hexyne

4.14 ALKADIENES

Compounds with two double bonds are **alkadienes,** commonly called dienes. When one single bond is located between the two double-bonded units, the compounds are chemically different from simple alkenes. These compounds, which are said to be **conjugated,** are the subject of this section. When more than one single bond is located between the two double-bonded units, the compounds are not chemically different from alkenes. The double bonds are said to be **isolated,** or **nonconjugated.** Double bonds that share a common atom are said to be **cumulated.** These compounds are relatively rare, and will not be discussed further.

$CH_3{-}CH{=}CH{-}CH{=}CH_2$
1,3-pentadiene (a conjugated diene)

$CH_2{=}CH{-}CH_2{-}CH{=}CH_2$
1,4-pentadiene (a nonconjugated diene)

$CH_2{=}C{=}CH{-}CH_2{-}CH_3$
1,2-pentadiene (a cumulated diene)

Electrophilic Conjugate Addition

Addition of electrophilic reagents to nonconjugated alkadienes can occur at one or both double bonds. The products are those predicted by Markovnikov's rule.

$$CH_2{=}CHCH_2CH{=}CH_2 \xrightarrow{HBr} CH_3CH(Br)CH_2CH{=}CH_2$$

4-bromo-1-pentene

$$CH_3\overset{\overset{Br}{|}}{C}HCH_2CH{=}CH_2 \xrightarrow{HBr} CH_3\overset{\overset{Br}{|}}{C}HCH_2\overset{\overset{Br}{|}}{C}HCH_3$$

2,4-dibromopentane

Addition of HBr to a conjugated diene is strikingly different. Two products are obtained when one molar equivalent of HBr reacts.

$$CH_2{=}CH{-}CH{=}CH_2 \xrightarrow{HBr} \overset{\overset{H}{|}}{C}H_2{-}\overset{\overset{Br}{|}}{C}H{-}CH{=}CH_2 + \overset{\overset{H}{|}}{C}H_2{-}CH{=}CH{-}\overset{\overset{Br}{|}}{C}H_2$$

3-bromo-1-butene (1,2-addition, 70%) 1-bromo-2-butene (1,4-addition, 30%)

The 3-bromo-1-butene is the product of direct addition to a double bond. This product is predicted by Markovnikov's rule. The 1-bromo-2-butene is an unusual product that results from the addition of HBr to the C-1 and C-4 atoms. Note that the double bond in the product is between the C-2 and C-3 atoms. This product results from a **1,4-addition reaction.**

Before considering the origin of the 1,4-addition product derived from a conjugated diene, let's examine the structure of the allylic carbocation. This cation can be represented by two contributing resonance structures.

$$CH_2{=}CH{-}CH_2^{+} \longleftrightarrow {}^{+}CH_2{-}CH{=}CH_2$$

Thus, the positive charge is distributed between both terminal carbon atoms. Either terminal carbon atom could react with a nucleophile, but the product of the reaction would be the same.

Now let's consider the electrophilic addition of a proton to a conjugated diene.

$$CH_2{=}CH{-}CH{=}CH_2 \xrightarrow{H^+} \underset{}{\overset{\overset{H}{|}}{C}H_2{-}\overset{+}{C}H{-}CH{=}CH_2} \longleftrightarrow \overset{\overset{H}{|}}{C}H_2{-}CH{=}CH{-}CH_2^{+}$$

(1) (2)

The carbocation intermediate is an allylic-type ion that can be represented by two contributing resonance structures. In the next step in the addition reaction, the nucleophilic bromide ion can bond to the secondary carbon atom (see structure 1) to give the 1,2 addition product. However, if the bromide ion bonds to the primary carbon atom (see structure 2), the 1,4-addition product results.

$$\overset{\overset{H}{|}}{C}H_2{-}\overset{+}{C}H{-}CH{=}CH_2 \longleftrightarrow \overset{\overset{H}{|}}{C}H_2{-}CH{=}CH{-}\overset{+}{C}H_2$$

(1) (2)

$$\downarrow Br^-$$

$$\overset{\overset{H}{|}}{C}H_2{-}\overset{\overset{Br}{|}}{C}H{-}CH{=}CH_2 + \overset{\overset{H}{|}}{C}H_2{-}CH{=}CH{-}\overset{\overset{Br}{|}}{C}H_2$$

Allylic Oxidation and the Metabolism of Marijuana

We have already seen several examples of metabolic oxidation reactions catalyzed by cytochrome P-450. This detoxifying agent also oxidizes allylic sites to produce allylic alcohols. Although the nature of the intermediates is not always well-known, the ready oxidation of allylic C—H bonds by cytochrome P-450 must involve intermediates with some resonance stabilization.

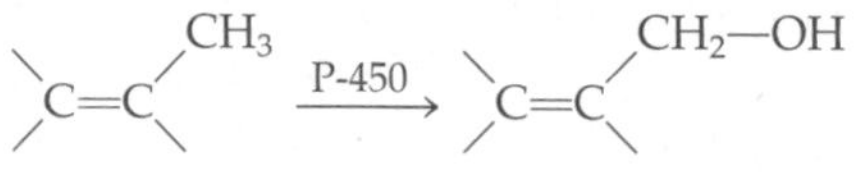

An example of such a reaction is the first step in the degradation of marijuana. Marijuana contains Δ^1-tetrahydrocannabinol (Δ^1-THC), which has allylic centers at C-3, C-6, and C-7. Allylic oxidation does not occur at C-3 because of steric hindrance caused by the methyl groups and the aromatic ring. Of the other possible sites, the C-7 product predominates over the C-6 product. Interestingly, the C-7 product has been shown to be even more psychoactive than Δ^1-THC.

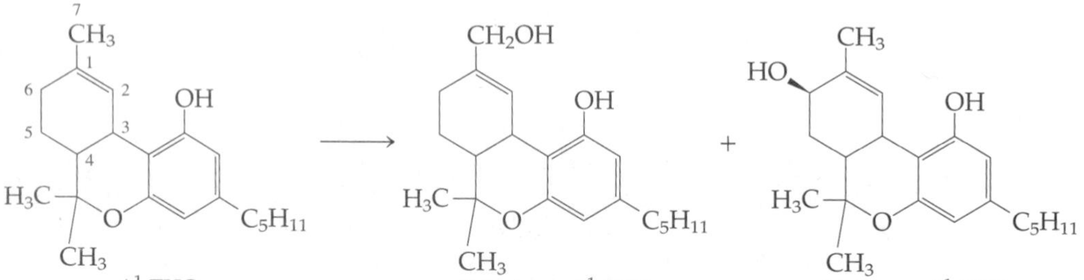

$$\mathrm{>C{=}C<^{CH_3}} \xrightarrow{\text{P-450}} \mathrm{>C{=}C<^{CH_2{-}OH}}$$

an allyl alcohol

Δ^1-THC → 7-hydroxy-Δ^1-THC + 6-hydroxy-Δ^1-THC

4.15 TERPENES

Terpenes, which are abundant in the oils of plants and flowers, have distinctive odors and flavors. They are responsible for the odors of pine trees and for the colors of carrots and tomatoes. Vitamins A, E, and K are terpenes.

Terpenes consist of two or more isoprene units joined together, usually head to tail. These compounds may have different degrees of unsaturation and a variety of functional groups. Nevertheless, the isoprene units are usually easy to identify.

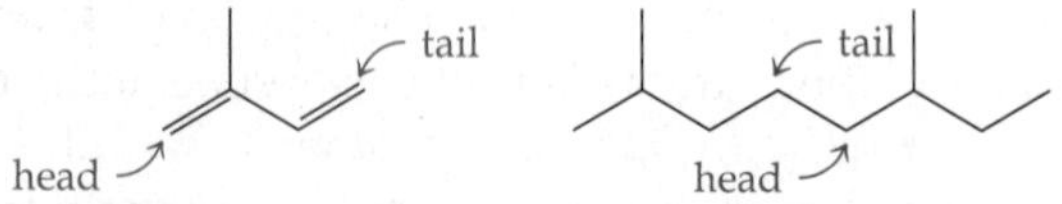

head, tail
isoprene
2-methyl-1,3-butadiene

head, tail
two isoprene units
linked head to tail

Farnesol contains three isoprene units joined head to tail. Dashed lines indicate where the three units are joined.

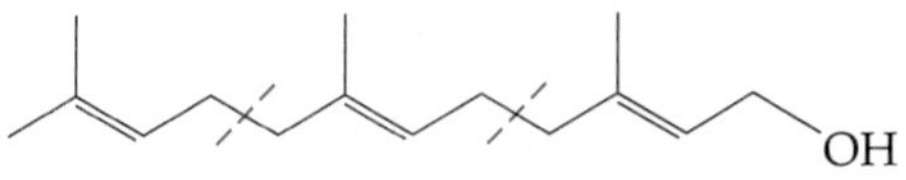

OH

isoprene units in farnesol

Many terpenes contain one or more rings that can be mentally dissected into isoprene units, as in the case of carvone.

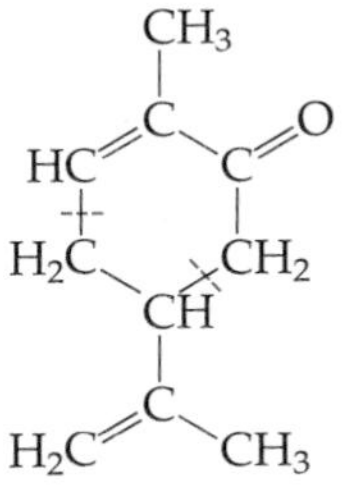

CH_3, HC, C, $C{=}O$, H_2C, CH_2, CH, C, H_2C, CH_3

carvone
(oil of caraway)

Terpenes are classified by the number of isoprene units they contain. The **monoterpenes,** the simplest terpene class, contain two isoprene units and **sesquiterpenes** have three isoprene units. Examples of these structures are shown in Table 4.3 using bond-line structures. **Diterpenes, triterpenes,** and **tetraterpenes** contain 4, 6, and 8 isoprene units, respectively.

EXAMPLE 4.11 Classify the following terpene and indicate the division into isoprene units.

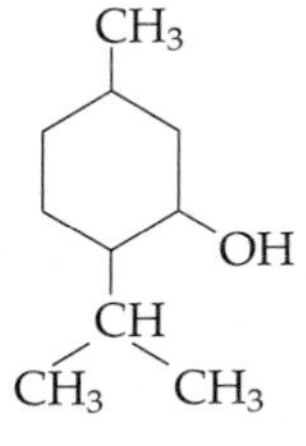

CH_3, OH, CH, CH_3, CH_3

Solution The compound is a monoterpene; it contains 10 carbon atoms or two isoprene units. The isopropyl group provides three of the necessary carbon atoms for one isoprene unit; only two other carbon atoms are required. One carbon atom is the atom to which the isopropyl group is attached. The other carbon atom may be either the ring —CH_2— group or the —CHOH group.

TABLE 4.3 Classification of Terpenes

Monoterpene

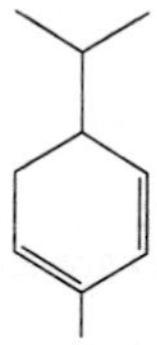

α-phellandrene
(eucalyptus)

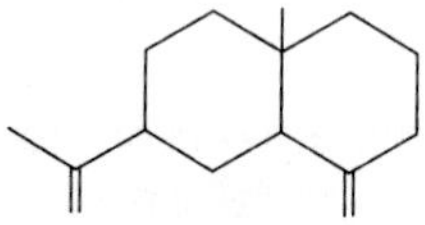

Sesquiterpene

α-selinene
(celery)

Diterpene

OH

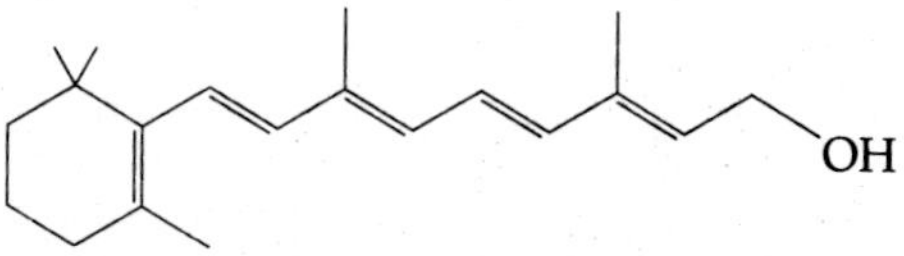

vitamin A
(present in mammalian tissue and fish oil)

Triterpene

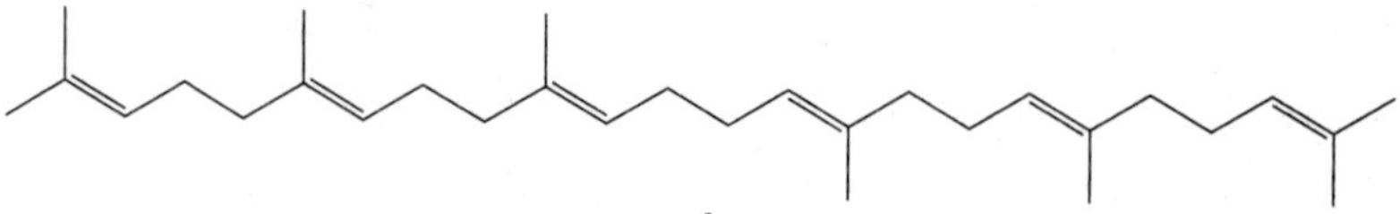

squalene
(shark liver oil)

Tetraterpene

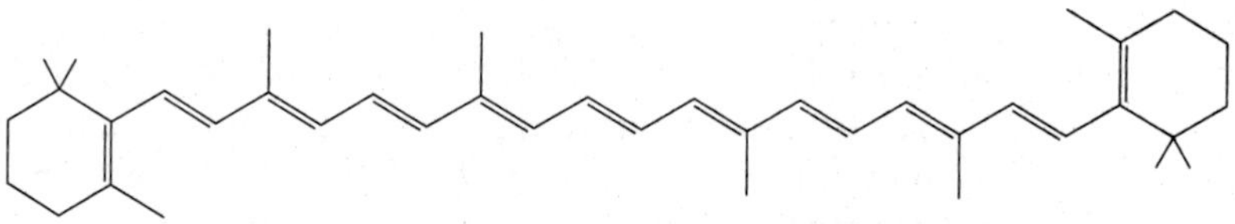

β-carotene
(present in carrots and other vegetables)

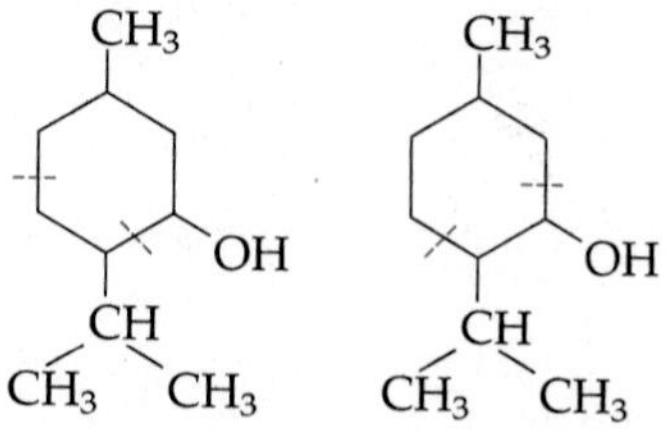

CH_3 ... OH ... CH ... CH_3 CH_3

CH_3 ... OH ... CH ... CH_3 CH_3

The two pairs of dashed lines indicate the possible divisions into isoprene units.

EXERCISES

Molecular Formulas

4.1 What is the molecular formula for a compound with each of the following structural features?

(a) six carbon atoms and one double bond

(b) five carbon atoms and two double bonds
(c) seven carbon atoms, a ring, and one double bond
(d) four carbon atoms and one triple bond

4.2 What is the molecular formula for a compound with each of the following structural features?
(a) four carbon atoms and two triple bonds
(b) four carbon atoms, a double bond, and a triple bond
(c) ten carbon atoms and two rings
(d) ten carbon atoms, two rings, and five double bonds

4.3 Write the molecular formula for each of the following compounds.

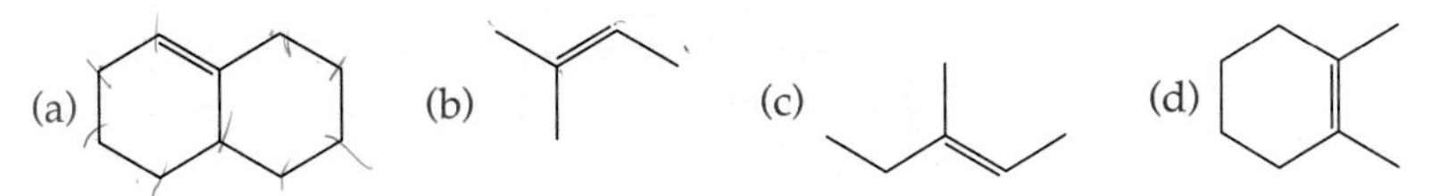

4.4 Write the molecular formula for each of the following compounds.

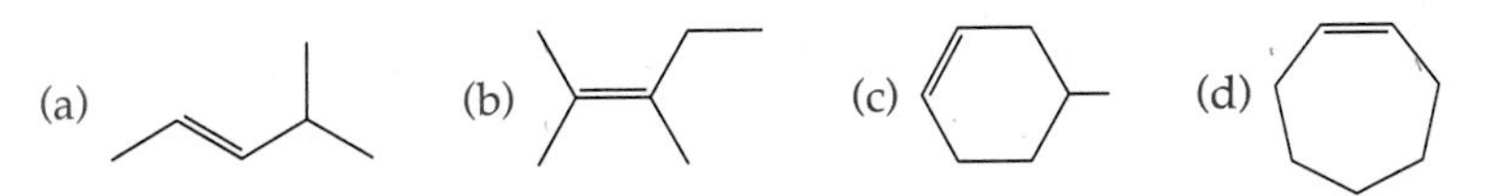

Classification of Alkenes and Alkynes

4.5 Classify the double bond in each alkene in Exercise 4.3 by its substitution pattern.

4.6 Classify the double bond in each alkene in Exercise 4.4 by its substitution pattern.

4.7 Indicate the degree of substitution of the double bond in each of the following compounds.
(a) cholesterol, a steroid required for growth in almost all life forms

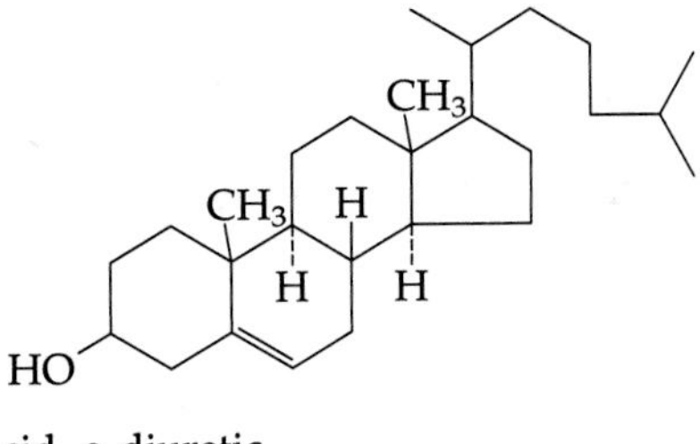

CH_3 CH_3 H H H HO

(b) ethacrynic acid, a diuretic

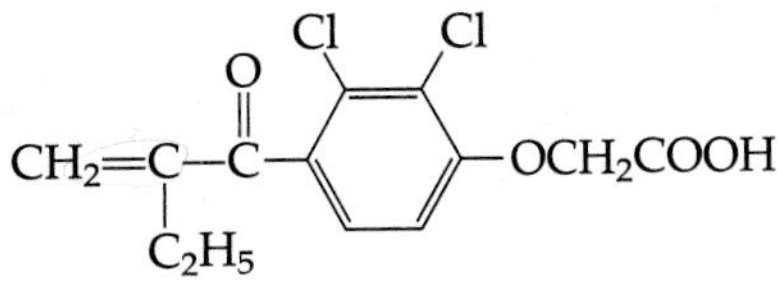

Cl Cl O

$CH_2=C(C_2H_5)-C(=O)-$ (dichlorophenyl) $-OCH_2COOH$

(c) safrole, a carcinogen found in sassafras root

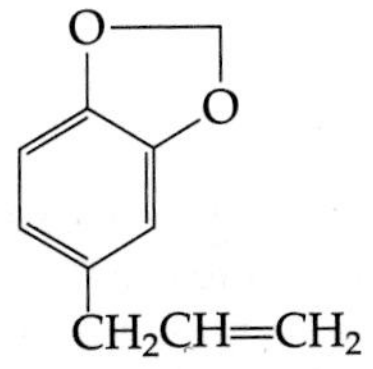

O O

$CH_2CH=CH_2$

(d) tamoxifen, a drug used in treatment of breast cancer

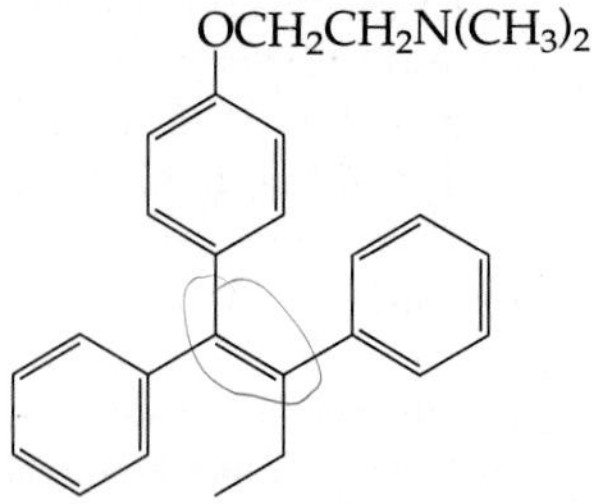

$OCH_2CH_2N(CH_3)_2$

4.8 Indicate the degree of substitution of the triple bond in each of the following compounds.

(a) ethinamate, a sedative and hypnotic drug

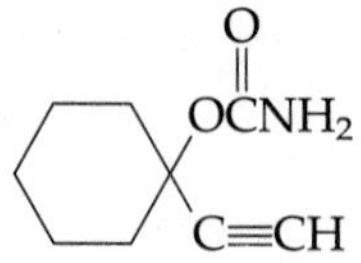

$OCNH_2$ (C=O), C≡CH

(b) tremorine, a drug used to treat Parkinson's disease

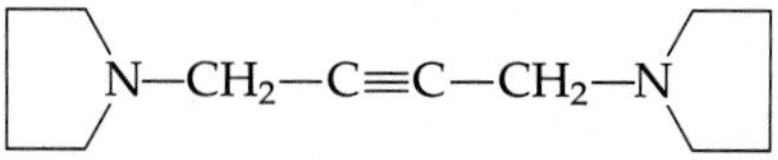

N—CH_2—C≡C—CH_2—N

(c) MDL, a drug used in breast cancer therapy

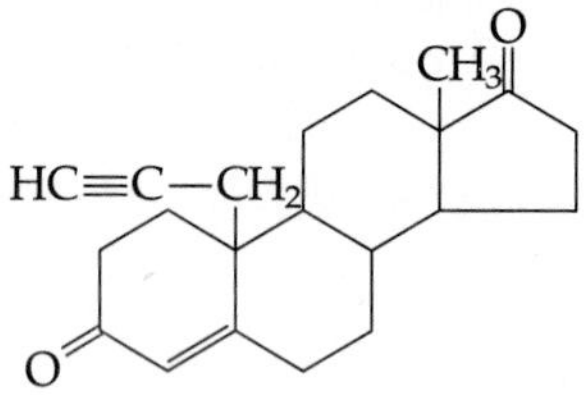

HC≡C—CH_2, CH_3, O, O

(d) RU-486, a drug used to induce abortion

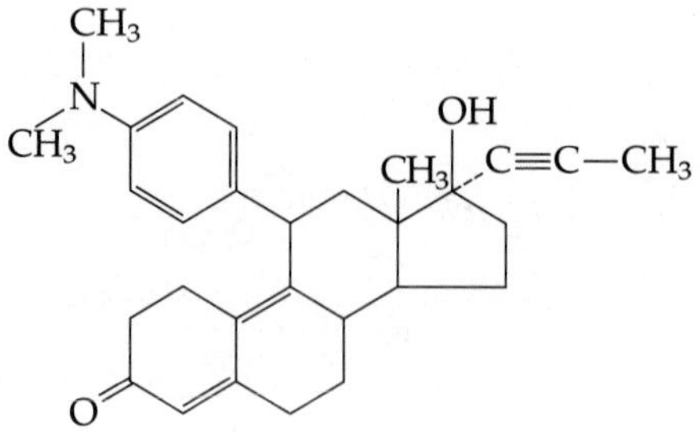

CH_3, N, CH_3, OH, CH_3, C≡C—CH_3, O

Isomers

4.9 Draw structural formulas for four isomeric alkenes with the molecular formula C_4H_8 and name each compound.

4.10 Draw structural formulas for six isomeric alkenes with the molecular formula C_5H_{10} and name each compound.

4.11 Which of the following molecules can exist as cis and trans isomers?
(a) 1-hexene (b) 3-heptene (c) 4-methyl-2-pentene (d) 2-methyl-2-butene

4.12 Which of the following molecules can exist as cis and trans isomers?
(a) 3-methyl-1-hexene (b) 3-ethyl-3-heptene (c) 2-methyl-2-pentene
(d) 3-methyl-2-pentene

E,Z System of Nomenclature

4.13 Select the group with the highest priority in each of the following sets.
(a) $-CH(CH_3)_2$ $-CHClCH_3$ CH_2CH_2Br
(b) $-CH_2CH{=}CH_2$ $-CH_2CH(CH_3)_2$ $-CH_2C{\equiv}CH$
(c) $-OCH_3$ $-N(CH_3)_2$ $-C(CH_3)_3$

4.14 Select the group with the highest priority in each of the following sets.
(a) $-C(=O)-CH_3$ $-C(=O)-OH$ $-C(=O)-F$
(b) $-C(=O)-NH_2$ $-C(=O)-O-CH_3$ $-C(=O)-N(CH_3)_2$
(c) $-C(=O)-S-CH_3$ $-C(=O)-O-CH_3$ $-C(=O)-Cl$

4.15 Assign the *E* or *Z* configuration to each of the following antihistamines.
(a) pyrrobutamine

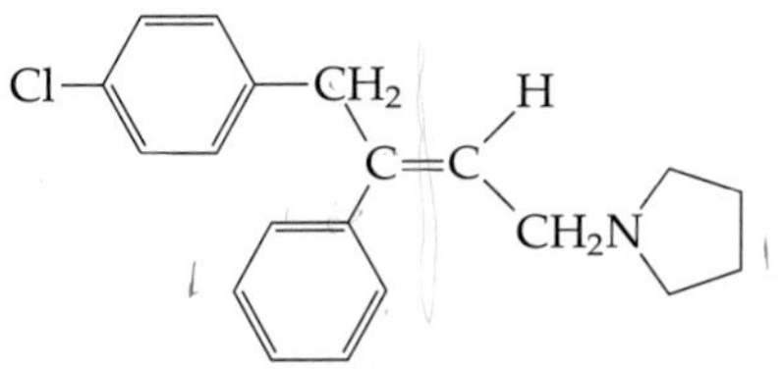

Cl CH₂ H C=C CH₂N

(b) triprolidine

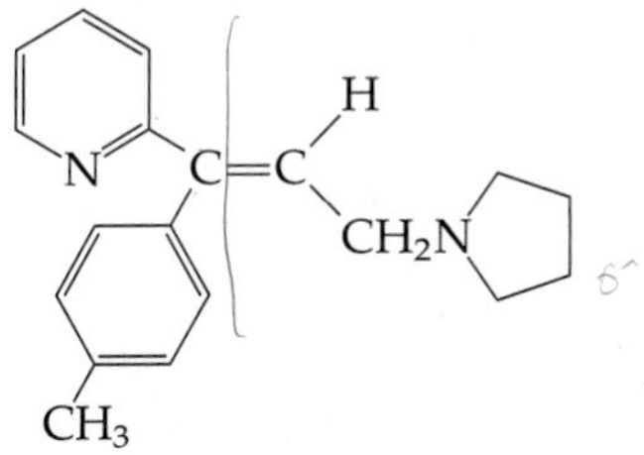

H N C=C CH₂N CH₃

(c) chloroprothixene

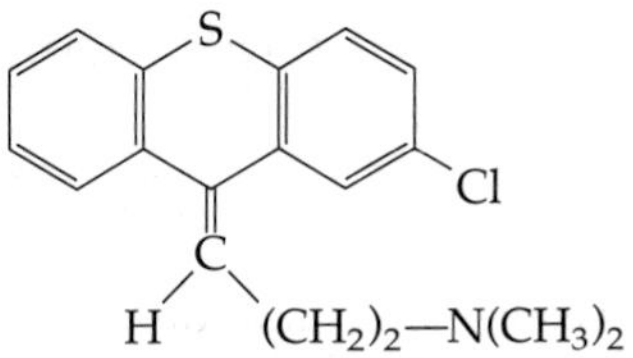

S Cl C H (CH₂)₂—N(CH₃)₂

4.16 Assign the *E* or *Z* configuration to each of the following hormone antagonists used to control cancer.

(a) clomiphene

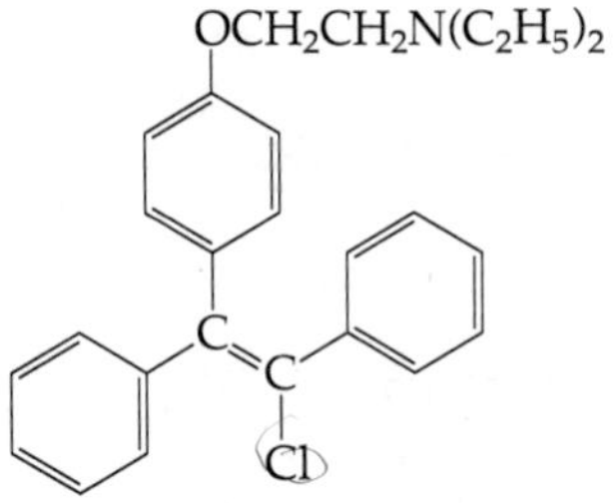

$OCH_2CH_2N(C_2H_5)_2$

C=C

Cl

(b) tamoxifen

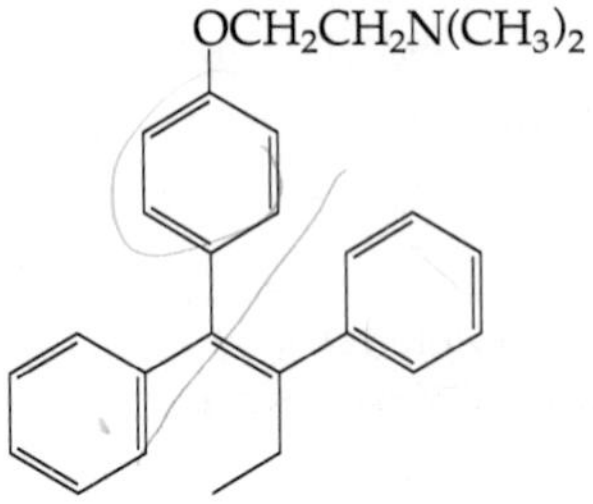

$OCH_2CH_2N(CH_3)_2$

(c) nitromifene

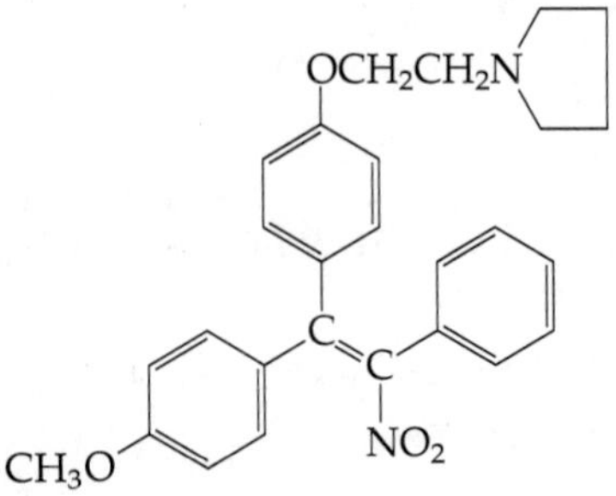

OCH_2CH_2N

C=C

NO_2

CH_3O

4.17 Draw the structural formula for each of the following pheromones with the indicated configuration.

(a) sex pheromone of Mediterranean fruit fly, *E* isomer

$$CH_3CH_2CH{=}CH(CH_2)_4CH_2OH$$

(b) sex pheromone of honeybee, *E* isomer

$$CH_3\overset{\overset{\large O}{\|}}{C}(CH_2)_4CH_2CH{=}CHCO_2H$$

(c) defense pheromone of termite, *E* isomer

$$CH_3(CH_2)_{12}CH{=}CHNO_2$$

4.18 Assign the configuration at all double bonds where geometrical isomerism is possible in each of the following sex pheromones.

(a) European vine moth

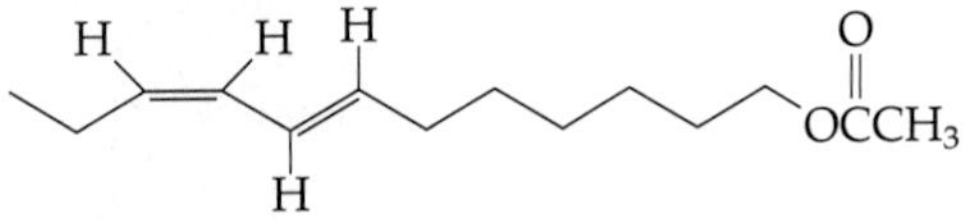

(b) pink bollworm moth

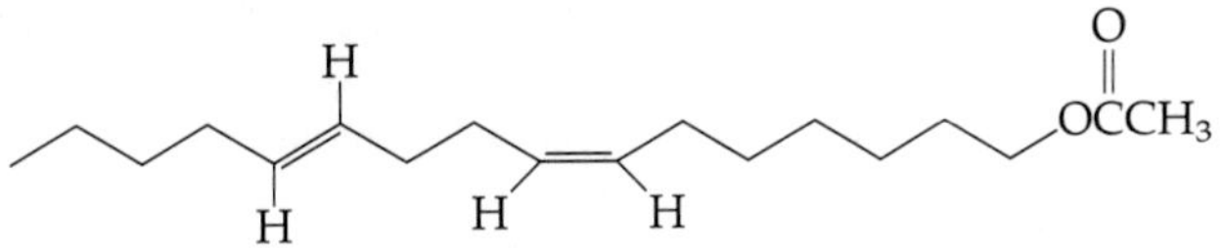

(c) Japanese beetle

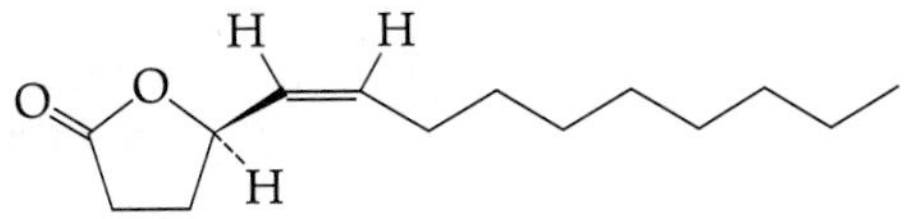

Nomenclature of Alkenes

4.19 Name each of the following compounds.

(a) 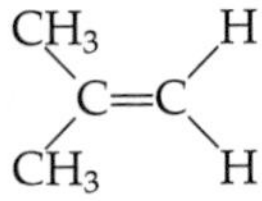 $(CH_3)_2C{=}CH_2$ (b) 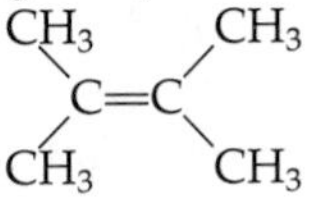 $(CH_3)_2C{=}C(CH_3)_2$

(c) $(CH_3)_2C{=}CHCH_3$ (d) 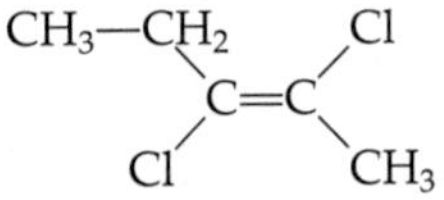 $CH_3CH_2(Cl)C{=}C(Cl)CH_3$

4.20 Name each of the following compounds.

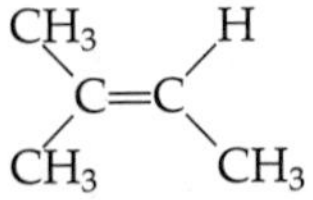

(a) 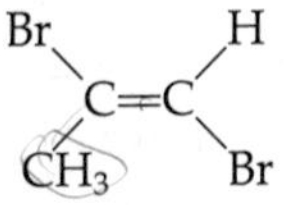 $Br(CH_3)C{=}CH(Br)$ (b) $(CH_3)_2C{=}CH{-}CH_2{-}Br$

(c) $(CH_3)_2C{=}C(Cl)CH_2{-}CH_3$ (d) 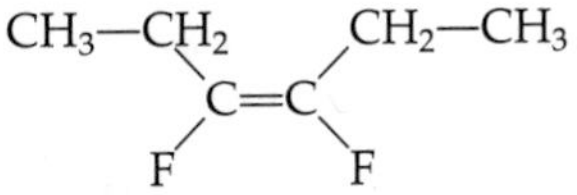 $CH_3{-}CH_2(F)C{=}C(F)CH_2{-}CH_3$

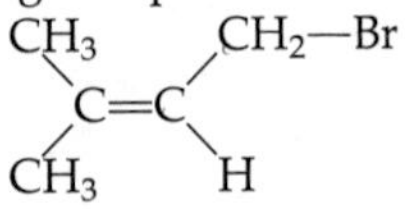

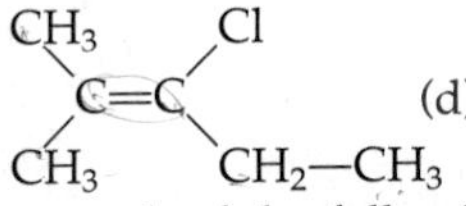

4.21 Name each of the following compounds.

(a) 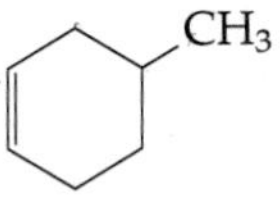 (b) 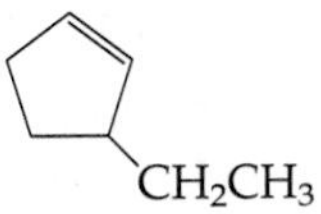 (c) 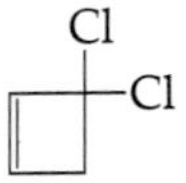 (d)

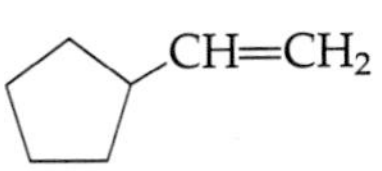

4.22 Name each of the following compounds.

(a) 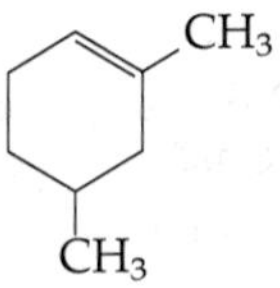 (b) 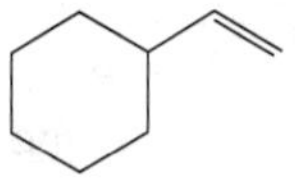 (c) 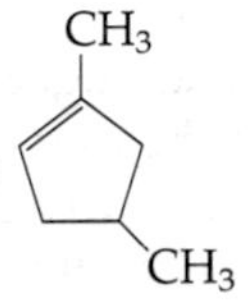 (d)

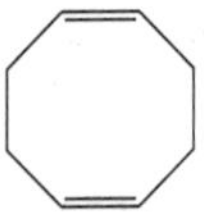

4.23 Draw a structural formula for each of the following compounds.
(a) 2-methyl-2-pentene (b) 1-hexene (c) (*Z*)-2-methyl-3-hexene
(d) (*E*)-5-methyl-2-hexene

4.24 Draw a structural formula for each of the following compounds.
(a) (*E*)-1-chloropropene (b) (*Z*)-2,3-dichloro-2-butene (c) 3-chloropropene
(d) 4-chloro-2,4-dimethyl-2-hexene

4.25 Draw a structural formula for each of the following compounds.
(a) cyclohexene (b) 1-methylcyclopentene (c) 1,2-dibromocyclohexene
(d) 4,4-dimethylcyclohexene

4.26 Draw a structural formula for each of the following compounds.
(a) cyclopentene (b) 3-methylcyclohexene (c) 1,3-dibromocyclopentene
(d) 3,3-dichlorocyclopentene

Nomenclature of Alkynes

4.27 Name each of the following compounds.
(a) $CH_3CH_2CH_2C{\equiv}CH$ (b) $(CH_3)_3CC{\equiv}CCH_2CH_3$
(c) $CH_3{-}C{\equiv}C{-}CH{-}CH_3$
$\qquad\qquad\quad |$
$\qquad\qquad CH_2{-}CH_3$

4.28 Name each of the following compounds.
(a) $CH_3CHBrCHBrC{\equiv}CCH_3$ (b) $Cl(CH_2)_2C{\equiv}C(CH_2)_3CH_3$
(c) $CH_3{-}CH{-}CH_2{-}C{\equiv}C{-}CH{-}CH_3$
$\qquad\quad |\qquad\qquad\qquad\qquad |$
$\qquad CH_2{-}CH_3\qquad\qquad\quad Cl$

4.29 Write the structural formula for each of the following compounds.
(a) 2-hexyne (b) 3-methyl-1-pentyne (c) 5-ethyl-3-octyne

4.30 Write the structural formula for each of the following compounds.
(a) 3-heptyne (b) 4-methyl-1-pentyne (c) 5-methyl-3-heptyne

Hydrogenation of Alkenes and Alkynes

4.31 How many moles of hydrogen gas will react with each of the following compounds?
(a) $CH_3{-}CH{=}CH{-}C{\equiv}CH$ (b) $HC{\equiv}C{-}C{\equiv}CH$
(c) $CH_2{=}CH{-}C{\equiv}C{-}CH{=}CH_2$ (d) $HC{\equiv}C{-}C{\equiv}C{-}C{\equiv}CH$

4.32 How many moles of hydrogen gas will react with each of the following compounds?
(a) ichthyothereol, a convulsant (see structure in Section 4.1)
(b) vitamin A (see structure in Section 4.1)
(c) mycomycin, an antibiotic

$$HC{\equiv}C{-}C{\equiv}C{-}CH{=}C{=}CH{-}CH{=}CH{-}CH{=}CH{-}CH_2COOH$$

(d) squalene, found in shark liver oil

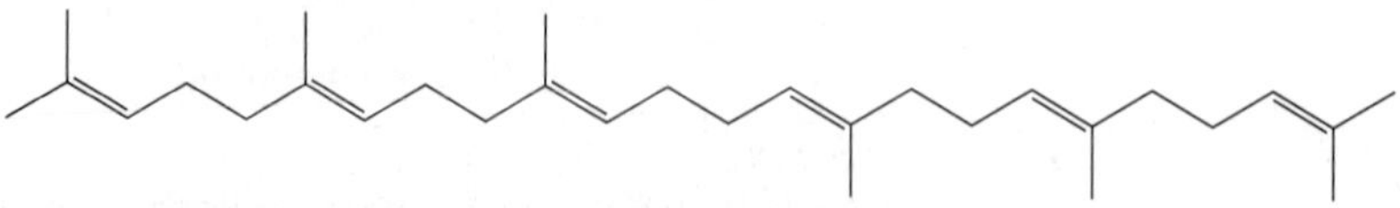

4.33 How could muscalure (Section 4.1) be prepared from a structurally related alkyne? What reagent is required?

4.34 How could the following compound, which is a constituent of the sex pheromone of the male oriental fruit moth, be prepared from a structurally related alkyne? What reagent is required?

Oxidation of Alkenes

4.35 Describe the observation that is made when *cis*-2-pentene reacts with potassium permanganate. How could this reagent be used to distinguish between *cis*-2-pentene and cyclopentane?

4.36 Write the products of the reactions of vinylcyclohexane and allylcyclopentane with potassium permanganate.

4.37 Write the product(s) of the ozonolysis of each of the following compounds.

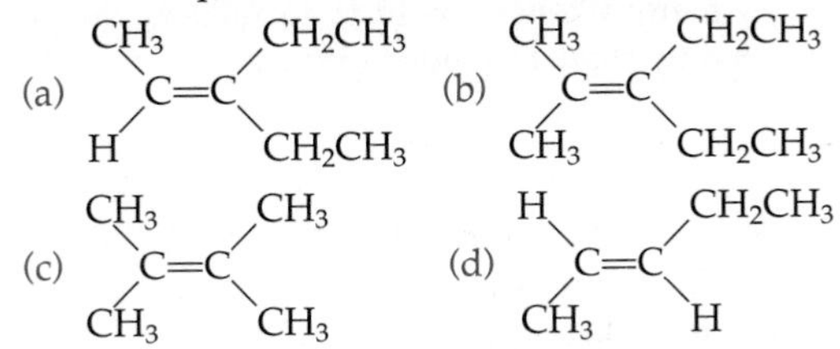

4.38 Write the product(s) of the ozonolysis of each of the following compounds.

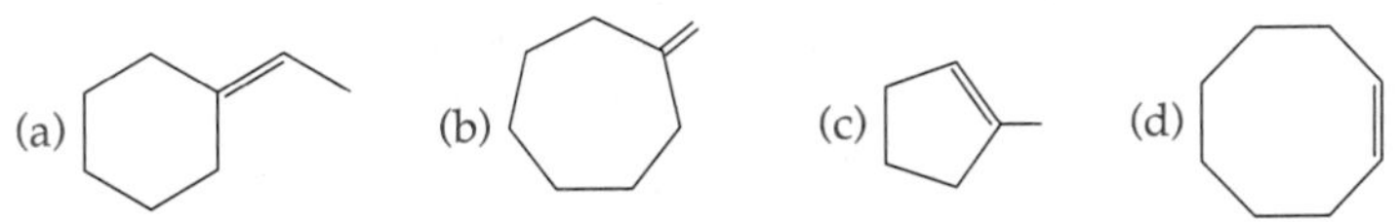

Addition Reactions

4.39 Four compounds with molecular formula C_4H_6 can react with excess bromine to give $C_4H_6Br_2$. Draw their structures.

4.40 Four compounds with molecular formula C_4H_6 can react with excess bromine to give $C_4H_6Br_4$. Write their structures.

4.41 Which of the compounds in Exercise 4.37 would give a single product when reacted with HBr? Why?

4.42 Write the product(s) of each of the following reactions.

(a) $CH_3CH_2C{\equiv}CH + HBr \longrightarrow$ (b) $CH_3CH_2C{\equiv}CH + 2\,HBr \longrightarrow$

(c) $CH_3C{\equiv}CCH_3 + HBr \longrightarrow$ (d) $CH_3C{\equiv}CCH_3 + 2\,HBr \longrightarrow$

4.43 Write the product of the reaction of HBr with each of the following compounds.

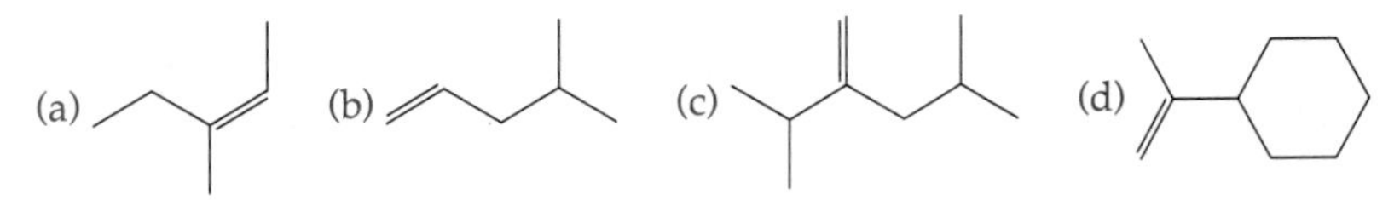

4.44 Write the product of the reaction of HBr with each of the compounds in Exercise 4.38.

4.45 Write the product of hydration of each of the compounds in Exercise 4.43.

4.46 Hydration of one of the following two compounds yields a single ketone product. The other compound yields a mixture of ketones. Which one yields only one ketone product? Why?

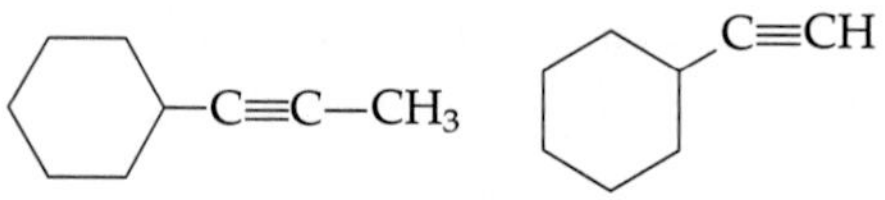

$-C{\equiv}C-CH_3$ $C{\equiv}CH$

Syn–Anti Addition

4.47 Write the product of the reaction of cyclopentene with potassium permanganate.

4.48 Write the product of the reaction of 1,2-dimethylcyclobutene with hydrogen in the presence of a platinum–carbon catalyst.

4.49 Write the product of the reaction of 1-methylcyclopentene with bromine in CCl_4 as solvent.

4.50 The reaction of cyclopentene with bromine in water as the solvent yields the alcohol *trans*-2-bromocyclohexanol. Explain why.

Preparation of Alkenes and Alkynes

4.51 How many alkenes would be formed by dehydrohalogenation of each of the following alkyl bromides? Which compound should be the major isomer?

(a) $CH_3CH_2CH(Br)CH_2CH_3$ (b) $(CH_3)_3CCHBrCH_3$

(c) $CH_3(CH_2)_3CH_2Br$ (d) $CH_3(CH_2)_3CHBrCH_3$

4.52 Write the structure of a bromo compound that gives exclusively each of the following alkenes by dehydrohalogenation.

(a) 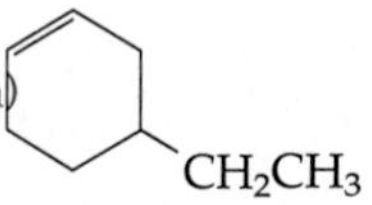 CH_2CH_3 (b) 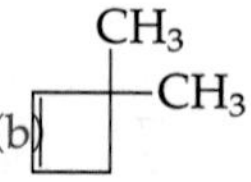 CH_3, CH_3 (c) 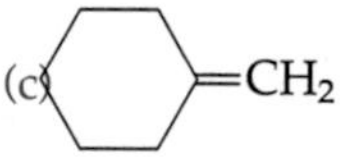 $=CH_2$ (d)

4.53 Write the structure of a compound that would yield the following alkyne upon dehydrohalogenation.

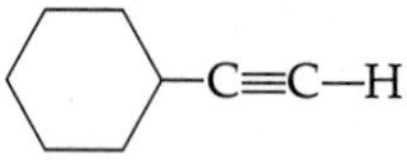

$-C{\equiv}C-H$

4.54 Would the following reaction provide a good yield of the indicated product? Explain.

$$CH_3CH_2CH_2CBr_2CH_3 \xrightarrow{NaNH_2} CH_3CH_2C{\equiv}CCH_3$$

Polyunsaturated Compounds

4.55 Which of the following compounds has conjugated double bonds?

(a) 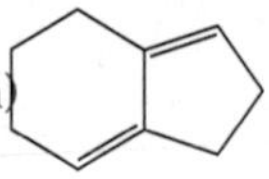 (b)

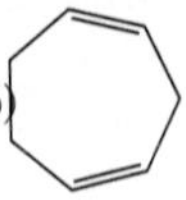

(c) 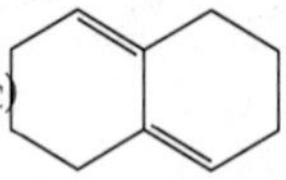 (d)

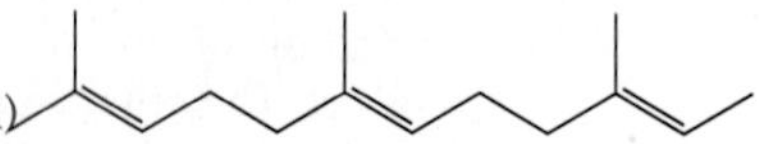

4.56 Which of the following compounds has conjugated double bonds?

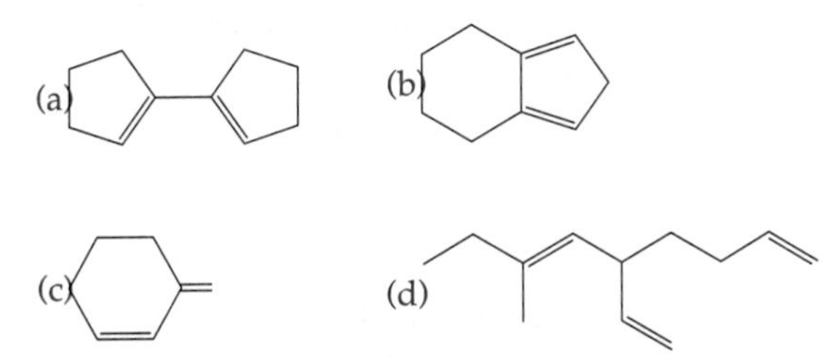

4.57 How many compounds in each of the following sets of isomeric compounds contain conjugated double bonds?

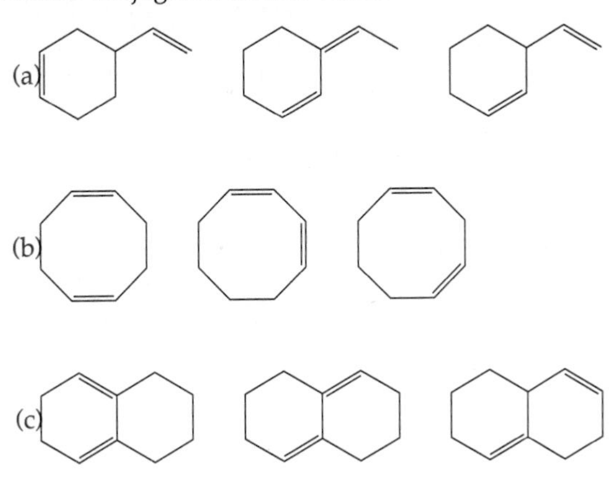

4.58 How many conjugated multiple bonds are contained in each of the following compounds?

(a) ichthyothereol, a convulsant (see structure in Section 4.1)

(b) natamycin, an antifungal drug

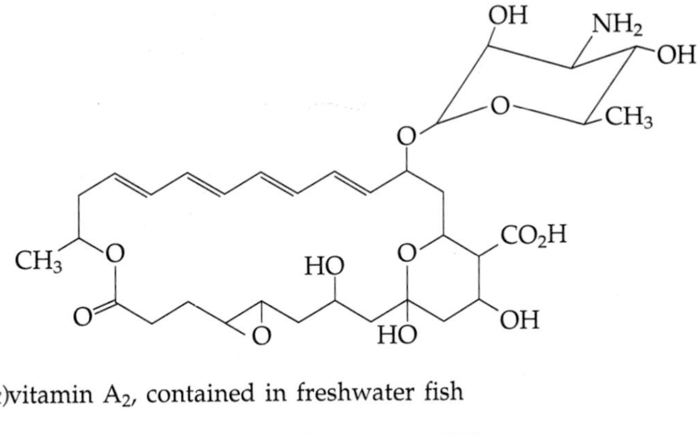

(c) vitamin A_2, contained in freshwater fish

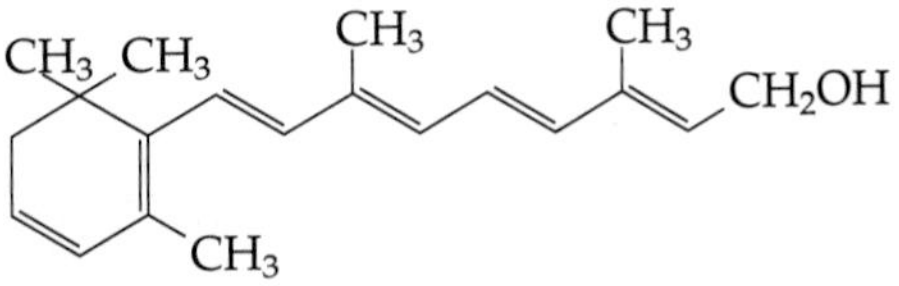

(d) zingiberene, contained in oil of ginger

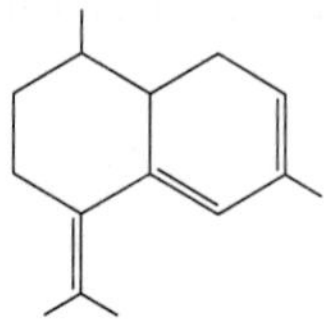

Polymers

4.59 Draw a representation of the polymer produced from $CH_2{=}CHCO_2CH_3$.

4.60 Draw a representation of the polymer produced from $CH_2{=}CCl_2$.

4.61 Write the structure of the intermediate formed when a radical formed from a peroxide reacts with styrene, $C_6H_5{-}CH{=}CH_2$.

4.62 A nucleophilic catalyst is used to polymerize methyl acrylate, $CH_2{=}CH{-}CO_2CH_3$. Write the structure of the intermediate.

Terpenes

4.63 Classify each of the following terpenes and divide it into isoprene units.

(a) 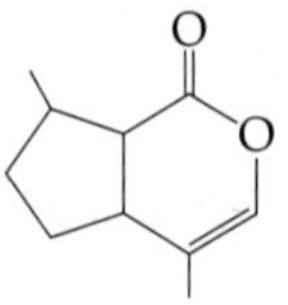 (b)

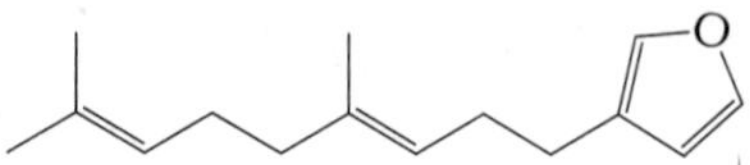

(c)

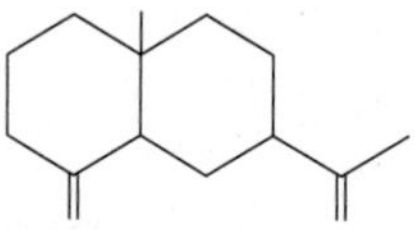

4.64 Classify each of the following terpenes and divide it into isoprene units.

(a) 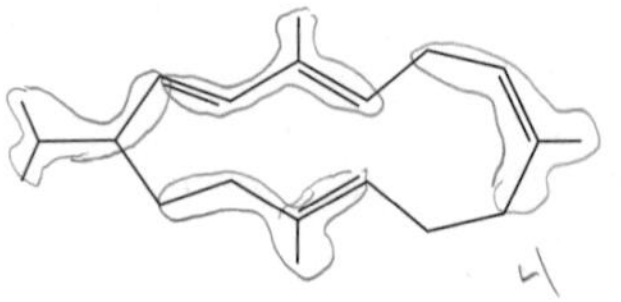 (b)

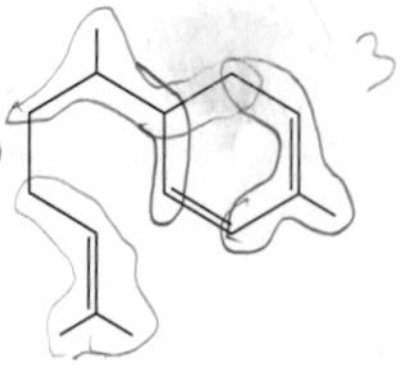

(c)

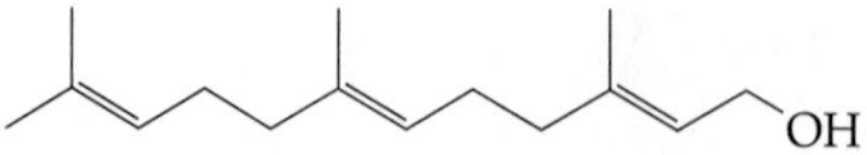

Metabolic Oxidations

4.65 Write the structures of two possible allylic oxidation products of safrole, a carcinogenic compound.

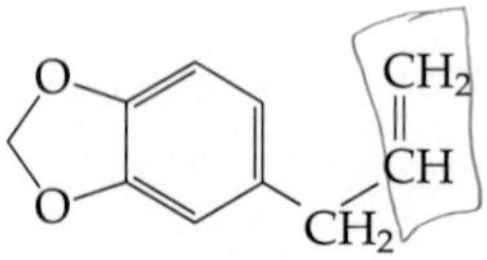

4.66 The anticonvulsant drug carbamazepine is metabolized to initially produce an epoxide. Write the structure of the product.

N
O NH$_2$

4.67 Secobarbital, a barbiturate, undergoes allylic oxidation when metabolized. Write the product of oxidation.

CH$_3$
O
H CHCH$_2$CH$_2$CH$_3$
N CH$_2$CH=CH$_2$
O N O
H

4.68 Write two possible allylic oxidation products of the barbiturate hexobarbital.

O
CH$_3$
HN
O N O
CH$_3$

or

OH
OH

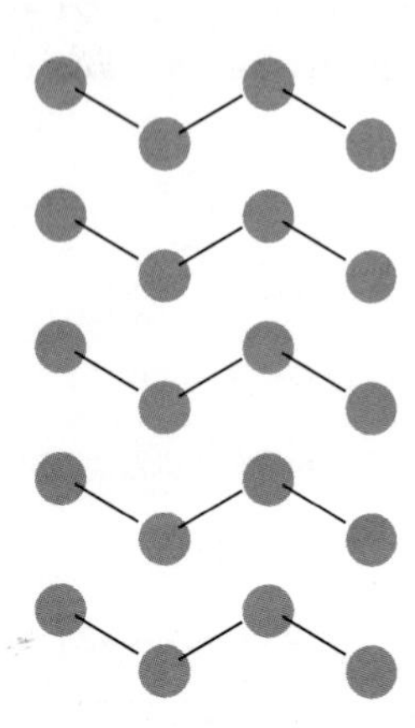

5

AROMATIC COMPOUNDS

5.1 BENZENE COMPOUNDS

The term *aromatic* means fragrant. For this reason many fragrant substances, known from earliest time, were classified as aromatic compounds. These compounds contain a benzene ring that is bonded to one or more substituents. The structures of a few fragrant compounds containing a benzene ring are shown below.

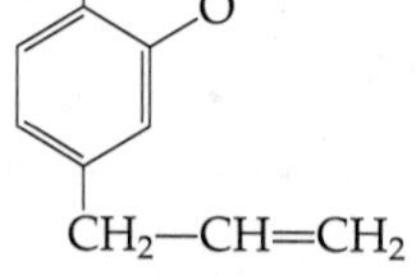

safrole (oil of sassafras) | methyl salicylate (oil of wintergreen) | vanillin (vanilla)

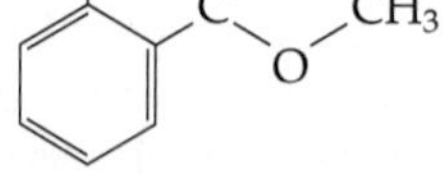

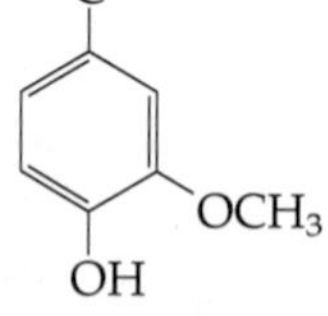

Today, the classification of aromatic compounds is no longer based on odor because most compounds containing a benzene ring are not in fact fragrant. Some aromatic compounds—including the pain relievers or analgesics aspirin, ibuprofen, and acetaminophen—are solids that have little or no odor.

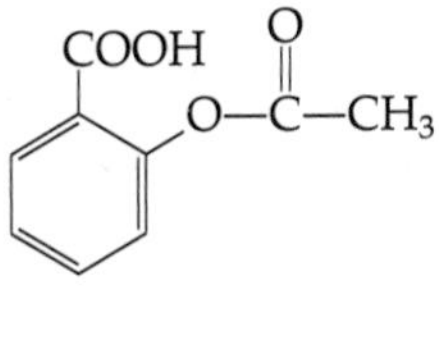

aspirin 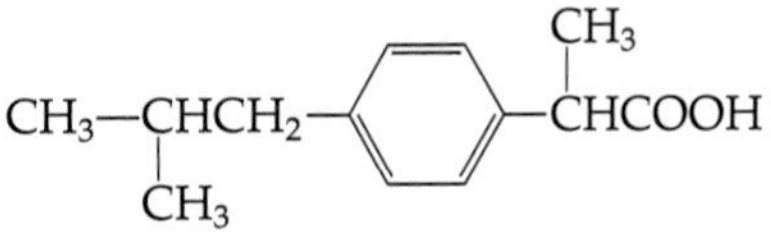　　ibuprofen 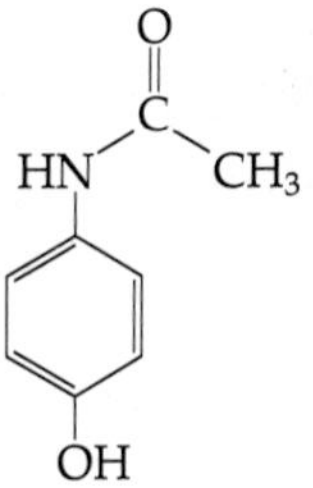　　acetaminophen

A benzene ring is found in many drugs with very complex structures. Examples include the antibiotics chloramphenicol and protosil.

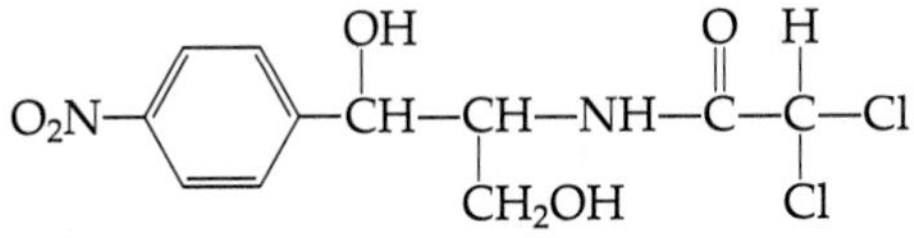

chloramphenicol 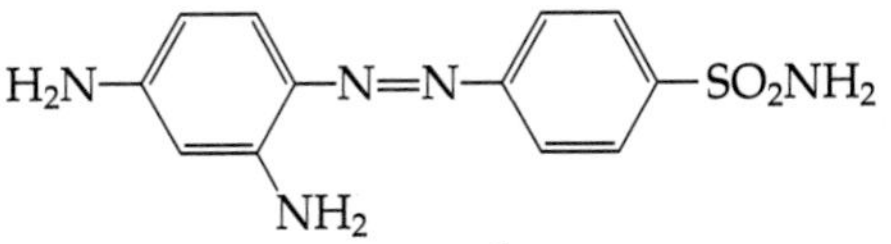　　protosil

5.2 AROMATICITY

Rather than classifying aromatic compounds by their odor, we now classify aromatic hydrocarbons by their inability to react in addition reactions. Benzene, C_6H_6, has eight fewer hydrogen atoms than the saturated hydrocarbon hexane, C_6H_{14}. Thus, benzene is highly unsaturated. It is represented by a hexagon containing three double bonds. However, unlike alkenes, benzene does not add bromine to give dibromo compounds. Furthermore, benzene does not add HBr, cannot be hydrated, and does not react with the powerful oxidizing agent potassium permanganate.

The reactivity of benzene contradicts what we know about unsaturated compounds. Benzene does not react with most reactants that attack π bonds to form addition products. That is to say, it does not behave like the "triene" depicted by its Lewis structure. It undergoes reactions that are different from those of alkenes. Thus, benzene reacts with bromine to give a substitution product in which a bromine atom replaces a hydrogen atom. The reaction requires iron(III) bromide as a catalyst. Only one compound, C_6H_5Br, is formed.

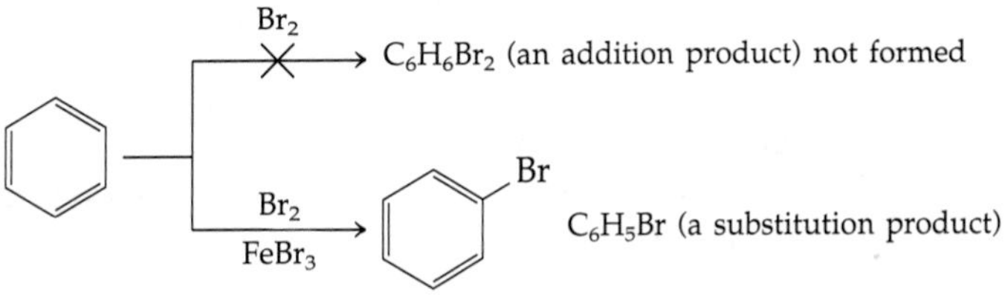

Br_2 → $C_6H_6Br_2$ (an addition product) not formed

Br_2, $FeBr_3$ → C_6H_5Br (a substitution product)

Kekulé's Concept of Benzene

In 1865, a German chemist, August Kekulé, suggested that benzene is a ring of six carbon atoms linked by alternating single and double bonds. He proposed that benzene actually exists as two structures differing only in the arrangement of the single and double bonds, which oscillate around the ring. Kekulé proposed that the rapid oscillation of single and double bonds somehow made benzene resist addition reactions.

(Kekulé structures of benzene, C_6H_6) or (Kekulé structures of benzene, C_6H_6)

Resonance Theory and Benzene

Benzene is a planar molecule in which all carbon–carbon bond lengths are the same; the bond angles of the ring are all 120°. Thus, the π bonds in benzene are made with sp^2-hybridized carbon atoms. Each carbon atom shares one electron in each of its three σ bonds: two σ bonds are to adjacent carbon atoms; the third σ bond is to a hydrogen atom. The fourth electron is in a $2p$ orbital perpendicular to the plane of the benzene ring (Figure 5.1). A set of six $2p$ orbitals (one from each carbon atom) and their six electrons overlap to share electrons in a π system that extends over the entire ring of carbon atoms. The electrons are located both above and below the plane of the ring. The sharing of electrons over many atoms is called **delocalization.** This delocalization of electrons accounts for the unique chemical stability of benzene.

FIGURE 5.1 **Bonding in the Benzene Ring**
The lines between carbon atoms represent the σ bonds of the benzene ring. In addition, each carbon atom has one p orbital that contributes one electron to the π system. Overlap of the six p orbitals that are maintained mutually parallel results in a delocalized system that distributes the electrons over the entire carbon framework.

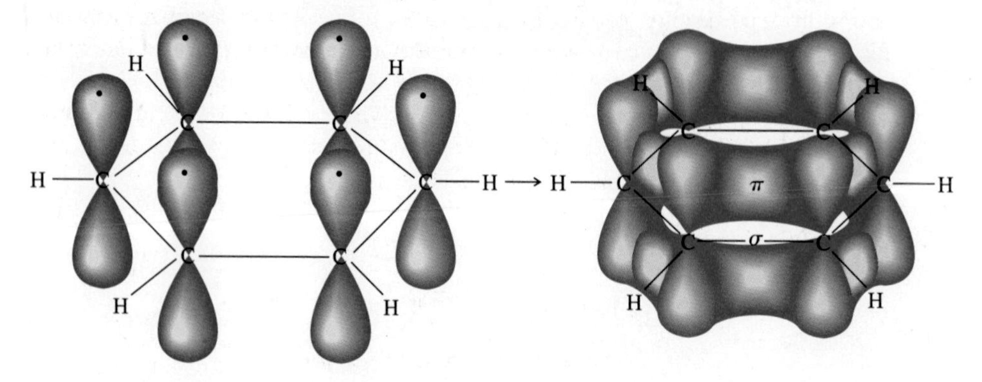

Two resonance structures are used to depict the structure of benzene. They are the Kekulé structures and differ only in the positions of the double bonds. Because the two resonance structures are otherwise identical, they are said to contribute equally to the actual structure of benzene. Kekulé structures are not "real", but the benzene molecule can be viewed as a resonance hybrid of these two nonexistent structures. We indicate the relationship between the contributing structures by a single double-headed arrow.

The structure of benzene is usually represented in chemical equations as one of the two possible Kekulé structures. Each corner of the hexagon represents a carbon atom with one attached hydrogen atom, which is often not shown.

The Hückel Rule

Most aromatic compounds contain a benzene ring or a related structure (Section 5.3). What is responsible for the characteristic stability of benzene and its unique reactivity? Three general criteria must be met if a molecule is to be aromatic. First, the molecule must be cyclic and planar. Second, the ring must contain only sp^2-hybridized atoms that can form a delocalized system of π electrons. (There can be no interruption by sp^3-hybridized atoms.) Third, the number of π electrons must equal $4n + 2$, where n is an integer. (Note that n is *not* the number of carbon atoms in the ring.) The "$4n + 2$ rule" was proposed by E. Hückel and is known as the **Hückel rule.** The theoretical basis for this rule is beyond the scope of this text. However, based on the Hückel rule, cyclic π systems with 6 ($n = 1$), 10 ($n = 2$), and 14 ($n = 3$) electrons will be aromatic.

Benzene meets the criteria for aromaticity. Other examples will be presented in the following section for various 6-, 10-, and 14-π-electron systems. We will also see that aromatic compounds can contain atoms other than carbon (Section 5.3).

Some cyclic polyenes with alternating single and double bonds are not aromatic. These compounds do not obey the Hückel rule. Two examples are cyclobutadiene and cyclooctatetraene.

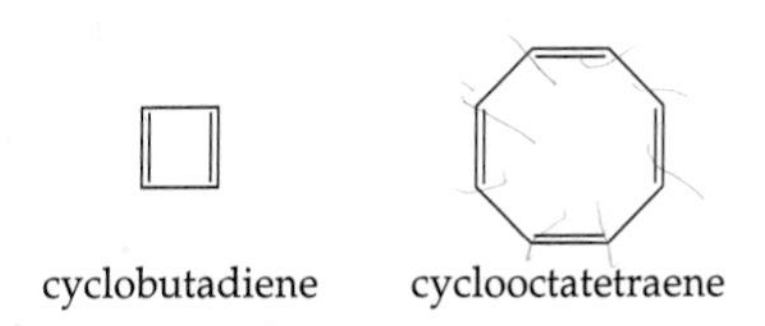

Cyclobutadiene is extremely unstable and has never been isolated, although its fleeting existence has been inferred from the products of its reactions. Cyclobutadiene has four π electrons, a number that does not satisfy the Hückel rule.

Cyclooctatetraene has eight π electrons, and eight is not a Hückel number; that is, there is no integer n for which $4n + 2 = 8$. Cyclooctatetraene is a stable molecule, but it reacts like an alkene. For example, it undergoes addition reactions with bromine and is easily hydrogenated. Also, cyclooctatetraene is not a planar molecule, so its $2p$ orbitals cannot overlap to form a π system. Cyclooctatetrene does not satisfy the general characteristics for aromaticity.

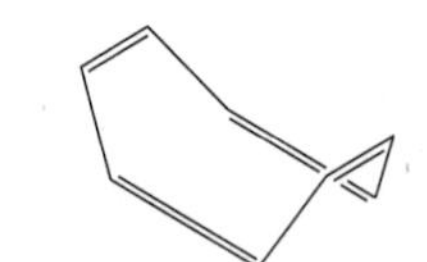

"tub" conformation of cyclooctatetraene

5.3 POLYCYCLIC AND HETEROCYCLIC AROMATIC COMPOUNDS

Some aromatic compounds contain two or more rings that are said to be "fused"; that is, two carbon atoms are common to two rings. Compounds of this type are called **polycyclic aromatic hydrocarbons.** These molecules are planar; that is, all atoms in the rings and those directly attached to the rings are in a plane. Several examples of polycyclic aromatic compounds are shown below. Note that all the carbon atoms have a bond to a hydrogen atom, except those at the points of fusion.

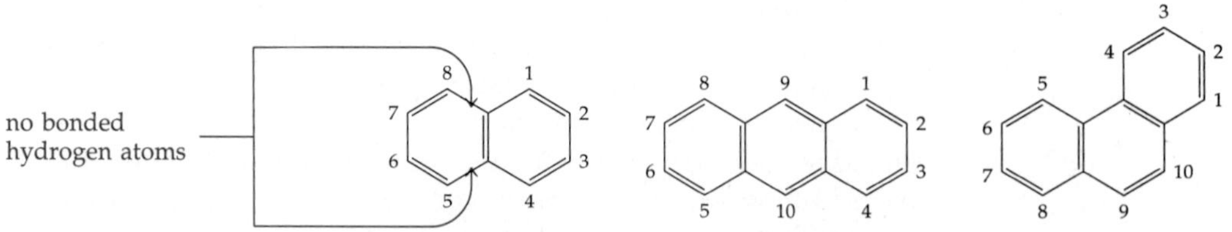

no bonded hydrogen atoms

naphthalene anthracene phenanthrene

EXAMPLE 5.1 What are the structural and electronic similarities of anthracene and phenanthrene?

Solution Both compounds have 14 carbon atoms and 10 hydrogen atoms; they are isomers. Both molecules consist of three fused rings which are depicted with a series of alternating single and double bonds, and each has

14 π electrons—a number that fits the Hückel rule for $n = 3$. Thus, both of these compounds are aromatic.

Cyclic compounds that have one or more atoms other than carbon within the ring are said to be **heterocyclic compounds;** those having $4n + 2$ π electrons are **heterocyclic aromatic compounds.** The hetero atoms most commonly encountered in naturally occurring compounds are nitrogen and oxygen; sulfur-containing compounds also exist. The structures of a few heterocyclic aromatic compounds are shown below.

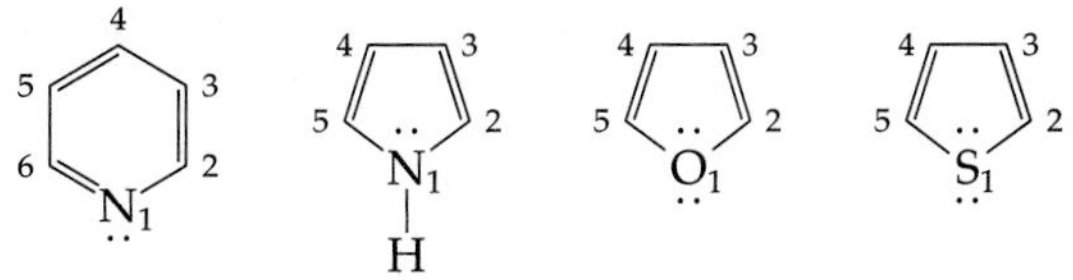

pyridine pyrrole furan thiophene

Pyridine closely resembles benzene: it is planar; each of its ring atoms, including the nitrogen atom, is sp^2-hybridized; and each ring atom contrib-

Carcinogenic Aromatic Compounds

Fused polycyclic aromatic hydrocarbons containing four or more rings that are not colinear—that is, there is an "angle" in the series of rings—are carcinogenic. These compounds structurally resemble phenanthrene. Three of the most potent carcinogens are 1,2-benzanthracene, 1,2,5,6-dibenzanthracene, and 3,4-benzpyrene. The angular area is shaded in the structures shown here.

Small amounts of these angular fused-ring aromatic hydrocarbons cause cancer in about a month when they are applied to the skin of a mouse. These compounds are present in the effluent from coal-burning power plants and in automobile exhaust. They are also present in tobacco smoke and in meat cooked over charcoal. The incidence of lung cancer among smokers and inhabitants of large urban areas may partially result from inhaling these airborne compounds in minute amounts over a period of time.

It was once common for chimney sweeps in England to develop cancer. Today the reason is understood. While they worked, the chimney sweeps became covered with chimney soot and inhaled sooty dust that contained angular, fused, aromatic hydrocarbons.

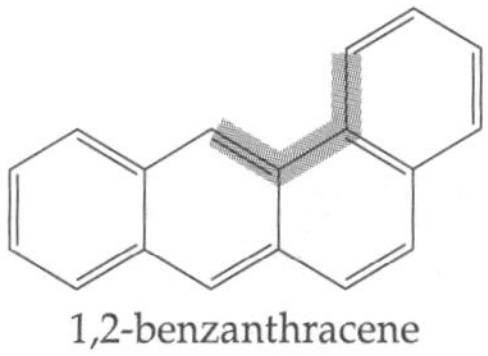

1,2-benzanthracene 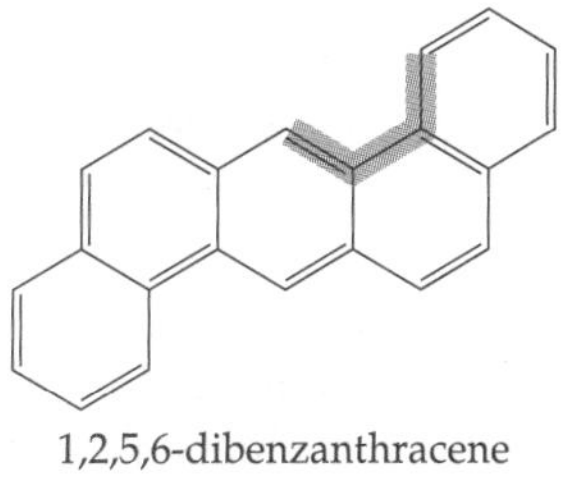 1,2,5,6-dibenzanthracene 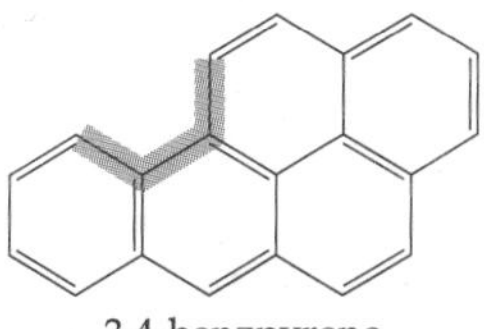 3,4-benzpyrene

utes one electron in a p orbital to an aromatic system of six π electrons. The sp^2-hybridized nitrogen atom has five valence electrons: one is contributed to the aromatic sextet. The remaining four valence electrons of nitrogen are distributed in the three sp^2 orbitals. Two valence electrons form σ bonds to two carbon atoms, and two valence electrons are present as a nonbonded pair in an sp^2-hybridized orbital. The nonbonded pair projects out from the plane of the ring in the same direction as the carbon–hydrogen bonding electrons of benzene (Figure 5.2). This nonbonded electron pair allows pyridine to act as a base.

The nitrogen atom in pyrrole is also sp^2-hybridized. However, the electrons are distributed differently than in pyridine. The nitrogen atom contributes one electron to each of the three sp^2 orbitals; two of them form σ bonds to carbon atoms; the third orbital forms a bond with the hydrogen atom (Figure 5.2). The nitrogen atom's remaining two valence electrons are lo-

FIGURE 5.2
Bonding in Heterocyclic Aromatic Hydrocarbons

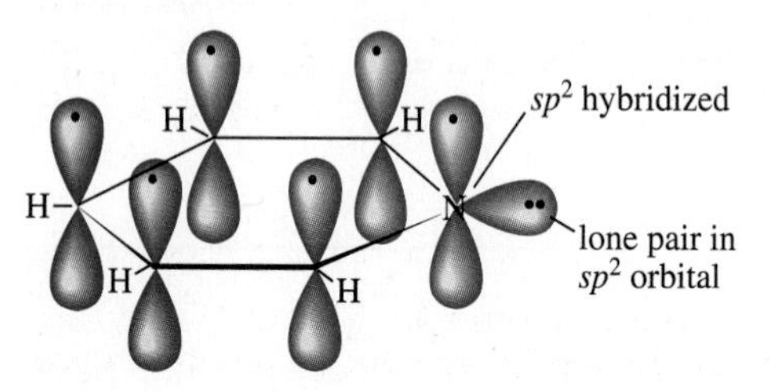

In pyridine two of the electrons of the nitrogen atom are located in an sp^2 hybrid orbital directed outward from the plane. These electrons are not involved in resonance with the π electrons.

pair in p orbital

sp^2 hybridized

In pyrrole two of the electrons of the nitrogen atom are located in a p orbital that is part of the π system. The other three electrons of nitrogen are in sp^2 hybrid orbitals which form three σ bonds—two with carbon atoms and one with a hydrogen atom.

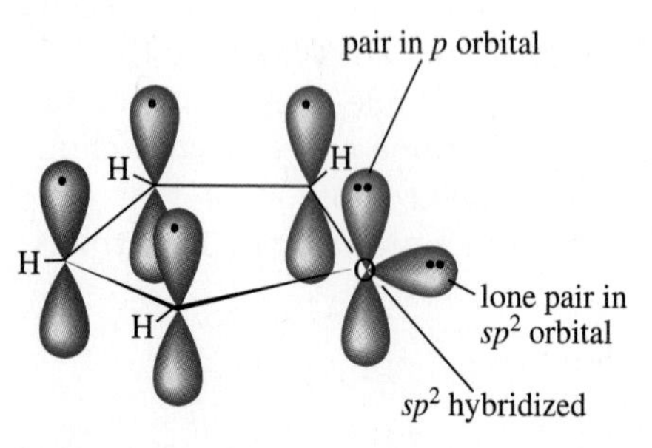

In furan two of the electrons of the oxygen atom are located in a p orbital that is part of the π system. The other four electrons of oxygen are in sp^2 orbitals. Two of the electrons form σ bonds with carbon atoms. The remaining two electrons are located in an sp^2 hybrid orbital directed outward from the plane of the ring.

cated in a $2p$ orbital. These two electrons are added to the four electrons in the $2p$ orbitals of the four carbon atoms in the ring to provide a six-electron π system—again an aromatic ring like benzene.

Furan and thiophene are heterocyclic aromatic compounds whose heteroatoms are oxygen and sulfur, respectively. The oxygen atom of furan and the sulfur atom of thiophene are both sp^2-hybridized. Oxygen has six valence electrons. Two are in a $2p$ orbital that can interact with the four π electrons of the carbon atoms in the ring to provide a six-electron π system. The remaining four valence electrons of oxygen are distributed in three sp^2 orbitals. Two of the orbitals have one electron each, and these form σ bonds to two carbon atoms. The remaining sp^2 orbital has two electrons, and, as in pyridine, this unshared pair of electrons has the same relationship to the ring as the carbon–hydrogen bonding electrons of benzene (Figure 5.2).

Many classes of naturally occurring biologically important compounds—such as vitamins B_1, B_3, and B_6—have one or more aromatic heterocyclic rings of five or six atoms. Vitamin B_1 has two hetero atoms in each ring.

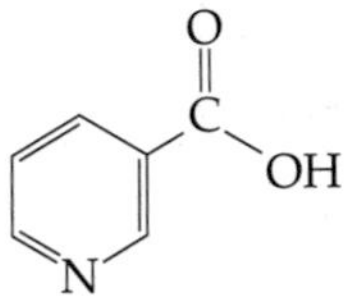

O C OH N
niacin (B_3)

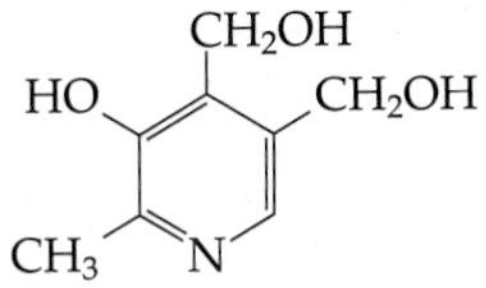

CH_2OH HO CH_2OH CH_3 N
pyridoxine (B_6)

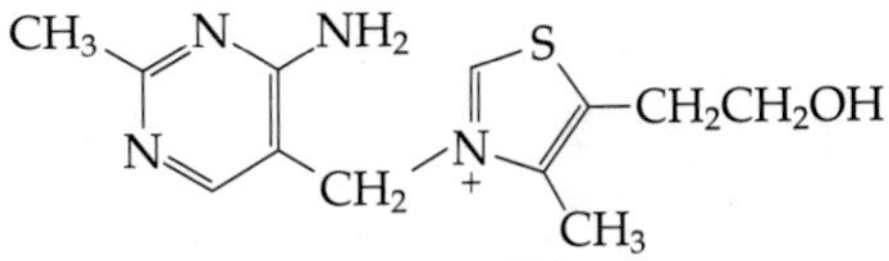

CH_3 N NH_2 S CH_2CH_2OH N CH_2 N + CH_3
thiamine (B_1)

Many pharmaceutical compounds contain heterocyclic rings that contribute to their effectiveness as drugs. One example is methotrexate, a chemotherapeutic agent, which contains a ring resembling naphthalene. There are four nitrogen atoms in the rings. Another drug containing a heterocyclic aromatic ring is Tagamet (generic name, cimetidine), an antiulcer drug. The heterocyclic ring system contains two nitrogen atoms.

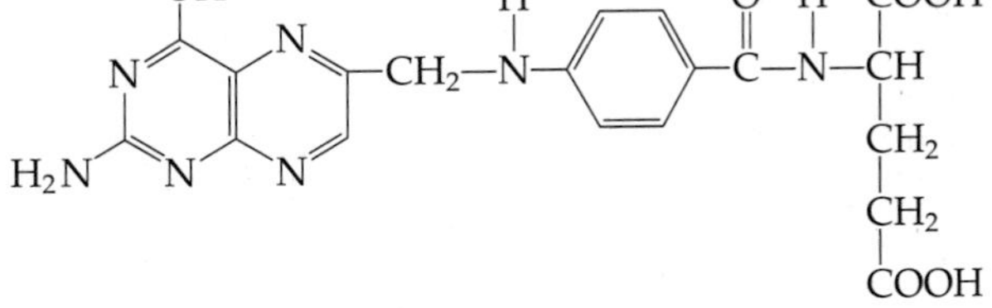

OH N N H_2N N N CH_2—N(H)— ... —C(=O)—N(H)—CH(COOH)—CH_2—CH_2—COOH
methotrexate

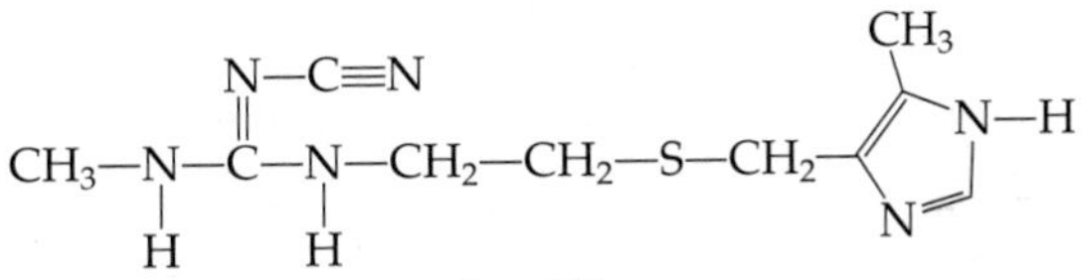

CH_3—N(H)—C(=N—C≡N)—N(H)—CH_2—CH_2—S—CH_2— (ring: CH_3, N—H, N)
cimetidine

EXAMPLE 5.2 Histamine is released in the body in persons with allergic hypersensitivities, such as hay fever. Distinguish between the locations of the electrons on the two nitrogen atoms in the heterocyclic ring.

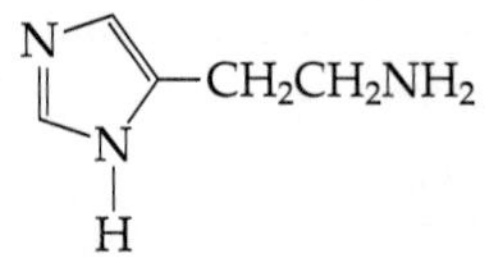

N
$CH_2CH_2NH_2$
N
H

Solution The nitrogen atom on the bottom is bonded to a hydrogen atom and resembles the nitrogen atom of pyrrole. Three σ bonds are shown. Each has one valence electron contributed from the nitrogen atom. Thus, the remaining two valence electrons of nitrogen are located in a $2p$ orbital. These two electrons, along with the four electrons of the two π bonds shown in the structure, account for six electrons of an aromatic system.

The nitrogen atom on the left has one single and one double bond as shown. This nitrogen atom resembles the nitrogen atom of pyridine. Two of its electrons are used to form two σ bonds; one electron is contributed to the π bond with a carbon atom. The remaining two electrons are in an sp^2 hybrid orbital projecting out from the plane of the ring.

5.4 NOMENCLATURE OF BENZENE COMPOUNDS

Many compounds with a benzene ring have well-established, nonsystematic names. Such names are often based on sources of the compounds and have been used for so long that they have become accepted by IUPAC. One example is toluene, which used to be obtained from the South American gum tree *Toluifera balsamum.*

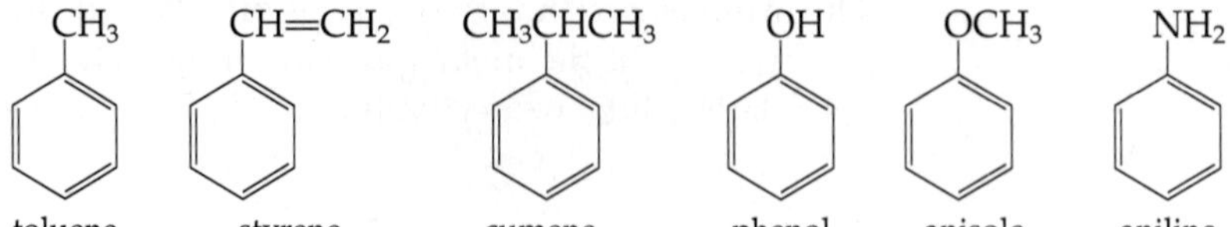

CH_3 toluene
$CH{=}CH_2$ styrene
CH_3CHCH_3 cumene
OH phenol
OCH_3 anisole
NH_2 aniline

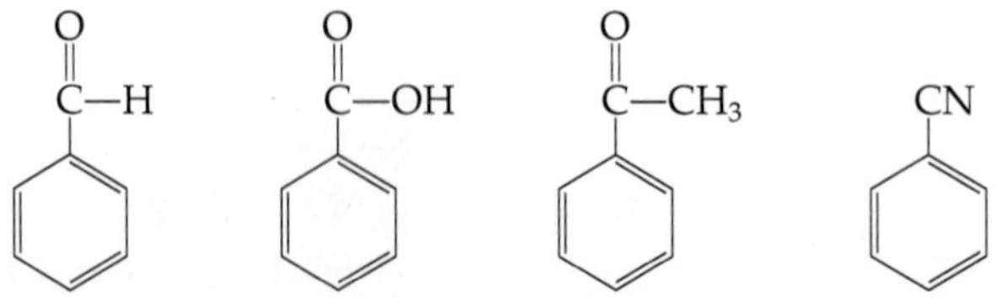

O ‖ C—H benzaldehyde
O ‖ C—OH benzoic acid
O ‖ C—CH_3 acetophenone
CN benzonitrile

The IUPAC system of naming substituted aromatic hydrocarbons uses the names of the substituents as prefixes to benzene. Examples include

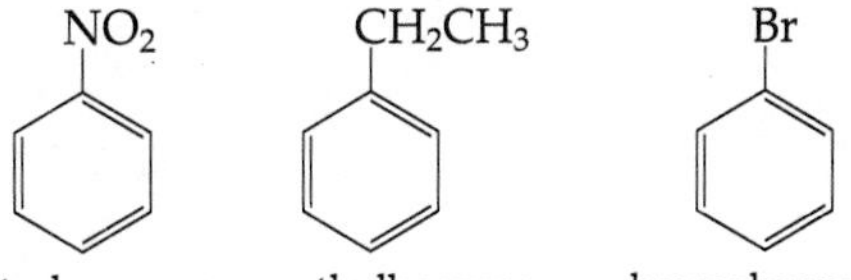

nitrobenzene ethylbenzene bromobenzene

Many compounds shown in this section have the substituent at a "12 o'clock" position. However, all six positions on the benzene ring are equivalent, and you should be able to recognize a compound, such as bromobenzene, no matter where the bromine atom is written.

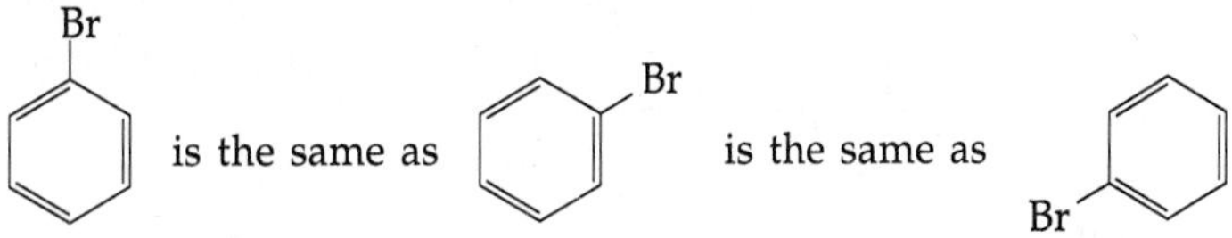

is the same as is the same as

Disubstituted compounds result when two hydrogen atoms of the benzene ring are replaced by other groups. Two substituents can be arranged in a benzene ring in three different ways to give three different isomers. These isomers are designated ortho, meta, and para. As prefixes these terms are abbreviated as *o*-, *m*-, and *p*-, respectively.

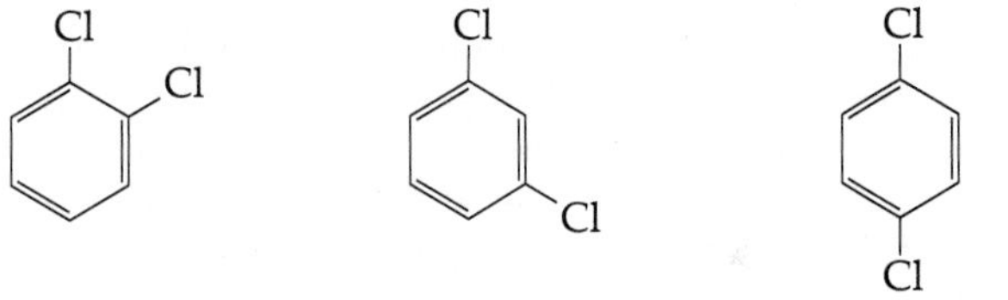

o-dichlorobenzene
1,2-dichlorobenzene

m-dichlorobenzene
1,3-dichlorobenzene

p-dichlorobenzene
1,4-dichlorobenzene

The ortho isomer has two groups on adjacent carbon atoms—that is, in a 1,2 relationship. In the meta and para isomers, the two groups are in a 1,3 and a 1,4 relationship, respectively. The IUPAC name of disubstituted aromatic compounds is obtained by numbering the benzene ring to give the lowest possible numbers to the carbon atoms bearing the substituents. When three or more substituents are present, the ring carbon atoms must be numbered.

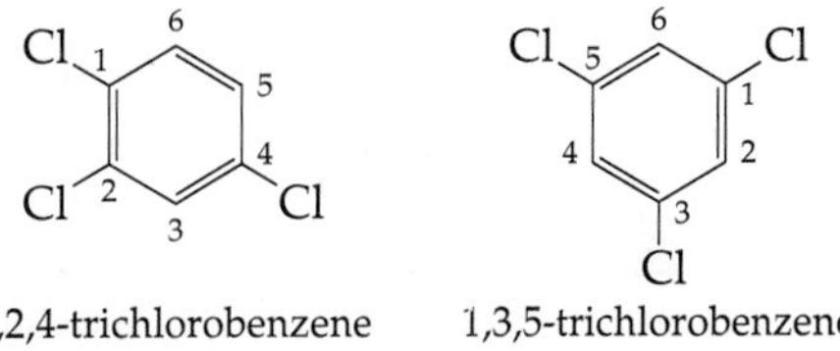

1,2,4-trichlorobenzene 1,3,5-trichlorobenzene

Many disubstituted compounds have common names. Examples include the xylenes, cresols, and toluidines—all of which can be ortho, meta, or para isomers.

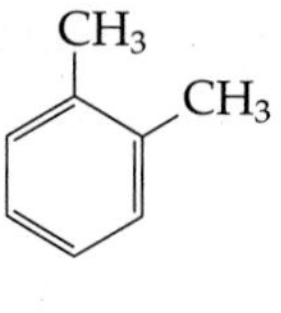

CH_3 CH_3 OH CH_3 NH_2 CH_3

o-xylene 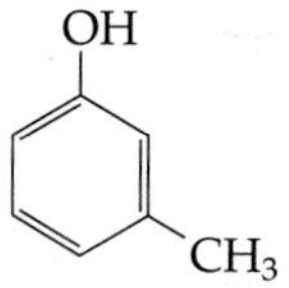 *m*-cresol 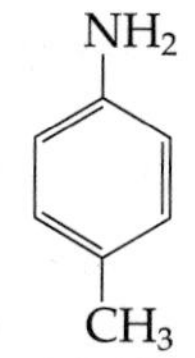 *p*-toluidine

Many derivatives of benzene are named with the common name of the monosubstituted aromatic compound as the parent. The position of the substituent of the "parent" is automatically designated 1, but the number is not used in the name. The remaining substituents are prefixed in alphabetical order to the parent name along with numbers indicating their locations.

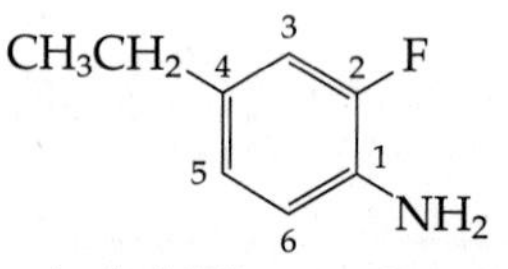

CH_3CH_2 4 3 2 F 1 NH_2 6 5 OCH_3 1 2 CH_3 3 CH_2CH_3 4 5 6

4-ethyl-2-fluoroaniline 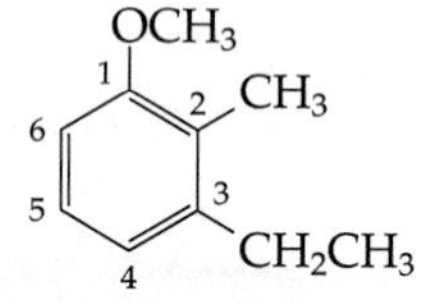 3-ethyl-2-methylanisole

---

EXAMPLE 5.3 Indicate how a name could be derived for the following trisubstituted compound, which is used as an antioxidant in some food products.

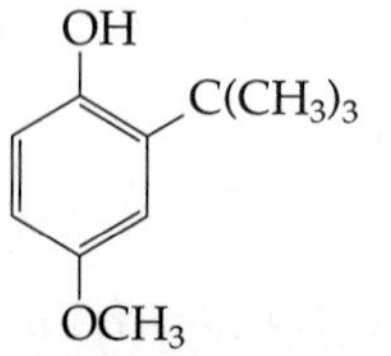

OH $C(CH_3)_3$ OCH_3

Solution Either the —OH group or the —OCH_3 group could provide the parent name, which would be phenol or anisole, respectively. Let's assume that the compound is a substituted anisole. When we assign the number 1 to the carbon atom bearing the —OCH_3 group at the "six o'clock" position, we must number the ring in a counterclockwise direction. A *t*-butyl group is located at the 3 position and a hydroxyl group at the 4 position, so the compound is 3-*tert*-butyl-4-hydroxyanisole.

Aromatic hydrocarbons belong to a general class called **arenes.** An aromatic ring residue attached to a larger parent structure is an **aryl group;** it is symbolized as Ar (not to be confused with argon), just as the symbol R is used for an alkyl group. Two groups whose names unfortunately do not make much sense are the phenyl (fen′-nil) and benzyl groups. We might reasonably expect that the aryl group derived from benzene (C_6H_5—) would be named benzyl. Alas, it is a phenyl group. A benzyl group, which is derived from toluene, has the formula $C_6H_5CH_2$—.

C_6H_5— or [benzene ring]—

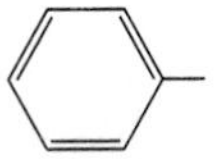

phenyl group

$C_6H_5CH_2$— or [benzene ring]—CH_2—

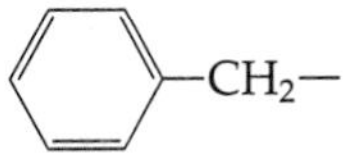

benzyl group

If alkyl groups containing fewer than six carbon atoms are bonded to a benzene ring, the compound is named as an alkyl-substituted benzene. For more complex molecules, the term *phenyl* is used to designate the aryl group on the parent chain of carbon atoms, as in 3-phenylheptane.

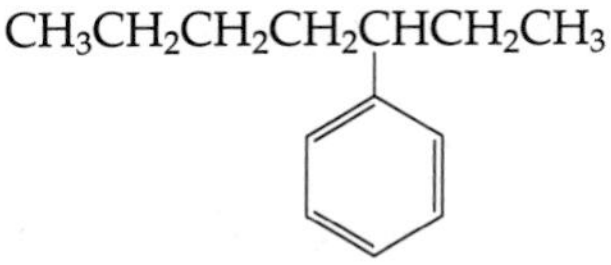

$CH_3CH_2CH_2CH_2CHCH_2CH_3$ (with a benzene ring on the CH)

3-phenylheptane

EXAMPLE 5.4 What is the name of the following compound?

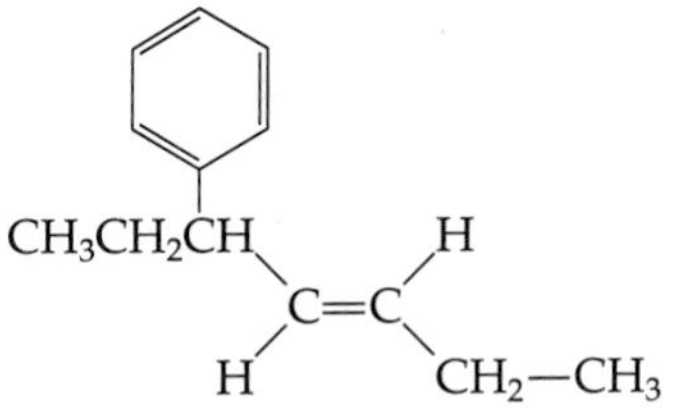

CH_3CH_2CH(phenyl)—CH=CH—CH_2—CH_3 (H on C=C trans)

Solution The compound is an alkene with an aromatic ring as a substituent. First, we determine that the chain has seven carbon atoms. Next, we number the chain from right to left so that the double bond is assigned to the C-3 atom. The phenyl group is then located on the C-5 atom. Also, we note that the compound is the *E* isomer. The complete name is (*E*)-5-phenyl-3-heptene.

5.5 ELECTROPHILIC AROMATIC SUBSTITUTION

The aromatic ring does not undergo the addition reactions we discussed in Chapter 4 for alkenes. Instead, aromatic rings react with electrophiles, in the presence of a proper catalyst, to give substitution products. In these reactions an electrophile (E^+) substitutes for H^+. The general process is shown below.

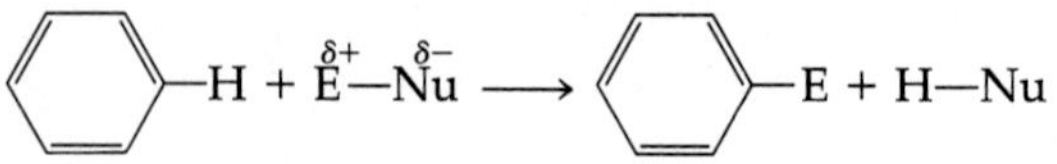

$$C_6H_5\text{—H} + \overset{\delta+}{E}\text{—}\overset{\delta-}{Nu} \longrightarrow C_6H_5\text{—E} + \text{H—Nu}$$

Mechanism of Electrophilic Aromatic Substitution

The first step of electrophilic aromatic substitution is similar to that discussed for the addition of electrophiles to alkenes. The electrophile accepts an electron pair from the aromatic ring. However, the electrons are delocalized in an aromatic ring. Therefore, aromatic compounds are significantly less reactive than alkenes. In fact, they are so much less reactive than alkenes that a strong Lewis acid—such as $AlCl_3$ or $FeBr_3$—is required as a catalyst to generate an electrophile sufficiently reactive to attack the aromatic ring. Details of the relationship between reagent and catalyst will be discussed in the next section.

In the first step of electrophilic aromatic substitution, an electrophile adds to the aromatic ring and produces a carbocation intermediate stabilized by resonance. However, this carbocation is not aromatic—it has only four π electrons and it has an sp^3-hybridized carbon atom. Therefore, it is less stable than the original aromatic ring, which had a full array of six π electrons.

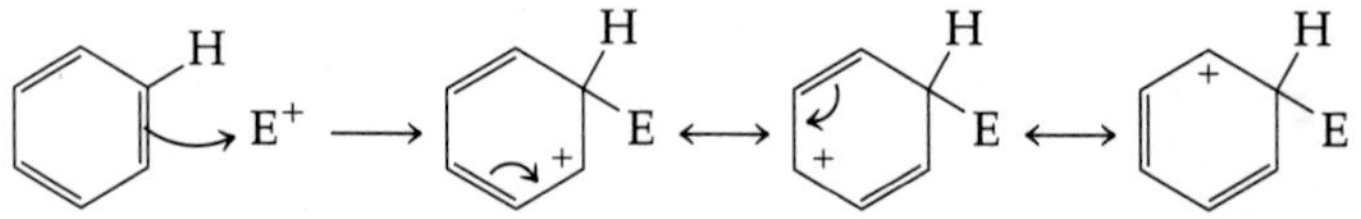

In the second step of the reaction, the proton bound to the same carbon atom as the newly arrived electrophile is lost, and the aromatic π system is restored. The proton that leaves in this step is extracted by a nucleophile acting as a base.

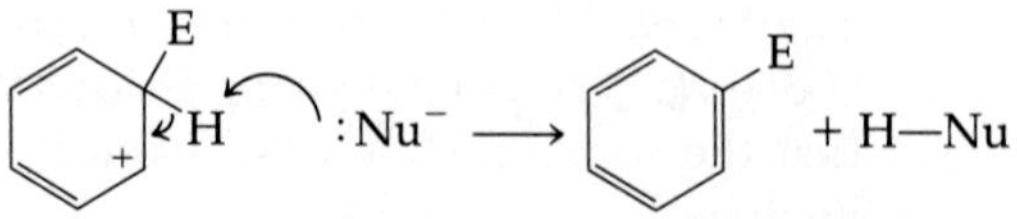

Typical Electrophilic Substitution Reactions

In the preceding discussion we used a generic electrophile, E^+. In this section we will consider some specific examples of electrophiles that react with aromatic rings. Our first examples are bromination and chlorination. Bromination requires both Br_2 and a Lewis acid catalyst, $FeBr_3$. The catalyst generates a bromine cation, Br^+, in a Lewis acid–base reaction as shown below. Chlorination requires both Cl_2 and a Lewis acid catalyst, $FeCl_3$. Chlorination proceeds in the same manner as bromination.

$$FeBr_3 + Br_2 \longrightarrow FeBr_4^- + Br^+$$

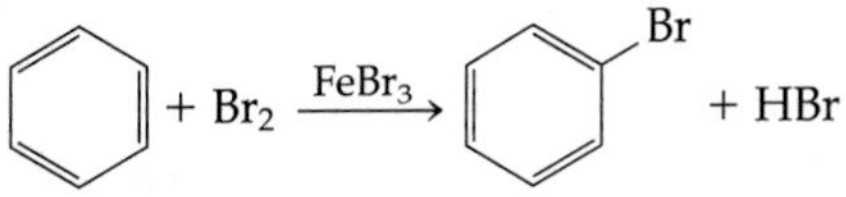

$$C_6H_6 + Br_2 \xrightarrow{FeBr_3} C_6H_5Br + HBr$$

In **nitration,** a nitro group, $—NO_2$, is introduced onto an aromatic ring. The electrophilic aromatic substitution is accomplished by using nitric acid, HNO_3, with sulfuric acid as the catalyst. The electrophile is the nitronium ion, NO_2^+, which is produced by the reaction of nitric acid with sulfuric acid.

$$HO—NO_2 + H—OSO_3H \longrightarrow NO_2^+ + HSO_4^- + H_2O$$

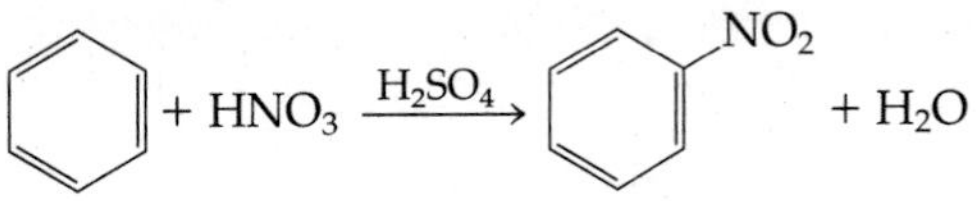

$$C_6H_6 + HNO_3 \xrightarrow{H_2SO_4} C_6H_5NO_2 + H_2O$$

nitrobenzene

A sulfonic acid group, $—SO_3H$, can also be introduced onto an aromatic ring by electrophilic aromatic substitution. The process is called **sulfonation.** In sulfonation, the reagent is SO_3. The reaction requires a mixture of SO_3 and sulfuric acid, called fuming sulfuric acid. The electrophile is $^+SO_3H$.

$$SO_3 + H_2SO_4 \longrightarrow {}^+SO_3H + HSO_4^-$$

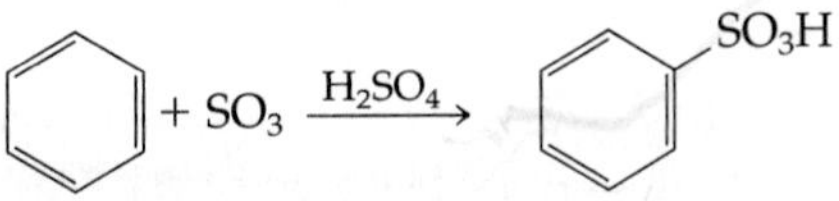

$$C_6H_6 + SO_3 \xrightarrow{H_2SO_4} C_6H_5SO_3H$$

benzenesulfonic acid

An alkyl group can be substituted for a hydrogen atom of an aromatic ring by a reaction called the **Friedel–Crafts alkylation.** This reaction requires an alkyl halide with an aluminum trihalide as the catalyst. The catalyst produces an electrophilic carbocation. The reaction is commonly done only with alkyl bromides or alkyl chlorides.

$$(CH_3)_2CHCl + AlCl_3 \longrightarrow (CH_3)_2CH^+ + AlCl_4^-$$

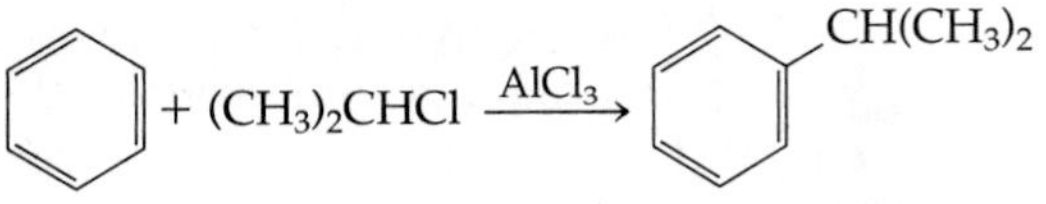

$$C_6H_6 + (CH_3)_2CHCl \xrightarrow{AlCl_3} C_6H_5CH(CH_3)_2$$

isopropylbenzene
(cumene)

An acyl group can replace hydrogen in an aromatic ring by a reaction called the **Friedel–Crafts acylation.** The reaction requires an acyl halide and the corresponding aluminum trihalide. The reaction is commonly done only with acyl chlorides. The electrophile is the acylium ion.

$$C_6H_6 + CH_3\overset{O}{\overset{\|}{C}}{-}Cl \xrightarrow{AlCl_3} C_6H_5{-}\overset{O}{\overset{\|}{C}}{-}CH_3 + HCl$$

acetophenone

Limitations of Friedel–Crafts Reactions

Neither Friedel–Crafts alkylation nor acylation occurs on aromatic rings that have one of the groups $-NO_2$, $-SO_3H$, $-C{\equiv}N$, and any carbonyl-containing group bonded directly to the aromatic ring. The carbonyl-containing compounds include aldehydes, ketones, carboxylic acids, and esters. All of these substituents make the benzene ring less reactive, a subject to be discussed in the next section.

The Friedel–Crafts alkylation is also seriously limited by the structural rearrangement of the alkyl carbocation generated from the alkyl halide. Such a rearrangement produces a different product from the one desired. For example, the reaction with 1-chloropropane in the presence of $AlCl_3$ yields a small amount of propylbenzene but a larger amount of the isomer, isopropylbenzene.

$$C_6H_6 + CH_3CH_2CH_2{-}Cl \xrightarrow{AlCl_3} C_6H_5{-}CH(CH_3)_2 + C_6H_5{-}CH_2CH_2CH_3$$

major product — minor product

Isomerization of carbocations occurs in the Friedel–Crafts reaction. The isomerization converts a less stable carbocation into a more stable one. (We recall from Chapter 4 that the order of carbocation stability is tertiary > secondary > primary.) Thus, if the Friedel–Crafts reaction is carried out with 1-chloropropane, a relatively unstable propyl carbocation forms initially. But this intermediate rearranges by shifting a hydrogen atom along with the bonding electron pair, $H:^-$, from the C-2 to the C-1 atom.

$$CH_3{-}\overset{H}{\overset{|}{C}}H{-}CH_2^+ \longrightarrow CH_3{-}\overset{+}{C}H{-}\overset{H}{\overset{|}{C}}H_2$$

Acylium ions produced in the Friedel–Crafts reaction do not rearrange. The acyl group in the product can be reduced by a zinc–mercury amalgam and HCl (a reaction called a **Clemmensen reduction**) to produce an alkylbenzene. By this means, the rearrangement of primary alkyl groups that occurs in the Friedel–Crafts alkylation reaction is circumvented. For example, propyl-benzene can be synthesized by acylation of benzene with propanoyl chloride followed by a Clemmensen reduction.

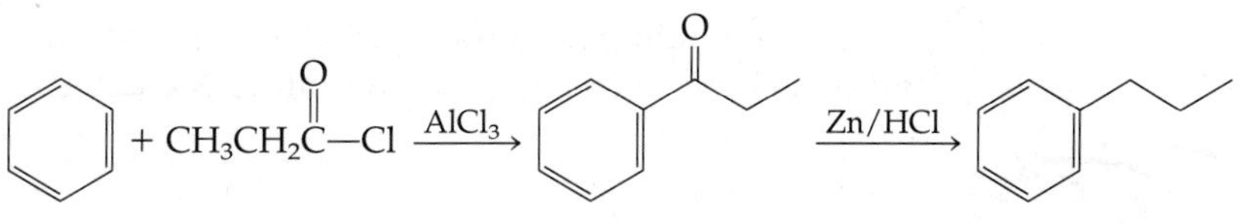

$$\text{C}_6\text{H}_6 + CH_3CH_2\overset{\overset{O}{\|}}{C}-Cl \xrightarrow{AlCl_3} C_6H_5COCH_2CH_3 \xrightarrow{Zn/HCl} C_6H_5CH_2CH_2CH_3$$

5.6 STRUCTURAL EFFECTS IN ELECTROPHILIC AROMATIC SUBSTITUTION

To this point, we have discussed only electrophilic substitution reactions of benzene itself. Now let's examine the effect that a substituent already bonded to the aromatic ring has on the introduction of a second substituent. For a substitution reaction on benzene, only one product is possible. But if a second substituent is placed onto a substituted benzene, any of three possible products—the ortho, meta, and para isomers—can be produced. We would like to know how the presence of the original substituent affects (a) the rate at which these products form and (b) the distribution of the products.

Effects on Reaction Rate

To examine the effect of a substituent on the rate of electrophilic aromatic substitution, let us consider the rates of nitration of benzene and several substituted benzenes. The difference in rate between substituting a nitro group onto phenol and substituting a nitro group onto nitrobenzene is a phenomenal 10^{10}. (For comparison, this rate difference is like the difference between the speed of light and the speed of walking!)

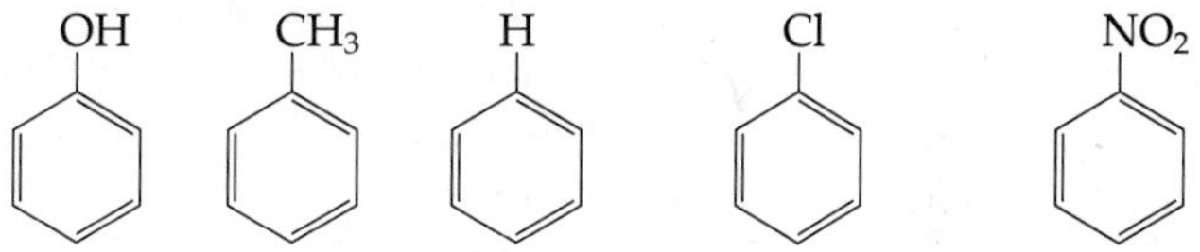

	OH	CH_3	H	Cl	NO_2
	phenol	toluene	benzene	chlorobenzene	nitrobenzene
Relative rate of nitration	10^3	25	1	3×10^{-2}	1×10^{-7}

Two of the above compounds undergo nitration faster than benzene: phenol has a hydroxyl group as a substituent, and toluene has a methyl group as a substituent. Thus, either a hydroxyl group or a methyl group makes the aromatic ring more reactive compared to benzene. These groups are called **activating groups** because they make the aromatic ring more reactive. On the other hand, two of the compounds shown—chlorobenzene and nitrobenzene—react more slowly than benzene. Thus, the chloro and nitro groups are called **deactivating groups** because they make the aromatic ring less reactive. Table 5.1 lists some common substituents and divides them into activating and deactivating groups toward electrophilic aromatic substitution.

TABLE 5.1 Properties of Substituents on an Aromatic Ring

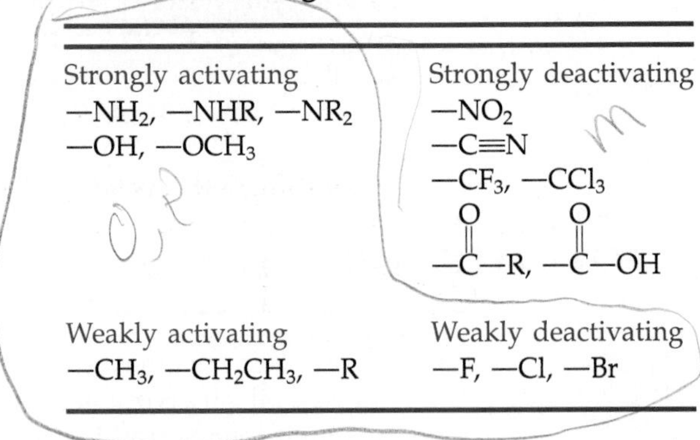

Strongly activating	Strongly deactivating
$-NH_2$, $-NHR$, $-NR_2$	$-NO_2$
$-OH$, $-OCH_3$	$-C{\equiv}N$
	$-CF_3$, $-CCl_3$
	$-C(=O)-R$, $-C(=O)-OH$
Weakly activating	**Weakly deactivating**
$-CH_3$, $-CH_2CH_3$, $-R$	$-F$, $-Cl$, $-Br$

EXAMPLE 5.5 Arrange the following compounds in order of increasing rate of reaction with bromine and $FeBr_3$.

$C_6H_5-CH_2-CH_3$ ethylbenzene; $C_6H_5-C(=O)-O-CH_3$ methyl benzoate; $C_6H_5-O-CH_2-CH_3$ ethoxybenzene

Solution Ethylbenzene contains an alkyl substituent and it is slightly more reactive than benzene. Methyl benzoate has a carbonyl carbon atom bonded to the aromatic ring. As a result, its rate of bromination will be significantly slower than that of benzene. Ethoxybenzene is an ether—it structurally resembles anisole. Thus the order of increasing rate of reaction in an electrophilic aromatic substitution reaction such as bromination is methyl benzoate < ethylbenzene < ethoxybenzene.

Orientation Effect of Substituents

Now let's consider the distribution of products formed in the nitration of toluene. The nitro group which enters the ring can bond at three nonequivalent sites to give *ortho-*, *meta-*, or *para-*nitrotoluene. When we examine the product distribution, we find that the ortho and para isomers predominate and that very little of the meta isomer forms. The methyl group is said to be

$$C_6H_5CH_3 \xrightarrow{HNO_3} o\text{-}CH_3C_6H_4NO_2 + m\text{-}CH_3C_6H_4NO_2 + p\text{-}CH_3C_6H_4NO_2$$

minor product (meta isomer)

an **ortho,para director.** That is, the methyl group directs or orients the incoming substituent into positions ortho and para to itself. All activating groups are ortho,para directors. Halogens, which are weakly deactivating, are also ortho,para directors.

A second class of ring substituents directs incoming substituents into the meta position. These groups, known as **meta directors,** include nitro, trifluoromethyl, cyano, sulfonic acid and all carbonyl-containing groups. For a nitration reaction, the trifluoromethyl group orients the incoming nitro group to a position meta to itself. Very small amounts of the ortho and para isomers are formed. All deactivating groups (except halogens) are meta-directing groups.

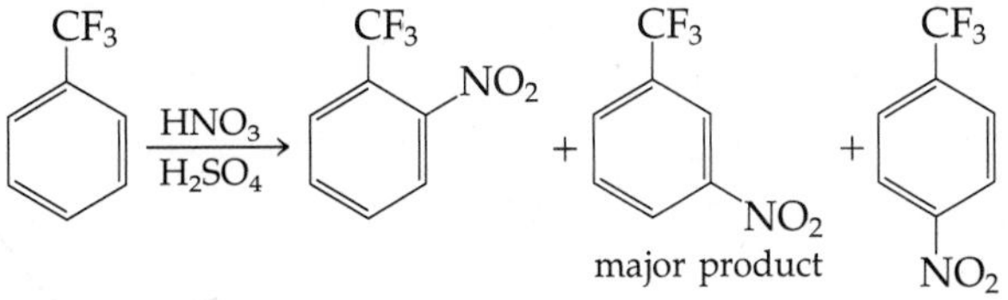

CF_3 $\xrightarrow[H_2SO_4]{HNO_3}$ CF_3 NO_2 + CF_3 NO_2 + CF_3 NO_2

major product

EXAMPLE 5.6 Predict the structure of the product(s) formed in the bromination of each of the following compounds.

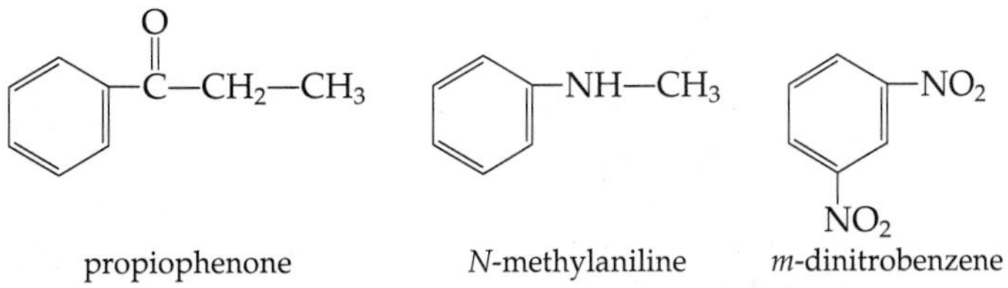

O ‖ C—CH_2—CH_3 —NH—CH_3 —NO_2 NO_2

propiophenone   *N*-methylaniline   *m*-dinitrobenzene

Solution Propiophenone has a carbonyl group bonded to the benzene ring. Therefore, the substituent should direct the bromine to the meta position. *N*-Methylaniline resembles aniline, and it should direct the bromine to the ortho or para position. Two isomeric compounds should result. The third compound has two nitro groups bonded to the ring. Each nitro group directs an electrophile onto the ring in a position meta to itself. Thus both groups direct the bromine into the same position. The product is 3,5-dinitrobromobenzene.

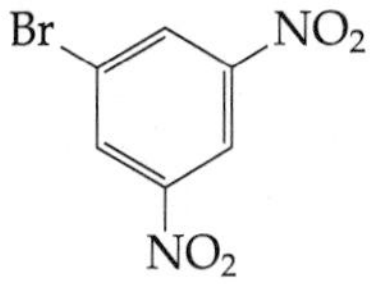

Br NO_2 NO_2

5.7 INTERPRETATION OF RATE EFFECTS

In the preceding section, we saw that a substituent influences both the rate and distribution of products in electrophilic aromatic substitution reactions. These two properties are directly related and can be understood with one model based on the ability of the substituents to either donate or withdraw electron density from the aromatic ring. Let us consider the effect of a group, G, on the electron density of the benzene ring.

If G is an electron donor, the ring gains electrons and becomes more reactive.

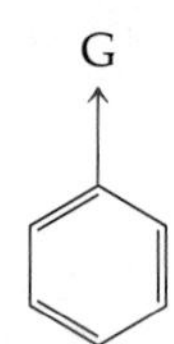

If G is an electron acceptor, the ring loses electrons and becomes less reactive.

Any substituent that donates electrons to the aromatic ring will activate the ring toward attack by an electrophile. Electron-withdrawing substituents decrease the electron density in the ring and make it less reactive toward an attacking electrophile. Therefore, all activating groups listed in Table 5.1 are electron-donating groups; deactivating groups are electron-withdrawing groups. Substituents can donate or withdraw electron density by inductive or resonance effects.

Inductive Effects of Substituents

Inductive effects are perhaps more easily visualized because they are related to the concept of electronegativity. As discussed in Chapter 4, alkyl groups tend to stabilize double bonds; they also stabilize carbocations. These alkyl groups transfer electron density through their σ bonds to electron-deficient carbon atoms or groups. That is, they donate electron density to the benzene ring by an inductive effect.

A methyl group is inductively electron-donating.

Electronegative groups, such as the halogens, have an attraction for electrons, and they withdraw electron density from a benzene ring. Also, any functional group that has a partial positive charge on the atom bonded to the aromatic ring withdraws electron density from the ring. Examples include the trifluoromethyl group, whose fluorine atoms pull electrons away from the carbon atom to which they are bound. To compensate, the carbon atom bearing the fluorine atoms withdraws electron density from the benzene ring.

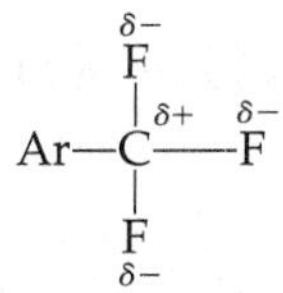

$$Ar{-}\overset{\overset{\delta-}{F}}{\underset{\underset{\delta-}{F}}{\overset{\delta+}{C}}}{-}\overset{\delta-}{F}$$

The trifluoromethyl group is inductively electron-withdrawing.

Other common electron-withdrawing substituents are the nitro, cyano, and any carbonyl-containing groups. The atom directly bonded to the ring in these substituents either has a formal positive charge or a partial positive charge.

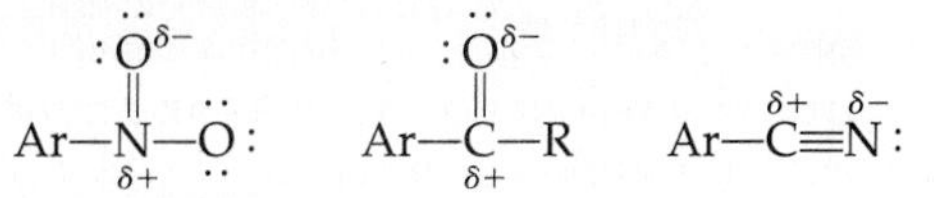

$$Ar{-}\underset{\delta+}{\overset{\overset{:\ddot{O}^{\delta-}}{\|}}{N}}{-}\ddot{\underset{..}{O}}: \qquad Ar{-}\underset{\delta+}{\overset{\overset{:\ddot{O}^{\delta-}}{\|}}{C}}{-}R \qquad Ar{-}\overset{\delta+}{C}{\equiv}\overset{\delta-}{N}:$$

Resonance Effects of Substituents

Next, let's consider how electron density is shifted into or out of a benzene ring by resonance effects. Resonance effects are depicted by moving electrons and drawing alternate resonance forms. Nitro and carbonyl-containing groups have sp^2-hybridized atoms bonded directly to the benzene ring. Thus, these groups are conjugated with the ring. First, consider the resonance effects of the nitro group. Oxygen is more electronegative than nitrogen. Hence, the electron pair in a nitrogen–oxygen double bond can "shift" onto the oxygen atom; simultaneously, an electron pair can "shift" out of the ring to make a carbon–nitrogen double bond and leave a positive charge on the aromatic ring. Because a positive charge develops in the ring, the ring is less reactive toward electrophiles. A similar effect is illustrated below for the acyl group, which also makes the ring less reactive toward electrophiles.

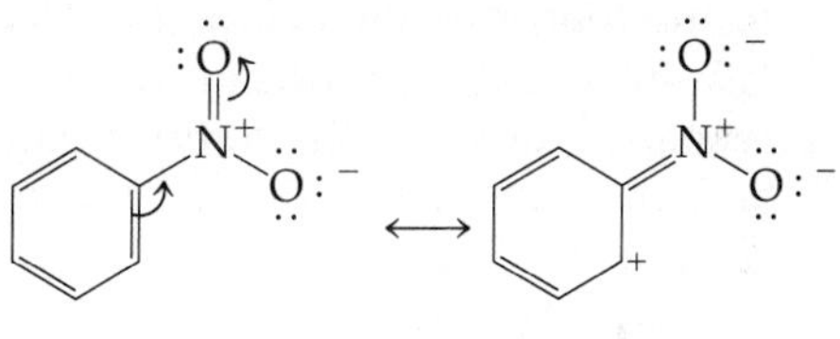

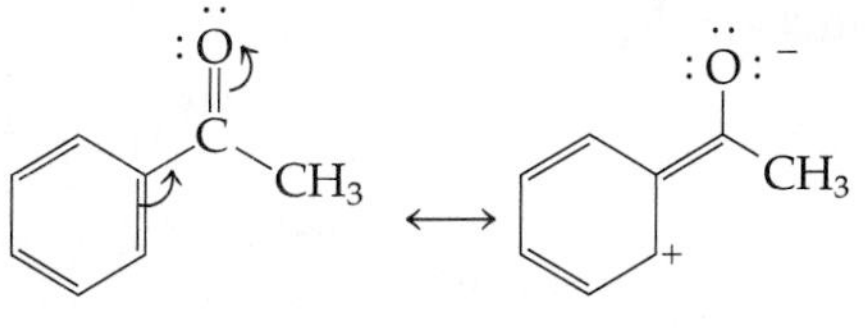

Now let's consider ring substituents that have atoms with lone-pair electrons. These electrons can be donated to the ring by resonance. As a consequence, the ring develops a negative charge and is more reactive toward electrophiles. Groups that have an unshared electron pair on the atom attached to the ring include the hydroxyl (—OH), any alkoxyl group (—OR), and the amino (—NH_2) or any substituted amino group (—NHR, —NR_2). These groups all donate electrons to the aromatic ring by resonance.

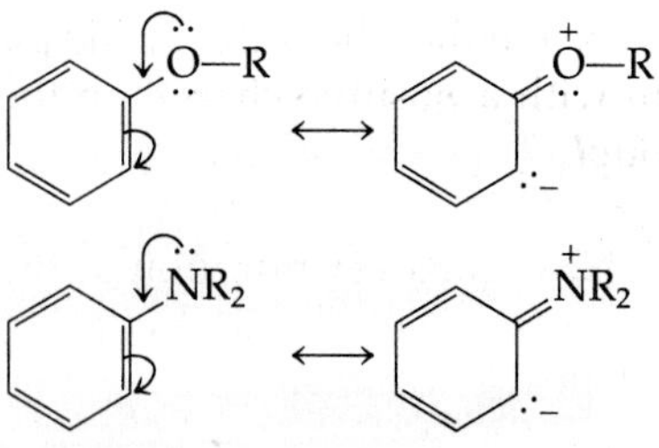

Groups that donate electrons by resonance are often electronegative. Therefore, they also withdraw electrons from the ring by an inductive effect. Thus, these substituents inductively take electron density from the ring while giving electron density back by resonance. A group that donates electrons by resonance, such as an amino or hydroxyl group, interacts with the ring through its 2*p* orbital, which very effectively overlaps with the 2*p* orbital of a ring carbon. Thus, donation of electrons by resonance is very effective and is more important than inductive electron withdrawal. This situation, however, does not hold true for chlorine or bromine. These electronegative atoms pull electron density out of the aromatic ring by an inductive effect. However, neither the 3*p* orbital of chlorine nor the 4*p* orbital of bromine overlaps effectively with the 2*p* orbital of carbon, and electron donation by resonance is less important.

5.8 INTERPRETATION OF DIRECTING EFFECTS

We noted earlier that, with the exception of the halogens, ortho,para directors activate the ring toward electrophilic substitution because they supply electron density to the ring. But why are the ortho and para positions especially susceptible to attack? To answer this question, we will examine the resonance forms of the intermediate carbocations resulting from attack at the ortho and para positions and compare them with those resulting when an electrophile attacks at the meta position.

First, we will consider the nitration of toluene at the ortho and para positions. The resulting resonance forms are shown below.

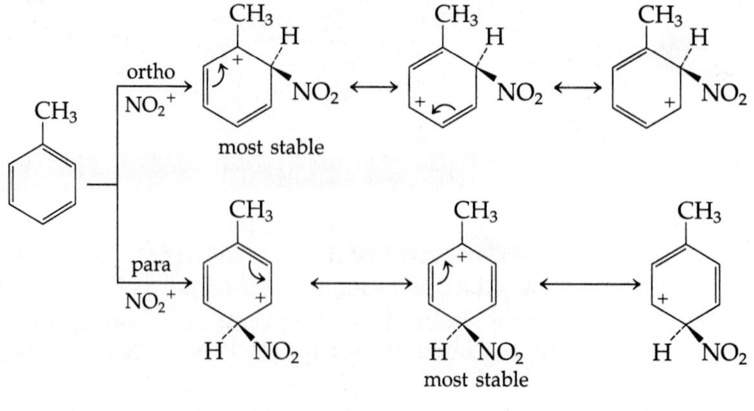

most stable

most stable

Attack at either the ortho or the para position results in one resonance structure with a positive charge on the ring carbon atom bonded to the methyl group. This form makes a major contribution to the stability of the resonance hybrid because it is a tertiary carbocation.

Now consider nitration at the meta position. The resonance structures for the carbocation intermediate are shown below.

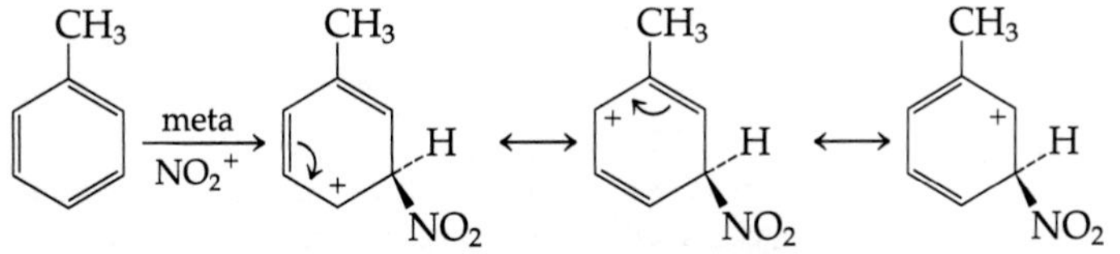

CH_3 $\xrightarrow[NO_2^+]{meta}$ CH_3 H NO_2 ⟷ CH_3 H NO_2 ⟷ CH_3 H NO_2

The structures show that positive charge cannot be placed on the carbon atom attached to the methyl group. Thus, only secondary carbocations are possible; they are less stable than the tertiary carbocation. As a consequence, attacks at the ortho and para positions are favored over an attack at the meta position.

Next, let's consider the ortho,para-directing effect of a hydroxyl group or any other group that can donate an unshared pair of electrons by resonance. An attack either ortho or para to the hydroxyl group leads to an intermediate that is resonance-stabilized by the substituent. As in the case of the methyl group, a contributing structure exists in which the positive charge is located on the carbon atom bonded to the substituent. This positive charge is stabilized by an electron pair provided by oxygen when the electrophile is located ortho or para to it. No such stabilization is possible for a group that attacks meta to the hydroxyl substituent. Hence, ortho,para substitution is preferred over meta.

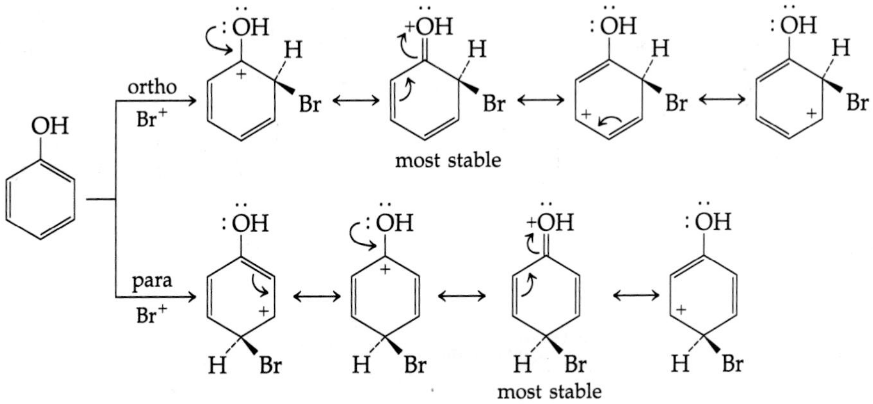

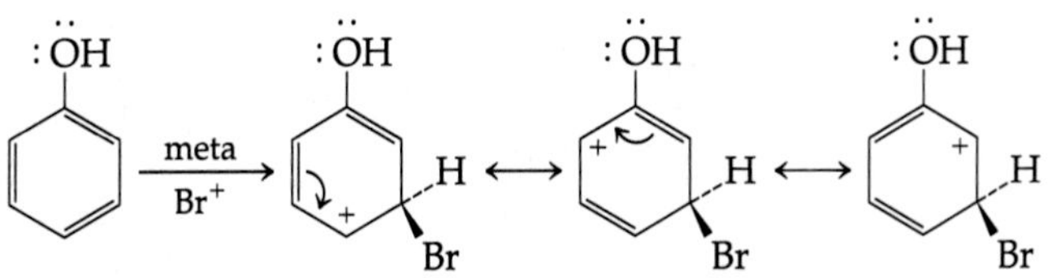

OH

ortho, Br^+: :OH H Br ⟷ +OH H Br (most stable) ⟷ :OH H Br ⟷ :OH H Br

para, Br^+: :OH H Br ⟷ :OH H Br ⟷ +OH H Br (most stable) ⟷ :OH H Br

:OH $\xrightarrow[Br^+]{meta}$:OH H Br ⟷ :OH H Br ⟷ :OH H Br

We saw in Table 5.1 that some substituents strongly deactivate the ring toward electrophilic aromatic substitution. All of the strong deactivating groups withdraw electrons from the ring and are meta directors. Why is there a preference for attack at the meta position? First, let's consider the possible nitration of nitrobenzene at the ortho and para positions. The resonance forms of the intermediate carbocations are shown below.

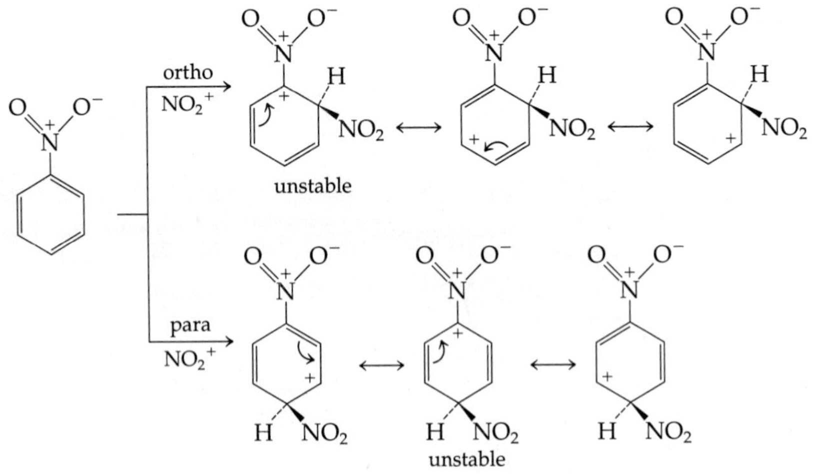

In one of the resonance forms for both ortho- and para-substituted intermediates, a positive charge is located at the carbon atom containing the original nitro group. The nitrogen atom of the nitro group has a formal positive charge. The presence of a positive charge on the carbon atom bearing the positively charged nitrogen atom is not favorable. Thus, these are unstable resonance forms.

Next, consider an attack at the meta position. The contributing resonance structures of the carbocation intermediate are shown below.

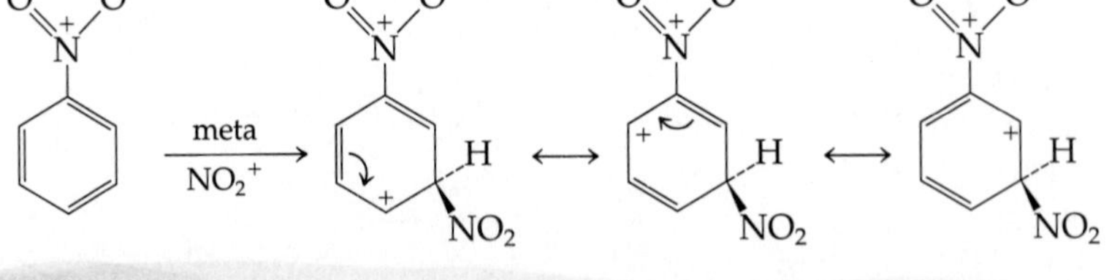

None of the resonance forms of the intermediate has a positive charge on the carbon atom bonded to a nitro group—whose nitrogen atom, we noted above, has a formal charge of +1. Thus, the resonance forms are more stable overall than the resonance forms of the intermediates formed from ortho,para substitution. As a consequence, meta substitution is favored over ortho,para substitution.

Biological Effects of Benzene and Its Halogenated Derivatives

We have seen that benzene is remarkably unreactive even under very strong reaction conditions. We might therefore expect benzene to be inert in cells at pH 7 and 37 °C, and it is. Benzene itself is not metabolized in most human cells, rather it accumulates in the liver, where it does great harm. Benzene is carcinogenic and extremely toxic. Oxidation of benzene and aromatic compounds by cytochrome P-450 often yields phenols. Although the process shown below appears to be aromatic hydroxylation, the reaction actually occurs via a three-membered heterocyclic ring called an epoxide. (The chemistry of epoxides will be discussed in Chapter 8.) The epoxide intermediates, called **arene oxides,** rearrange to phenols.

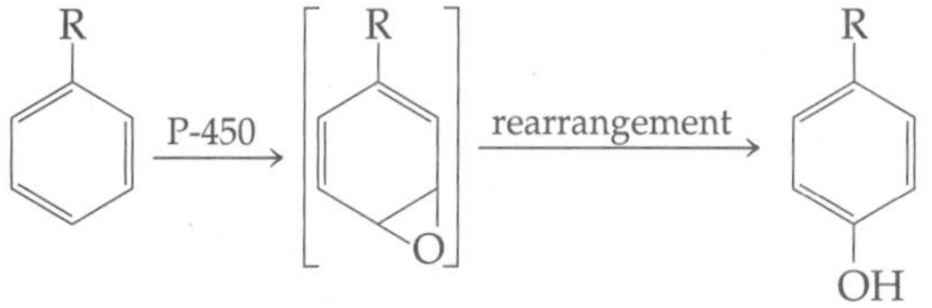

arene —P-450→ [arene oxide] —rearrangement→ a phenol

All arene oxide intermediates are very reactive and undergo several types of reactions besides the rearrangement reaction to form phenols. Arene oxides react with proteins, RNA, and DNA. As a consequence, serious cellular disruptions can occur. These processes will be discussed in Chapter 9, when the chemistry of epoxides is presented.

Some drugs contain aromatic rings that are hydroxylated when metabolized. Hydroxylation most commonly occurs at the para position. Phenytoin, an anticonvulsant, is an example. The phenolic compounds react further to form water-soluble derivatives.

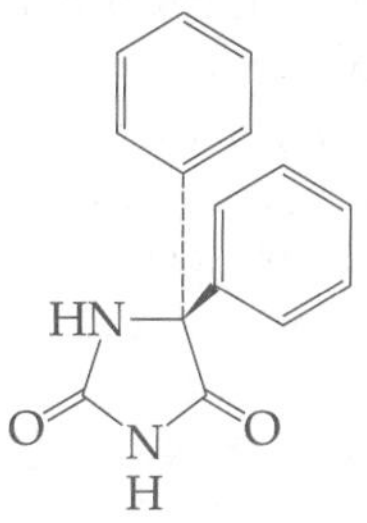

phenytoin

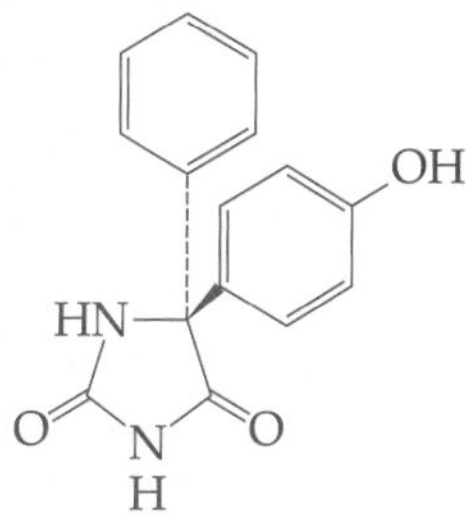

p-hydroxyphenytoin

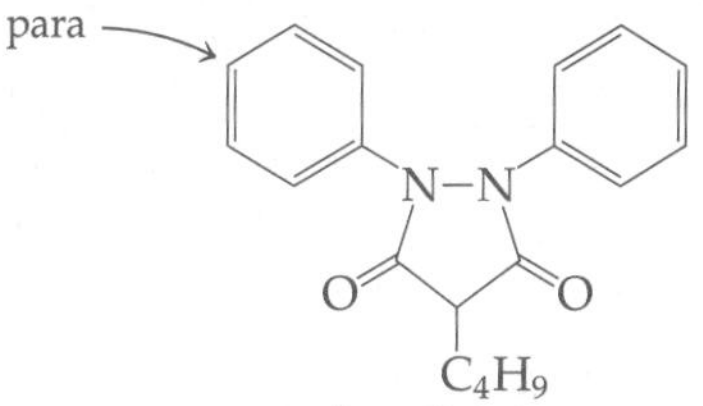

phenylbutazone

When some drugs are hydroxylated in the liver, they are also pharmacologically active. The site of hydroxylation of phenylbutazone, an anti-inflammatory agent, is at the indicated para position. This hydroxylated product has been produced in the laboratory and is now marketed under the trade names Tandearil and Oxalid.

We described the deactivating effect of halogens on the reactivity of aromatic rings. The environmental pollutants dioxin (2,3,7,8-tetrachlorodibenzo-*p*-dioxin) and PCBs (polychlorinated biphenyls) have many deactivating groups. This effect has grave environmental consequences. The substituents on the aromatic rings affect the ease of hydroxylation; electron-withdrawing groups deactivate the ring toward the initial epoxidation. Because the aromatic rings lose electrons in the oxidation process, the electron-withdrawing groups slow the rate of biological oxidation. These halogenated compounds are nonpolar and quite soluble in fatty tissue (they are sometimes said to be "lipophilic"). Thus, they tend to persist in the bodies of organisms that inadvertently ingest them.

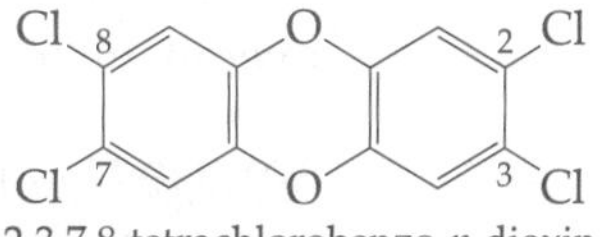

2,3,7,8-tetrachlorobenzo-*p*-dioxin

Table 5.1 shows that halogens are weakly deactivating. Yet, these deactivating groups are ortho,para directors. Why is that? The answer lies in their electronegativity. Because the halogens are more electronegative than a benzene ring, they withdraw electron density from the ring by an inductive effect. But because halogens have lone-pair electrons, they can donate electrons to a carbocation intermediate. However, this resonance effect only comes into play if the entering electrophile attacks ortho or para to the halogen atom. Thus, although weakly deactivating, the halogens are ortho,para directors.

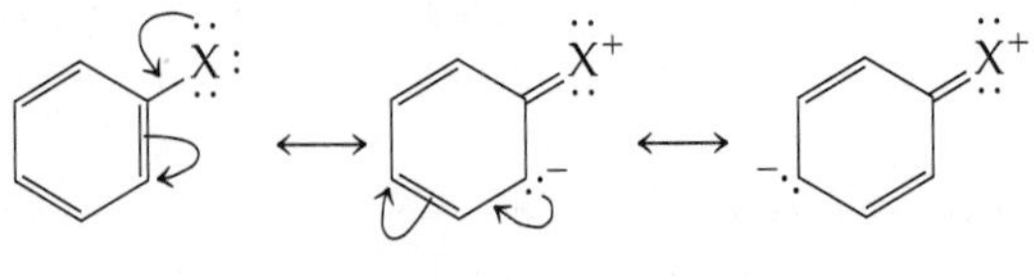

As we noted earlier, the 3*p* orbital of chlorine and the 4*p* orbital of bromine do not overlap effectively with the 2*p* orbital of carbon, so electrons cannot be donated efficiently to the ring.

5.9 REACTIONS OF SIDE CHAINS

A group of carbon atoms bonded to an aromatic ring constitutes a **side chain.** Side-chain carbon atoms, which are separated from the aromatic ring by two or more σ bonds, behave independently of the aromatic ring. However, carbon atoms directly bonded to the aromatic ring are influenced by the ring. For example, any reaction that forms a carbocation at the carbon atom bonded to the aromatic ring is especially favored. A carbocation in which the positively charged carbon atom is directly attached to a benzene ring is a **benzylic carbocation.** We recall that the formula of a benzyl group is $C_6H_5CH_2$—, so a benzyl carbocation has the formula $C_6H_5CH_2^+$. A benzyl carbocation is resonance-stabilized in the same manner as the allylic carbocation we discussed in the previous chapter. That is, the positive charge at the benzylic carbon atom can be delocalized among the carbon atoms of the benzene ring.

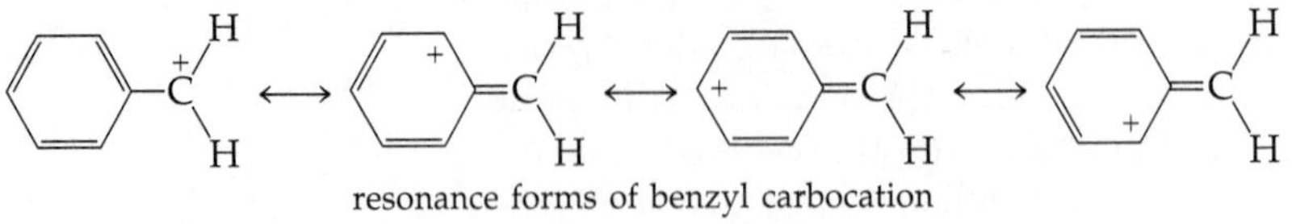

resonance forms of benzyl carbocation

Let's consider the addition reaction of HBr to indene, which has a double bond conjugated with the benzene ring. Electrophilic attack of H^+ at the carbon atom two atoms away from the ring gives a stable secondary benzyl carbocation. Attack at the carbon atom directly attached to the ring would produce a much less stable secondary carbocation.

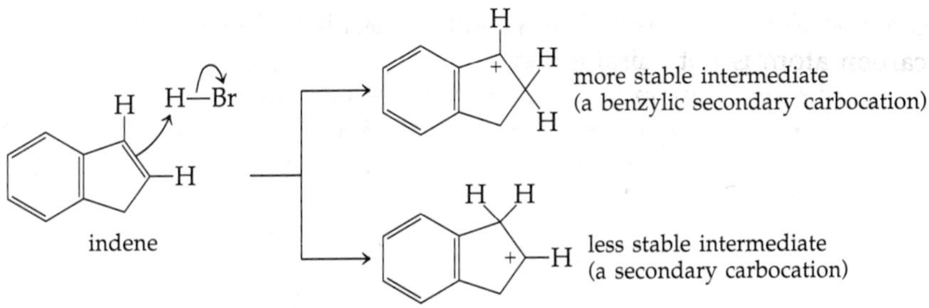

H H—Br

H

indene

H

H

more stable intermediate
(a benzylic secondary carbocation)

H

H H

H

less stable intermediate
(a secondary carbocation)

The only product in the reaction is derived from the more stable carbocation, which then reacts with the bromide ion.

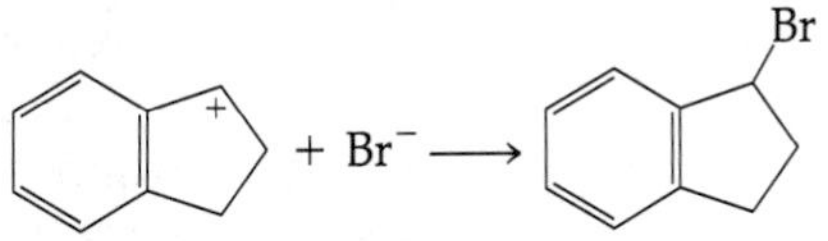

$$+ \ Br^- \longrightarrow$$

Br

Although the aromatic ring causes special reactivity on the side chain, the ring itself is quite unreactive toward many reagents and remains intact. An illustration of the special stability of the aromatic ring is provided by oxidation of the side-chain alkyl groups of alkyl benzenes. We recall that potassium permanganate (Section 4.7) reacts with the π bonds of an alkene under conditions where the σ bonds of the saturated part of the molecule do not react. The benzene ring, in spite of being "unsaturated", also is not oxidized by potassium permanganate. However, under vigorous conditions, the alkyl side chain is totally oxidized to produce a carboxylic acid—that is, benzoic acid—at the site of the alkyl group. The benzene ring itself remains unscathed!

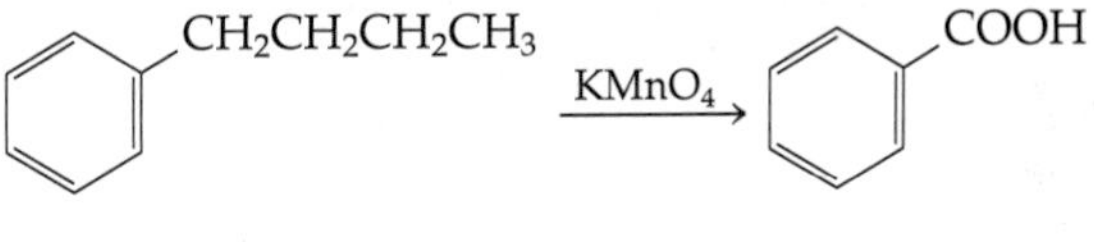

$CH_2CH_2CH_2CH_3$ $\xrightarrow{KMnO_4}$ COOH

EXAMPLE 5.7 Predict the product of the reaction of the following compound with potassium permanganate.

CH_3

Br— —CH—CH_2—CH_3

NO_2

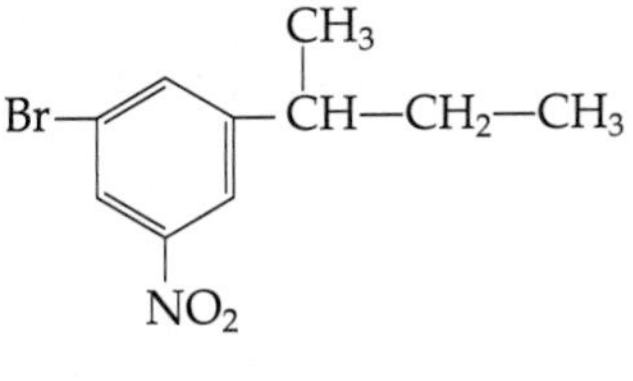

Solution Potassium permanganate oxidizes the side chain of substituted aromatic compounds completely and forms a carboxylic acid. The *sec*-butyl

Metabolic Oxidation of Aromatic Side Chains

In our discussion of the biological effects of benzene and its halogenated derivatives we noted that benzene is quite toxic. At one time benzene was widely used in the chemical industry and in many commercial products as a solvent. However, it is no longer used for these purposes; it has been supplanted by toluene because the side-chain methyl group can be metabolized to produce a nontoxic byproduct. The oxidation of side chains in alkylbenzenes occurs in the liver and is catalyzed by cytochrome P-450.

Oxidation of methyl groups bonded to aromatic rings occurs to give a primary alcohol called a benzylic alcohol, Ar—CH_2OH. Benzyl alcohols are subsequently eliminated from the body by reaction via the oxygen atom with glucuronic acid, a compound produced by the oxidation of glucose (Chapter 11).

The resultant product—a glucuronide—is very polar and hence it tends to be water-soluble. The details of how this reaction occurs will be discussed in Chapter 10 when the chemistry of hemiacetals and hemiketals is discussed.

A benzyl alcohol derived from the oxidation of methyl-substituted aromatic compounds can also be oxidized further in metabolic reactions to produce aromatic aldehydes, Ar—CHO, and finally to carboxylic acids, Ar—CO_2H. These acids are more water-soluble than the original methyl-substituted compound and can be excreted. The oral hypoglycemic drug tolbutamide (Orinase) is oxidized in several steps to the corresponding carboxylic acid. The metabolite is excreted as the conjugate base of the carboxylic acid at urinary pH.

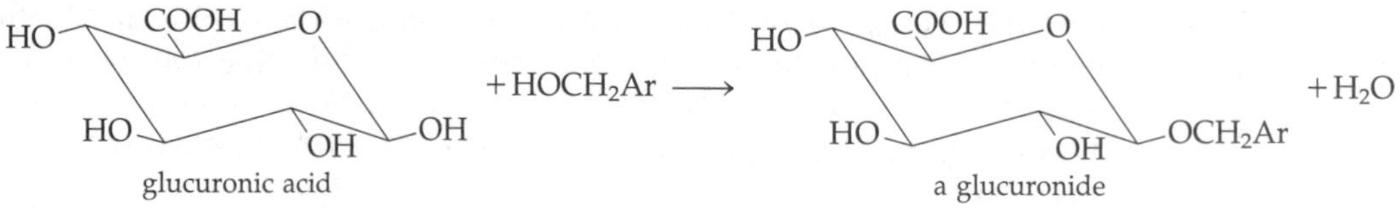

glucuronic acid + $HOCH_2Ar$ ⟶ a glucuronide + H_2O

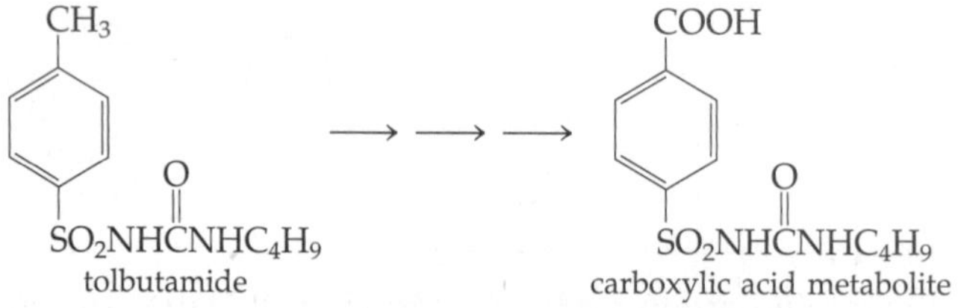

tolbutamide ⟶ ⟶ ⟶ carboxylic acid metabolite

group is oxidized, and a —CO_2H group results. The product of the reaction is 3-bromo-5-nitrobenzoic acid.

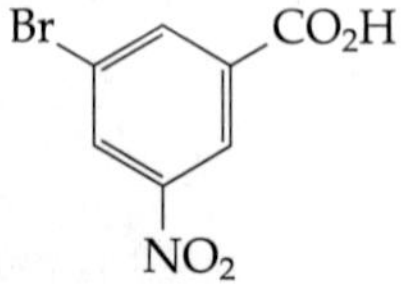

5.10 FUNCTIONAL GROUP MODIFICATION

The oxidation of alkyl side chains is one example of a reaction that modifies a substituent bonded to an aromatic ring. Functional group modifications are important because only a few functional groups can be placed directly on an aromatic ring by electrophilic aromatic substitution. The remaining groups are obtained by modifying a group already bonded to the aromatic ring. When one functional group is changed to another, its ortho,para- or meta-directing properties can also change. For example, we just saw that a methyl group can be changed to a carboxylic acid group. As a result, an ortho,para-directing methyl group is changed into a meta-directing carboxylic acid group, $—CO_2H$.

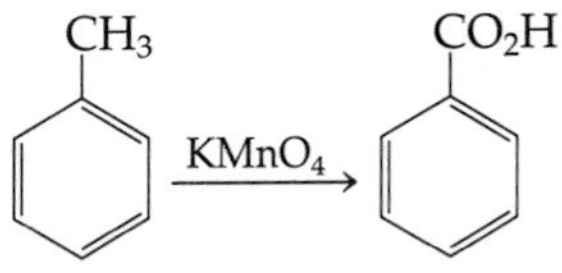

CH_3 $\xrightarrow{KMnO_4}$ CO_2H

We have also seen that an acyl group bonded to a benzene ring can be converted into an alkyl group by reduction with a zinc–mercury amalgam and HCl (Section 5.5). An acyl group has a carbonyl carbon atom directly attached to the ring. It is a deactivating, meta-directing substituent. However, an alkyl group is an activating, ortho,para-directing group.

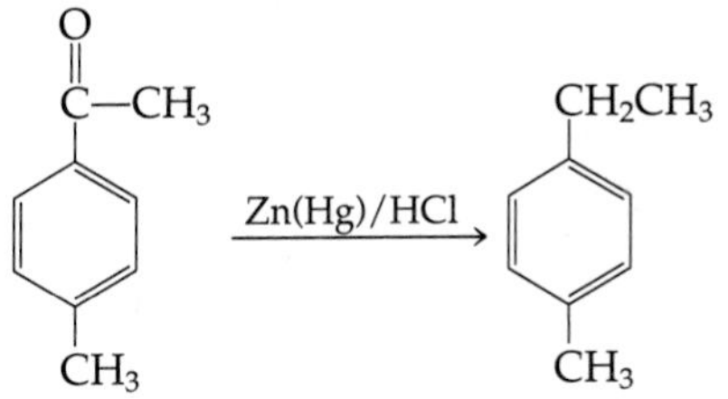

O
‖
C—CH_3 $\xrightarrow{Zn(Hg)/HCl}$ CH_2CH_3

CH_3 CH_3

A nitro group can be attached to a benzene ring by electrophilic aromatic substitution, but an amino group cannot be attached to an aromatic ring in one step. However, after a nitro group is introduced, it can easily be reduced to an amino group; the product is an aniline. This reaction transforms a strongly deactivating, meta-directing nitro group into a strongly activating, ortho,para-directing amino group.

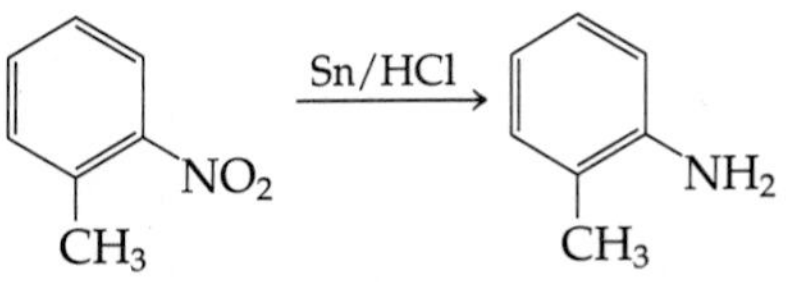

$\xrightarrow{Sn/HCl}$

NO_2 NH_2

CH_3 CH_3

Once an amino group has been introduced onto an aromatic ring, the possibilities for further functional group modifications are vastly increased. The amino groups of anilines can be converted into many other groups. The door to other functional groups is opened by converting the amino group into a diazonium ion, $Ar—N_2^+$. A diazonium ion is made by treating aniline with nitrous acid, HNO_2, prepared by reaction of sodium nitrite with sulfuric acid. This step, which produces a diazonium ion, is called **diazotization.**

$$Ar—NH_2 \xrightarrow{HNO_2} \underset{\text{a diazonium ion}}{Ar—N{\equiv}N^+}$$

Aromatic diazonium ions are extremely reactive. They react with electron-pair donors—that is, nucleophiles—that replace the diazonium group. Nitrogen gas is liberated in the process.

$$Ar—N{\equiv}N^+ + :Nu^- \longrightarrow Ar—Nu + N_2$$

In 1884, the German chemist T. Sandmeyer found that diazonium ions react with nucleophiles supplied in the form of a Cu(I) salt. Thus, a solution of an aromatic diazonium ion can be treated with Cu_2Cl_2 or Cu_2Br_2 to yield chlorobenzene or bromobenzene, respectively. These reactions are known collectively as the **Sandmeyer reaction.**

$$Ar—N{\equiv}N^+ \xrightarrow{Cu_2Br_2} Ar—Br + N_2$$

$$\text{2-methylaniline } (NH_2, CH_3) \xrightarrow{HNO_2} (N_2^+, CH_3) \xrightarrow{Cu_2Br_2} (Br, CH_3)$$

Cuprous salts of the cyanide ion result in the formation of aryl nitriles. (The chemistry of nitriles will be discussed in Chapters 12 and 14.)

$$Ar—N{\equiv}N^+ \xrightarrow{Cu_2(CN)_2} Ar—C{\equiv}N + N_2$$

$$\text{4-methylaniline } (NH_2, CH_3) \xrightarrow{HNO_2} (N_2^+, CH_3) \xrightarrow{Cu_2(CN)_2} (CN, CH_3)$$

Phenols can be synthesized by reaction of the aryldiazonium compound with hot aqueous acid. This reaction is the best way to introduce an —OH group onto an aromatic ring.

$$Ar{-}N{\equiv}N^+ \xrightarrow{H_3O^+} Ar{-}OH + N_2$$

NH_2 / $C(CH_3)_3$ $\xrightarrow{HNO_2}$ N_2^+ / $C(CH_3)_3$ $\xrightarrow{H_3O^+}$ OH / $C(CH_3)_3$

A diazonium group can also be replaced by hydrogen. Hypophosphorous acid, H_3PO_2, is the reagent used to replace the diazonium group.

$$Ar{-}N{\equiv}N^+ \xrightarrow{H_3PO_2} Ar{-}H + N_2$$

NH_2, Br / CH_3 $\xrightarrow{HNO_2}$ N_2^+, Br / CH_3 $\xrightarrow{H_3PO_2}$ Br / CH_3

5.11 SYNTHESIS OF SUBSTITUTED AROMATIC COMPOUNDS

Chemists often want to design benzene derivatives with two or more substituents strategically placed around the ring. A project of this type begins with an analysis of the ortho,para- or meta-directing characteristics of the substituents. For example, consider the problem of synthesis of *m*-chloronitrobenzene. A nitro group is meta-directing; a chlorine atom is ortho,para-directing. The order in which we add these groups is clearly important. If chlorination precedes nitration, the entering nitro group will be largely directed to form *o*-chloronitrobenzene and *p*-chloronitrobenzene. Very little of the desired meta isomer will be formed.

$\xrightarrow[FeCl_3]{Cl_2}$ Cl $\xrightarrow[H_2SO_4]{HNO_3}$ Cl, NO_2 + Cl / NO_2

However, we can obtain the desired compound by introducing the nitro group first and the chlorine atom second. Because the nitro group is a meta director, the entering chlorine is directed to the desired meta position.

$$\text{C}_6\text{H}_6 \xrightarrow[H_2SO_4]{HNO_3} \text{C}_6\text{H}_5\text{NO}_2 \xrightarrow[FeCl_3]{Cl_2} m\text{-ClC}_6\text{H}_4\text{NO}_2$$

EXAMPLE 5.8 Devise a synthesis of *m*-bromoaniline starting from benzene.

Solution Bromine, which is an ortho,para director, can be introduced directly onto the benzene ring by reaction with bromine and $FeBr_3$. The amino group, $—NH_2$, of aniline is also an ortho,para director. It can be introduced indirectly by first nitrating benzene and then reducing the nitro compound. Recall that the nitro group is a meta director; the amino group is an ortho,para director.

Bromination of benzene followed by nitration gives a mixture of *ortho-* and *para*-bromonitrobenzene. The desired meta isomer is not formed. Nitration of benzene gives nitrobenzene—a compound that can direct subsequent electrophilic substitution reactions to produce the meta isomer. Thus, bromination of nitrobenzene followed by reduction of the product gives the desired *m*-bromoaniline.

$$\text{C}_6\text{H}_6 \xrightarrow[H_2SO_4]{HNO_3} \text{C}_6\text{H}_5\text{NO}_2 \xrightarrow[FeBr_3]{Br_2} m\text{-BrC}_6\text{H}_4\text{NO}_2 \xrightarrow{Sn/HCl} m\text{-BrC}_6\text{H}_4\text{NH}_2$$

Now, we consider a task that appears at first glance to be impossible; namely, the synthesis of *m*-dibromobenzene. Why "impossible"? Because the bromo groups are meta to each other, but bromine is an ortho,para director! Direct bromination of benzene would place on the ring one bromine atom that would then direct the second bromine atom into the ortho or para position.

$$\text{C}_6\text{H}_6 \xrightarrow[FeBr_3]{Br_2} \text{C}_6\text{H}_5\text{Br} \xrightarrow[FeBr_3]{Br_2} o\text{-C}_6\text{H}_4\text{Br}_2 + p\text{-C}_6\text{H}_4\text{Br}_2$$

However, we know that a nitro group is a meta director. So, we first make nitrobenzene, then brominate it to obtain *m*-bromonitrobenzene.

benzene $\xrightarrow[H_2SO_4]{HNO_3}$ nitrobenzene (NO_2) $\xrightarrow[FeBr_3]{Br_2}$ m-bromonitrobenzene (NO_2, Br)

In the preceding section we saw that a nitro group can be converted to a bromo group by (a) reducing the nitro group to an amino group, (b) converting the amino group to a diazonium group, and (c) treating the diazonium compound with copper(I) bromide. The procedure requires several steps, but it accomplishes the apparently "impossible" task of preparing *m*-dibromobenzene.

(NO_2, Br) $\xrightarrow{Sn/HCl}$ (NH_2, Br) $\xrightarrow{HNO_2}$ (N_2^+, Br) $\xrightarrow{Cu_2Br_2}$ (Br, Br)

EXERCISES

Aromaticity

5.1 Determine whether each of the following is an aromatic compound.

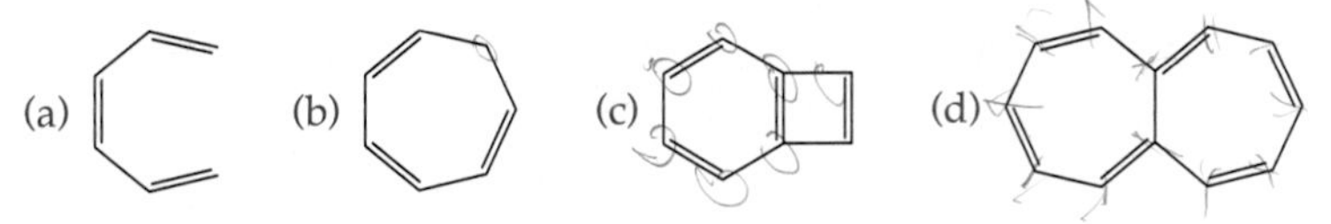

5.2 Determine whether each of the following is an aromatic compound.

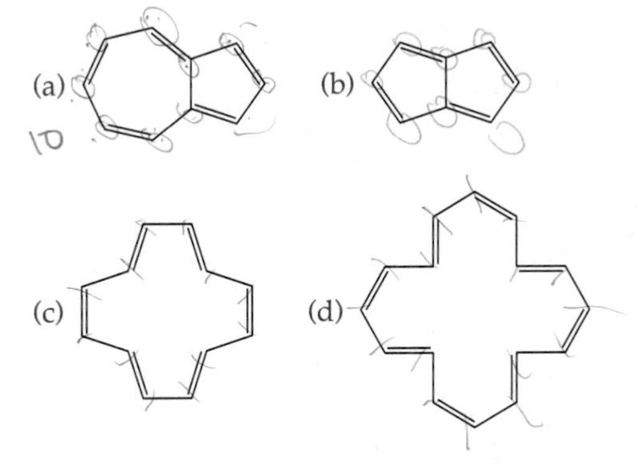

Polycyclic and Heterocyclic Aromatic Compounds

5.3 There are two isomeric bromonaphthalenes. Draw their structures.

5.4 There are three isomeric bromoanthracenes. Draw their structures.

5.5 There are three isomeric diazines $C_4N_2H_4$ that resemble benzene but have two nitrogen atoms in place of carbon atoms in the ring. Draw their structures. Which of the isomers should have no dipole moment?

5.6 There are three isomeric triazines, $C_3N_3H_3$, that resemble benzene but have three nitrogen atoms in place of carbon atoms in the ring. Draw their structures. Which of the isomers should have no dipole moment?

5.7 How many electrons does each heteroatom contribute to the π system in each of the following compounds?

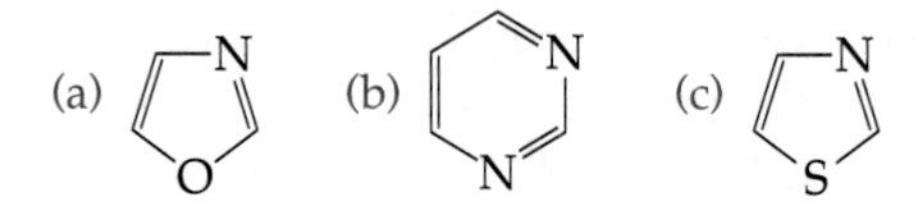

5.8 How many electrons does each heteroatom contribute to the π system in each of the following compounds?

(a) N, N–H (b) N, N, N, N–H (c) N=N

5.9 Identify the heterocyclic ring structure contained in each of the following compounds, which have been investigated as possible male contraceptives.

(a) O_2N–(furan, O)–CH=N—NHCNH$_2$ (with C=O)

(b) NO_2, NO_2, N–$CH_2CON(C_2H_5)_2$

(c) Cl–(thiophene, S)–C(=O)–CH_3

5.10 Identify the aromatic heterocyclic ring structure contained in each of the following compounds.

(a) tolmetin, a drug used to lower blood sugar levels

H_3C–(benzene)–C(=O)–(pyrrole, N–CH_3)–CH_2COOH

(b) cephalothin sodium, a broad spectrum antibacterial

(thiophene, S)–CH_2CONH—CH—CH(S)CH_2; CO—N; C=C–CH_2OCOCH_3; COO^-Na^+

(c) dantrolene, a muscle relaxant

(d) ethionamide, an antitubercular agent

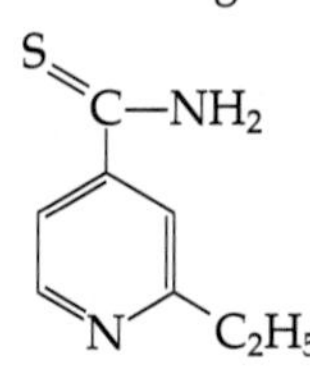

Isomeric Benzene Compounds

5.11 There are three isomeric dichlorobenzenes. One compound is nonpolar. Which one?

5.12 There are three isomeric trichlorobenzenes. One compound is nonpolar. Which one?

5.13 There are four isomeric aromatic compounds with the molecular formula C_8H_{10}. Name each compound.

5.14 There are three isomeric aromatic compounds with the molecular formula $C_6H_3Br_3$. Name each compound.

5.15 The boiling points of benzyl alcohol and anisole are 205 and 154 °C, respectively. Explain this difference.

5.16 The boiling points of *ortho-*, *meta-*, and *para*-chlorophenol are 175, 214, and 220 °C, respectively. Considering the phenomenon of hydrogen bonding, explain why the ortho isomer has a significantly lower boiling point that the other isomers.

Nomenclature of Aromatic Compounds

5.17 Identify each of the following as an ortho-, meta-, or para-substituted compound.

(a) methylparaben, a food preservative to protect against molds

(b) crotamiton, used in creams for topical treatment of scabies

(c) diethyltoluamide, an insect repellent

5.18 Identify each of the following as an ortho-, meta-, or para-substituted compound.

(a) resorcinol monoacetate, a germicide used to treat skin conditions

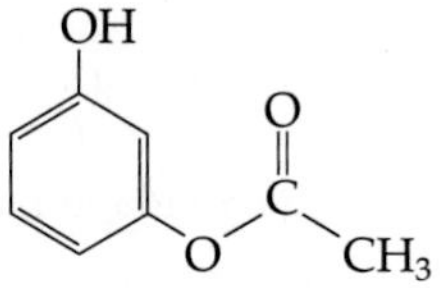

OH, O, C, O, CH_3

(b) halazone, used to disinfect water

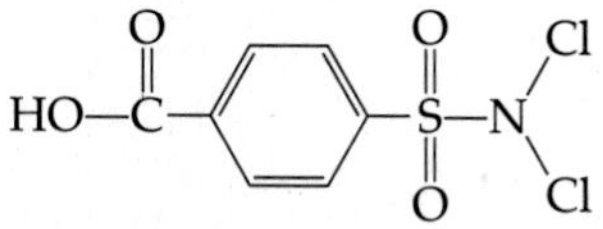

O, HO—C—, O, S—N, O, Cl, Cl

(c) salicylamide, an analgesic

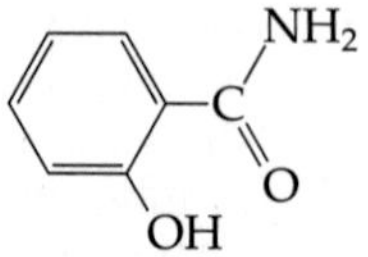

NH_2, C, O, OH

5.19 Name each of the following compounds.

(a) 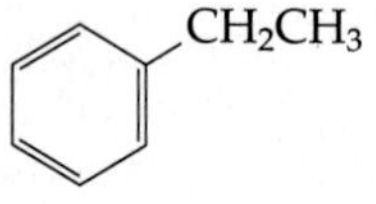 CH_2CH_3

(b) 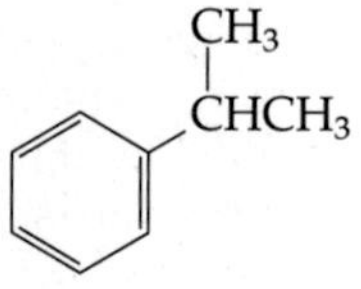 CH_3, $CHCH_3$

(c) 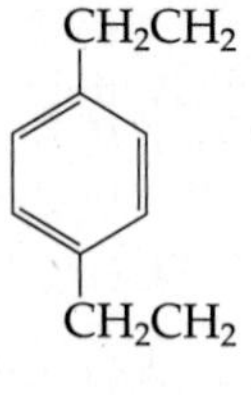 CH_2CH_2, CH_2CH_2

(d) 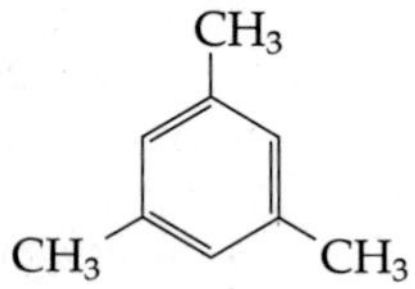 CH_3, CH_3, CH_3

5.20 Name each of the following compounds.

(a) 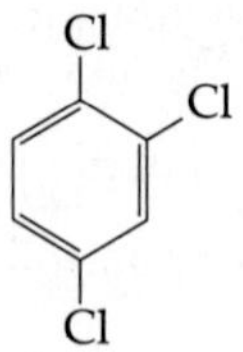 Cl, Cl, Cl

(b) 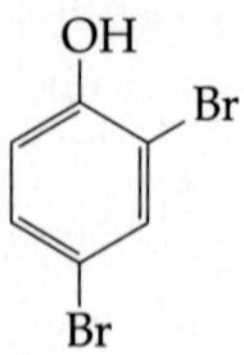 OH, Br, Br

(c) 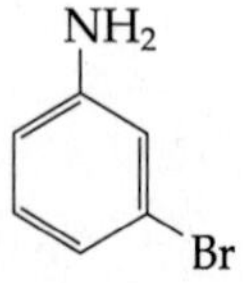 NH_2, Br

(d) 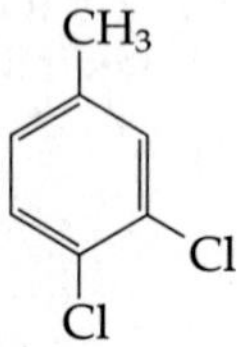 CH_3, Cl, Cl

5.21 Name each of the following compounds.
(a) an antiseptic agent used to treat athlete's foot and jock itch

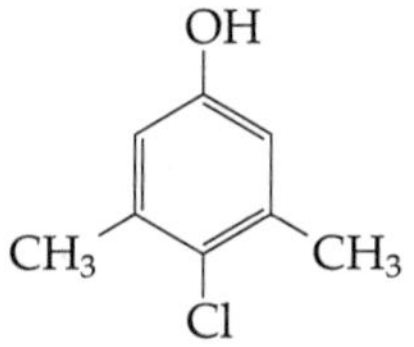

(b) a compound used to make a local anesthetic

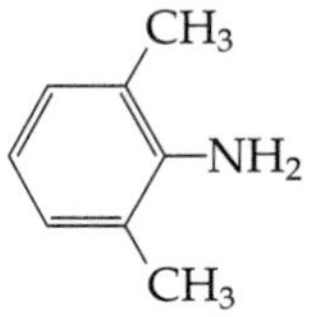

(c) a disinfectant

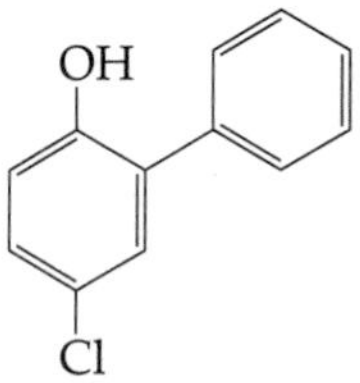

5.22 Draw the structure of each of the following compounds.
(a) 5-isopropyl-2-methylphenol, found in oil of marjoram
(b) 2-isopropyl-5-methylphenol, found in oil of thyme
(c) 2-hydroxybenzyl alcohol, found in the bark of the willow tree

5.23 Draw the structure of 3,4,6-trichloro-2-nitrophenol, a lampricide used to control sea lampreys in the Great Lakes.

5.24 *N,N*-Dipropyl-2,6-dinitro-4-trifluoromethylaniline is the IUPAC name for Treflan, a herbicide. Draw its structure. (The prefix *N* signifies the location of substituents on a nitrogen atom.)

Electrophiles and Ring Substituents

5.25 Some activated rings may be hydroxylated by reacting hydrogen peroxide (H_2O_2) with acid. What is the formula of the electrophile? How is it formed?

5.26 An aromatic ring can be alkylated with a tertiary butyl group by treating *t*-butyl alcohol, $(CH_3)_3COH$, with acid. What is the formula of the electrophile? How is it formed?

5.27 Consider the thiomethyl group, —S—CH_3. Predict whether it is an activating or deactivating group. Will it be an ortho,para-directing group or a meta-directing group?

5.28 The sulfonamide group is found in sulfa drugs. Consider its structure and determine if it is an activating or deactivating group. Will it be an ortho,para- or a meta-directing group?

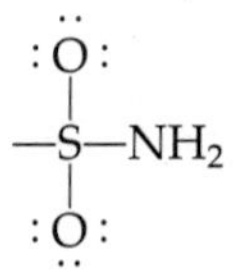

5.29 What product will result from the Friedel–Crafts alkylation of benzene using 2-methyl-1-chloropropane and aluminum trichloride?

5.30 Alkylation of benzene can be accomplished by using an alkene such as propene and an acid catalyst. What is the electrophile and what is the product?

5.31 Indicate on which ring and at what position bromination of the following compound will occur.

H O
N—C

5.32 Indicate on which ring and at what position nitration of the following compound will occur.

CCl_3

Reactions of Side Chains of Aromatic Compounds

5.33 Draw the oxidation product of each compound in Exercise 5.19 when reacted with potassium permanganate.

5.34 Predict the oxidation product of each of the following compounds when reacted with potassium permanganate.

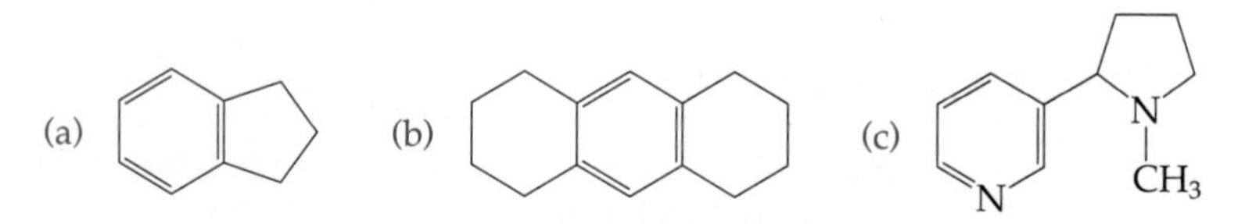

5.35 Free-radical bromination of ethylbenzene yields essentially one product. What is its structure? Why aren't other isomers formed?

5.36 Treatment of allyl benzene with dilute acid causes isomerization to an isomeric compound. Suggest its structure and propose a mechanism for its formation.

Synthesis of Aromatic Compounds

5.37 What reagent is required for each of the following reactions? Will an ortho,para mixture of products or the meta isomer predominate?
(a) nitration of bromobenzene (b) sulfonation of nitrobenzene
(c) bromination of ethylbenzene (d) methylation of anisole

5.38 What reagent is required for each of the following reactions? Will an ortho,para mixture of products or the meta isomer predominate?
(a) bromination of benzoic acid (b) acetylation of isopropylbenzene
(c) nitration of acetophenone (d) nitration of phenol

5.39 Starting with benzene, describe the series of reagents and reactions required to produce each of the following compounds.
(a) *p*-bromonitrobenzene (b) *m*-bromonitrobenzene
(c) *p*-bromoethylbenzene (d) *m*-bromoethylbenzene

5.40 Starting with benzene, describe the series of reagents and reactions required to produce each of the following compounds.
(a) *m*-bromobenzenesulfonic acid (b) *p*-bromobenzenesulfonic acid
(c) *p*-nitrotoluene (d) *p*-nitrobenzoic acid

5.41 Starting with either benzene or toluene, describe the series of reagents and reactions required to produce each of the following compounds.
(a) 3,5-dinitrochlorobenzene (b) 2,4,6-trinitrotoluene
(c) 2,6-dibromo-4-nitrotoluene

5.42 Starting with either benzene or toluene, describe the series of reagents and reactions required to produce each of the following compounds.
(a) 2,4,6-tribromobenzoic acid (b) 2-bromo-4-nitrotoluene
(c) 1-bromo-3,5-dinitrobenzene

5.43 Starting with either benzene or toluene, describe the series of reagents and reactions required to produce each of the following compounds.
(a) *m*-bromophenol (b) *m*-bromoaniline (c) *p*-methylphenol

5.44 Starting with either benzene or toluene, describe the series of reagents and reactions required to produce each of the following compounds.
(a) *m*-bromochlorobenzene (b) *p*-methylbenzonitrile
(c) 1-bromo-3,5-dichlorobenzene

Metabolic Oxidation of Aromatic Compounds

5.45 Toluene is easily oxidized in the liver but benzene is not. Why?

5.46 Why are polychlorinated biphenyls (PCBs) such as the following compound difficult to oxidize, causing them to accumulate in fatty tissue?

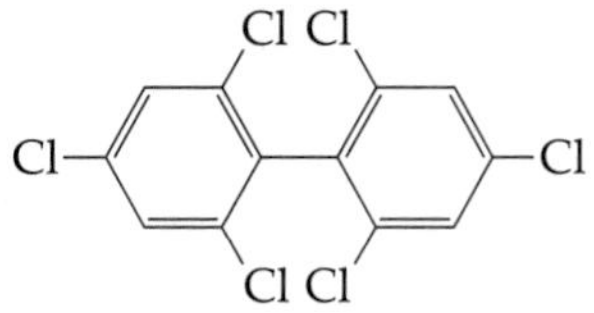

5.47 Explain why aromatic hydroxylation of chlorpromazine, an antipsychotic drug, occurs at the indicated position and in that ring.

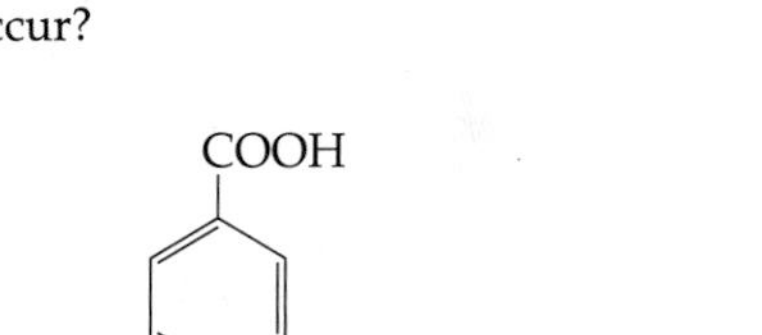

5.48 Why doesn't aromatic hydroxylation of probenecid, a drug used to treat chronic gout, occur?

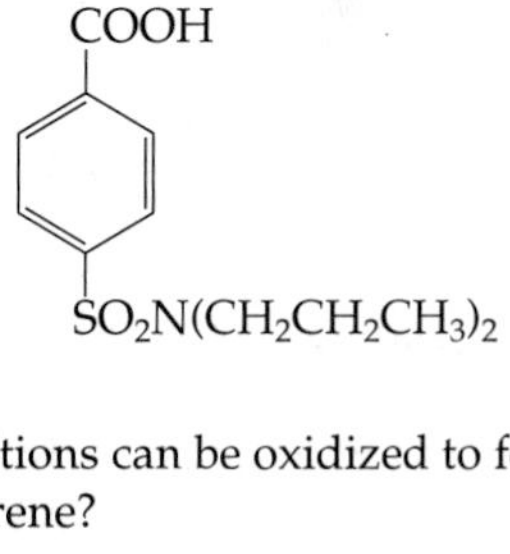

5.49 How many allylic positions can be oxidized to form an alcohol in the carcinogen 3-methylcholanthrene?

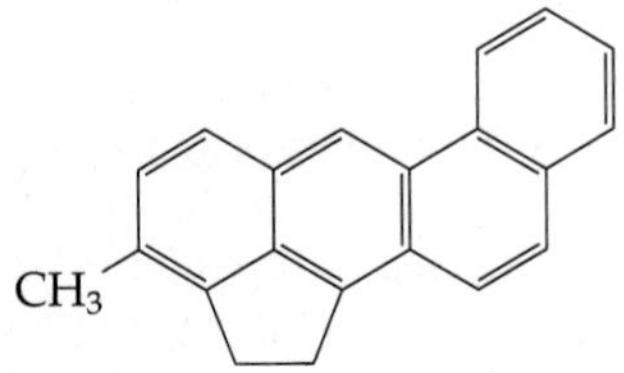

CH_3

5.50 Oxidation of the sedative–hypnotic methaqualone occurs to produce a benzyl alcohol. Draw the structure of the product.

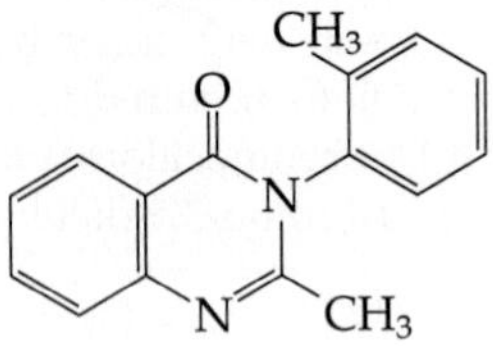

CH_3 O N N CH_3

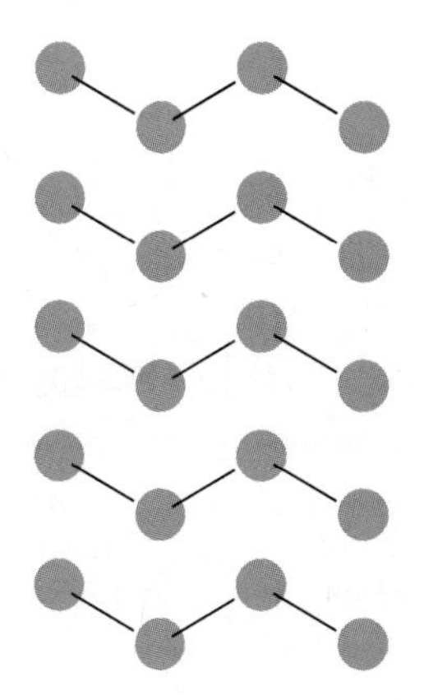

6

STEREOCHEMISTRY

6.1 CONFIGURATIONS OF MOLECULES

Until now, we have primarily drawn structures of organic compounds in two dimensions. These planar representations enable us to explain many chemical properties of compounds. In Chapters 3 and 4 we considered the structures of geometrical isomers, which are one of a general class of **stereoisomers.** Stereoisomers have the same connectivity—the same sequence of bonded atoms—but different configurations. The different three-dimensional arrangements of atoms in molecules in space determine their **configurations.** Geometric isomers have different configurations.

The configuration of a molecule plays a major role in its biological function. Two molecules that have the same connectivity of atoms but differ in configuration often have entirely different biological properties. There are many examples of the effect of geometric isomerism on biological processes. Geometric isomers invariably elicit different responses in organisms. For example, bombykol, the sex attractant of the male silkworm moth (Chapter 4), has a trans–cis arrangement about the double bonds at the C-10 and C-12 positions. It is 10^9 to 10^{13} times more potent than the other three possible geometric isomers. Disparlure, the sex attractant of the female gypsy moth (Chapter 3), is active only if the large alkyl groups bonded to the three-membered ring are cis.

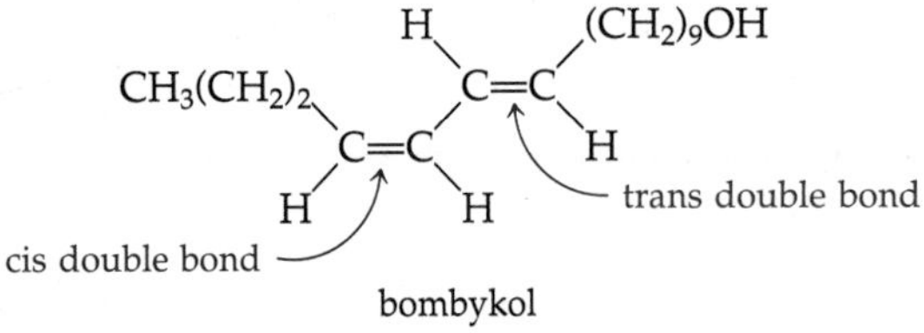

$CH_3(CH_2)_2$, H, $C=C$, H, $C=C$, $(CH_2)_9OH$, H, H
cis double bond
trans double bond
bombykol

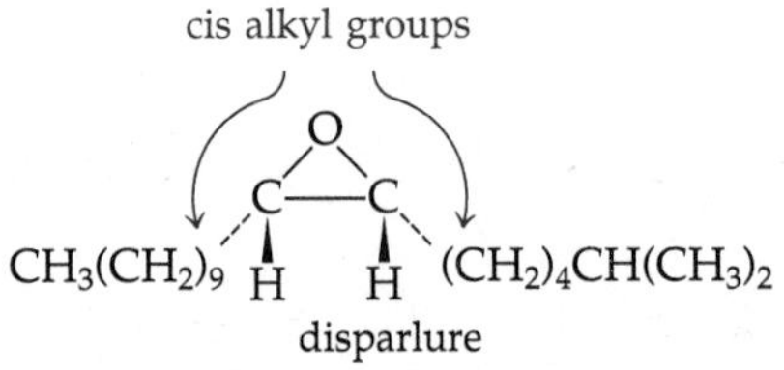

cis alkyl groups
$CH_3(CH_2)_9$, H, O, C—C, H, $(CH_2)_4CH(CH_3)_2$
disparlure

Geometrical isomers are only one type of a general class of stereoisomers. Another type of stereoisomerism is based on mirror-image relationships between molecules and is the subject of this chapter. This phenomenon is not as easily visualized as geometrical isomers, but its consequences are even more vital to life processes. In this chapter we will see that the configuration about a tetrahedral carbon atom containing four different groups of atoms significantly affects its chemical reactions, especially those occurring in living organisms.

6.2 MIRROR IMAGES AND CHIRALITY

The fact that we live in a three-dimensional world has important personal consequences. In the simple act of looking into a mirror, you see someone who does not actually exist, namely your mirror image. Every object has a mirror image, but this reflected image need not be identical to the actual object. One such example is lettering in two-dimensional space, such as the reverse lettering on the front of an ambulance. The letters are painted as their mirror image, so that the lettering we see in the rear-view mirror of our automobiles clearly identifies the ambulance.

Now let's think about the mirror images of a few common three-dimensional objects. A simple wooden chair looks exactly like its mirror image (Figure 6.1). Similarly, the mirror images of items such as a teacup or a hammer are identical to the objects themselves. When an object and its mirror image exactly match, we say that they are **superimposable.** Superimposable objects can be "placed" on each other so that each three-dimensional feature of one object coincides with an equivalent three-dimensional feature in the mirror image.

Now let's consider some objects that cannot be superimposed upon their mirror images; these images are said to be **nonsuperimposable.** One example is the side-arm chair found in many classrooms. When a chair with a "right-handed arm" is reflected in a mirror, it becomes a chair with a "left-handed arm" (Figure 6.1). We can convince ourselves of this by imagining sitting in the chair or its mirror image.

Now consider the nonsuperimposability of our hands, which are also related as mirror images. The mirror image of a left hand looks like a right hand. But when we try to superimpose our hands, we find that it can't be done (Figure 6.2). Thus, our hands are related as nonsuperimposable mirror images.

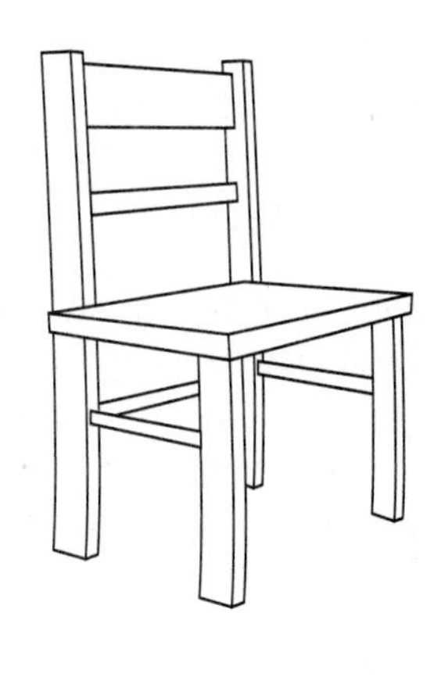

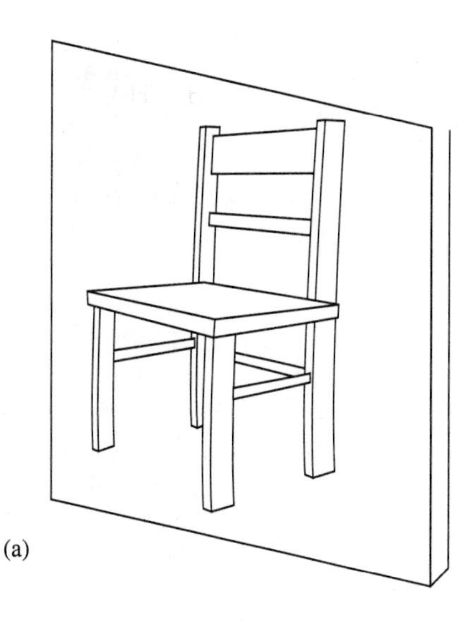

(a)

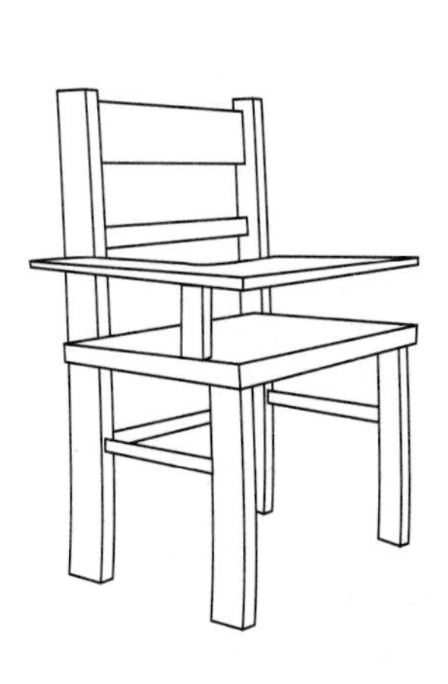

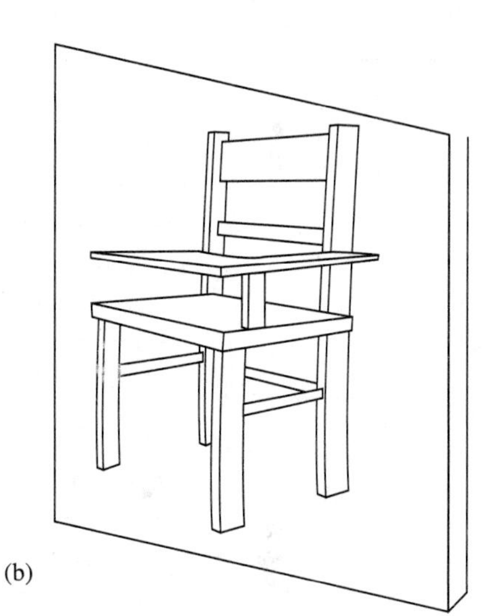

(b)

FIGURE 6.1
Objects and Their Mirror Images
In (a) the simple chair and its mirror image are identical. The mirror image can be superimposed on the original chair. In (b) the sidearm chair has a mirror image that is different. The chair is a right-handed object; the mirror image is a left-handed object. The mirror image cannot be superimposed on the original chair.

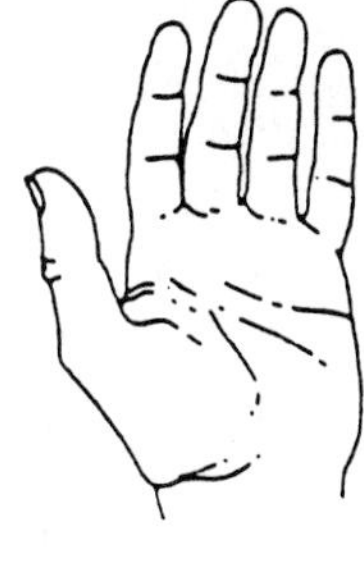

left hand

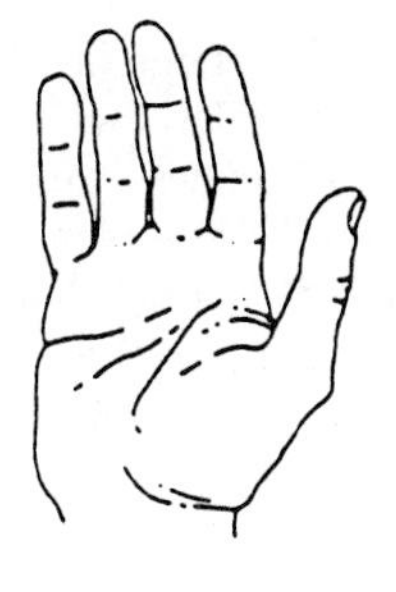

right hand

(a)

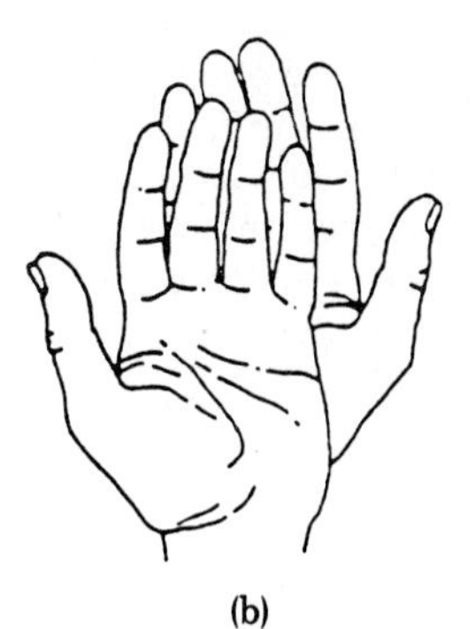

(b)

FIGURE 6.2
Chiral Objects Are Nonsuperimposable
The hands in (a) are mirror images and resemble each other. However, as shown in (b), the hands cannot be superimposed. Hands are nonsuperimposable mirror images; they are chiral.

An object that is not superimposable on its mirror image is said to be **chiral** (Gk. *chiron*, hand). Thus, objects such as gloves and shoes that have a "handedness" are chiral. An object that is superimposable on its mirror image is **achiral.** Objects such as a cup or a hammer are achiral. We can determine whether an object is chiral or achiral without trying to superimpose its mirror image. An object, such as a cup, that has a plane of symmetry that divides it so that one-half of it is the mirror image of the other half is achiral (Figure 6.3). However, if an object, such as your hand, has no such plane of symmetry, it is chiral. Thus, a cup is achiral, but a hand is chiral.

EXAMPLE 6.1 If you ignore the motor and the other parts underneath a car, is a car chiral or achiral?

Solution Consider the mirror image of the car. Usually the exterior of the car carries name plates identifying the specific model and the dealership that sold the car. Any item on the right would be on the left in the mirror image and vice versa. Thus, the car is chiral. However, even if these exterior items were symmetrically placed, the car would still be chiral. Consider possible planes of symmetry through the car. One possible plane would be the one perpendicular to the road, which splits the car along its length into two halves. But they are not equal halves. Everything on the dashboard is located on one side or the other, not in a symmetrical manner. The most obvious difference is the location of the steering wheel, which is on the left in an American car but on the right in an English car.

Molecules Can Be Chiral

The concepts of chirality can be extended from macroscopic objects to molecules. Most molecules produced in living organisms are chiral. A molecule is chiral if it contains a tetrahedral carbon atom attached to four different atoms or groups of atoms. These four atoms or groups of atoms can be

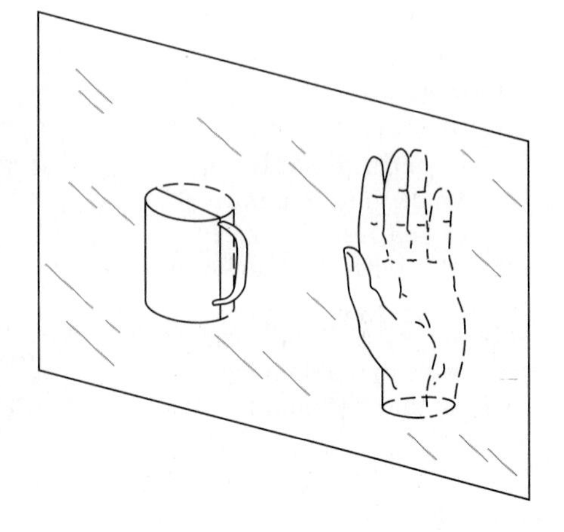

FIGURE 6.3
Plane of Symmetry
Any object with a plane of symmetry is achiral. The half on one side of the plane is the mirror image of the half on the other side of the plane. The cup shown can be divided into two equal halves that are mirror images of each other. It is achiral. Any object that does not have a plane of symmetry is chiral. The plane shown does not split a hand into two equal halves. A hand is chiral.

arranged in two different ways to correspond to two different stereoisomers. Let's consider the stereoisomers of bromochlorofluoromethane. One stereoisomer of this molecule and its mirror image are illustrated in Figure 6.4. The two structures cannot be superimposed; therefore, bromochlorofluoromethane is chiral.

In our discussion of macroscopic achiral objects we noted that they contain a plane of symmetry. This generalization applies equally to molecules. Figure 6.5 shows ball-and-stick models of dichloromethane and bromochloromethane. Dichloromethane has two planes of symmetry; bromochloromethane has one plane of symmetry. Each molecule can be superimposed on its mirror image and is therefore achiral. In contrast, bromochlorofluoromethane does not have a plane of symmetry and is chiral.

Enantiomers Are Mirror-Image Isomers

Stereoisomers that are related as nonsuperimposable mirror images are called **enantiomers** (Gk. *enantios*, opposite + *meros*, part). We can tell that a substance is chiral and predict that two enantiomers exist by identifying the substituents on each carbon atom. A carbon atom with four different substituents is chiral, and the molecule containing this carbon atom can exist as a pair of enantiomers. For example, 2-butanol is chiral because the C-2 atom is attached to four different groups (CH_3, CH_3CH_2, H, and OH). In contrast, 2-propanol is achiral because it does not contain any carbon atom attached to four different groups. The C-2 atom has two methyl groups attached to it.

FIGURE 6.4
Criteria for Chirality in Molecules
The two molecular models representing the chiral molecule bromochlorofluoromethane cannot be superimposed. The F and Br spheres of structure I held in the hand do not line up with those of structure II, which has been rotated 180° to line up the chlorine atoms. The mirror image cannot be superimposed on the original structure.

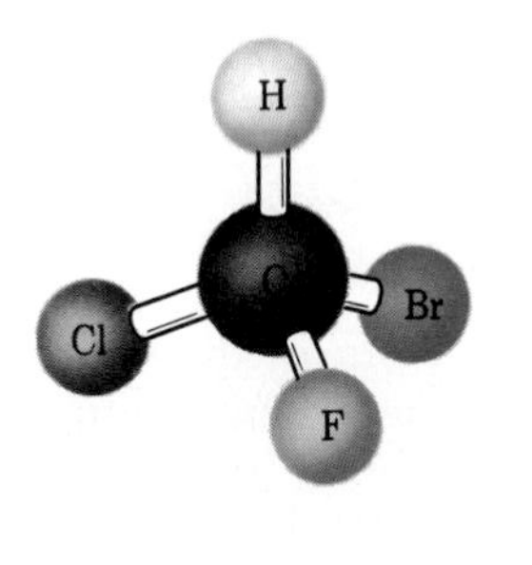

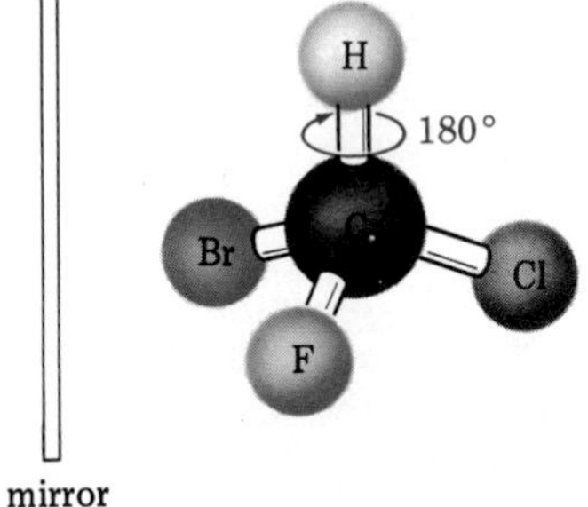

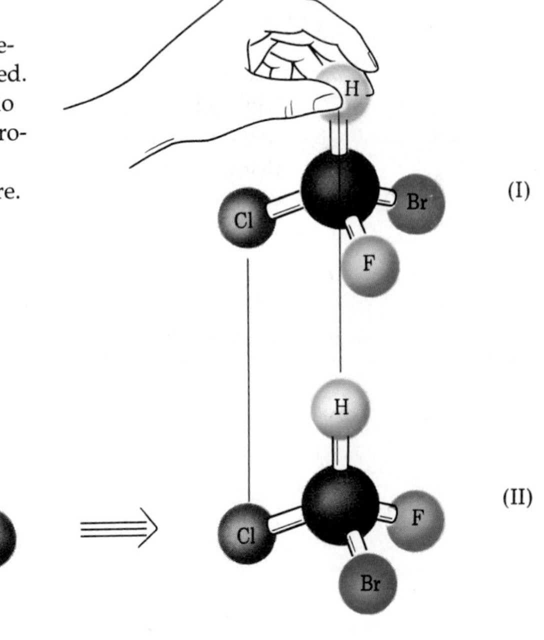

FIGURE 6.5 Planes of Symmetry for Molecules
Dichloromethane is bisected by two planes of symmetry. For each plane, one part of the structure mirrors the part of the molecule lying on the other side of the plane. Bromochloromethane has one plane of symmetry. Bromochlorofluoromethane has no plane of symmetry.

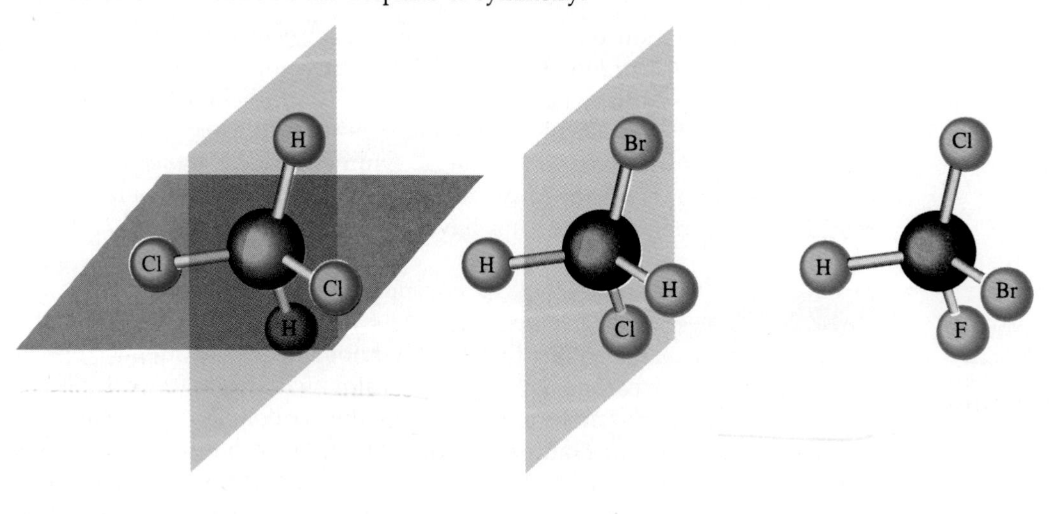

chiral carbon atom

$$CH_3{-}CH_2{-}\overset{\overset{\textstyle OH}{|}}{\underset{\underset{\textstyle H}{|}}{C}}{-}CH_3$$

not a chiral carbon atom

$$CH_3{-}\overset{\overset{\textstyle OH}{|}}{\underset{\underset{\textstyle H}{|}}{C}}{-}CH_3$$

EXAMPLE 6.2 Consider phenytoin, a compound with anticonvulsant activities. Is the molecule chiral or achiral? Determine your answer by identifying the number of different groups bonded to tetrahedral carbon atoms and by determining whether or not the molecule has a plane of symmetry.

HN, O, N, H, O

Solution Phenytoin has only one tetrahedral carbon atom in the entire molecule! That carbon atom is bonded to a nitrogen atom, a carbonyl group, and two phenyl groups. Because the tetrahedral carbon atom is attached to two identical phenyl groups, the molecule is achiral.

Phenytoin has a plane of symmetry that lies in the plane of the page. The two phenyl groups of phenytoin lie above and below the symmetry plane. Note that the other atoms of phenytoin are bisected by this plane.

EXAMPLE 6.3 Consider the following structural formula for nicotine. Is the molecule chiral?

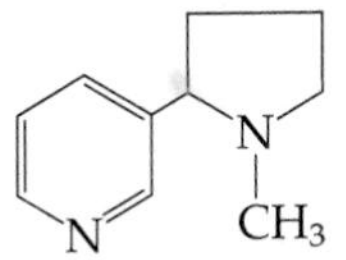

N N CH_3

Solution It is often difficult to visualize a three-dimensional structure based on a two-dimensional drawing. Nicotine consists of a six-membered pyridine ring linked to a five-membered ring. The pyridine ring has no tetrahedral carbon atoms, and all of its atoms lie in one plane. The five-membered ring, however, contains four tetrahedral carbon atoms. Atoms bonded to the ring are located both above and below the plane of the ring. Three carbon atoms of the ring have two identical groups—namely, two hydrogen atoms—so they are not chiral. However, the carbon atom bonded to the pyridine ring is bonded to four different groups—the pyridine ring, a carbon atom of the ring, a nitrogen atom, and a hydrogen atom that is not indicated in the bond-line structure. Thus, the molecule is chiral and can exist in two enantiomeric forms.

Physical Properties of Enantiomers

Enantiomers have many identical physical properties, such as density, melting point, and boiling point. They also have the same chemical properties in a symmetrical environment. However, they can be distinguished in a chiral environment. This difference is important in many processes in living cells.

We can regard our hands as analogous to the enantiomers of a chiral molecule. Let's consider the interaction of our hands with a symmetrical object such as a pair of tweezers. The tweezers are symmetrical because they have a plane of symmetry. Therefore, they can be used equally well with either hand by an ambidextrous person. That is, there is no preferred way to pick up or manipulate a pair of tweezers. However, you can easily use your hands to distinguish right- and left-handed gloves. Your hands are "a chiral environment", and in this environment, mirror-image gloves do not have identical properties. The right glove will fit only the right hand. We can distinguish chiral objects only because we, ourselves, are chiral.

In much the same way, only one of a pair of enantiomers fits into a specific site in a biological molecule such as an enzyme catalyst, because the site on the enzyme that binds the enantiomer is chiral. The binding of this enantiomer is said to be **stereospecific.** An example of such a stereospecific process is the conversion of the drug levodopa to dopamine, a neurotransmitter in the brain. Levodopa, the precursor of dopamine, is administered to treat Parkinson's disease. Levodopa has one chiral carbon atom and can thus exist as either of two enantiomers. Only the enantiomer with the configuration shown below is transformed into dopamine.

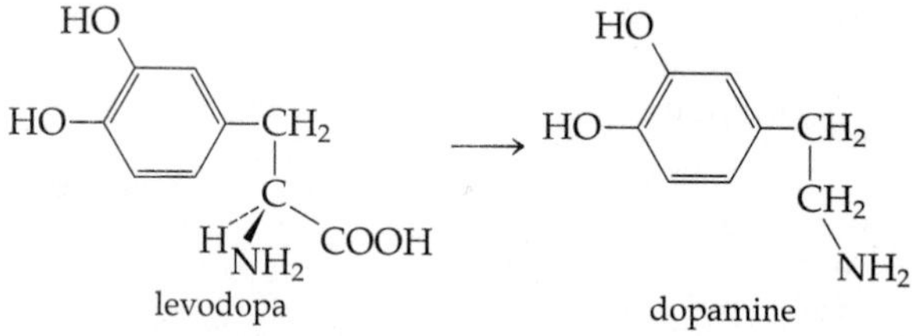

HO, HO, CH_2, C, H, NH_2, COOH → HO, HO, CH_2, CH_2, NH_2

levodopa   dopamine

The reaction that occurs is the loss of a carboxyl group by formation of carbon dioxide (decarboxylation), and the biological catalyst that causes decarboxylation is a stereospecific decarboxylase. This enzyme has a chiral binding site for levodopa; it does not bind the enantiomer of levodopa, which is toxic to humans.

6.3 OPTICAL ACTIVITY

Although enantiomers have many identical physical properties, they are not identical in every respect: enantiomers behave differently toward plane-polarized light. This one difference in physical properties is used to distinguish a chiral molecule from its enantiomer.

Plane-Polarized Light

Light consists of waves oscillating in an infinite number of planes that lie at right angles to the direction of propagation of the light. When a beam of "ordinary" light passes through a *polarizing filter*, **plane-polarized light,** which vibrates in a single plane, emerges. We are familiar with this phenomenon in everyday life: plane-polarized light can be produced by Polaroid sunglasses, which reduce glare by acting as a polarizing filter.

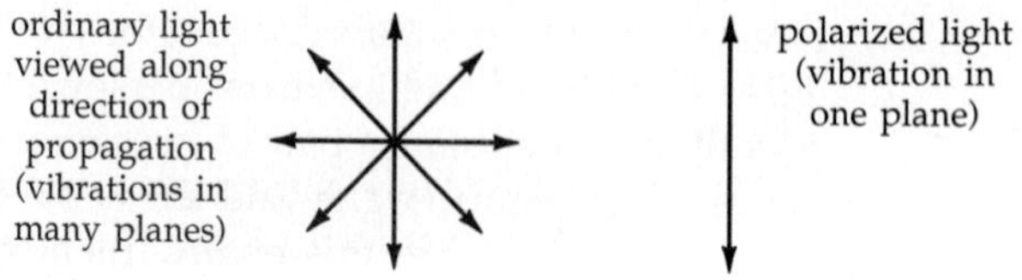

ordinary light viewed along direction of propagation (vibrations in many planes)

polarized light (vibration in one plane)

The Polarimeter

Plane-polarized light interacts with chiral molecules, and this interaction can be measured by an instrument called a **polarimeter.** A diagram of a polarimeter is shown in Figure 6.6. In a polarimeter, light having a single frequency of vibration—that is, monochromatic light—passes through a polarizing filter. The polarized light then traverses a tube containing a solu-

FIGURE 6.6 Representation of a Polarimeter
Plane-polarized light is obtained by the passage of light through the polarizer. Any chiral compound contained in the sample tube rotates the plane of light. The direction and magnitude of the rotation is determined by rotating the analyzer to allow the light to emerge.

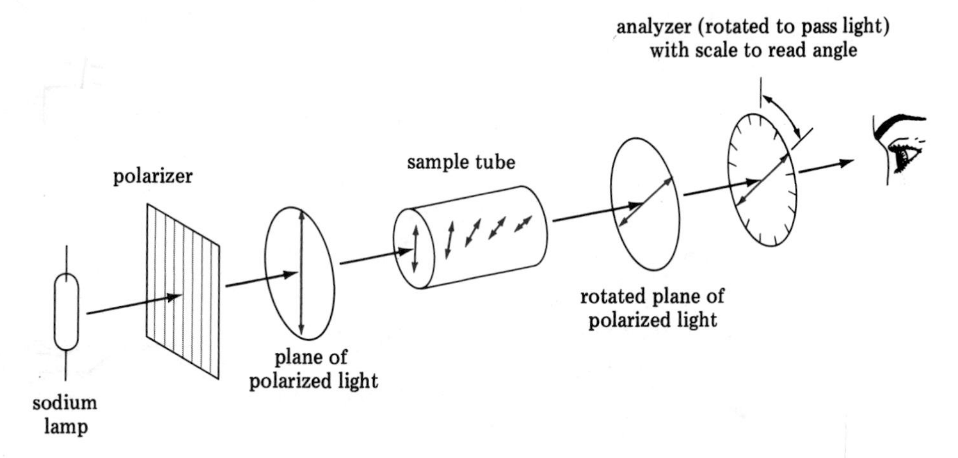

tion of the compound to be examined. During this passage, the plane of polarized light is rotated by an interaction with the chiral compound. After the plane-polarized light leaves the sample tube, it passes through a second polarizing filter called an analyzer. The analyzer is rotated either clockwise or counterclockwise to allow the light to escape. The angle of rotation of the analyzer equals the angle by which the light has been rotated by the chiral compound.

Because chiral molecules rotate plane-polarized light, they are **optically active.** Achiral molecules do not rotate plane-polarized light and are optically inactive.

Specific Rotation

The amount of rotation observed in a polarimeter depends on the structure of the substance and the number of molecules encountered by the light. The optical activity of a substance is reported as its **specific rotation,** symbolized by $[\alpha]_D$. It is the number of degrees of rotation of a solution at a concentration of 1 g/mL in a tube 1 dm long. The standard conditions selected for these experiments are 25 °C and the yellow light (D line, 589 nm) of the sodium vapor lamp.

If a chiral substance rotates plane-polarized light to the right, in a positive (+) clockwise direction, the substance is **dextrorotatory** (L., *dextro,* right). If a chiral substance rotates plane-polarized light to the left, in a negative (−) counterclockwise direction, the substance is **levorotatory** (L., *laevus,* left). The dextrorotatory and levorotatory isomers of a given substance rotate polarized light the same number of degrees but in opposite directions. Therefore, they are sometimes called *optical isomers.* The (+) isomer is sometimes called the **d form,** and the (−) isomer is called the ***l* form.**

TABLE 6.1 Specific Rotations of Common Compounds

Compound	$[\alpha]_D$
azidothymidine (AZT)	+99
cefotaxine (a cephalosporin)	+55
cholesterol	−31.5
cocaine	−16
codeine	−136
epinephrine (adrenaline)	−5.0
heroin	−107
levodopa	−13.1
monosodium glutamate (MSG)	+25.5
morphine	−132
oxacillin (a penicillin)	+201
progesterone (female sex hormone)	+172
sucrose (table sugar)	+66.5
testosterone (male sex hormone)	+109

Earlier, we encountered levodopa, which is so named because it is levorotatory; it is also called L-dopa and (−)-dopa. The specific rotation of L-dopa is −13.1. The specific rotations of some common substances are listed in Table 6.1.

6.4 FISCHER PROJECTION FORMULAS

Drawing molecules in three dimensions is time-consuming and the resulting perspective structural formulas are difficult to use, especially in the case of compounds containing several chiral carbon atoms (Section 6.6). The enantiomers of chiral substances are conveniently drawn by a convention proposed by the German chemist Emil Fischer over a century ago. The configurations of substances are indicated by comparing them to the configuration of a reference compound called glyceraldehyde.

Glyceraldehyde contains a carbon atom bonded to four different groups; thus, it exists as two enantiomeric forms (Figure 6.7). The enantiomers of glyceraldehyde are written according to the projection method proposed by Fischer. The carbonyl group, —CHO, the hydroxymethyl group, —CH_2OH, and the chiral carbon atom are arranged vertically with the most oxidized group, —CHO, at the "top". The chiral carbon atom is placed in the plane of the paper. Because this carbon atom is tetrahedral, the —CHO group and the —CH_2OH group extend behind the plane of the page, and the hydrogen atom and the hydroxyl group extend up and out of the plane. When these four groups are projected onto a plane, the projection is called the **Fischer projection formula.** In this convention, the chiral carbon atom is

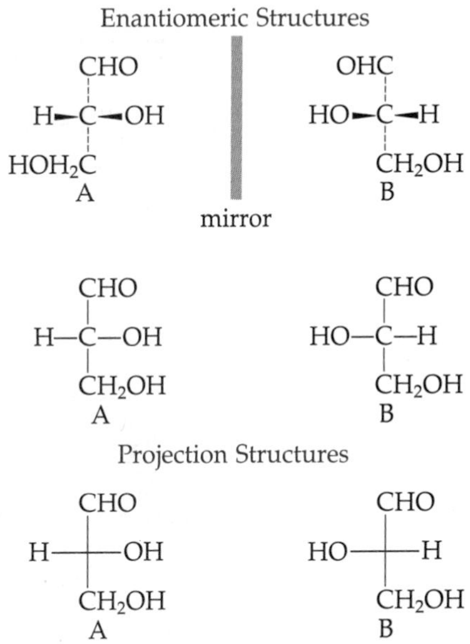

FIGURE 6.7

Projection Formulas of Enantiomers of Glyceraldehyde

The Fischer projection of the two enantiomers of glyceraldehyde has crossed lines at a point where the chiral carbon atom would be. However, the carbon atom is not usually shown. The vertical lines are assumed to project away from the viewer. The horizontal lines project out toward the viewer.

Enantiomeric Structures

CHO / H—C—OH / HOH_2C — A | mirror | OHC / HO—C—H / CH_2OH — B

CHO / H—C—OH / CH_2OH — A | CHO / HO—C—H / CH_2OH — B

Projection Structures

CHO / H—+—OH / CH_2OH — A | CHO / HO—+—H / CH_2OH — B

usually not shown. It is located at the point where the bond lines cross. The vertical lines are assumed to project away from the viewer. The horizontal lines project out toward the viewer.

A Fischer projection formula is a two-dimensional representation. It might appear that if we lifted one formula out of the plane and flipped it over, we would obtain the formula of the enantiomer. However, if the Fischer projection formula of molecule A were flipped over, the carbonyl group and the hydroxymethyl group, originally behind the plane, would be in front of the plane. These groups would not occupy identical positions with respect to the carbonyl group and hydroxymethyl group of molecule B, which are behind the plane. Therefore, in order to avoid the error of apparently achieving a two-dimensional equivalence of nonequivalent three-dimensional molecules, it is important not to lift two-dimensional representations out of the plane of the paper.

Fischer projection formulas can be drawn to depict any pair of enantiomers. These formulas imply that we "know" the configuration at the chiral carbon atom. However, the true configuration could not be determined by early chemists because there was no way to "see" the arrangement of the atoms in space. Therefore, Fischer arbitrarily assigned a configuration to one member of an enantiomeric pair of glyceraldehydes. The dextrorotatory enantiomer of glyceraldehyde, which rotates plane-polarized light in a clockwise direction (+13.5), was assigned to the Fischer projection with the hydroxyl group on the right side. The mirror image compound, (−)-glyceraldehyde, corresponds to the structure in which the hydroxyl group is on the left. It rotates plane-polarized light in a counterclockwise direction.

$$\begin{array}{c} CHO \\ | \\ H-C-OH \\ | \\ CH_2OH \end{array} \qquad \begin{array}{c} CHO \\ | \\ HO-C-H \\ | \\ CH_2OH \end{array}$$

(+)-glyceraldehyde $[\alpha] = +13.5$ (−)-glyceraldehyde $[\alpha] = -13.5$

6.5 ABSOLUTE CONFIGURATION

We began this chapter by saying that the arrangement of atoms in space determines the configuration of a molecule. When we know the exact positions of these atoms in space, we know the **absolute configuration** of the molecule. The absolute configuration of a molecule cannot be established by measuring the direction or magnitude of the optical rotation of an enantiomer. Optical rotation is a physical property. It depends on both the configuration and the identity of the four groups about the central carbon atom. One "right-handed" molecule could be dextrorotatory, whereas another "right-handed" molecule with different groups could be levorotatory.

To determine the absolute configuration, we require a method that can specify the positions of all atoms in the molecule. The best way to do this is by X-ray crystallography. The absolute configuration of an optically active substance was determined in 1950. The arrangement of its atoms in space corresponds to the arrangement of atoms in (+)-glyceraldehyde arbitrarily assigned by Fischer; his original choice was correct. As a result, all the configurations deduced by using (+)-glyceraldehyde as the reference compound are also correct.

R,S Configurations

The configurations of some molecules, such as amino acids and carbohydrates, can easily be compared to that of (+)-glyceraldehyde. But this procedure is not easily applied to molecules whose structures differ considerably from the reference compound. To circumvent this difficulty, three chemists—R. S. Cahn, K. C. Ingold, and V. Prelog—established a set of rules that define the absolute configuration of any chiral molecule. Once established, the configuration is designated by placing the symbol ***R*** or ***S*** within parentheses in front of the name of the compound.

The ***R,S*** **system of configurational nomenclature** for describing absolute configurations is related to the method we introduced in Chapter 4 to describe the location of groups in geometrical isomers of alkenes. In the *R,S* system, the four groups bonded to each chiral carbon atom are arranged from highest to lowest priority. The highest priority group is assigned the number 1, the lowest priority group the number 4. Then the molecule is oriented so that the bond from the carbon atom to the group of lowest priority is arranged directly along our line of sight (Figure 6.8). When this has been done, the highest three priority groups point up and lie on the circumference of a circle. (It may help to imagine holding the lowest priority group in your hand like the stem of a flower as you then examine the petals.)

FIGURE 6.8
The Cahn–Ingold–Prelog System
Place the atom of lowest priority away from your eye and view the chiral site along the carbon bond to that atom. To do this, reorient the molecule so that the low priority group is pointed behind the page. Determine the sequence of the three remaining groups that are pointed toward you.

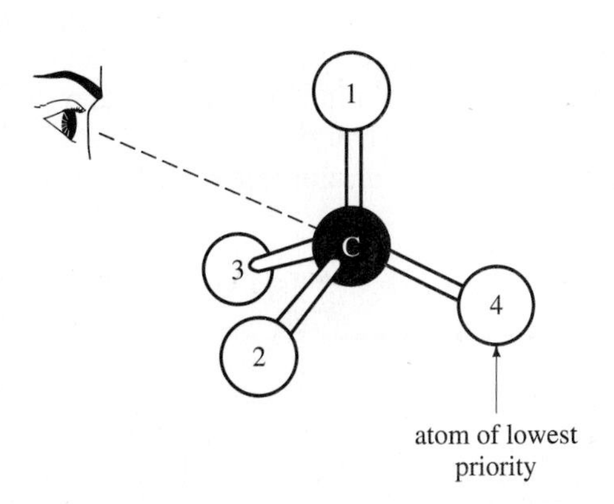

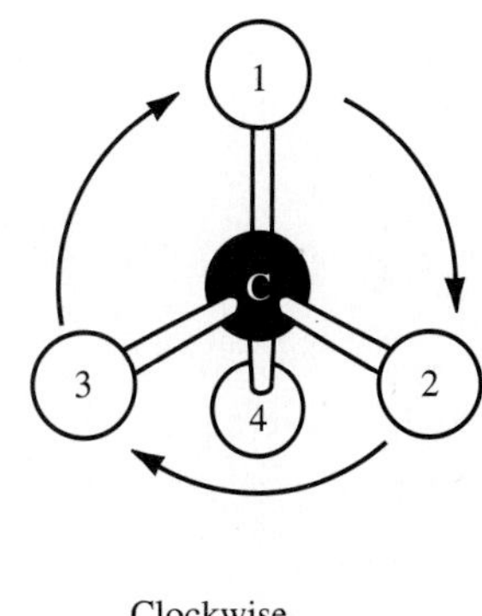

Consider the path taken as we trace the groups ranked 1 to 3. In Figure 6.8 this direction is clockwise. Therefore, the configuration is designated *R* (L. *rectus,* right). If we trace a counterclockwise path from groups ranked 1 to 3, the configuration is designated *S* (L. *sinister,* left).

Priority Rules

The priority rules we defined in Chapter 4 for determining the configuration of geometric isomers also apply for chiral compounds.

1. *Atoms.* We rank the four *atoms* bonded to a chiral carbon atom in order of decreasing atomic number; the higher the atomic number, the higher the priority.

$$I > Br > Cl > F > O > N > C > {}^2H > {}^1H$$

highest priority   lowest priority

   As the priority order ^2H (deuterium) > ^1H indicates, isotopes are ranked in order of decreasing mass.

2. *Groups of atoms.* If a chiral atom is attached to two or more identical atoms, move down the chain until a difference is encountered. Then apply rule 1. Using this rule, we find that the priority of alkyl groups is

$$(CH_3)_3C\text{—} > (CH_3)_2CH\text{—} > CH_3CH_2\text{—} > CH_3\text{—}$$

3. *Multiple Bonds.* If a group contains a double bond, both atoms are doubled. That is, a double bond is counted as two single bonds to each of the atoms of the double bond. The same principle is used for a triple bond. Thus, the order $HC{\equiv}C\text{—} > CH_2{=}CH\text{—} > CH_3CH_2\text{—}$ is obtained. The priority order for common functional groups containing oxygen is $\text{—}CO_2H$ (carboxylic acid) > $\text{—}CHO$ (aldehyde) > $\text{—}CH_2OH$ (alcohol).

Let's use the *R,S* system to determine the configuration of one of the enantiomers of alanine, an amino acid isolated from proteins. Alanine has one chiral carbon atom; it is bonded to a hydrogen atom, a methyl group, a carboxyl group, $—CO_2H$, and an amino group, $—NH_2$. A perspective drawing of this enantiomer of alanine is shown below.

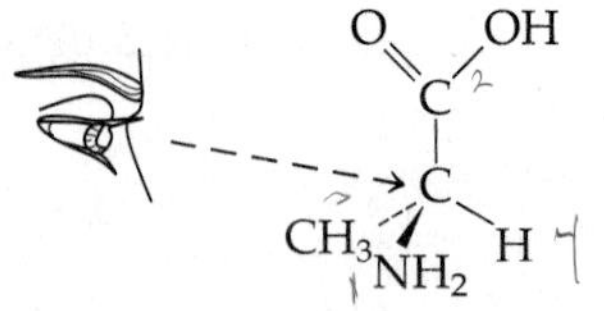

alanine

First, we rank the four groups attached to the chiral carbon atom in order of their priority from highest to lowest. The lowest priority (4) is given to the atom with the lowest atomic number directly attached to the chiral carbon atom; namely, hydrogen. The highest priority (1) is given to the atom with the highest atomic number directly attached to the chiral carbon atom; namely, nitrogen. The chiral carbon atom is attached to two other carbon atoms: one is in a methyl group, the other is a carboxyl group. The carboxyl group has the higher priority (2) because the carbon atom is attached to two oxygen atoms, whereas the carbon atom in the methyl group is attached only to hydrogen atoms; it has priority (3). Having assigned priorities, we next look into the molecule along the C—H bond and align the molecule so that the hydrogen atom points away from us. When this is done, we trace a path from the group with priority 1 to the group with priority 2, to the group with priority 3. The highest priority amino group lies at "4 o'clock", the next highest group (COOH) is at the "12 o'clock" position, and the methyl is at the "6 o'clock" position. Tracing a path from the amino group to the carboxyl group to the methyl group requires a counterclockwise motion. Therefore, the configuration of this enantiomer of alanine in proteins is *S*.

EXAMPLE 6.4 Warfarin is a drug that prevents blood clotting—that is, it is an anticoagulant drug. Warfarin is used both to treat thromboembolic disease and, in larger doses, as a rat poison. Assign its configuration.

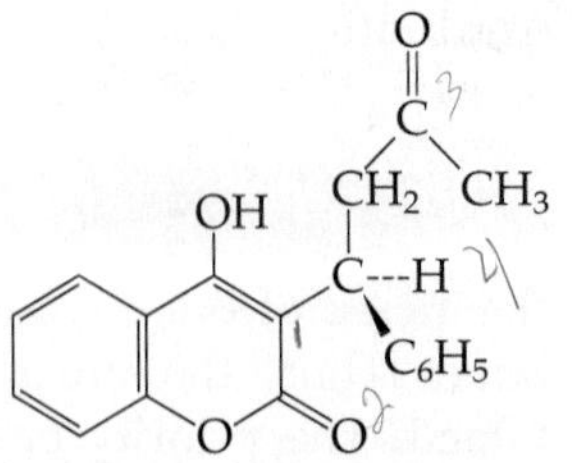

Solution Warfarin contains only one tetrahedral carbon atom that is attached to four different groups. That chiral carbon atom is attached to a hydrogen atom, a —C_6H_5 group (a phenyl group), and two other more complex groups. The lowest priority group is the hydrogen atom. The remaining three groups are linked through carbon atoms. One of them, the methylene group at the 12 o'clock position, has the next lowest priority (3) because it is attached to two hydrogen atoms. Next, we assign the priorities of the phenyl group and the ring system to the left. Both groups are attached to the chiral carbon atom by a carbon atom with a single and a double bond. Therefore, we must move to the next atom. When we do this in the complex ring, we find a carbon atom bonded to oxygen, which has a higher priority than the carbon atom bonded to hydrogen at the equivalent position in the phenyl ring. Therefore, the complex ring has a higher priority (1) than phenyl (2). Looking into the carbon–hydrogen bond at the chiral carbon atom, so that the hydrogen atom points away from us, we trace a counterclockwise path from group 1 to group 2 to group 3: this enantiomer of warfarin has the *S* configuration.

6.6 COMPOUNDS WITH MULTIPLE CHIRAL CENTERS

So far, we have considered only molecules with a single chiral center. However, many compounds contain several chiral centers. For example, the antibiotic erythromycin (Figure 6.9) contains 18 chiral centers! Erythromycin is effective against many venereal diseases, including chlamydia, syphilis, and gonorrhea, and is particularly useful for individuals who are allergic to penicillins. It has also been used to treat Legionnaires' disease.

How is the number of stereoisomers in a molecule with two or more chiral centers related to the number of chiral carbon atoms? The chirality of a molecule with two or more chiral centers depends on whether the centers are equivalent or nonequivalent. The term *nonequivalent* means that the chiral carbon atoms are not bonded to identical sets of substituents. We will first consider molecules with nonequivalent chiral centers. The number of

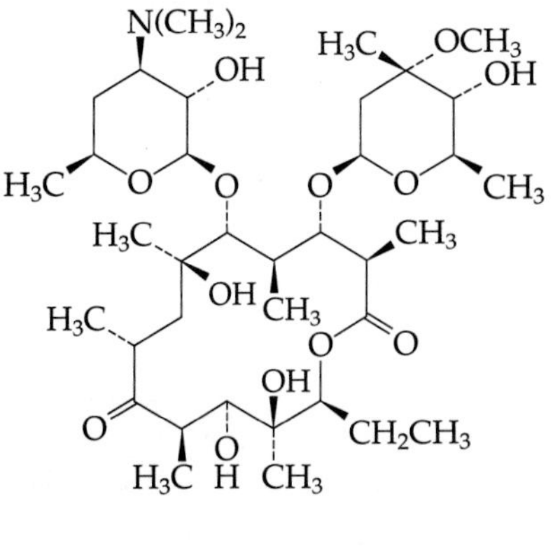

FIGURE 6.9
Erythromycin—A Chiral Antibiotic
Erythromycin has 18 chiral centers. Each center is designated with wedge-shaped or dashed lines.

$N(CH_3)_2$ OH H_3C OCH_3 OH H_3C O O O O CH_3 H_3C CH_3 OH CH_3 H_3C O O OH O CH_2CH_3 H_3C H CH_3

optically active isomers for a molecule containing n nonequivalent chiral carbon atoms is 2^n. Consider the following example.

$$\overset{4}{C}H_2(OH)-\overset{3}{C}H(OH)-\overset{2}{C}H(OH)-\overset{1}{C}HO$$

The C-2 and C-3 atoms are both chiral. They are also nonequivalent, because they are not bonded to identical sets of substituents. Therefore, the indicated structure has four stereoisomers (Figure 6.10).

In Figure 6.10, structures I and II are mirror images, not superimposable on each other (they are chiral), and enantiomers. This can be verified in two dimensions by imagining a mirror placed between I and II. Structures III and IV are also mirror images and are nonsuperimposable. Like all enantiomers, they rotate plane-polarized light in opposite directions but by the same absolute value.

Now consider structures I and III. These stereoisomers do not have a mirror-image relationship. Stereoisomers that are not enantiomers are called **diastereomers.** The pairs II and III, I and IV, and II and IV are all diastereomeric pairs. In contrast to enantiomers, which have the same chemical and physical properties, diastereomers have different chemical and physical properties. A comparison of some physical properties of enantiomers and diastereomers is given in Figure 6.10.

Nomenclature of Diastereomers

The name of a compound with two or more chiral centers must indicate the configuration of every chiral center. The configuration of each chiral carbon atom is indicated by a number that corresponds to its position in the carbon chain, and the letter R or S, separated by commas. Figure 6.10 shows the structures of the four stereoisomers of 2,3,4-trihydroxybutanal. Each struc-

FIGURE 6.10 Enantiomers and Diastereomers
There are four stereoisomers of a compound containing two nonequivalent chiral centers. There are two sets of enantiomers. Any combination of stereoisomers that are not enantiomers are diastereomers.

	I	II	III	IV
	CHO	CHO	CHO	CHO
	H—C—OH	HO—C—H	H—C—OH	HO—C—H
	H—C—OH	HO—C—H	HO—C—H	H—C—OH
	CH_2OH	CH_2OH	CH_2OH	CH_2OH
	enantiomers		enantiomers	
configuration	(2*R*,3*R*)	(2*S*,3*S*)	(2*S*,3*R*)	(2*R*,3*S*)
$[\alpha]_D$	−21.5	+21.5	−29.1	+29.1
melting point	(liquid)	(liquid)	130 °C	130 °C
solubility in ethanol	very soluble	very soluble	slightly soluble	slightly soluble

ture has two chiral carbon atoms: C-2 and C-3. Each of these chiral carbon atoms can be *R* or *S*. Thus, the four possibilities are (2*R*,3*R*), (2*S*,3*S*), (2*S*,3*R*), and (2*R*,3*S*). The enantiomer of the (2*R*,3*R*) compound is the (2*S*,3*S*) isomer, which has the opposite configuration at each chiral center. Compounds whose configurations differ at only one of the two chiral centers are diastereomers. For example, the (2*R*,3*R*) compound is a diastereomer of the (2*S*,3*R*) compound.

EXAMPLE 6.5 Threonine, an amino acid isolated from proteins, has the following condensed molecular formula. Write the Fischer projections of the possible stereoisomers.

$$\underset{4}{CH_3}\underset{3}{CH}(OH)\underset{2}{CH}(NH_2)\underset{1}{CO_2H}$$

Solution Both the C-2 and C-3 atoms are attached to four different substituents. Therefore, the compound has two chiral centers. Because the chiral centers are nonequivalent, four diastereomers are possible. The Fischer projections are written by placing the carboxyl group at the top of the vertical chain. The amino and hydroxyl groups may be on the right or left sides of the projection formulas.

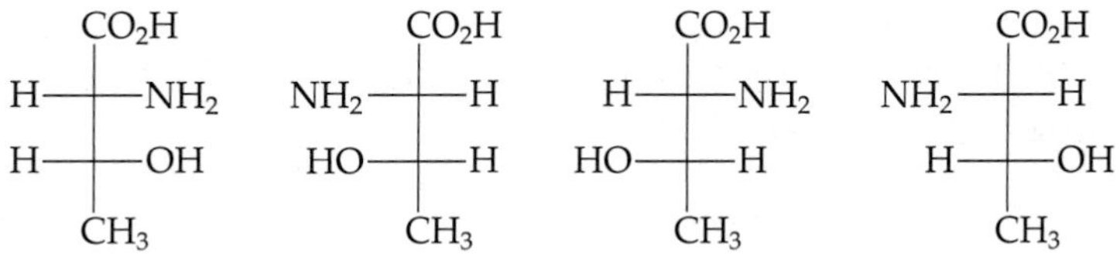

CO_2H	CO_2H	CO_2H	CO_2H
H—+—NH_2	NH_2—+—H	H—+—NH_2	NH_2—+—H
H—+—OH	HO—+—H	HO—+—H	H—+—OH
CH_3	CH_3	CH_3	CH_3

The structure of threonine isolated from proteins is given by the Fischer projection at the right.

EXAMPLE 6.6 Determine the number of chiral centers in vitamin K_1. How many stereoisomers are possible?

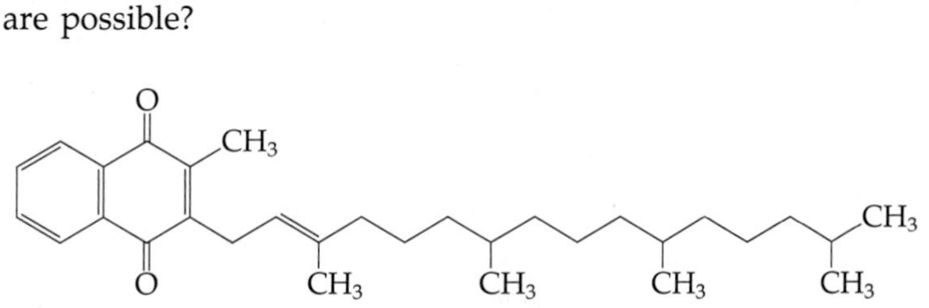

O, O, CH_3, CH_3, CH_3, CH_3, CH_3, CH_3

Solution The carbon atoms in the two rings are not chiral because they are not tetrahedral. The long alkyl chain contains 10 methylene units, none of which is chiral because a carbon atom in a methylene group is bonded to two hydrogen atoms. Similarly, the end of the alkyl chain is $CH(CH_3)_2$, and this tertiary carbon atom is not chiral either.

Next, consider the positions in the middle of the alkyl chain that have methyl-group branches. The methyl group on the left is bonded to a double-bonded carbon atom, which does not have four groups bonded to it: it is not chiral. The next two methyl groups are located on chiral centers. Because there are two chiral carbon atoms, $2^2 = 4$ stereoisomers are possible.

Some Molecules with Chiral Centers Are Achiral

We began our discussion of molecules with two or more chiral centers by noting that these chiral centers could be either equivalent or nonequivalent. Now let's consider compounds with equivalently substituted chiral centers as in the tartaric acids given in Figure 6.11. In each structure, the C-2 and C-3 atoms are connected to four different groups. But instead of the four diastereomers that would exist if the chiral centers were nonequivalent, only three stereoisomers exist and one is optically inactive! The compounds labeled (2*S*,3*S*) and (2*R*,3*R*) are enantiomers; therefore, they are optically active. But

Chirality and Your Senses

Our senses are sensitive to the configurations of molecules. Both the sense of taste and the sense of smell result from changes in specific sensory receptors when a receptor binds a specific molecule. The receptor's conformation changes and triggers a sequence of events that transmits a nerve impulse to the brain by sensory neurons. The brain interprets the input from sensory neurons as the "odor" of, say, spearmint.

Both enantiomeric and diastereomeric molecules interact differently in living systems. Differences in biological response to diastereomeric compounds are perhaps easier to understand. After all, diastereomeric compounds have different physical properties. Mannose, a carbohydrate, exists in two diastereomeric forms that differ in configuration at one center. The α form tastes sweet, but the β form tastes bitter.

Sensory receptors also readily distinguish enantiomers. The specificity of response is similar to the relationship between our hands and how they fit into gloves. Because sensory receptors are chiral, they stereospecifically interact with only one of a pair of enantiomers.

The two enantiomeric forms of carvone have very different odors. (+)-Carvone is present in spearmint oil and has the odor of spearmint. In contrast, its enantiomer, (−)-carvone, is present in caraway seed and has the odor associated with rye bread.

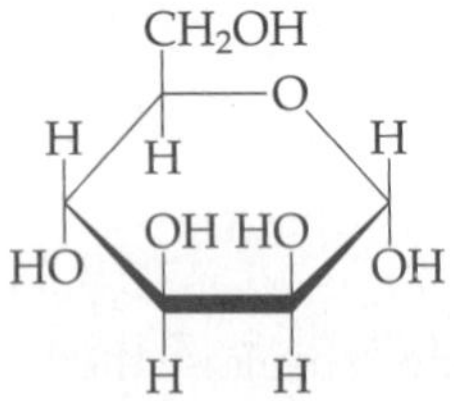

α-D-mannose 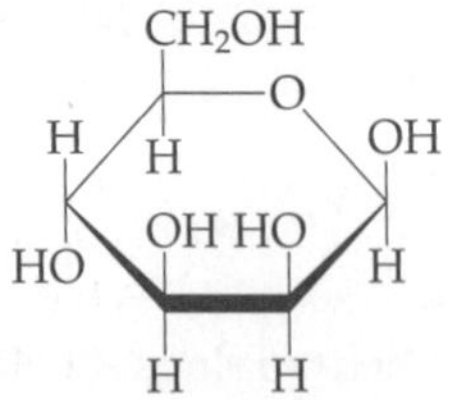 β-D-mannose

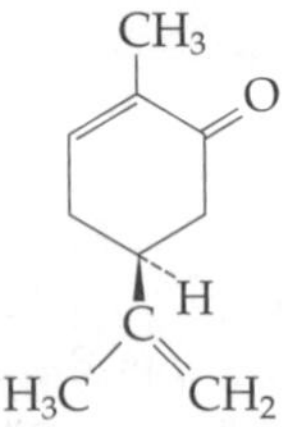

(+)-carvone (spearmint) 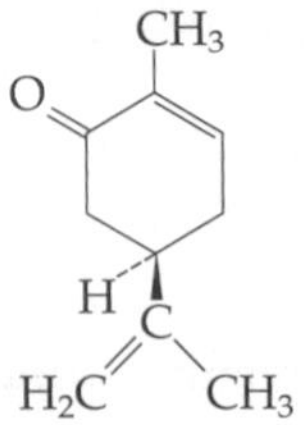 (−)-carvone (caraway)

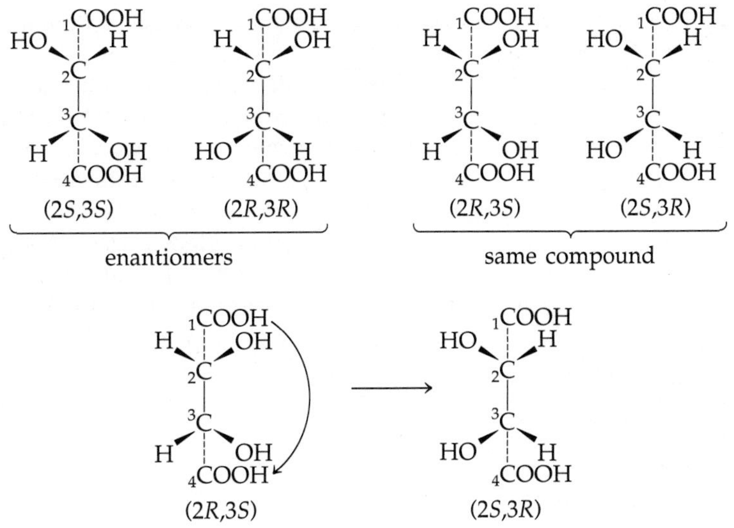

FIGURE 6.11
Tartaric Acids—Optically Active and Meso Compounds
There are only three stereoisomers of a compound with two equivalent chiral centers. Two compounds are enantiomers. The third compound has a plane of symmetry. It is optically inactive and is called a meso compound.

(2*S*,3*S*) (2*R*,3*R*)
enantiomers

(2*R*,3*S*) (2*S*,3*R*)
same compound

(2*R*,3*S*) → (2*S*,3*R*)

look at the structures labeled (2*R*,3*S*) and (2*S*,3*R*). Although the structures are drawn as "mirror images", these mirror images are, in fact, superimposable and identical. To show that this is so, rotate one structure 180° in the plane of the paper: the resulting structure superimposes on the original structure. Thus, these two structures represent the same molecule, which is not optically active.

The structures labeled (2*R*,3*S*) and (2*S*,3*R*) have two equivalent chiral carbon atoms. Each structure has a plane of symmetry. We recall from Section 6.2 that a structure with a plane of symmetry is achiral and that it is superimposable on its mirror image. In the case of this structure, the plane of symmetry is between the C-2 and C-3 atoms, so that the top half of the molecule is the mirror image of the bottom half.

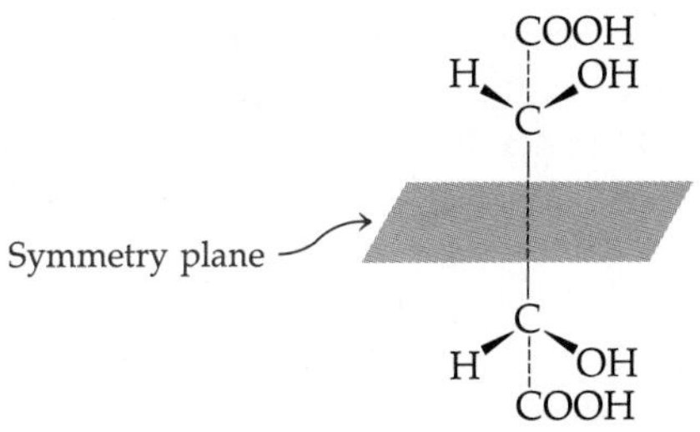

Symmetry plane

Compounds such as tartaric acid that have two or more chiral centers, but are nevertheless achiral, are called **meso compounds** (Gk., *meso*, middle). Meso compounds are not optically active.

6.7 SYNTHESIS OF STEREOISOMERS

If we attempt to synthesize a chiral compound by reacting a symmetrical (achiral) reactant with a symmetrical (achiral) reagent, the result will be a 50:50 mixture of enantiomers called a **racemic mixture.** Consider the reduction of pyruvic acid with $NaBH_4$. The C-2 atom is a carbonyl carbon atom. The atoms directly bonded to the carbonyl carbon atom are arranged in a trigonal plane. Addition of hydrogen to the carbon atom can occur from either face of the molecular plane. Thus, the tetrahedral carbon atom of lactic acid formed can have two possible configurations.

Metabolism May Vary by Species and Within Species

The metabolism of drugs is often species-dependent—a fact that must be considered because drugs are usually tested on animals prior to human trials. Even within the same species there are often strain differences, which are common among inbred test animals such as mice and rabbits.

Genetic differences in drug metabolism have been clearly established in humans. The differences between Northern Europeans, Eskimos, Mediterraneans, and Orientals must be considered in prescribing certain drugs.

Metabolism of drugs also varies by sex. Some oxidative processes are controlled by sex hormones, particularly the androgens (Section 3.7). This factor has become a matter of some concern recently because it has been common practice to test drugs on men rather than on women. Metabolism in men is more easily studied because of smaller hormonal changes day to day.

The anticonvulsant phenytoin shows a dramatic difference in metabolism depending on species. This achiral compound is oxidized to a chiral phenol. In humans, the hydroxylation occurs at the para position of one ring, and the compound has the *S* configuration. In dogs, the hydroxylation occurs at the meta position of the other ring, and the compound has the *R* configuration.

The processes by which metabolites are converted into water-soluble compounds to be excreted is also species-dependent. For example, it is not advisable to feed animals certain foods that humans may enjoy such as chocolate. The metabolites from chocolate can accumulate and poison a dog.

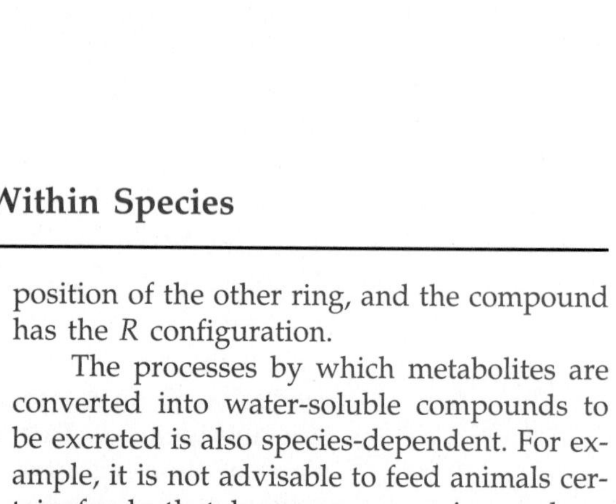

phenytoin

(man)

S(−)-*p*-hydroxyphenytoin

(dog)

R(+)-*m*-hydroxyphenytoin

$$\begin{array}{c} CO_2H \\ | \\ C{=}O \\ | \\ CH_3 \end{array} \xrightarrow{NaBH_4} H{-}\overset{\displaystyle CO_2H}{\underset{\displaystyle CH_3}{C}}{-}OH + HO{-}\overset{\displaystyle CO_2H}{\underset{\displaystyle CH_3}{C}}{-}H$$

pyruvic acid (*R*)-lactic acid (*S*)-lactic acid

racemic mixture of lactic acids

The individual lactic acid molecules produced are optically active. But a solution containing the products of the reaction is not optically active because the rotation of plane-polarized light by the (*R*)-lactic acid is canceled by the opposite optical rotation of (*S*)-lactic acid, which is formed in equal amounts. Note that there is a difference between a racemic mixture and a meso compound. A racemic mixture contains optically active components; the meso compound is a single achiral substance.

If we wish to synthesize a chiral product from an achiral reactant, the reaction must occur in a chiral environment. Protein catalysts called enzymes are examples of chiral reagents. Reduction of pyruvic acid using the liver enzyme lactate dehydrogenase yields exclusively (*S*)-lactic acid. The reducing agent for the reaction is nicotinamide adenine dinucleotide, NADH.

$$\begin{array}{c} CO_2H \\ | \\ C{=}O \\ | \\ CH_3 \end{array} \xrightarrow[\text{enzyme}]{NADH} HO{-}\overset{\displaystyle CO_2H}{\underset{\displaystyle CH_3}{C}}{-}H$$

pyruvic acid (*S*)-lactic acid

Cytochrome P-450, a widely distributed oxidizing enzyme system, very selectively forms one compound when a new chiral center is created. This selectivity occurs regardless of whether the compound is a natural part of metabolic processes, derived from a drug, or some substance that should not have been ingested.

Polycyclic aromatic hydrocarbons (Section 5.3) are formed in combustion processes and are found in auto emissions, cigarette smoke, and on meat cooked over a barbecue. Some of these compounds are carcinogens (cause cancer). The formation of an arene oxide (Section 5.8) from the carcinogen benzo[*a*]pyrene occurs at the C-7 and C-8 centers to produce the (7*R*,8*S*) epoxide (Figure 6.12). Further reaction with water occurs to give the (7*R*,8*R*) diol, which in turn is oxidized further at the 9,10 position to give the (9*S*,10*R*) epoxide—the ultimate carcinogenic species. This diol epoxide is very reactive; it combines with nitrogen atoms of amino groups in DNA to form compounds that disrupt the genetic code.

6.8 REACTIONS OF CHIRAL COMPOUNDS

As noted in the previous section, chiral compounds produced in reactions catalyzed by enzymes have a single configuration. Enzymes can also distinguish between enantiomers and bind only one of the two isomers to cause a

FIGURE 6.12 **Metabolic Sequence for Benzo[*a*]pyrene**

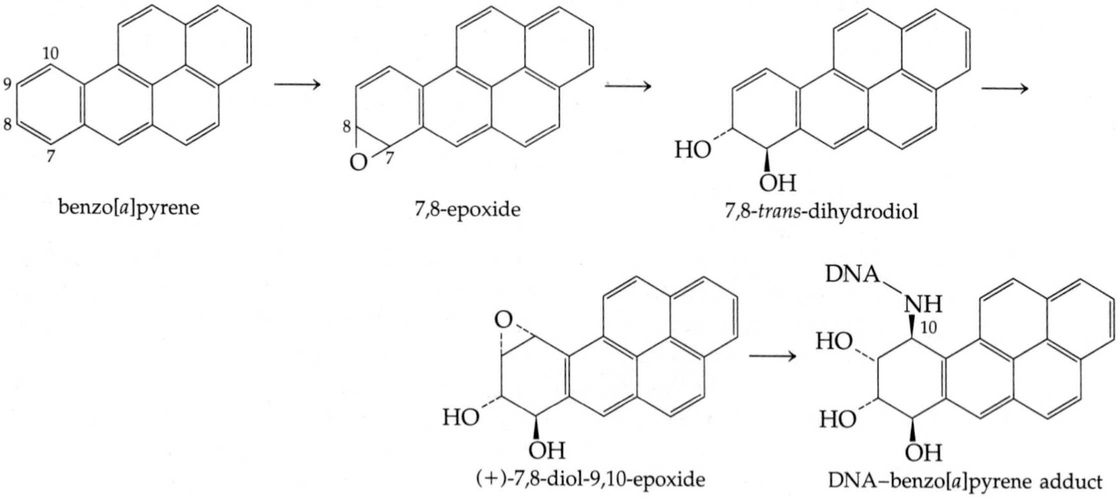

selective reaction. Enzymes themselves contain many chiral centers and have a "handedness". They will bind with a particular molecule but not with its enantiomer (Figure 6.13). For example, the enzyme tryptophan synthetase catalyzes the formation of the amino acid tryptophan from the amino acid serine and indole. Serine contains one chiral carbon atom and can exist as the mirror-image pair (*R*)- and (*S*)-serine. Of the two possible enantiomers, tryptophan synthetase binds only (*S*)-serine and produces only (*S*)-tryptophan.

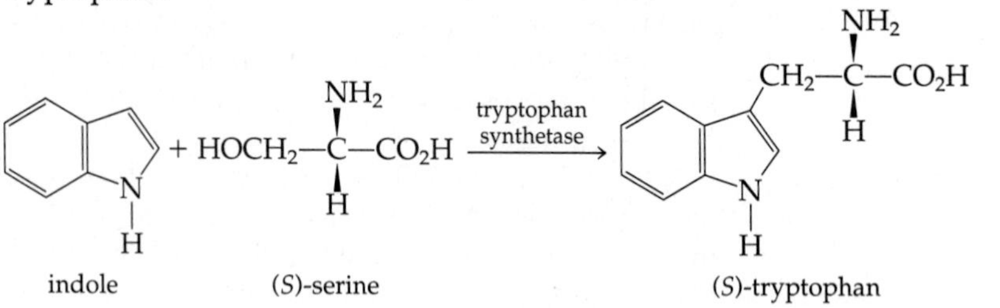

FIGURE 6.13
Specificity of an Enzyme for an Enantiomer
The enantiomer on the left fits into the template created by the enzyme, and the functionality is available for reaction. The mirror-image enantiomer on the right does not fit the enzyme, and its reactions are not catalyzed.

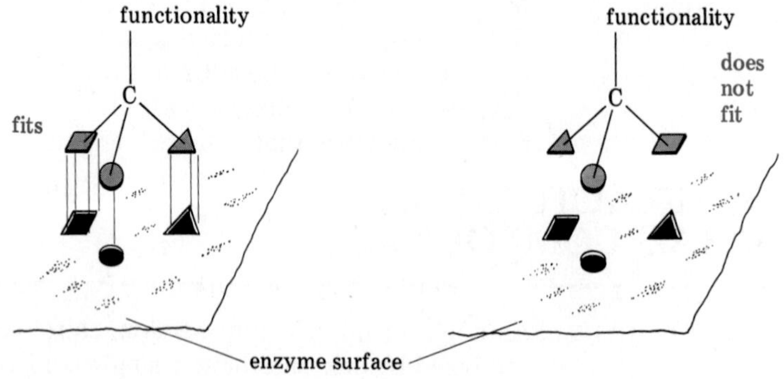

Reactions in the Chemistry Laboratory

We will see in subsequent chapters that the chirality of the products of reactions provides chemists with information about how bonds are broken and formed. Three types of reactions will be encountered.

1. Reactions that do not involve the chiral carbon atom.
2. Reactions that create a new chiral carbon atom.
3. Reactions that take place at the chiral carbon atom.

Any laboratory synthesis using chiral reactants will give chiral products if the bonds at the chiral center are not affected by the reagents. Formation of an ester (esterification) of (*R*)-lactic acid gives exclusively (*R*)-methyl lactate because the chiral carbon atom is not involved.

$$\begin{array}{c} CO_2H \\ | \\ H-C-OH \\ | \\ CH_3 \end{array} \xrightarrow{CH_3OH} \begin{array}{c} CO_2CH_3 \\ | \\ H-C-OH \\ | \\ CH_3 \end{array} + H_2O$$

(*R*)-lactic acid $[\alpha] = -3.8$ (*R*)-methyl lactate $[\alpha] = +7.5$

Reactions that break and make bonds at a chiral carbon atom will be the subject of the next chapter.

EXERCISES

Chirality

6.1 Which of the following isomeric methylheptanes has a chiral center?
(a) 2-methylheptane (b) 3-methylheptane (c) 4-methylheptane

6.2 Which of the isomeric bromohexanes has a chiral center?
(a) 1-bromohexane (b) 2-bromohexane (c) 3-bromohexane

6.3 Which of the compounds with molecular formula $C_5H_{11}Cl$ has a chiral center?

6.4 Which of the compounds with molecular formula $C_3H_6Cl_2$ has a chiral center?

6.5 How many chiral centers does each of the following alkanes have?

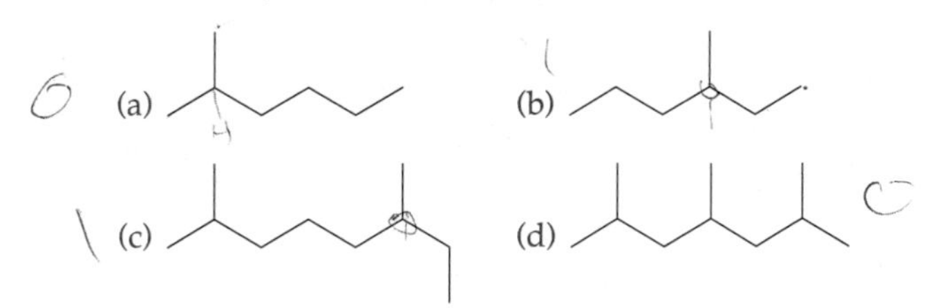

6.6 How many chiral centers does each of the following cyclic compounds have?

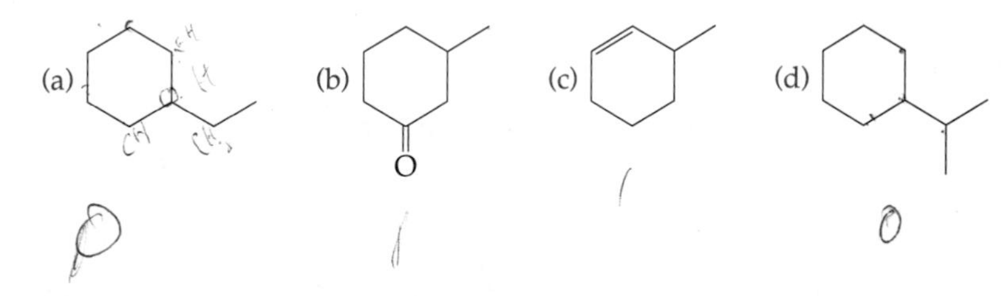

6.7 How many chiral centers does each of the following barbiturates have?

(a) phenobarbital (b) secobarbital

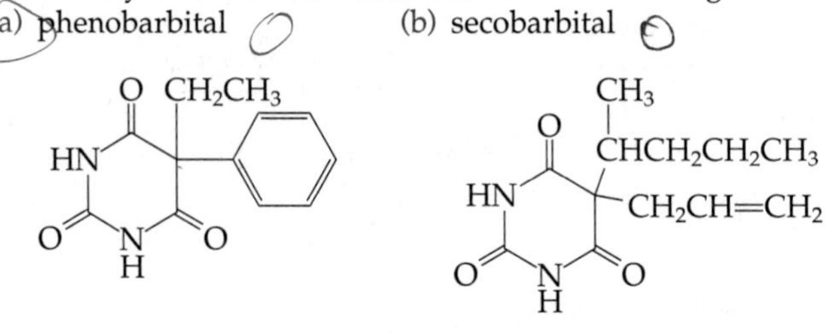

(c) hexobarbital (d) amobarbital

6.8 How many chiral centers does each of the following drugs have?

(a) phenylbutazone, used to treat gout

(b) ibuprofen, an analgesic

(c) chlorphentermine, a nervous system stimulant

(d) chloramphenicol, an antibiotic

6.9 How many chiral carbon atoms are in each of the following synthetic anabolic steroids?

(a) dianabol (b) stanozolol

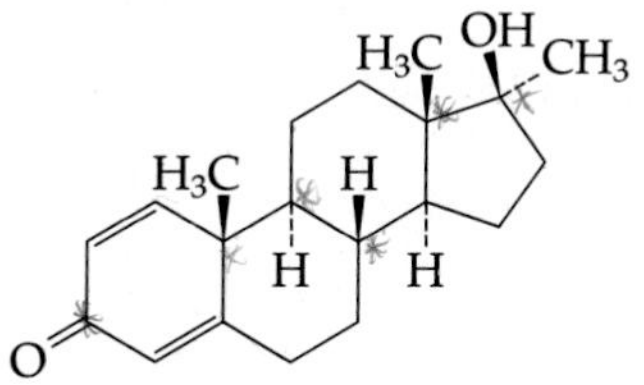

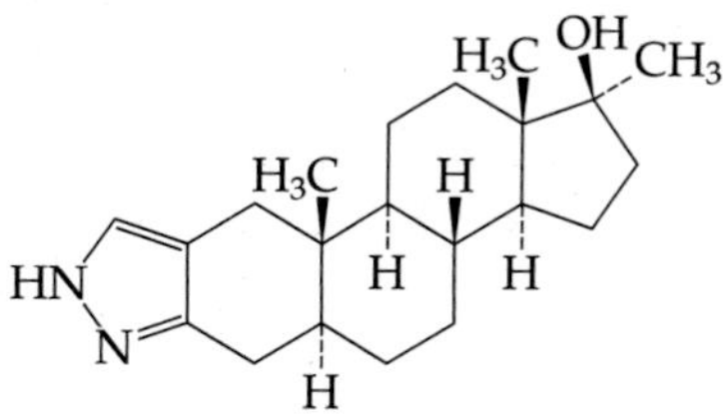

6.10 Determine the number of chiral centers in the male sex hormone testosterone and in the female sex hormone estradiol.

(a) testosterone (b) estradiol

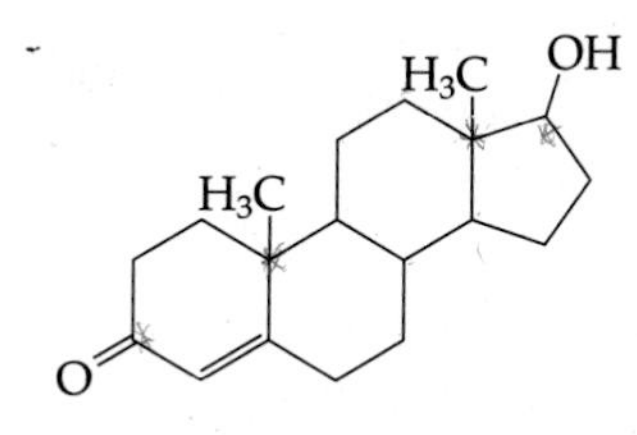

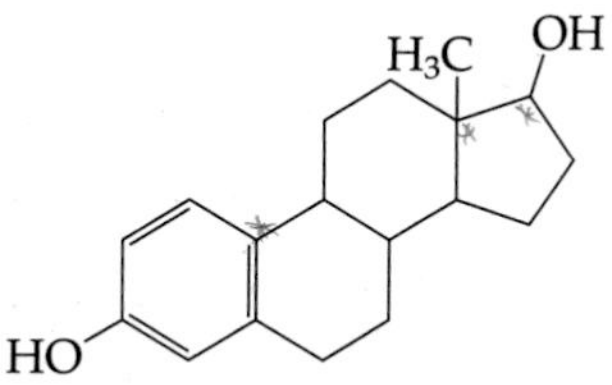

Priority Rules

6.11 Arrange the following groups in order of increasing priority.

(a) —OH, —SH, —SCH_3, —OCH_3

(b) —CH_2Br, —CH_2Cl, —Cl, —Br

(c) —CH_2—CH=CH_2, —CH_2—O—CH_3, —CH_2—C≡CH, —C≡C—CH_3

(d) —CH_2CH_3, —CH_2OH, —CH_2CH_2Cl, —OCH_3

6.12 Arrange the following groups in order of increasing priority.

(a) —O—C(=O)—CH_3 —C(=O)—CH_3 —C(=O)—OH

(b) —O—C(=O)—CH_3 —NH—C(=O)—CH_3 —C(=O)—NH_2

(c) —S—C(=O)—CH_3 —O—C(=O)—CH_2Br —C(=O)—Cl

(d) —C≡CH —C≡N —N≡C

6.13 Consider the following drugs and arrange the groups in order from low to high priority.

(a) ethchlorvynol, a sedative–hypnotic

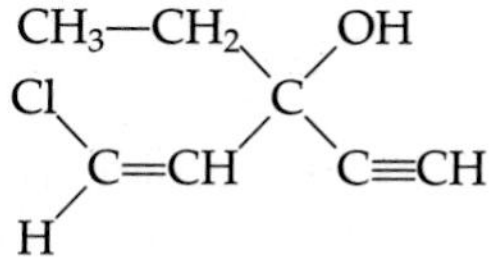

CH_3—CH_2—C(OH)(C≡CH)—CH=C(Cl)H

(b) chlorphenesin carbamate, a muscle relaxant

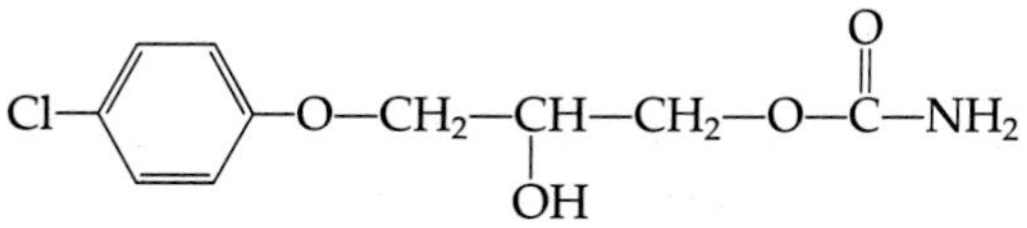

Cl—C_6H_4—O—CH_2—CH(OH)—CH_2—O—C(=O)—NH_2

(c) mexiletine, an antiarrhythmic

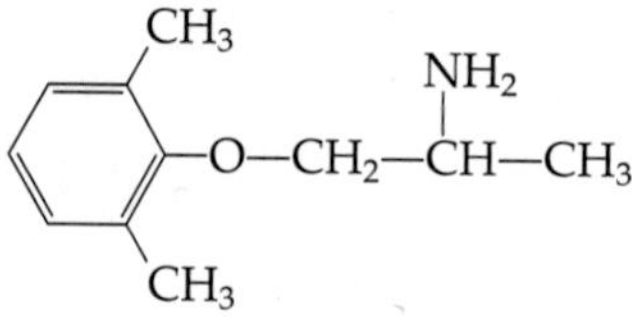

CH_3 / NH_2 / $-O-CH_2-CH-CH_3$ / CH_3

6.14 Consider the following drugs and arrange the groups in order of increasing priority.

(a) brompheniramine, an antihistamine

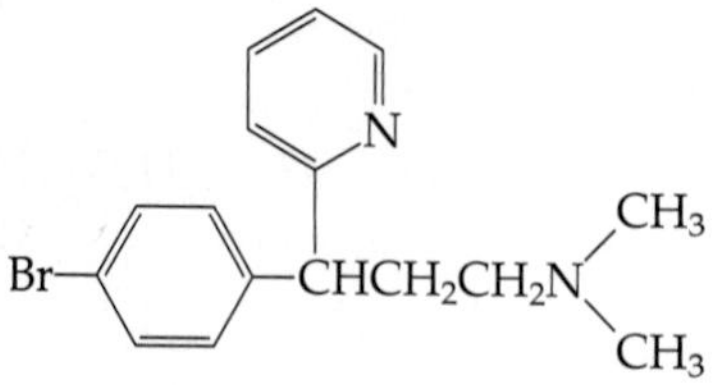

Br— —$CHCH_2CH_2N(CH_3)_2$

(b) fluoxetine, an antidepressant

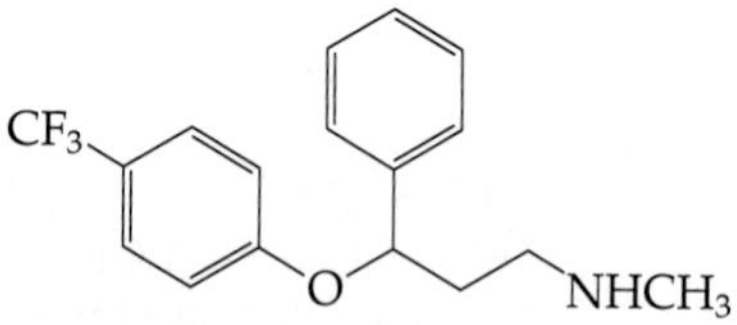

CF_3 O $NHCH_3$

(c) bachlofen, an antispastic

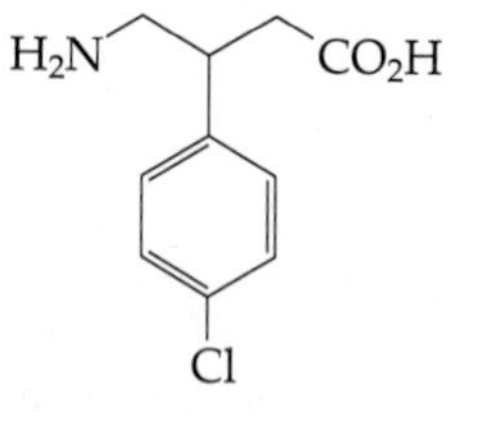

H_2N CO_2H Cl

R,S Configuration

6.15 Draw the structure of each of the following compounds.
(a) (*R*)-2-chloropentane (b) (*R*)-3-chloro-1-pentene
(c) (*S*)-3-chloro-2-methylpentane

6.16 Draw the structure of each of the following compounds.
(a) (*S*)-2-bromo-2-phenylbutane (b) (*S*)-3-bromo-1-hexyne
(c) (*R*)-2-bromo-2-chlorobutane

6.17 Assign the configuration of each of the following compounds.

(a) CH_3 / C / CH_3CH_2 H OH

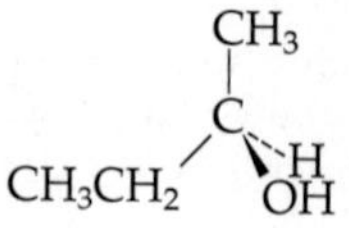

(b) CH_2CH_3 / C / H Cl CH_2CH_2Cl

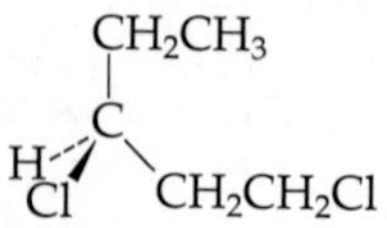

(c) 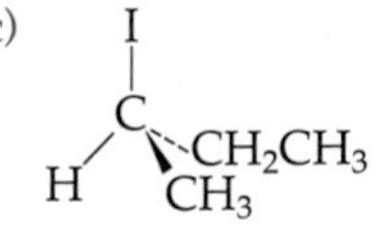 (d)

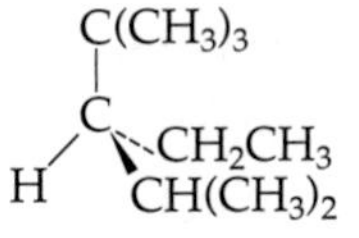

6.18 Assign the configuration of each of the following compounds.

(a) 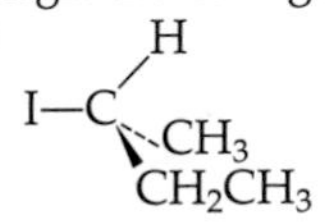 (b)

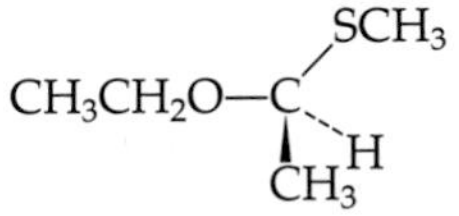

(c) 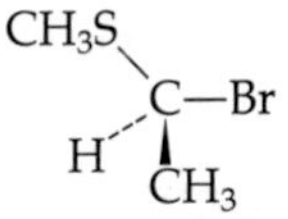 (d)

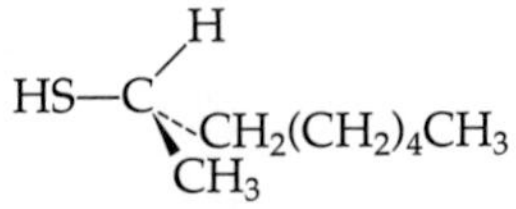

6.19 Assign the configuration of terbutaline, a drug used to treat bronchial asthma, and warfarin, an anticoagulant drug.

(a) terbutaline (b) warfarin

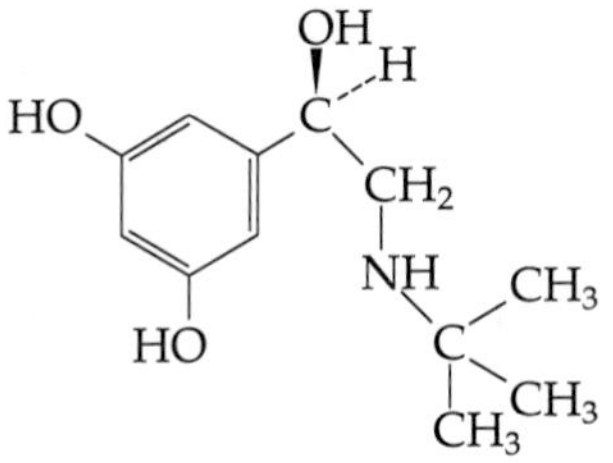

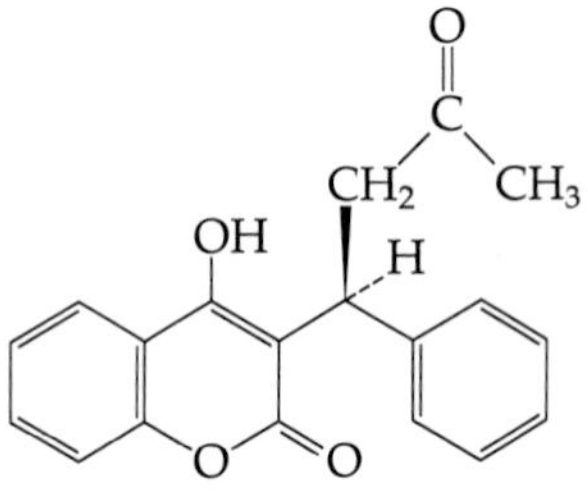

6.20 Assign the configuration of the following hydroxylated metabolite of diazepam, a sedative.

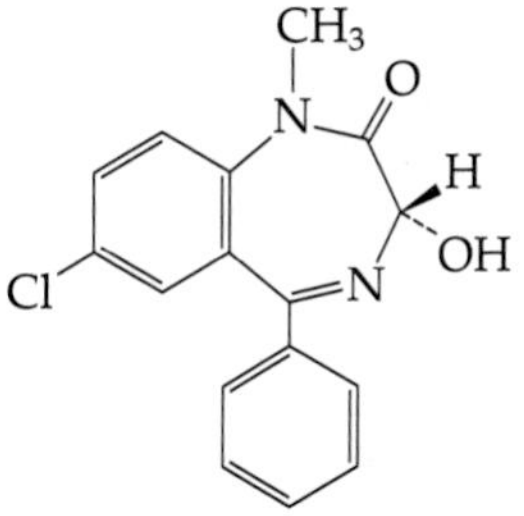

Optical Activity

6.21 The naturally occurring form of glucose has a specific rotation of +53. What is the specific rotation of its enantiomer?

6.22 The naturally occurring form of the amino acid threonine has a specific rotation of +28.3. What is the specific rotation of its enantiomer?

6.23 What do the various prefixes in (*R*)-(+)-glyceraldehyde and (*S*)-(−)-lactic acid mean?

6.24 (*R*)-(−)-Lactic acid is converted into a methyl ester when it reacts with methanol. What is the configuration of the ester? Can you predict its sign of rotation?

6.25 Consider the following four projection formulas. Determine the two missing rotations.

$$
\begin{array}{c}
\mathrm{CHO} \\
| \\
\mathrm{H{-}C{-}OH} \\
| \\
\mathrm{H{-}C{-}OH} \\
| \\
\mathrm{CH_2OH} \\
[\alpha] = -14.8
\end{array}
\qquad
\begin{array}{c}
\mathrm{CHO} \\
| \\
\mathrm{HO{-}C{-}H} \\
| \\
\mathrm{H{-}C{-}OH} \\
| \\
\mathrm{CH_2OH} \\
[\alpha] = +19.6
\end{array}
\qquad
\begin{array}{c}
\mathrm{CHO} \\
| \\
\mathrm{H{-}C{-}OH} \\
| \\
\mathrm{HO{-}C{-}H} \\
| \\
\mathrm{CH_2OH}
\end{array}
\qquad
\begin{array}{c}
\mathrm{CHO} \\
| \\
\mathrm{HO{-}C{-}H} \\
| \\
\mathrm{HO{-}C{-}H} \\
| \\
\mathrm{CH_2OH}
\end{array}
$$

6.26 What relationships between the rotations would exist for the following four compounds?

$$
\begin{array}{c}
\mathrm{CH_2OH} \\
| \\
\mathrm{C{=}O} \\
| \\
\mathrm{H{-}C{-}OH} \\
| \\
\mathrm{H{-}C{-}OH} \\
| \\
\mathrm{CH_2OH} \\
\mathrm{(I)}
\end{array}
\qquad
\begin{array}{c}
\mathrm{CH_2OH} \\
| \\
\mathrm{C{=}O} \\
| \\
\mathrm{HO{-}C{-}H} \\
| \\
\mathrm{HO{-}C{-}H} \\
| \\
\mathrm{CH_2OH} \\
\mathrm{(II)}
\end{array}
\qquad
\begin{array}{c}
\mathrm{CH_2OH} \\
| \\
\mathrm{C{=}O} \\
| \\
\mathrm{H{-}C{-}OH} \\
| \\
\mathrm{HO{-}C{-}H} \\
| \\
\mathrm{CH_2OH} \\
\mathrm{(III)}
\end{array}
\qquad
\begin{array}{c}
\mathrm{CH_2OH} \\
| \\
\mathrm{C{=}O} \\
| \\
\mathrm{HO{-}C{-}H} \\
| \\
\mathrm{H{-}C{-}OH} \\
| \\
\mathrm{CH_2OH} \\
\mathrm{(IV)}
\end{array}
$$

Diastereomers

6.27 Consider the structure of 5-hydroxylysine, and determine the number of stereoisomers possible.

$$
\mathrm{NH_2CH_2\underset{\displaystyle OH}{\underset{|}{C}}HCH_2CH_2\underset{\displaystyle NH_2}{\underset{|}{C}}H\overset{\displaystyle O}{\overset{\|}{C}}OH}
$$

6.28 Consider the structure of pantothenic acid (vitamin B_3), and determine the number of stereoisomers possible.

$$
\mathrm{HOCH_2\overset{\displaystyle CH_3}{\underset{\displaystyle CH_3}{\overset{|}{\underset{|}{C}}}}{-}\underset{\displaystyle OH}{\underset{|}{C}}H\overset{\displaystyle O}{\overset{\|}{C}}NHCH_2CH_2CO_2H}
$$

6.29 Ribose is optically active but ribitol, its reduction product, is optically inactive. Why?

$$
\begin{array}{c}
\mathrm{CHO} \\
| \\
\mathrm{H{-}C{-}OH} \\
| \\
\mathrm{H{-}C{-}OH} \\
| \\
\mathrm{H{-}C{-}OH} \\
| \\
\mathrm{CH_2OH} \\
\text{ribose}
\end{array}
\qquad
\begin{array}{c}
\mathrm{CH_2OH} \\
| \\
\mathrm{H{-}C{-}OH} \\
| \\
\mathrm{H{-}C{-}OH} \\
| \\
\mathrm{H{-}C{-}OH} \\
| \\
\mathrm{CH_2OH} \\
\text{ribitol}
\end{array}
$$

6.30 Which of the following carbohydrate derivatives are meso compounds?

(a) 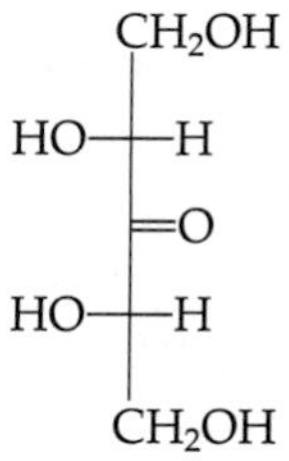
CH_2OH
HO—H
=O
HO—H
CH_2OH

(b) 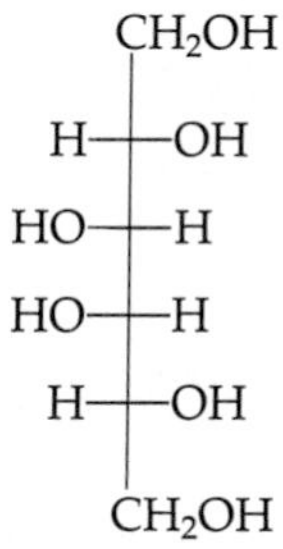
CH_2OH
H—OH
HO—H
HO—H
H—OH
CH_2OH

(c) 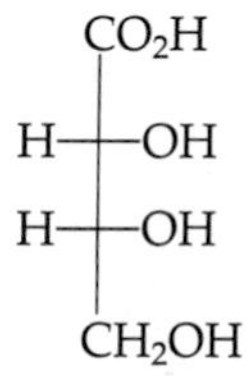
CO_2H
H—OH
H—OH
CH_2OH

6.31 There are four isomeric 2,3-dichloropentanes but only three isomeric 2,4-dichloropentanes. Explain why.

6.32 Which of the following compounds has a plane of symmetry?
(a) *cis*-1,2-dibromocyclobutane
(b) *trans*-1,2-dibromocyclobutane
(c) *cis*-1,3-dibromocyclobutane
(d) *trans*-1,3-dibromocyclobutane

Chemical Reactions

6.33 Addition of HBr to 1-butene yields a racemic mixture of 2-bromobutane. Explain why.

6.34 Reduction of acetophenone with $NaBH_4$ produces a racemic mixture of 1-phenyl-1-ethanol. Explain why.

6.35 How many products are possible when HBr adds to the double bond of (*R*)-3-bromo-1-butene? Which are optically active?

6.36 How many products are possible when HBr adds to the double bond of 4-methylcyclohexene? Which are optically active?

Stereoisomers in Biochemistry

6.37 Natural glucose is a sugar that the body can metabolize. Suggest what would happen if one were to eat its enantiomer.

6.38 The mold *Penicillium glaucum* can metabolize one enantiomer of optically active tartaric acid. Explain what would happen if a racemic mixture of tartaric acid were "fed" to the mold.

6.39 Natural adrenaline is levorotatory. The enantiomer has about 5% of the biological activity of the natural compound. Explain why.

6.40 The isomer of hydroxycitronellal shown has the odor of lily of the valley. Its mirror image has a minty odor. Explain why.

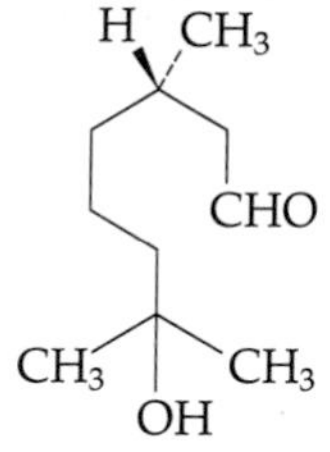

H CH_3
CHO
CH_3 CH_3
OH

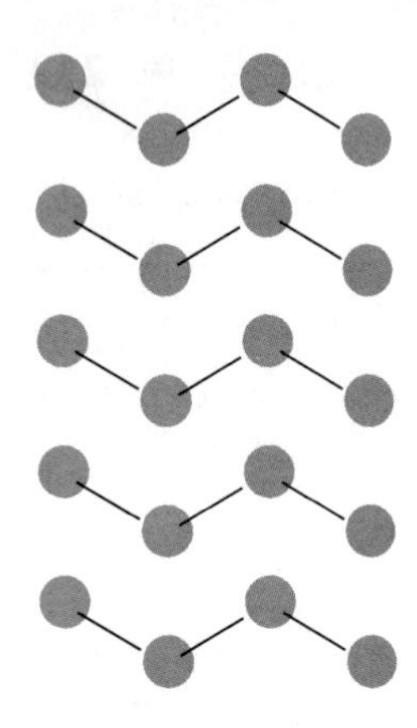

7

HALOALKANES

7.1 USES OF HALOALKANES

In this chapter we will consider compounds in which a halogen atom is bonded to an sp^3-hybridized carbon atom, hence the title haloalkanes. The chemistry of compounds in which a halogen atom is bonded to an sp^2-hybridized carbon atom—that is, aryl halides and vinyl halides—is very different. Compounds with a halogen atom bonded to an *sp*-hybridized carbon atom are very seldom encountered and are very unstable.

Haloalkanes, also commonly called alkyl halides, rarely occur in terrestrial plants and animals. Most of the haloalkanes encountered in everyday life were initially synthesized in a chemical laboratory and ultimately produced in large quantities by chemical industry. Many haloalkanes are potentially dangerous substances. For example, although trichloromethane (chloroform) and tetrachloromethane (carbon tetrachloride) are still used as solvents in the laboratory, they are no longer used for general commercial applications because they are on the Environmental Protection Agency's list of suspected carcinogens.

Another group of haloalkanes, collectively known as freons, contain fluorine and chlorine. Thus, freons are also called chlorofluorocarbons (CFCs). Freons are used as refrigerants and until recently as propellants in aerosol cans. The CFCs are unreactive, nontoxic, and nonflammable. How-

Halogen Compounds and Life in the Ocean

Although halogen-containing compounds are not common in terrestrial plants and animals, they do occur in marine organisms such as algae, sponges, and mollusks. The compounds made by these species have unusual structures, and some of them are clinically useful as antimicrobial, antifungal, and antitumor agents.

Some marine organisms use haloalkanes as part of a chemical defense mechanism to avoid predators. Thus, they can survive in an environment where virtually every organism is simultaneously predator and prey.

Marine algae produce halogen-containing metabolites that defend them against intense feeding by herbivores (animals that eat plants.) For example, red algae produce halogen compounds that repel most herbivores and provide a chemical "shield". However, the sea hare, a soft-bodied, shell-less mollusk, is not repelled by compounds from the red algae, which are a source of its food.

Most mollusks are protected from predators by a hard shell. Thus, the shell-less sea hare might seem to have little prospect of survival in the face of large carnivores. However, the sea hare converts the haloalkanes in red algae into closely related substances that it uses for its own chemical defense. The sea hare coats itself with a mucus that contains the halogenated compounds. These compounds protect its soft body against carnivorous fish. The structures of two of the closely related haloalkanes produced by red algae and the sea hare are shown below.

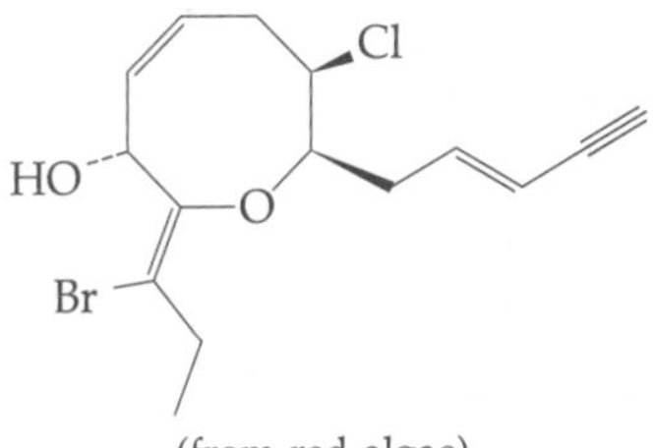

(from red algae)

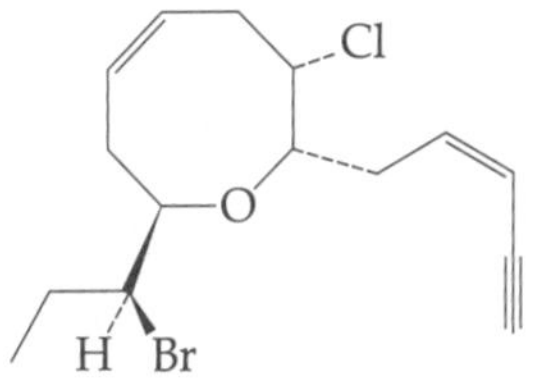

(from sea hare)

ever, it is now apparent that CFCs do reach the upper atmosphere, where they react to deplete the ozone layer, which protects us from ultraviolet radiation. Freon use is now severely restricted, and a search for substitute compounds is underway.

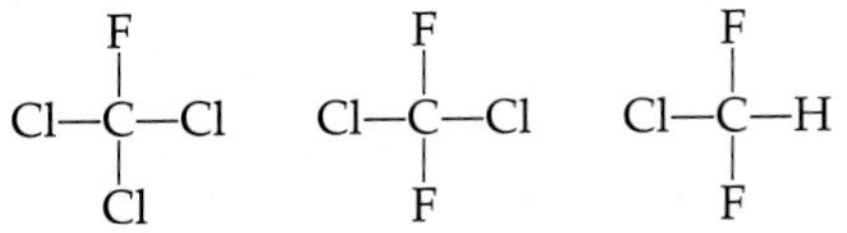

Cl—C(F)(Cl)—Cl
CFC-11

Cl—C(F)(F)—Cl
CFC-12

Cl—C(F)(F)—H
CFC-22

Certain chlorinated hydrocarbons are effective insecticides. DDT was introduced during World War II to control mosquitoes, which carry the malarial parasite responsible for millions of deaths each year. But DDT is only slowly degraded in nature; that is, DDT is a **persistent pesticide.**

FIGURE 7.1 **Chlorine-Containing Insecticides**

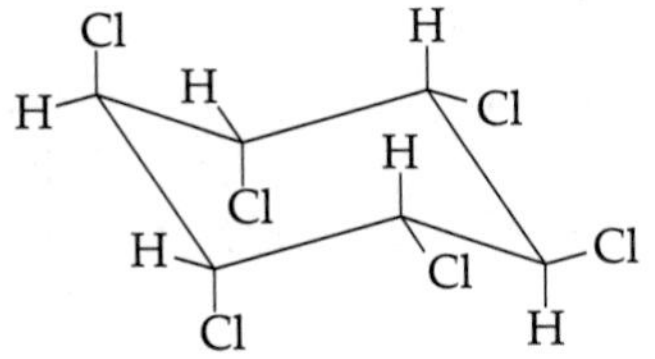

lindane

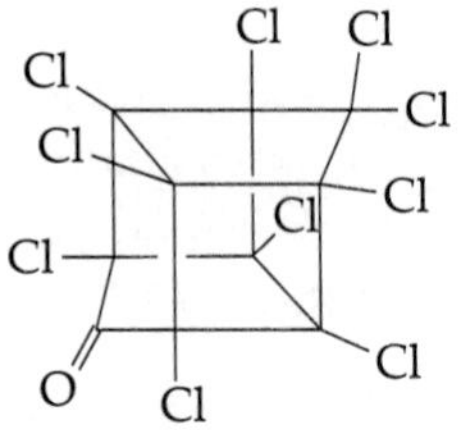

kepone

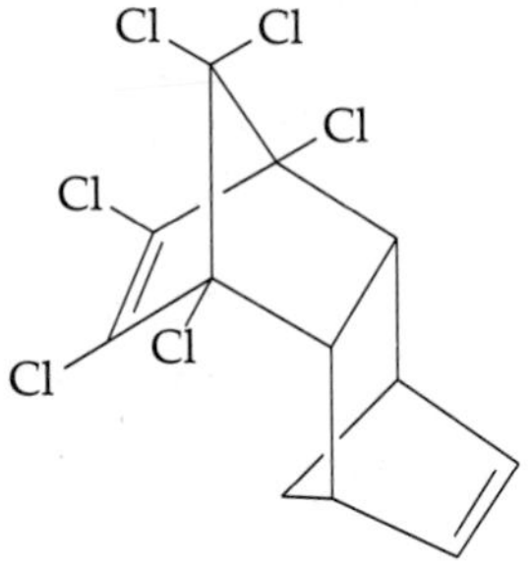

aldrin

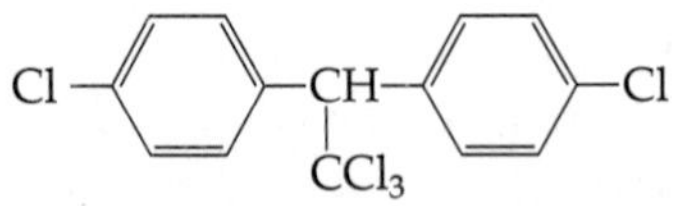

DDT [1,1-bis(*p*-chlorophenyl)-2,2,2-trichloroethane]

Although DDT is not toxic to humans, it is not easily metabolized in animals and tends to accumulate in fatty tissue. As a consequence, birds of prey—such as the eagle, osprey, and peregrine falcon—build up substantial quantities of DDT in their tissues. DDT alters calcium metabolism in these birds, and they produce eggs with thin, fragile shells that break before the baby bird hatches. As a result, the populations of eagle, osprey, and falcon have plummeted. In fact, DDT has threatened the survival of these species and is now banned in the United States in order to protect these birds.

Many other chlorinated hydrocarbons were once used in agriculture, but these have also been banned because they are environmentally unsafe. Figure 7.1 shows the structures of some of these compounds. Chlorinated hydrocarbons used as insecticides have been replaced by other classes of compounds, which are more polar and tend to have short lives in the environment. Among these insecticides are parathion and malathion, which contain phosphorus in P—O and P—S bonds. However, they are very toxic to human beings and act as neurotoxins against the central nervous system.

7.2 NOMENCLATURE OF HALOALKANES

Low molecular weight haloalkanes are often named by the name of the alkyl group followed by the name of the halide.

$CH_3—CH_2—Br$ — ethyl bromide

$CH_3—CH(CH_3)—F$ — isopropyl fluoride

$CH_3—C(CH_3)_2—Cl$ — *t*-butyl chloride

Haloalkanes are named in the IUPAC system by an extension of the rules outlined in Chapters 3 and 4.

1. The longest continuous chain of carbon atoms is chosen as the parent. If the parent chain has no branching alkyl groups, the position of the halogen atom is given by a number that corresponds to the carbon atom to which it is attached. The carbon chain is numbered so that the carbon atom bearing the halogen atom has the lowest number.

```
          H   H
          |   |
    CH3—C—C—CH3
    4     |3  |2  1
          H   Cl
```

This is 2-chlorobutane, not 3-chlorobutane.

2. If the parent chain has branching alkyl groups, number the chain from the end nearer the first substituent, regardless of whether it is an alkyl group or a halogen atom.

```
       CH3  H                          CH3  Br
        |   |                           |   |
  CH3—C——C—CH2—CH3       CH3—CH2—C——C—CH3
  1     |2  |3  4    5        5    4    |3  |2  1
        H   Cl                          H   H
```

3-chloro-2-methylpentane

2-bromo-3-methylpentane

3. If the compound contains two or more halogen atoms of the same type, they are indicated with the prefixes *di-*, *tri-*, etc. Each halogen atom is given a number that corresponds to its position in the parent chain.

```
        Cl  H                          Br  Br
        |   |                          |   |
  CH3—C—C—CH2—CH3        CH3—CH2—C—C—CH2—CH3
  1     |2  |3  4    5        6    5    |4  |3   2    1
        H   Cl                         H   Br
```

2,3-dichloropentane

3,4,4-tribromohexane

4. If a compound contains different halogen atoms, they are numbered according to their positions on the chain and listed in alphabetical order.

```
       1     2     3     4    5
  Cl—CH2—CH—CH2—CH—CH3
             |           |
             Br          CH3
```

2-bromo-1-chloro-4-methylpentane

5. If the chain can be numbered from either end based on the location of the substituents, begin at the end nearer the substituent that has alpha-

betical precedence, whether it is an alkyl group or a halogen atom.

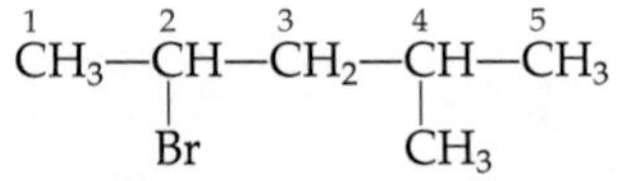

$$\overset{1}{CH_3}-\overset{2}{\underset{Br}{CH}}-\overset{3}{CH_2}-\overset{4}{\underset{CH_3}{CH}}-\overset{5}{CH_3}$$

2-bromo-4-methylpentane

6. Halocycloalkanes are numbered from the carbon atom bearing the halogen atom unless another group such as a double bond takes precedence. Carbon atoms in the ring are numbered to give the lower numbers to the substituents.

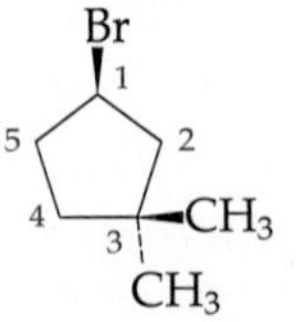

1-bromo-3,3-dimethylcyclopentane 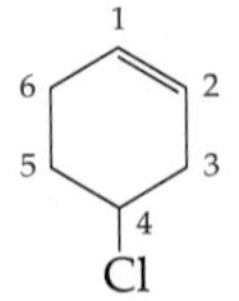   4-chlorocyclohexene

EXAMPLE 7.1 (*E*)-8-Bromo-3,7-dichloro-2,6-dimethyl-1,5-octadiene is produced by a species of red algae. Draw its structure.

Solution Draw the eight-carbon-atom parent chain, and select a direction for numbering it. Place double bonds between the C-1 and C-2 atoms and between the C-5 and C-6 atoms.

$$\underset{1}{C}=\underset{2}{C}-\underset{3}{C}-\underset{4}{C}-\underset{5}{C}=\underset{6}{C}-\underset{7}{C}-\underset{8}{C}$$

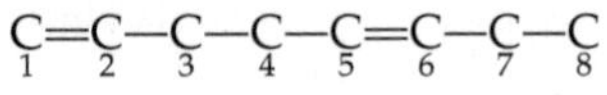

Next, place a bromine atom at the C-8 atom, chlorine atoms at the C-3 and C-7 atoms, and methyl groups at C-2 and C-6 atoms.

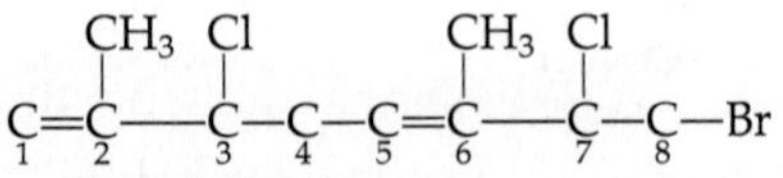

$$\underset{1}{C}=\overset{CH_3}{\underset{2}{C}}-\overset{Cl}{\underset{3}{C}}-\underset{4}{C}-\underset{5}{C}=\overset{CH_3}{\underset{6}{C}}-\overset{Cl}{\underset{7}{C}}-\underset{8}{C}-Br$$

We now have attached all the substituents to the chain. Next, we must arrange them so that the configuration about the double bond between the C-5 and C-6 atoms is *E*. To do this, we must assign priorities to each group at the C-5 and C-6 atoms. The higher priority groups at both the C-5 and C-6 atoms are the alkyl groups that are part of the parent carbon chain. Indicate the (*E*) configuration by placing the atoms of the carbon chain on the opposite sides of the double bond. Finally, fill in the requisite hydrogen atoms.

```
                                    Cl
                                    |
                     H             CH—CH2—Br
                      \           /
           Cl          C=C
           |          /    \
H2C=C—CH—CH2            CH3
      |
      CH3
```

7.3 STRUCTURE AND PROPERTIES OF HALOALKANES

We recall from Chapter 1 that the halogens are more electronegative than carbon. As a result, a carbon–halogen bond is polar. The carbon atom bears a partial positive charge, and the halogen atom has a partial negative charge.

$$\overset{\delta+}{\text{—C}}\text{—}\overset{\delta-}{\text{X}} \quad \text{where X = F, Cl, Br, I}$$

Because the carbon atom in a C—X bond has a partial positive charge, it is electrophilic and reacts with nucleophiles, which have a full or partial negative charge.

The atomic radii of the halogens increase going from top to bottom in the periodic table. This trend is reflected in the bond lengths of the carbon–halogen bond.

	CH_3—F	CH_3—Cl	CH_3—Br	CH_3—I
bond length (Å)	1.39	1.78	1.93	2.14

We recall from Chapter 1 that the polarizability of an atom is a measure of the ease with which its electrons can be distorted in an electric field. The polarizability of the halogen atoms increases as we move down the periodic table: F < Cl < Br < I. Highly polarizable atoms interact more strongly by London forces than less polarizable atoms. Therefore, intermolecular forces for haloalkanes increase in the order RF < RCl < RBr < RI. The effect of intermolecular forces is reflected in the boiling points of haloalkanes, which increase in the same order as the polarizability of their halogen components.

	CH_3CH_2—F	CH_3CH_2—Cl	CH_3CH_2—Br	CH_3CH_2—I
bp (°C)	−37.7	12.2	38.4	72

Halogen atoms, except for fluorine, are more massive than the other atoms most often found in organic compounds. The relatively high masses of halogens are reflected in the densities of haloalkanes. Haloalkanes are

TABLE 7.1 Boiling Points and Densities of Haloalkanes

Compound	Boiling point (°C)	Density (g/mL)
CH_3F	−78	
CH_3Cl	−24	
CH_3Br	4	
CH_3I	42	2.28
CH_2Cl_2	40	1.34
$CHCl_3$	61	1.50
CCl_4	77	1.60
CH_3CH_2F	−38	
CH_3CH_2Cl	12	
CH_3CH_2Br	38	1.46
CH_3CH_2I	72	1.94
$CH_3CH_2CH_2F$	3	
$CH_3CH_2CH_2Cl$	47	
$CH_3CH_2CH_2Br$	71	1.35
$CH_3CH_2CH_2I$	102	1.75

more dense than alkanes and, in fact, more dense than water (Table 7.1). Haloalkanes sink to the bottom of a container of water. In contrast, we recall that alkanes are less dense than water and, therefore, they float.

7.4 FORMATION OF ORGANOMETALLIC REAGENTS

Most reactions of haloalkanes described in this chapter involve displacement of the halogen by a nonmetal such as oxygen, nitrogen, or sulfur. However, haloalkanes also react with certain metals to produce compounds with a carbon to metal bond. These are **organometallic compounds.** When the metal in an organometallic compound is magnesium, the compound is called a **Grignard reagent,** because such compounds were synthesized by the French chemist Victor Grignard, who received the Nobel Prize in 1912.

$$R{-}X \xrightarrow[\text{ether}]{\text{Mg}} R{-}Mg{-}X$$

In a Grignard reagent, the R group may be a 1°, 2°, or 3° alkyl group as well as a vinyl or aryl group. The halogen may be Cl, Br, or I; fluorine compounds do not form Grignard reagents.

Grignard reagents are very reactive and versatile reactants that are used synthetically to form new carbon–carbon bonds. Examples of these reactions will be discussed in Chapter 10.

A Grignard reagent has a very polar carbon–magnesium bond in which the carbon atom has a partial negative charge and the metal a partial positive charge.

$$\overset{\delta-}{\text{—C}}\text{—}\overset{\delta+}{\text{MgX}}$$

This bond polarity is opposite that of the carbon–halogen bond of haloalkanes. Because the carbon atom in a Grignard reagent has a partial negative charge, it resembles a carbanion and reacts with electrophiles. Grignard reagents react rapidly with acidic hydrogen atoms in molecules such as alcohols and water. When a Grignard reagent reacts with water, the product is an alkane. Thus, the Grignard reagent provides a pathway for converting a haloalkane to an alkane in two steps.

$$\text{R—X} \xrightarrow[\text{ether}]{\text{Mg}} \text{R—Mg—X} \xrightarrow{\text{H}_2\text{O}} \text{R—H} + \text{HOMgX}$$

EXAMPLE 7.2 Devise a synthesis of $CH_3CH_2CHDCH_3$ starting from 1-butene and heavy water (D_2O).

Solution Reaction of a Grignard reagent R—MgBr with D_2O will yield R—D. The necessary Grignard reagent is obtained from the corresponding bromoalkane R—Br.

$$\text{CH}_3\text{CH}_2\underset{}{\overset{\text{Br}}{\text{C}}}\text{HCH}_3 \xrightarrow{\text{Mg}} \text{CH}_3\text{CH}_2\overset{\text{MgBr}}{\text{C}}\text{HCH}_3 \xrightarrow{\text{D}_2\text{O}} \text{CH}_3\text{CH}_2\overset{\text{D}}{\text{C}}\text{HCH}_3$$

The required 2-bromobutane can be prepared from 1-butene by adding HBr. This reaction occurs according to Markovnikov's rule, so that a hydrogen atom adds to the less substituted carbon atom of the double bond.

$$\text{CH}_3\text{CH}_2\text{CH}{=}\text{CH}_2 \xrightarrow{\text{HBr}} \text{CH}_3\text{CH}_2\overset{\text{Br}}{\text{C}}\text{HCH}_3$$

7.5 REACTIVITY OF HALOALKANES

A haloalkane has two sites of reactivity. One is at the carbon atom bonded to the halogen atom. This carbon atom is electropositive and attracts reactants with a negative or partial negative charge. That is, the electropositive carbon atom reacts with nucleophiles. The second site of reactivity in a haloalkane is the hydrogen atom bonded to the carbon atom adjacent to the carbon atom bonded to the halogen atom. This C—H bond is more acidic than the C—H bonds in alkanes because the halogen atom on the adjacent carbon atom withdraws electron density by an inductive effect.

$$-\underset{\underset{H^{\delta+}}{\uparrow}}{\overset{|}{C}}\rightarrow\underset{|}{\overset{|\,\delta+}{C}}\longrightarrow\overset{\delta-}{X}$$

First, let's consider the reaction of the nucleophile hydroxide ion with an electropositive carbon atom bonded to a halogen atom. Hydroxide ion can displace the halide ion in a substitution reaction. However, the hydroxide ion is not only a nucleophile but also a strong base that can remove a proton from the carbon atom adjacent to the one bonded to the halogen atom. When the proton is extracted, the halide ion simultaneously departs, and a double bond forms in an elimination reaction.

$$H—\ddot{\underset{..}{O}}:^- + —\underset{H}{\overset{|}{C}}—\underset{|}{\overset{|}{C}}—\ddot{\underset{..}{Br}}: \longrightarrow —\underset{H}{\overset{|}{C}}—\underset{|}{\overset{|}{C}}—\ddot{\underset{..}{O}}—H + :\ddot{\underset{..}{Br}}:^-$$

$$H—\ddot{\underset{..}{O}}:^- + —\underset{H}{\overset{|}{C}}—\underset{|}{\overset{|}{C}}—\ddot{\underset{..}{Br}}: \longrightarrow \rangle C{=}C\langle + H_2O + :\ddot{\underset{..}{Br}}:^-$$

The substitution and elimination reactions usually occur concurrently, and mixtures of products result. In the following sections, we will evaluate the conditions that cause one reaction to be favored over the other.

7.6 NUCLEOPHILIC SUBSTITUTION REACTIONS

In a nucleophilic substitution reaction, the nucleophile donates an electron pair to the electrophilic carbon atom to form a "carbon–nucleophile" bond. The nucleophile may be either negatively charged, as in the case of OH^-, or neutral, as in the case of NH_3. These two types of nucleophiles are commonly represented as Nu:$^-$ and Nu:, respectively. If the nucleophile is negatively charged, the product has no net charge. If the nucleophile is neutral, the product is positively charged. The nucleophile reacts with a haloalkane, which is called the **substrate,** that is, the compound upon which the reaction occurs.

$$Nu:^- + R—X \longrightarrow R—Nu + X:^-$$
$$Nu: + R—X \longrightarrow R—Nu^+ + X:^-$$

The nucleophile displaces an atom or group of atoms in the reaction, and this group is called the **leaving group.** It has an electron pair that was originally part of the C—X bond. The leaving group is usually negatively charged, as in the case of halide ions. Some neutral leaving groups such as H_2O will be examined in later chapters.

Haloalkanes are substrates in an extremely broad range of nucleophilic substitution reactions. Haloalkanes react with nucleophilic anions derived from the halogens, oxygen, sulfur, and even carbon. Haloalkanes also react with neutral nucleophiles that contain nitrogen, such as NH_3 or amines. We will discuss nitrogen-containing nucleophiles in Chapter 14. Some simple examples of nucleophilic substitution reactions using bromomethane as the substrate are described below.

First, let's consider the reaction of a haloalkane with a halide anion. One example is nucleophilic substitution of iodide ion for chloride or bromide ion in a haloalkane such as methyl chloride or methyl bromide.

$$:\ddot{\underset{..}{I}}:^- + CH_3-\ddot{\underset{..}{Br}}: \longrightarrow CH_3-\ddot{\underset{..}{I}}: + :\ddot{\underset{..}{Br}}:^-$$

A similar reaction occurs when the hydroxide ion of an alkali metal hydroxide replaces the halide ion to produce an alcohol. When the oxygen-containing nucleophile is an alkoxide ion (RO^-), the product is an ether. These two reactions will be discussed in Chapters 8 and 9.

$$H-\ddot{\underset{..}{O}}:^- + CH_3-\ddot{\underset{..}{Br}}: \longrightarrow \underset{\text{an alcohol}}{CH_3-\ddot{\underset{..}{O}}-H} + :\ddot{\underset{..}{Br}}:^-$$

$$CH_3CH_2-\ddot{\underset{..}{O}}:^- + CH_3-\ddot{\underset{..}{Br}}: \longrightarrow \underset{\text{an ether}}{CH_3CH_2-\ddot{\underset{..}{O}}-CH_3} + :\ddot{\underset{..}{Br}}:^-$$

Haloalkanes also undergo nucleophilic substitution reactions with sulfur-containing nucleophiles such as hydrogen sulfide ion, HS^-, and thiolate ions, RS^-. These reactions yield sulfur analogs of alcohols and ethers—namely, thiols and thioethers (Chapter 8).

$$H-\ddot{\underset{..}{S}}:^- + CH_3-\ddot{\underset{..}{Br}}: \longrightarrow \underset{\text{a thiol}}{CH_3-\ddot{\underset{..}{S}}-H} + :\ddot{\underset{..}{Br}}:^-$$

$$CH_3CH_2-\ddot{\underset{..}{S}}:^- + CH_3-\ddot{\underset{..}{Br}}: \longrightarrow \underset{\text{a thioether}}{CH_3CH_2-\ddot{\underset{..}{S}}-CH_3} + :\ddot{\underset{..}{Br}}:^-$$

Haloalkanes also react with carbon-containing nucleophiles resulting in the formation of carbon–carbon bonds. These reactions increase the length of the carbon chain. Thus, carbon chains that have different carbon skeletons than the substrate can be synthesized. One carbon-containing nucleophile is cyanide ion, CN^-, which produces a nitrile with the formula R—CN. Nitriles can be transformed into carboxylic acids (Chapter 12) and amines (Chapter 14). Carbon-containing nucleophiles derived from alkynes are called **alkynide** ions. These nucleophiles, the conjugate bases of alkynes (Chapter 4), react to form alkynes containing the carbon atoms of both the haloalkane and the alkynide.

$$:N\equiv C:^- + CH_3-\ddot{\underset{..}{Br}}: \longrightarrow CH_3-C\equiv N: + :\ddot{\underset{..}{Br}}:^-$$

$$R-C\equiv C:^- + CH_3-\ddot{\underset{..}{Br}}: \longrightarrow R-C\equiv C-CH_3 + :\ddot{\underset{..}{Br}}:^-$$

EXAMPLE 7.3 Using compounds containing no more than three carbon atoms, propose two ways to prepare $CH_3CH_2-S-CH_2CH_2CH_3$.

Solution A thioether can be prepared by reaction of a thiolate with a haloalkane. This thioether has two different alkyl groups bonded to sulfur. One alkyl group can be bonded to the sulfur atom in the thiolate and the other can be in the haloalkane. Thus, two possible combinations of reactants can yield the product.

$$\underset{\text{ethylthiolate}}{CH_3CH_2-S^-} + \underset{\text{1-bromopropane}}{Br-CH_2CH_2CH_3} \longrightarrow CH_3CH_2-S-CH_2CH_2CH_3$$

$$\underset{\text{bromoethane}}{CH_3CH_2-Br} + \underset{\text{propylthiolate}}{^-S-CH_2CH_2CH_3} \longrightarrow CH_3CH_2-S-CH_2CH_2CH_3$$

7.7 NUCLEOPHILICITY VERSUS BASICITY

In Section 7.5 we saw that haloalkanes undergo two kinds of reactions: (1) substitution of a halogen atom by another group and (2) elimination of a halogen (as X^-) and a hydrogen ion (as H^+) from adjacent carbon atoms to give an alkene. The type of reaction depends upon two properties of the nucleophile: (1) its nucleophilicity and (2) its basicity.

Nucleophilic substitution occurs at a rate that depends upon the ability of the nucleophile to displace the leaving group. This property of the nucleophile is called **nucleophilicity.** Because nucleophiles are electron-pair donors, they are also bases. When an elimination reaction occurs, a basic nucleophile extracts a proton from the carbon adjacent to the one bonded to the halogen atom. Therefore, the elimination reaction depends upon the basicity of the nucleophile. Hence, the terms *nucleophilicity* and *basicity* describe different phenomena. Nucleophilicity affects the rate of a substitution reaction at a carbon center. Basicity affects the equilibrium constant for an acid–

base reaction between the hydrogen atom of the haloalkane and the basic nucleophile in an elimination reaction.

The nucleophilicities and basicities of a series of structurally or chemically related nucleophiles—such as halogens, oxygen-containing anions, and sulfur-containing anions—are not always related in a simple way. However, trends are evident based on periodic properties of the elements. The relative rates of substitution reactions of various nucleophiles with iodomethane are given in Table 7.2.

Trends Within a Period

When a group of nucleophiles have the same charge and the nucleophilic ions are in the same period of the periodic table, nucleophilicity and basicity parallel each other and decrease from left to right in the period. Thus, the hydroxide ion is both more basic and more nucleophilic than the fluoride ion. The oxygen atom of the hydroxide ion is less electronegative and holds its electrons less firmly than the fluoride ion. As a consequence the nonbonding electrons of the oxygen atom can be more easily donated to carbon in a nucleophilic substitution reaction. The same periodic trend is observed for organic anions.

$R_3C:^-$	$R_2\ddot{N}:^-$	$R\ddot{\underset{..}{O}}:^-$	$:\ddot{\underset{..}{F}}:^-$
carbanion	amide ion	alkoxide ion	fluoride ion

most basic ⟶ least basic

most nucleophilic ⟶ least nucleophilic

Effect of Charge

When a nucleophile can exist either as an anion or as its uncharged conjugate acid, the anion is more nucleophilic than the neutral conjugate acid. Thus, alkoxide ions (RO^-) are better nucleophiles than alcohols (ROH), and hydroxide ion is a better nucleophile than water. Similarly, amide ions, RNH^-, are better nucleophiles than amines, RNH_2.

TABLE 7.2 Relative Rates of Reaction of Nucleophiles with Iodomethane

Nucleophile	Relative rate
CH_3OH	1
F^-	5×10^2
Cl^-	2×10^4
NH_3	3×10^5
Br^-	6×10^5
CH_3O^-	2×10^6
CN^-	5×10^6
I^-	3×10^7
CH_3S^-	1×10^8

Biological Substitution Reactions by Sulfur-Containing Nucleophiles

Many cellular molecules possess a nucleophilic sulfur atom. Of these, one of the most important is glutathione, which contains a sulfhydryl group (SH). Glutathione is present in a concentration of about 1–5 mM in most animal cells. It participates in several enzyme-catalyzed reactions. In some reactions, glutathione acts as a reducing agent. In other reactions, its sulfhydryl group reacts with various toxic intermediates that are produced when drugs are metabolized in liver cells. The type of reaction that occurs depends on the cell type and the nature of the enzyme catalyzing the reaction.

The sulfhydryl group of glutathione, often represented as GSH, is a nucleophile that displaces substituents bound to carbon. The various leaving groups of reactive metabolites are each represented as L. They are all strongly electron-withdrawing groups, and they make the carbon atom to which they are bound partially positive and susceptible to nucleophilic attack by an essential macromolecule with a nucleophilic center (M—Nu:) or by glutathione (GSH). Glutathione protects cells by reacting with toxic metabolites, represented below by R—L, before they react with cellular macromolecules (M—Nu:).

Glutathione not only protects cells against toxic byproducts of drug metabolism but also provides some degree of protection against toxic industrial chemicals, such as benzyl, allyl, and methyl halides. However, long-term exposure to these chemicals eventually overwhelms the protection provided by glutathione and damages the organism.

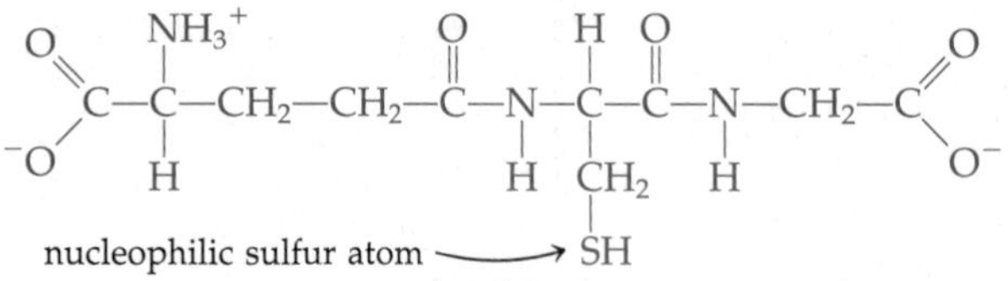

$$^{-}O_2C{-}CH(NH_3^+){-}CH_2{-}CH_2{-}C(=O){-}NH{-}CH(CH_2SH){-}C(=O){-}NH{-}CH_2{-}CO_2^{-}$$

nucleophilic sulfur atom → SH

glutathione

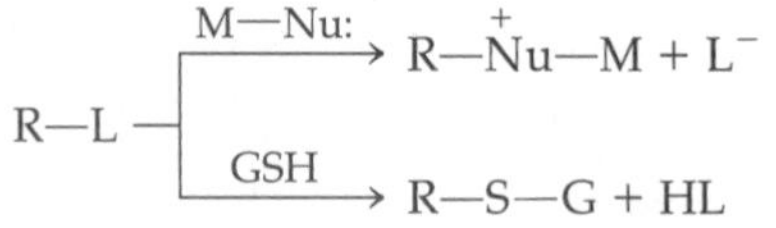

$$R{-}L \xrightarrow{M{-}Nu:} R{-}\overset{+}{N}u{-}M + L^-$$

$$R{-}L \xrightarrow{GSH} R{-}S{-}G + HL$$

$R{-}\ddot{O}:^-$ > $R{-}\ddot{O}{-}H$

better nucleophile   poorer nucleophile

$R{-}\ddot{N}H:^-$ > $R{-}NH_2:$ (R—N(H)(H):)

better nucleophile   poorer nucleophile

Trends Within a Group

When a group of nucleophiles contain atoms in the same group of the periodic table, the order of nucleophilicity runs opposite to the order of basicity. First, consider the nucleophilicities of thiolate and alkoxide. Thiolate is more nucleophilic but less basic than alkoxide ion (Table 7.2).

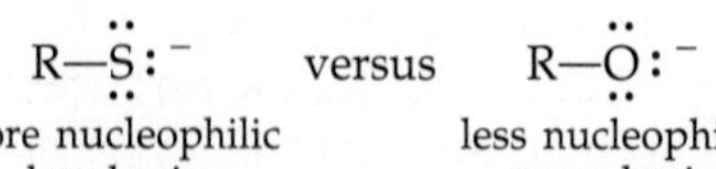

$R{-}\ddot{S}:^-$ versus $R{-}\ddot{O}:^-$

more nucleophilic, less basic   less nucleophilic, more basic

Similarly, when we compare the nucleophilicities and basicities of the halides, we find that the least basic one, iodide, is the most nucleophilic, whereas the most basic one, fluoride, is the least nucleophilic.

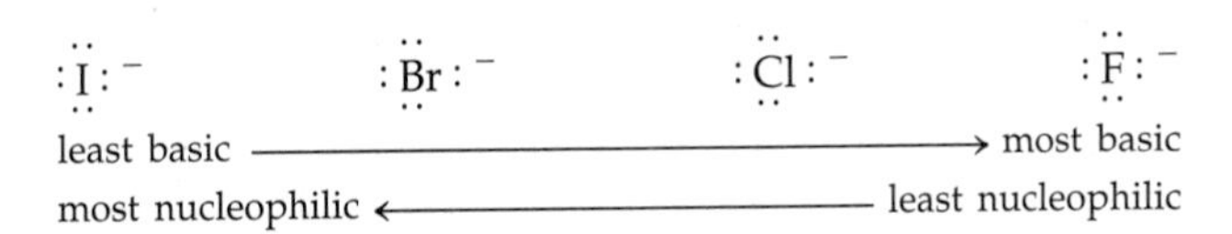

The order of nucleophilicities can be explained in terms of the polarizability of the atoms. We recall that the atomic radii of elements increase going down a family in the periodic table. As a consequence, the electrons are more polarizable. Thus, iodide ion is more polarizable and more nucleophilic than bromide ion. The polarizability of a nucleophile is important in a nucleophilic substitution reaction because an electron pair in the nucleophile forms a bond to the electrophilic carbon atom during the reaction. Basicity, which is a measure of the ability to bond to a proton, a center of highly concentrated positive charge, is not as sensitive to polarizability.

7.8 MECHANISMS OF SUBSTITUTION REACTIONS

Nucleophilic substitution reactions can occur by either of two mechanisms. These mechanisms are described by the symbols S_N2 and S_N1, where the term S_N means: substitution, nucleophilic. The numbers 2 and 1 refer to the number of reactants present in the transition state for the rate-determining step.

The S_N2 Mechanism

The S_N2 mechanism is a one-step process in which the nucleophile attacks the substrate and the leaving group, L, departs simultaneously. The reaction occurs in one step and is said to be **concerted.** The substrate and the nucleophile are both present in the transition state for this step. That is, two molecules are present, and the reaction is **bimolecular** as indicated by the number 2 in the S_N2 symbol. As a consequence, the reaction rate depends on the concentrations of both the nucleophile and the substrate. If the substrate concentration is doubled, the reaction rate is doubled. Similarly, if the concentration of the nucleophile is doubled, the rate again doubles. This relationship between the rate and the concentration of the reactants exists because the reactants must collide for reaction to occur. The probability that the nucleophile will collide with the substrate increases if the concentration of either or both species is increased.

Let's consider the S_N2 reaction of hydroxide ion with chloromethane to give methanol and chloride ion. This reaction is shown with the energy diagram in Figure 7.2. In this plot of energy versus progress of the reaction,

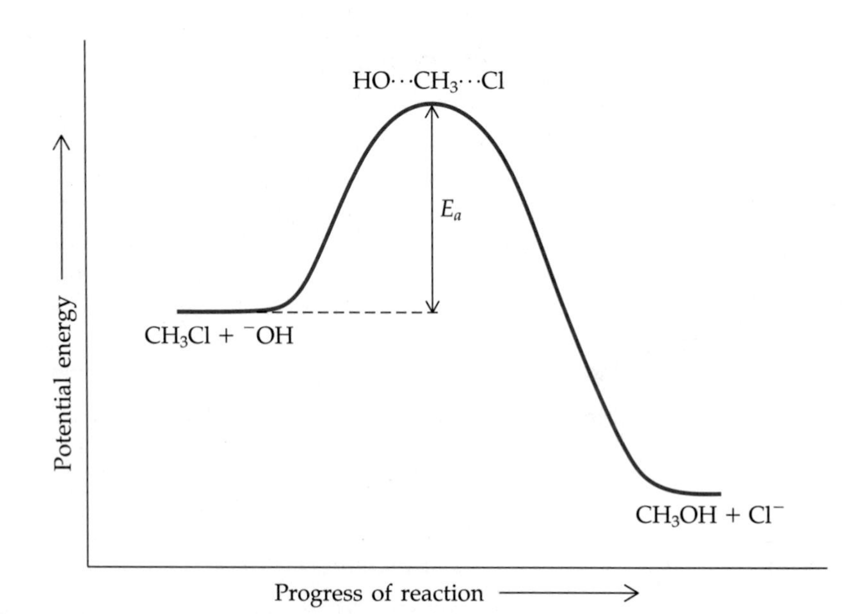

FIGURE 7.2
Activation Energy and the S_N2 Reaction Mechanism
The reaction of chloromethane with hydroxide ion occurs in a single-step process. The activation energy reflects the stability of the transition state, which in turn is related to the structures of the alkyl group, the nucleophile, and the leaving group.

we see that the point of highest energy—that is, the transition state—contains two species: hydroxide ion and the substrate. In the transition state neither the nucleophile nor the leaving group is fully bonded to carbon. But as the reaction proceeds through the transition state, a bond between carbon and hydroxide ion forms and the bond between carbon and chlorine breaks.

The energy of the transition state is related to the rate of the reaction. If the energy of the transition state is low, the reaction is fast. The rate of reaction for haloalkanes via the S_N2 mechanism is primary > secondary > tertiary. This trend is observed because adding alkyl groups to the carbon atom bearing the halogen atom shields the carbon atom from attack by

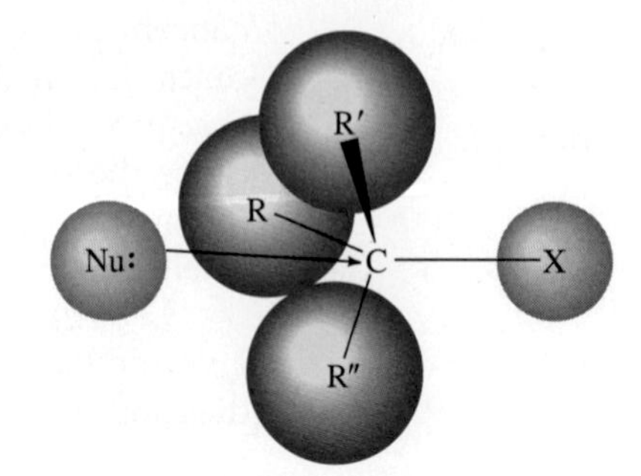

FIGURE 7.3
Effect of Steric Hindrance on the S_N2 Reaction
When R, R′, and R″ are all alkyl groups, the reaction rate is slow because the nucleophile cannot easily attack the carbon atom. As alkyl groups are replaced by hydrogen atoms the reaction rate becomes faster. The rate of reaction is 1° > 2° > 3°.

nucleophiles (Figure 7.3). This crowding of alkyl groups around the reactive carbon atom is called **steric hindrance.** Alkyl groups block the approach of the nucleophile and slow the rate of the reaction.

When (*S*)-2-bromobutane reacts with sodium hydroxide, (*R*)-2-butanol is formed. The reaction occurs with inversion of configuration. Thus, the nucleophile approaches the electrophilic carbon atom from the back and the leaving group simultaneously departs from the front of the substrate. This process explains why the carbon atom undergoes inversion of configuration in an S_N2 mechanism.

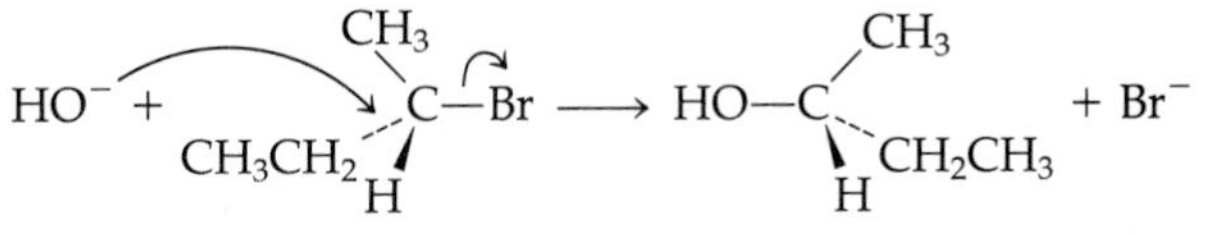

$$HO^- + CH_3(CH_3CH_2)(H)C{-}Br \longrightarrow HO{-}C(CH_3)(H)(CH_2CH_3) + Br^-$$

Biological Methylation by an S_N2 Reaction

An S_N2 reaction occurs in all living cells in which a methyl group is transferred from a methylating agent called *S*-adenosylmethionine (SAM) to various biological substrates.

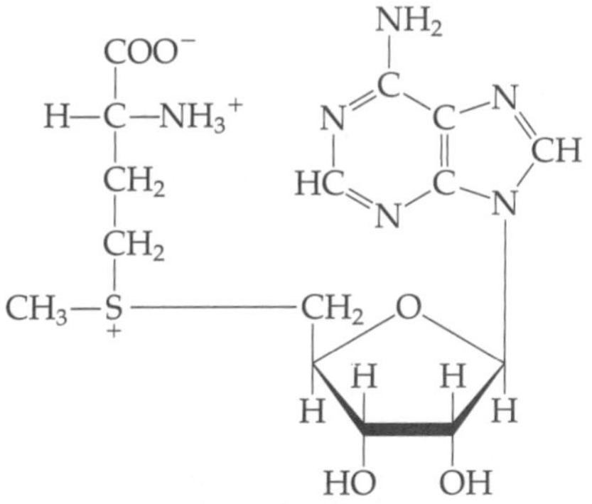

S-adenosylmethionine

Three carbon atoms are bonded to the positively charged sulfur atom, known as a sulfonium ion, which is part of a leaving group called *S*-adenosylhomocysteine. In short, SAM reacts with nucleophiles, which displace *S*-adenosylhomocysteine so that the methyl group is transferred from SAM to the nucleophile. This nucleophilic substitution reaction is shown below with a generic nucleophile (Nu$:^-$) and an abbreviated representation of SAM.

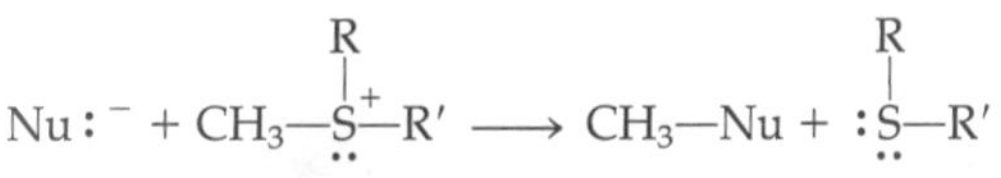

$$Nu:^- + CH_3{-}\overset{+}{S}(R){-}R' \longrightarrow CH_3{-}Nu + :S(R){-}R'$$

An important example of methyl-group transfer from SAM to a nucleophile occurs in the biosynthesis of the neurotransmitter epinephrine. In this reaction, an amino group of norepinephrine attacks the methyl-group carbon atom of *S*-adenosylmethionine in an S_N2 reaction to produce epinephrine. The leaving group is *S*-adenosylhomocysteine—a compound that results from the loss of a methyl group from *S*-adenosylmethionine.

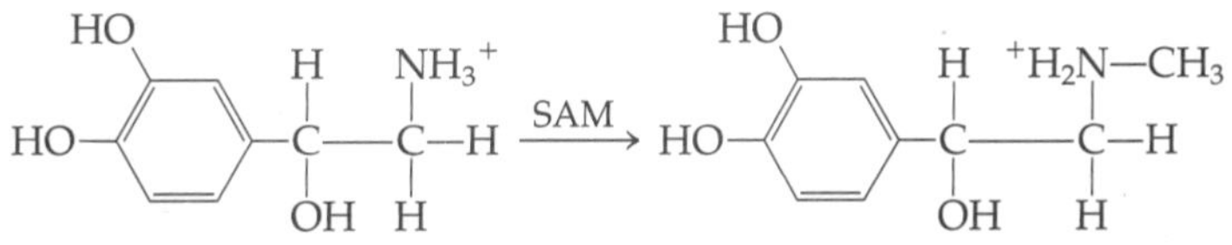

$$(HO)_2C_6H_3{-}CH(OH){-}CH_2{-}NH_3^+ \xrightarrow{SAM} (HO)_2C_6H_3{-}CH(OH){-}CH_2{-}{}^+H_2N{-}CH_3$$

If the hydroxide ion had bonded to the carbon atom on the side originally occupied by the leaving group, the product would have the same configuration as the reactant.

The S_N1 Mechanism

A nucleophilic substitution reaction that occurs by an S_N1 mechanism proceeds in two steps. In the first step, the bond between the carbon atom and the leaving group breaks to produce a carbocation and, most commonly, an anionic leaving group. In the second step, the carbocation reacts with the nucleophile to form the product. The two-step process is shown below.

$$\underset{\text{substrate}}{\text{C—L}} \xrightarrow{\text{slow}} \underset{\text{carbocation}}{\text{C}^+} + :\text{L}^-$$

$$\underset{\text{carbocation}}{\text{C}^+} + \underset{\text{nucleophile}}{:\text{Nu}^-} \xrightarrow{\text{fast}} \text{C—Nu}$$

The first step in an S_N1 reaction—formation of a carbocation—is the slow or rate-determining step. The second step—formation of a bond between the nucleophile and the carbocation—occurs very rapidly. The slow step of the reaction involves only the substrate, so the reaction is **unimolecular.** Because only one molecular species is present in the transition state, the rate of the reaction depends only on the concentration of the substrate and not on the concentration of the nucleophile.

An energy diagram showing the progress of the reaction by an S_N1 mechanism is shown in Figure 7.4. The rate of the reaction depends on the

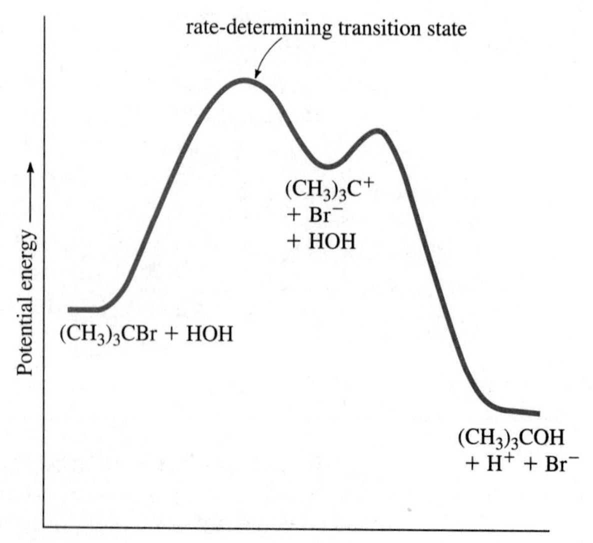

FIGURE 7.4
Activation Energy and the S_N1 Reaction Mechanism
The reaction of 2-bromo-2-methylpropane occurs via a tertiary carbocation formed in a rate-determining step that does not involve a nucleophile. In the second, faster step the carbocation reacts with water to form the product alcohol.

FIGURE 7.5
Stereochemistry of an S_N1 Reaction
The reaction of (*S*)-3-bromo-3-methylhexane with water occurs via a tertiary carbocation that is planar and, thus, achiral. Subsequent attack of the nucleophile water can occur from either side of the plane.

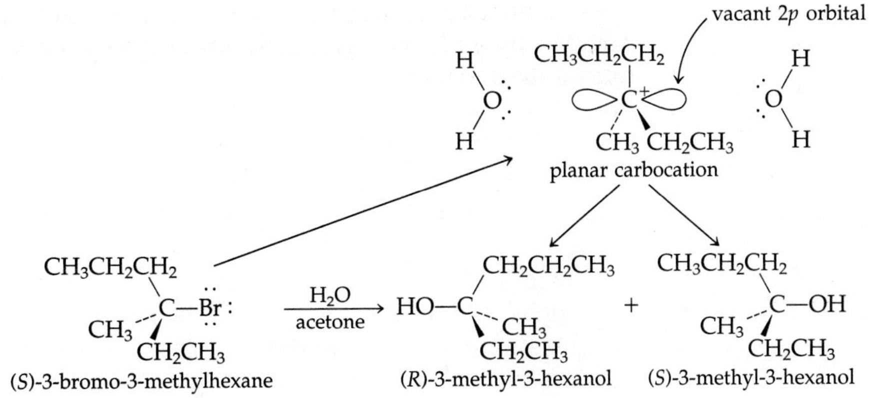

activation energy (Section 2.10) required to form the carbocation intermediate. The activation energy required for the second step, addition of the nucleophile to the carbocation, is much smaller, so step 2 is very fast. The rate of the second step has no effect on the net rate of the reaction.

The rate of S_N1 type reactions decreases in the order 3° > 2° > 1°. This trend is exactly the reverse of the trend observed in S_N2 reactions. The relative reactivity of haloalkanes in S_N1 reactions corresponds to the relative stability of carbocation intermediates that form during the reaction. We recall from Chapter 4 that the order of stability of carbocations is tertiary > secondary > primary. A relatively stable tertiary carbocation forms faster than a less-stable secondary carbocation, which in turn forms very much faster than a highly unstable primary carbocation. However, S_N1 mechanisms are favored by resonance-stabilized primary carbocations such as benzyl and allyl.

In the preceding section we saw that S_N2 reactions at chiral centers occur with inversion of configuration. In contrast, an S_N1 reaction gives a racemic mixture of enantiomers that has no optical rotation. For example, (*S*)-3-bromo-3-methylhexane reacts with water to give a racemic mixture of 3-methyl-3-hexanols. The reaction occurs via an achiral carbocation intermediate with a plane of symmetry (Figure 7.5). The carbocation intermediate's plane of symmetry allows the nucleophile to attack equally well from either side. The product is then a racemic mixture of enantiomers. Thus, a chiral substrate loses chirality in a reaction that occurs by an S_N1 mechanism.

7.9 S_N2 VERSUS S_N1 REACTIONS

Now that we have outlined the general properties of S_N2 and S_N1 reactions, let's see if we can predict which one is likely to occur. We will consider (1) the structure of the substrate, (2) the nucleophile, and (3) the nature of the solvent.

Structure of the Substrate

Primary haloalkanes almost always react in nucleophilic substitution reactions by the S_N2 mechanism, whereas tertiary haloalkanes react by the S_N1 mechanism. Secondary haloalkanes may react by either mechanism depending on the nature of the nucleophile and the solvent.

relative rate, S_N2: primary > secondary > tertiary
relative rate, S_N1: tertiary > secondary > primary

Effect of the Nucleophile

The nature of the nucleophile sometimes determines the mechanism of a nucleophilic substitution reaction. If the nucleophile is a highly polarizable species such as thiolate ion, RS^-, it tends to react by an S_N2 reaction. If the nucleophile is an uncharged species such as H_2O or CH_3OH, an S_N1 mechanism is more likely.

Effect of Solvent

Until now, we have neglected the role of solvent in nucleophilic substitution reactions, but the choice of solvent can tip the balance in favor of a particular mechanism. We noted that secondary haloalkanes can react by either an S_N1 or an S_N2 mechanism. In these cases, the polarity of the solvent plays an important role. The S_N1 process forms a carbocation intermediate. Because a polar solvent stabilizes charged species better than a nonpolar solvent, it increases the rate of S_N1 reactions much more than it affects the rate of S_N2 reactions.

The solvent also affects nucleophilicity. Solvents that have proton-donating ability, such as alcohols, are called **protic solvents.** A protic solvent interacts strongly with nucleophilic anions by forming hydrogen bonds with the unshared pairs of electrons on the nucleophiles. When the nucleophile is hydrogen-bonded to the solvent, its nucleophilicity decreases.

Solvents that do not have protons available for hydrogen bonding to nucleophiles are called **aprotic solvents.** Examples of aprotic solvents include dimethylformamide (DMF) and dimethyl sulfoxide (DMSO).

$$CH_3{-}\underset{}{N}(CH_3){-}C({=}\ddot{O}{:}){-}H \qquad CH_3{-}\ddot{S}(\ddot{O}{:}){-}CH_3$$

dimethylformamide   dimethyl sulfoxide

The electron pairs of the oxygen atoms of these aprotic solvents solvate cations but not anions. For example, these solvents tie up the cation of KCN, but leave the CN^- ion free. Consequently, the nucleophilicity of CN^- is greater in dimethyl sulfoxide than in ethanol (CH_3CH_2OH). Thus, an aprotic solvent such as dimethyl sulfoxide favors an S_N2 reaction.

EXAMPLE 7.4 Explain why the reaction of 3-bromocyclohexene with methanol (CH_3OH) is faster than the reaction of bromocyclohexane with methanol.

Br Br

bromocyclohexane 3-bromocyclohexene

Solution Both substrates are secondary bromides. The reaction with methanol, a neutral nucleophile, will tend to occur by an S_N1 process. Although both carbocations are secondary, the one derived from 3-bromocyclohexene is also a resonance-stabilized allylic carbocation. Resonance stabilization enhances the rate of its formation. No such stabilization occurs in the reaction of bromocyclohexane.

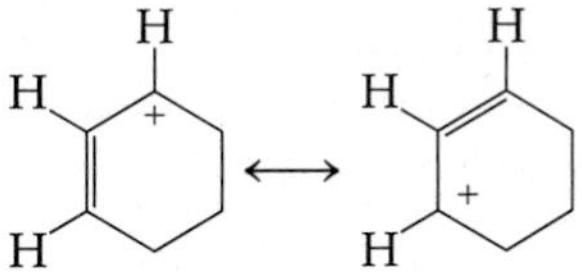

resonance-stabilized allylic carbocation

7.10 MECHANISMS OF ELIMINATION REACTIONS

We noted in Section 7.5 that when a haloalkane undergoes a nucleophilic substitution reaction, a competing elimination reaction may also occur because a nucleophile is also a base. In a substitution reaction, the nucleophile attacks an electrophilic carbon atom in the substrate. In an elimination reaction, the nucleophile acts as a base and removes a proton from a carbon atom adjacent to the carbon atom bearing the halogen. Both the proton and the leaving group are eliminated and a carbon–carbon π bond is formed. When a proton and a halogen are lost by a substrate, the elimination reaction is called **dehydrohalogenation.**

An elimination reaction is denoted by the symbol E. Elimination reactions can occur by two mechanisms, designated E1 and E2. An E1 or an E2 process may compete with S_N1 and S_N2 reactions.

The E2 Mechanism

Like the S_N2 reaction mechanism, the E2 mechanism is a one-step, concerted process. In an E2 dehydrohalogenation reaction, the base (nucleophile) removes a proton on the carbon atom adjacent to the carbon atom containing the leaving group. As the proton is removed, the leaving group departs and a double bond forms.

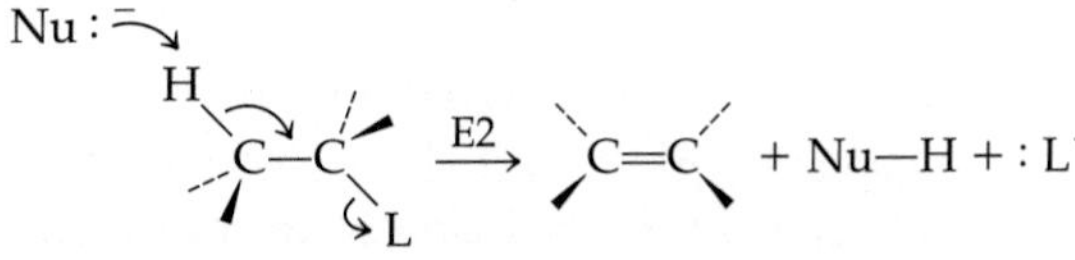

$$\text{Nu:}^- + \text{H–C–C–L} \xrightarrow{E2} \text{C=C} + \text{Nu–H} + :\text{L}^-$$

Like the S_N2 reaction, an E2 reaction requires a precise molecular arrangement. The anti conformation of the hydrogen and halide atoms is preferred for the reaction because it aligns the orbitals properly for the formation of the π bond. In terms of "electron pushing", we can visualize the process as the removal of the proton to provide an electron pair that attacks the neighboring carbon atom from the back to displace the leaving group.

An E2 reaction occurs at a rate that depends on the concentrations of both the substrate and the base. If the substrate concentration is doubled, the reaction rate also doubles, as in S_N2 processes. Thus, both E2 and S_N2 mechanisms are affected in the same way, and the two mechanisms compete with each other.

The E1 Mechanism

We recall that an S_N1 reaction proceeds in two steps, and that the rate-determining step is formation of a carbocation intermediate. Similarly, an E1 mechanism occurs in two steps, and the rate-determining step is the formation of a carbocation. Thus, just as E2 and S_N2 mechanisms compete with each other, an E1 mechanism competes with an S_N1 mechanism. Because the rate-determining step of an E1 reaction involves only the substrate, the formation of the carbocation is a unimolecular reaction. If the carbocation reacts with a nucleophile at the positively charged carbon atom, the result is substitution. But if the nucleophile acts as a base and removes a proton from the carbon atom adjacent to the cationic center, the result is the loss of HX and the formation of a π bond. These possibilities are outlined below.

$$\text{H}-\overset{|}{\underset{|}{\text{C}}}-\overset{|}{\underset{|}{\text{C}}}-\text{X} \xrightarrow{\text{slow}} \text{H}-\overset{|}{\underset{|}{\text{C}}}-\overset{|}{\underset{|}{\text{C}}}^{+} + \text{X}^{-}$$

$$\text{H}-\overset{|}{\underset{|}{\text{C}}}-\overset{|}{\underset{|}{\text{C}}}^{+} \xrightarrow[\text{Nu}^{-}]{\text{fast}} \text{H}-\overset{|}{\underset{|}{\text{C}}}-\overset{|}{\underset{|}{\text{C}}}-\text{Nu}$$

$$\text{H}-\overset{|}{\underset{|}{\text{C}}}-\overset{|}{\underset{|}{\text{C}}}^{+} \xrightarrow[\text{Nu}^{-}]{\text{fast}} \gt\text{C}=\text{C}\lt + \text{H}-\text{Nu}$$

7.11 EFFECT OF STRUCTURE ON COMPETING REACTIONS

Let's now examine the variety of product mixtures that result from competing substitution and elimination processes. We will divide our discussion according to the type of haloalkane.

Tertiary Haloalkanes

Tertiary haloalkanes can undergo substitution reactions only by an S_N1 mechanism because there is too much steric hindrance for an S_N2 reaction to occur. A tertiary haloalkane can undergo an elimination reaction by either an E2 or E1 process. The mechanism depends on the basicity of the nucleophile and the polarity of the solvent. If the nucleophile is a weak base, S_N1 and E1 processes compete, and the amounts of the two types of products depend only on the carbocation formed. For example, 2-bromo-2-methylbutane reacts in ethanol to give about 64% of the product from an S_N1 process.

$$CH_3CH_2C(CH_3)_2Br \xrightarrow{CH_3CH_2OH} CH_3CH_2C(CH_3)_2OCH_2CH_3 + CH_3CH{=}C(CH_3)CH_3 + CH_3CH_2C(CH_3){=}CH_2$$

2-bromo-2-methylbutane; 64% (S_N1 product); 30%, 6% (E1 products)

However, if sodium ethoxide, a strongly basic nucleophile, is added to the ethanol, an E2 process competes with the substitution reaction. The amount of elimination product is increased to a total of about 93% of the product mixture; only 7% of the ether product is formed.

$$CH_3CH_2C(CH_3)_2Br \xrightarrow[CH_3CH_2O^-]{CH_3CH_2OH} CH_3CH_2C(CH_3)_2OCH_2CH_3 + CH_3CH{=}C(CH_3)CH_3 + CH_3CH_2C(CH_3){=}CH_2$$

2-bromo-2-methylbutane; 7% (S_N1 product); 93% (E2 products)

Primary Haloalkanes

Primary haloalkanes can undergo either S_N2 or E2 reactions. They do not undergo S_N1 or E1 reactions because a primary carbocation is very unstable. Primary haloalkanes react with strongly nucleophilic, weakly basic reactants such as ethyl thiolate, $CH_3CH_2S^-$, exclusively by an S_N2 process. However, a primary haloalkane reacts with ethoxide ion, which is a weaker nucleophile but a stronger base than ethyl thiolate, to give some elimination product.

$$CH_3CH_2CH_2CH_2{-}Br \xrightarrow{CH_3CH_2S^-} CH_3CH_2CH_2CH_2{-}S{-}CH_2CH_3$$

1-bromobutane; exclusive product (S_N2 mechanism)

$$CH_3CH_2CH_2CH_2{-}Br \xrightarrow{CH_3CH_2O^-} CH_3CH_2CH_2CH_2{-}O{-}CH_2CH_3 + CH_3CH_2CH{=}CH_2$$

1-bromobutane; 90% (S_N2 mechanism); 10% (E2 mechanism)

If a primary haloalkane is treated with *t*-butoxide ion, $(CH_3)_3CO^-$, instead of ethoxide, the amount of elimination product increases significantly. The *t*-butoxide ion is not only more basic than the ethoxide ion but also much more sterically hindered. The combination of these two factors favors elimination by an E2 process over substitution by an S_N2 process.

$$CH_3CH_2CH_2CH_2{-}Br \xrightarrow{(CH_3)_3CO^-} CH_3CH_2CH_2CH_2{-}O{-}C(CH_3)_3 + CH_3CH_2CH{=}CH_2$$

1-bromobutane; 15% (S_N2 mechanism); 85% (E2 mechanism)

Secondary Haloalkanes

Secondary haloalkanes can react by S_N2, E2, S_N1, and E1 mechanisms, and it is sometimes difficult to predict which will occur in a given reaction. However, secondary haloalkanes tend to react with strong nucleophiles that are weak bases, such as thiolates or cyanide ion, by an S_N2 process.

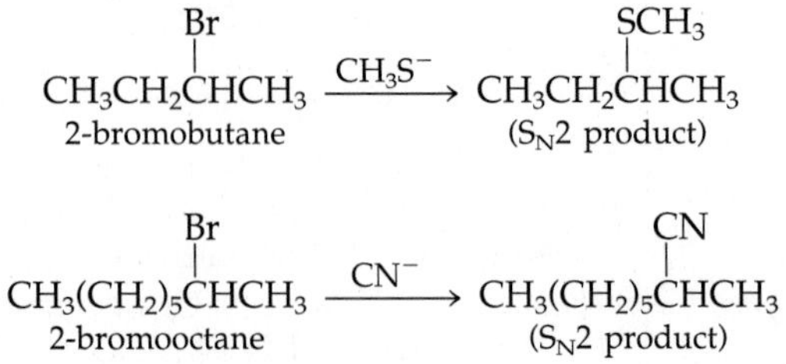

$$\underset{\text{2-bromobutane}}{CH_3CH_2\overset{\overset{Br}{|}}{C}HCH_3} \xrightarrow{CH_3S^-} \underset{(S_N2\text{ product})}{CH_3CH_2\overset{\overset{SCH_3}{|}}{C}HCH_3}$$

$$\underset{\text{2-bromooctane}}{CH_3(CH_2)_5\overset{\overset{Br}{|}}{C}HCH_3} \xrightarrow{CN^-} \underset{(S_N2\text{ product})}{CH_3(CH_2)_5\overset{\overset{CN}{|}}{C}HCH_3}$$

On the other hand, a secondary haloalkane tends to react with weak nucleophiles that are also weak bases, such as ethanol, by an S_N1 process with some accompanying E1 product. In contrast, we can tip the scales in the other direction by adding sodium ethoxide to ethanol. By adding a strong base, we find that the product of the S_N1 reaction drops to 18% of the total, and E2 products account for the rest.

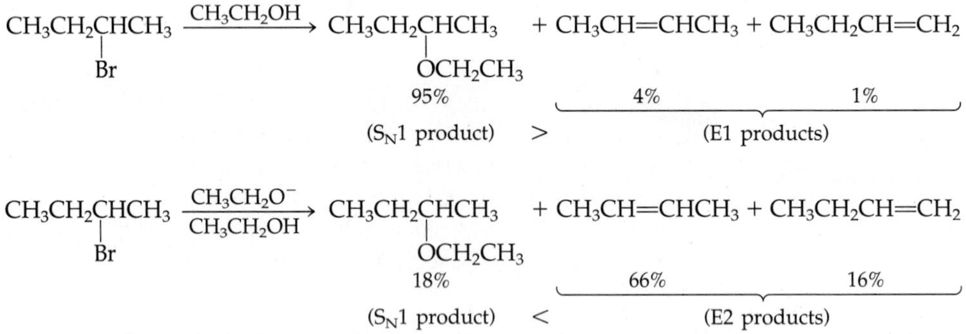

$$\underset{\underset{Br}{|}}{CH_3CH_2CHCH_3} \xrightarrow{CH_3CH_2OH} \underset{\underset{OCH_2CH_3}{|}}{CH_3CH_2CHCH_3} + CH_3CH{=}CHCH_3 + CH_3CH_2CH{=}CH_2$$

95% (SN1 product) > 4%, 1% (E1 products)

$$\underset{\underset{Br}{|}}{CH_3CH_2CHCH_3} \xrightarrow[CH_3CH_2OH]{CH_3CH_2O^-} \underset{\underset{OCH_2CH_3}{|}}{CH_3CH_2CHCH_3} + CH_3CH{=}CHCH_3 + CH_3CH_2CH{=}CH_2$$

18% (SN1 product) < 66%, 16% (E2 products)

EXAMPLE 7.5 Which of following two methods of preparing $CH_3CH_2OCH(CH_3)_2$ will give the better yield?

$$\underset{\text{ethoxide}}{CH_3CH_2{-}O^-} + \underset{\text{2-bromopropane}}{Br{-}CH(CH_3)_2} \longrightarrow CH_3CH_2{-}O{-}CH(CH_3)_2$$

$$\underset{\text{bromoethane}}{CH_3CH_2{-}Br} + \underset{\text{isopropoxide}}{{}^-O{-}CH(CH_3)_2} \longrightarrow CH_3CH_2{-}O{-}CH(CH_3)_2$$

Solution An ether can be prepared by treating a haloalkane with an alkoxide in an S_N2 reaction. This ether has two different alkyl groups bonded to oxygen: one from the alkoxide and the other from the haloalkane.

The first reaction is a nucleophilic displacement at a secondary center by a nucleophile that is also a strong base. A competing elimination reaction to yield propene will also occur. The second reaction occurs by an S_N2 reaction at a primary center, which tends to occur with little competition from an elimination reaction. Therefore, the second reaction is the better way to make the desired product.

EXERCISES

Nomenclature

7.1 What is the IUPAC name for each of the following compounds?
(a) vinyl fluoride (b) allyl chloride (c) propargyl bromide

7.2 What is the IUPAC name for each of the following compounds?
(a) $(CH_3)_3CCH_2Cl$ (neopentyl chloride)
(b) $(CH_3)_2CHCH_2CH_2Br$ (isoamyl bromide)
(c) $C_6H_5CH_2CH_2F$ (phenethyl fluoride)

7.3 Draw the structure of each of the following compounds.
(a) *cis*-1-bromo-2-methylcyclopentane (b) 3-chlorocyclobutene
(c) (*E*)-1-fluoro-2-butene (d) (*Z*)-1-bromo-1-propene

7.4 What is the IUPAC name for each of the following compounds?

(a) 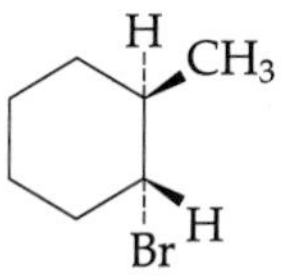 H, CH₃ / Br, H (cyclohexane)

(b) 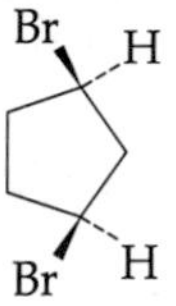 Br, H / Br, H (cyclopentane)

(c) 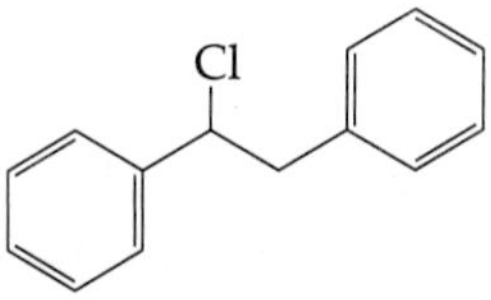 Cl

(d) 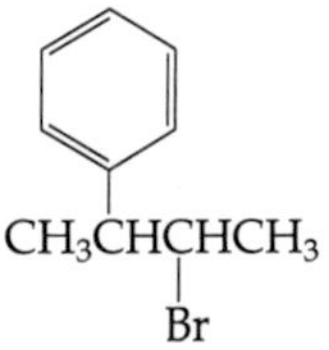 $CH_3CHCHCH_3$ with Br

Properties of Haloalkanes

7.5 Which compound is more polar, methylene chloride (CH_2Cl_2) or carbon tetrachloride (CCl_4)?

7.6 Tribromomethane is more polar than tetrabromomethane, but their boiling points are 150 and 189 °C, respectively. Explain why the more polar compound has the lower boiling point.

7.7 The densities of chloroiodomethane and dibromomethane are 2.42 and 2.49 g/mL, respectively. Why are these values similar?

7.8 The density of 1,2-dichloroethane is 1.26 g/mL. Predict the density of 1,1-dichloroethane.

7.9 The dipole moment of (*Z*)-1,2-dichloroethene is 1.90 Debye. Predict the dipole moment of the E isomer.

7.10 The dipole moment of 1,2-dichloroethane is 1.19 Debye. What does this value indicate about the conformational equilibrium of this compound?

Grignard Reagents

7.11 Devise a synthesis of 1-deutero-1-methylcyclohexane starting from 1-methylcyclohexene.

7.12 Devise a synthesis of 1,4-dideuterobutane starting from any organic compound that does not contain deuterium.

7.13 Suggest a reason why a Grignard reagent cannot be formed using ethanol (CH_3CH_2OH) as a solvent.

7.14 Reaction of 1,2-dibromoethane with magnesium yields ethylene and magnesium bromide. Write a mechanism that accounts for this reaction.

Nucleophilic Substitution Reactions

7.15 Write the structure of the product obtained for each of the following combinations of reactants.
(a) 1-chloropentane and sodium iodide
(b) 1,3-dibromopropane and excess sodium cyanide
(c) *p*-methylbenzylchloride and sodium acetylide
(d) 2-bromobutane and sodium hydrosulfide (NaSH)

7.16 What haloalkane and nucleophile are required to produce each of the following compounds?
(a) $CH_3CH_2CH_2C{\equiv}CH$ (b) $(CH_3)_2CHCH_2CN$ (c) $CH_3CH_2SCH_2CH_3$
(d) $C_6H_5CH_2CH_2OH$

7.17 Alcohols (ROH) are converted into alkoxides (RO^-) by reaction with NaH. Treatment of the following compound with sodium hydride yields C_4H_8O. What is the structure of the product? How is it formed?

$$HO{-}CH_2{-}CH_2{-}CH_2{-}CH_2{-}Br + NaH \longrightarrow C_4H_8O + H_2 + NaBr$$

7.18 Treatment of the following compound with sodium sulfide yields C_4H_8S. What is the structure of the product? How is it formed?

$$Cl{-}CH_2{-}CH_2{-}CH_2{-}CH_2{-}Cl + Na_2S \longrightarrow C_4H_8S + 2\,NaBr$$

Structure and Rates of Substitution Reactions

7.19 Which compound of the following pairs reacts with sodium iodide at the faster rate in an S_N2 reaction?
(a) 1-bromobutane or 2-bromobutane
(b) 1-chloropentane or chlorocyclopentane
(c) 2-bromo-2-methylpentane or 2-bromo-4-methylpentane

7.20 The rate of reaction of *cis*-1-bromo-4-*t*-butylcyclohexane with methylthiolate (CH_3S^-) is faster than for the trans isomer. Suggest a reason for this difference.

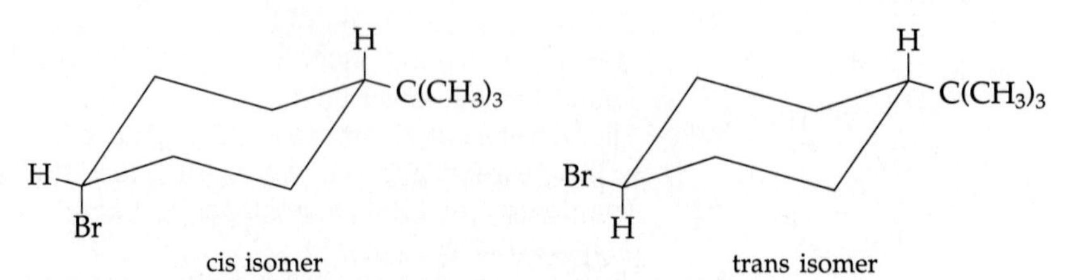

7.21 Predict which compound in each of the following pairs reacts with methanol (CH_3OH) at the faster rate in an S_N1 reaction. Explain why.
(a) bromocyclohexane or 1-bromo-1-methylcyclohexane
(b) isopropyl iodide or isobutyl iodide
(c) 3-bromo-1-pentene or 4-bromo-1-pentene

7.22 Predict which compound in each of the following pairs reacts with ethanol (CH_3CH_2OH) at the faster rate in an S_N1 reaction. Explain why.

(a) 1-bromo-1-phenylpropane or 2-bromo-1-phenylpropane
(b) 3-chlorocyclopentene or 4-chlorocyclopentene
(c) benzyl bromide or *p*-methylbenzyl bromide

7.23 *p*-Methylbenzyl bromide reacts faster than *p*-nitrobenzyl bromide with ethanol to form an ether product. Suggest a reason for this observation.

7.24 Although 1-bromo-2,2-dimethylpropane is a primary halide, S_N2 reactions of this compound are about 10,000 times slower than those of 1-bromopropane. Suggest a reason.

Stereochemistry of Substitution Reactions

7.25 Write the structure of the product of the reaction of (*R*)-2-bromobutane with sodium cyanide.

7.26 Write the structure of the product of the reaction of *cis*-1-bromo-2-methylcyclopentane with methylthiolate.

7.27 Write the structure of the product of the reaction of (*R*)-1-phenyl-1-bromobutane with water.

7.28 Write the structure of the product of the reaction of (*S*)-3-bromo-3-methylhexane with ethanol (CH_3CH_2OH).

7.29 Optically active 2-iodooctane slowly becomes racemic when treated with sodium iodide in an inert polar solvent. Explain why.

7.30 Optically active 2-butanol, $CH_3CH_2CH(OH)CH_3$, slowly becomes racemic when treated with dilute acid. Explain why.

Elimination Reactions

7.31 What alkene will be formed in the E2 reaction of the following compound with sodium methoxide ($CH_3O^-Na^+$)?

H CH3
Br H

7.32 How many isomeric alkenes could be formed by the elimination of HBr from the following compound by an E1 process? Which one will predominate?

$$CH_3-CH(CH_3)-C(Br)(CH_3)-CH_2-CH_3$$

7.33 The rate of elimination of hydrogen bromide in the reaction of *cis*-1-bromo-4-*t*-butylcyclohexane with hydroxide ion is faster than for the trans isomer. Suggest a reason for this difference.

H C(CH3)3 H Br

cis isomer

H C(CH3)3 Br H

trans isomer

7.34 Eight diastereomers of 1,2,3,4,5,6-hexachlorocyclohexane are possible. The following isomer undergoes an E2 reaction about 1,000 times slower than any of the others. Why?

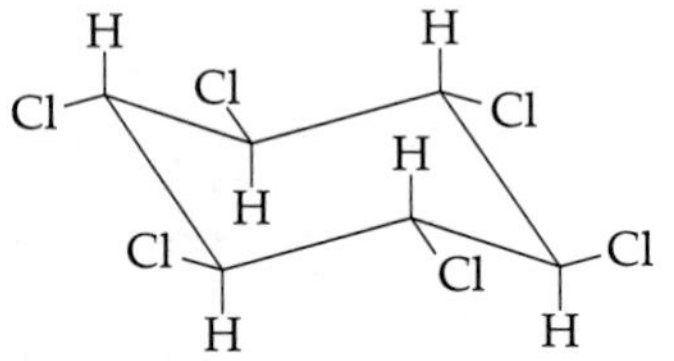

7.35 What is the configuration of the alkene formed by the elimination of one molar equivalent of HBr from the following compound?

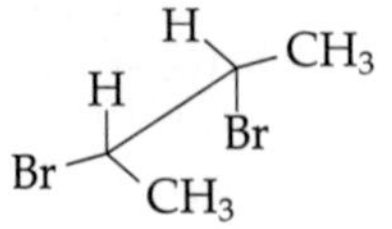

7.36 How many isomers can result from the elimination of HBr from the following compound. Which one will predominate?

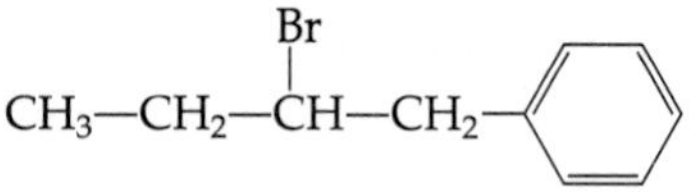

$CH_3-CH_2-CH(Br)-CH_2-C_6H_5$

Biological Molecules and Reactions

7.37 Draw the structure for the following compound, which inhibits feeding by herbivorous fish, with an E configuration at the double bond and a trans relationship for the bromine and chlorine atoms.

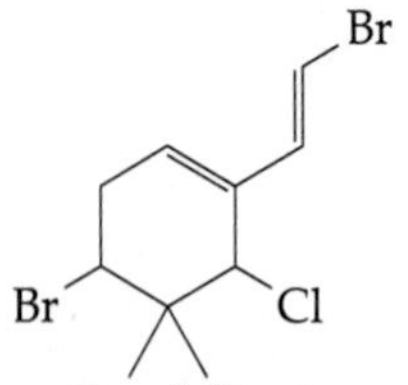

7.38 List the similarities between the following compound released by the Guamania mollusk and the compound released by the sea hare (page 251).

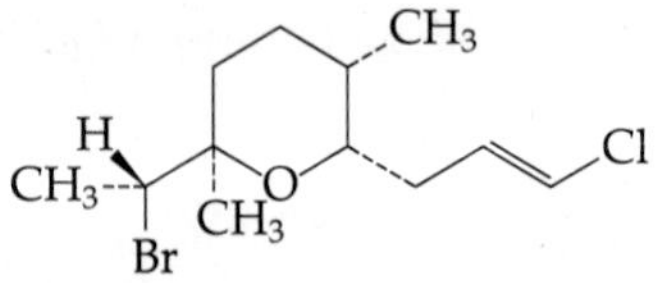

7.39 Glutathione is rapidly methylated by organophosphate insecticides such as methyl parathion. Draw the structure of the leaving group.

$$CH_3{-}O{-}\overset{\overset{\displaystyle S}{\|}}{\underset{\underset{\displaystyle OCH_3}{|}}{P}}{-}O{-}C_6H_4{-}NO_2$$

methyl parathion

7.40 Glutathione will substitute at heteroatoms such as oxygen. It reacts with nitroglycerine, a coronary vasodilator, as follows. What is the leaving group?

$$\begin{array}{l} CH_2{-}ONO_2 \\ | \\ CH{-}ONO_2 \\ | \\ CH_2{-}ONO_2 \end{array} \xrightarrow{GSH} \begin{array}{l} CH_2{-}OSG \\ | \\ CH{-}ONO_2 \\ | \\ CH_2{-}ONO_2 \end{array}$$

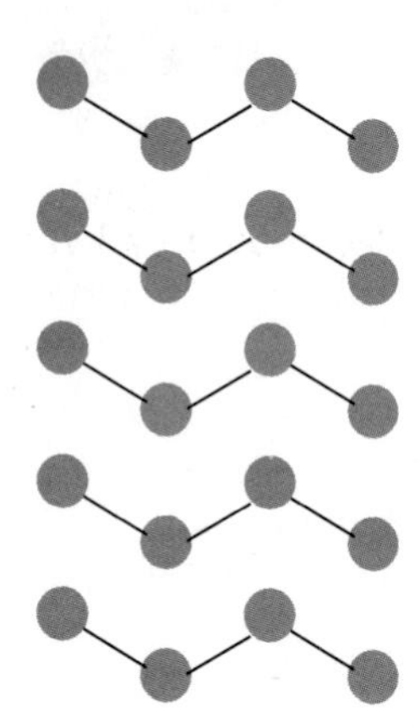

8

ALCOHOLS AND PHENOLS

8.1 THE HYDROXYL GROUP

Several families of organic compounds have functional groups that contain oxygen. These compounds include alcohols, phenols, ethers, aldehydes, ketones, acids, esters, and amides. Alcohols and phenols both contain a hydroxyl group, —OH. A hydroxyl group is also present in carboxylic acids, but it is bonded to a carbonyl carbon atom. Carboxylic acids are the subject of another chapter because most of the chemistry of the carboxyl group is different from that of the hydroxyl group. Compounds in which an —OH is bonded to the sp^2-hybridized carbon atom of an alkene are called **enols.** An enol is unstable and exists in equilibrium with a carbonyl compound (Chapter 10).

$$\underset{\text{carboxylic acid}}{R{-}\overset{\overset{\large O}{\|}}{C}{-}O{-}H} \qquad \underset{\text{enol (unstable)}}{H_2C{=}CH{-}O{-}H} \;\rightleftharpoons\; \underset{\text{aldehyde (stable)}}{H{-}\underset{H}{\overset{H}{C}}{-}\overset{\overset{\large O}{\|}}{C}{-}H}$$

Alcohols contain a hydroxyl group bonded to an sp^3-hybridized carbon atom. **Phenols** have a hydroxyl group bonded to an sp^2-hybridized carbon atom of an aromatic ring. Alcohols and phenols can be viewed as the organic "relatives" of water in which one hydrogen atom is replaced by an alkyl group or an aryl group. The two isomeric structures shown below illustrate the distinction between an alcohol and a phenol.

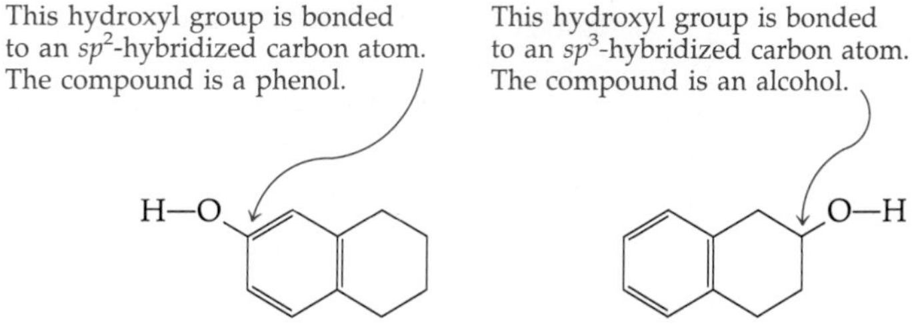

This hydroxyl group is bonded to an sp^2-hybridized carbon atom. The compound is a phenol.

This hydroxyl group is bonded to an sp^3-hybridized carbon atom. The compound is an alcohol.

H—O   O—H

Alcohols and phenols react differently at the carbon atom bearing the hydroxyl group because the hydridization of the carbon atoms differs. However, alcohols and phenols both have oxygen–hydrogen bonds and undergo many similar reactions at the oxygen atom. For example, because alcohols and phenols are organic analogs of water, they react with alkali metals to produce hydrogen gas, alkali metal ions, and an oxyanion derived from the alcohol or phenol. The oxyanions are called **alkoxides** and **phenoxides,** respectively. The evolution of hydrogen gas is a good qualitative test for the presence of an —OH group in an organic molecule.

$$2\,Na + 2\,H_2O \longrightarrow H_2 + \underset{\text{hydroxide}}{2\,HO^-} + 2\,Na^+$$

$$2\,Na + 2\,ROH \longrightarrow H_2 + \underset{\text{alkoxide}}{2\,RO^-} + 2\,Na^+$$

$$2\,Na + 2\,ArOH \longrightarrow H_2 + \underset{\text{phenoxide}}{2\,ArO^-} + 2\,Na^+$$

Some alcohols contain two or more hydroxyl groups. For example, ethylene glycol has two carbon atoms and two hydroxyl groups; glycerol has three carbon atoms and three hydroxyl groups.

$$\underset{\text{ethylene glycol}}{\underset{OH\quad OH}{CH_2{-}CH_2}} \qquad \underset{\text{glycerol}}{\underset{OH\quad OH\quad OH}{CH_2{-}CH{-}CH_2}}$$

Ethylene glycol is the major ingredient in automobile antifreeze. It is highly toxic; a dose of 50 mL is fatal. Ethylene glycol has important industrial uses in the manufacture of the synthetic fiber Dacron and Mylar film, which is used in cassette recorder tapes.

Glycerol, also known as glycerin, is the backbone of glycerophospholipids, which are the major components of cell membranes. Glycerol is also a component of triacylglycerols (fats), which are a major source of cellular energy (Chapter 13).

8.2 CLASSIFICATION AND NOMENCLATURE OF ALCOHOLS

Classification

Alcohols are classified by the number of alkyl groups bonded to the carbon atom bearing the hydroxyl group. You may wish to review the classification of carbon atoms in Chapter 3. Alcohols are classified as primary (1°), secondary (2°), or tertiary (3°) based on the number of alkyl groups bonded to the carbon atom to which the hydroxyl group is attached.

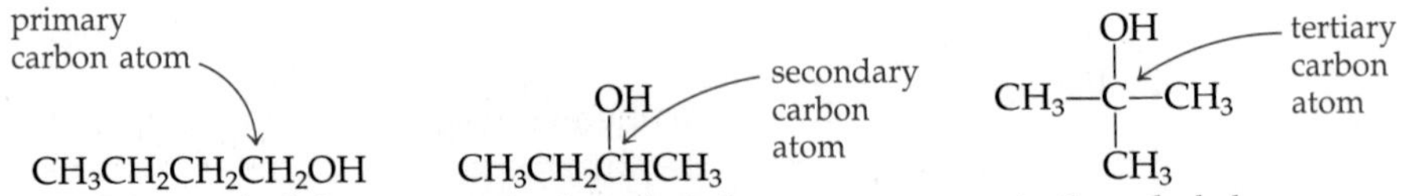

primary carbon atom → $CH_3CH_2CH_2CH_2OH$
a primary alcohol

$CH_3CH_2CH(OH)CH_3$ ← secondary carbon atom
a secondary alcohol

$CH_3—C(OH)(CH_3)—CH_3$ ← tertiary carbon atom
a tertiary alcohol

EXAMPLE 8.1 Classify the broad-spectrum antibiotic chloramphenicol as an alcohol.

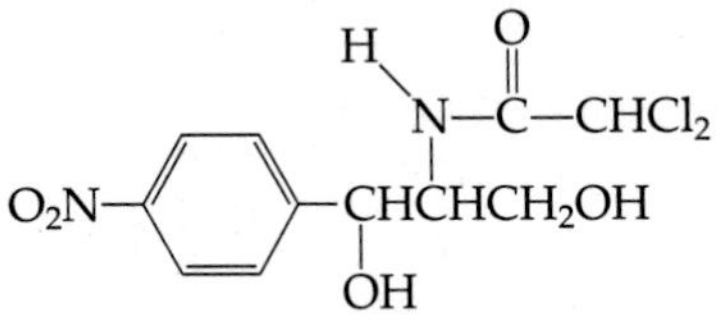

$O_2N—C_6H_4—CH(OH)CH(NH—C(=O)—CHCl_2)CH_2OH$

Solution First, we locate the oxygen atoms in the structure. The five oxygen atoms appear at four sites in the structure. Two oxygen atoms are bonded to a nitrogen atom and are part of the nitro group; the double-bonded oxygen atom is part of a carbonyl group of an amide. The hydroxyl group on the right is bonded to a carbon atom that has two hydrogen atoms and a carbon atom bonded to it; this is a primary alcohol. The hydroxyl group in the middle of the molecule is bonded to a carbon atom with two other carbon atoms bonded to it; one carbon atom is part of a substituted alkyl group, the other carbon atom is part of an aryl group. This alcohol is secondary.

Common Names

Alcohols containing one to four carbon atoms have common names that consist of the name of the alkyl group (Section 3.3) followed by the term *alcohol*. For example, CH_3CH_2OH is ethyl alcohol and $CH_3CH(OH)CH_3$ is isopropyl alcohol. Other common names are benzyl alcohol and allyl alcohol, whose structures are shown below.

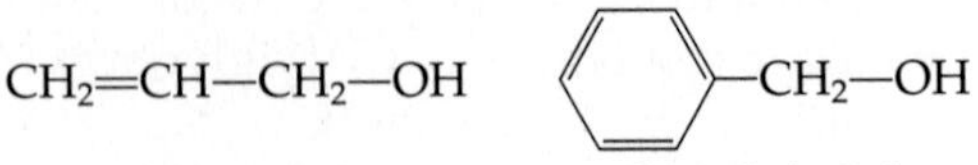

$CH_2{=}CH—CH_2—OH$
allyl alcohol

$C_6H_5—CH_2—OH$
benzyl alcohol

IUPAC Names

The IUPAC system of naming alcohols is as follows.

1. The longest continuous chain of carbon atoms that includes the hydroxyl group is the parent chain.

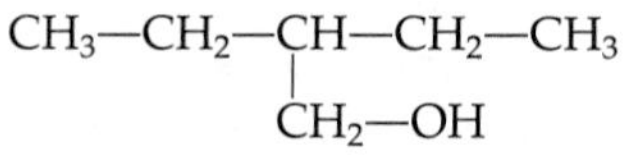

$$CH_3{-}CH_2{-}\underset{\displaystyle CH_2{-}OH}{\underset{|}{CH}}{-}CH_2{-}CH_3$$

The longest chain that contains the hydroxyl group has 4 carbon atoms, although the longest chain has 5 carbon atoms.

2. The parent name is obtained by substituting the suffix *-ol* for the final *-e* of the corresponding alkane.

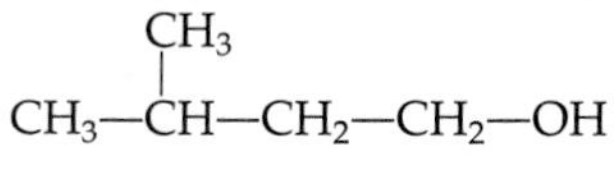

$$CH_3{-}\overset{\displaystyle CH_3}{\overset{|}{CH}}{-}CH_2{-}CH_2{-}OH$$

The parent alkane is butane. This is a substituted butanol. A methyl branch is attached to the butanol chain.

3. The position of the hydroxyl group is indicated by the number of the carbon atom to which it is attached. The chain is numbered so that the carbon atom bearing the hydroxyl group has the lower number.

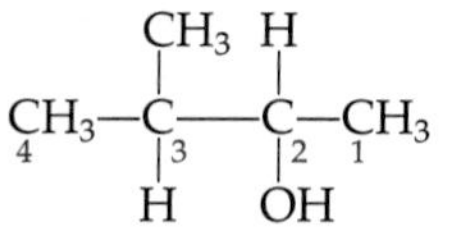

$$\underset{4}{CH_3}{-}\underset{\displaystyle H}{\underset{|3}{\overset{\displaystyle CH_3}{\overset{|}{C}}}}{-}\underset{\displaystyle OH}{\underset{|2}{\overset{\displaystyle H}{\overset{|}{C}}}}{-}\underset{1}{CH_3}$$

This is 3-methyl-2-butanol, not 2-methyl-3-butanol.

4. When the hydroxyl group is attached to a ring, the ring is numbered starting with the carbon atom bearing the hydroxyl group. Numbering continues in the direction that gives the lowest numbers to carbon atoms with substituents such as alkyl groups. The number 1 is not used in the name to indicate the position of the hydroxyl group.

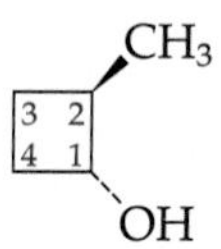

trans-2-methylcyclobutanol 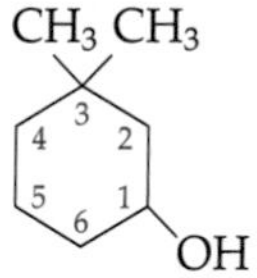   3,3-dimethylcyclohexanol

5. Alcohols containing two or more hydroxyl groups are called diols, triols, and so on. The terminal *-e* in the name of the parent alkane is retained, and the suffix *-diol* or *-triol* is added. The positions of the hydroxyl groups in the parent chain are indicated by numbers.

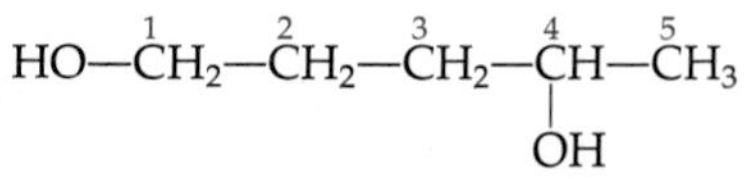

$$HO{-}\overset{1}{C}H_2{-}\overset{2}{C}H_2{-}\overset{3}{C}H_2{-}\underset{\displaystyle OH}{\underset{|}{\overset{4}{C}H}}{-}\overset{5}{C}H_3$$

1,4-pentanediol

6. When an alcohol contains a double or triple bond, the hydroxyl group takes precedence in numbering the carbon chain. The number that indicates the position of the multiple bond is located in front of the name of the alkene (or alkyne). The number that indicates the position of the hydroxyl group is appended to the name of the alkene (or alkyne) along with the suffix *-ol.*

$$\overset{5}{C}H_2{=}\overset{4}{C}H{-}\overset{3}{C}H_2{-}\overset{2}{C}H_2{-}\overset{1}{C}H_2{-}OH$$

4-penten-1-ol

$$\overset{6}{C}H_3{-}\overset{5}{C}{\equiv}\overset{4}{C}{-}\overset{3}{C}H_2{-}\underset{\displaystyle OH}{\overset{2}{C}H}{-}\overset{1}{C}H_2{-}OH$$

4-hexyne-1,2-diol

EXAMPLE 8.2 Assign the IUPAC name for citronellol, a compound found in geranium oil, which is used in perfumes.

OH

Solution The longest carbon chain that contains the hydroxyl group has eight carbon atoms. The hydroxyl group is on the carbon atom located on the right side of the chain. This carbon atom is C-1. Numbering the chain from right to left, we see that the methyl groups are at the C-3 and C-7 atoms. The double bond is located at the C-6 atom. The name is 3,7-dimethyl-6-octene-1-ol.

EXAMPLE 8.3 Ethchlorvynol is a sedative–hyponotic. The IUPAC name is (*E*)-1-chloro-3-ethyl-1-penten-4-yne-3-ol. Draw its structure.

Solution The term *1-penten* tells us that the parent chain contains five carbon atoms and has a double bond between the C-1 and C-2 atoms. The term *4-yne* informs us that there is a triple bond between the C-4 and C-5 atoms. Using this information, we draw the five-carbon-atom chain and select a direction for numbering it. Then, we place the multiple bonds in the proper places.

$$\underset{5}{C}{\equiv}\underset{4}{C}{-}\underset{3}{C}{-}\underset{2}{C}{=}\underset{1}{C}$$

The IUPAC name also tells us that a chlorine atom is located at the C-1 atom, an ethyl group at the C-3 atom, and a hydroxyl group at the C-3 atom.

Ethanol—Medical Considerations

Ethanol, CH_3CH_2OH, popularly known as alcohol, is a depressant and acts like a general anesthetic. In fact, in the days before modern anesthetics, alcohol was used to deaden the pain of persons undergoing surgery. Ethanol acts in the brain by inhibiting the firing of certain neurons, which are also inhibited by tranquilizers and sedatives such as Valium. Valium and ethanol bind to the same protein, and there is a strong cooperative interaction between them. Consuming Valium and ethanol at the same time is potentially lethal.

Ethanol is rapidly absorbed into the blood, and only about 1 ounce of pure alcohol per hour can be removed from the human body by oxidation in the liver. Even in moderate amounts, alcohol causes drowsiness and depresses brain functions. A blood alcohol level of 0.4% is fatal.

Several alcohols are antiseptics and disinfectants that act by denaturing bacterial proteins. Their effectiveness is related to structure. Primary alcohols are more effective antiseptics than secondary alcohols; tertiary alcohols are least effective. The most common alcoholic antiseptic is ethanol; a 70% solution of ethanol is often used for medical procedures such as preoperative sterilization of the skin. Another disinfectant is 2-propanol (also called isopropyl alcohol or rubbing alcohol). This secondary alcohol is a less effective disinfectant than 1-propanol, but it is less expensive to produce and is used for that reason.

Ethanol is used for pharmaceutical preparations such as tinctures and spirits. Tinctures are solutions of drugs in a water–alcohol mixture; spirits are solutions containing ethanol as the sole solvent.

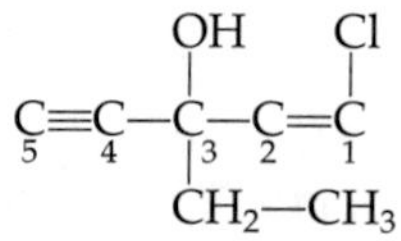

```
          OH    Cl
          |     |
C≡C—C—C=C
5   4   |3  2   1
        CH2—CH3
```

Next, we draw the (*E*) configuration by placing the chlorine atom and the C-3 atom on opposite sides of the double bond. Finally, we add the necessary hydrogen atoms.

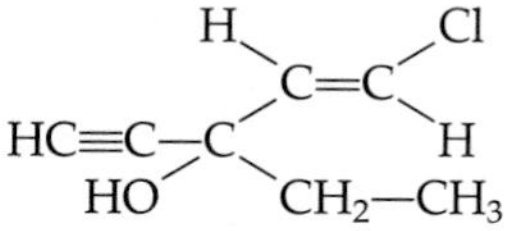

```
            H         Cl
             \       /
              C=C
             /     \
HC≡C—C          H
       /    \
    HO      CH2—CH3
```

8.3 PHYSICAL PROPERTIES OF ALCOHOLS

We noted earlier that alcohols can be viewed as the organic analogs of water, in which one hydrogen atom is replaced by an alkyl group. The oxygen atom of an alcohol is sp^3-hydridized. The C—O—H bond angle is approximately the tetrahedral bond angle. The bond angle in methyl alcohol is 108.9° (Figure 8.1). The lone-pair electrons in the remaining two sp^3 hybrid

FIGURE 8.1
Structure of Methanol
The oxygen atom in methanol is sp^3-hybridized. The C—O—H bond angle is close to the tetrahedral angle. The two sets of lone-pair electrons are in sp^3 hybrid orbitals that are directed to two of the corners of a tetrahedron.

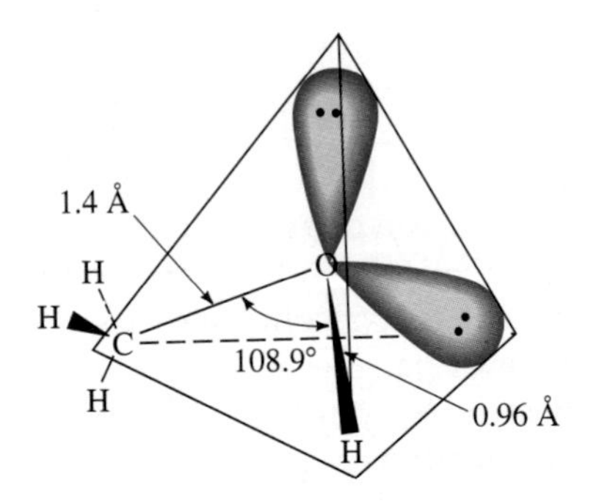

orbitals are directed to the remaining "corners" of a tetrahedron. The radius of the oxygen atom is smaller than that of a carbon atom. As a result, the O—H bond length (0.96 Å) is shorter than the C—H bond length (1.10 Å), and the C—O bond length (1.40 Å) is shorter than the average C—C bond length (1.54 Å).

The dipole moments of ethanol and propane are 1.69 and 0.08 D, respectively. Alcohols are much more polar than alkanes because they have both a polar C—O bond and a polar O—H bond. Alcohols form strong intermolecular hydrogen bonds, which have an enormous influence on their physical properties.

Boiling Points

When we compare the boiling points of alcohols with those of alkanes of comparable molecular weight, we find that alcohols boil at dramatically higher temperatures. For example, propane boils at −42 °C, whereas ethanol boils at 78 °C. These two compounds have approximately the same London forces, but ethanol also has strong dipole–dipole forces of attraction. However, the dramatic difference in boiling points is largely due to hydrogen bonding between alcohol molecules. The hydroxyl group of an alcohol can serve as both a hydrogen-bond donor and a hydrogen-bond acceptor. The hydrogen bonds of alcohols stabilize the liquid state, and much more energy is needed to separate hydrogen-bonded alcohol molecules than is required to disrupt the relatively weak London forces in alkanes.

hydrogen-bond acceptor
A hydrogen bond is formed here.
hydrogen-bond donor

The boiling points of the 1-alkanols and alkanes of approximately the same molecular weight are compared in Figure 8.2. Note the dramatic differences in boiling points for compounds at lower molecular weights. As the molecular weights of alkanes and alcohols increase, the two curves ap-

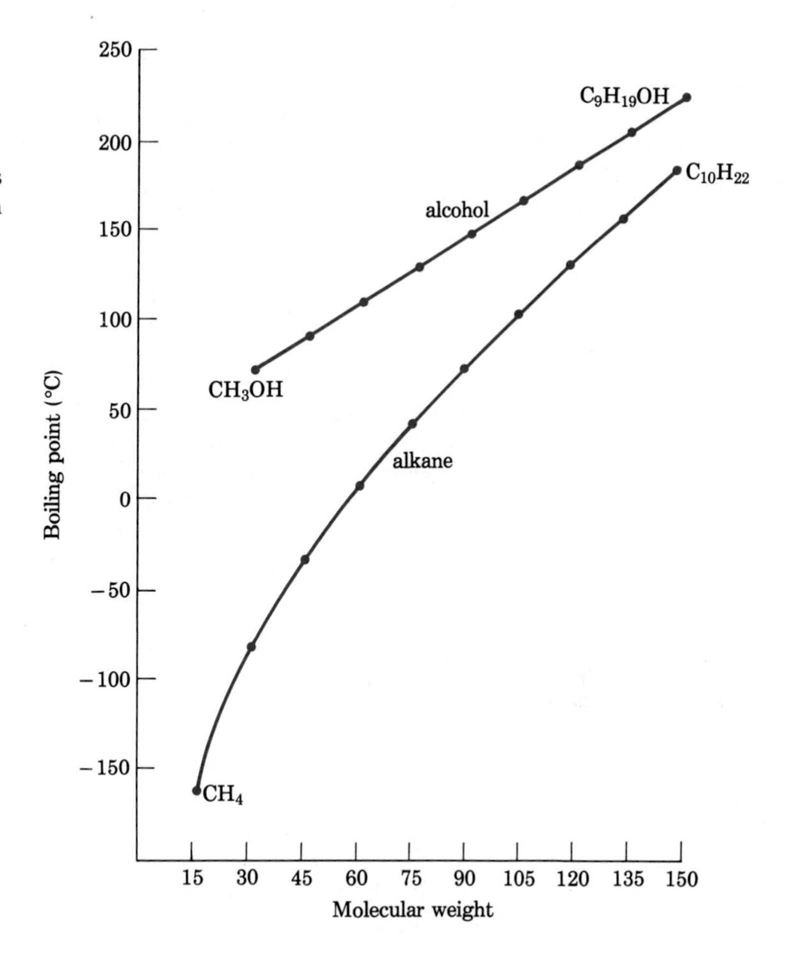

FIGURE 8.2
Comparison of the Boiling Points of Alcohols and Alkanes
The boiling points of both alkanes and normal alcohols increase with increasing chain length. Alcohols have higher boiling points than alkanes of similar molecular weight.

proach each other. In alcohols with high molecular weights, hydrogen bonding is still possible, but interactions due to London forces increase due to the longer carbon chain. As a consequence, the difference in boiling point between an alcohol and an alkane of comparable molecular weight decreases.

Solubility of Alcohols in Water

The ability of alcohols to form hydrogen bonds has an important effect on their solubilities in water. Table 8.1 lists the solubilities of some alcohols that contain normal alkyl groups. The lower molecular weight alcohols are completely soluble in water. These molecules, like water, are highly polar, and we know that "like dissolves like". Furthermore, the hydroxyl group can form three hydrogen bonds to water. Two can form between the two sets of lone-pair electrons of the alcohol oxygen atom, which are hydrogen-bond acceptors, and the hydrogen atoms of water. The third can form between the hydrogen atom of the hydroxyl group, which is a hydrogen-bond donor, and the lone-pair electrons on the oxygen atom in water. However, as the size of the alkyl group increases, alcohols more closely resemble alkanes,

TABLE 8.1 Boiling Points and Solubilities of Alcohols

Name	Formula	Boiling point (°C)	Solubility (g/100 mL in water)
methanol	CH_3OH	65	miscible
ethanol	CH_3CH_2OH	78	miscible
1-propanol	$CH_3CH_2CH_2OH$	97	miscible
1-butanol	$CH_3CH_2CH_2CH_2OH$	117	7.9
1-pentanol	$CH_3CH_2CH_2CH_2CH_2OH$	137	2.7
1-hexanol	$CH_3(CH_2)_4CH_2OH$	158	0.59
1-heptanol	$CH_3(CH_2)_5CH_2OH$	176	0.09
1-octanol	$CH_3(CH_2)_6CH_2OH$	194	insoluble
1-nonanol	$CH_3(CH_2)_7CH_2OH$	213	insoluble
1-decanol	$CH_3(CH_2)_8CH_2OH$	229	insoluble

and the hydroxyl group has less effect on their physical properties. Water can still form hydrogen bonds to the hydroxyl group. However, the long chain interferes with other water molecules and prevents them from hydrogen bonding to each other. The formation of a hydrogen bond between an alcohol and water releases energy. However, the energy released is not sufficient to compensate for disrupting the extensive hydrogen-bonding network of water. As a result, the solubility of alcohols decreases with increasing size of the alkyl group.

Alcohols as Solvents

Ethanol is an excellent solvent for many organic compounds, especially those with lone-pair electrons that are hydrogen-bond acceptors. Polar compounds dissolve readily in a "like" polar solvent. Nonpolar compounds dissolve in alcohols to some extent, but the solubility is often limited because the extensive hydrogen-bonding network of the alcohol must be broken to accommodate the solute.

Strongly basic compounds react with alcohols, thus destroying them. For example, Grignard reagents react with the acidic proton of an alcohol.

$$R{-}Mg{-}Br + R'{-}O{-}H \longrightarrow R{-}H + R'{-}O{-}Mg{-}Br$$

8.4 REACTIONS OF ALCOHOLS

Alcohols undergo various reactions in which different bonds can break, depending on experimental conditions. In some reactions the O—H bond breaks; in others the C—O bond breaks. In addition, the C—H bond on the carbon atom bearing the hydroxyl group or the C—H bond on the carbon atom adjacent to the —OH group may also be reactive under some conditions. We will divide our discussion of the reactions of alcohols into four classes based on the bonds that break.

1. Breaking the oxygen–hydrogen bond.

This bond is broken in an acid–base reaction.

2. Breaking the carbon–oxygen bond.

This bond is broken in a substitution reaction with a nucleophile.

3. Breaking both the carbon–oxygen bond and the carbon–hydrogen bond at a carbon atom adjacent to the carbon atom bearing the hydroxyl group.

These bonds are broken in an elimination reaction.

4. Breaking both the oxygen–hydrogen bond and the carbon–hydrogen bond at the carbon atom bearing the hydroxyl group.

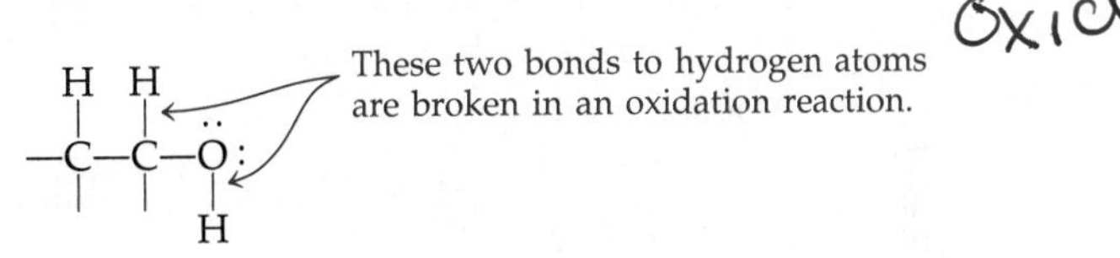

8.5 ACIDITY AND BASICITY OF ALCOHOLS

We know that water can act as a proton donor (an acid) in some reactions and as a proton acceptor (a base) in other reactions depending on conditions. Alcohols can also act as acids or bases. That is, alcohols are amphoteric substances.

Alcohols are slightly weaker acids than water; the K_a of ethanol is 1.3×10^{-16} ($pK_a = 16$) and the K_a of water is 1.8×10^{-16} ($pK_a = 15.7$). The pK_a values of some common alcohols are listed in Table 8.2. We recall that a strong acid has a large K_a and a small pK_a.

TABLE 8.2 Effect of Structure on the Acidity of Alcohols

Compound	K_a	pK_a
CH_3OH	3.2×10^{-16}	15.5
CH_3CH_2OH	1.3×10^{-16}	15.9
$(CH_3)_2CHOH$	1×10^{-18}	18.0
$(CH_3)_3COH$	1×10^{-19}	19.0
$ClCH_2CH_2OH$	5×10^{-15}	14.3
CF_3CH_2OH	4×10^{-13}	12.4
$CF_3CH_2CH_2OH$	2.5×10^{-15}	14.6
$CF_3CH_2CH_2CH_2OH$	4×10^{-16}	15.4

$$CH_3CH_2OH + H_2O \rightleftharpoons \underset{\text{ethoxide ion}}{CH_3CH_2O^-} + H_3O^+ \qquad K_a = 1.3 \times 10^{-16}$$

$$H_2O + H_2O \rightleftharpoons HO^- + H_3O^+ \qquad K_a = 1.8 \times 10^{-16}$$

The acidity of alcohols increases when electronegative substituents are added to the carbon atoms near the hydroxyl group. Such substitutents withdraw electron density from the oxygen atom by an inductive effect that acts through the network of σ bonds. This inductive effect weakens the O—H bond and destabilizes the alcohol. The inductive effect stabilizes the negative charge of the conjugate base. Table 8.2 shows this effect. Replacing a hydrogen atom at the C-2 atom of ethanol with a chlorine atom decreases the pK_a from 15.9 to 14.3, which means that K_a increases by a factor of 50. Replacing all three hydrogen atoms at the C-2 atom of ethanol with the more electronegative fluorine atoms substantially decreases the pK_a. The effect of the electron-withdrawing —CF_3 group decreases with distance from the oxygen atom. The pK_a of 4,4,4-trifluorobutanol is similar to the pK_a of a primary alcohol such as ethanol.

When an alcohol loses a proton, a conjugate base called an **alkoxide ion** is produced. Because alcohols are weaker acids than water, alkoxides are somewhat stronger bases than hydroxide ion. Alkoxides are used as bases in organic solvents because they are more soluble than hydroxide salts. Alkoxide anions can easily be prepared by adding an alkali metal to an alcohol.

$$2\,Na + 2\,CH_3OH \longrightarrow H_2 + 2\,CH_3O^- + 2\,Na^+$$

$$2\,K + 2\,(CH_3)_3COH \longrightarrow H_2 + 2\,(CH_3)_3CO^- + 2\,K^+$$

Alcohols can act as bases because they have two sets of lone-pair electrons on the oxygen atom. But alcohols are very weak bases and can only be protonated to form the conjugate acid, an **oxonium ion,** by a strong acid. The formation of an oxonium ion is analogous to the reaction of water with a strong acid to give the hydronium ion. Oxonium ions are intermediates in many reactions catalyzed by strong acids.

$$CH_3CH_2OH + HA \rightleftharpoons \underset{\text{an oxonium ion}}{CH_3CH_2OH_2^+} + A^-$$

$$H_2O + HA \rightleftharpoons \underset{\text{hydronium ion}}{H_3O^+} + A^-$$

In summary, alcohols are amphoteric. They exist in strongly basic solution as alkoxide ions, in neutral solution as alcohols, and in strongly acidic solution as oxonium ions.

8.6 SUBSTITUTION REACTIONS OF ALCOHOLS

The hydroxyl group of an alcohol can be replaced by a halogen in either an S_N2 or S_N1 reaction (Chapter 7). For example, treating a primary alcohol with hydrogen bromide, HBr, produces an alkyl bromide. Similarly, treating a primary alcohol with HCl in the presence of $ZnCl_2$, which is required as a catalyst, produces an alkyl chloride.

$$CH_3CH_2CH_2CH_2OH + HBr \longrightarrow CH_3CH_2CH_2CH_2Br + H_2O$$

$$CH_3CH_2CH_2OH + HCl \xrightarrow{ZnCl_2} CH_3CH_2CH_2Cl + H_2O$$

These reactions also occur when secondary and tertiary alcohols are the substrates. The relative reaction rates depend on the type of alcohol and decrease in the order tertiary > secondary > primary alcohols.

The reaction mechanism varies with the structure of the alkyl group. Primary alcohols react by an S_N2 mechanism, whereas secondary and tertiary alcohols react by an S_N1 mechanism. However, in each case the leaving group is not hydroxide ion but a water molecule. The acid catalyst is required to form the conjugate acid of the alcohol. The departure of a neutral leaving group from a developing carbocation center requires less energy than the departure of a negatively charged leaving group such as the hydroxide ion.

$$R{-}\ddot{O}{:}{-}H \xrightarrow{H^+} R{-}\ddot{O}^+{-}H \; (\text{with } H \text{ on } O)$$

S_N1: $R{-}\ddot{O}^+(H){-}H \longrightarrow R^+ + H_2O \xrightarrow{Nu:^-} R{-}Nu$

S_N2: $R{-}\ddot{O}^+(H){-}H \xrightarrow{Nu:^-} R{-}Nu + H_2O$

Pyrophosphate—Nature's Leaving Group

Many reactions that occur in cells require displacement of a hydroxyl group by a nucleophile, but the hydroxyl group of an alcohol is not a good leaving group. However, a hydroxyl group can be converted into a pyrophosphate group. Pyrophosphoric acid is a stronger acid than water. Thus, its conjugate base—pyrophosphate ion—is a weaker base than the conjugate base of water—hydroxide ion. Pyrophosphate ion is an excellent leaving group.

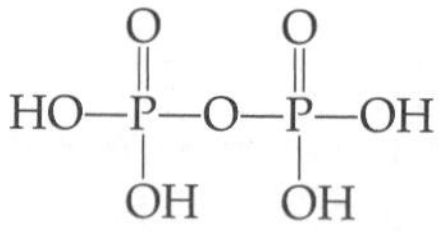

HO—P(=O)(OH)—O—P(=O)(OH)—OH

pyrophosphoric acid

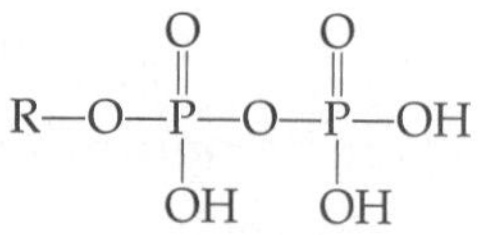

R—O—P(=O)(OH)—O—P(=O)(OH)—OH

a pyrophosphate

The concentration of pyrophosphate in cells is very low, and the conversion of alcohols to pyrophosphates essentially never occurs by itself. So, how are alcohols converted to pyrophosphates in cells? The answer is twofold. First, the source of pyrophosphate is not "free pyrophosphate" but a molecule called adenosine triphosphate (ATP), which contains a pyrophosphate group.

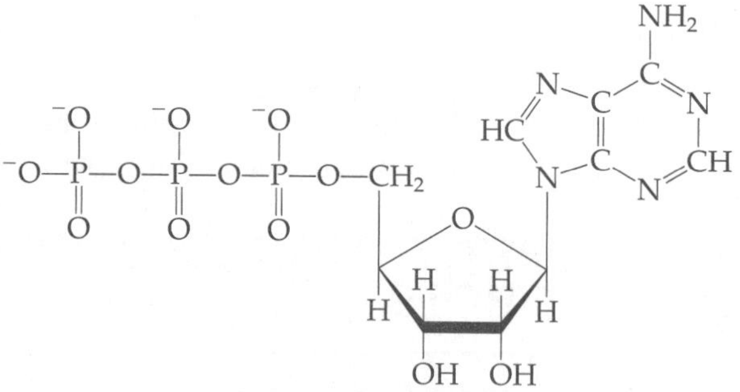

adenosine triphosphate (ATP)

Second, a pyrophosphate group is transferred from ATP to an alcohol by an enzyme, and enzyme-catalyzed reactions are very fast.

$$\text{ROH} + \text{ATP} \longrightarrow \text{R—OPP} + \text{AMP}$$

The pyrophosphate group is represented as —OPP in biochemical reactions.

Water is also a better leaving group than hydroxide ion because water is the weaker base. There is a general correlation between basicity and leaving-group ability. A weak base is a better leaving group than a stronger base in a substitution reaction in both S_N1 and S_N2 reactions.

Primary and secondary alcohols, which react only slowly with HBr and HCl, react readily with thionyl chloride and phosphorus trihalides such as phosphorus tribromide to give the corresponding alkyl halides. The products of these reactions are easily separated from the inorganic byproducts. Thionyl chloride produces hydrogen chloride and sulfur dioxide, which are released from the reaction as gases; the chloroalkane remains in solution.

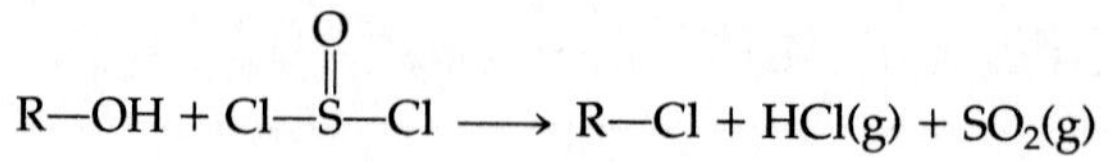

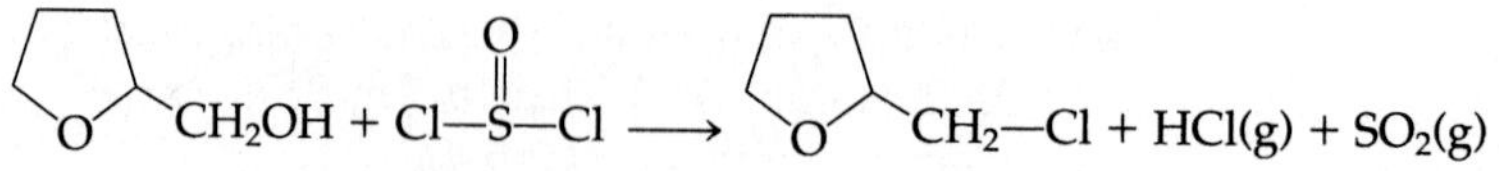

$$\text{R—OH} + \text{Cl—S(=O)—Cl} \longrightarrow \text{R—Cl} + \text{HCl}(g) + \text{SO}_2(g)$$

$$\text{(tetrahydrofuran-2-yl)CH}_2\text{OH} + \text{Cl—S(=O)—Cl} \longrightarrow \text{(tetrahydrofuran-2-yl)CH}_2\text{—Cl} + \text{HCl}(g) + \text{SO}_2(g)$$

Pyrophosphate derivatives of alcohols participate in the biosynthesis of many molecules, including cholesterol and steroid hormones. The biosynthesis of these molecules begins with a reaction between the five-carbon pyrophosphates isopentenyl pyrophosphate and dimethylallyl pyrophosphate.

OPP OPP
isopentenyl pyrophosphate dimethylallyl pyrophosphate

The π electrons of isopentenyl pyrophosphate act as a nucleophilic center to displace a pyrophosphate group from dimethylallyl pyrophosphate. The nucleophilic substitution reaction produces a new carbon–carbon bond and leaves a carbocation in the isopentenyl group. Loss of a proton at the C-2 atom yields geraniol pyrophosphate. When this product reacts with water, the pyrophosphate group is displaced, and the product is the alcohol geraniol.

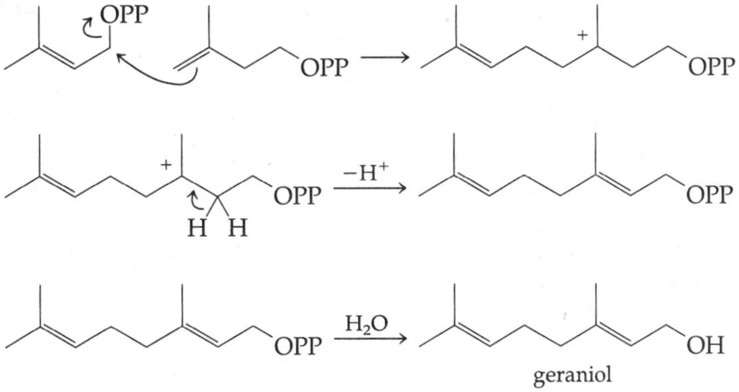

OPP OPP ⟶ + OPP

+ OPP H H $\xrightarrow{-H^+}$ OPP

OPP $\xrightarrow{H_2O}$ OH
geraniol

The reaction with phosphorus tribromide produces phosphorous acid, which has a high boiling point. Thus, the bromoalkane can be separated from the reaction mixture by distillation.

$$3\ R{-}OH + PBr_3 \longrightarrow 3\ R{-}Br + \underset{\text{phosphorous acid}}{H_3PO_3}$$

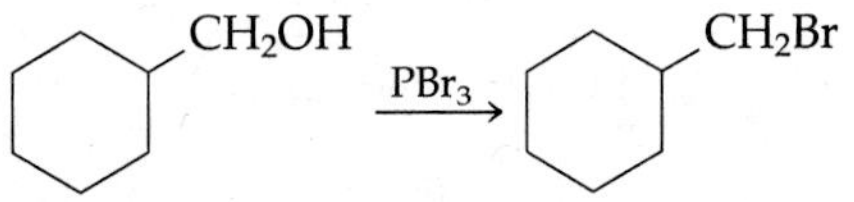

CH_2OH $\xrightarrow{PBr_3}$ CH_2Br

8.7 DEHYDRATION OF ALCOHOLS

The removal of a water molecule from an alcohol is a **dehydration reaction,** which is an example of an elimination reaction. This reaction requires an acid catalyst such as sulfuric acid or phosphoric acid and is illustrated by the formation of ethylene from ethyl alcohol.

These atoms are eliminated to form water.

$$H-\underset{H}{\overset{H}{C}}-\underset{H}{\overset{OH}{C}}-H \xrightarrow{H_2SO_4} \begin{matrix}H & & H\\ & C{=}C & \\ H & & H\end{matrix} + H_2O$$

Mechanism of Dehydration

The dehydration of alcohols occurs by mechanisms that depend on the structure of the alcohol. Tertiary alcohols undergo acid-catalyzed dehydration by an E1 mechanism; primary alcohols are dehydrated by an E2 mechanism. In either mechanism, the first step is the protonation of the lone-pair electrons of the oxygen atom (an acid–base reaction) to produce an oxonium ion. The acid is represented as HA.

$$H-\underset{H}{\overset{H}{C}}-\underset{CH_3}{\overset{CH_3}{C}}-\overset{H}{\ddot{O}}: + H-A \longrightarrow H-\underset{H}{\overset{H}{C}}-\underset{CH_3}{\overset{CH_3}{C}}-\overset{H}{\ddot{O}^{+}}-H + :A^{-}$$

an oxonium ion

A tertiary alcohol loses water by an S_N1 process to produce a tertiary carbocation. The tertiary carbocation then loses a proton from the carbon atom adjacent to the carbon atom bearing the positive charge, and an alkene is produced.

$$H-\underset{H}{\overset{H}{C}}-\underset{CH_3}{\overset{CH_3}{C}}-\overset{H}{\ddot{O}^{+}}-H \longrightarrow H-\underset{H}{\overset{H}{C}}-\underset{CH_3}{\overset{CH_3}{C^{+}}} + H_2O$$

a tertiary carbocation

$$H-\underset{H}{\overset{H}{C}}-\underset{CH_3}{\overset{CH_3}{C^{+}}} \longrightarrow \begin{matrix}H & & CH_3\\ & C{=}C & \\ H & & CH_3\end{matrix} + H^{+}$$

Primary alcohols are dehydrated by an E2 mechanism. The first step again is protonation of the OH group to give an oxonium ion. However, in the second step, the conjugate base of the acid removes a proton from the carbon atom adjacent to the carbon atom bearing the oxygen atom. In this step, the electron pair in the C—H bond "moves" to form a carbon–carbon double bond, whereas the electron pair of the C—O bond is retained by the oxygen atom. Formation of the double bond and loss of water occur in a single, concerted step.

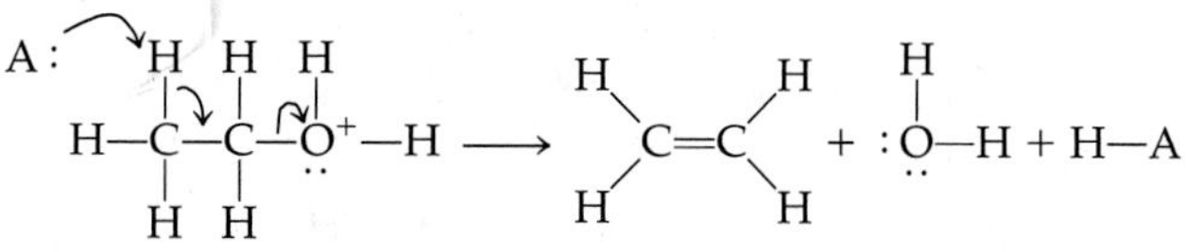

A: H H H / H—C—C—O⁺—H → H₂C=CH₂ + :O—H (H) + H—A

Note that in both the E1 process and the E2 process, the acid HA serves only as a catalyst; it is regenerated in the last step of the reaction.

Multiple Dehydration Products

In a dehydration reaction, the carbon–oxygen bond of one carbon atom and a carbon–hydrogen bond of an adjacent carbon atom both break. We recall from Chapter 4 that alcohols with two or more adjacent carbon atoms such as 2-butanol yield a mixture of products. Elimination can occur in either direction.

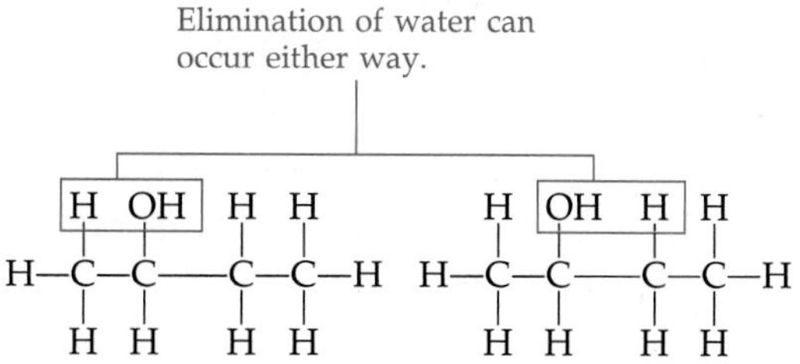

Elimination of water can occur either way.

H OH H H / H—C—C—C—C—H / H H H H   H OH H H / H—C—C—C—C—H / H H H H

The product formed in higher yield is the isomer that contains the greater number of alkyl groups attached to the double bond (the more substituted alkene). This more stable alkene is often called the Zaitsev product (Section 4.13). If the products are a mixture of geometric isomers, the more stable trans isomer predominates.

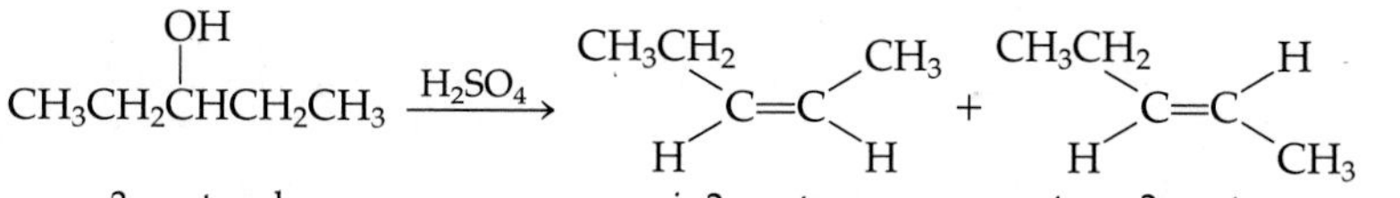

$$CH_3CH_2CH(OH)CH_2CH_3 \xrightarrow{H_2SO_4} \text{cis-}CH_3CH_2CH{=}CHCH_3 + \text{trans-}CH_3CH_2CH{=}CHCH_3$$

3-pentanol → *cis*-2-pentene (25%) + *trans*-2-pentene (75%)

EXAMPLE 8.4 Predict the product(s) of the dehydration of 1-methylcyclohexanol.

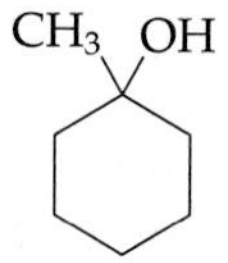

CH_3 OH (on cyclohexane ring)

Solution This tertiary alcohol has three carbon atoms adjacent to the carbon atom bearing the hydroxyl group. Each carbon atom can lose a hydrogen atom in the dehydration reaction. However, loss of a hydrogen atom from either C-2 or C-6 atoms results in the same product. Thus, only two isomers are formed.

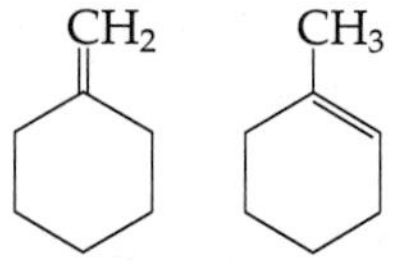

CH_2 CH_3

The second structure, with the double bond within the six-membered ring, predominates because it is the more highly substituted alkene.

8.8 OXIDATION OF ALCOHOLS

Like hydrocarbons, alcohols burn when heated in the presence of oxygen. Lower molecular weight alcohols such as methanol and ethanol burn with clean flames.

$$2\,CH_3OH + 3\,O_2 \longrightarrow 2\,CO_2 + 4\,H_2O$$
$$CH_3CH_2OH + 3\,O_2 \longrightarrow 2\,CO_2 + 3\,H_2O$$

Burning is a "brute force" oxidation method that converts an alcohol to carbon dioxide and water. Alcohols can be oxidized selectively by various chemical reagents in the laboratory to produce compounds in intermediate oxidation states. Primary and secondary alcohols react differently with oxidizing agents.

Primary alcohols, which have the general formula RCH_2OH, can be oxidized to aldehydes, which have the general formula RCHO. Note that this oxidation occurs with the loss of two hydrogen atoms. Aldehydes are easily oxidized and react further to produce carboxylic acids, RCOOH. When an aldehyde is oxidized to a carboxylic acid, the oxidized carbon atom gains an oxygen atom.

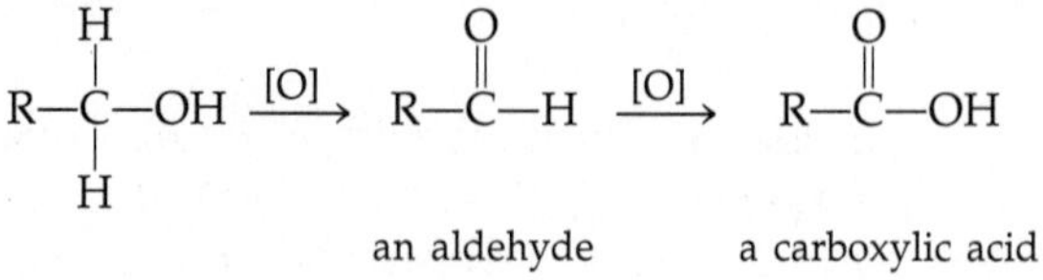

$$R\text{—}\overset{H}{\underset{H}{\overset{|}{\underset{|}{C}}}}\text{—}OH \xrightarrow{[O]} R\text{—}\overset{O}{\overset{\|}{C}}\text{—}H \xrightarrow{[O]} R\text{—}\overset{O}{\overset{\|}{C}}\text{—}OH$$

an aldehyde a carboxylic acid

Secondary alcohols are oxidized to form ketones, RCOR, which cannot be further oxidized because there is no hydrogen atom on the oxygen-bearing carbon atom of the ketone. Tertiary alcohols are not oxidized because the carbon atom bearing the —OH group has no hydrogen atom.

R–

a tertia

Oxidizing Agents for Alcohols

Alcohols are oxidized
trioxide in aqueous su
hols to aldehydes, wh
acids. This reagent als

$$\text{(CH}_3)_2\text{C=CHCH}_2\text{CH}_2\text{CH(CH}_3)\text{CH}_2\text{CH}_2\text{OH} \xrightarrow[\text{H}_2\text{SO}_4]{\text{CrO}_3} \text{(CH}_3)_2\text{C=CHCH}_2\text{CH}_2\text{CH(CH}_3)\text{CH}_2\text{COOH}$$

$$\text{cyclooctanol} \xrightarrow[\text{H}_2\text{SO}_4]{\text{CrO}_3} \text{cyclooctanone}$$

Alcohols are also oxidized with a milder oxidizing agent that consists of pyridinium chlorochromate (PCC) in methylene chloride (CH_2Cl_2) as solvent.

$$CrO_3 + HCl + C_5H_5N: \longrightarrow C_5H_5N^+\text{—H}\ CrO_3Cl^-$$

pyridine → pyridinium chlorochromate (PCC)

Oxidation reactions with PCC are carried out at a lower temperature than that needed for Jones's reagent. But the principal advantage of PCC is that primary alcohols are converted to aldehydes without continued oxidation to carboxylic acids.

$$\text{(CH}_3)_2\text{C=CHCH}_2\text{CH}_2\text{CH(CH}_3)\text{CH}_2\text{CH}_2\text{OH} \xrightarrow{\text{PCC}} \text{(CH}_3)_2\text{C=CHCH}_2\text{CH}_2\text{CH(CH}_3)\text{CH}_2\text{CHO}$$

es. Methanol ttle as 15 mL (1 ol can cause blind- se death. Prolonged l vapors is also a serious

ethanol is the least toxic of the nols, it is still a poisonous sub- at must be oxidized in the body to t high blood alcohol levels, which can ison" the brain. Alcohol is oxidized in the ver by the enzyme alcohol dehydrogenase (ADH). ADH requires a coenzyme nicotinamide adenine dinucleotide (NAD^+) as an oxidizing agent. The coenzyme can exist in an oxidized form, NAD^+, and a reduced form, NADH. Ethanol is oxidized to ethanal (acetaldehyde) by NAD^+-dependent liver ADH. Subsequent oxidation of ethanal yields ethanoic acid (acetic acid), which is nontoxic.

$$CH_3\text{—}CH_2\text{—}OH \xrightarrow[NAD^+]{ADH} CH_3\text{—}C(=O)\text{—}H$$

ethanal
(acetaldehyde)

$$CH_3\text{—}C(=O)\text{—}H \xrightarrow[NAD^+]{ADH} CH_3\text{—}C(=O)\text{—}OH$$

ethanoic acid
(acetic acid)

The oxidation products of some other alcohols are toxic. In the case of methanol, oxidation catalyzed by ADH gives methanal (formaldehyde) and then methanoic acid (formic acid).

$$H\text{—}CH_2\text{—}OH \xrightarrow[NAD^+]{ADH} H\text{—}C(=O)\text{—}H$$

methanal
(formaldehyde)

$$H\text{—}C(=O)\text{—}H \xrightarrow[NAD^+]{ADH} H\text{—}C(=O)\text{—}OH$$

methanoic acid
(formic acid)

Formaldehyde is transported in the blood thoughout the body and reacts very rapidly with proteins and destroys their biological function. For example, it reacts with an amine functional group of lysine contained in a protein required for vision called rhodopsin (Chapter 10), which is why the ingestion of methanol causes blindness. Formaldehyde also reacts with amino groups in other proteins, including many enzymes, and the loss of the function of these biological catalysts causes death.

Ethylene glycol is also toxic. This sweet-tasting substance is the primary component of antifreeze. Dogs are poisoned when they ingest antifreeze left in open containers. Oxidation occurs to give oxalic acid, which causes kidney failure.

$$H\text{—}CH(OH)\text{—}CH(OH)\text{—}H \longrightarrow HO\text{—}C(=O)\text{—}C(=O)\text{—}OH$$

ethylene glycol → oxalic acid

Intravenous injections of ethanol are used to treat methanol or ethylene glycol poisoning before substantial oxidation has occurred. ADH binds more tightly to ethanol than to methanol or ethylene glycol, and the rate of oxidation of ethanol is about six times faster than that of methanol. The ethanol is also present in higher concentration because it is directly injected. As a result, neither methanol nor ethylene glycol are oxidized to toxic products. These "toxic" alcohols are then slowly excreted by the kidneys.

EXAMPLE 8.5 Which of the isomeric $C_4H_{10}O$ alcohols react with Jones's reagent to produce a ketone, C_4H_8O?

Solution There are four isomeric alcohols because there are four C_4H_9 alkyl groups (Section 3.3). Two alkyl groups are primary—*n*-butyl and isobutyl; one is tertiary—*t*-butyl. Only the *sec*-butyl group provides a secondary alcohol, and only secondary alcohols yield ketones when oxidized. Thus, only 2-butanol (*sec*-butyl alcohol) yields a ketone.

$$CH_3-\underset{\displaystyle H}{\overset{\displaystyle OH}{\underset{|}{\overset{|}{C}}}}-CH_2-CH_3 \xrightarrow{\text{Jones's reagent}} CH_3-\overset{\displaystyle O}{\overset{\|}{C}}-CH_2-CH_3$$

8.9 SYNTHESIS OF ALCOHOLS

We have already encountered two methods of preparing alcohols: (1) substitution of a halide by hydroxide and (2) hydration of an alkene. These reactions often have low yields. The yield in the substitution reaction is diminished by the competing elimination reaction (Chapter 7). The yield in the hydration of an alkene is somewhat limited because the reaction is reversible (Chapter 4).

In this section we will discuss two types of reactions that give excellent yields of alcohols. The first type is the reduction of carbonyl compounds; the second is an "indirect" hydration of alkenes. In both reactions, the functional group converted to the hydroxyl group is located on the proper hydrocarbon skeleton. In Chapter 10 we will examine reactions that simultaneously form alcohols and build new hydrocarbon skeletons.

Reduction of Carbonyl Compounds

Alcohols can be produced by reducing the carbonyl group of aldehydes and ketones with hydrogen gas in the presence of a metal catalyst such as palladium, platinum, or a special reactive form of nickel called Raney nickel. Aldehydes yield primary alcohols; ketones yield secondary alcohols.

$$R-\overset{\displaystyle O}{\overset{\|}{C}}-H + H_2 \xrightarrow{Ni} R-\underset{\displaystyle H}{\overset{\displaystyle OH}{\underset{|}{\overset{|}{C}}}}-H$$

an aldehyde → a primary alcohol

$$R-\overset{\displaystyle O}{\overset{\|}{C}}-R + H_2 \xrightarrow{Ni} R-\underset{\displaystyle H}{\overset{\displaystyle OH}{\underset{|}{\overset{|}{C}}}}-R$$

a ketone → a secondary alcohol

The reduction reaction occurs by the transfer of hydrogen atoms bound to the surface of the metal catalyst to the carbonyl oxygen and carbon atoms. Note that the same catalysts are used for the hydrogenation of alkenes, which is a much faster reaction. Thus, both the carbon–carbon double bond and the carbonyl group are reduced in compounds containing both functional groups.

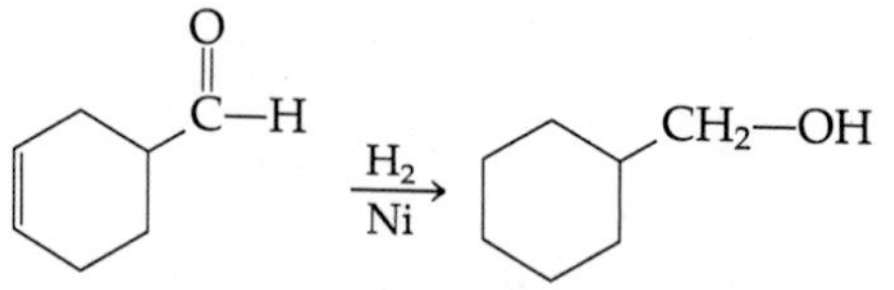

$$\text{cyclohexenyl–CHO} \xrightarrow[\text{Ni}]{H_2} \text{cyclohexyl–}CH_2\text{–OH}$$

A carbonyl group can be reduced to an alcohol selectively by reagents attracted to the highly polarized carbonyl group. The carbonyl carbon atom has a partial positive charge and tends to react with nucleophiles.

$$\overset{\delta^+}{C}=\overset{\delta^-}{\ddot{O}}:$$

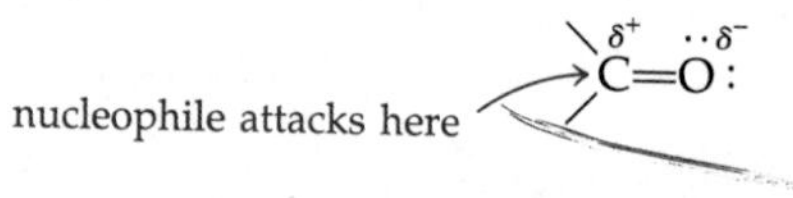

nucleophile attacks here

The carbon–carbon double bond of alkenes is not polar and does not react with nucleophiles. This difference in reactivity is the basis for the reduction of carbonyl compounds by metal hydrides, such as sodium borohydride, $NaBH_4$, and lithium aluminum hydride, $LiAlH_4$, which do not react with alkenes. Both reagents are a source of a nucleophilic hydride ion. Sodium borohydride is less reactive than lithium aluminum hydride, but both easily reduce both aldehydes and ketones. Sodium borohydride can be used in water or ethanol as the solvent.

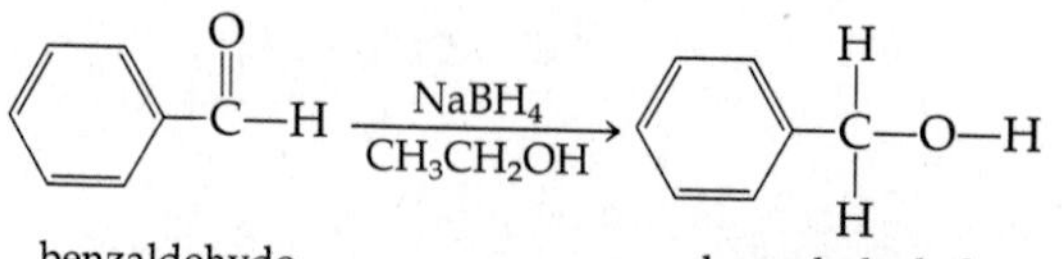

$$C_6H_5\text{–}\overset{O}{\overset{\|}{C}}\text{–H} \xrightarrow[CH_3CH_2OH]{NaBH_4} C_6H_5\text{–}CH_2\text{–O–H}$$

benzaldehyde → benzyl alcohol

In reduction by sodium borohydride, a hydride ion of the borohydride ion, BH_4^-, is transferred to the carbonyl carbon atom from boron and the carbonyl oxygen atom is protonated by the ethanol.

$$H_3\bar{B}\text{—H} + \text{>C}=\ddot{O}: + \text{H—}OCH_2CH_3 \longrightarrow \text{H—}\overset{|}{\underset{|}{C}}\text{—O—H} + CH_3CH_2OBH_3^-$$

ethoxyborohydride

The ethoxyborohydride product in the above reaction has three remaining hydride ions available for further reduction reactions, and the ultimate boron product is tetraethoxyborohydride, $(RO)_4B^-$. Thus, one mole of $NaBH_4$ reduces four moles of a carbonyl compound.

When lithium aluminum hydride is used to reduce carbonyl compounds, an ether such as diethyl ether, $(CH_3CH_2)_2O$, is used as the solvent.

$$\text{cyclohexanone} \xrightarrow[\text{2. } H_3O^+]{\text{1. } LiAlH_4 \text{ (ether)}} \text{cyclohexanol}$$

The reduction of a carbonyl group by lithium aluminum hydride occurs by the transfer of a hydride ion from AlH_4^- to the carbonyl carbon atom. The carbonyl oxygen atom forms a salt with aluminum.

$$\text{:}\ddot{O}\text{=C<} \quad \bar{Al}H_3\text{—H} \longrightarrow \text{:}\ddot{O}\text{—}AlH_3^- \text{ on —C—H}$$

an alkoxyaluminate

The initial alkoxyaluminate has three remaining hydride ions available for further reduction reactions, and the ultimate aluminum product is tetraalkoxyaluminate, $(RO)_4Al^-$. Thus, one mole of $LiAlH_4$ reduces four moles of a carbonyl compound. The tetraalkoxyaluminate is hydrolyzed with aqueous acid in a separate, second step.

$$(H\text{—}\overset{|}{\underset{|}{C}}\text{—}O)_4Al^-Li^+ + 4\,H_2O \longrightarrow 4\,H\text{—}\overset{|}{\underset{|}{C}}\text{—}OH + Al(OH)_3 + LiOH$$

Indirect Hydration of Alkenes

In Chapter 4 we discussed the electrophilic addition of water to alkenes to give alcohols. In this section we will consider two "indirect" ways to add the elements of water to a double bond. These methods are indirect because the hydroxyl group, the hydrogen atom, or both originate in reagents other than water. One such reaction is oxymercuration–demercuration. Oxymercuration–demercuration of an alkene gives a product that corresponds to Markovnikov addition of water. Another indirect reaction is hydroboration–oxidation. Hydroboration–oxidation is equivalent to an anti-Markovnikov addition of water to the alkene.

In an oxymercuration–demercuration reaction, an alkene is treated with mercuric acetate, $Hg(OAc)_2$, and the product is treated with sodium borohydride. The net result is a **Markovnikov addition product** in which the —OH group is bonded to the more substituted carbon atom of the alkene.

$$CH_3(CH_2)_3CH\text{=}CH_2 \xrightarrow[\text{2. } NaBH_4]{\text{1. } Hg(OAc)_2} CH_3(CH_2)_3\overset{HO}{\overset{|}{C}}H\text{—}\overset{H}{\overset{|}{C}}H_2$$

In the first step, an electrophilic $HgOAc^+$ ion adds to the double bond to give a mercurinium ion whose structure is similar to that of the bromonium ion (Section 4.11). This species subsequently reacts with a nucleophilic water molecule. The net result is the bonding of —HgOAc and a hydroxyl group on adjacent carbon atoms. The product corresponds to a Markovnikov addition because the water attacks the more positive carbon atom of the intermediate, which is the more substituted center.

$$\text{>C=C<} + {}^{+}HgOAc \xrightarrow{H_2O} -\underset{AcOHg}{\underset{|}{\overset{|}{C}}}-\underset{|}{\overset{\overset{OH}{|}}{C}}- + H^+$$

The organomercury compound is reduced with sodium borohydride, and the —HgOAc group is replaced by a hydrogen atom.

$$-\underset{AcOHg}{\underset{|}{\overset{|}{C}}}-\underset{|}{\overset{\overset{OH}{|}}{C}}- \xrightarrow{NaBH_4} -\underset{H}{\underset{|}{\overset{|}{C}}}-\underset{|}{\overset{\overset{OH}{|}}{C}}- + Hg + OAc^-$$

Oxymercuration–demercuration gives the product that would result from direct hydration of an alkene. However, the reactions occur with a higher yield because the competing reverse reaction—dehydration—does not occur.

Hydroboration–oxidation of alkenes, which was developed by the American chemist H. C. Brown, also requires two steps. The sequence of reactions adds the elements of water to a double bond to give a product that corresponds to an **anti-Markovnikov product.**

$$CH_3(CH_2)_3CH{=}CH_2 \xrightarrow[2.\ H_2O_2,\ OH^-]{1.\ B_2H_6} CH_3(CH_2)_3\overset{\overset{H}{|}}{C}H-\overset{\overset{OH}{|}}{C}H_2$$

In the first step—hydroboration—an alkene is treated with diborane, $(BH_3)_2$ or B_2H_6. Diborane acts as if it were the monomeric species called borane, BH_3. The compound is usually prepared in an ether solvent such as diethyl ether or tetrahydrofuran (Chapter 9). It adds to the carbon–carbon double bond of one alkene and then adds successively to two additional alkenes to produce a trialkylborane, R_3B. These steps are hydroboration reactions.

$$\text{>C=C<} + \text{H}_2\text{B–H} \longrightarrow -\underset{|}{\overset{\overset{H}{|}}{C}}-\underset{|}{\overset{\overset{BH_2}{|}}{C}}- \longrightarrow \longrightarrow (-\underset{|}{\overset{\overset{H}{|}}{C}}-\underset{|}{\overset{|}{C}})_3B$$

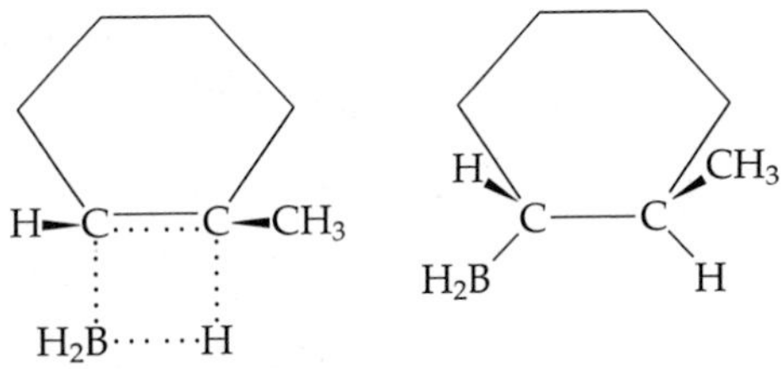

FIGURE 8.3
Mechanism of Borane Addition to Double Bonds
The boron and hydrogen atoms add to the same side of the double bond. The resulting product has the groups cis to each other.

In the oxidation step, the trialkyl borane is treated with hydrogen peroxide and base to oxidize the organoborane to form an alcohol.

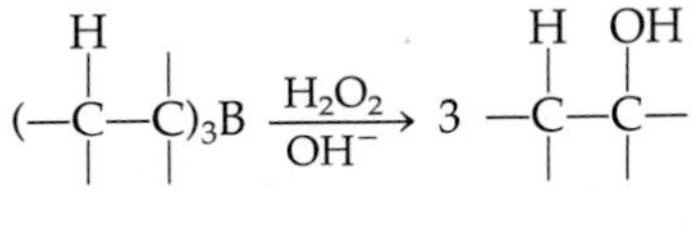

$$(\text{—}\overset{\text{H}}{\underset{|}{\overset{|}{\text{C}}}}\text{—}\underset{|}{\overset{|}{\text{C}}})_3\text{B} \xrightarrow[\text{OH}^-]{\text{H}_2\text{O}_2} 3\ \text{—}\overset{\text{H}}{\underset{|}{\overset{|}{\text{C}}}}\text{—}\overset{\text{OH}}{\underset{|}{\overset{|}{\text{C}}}}\text{—}$$

Let's consider the result of the hydroboration–oxidation of 1-methylcyclohexene.

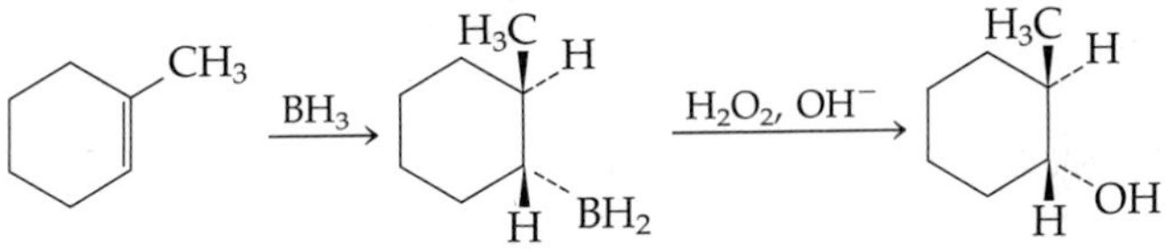

The overall addition of water is anti-Markovnikov; the hydrogen atom is added to the more substituted carbon atom, and the hydroxyl group is on the less substituted carbon atom. The hydrogen atom and hydroxyl group are introduced from the same side of the double bond; that is, the addition is cis. This mode of addition is observed because hydroboration is a concerted syn process (Figure 8.3). That is, the carbon–boron and carbon–hydrogen bonds are formed at the same time that the boron–hydrogen bond is broken. In the oxidation step a hydroxyl group replaces the boron with retention of configuration.

Borane reacts with alkenes for two reasons. First, the boron atom in borane is an electron-deficient species—it has only six electrons. Thus, the boron atom has a vacant 2*p* orbital and is an electrophilic Lewis acid. Because boron is electrophilic, it bonds to the least substituted carbon atom—much like a proton. Second, boron is more electropositive than hydrogen. Therefore, the hydrogen atom of the boron–hydrogen bond has a partial negative charge. This hydrogen atom behaves like a hydride ion, not like a proton. In summation, two properties of BH_3, the electrophilic character of the boron atom and the hydride character of the hydrogen atom, account for the anti-Markovnikov addition of BH_3 to alkenes.

EXAMPLE 8.6 What product is formed from the following alkene by oxymercuration–demercuration? What product is formed by hydroboration–oxidation?

CH_2

Solution The alkene is disubstituted, and both alkyl groups are bonded to the same carbon atom. The double-bonded CH_2 is the less substituted carbon atom; the ring carbon atom is the more substituted carbon atom. An oxymercuration–demercuration reaction places a hydrogen atom at the CH_2 site and a hydroxyl group on the ring carbon atom. This is equivalent to Markovnikov addition to water to the alkene.

CH_2 —(1. $Hg(OAc)_2$; 2. $NaBH_4$)→ HO CH_3

The hydroboration–oxidation product has a hydroxyl group at the CH_2 site and a hydrogen atom at the ring carbon atom. This process is equivalent to anti-Markovnikov addition of water to the double bond.

CH_2 —(1. BH_3; 2. H_2O_2, OH^-)→ H CH_2OH

8.10 PHENOLS

Many of the common reactions of alcohols do not occur with phenols. The chemistry of phenols is quite different from the chemistry of alcohols because it is very difficult to break the C—O bond in phenols. The carbon atom is sp^2-hybridized, and the C—O bond in phenols is shorter and stronger than the C—O bond of alcohols, where the carbon atom is sp^3-hybridized. The phenol oxygen atom is readily protonated to give an oxonium ion, but an S_N1 process in which water is a leaving group does not occur.

$$C_6H_5-\overset{+}{O}H_2 \not\rightarrow C_6H_5^+ + H_2O$$

Phenols do not react in S_N2 reactions because the geometry of the ring prevents nucleophilic attack at the ring carbon atom from the back side.

Acidity

Phenols are stronger acids than alcohols, but they are still quite weak acids. A typical alcohol has a pK_a of 16–17. In contrast, phenol is ten million times more acidic: its pK_a is 10.

$$R{-}\ddot{\underset{..}{O}}{-}H + H_2O \rightleftharpoons R{-}\ddot{\underset{..}{O}}:^- + H_3O^+ \qquad pK_a = 17$$

$$Ar{-}\ddot{\underset{..}{O}}{-}H + H_2O \rightleftharpoons Ar{-}\ddot{\underset{..}{O}}:^- + H_3O^+ \qquad pK_a = 10$$

Phenol is more acidic than cyclohexanol and acyclic alcohols because the phenoxide ion is more stable than the alkoxide ion. In an alkoxide ion, the negative charge is localized at the oxygen atom. However, in a phenoxide ion, the negative charge is stabilized by resonance in which the charge is delocalized over the benzene ring.

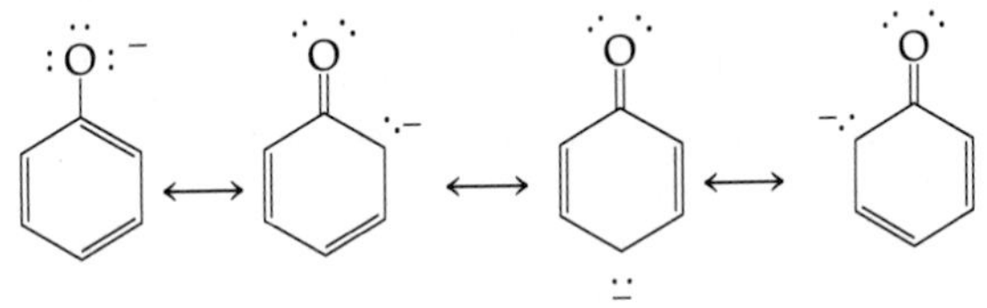

The acidity of phenols increases when the ring is substituted with electron-withdrawing groups. These substituents stabilize the phenoxide ion by further delocalizing the negative charge. Phenols substituted with electron-donating groups are less acidic than phenol.

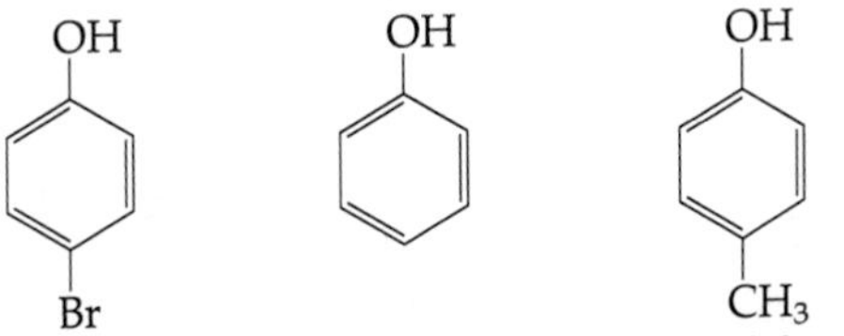

OH
Br
p-bromophenol
($pK_a = 9.35$)

OH
phenol
($pK_a = 10.00$)

OH
CH_3
p-methylphenol
($pK_a = 10.26$)

Phenols react with hydroxide ions to produce phenoxide ions. Thus, phenols dissolve in basic solution, whereas high molecular weight alcohols are insoluble. Phenol is not a strong enough acid to react with aqueous sodium bicarbonate. In contrast, stronger organic acids such as carboxylic acids ($pK_a = 5$) react with aqueous sodium bicarbonate to form carbon dioxide gas and a carboxylate anion.

Oxidation

Phenols are oxidized to give conjugated 1,4-diketones called quinones.

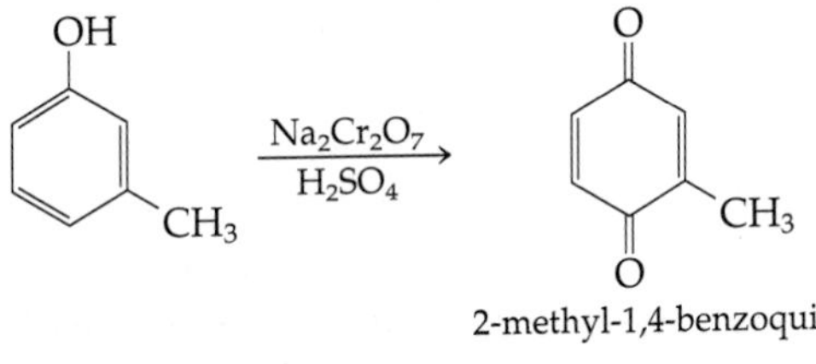

OH, CH_3 $\xrightarrow[H_2SO_4]{Na_2Cr_2O_7}$ O, O, CH_3

2-methyl-1,4-benzoquinone

Hydroquinone, a phenol with two hydroxyl groups, is very easily oxidized.

Phenols Are Germicides

Disinfectants are compounds that decrease the bacterial count on objects such as medical instruments. Antiseptics also inhibit bacterial growth but are used on living tissue. The English surgeon Joseph Lister used phenol itself as a hospital disinfectant in the late nineteenth century. But phenol is no longer used as an antiseptic because it causes severe burns to skin. A 2% solution was formerly used to decontaminate medical instruments. However, this use has also been largely discontinued because substituted phenols and other compounds are more effective.

The efficiency of a germicide is measured in terms of its **phenol coefficient (PC).** Phenol itself has a PC value of 1. If a 1% solution of a germicide is as effective as a 10% solution of phenol, it has a PC of 10. If a phenol is substituted with alkyl groups, its germicidal action increases. For example, the methyl-substituted phenols called *ortho-*, *meta-*, and *para-*cresol are used in the commercial disinfectant, Lysol. Another phenol, called thymol, is used by dentists to sterilize a tooth before filling it.

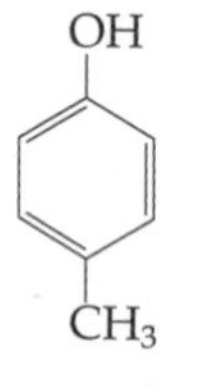

p-cresol 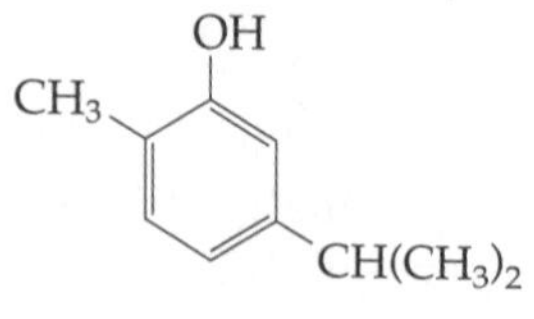　　thymol

The phenol coefficient increases when the phenol is halogenated, particularly if the halogen is para to the hydroxyl group. The structures of a few halogenated phenols are shown here. Chlorophene is a more effective germicide than o-phenylphenol. *p*-Chloro-*m*-xylenol is more effective than the cresols; it is used in topical preparations for athlete's foot and jock itch. Hexachlorophene (PC = 120) has been used in some toothpastes, deodorants, and soaps. However, it is toxic to infants and is no longer used in commercial products, although it is still used as a surgical scrub.

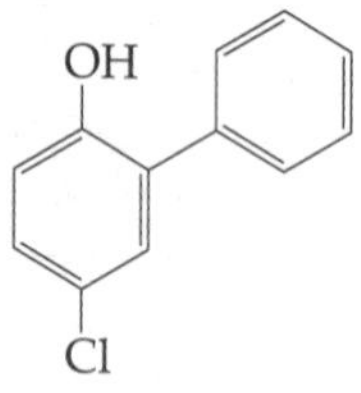

chlorophene 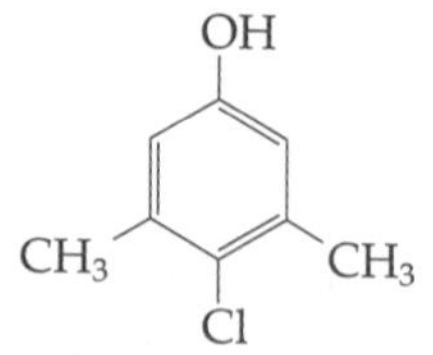　　*p*-chloro-*m*-xylenol

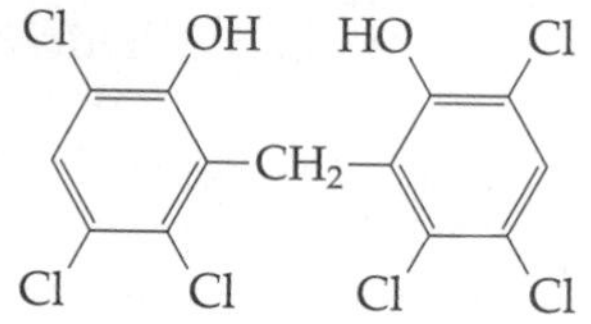

hexachlorophene

Phenols that contain two hydroxyl groups are called bis-phenols (bis meaning two). When the two hydroxyl groups are meta, the phenols are called resorcinols. Resorcinol and hexylresorcinol are effective germicides. Resorcinol has a PC of only 0.4, but it is useful in the treatment of psoriasis and seborrhea. As in the case of phenols, alkyl substitution increases the PC. Hexylresorcinol has PC = 93; it is used in throat lozenges.

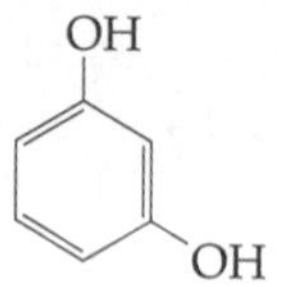

resorcinol 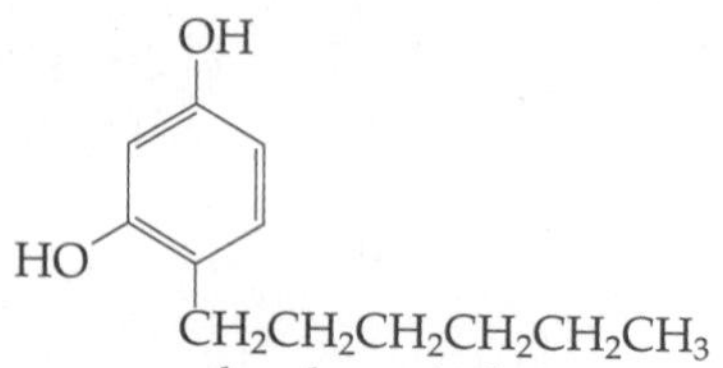　　hexylresorcinol

It will reduce silver bromide that has been activated by exposure to light in photographic film emulsions.

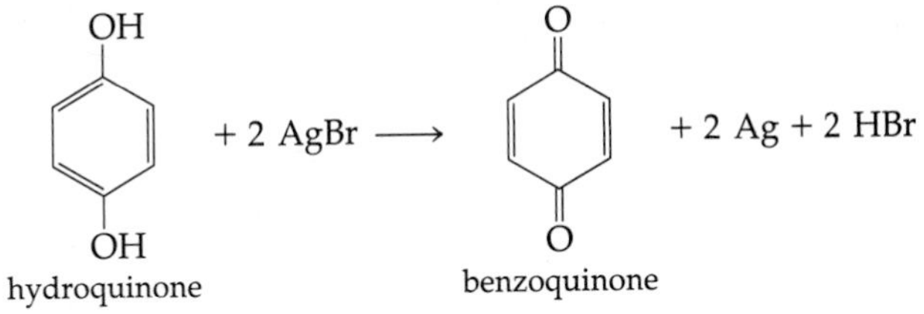

hydroquinone + 2 AgBr ⟶ benzoquinone + 2 Ag + 2 HBr

Hydroquinone is used as a developer. It reacts faster with the light-activated silver bromide than with unexposed grains of the salt. As a consequence, silver deposits in the film at points where exposure to light occurred. The result is a "negative" image. Areas in the film negative that are clear correspond to unlit areas in the real world; areas that are very dark in the negative correspond to well-lit areas in the scene photographed.

Quinones are widely distributed in nature, where along with the reduced hydroquinone form they serve as oxidizing and reducing agents. Coenzyme Q, also called ubiquinone, is found within the mitochondria of eucaryotic cells. Coenzyme Q, in its oxidized form, oxidizes NADH to regenerate NAD^+, a common oxidizing agent in biological reactions. The side chain represented by R in the following structure is a polyunsaturated unit that consists of isoprene units.

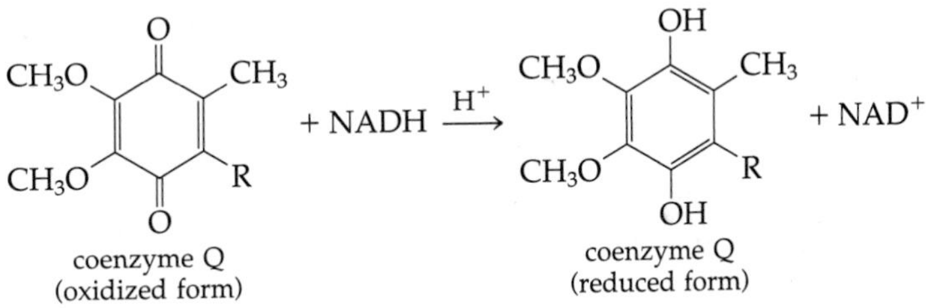

coenzyme Q (oxidized form) + NADH $\xrightarrow{H^+}$ coenzyme Q (reduced form) + NAD^+

8.13 SULFUR COMPOUNDS

Sulfur is in the same group of the periodic table as oxygen. It forms compounds that are structurally similar to alcohols. The —SH group is called the **sulfhydryl group.** Compounds containing an —SH group are called **mercaptans** or **thiols.** The nomenclature of these compounds resembles that of alcohols, except that the suffix *-thiol* replaces the suffix *-ol.* The *-e* of the alkane name is retained.

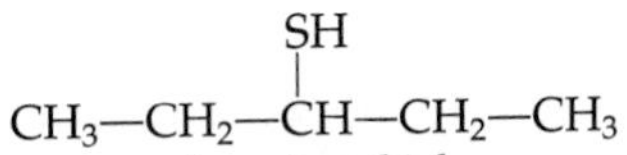

$$CH_3{-}CH_2{-}CH(SH){-}CH_2{-}CH_3$$

3-pentanethiol

Alcohols and thiols resemble each other in many ways, but they also differ in some significant respects. For example, thiols have lower boiling points than the corresponding alcohols because sulfur does not form hydrogen bonds. We recall from Chapter 2 that only nitrogen, oxygen, and fluorine form hydrogen bonds.

$$\underset{\text{bp} = 98\ ^\circ\text{C}}{CH_3CH_2CH_2CH_2{-}SH} \qquad \underset{\text{bp} = 117\ ^\circ\text{C}}{CH_3CH_2CH_2CH_2{-}OH}$$

Some alcohols have rather sweet odors, but one of the distinguishing properties of thiols is their strong, disagreeable odor. The odor of the striped skunk *(Memphitis mephitis)* is due to 3-methyl-1-butanethiol. Thiols can be detected by the human nose at the level of parts per billion in air. Small amounts of thiols are added to natural gas so that leaks can easily be detected. A skunk takes more drastic measures to fend off predators.

Although thiols are weak acids, they are far stronger than alcohols.

$$R{-}S{-}H + H_2O \rightleftharpoons R{-}S^- + H_3O^+ \qquad pK_a = 8$$

$$R{-}O{-}H + H_2O \rightleftharpoons R{-}O^- + H_3O^+ \qquad pK_a = 17$$

The sulfhydryl group is acidic enough to react with hydroxide ions to form thiolate salts.

$$R{-}S{-}H + NaOH \longrightarrow \underset{\text{a thiolate}}{R{-}S^-} + Na^+ + H_2O$$

Thiols can be obtained from haloalkanes by nucleophilic displacement of halide ion with the sulfhydryl ion (HS^-). The sulfur analogs of ethers—that is, thioethers—are obtained from haloalkanes by nucleophilic displacement with thiolates. We recall that sulfur anions are excellent nucleophiles (Chapter 7).

$$R{-}Br + HS^- \longrightarrow R{-}SH + Br^-$$

$$R{-}Br + R'{-}S^- \longrightarrow \underset{\text{a thioether}}{R'{-}S{-}R} + Br^-$$

Thiols are easily oxidized but yield disulfides rather than the structural analogs of aldehydes and ketones. In the following equation, the symbol [O] represents an unspecified oxidizing agent that removes the hydrogen atoms.

$$2\,R{-}S{-}H \xrightarrow{[O]} \underset{\text{a disulfide}}{R{-}S{-}S{-}R}$$

This reaction is of great biological importance because many proteins contain the amino acid cysteine (Chapter 15), which contains a sulfhydryl group. Oxidation of the —SH group of cysteine gives a disulfide bond in cystine.

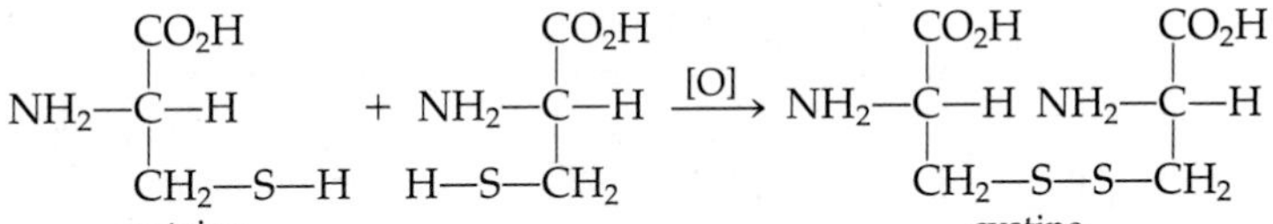

$$\underset{\text{cysteine}}{NH_2-\overset{CO_2H}{\underset{CH_2-S-H}{C}}-H} + NH_2-\overset{CO_2H}{\underset{H-S-CH_2}{C}}-H \xrightarrow{[O]} \underset{\text{cystine}}{NH_2-\overset{CO_2H}{C}-H \quad NH_2-\overset{CO_2H}{C}-H \atop CH_2-S-S-CH_2}$$

In some cases the sulfhydryl groups in an enzyme must be maintained in the reduced state for proper biological function. If an essential cysteine sulfhydryl group is oxidized, the enzyme becomes inactive.

Sulfhydryl groups in enzymes (E—S—H) also react with salts of lead and mercury. Because these reactions render the enzymes inactive, lead and mercury salts are highly toxic.

$$2\,E-S-H + Hg^{2+} \longrightarrow E-S-Hg-S-E + 2\,H^+$$

EXERCISES

Nomenclature of Alcohols

8.1 Write the structural formula of each of the following.
(a) 2-methyl-2-pentanol (b) 2-methyl-1-butanol (c) 2,3-dimethyl-1-butanol

8.2 Write the structural formula of each of the following.
(a) 2-methyl-3-pentanol (b) 3-ethyl-3-pentanol (c) 4-methyl-2-pentanol

8.3 Write the structural formula of each of the following.
(a) cyclopentanol (b) 1-methylcyclohexanol (c) *trans*-2-methylcyclohexanol
(d) *cis*-3-ethylcyclopentanol

8.4 Write the structural formula of each of the following.
(a) 1,2-hexanediol (b) 1,3-propanediol (c) 1,2,4-butanetriol
(d) 1,2,3,4,5,6-hexanehexol

8.5 Name each of the following compounds.

(a) OH (b) OH

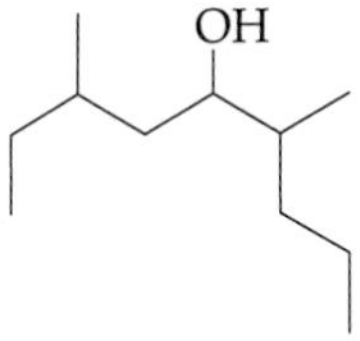

(c) OH (d) OH

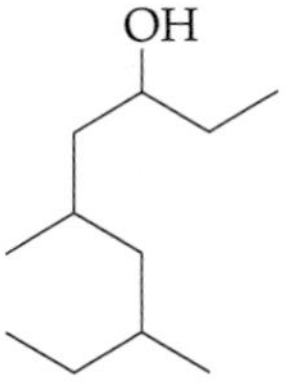

8.6 Name each of the following compounds.

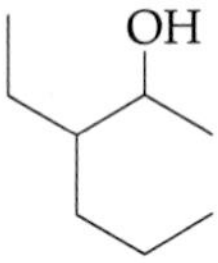

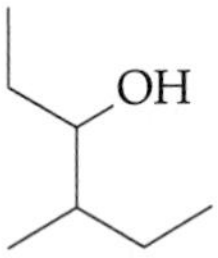

(a) CH_3—⬡---OH (b) ⬠—$CH_2CH_2CH_2CH_2OH$

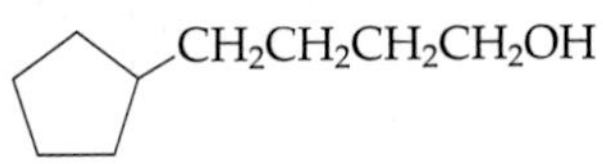

(c) 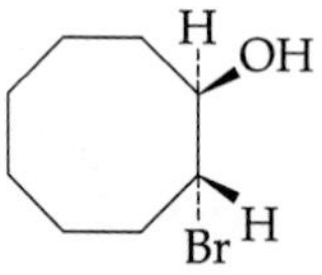 [cyclooctane ring bearing H, OH (wedge) and Br, H (wedge) on adjacent carbons]

(d) 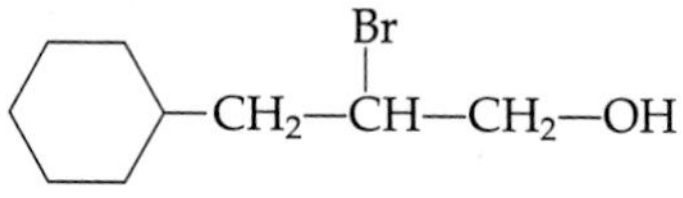 [cyclohexyl]$-CH_2-\underset{|}{\overset{Br}{CH}}-CH_2-OH$

8.7 Name the sex attractant of the Mediterranean fruit fly.

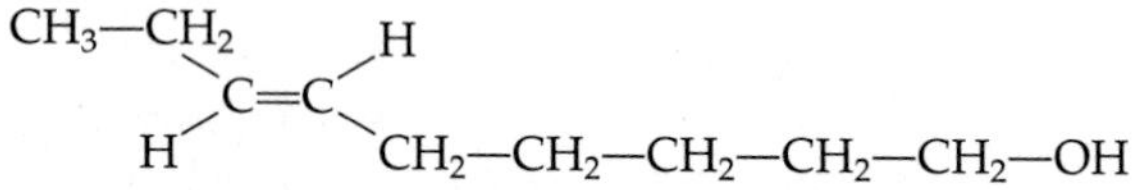

$$\begin{matrix} CH_3-CH_2 & & H \\ & C=C & \\ H & & CH_2-CH_2-CH_2-CH_2-CH_2-OH \end{matrix}$$

8.8 Name the following compound that is used as a mosquito repellant.

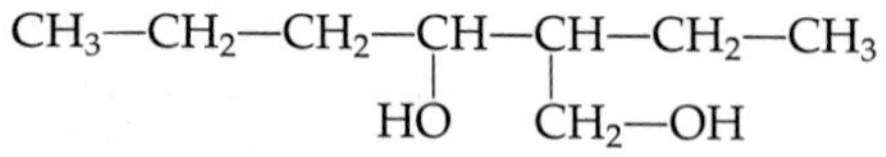

$$CH_3-CH_2-CH_2-\underset{HO}{CH}-\underset{CH_2-OH}{CH}-CH_2-CH_3$$

Classification of Alcohols

8.9 Classify each of the following compounds.

(a) 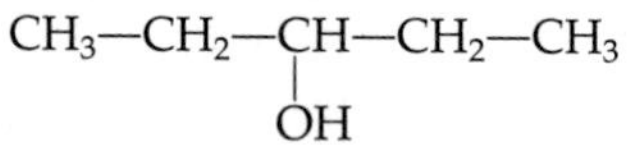 $CH_3-CH_2-\underset{OH}{CH}-CH_2-CH_3$ (b) 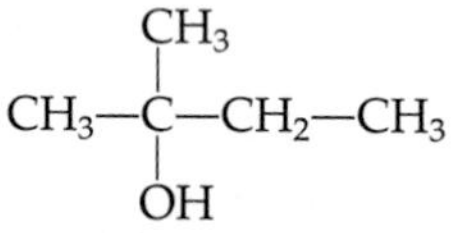 $CH_3-\overset{CH_3}{\underset{OH}{C}}-CH_2-CH_3$

(c) $CH_3-CH_2-\overset{OH}{CH}-CH_3$ (d) 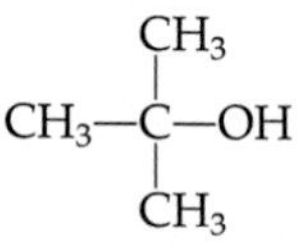 $CH_3-\overset{CH_3}{\underset{CH_3}{C}}-OH$

8.10 Classify each of the following compounds.

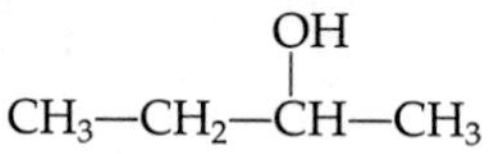

(a) 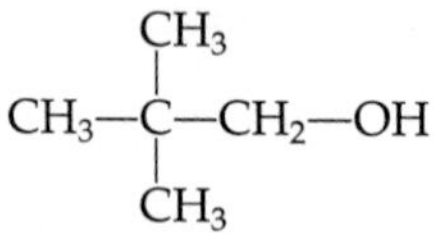 $CH_3-\overset{CH_3}{\underset{CH_3}{C}}-CH_2-OH$ (b) $CH_3-\underset{OH}{CH}-CH_2-\underset{CH_3}{CH}-CH_2-CH_3$

(c) $CH_3-\underset{OH}{CH}-CH_2-CH_2-CH_3$ (d) 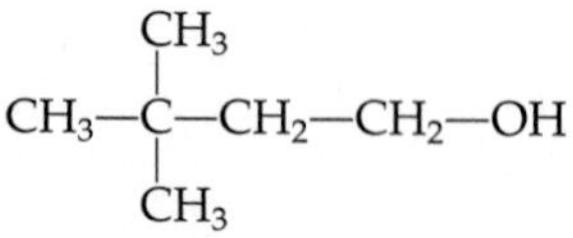 $CH_3-\overset{CH_3}{\underset{CH_3}{C}}-CH_2-CH_2-OH$

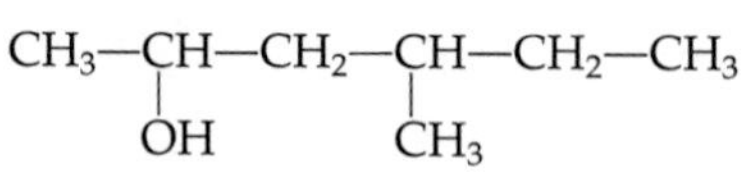

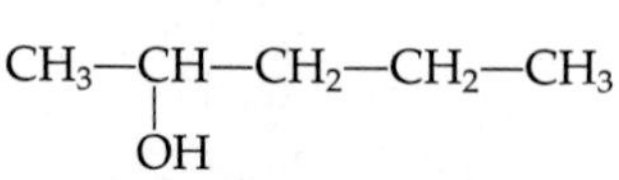

8.11 Classify each of the hydroxyl groups in the following vitamins.
(a) pyridoxal (vitamin B_6) (b) thiamine (vitamin B_1)

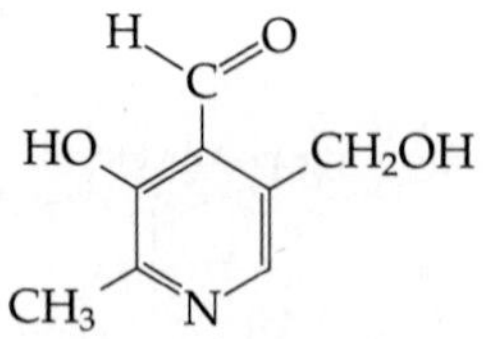

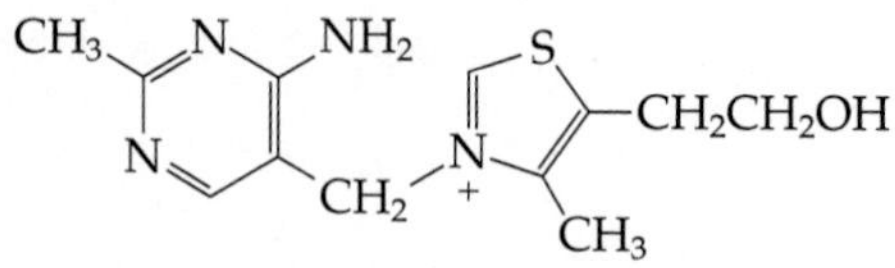

[pyridoxal: pyridine ring with H–C=O, HO, CH_2OH, CH_3, N]

[thiamine: pyrimidine ring with CH_3, N, NH_2, N, linked by CH_2 to thiazolium ring with S, N^+, $-CH_2CH_2OH$, CH_3]

(c) riboflavin (vitamin B_2)

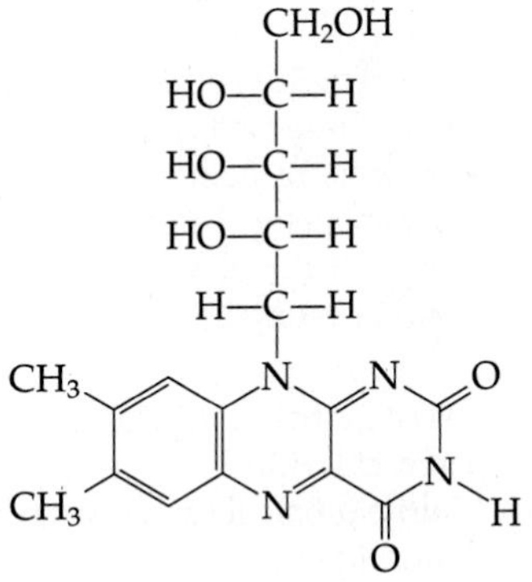

8.12 Classify each of the hydroxyl groups in the following steroids.
(a) digitoxigenin, a cardiac glycoside

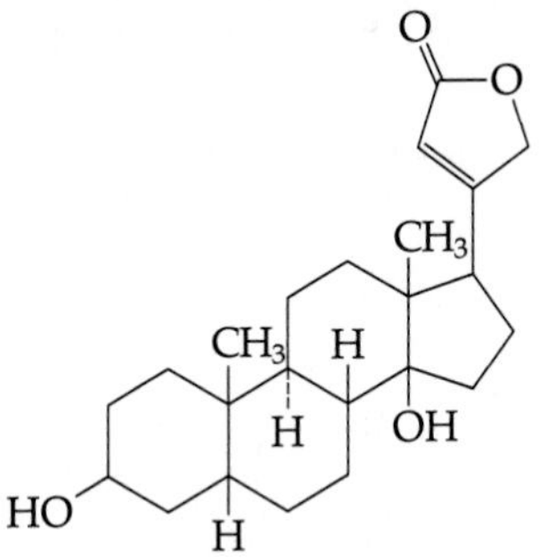

(b) hydrocortisone, an anti-inflammatory drug

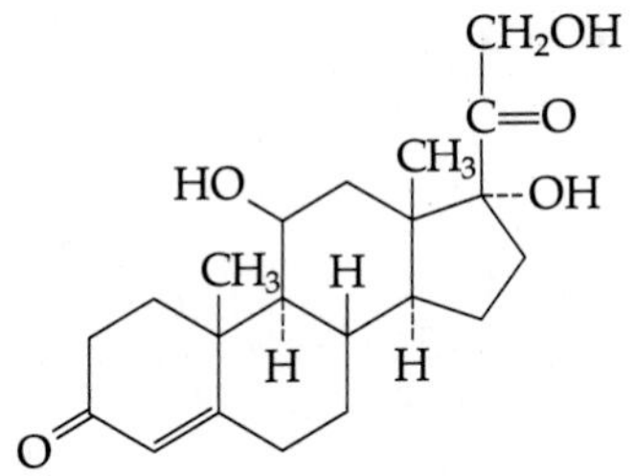

(c) norethindrone, an oral contraceptive

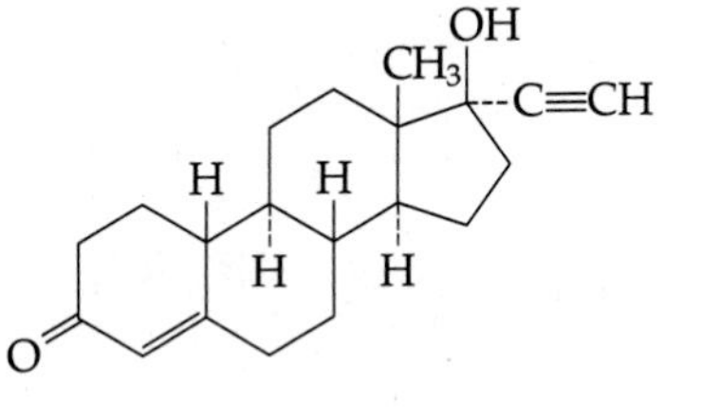

Physical Properties of Alcohols

8.13 1,2-Hexanediol is very soluble in water but 1-heptanol is not. Explain why these two compounds with similar molecular weights have different solubilities.

8.14 Ethylene glycol and 1-propanol boil at 198 and 97 °C, respectively. Explain why these two compounds with similar molecular weights have such different boiling points.

8.15 Explain why 1-butanol is less soluble in water than 1-propanol.

8.16 Suggest a reason why 2-methyl-1-propanol is much more soluble in water than 1-butanol.

Acid–Base Properties of Alcohols

8.17 Write a balanced equation for the reaction of sodium methoxide and hydrogen chloride.

8.18 Write a balanced equation for the reaction of lithium ethoxide and hydrogen bromide.

8.19 1,1,1-Trichloro-2-methyl-2-propanol is used as a bacteriostatic agent. Compare its K_a to that of 2-methyl-2-propanol.

8.20 Based on the data in Table 8.2, estimate the K_a of 2-bromoethanol.

8.21 Based on the data in Table 8.2, estimate the K_a of cyclohexanol.

8.22 Which base is stronger, methoxide ion or *t*-butoxide ion?

Substitution Reactions

8.23 Rank the following isomeric compounds according to reactivity with HBr.

(I) C_6H_{11}—$CHOHCH_3$ (II) $C_6H_{10}(OH)$—CH_2CH_3 (III) C_6H_{11}—CH_2CH_2OH

8.24 Rank the following isomeric compounds according to reactivity with HCl and $ZnCl_2$.

(I) $CH_3—CH_2—C(CH_3)_2—OH$

(II) $CH_3—CH_2—CH(CH_3)—CH_2—OH$

(III) $CH_3—CH(OH)—C(CH_3)(H)—CH_3$

8.25 Write the structure of the product of reaction for each of the following compounds with PBr_3.

(a) $CH_3—C(CH_3)_2—CH_2—CH_2—OH$ (b) $CH_3—C(CH_2—CH_3)(OH)—CH_2—CH_3$

(c) $CH_3-C(H)(CH_3)-CH_2-CH(OH)-CH_3$

8.26 Write the structure of the product of reaction for each of the following compounds with $SOCl_2$.

(a) $C_6H_5-CH(OH)CH_3$ (b) cyclopentanol (c) $C_6H_5-CH_2CH_2OH$

8.27 Reaction of 3-buten-2-ol with HBr yields a mixture of two products: 3-bromo-1-butene and 1-bromo-2-butene. Explain why. (*Hint:* The reaction of this allyl alcohol occurs via an S_N1 process.)

8.28 The rate of reaction of the following unsaturated alcohol with HBr is faster than the rate of reaction of the saturated alcohol. Explain why.

1-methyl-2-cyclohexen-1-ol (H_3C, OH) and 1-methylcyclohexanol (H_3C, OH)

Dehydration of Alcohols

8.29 Draw the structure of the dehydration product(s) when each of the following compounds reacts with sulfuric acid. If more than one product is formed, predict the major isomer.

(a) $CH_3-C(CH_3)(OH)-CH_2-CH_3$ (b) $CH_3-CH_2-CH(CH_3)-CH_2-CH_2-OH$

(c) $CH_3-CH_2-CH(OH)-CH_2-CH_2-CH_3$

8.30 Draw the structure of the dehydration product(s) when each of the following compounds reacts with sulfuric acid. If more than one product is formed, predict the major isomer.

(a) cyclopentane ring with OH and CH_3 on adjacent carbons (wedge bonds) (b) 1-methylcyclopentanol (CH_3, OH) (c) bicyclic (cyclohexane fused to cyclopentane) with OH on the five-membered ring

8.31 Write the expected product of the acid-catalyzed dehydration of 1-phenyl-2-propanol. The reaction is more rapid than the rate of dehydration of 2-propanol. Explain why.

8.32 The dehydration of *trans*-2-methylcyclopentanol occurs via an E2 process to give predominately 3-methylcyclopentene rather than 1-methylcyclopentene.

Stop

What does this information indicate about the stereochemistry of the elimination reaction?

Oxidation of Alcohols

8.33 Write the product formed from the oxidation of each of the compounds in Exercise 8.23 with the Jones reagent.

8.34 Write the product formed from the oxidation of each of the compounds in Exercise 8.24 by the Jones reagent.

8.35 Write the product formed from the oxidation of the sex attractant of the Mediterranean fruit fly (see Exercise 8.7) with PCC.

8.36 Write the product formed by oxidation with PCC of the mosquito repellant in Exercise 8.8.

8.37 Which of the compounds in Exercises 8.25 and 8.26 will give a ketone when oxidized by Jones reagent?

8.38 Which of the compounds in Exercise 8.25 and 8.26 will give an acid product when oxidized by Jones reagent?

8.39 1,2-Propanediol is oxidized in the liver to pyruvic acid ($C_3H_4O_3$), which can be metabolized by the body. Draw the structure of pyruvic acid.

8.40 Why is the ingestion of isopropyl alcohol not advisable?

Preparation of Alcohols

8.41 Which of the isomeric $C_4H_{10}O$ alcohols can be produced by an acid-catalyzed hydration reaction of an alkene?

8.42 2-Propanol is produced in industry by a hydration reaction of propene but 1-propanol is produced by other methods. Why?

8.43 Name the final product of oxymercuration–demercuration of each of the following compounds.

(a) $(CH_3)_2C{=}CH_2$ (b) $(CH_3)_2C{=}C(CH_3)_2$

(c) $(CH_3)_2C{=}CHCH_3$ (d) $CH_3{-}CH_2{-}CH{=}CH_2$

8.44 Name the final product of hydroboration–oxidation of each of the compounds in Exercise 8.43.

8.45 Draw the structure of the hydroboration–oxidation product of each of the following compounds.

(a) 1-methylcyclodecene (CH_3 on ring double bond) (b) cyclobutyl–$CH{=}CH_2$ (c) cyclohexylidene$=CHCH_2CH_3$

8.46 Draw the structure of the oxymercuration–demercuration product of each compound in Exercise 8.45.

Synthesis Sequences

8.47 Draw the structure of the final product of each of the following sequences of reactions.

(a) cyclopentyl$-CH_2CH{=}CH_2$ $\xrightarrow[\text{2. } NaBH_4]{\text{1. } Hg(OAc)_2}$ $\xrightarrow[H_2SO_4]{Cr_2O_3}$

(b) 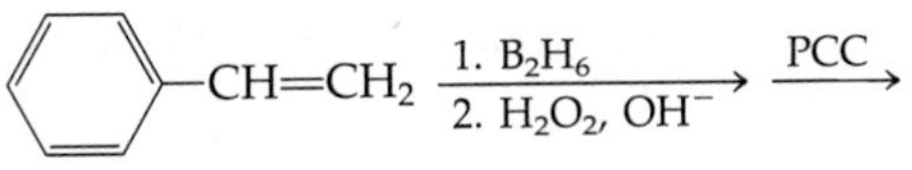 $C_6H_5-CH{=}CH_2 \xrightarrow[2.\ H_2O_2,\ OH^-]{1.\ B_2H_6} \xrightarrow{PCC}$

(c) $\square-CH_2CH{=}CH_2 \xrightarrow[2.\ H_2O_2,\ OH^-]{1.\ B_2H_6} \xrightarrow[H_2SO_4]{Cr_2O_3}$

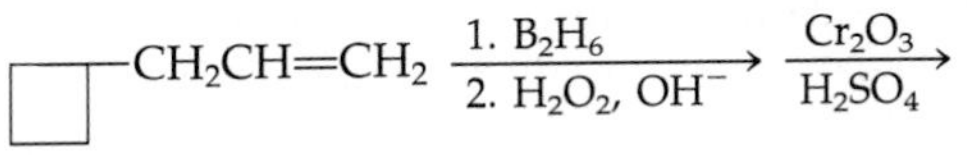

8.48 Draw the structures of the final product of each of the following sequences of reactions.

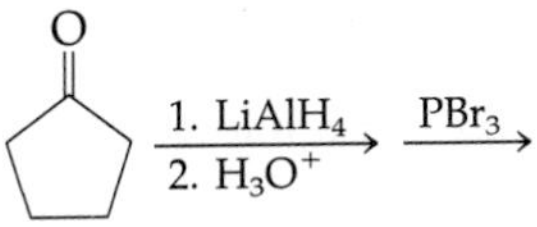

(a) 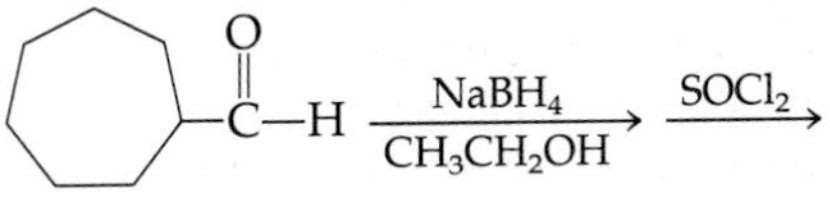 cycloheptyl$-\overset{O}{\overset{\|}{C}}-H \xrightarrow[CH_3CH_2OH]{NaBH_4} \xrightarrow{SOCl_2}$

(b) cyclopentanone $\xrightarrow[2.\ H_3O^+]{1.\ LiAlH_4} \xrightarrow{PBr_3}$

(c) $C_6H_5-\overset{O}{\overset{\|}{C}}-CH_3 \xrightarrow[CH_3CH_2OH]{NaBH_4} \xrightarrow{HBr}$

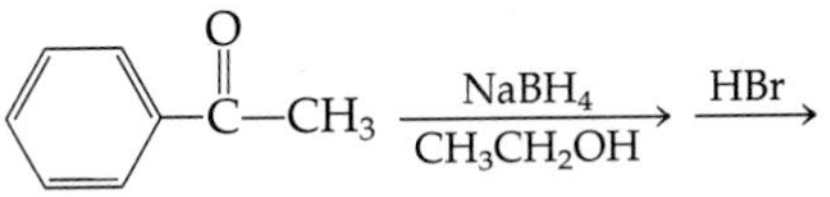

Phenols

8.49 Which of the following compounds is a phenol?

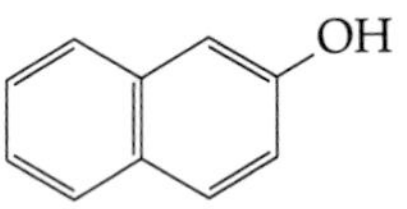

(I) 2-naphthol (naphthalene–OH)

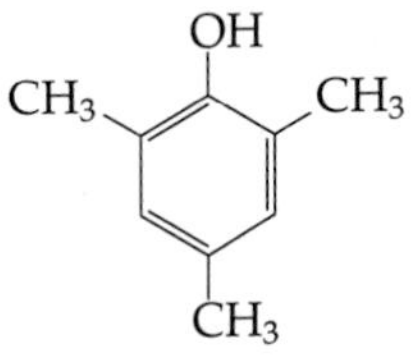

(II) OH, CH_3, CH_3, CH_3 on benzene ring

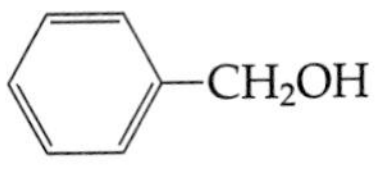

(III) $C_6H_5-CH_2OH$

8.50 Which of the following compounds is a phenol?

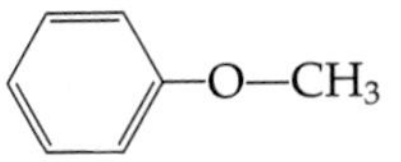

(I) $C_6H_5-O-CH_3$

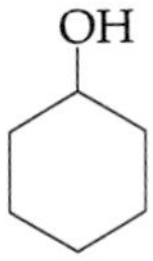

(II) cyclohexane–OH

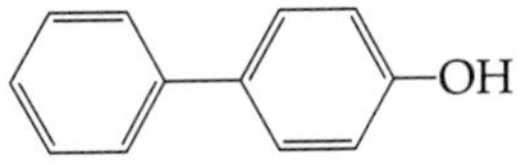

(III) $C_6H_5-C_6H_4-OH$

8.51 *p*-Nitrophenol is a much stronger acid than phenol. Explain why.

8.52 Which phenoxide is the stronger base, *p*-ethylphenoxide or *p*-chlorophenoxide?

8.53 Draw the structure of the quinone obtained from the oxidation of the following substituted naphthalene.

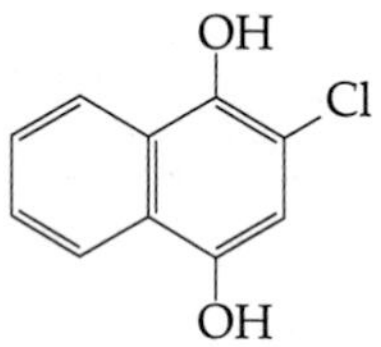

OH

Cl

OH

8.54 2-Methylhydroquinone is more easily oxidized to a quinone than is 2-chlorohydroquinone. Explain why.

Sulfur Compounds

8.55 There are four isomeric compounds $C_4H_{10}S$ with an —SH group. Draw the structures of the compounds.

8.56 There are three isomeric compounds C_3H_8S. Draw their structures.

8.57 Draw the structure of each of the following compounds.
(a) 1-propanethiol (b) 2-methyl-3-pentanethiol (c) cyclopentanethiol

8.58 Draw the structure of each of the following compounds.
(a) 2-propanethiol (b) 2-methyl-1-propanethiol (c) cyclobutanethiol

8.59 Addition of sodium hydroxide to an aqueous solution of $CH_3CH_2CH_2SH$ eliminates the odor. Explain why.

8.60 The boiling points of ethanethiol and dimethyl sulfide are 35 and 37 °C, respectively. Why are the boiling points similar? What types of intermolecular forces are responsible for this similarity?

$CH_3—CH_2—SH$ ethanethiol $CH_3—S—CH_3$ dimethyl sulfide

8.61 Indicate two methods to produce the scent marker of the red fox using a thiol as one of the reactants.

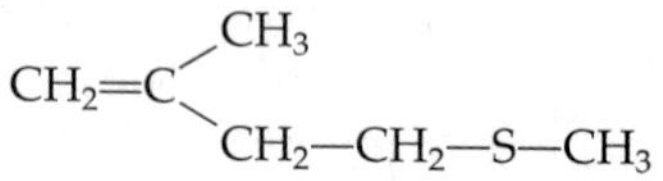

$CH_2{=}C(CH_3)—CH_2—CH_2—S—CH_3$

8.62 Outline a series of reactions to produce the compound used for defense by the skunk starting with 3-methyl-1-butene.

$CH_3—CH(CH_3)—CH_2—CH_2—SH$

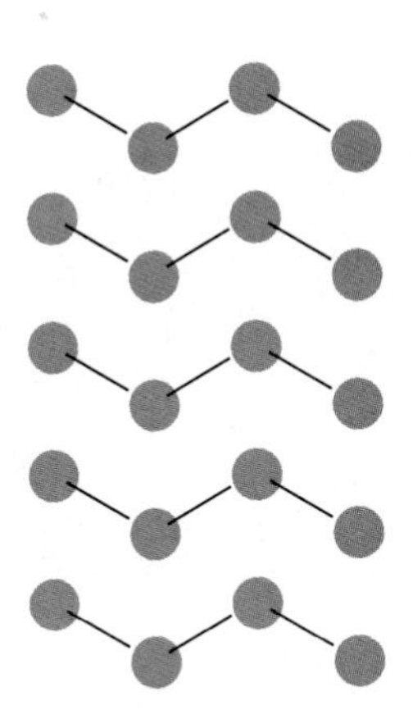

9

ETHERS AND EPOXIDES

9.1 ETHERS

Like alcohols, ethers can be viewed as organic "cousins" of water. **Ethers** contain two groups, which may be alkyl or aryl groups, bonded to an oxygen atom. The groups are the same in a **symmetrical ether** and different in an **unsymmetrical ether.**

$CH_3CH_2—O—CH_2CH_3$ diethyl ether (symmetrical ether)

$CH_3CH_2CH_2—O—C_6H_5$ phenyl propyl ether (unsymmetrical ether)

$C_6H_5—O—C_6H_5$ diphenyl ether (symmetrical ether)

Diethyl ether, often called ethyl ether or just "ether", is widely used as a solvent and also as a general anesthetic that is administered as a vapor. It acts as a depressant on the central nervous system and causes unconsciousness. Diethyl ether is used with oxygen, and the mixture is potentially explosive. An explosion could easily be caused by static electricity.

Many naturally occurring cyclic ethers are known. Tetrahydrocannabinol (THC), the principal active ingredient in marijuana, has a six-membered ring ether. Other cyclic ethers include drugs such as codeine, morphine, and heroin.

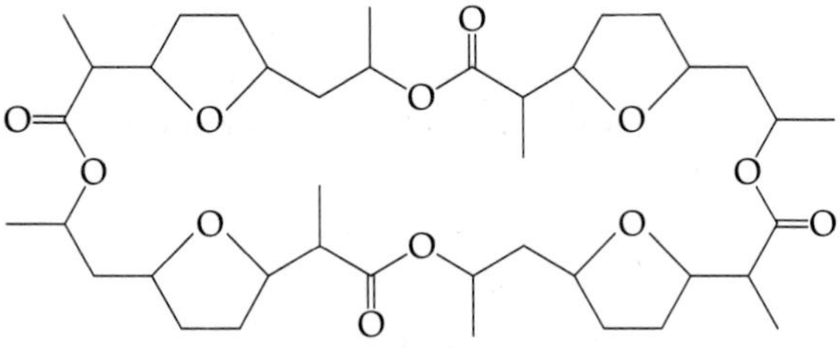

FIGURE 9.1
Antibiotic Ether Compounds
The cyclic polyethers can coordinate with ions such as the alkali metal ions. The selectivity of the ether for one metal ion over another depends on the geometry of the molecule and the location of the ether oxygen atoms.

nonactin

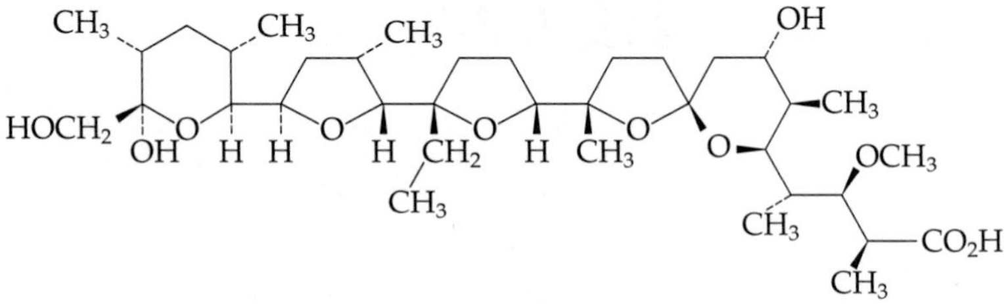

monensin

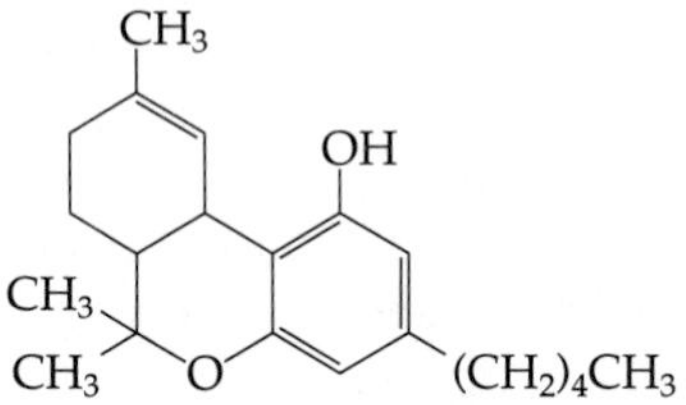

tetrahydrocannabinol

Certain cyclic ethers that contain several oxygen atoms can transport ions across biological membranes. One is nonactin (Figure 9.1), an antibiotic that selectively transports potassium out of bacterial cells. This compound, which contains four five-membered ring ethers linked by ester units, binds potassium about 10 times better than it binds sodium. Because cells must maintain a higher concentration of potassium ions than of sodium ions, the selective removal of potassium kills bacteria. Monensin (Figure 9.1), a more complex polyether, forms complexes with sodium ions, and the complex transports sodium ions into cells. The increase in concentration of sodium ions within the cell increases the osmotic pressure, and the cell membrane ruptures, killing the cells. Monensin is added to poultry feed to kill intestinal parasites.

9.2 NOMENCLATURE OF ETHERS

Common Names

Simple ethers are named as *alkyl alkyl ethers*. The name is constructed by listing the alkyl (or aryl) groups in alphabetical order and appending the name *ether*. For example, an unsymmetrical ether that has a butyl group and a methyl group bonded to an oxygen atom is called butyl methyl ether.

Symmetrical ethers are named by using the prefix *di-* in conjunction with the name of the alkyl group. For example, an ether with two isopropyl groups bonded to an oxygen atom is called diisopropyl ether.

$$CH_3—CH_2—CH_2—CH_2—O—CH_3$$

butyl methyl ether

$$CH_3—\underset{}{\overset{CH_3}{\overset{|}{C}}}H—O—\overset{CH_3}{\overset{|}{C}}H—CH_3$$

diisopropyl ether

IUPAC Names

Ethers are named according to IUPAC nomenclature as *alkoxy alkanes,* where the smaller alkyl group and the oxygen atom constitute an **alkoxy group.** An alkoxy group is treated as a substituent on the larger parent alkane chain. For example, a five-carbon chain (pentane) with a $—OCH_3$ group at the C-2 atom is named 2-methoxypentane. Other examples of ether nomenclature are given in Figure 9.2.

$$CH_3CH_2CH_2\underset{\underset{OCH_3}{|}}{C}HCH_3$$

$CH_3CH_2CH_2CHCH_3$ ⟵ The larger group is the parent chain.

OCH_3 ⟵ The smaller group is the substituent.

2-methoxypentane

Cyclic Ethers

The three- through six-membered cyclic ethers have common names. In all ring systems, the oxygen atom is assigned the number 1, and the rings are numbered in the direction that gives the first substituent the lower number. Cyclic ethers with three-ring atoms are called epoxides. Because these compounds are formed from the oxidation of an alkene, the common name of an epoxide is formed by adding *oxide* to the name of the alkene.

FIGURE 9.2
Nomenclature of Ethers

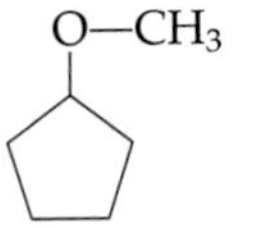

$O—CH_3$

methoxycyclopentane

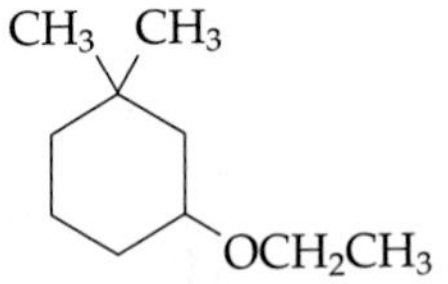

CH_3 CH_3 OCH_2CH_3

1-ethoxy-3,3-dimethylcyclohexane

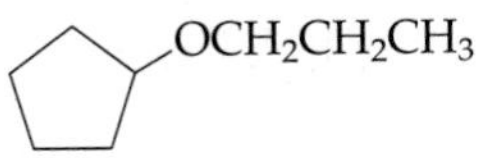

$OCH_2CH_2CH_3$

propoxycyclopentane

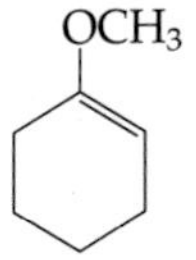

OCH_3

1-methoxycyclohexene

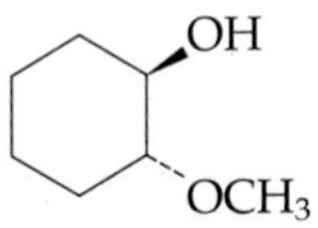

OH OCH_3

trans-2-methoxycyclohexanol

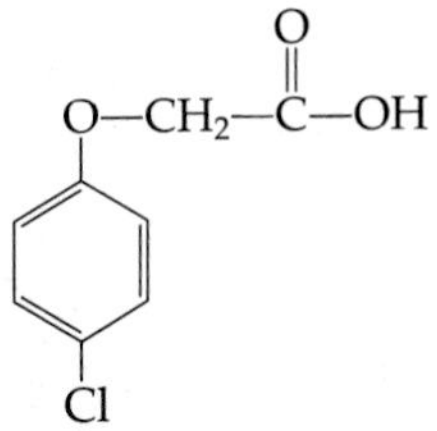

$O—CH_2—\overset{O}{\overset{\|}{C}}—OH$

Cl

p-chlorophenoxyacetic acid

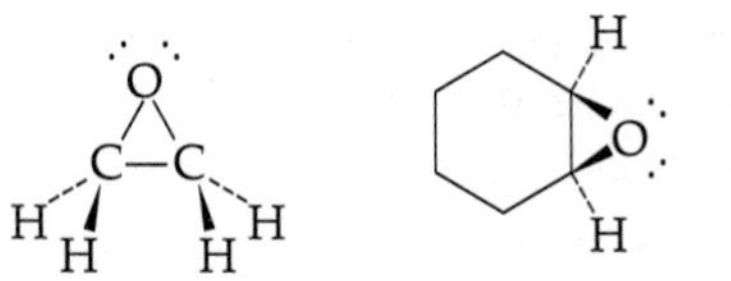

ethylene oxide cyclohexene oxide

The four-membered ether ring compounds, called trimethylene oxides, are not common substances. The name of the five-membered ring ether is tetrahydrofuran (THF), based on its relationship to the aromatic compound furan. Similarly tetrahydropyran (THP), a six-membered ring ether, is related to pyran, an unsaturated ether. Six-membered ring ethers with two oxygen atoms in a 1,4 relationship are 1,4-dioxanes.

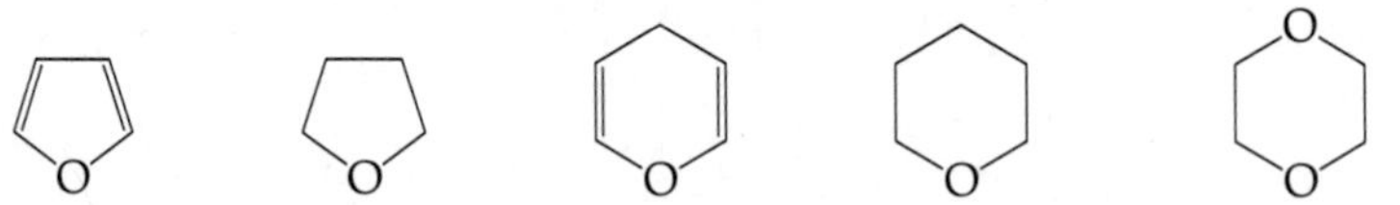

furan tetrahydrofuran pyran tetrahydropyran 1,4-dioxane

Cyclic ethers can also be named by the IUPAC system. In this system, each ring size receives a specific name. The names for cyclic ethers having three-, four-, five-, and six-membered rings are oxirane, oxetane, oxolane, and oxane, respectively. The oxygen atom in each of these rings receives the number 1. The ring is numbered in the direction that gives the lowest numbers to substituents.

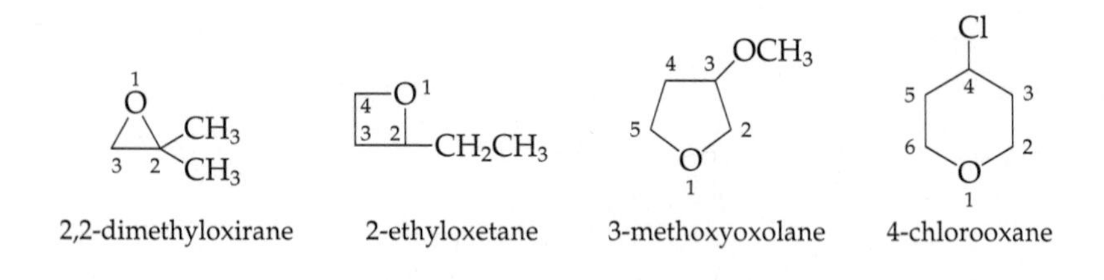

2,2-dimethyloxirane 2-ethyloxetane 3-methoxyoxolane 4-chlorooxane

---

EXAMPLE 9.1 What are the common and IUPAC names of the following compound?

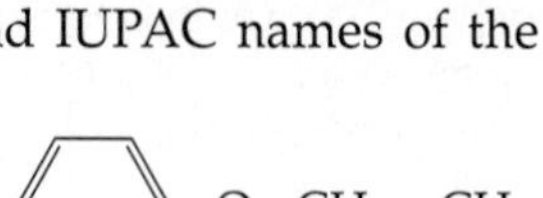

$-O-CH_2-CH_3$

Solution The common name of an unsymmetrical ether is based on the names of the two alkyl (or aryl groups). In this case they are phenyl and ethyl. The names of the two groups are arranged alphabetically, and the word ether is added to give the name ethyl phenyl ether.

The IUPAC name is obtained by first identifying the smaller of the two possible alkoxy groups and then indicating the location of the alkoxy group on the larger parent alkane chain. In this case, the benzene ring is selected as the parent and the ethoxy group is the substituent. The IUPAC name is ethoxybenzene.

9.3 PHYSICAL PROPERTIES OF ETHERS

The oxygen atom of an ether is sp^3-hybridized. The C—O—C bond angle is approximately the tetrahedral bond angle—the bond angle in dimethyl ether is 112° (Figure 9.3). The two O—C bonds are directed to two of the corners of a tetrahedron. The lone pair electrons in the remaining two sp^3 hybrid orbitals are directed to the remaining "corners" of a tetrahedron.

The geometry of the C—O bonds of ethers allows us to make predictions about their most stable conformations. We can imagine creating an ether by replacing a CH_2 group of an alkane with an oxygen atom. For example, replacing the C-3 methylene group of pentane with an oxygen atom would give diethyl ether. Diethyl ether has an anti arrangement of all carbon atoms and the oxygen atom in its most stable conformation (Figure 9.4). A similar situation prevails for the conformations of cyclic ethers

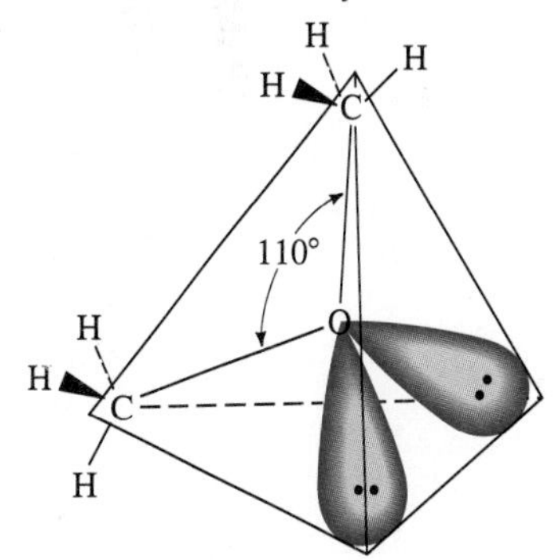

FIGURE 9.3
Structure of Dimethyl Ether
The oxygen atom of ethers is sp^3-hybridized. The C—O—C bond angle is close to the tetrahedral angle. The two sets of lone-pair electrons are in sp^3 hybrid orbitals that are directed to two of the corners of a tetrahedron.

most stable conformation of diethyl ether

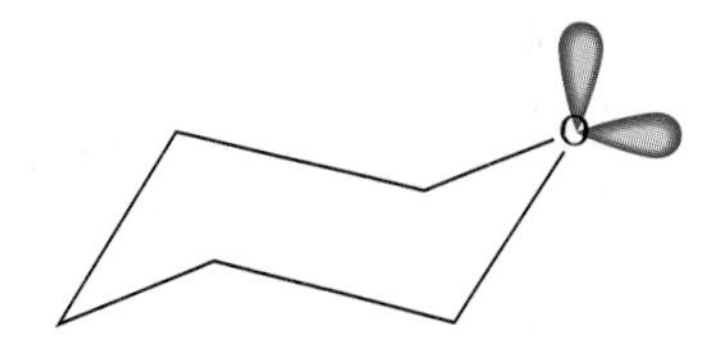

chair conformation of tetrahydropyran

FIGURE 9.4
Structures of Ethers
The lone-pair electrons of the oxygen atom in both acyclic and cyclic ethers occupy positions where the C—H bonds are located in the analogous hydrocarbons. The conformations of ethers are similar to those of the hydrocarbon analogs.

compared to those of the corresponding cycloalkanes. For example, tetrahydropyran, the ether analog of cyclohexane, exists in a chair conformation similar to that of cyclohexane. The tetrahedral oxygen atom has two lone pairs that occupy positions in space corresponding to the axial and equatorial C—H bonds of cyclohexane (Figure 9.4). The conformation of tetrahydropyran is particularly important because many sugars exist in solution as cyclic, six-membered, tetrahydropyran rings (Chapter 11).

Ethers have two polar C—O bonds and as a result, substantial dipole moments. Ethers are more polar than alkanes, but less polar than alcohols.

$CH_3—CH_2—CH_2—CH_2—CH_3$	$CH_3—CH_2—O—CH_2—CH_3$	$CH_3—CH_2—CH_2—CH_2—OH$
pentane (0.1 Debye)	diethyl ether (1.2 D)	1-butanol (1.7 D)
bp = 35 °C	bp = 36 °C	bp = 117 °C

Although alcohols and ethers are polar compounds, ethers do not have an O—H bond and cannot serve as hydrogen-bond donors. (We recall that alcohols can serve as both hydrogen-bond donors and acceptors). As a consequence, ether molecules do not hydrogen bond to each other. Thus, ethers have boiling points substantially lower than those of alcohols of comparable molecular weight. The boiling points of ethers are very close to those of alkanes of similar molecular weight.

Solubility in Water

Because ethers are polar, they are more soluble in water than alkanes of similar molecular weight. The slight solubility of ethers in water results from hydrogen bonds between the water molecules, hydrogen-bond donors, and the lone-pair electrons of ether molecules, hydrogen-bond acceptors. Tetrahydrofuran is miscible in water. The solubility of diethyl ether is about 10 g per 100 mL of water, much higher than that of pentane, which dissolves only slightly in water. The solubilities of ethers and alkanes approach one another as their molecular weights increase because the functional group contributes less to the overall structure of the compound.

Ethers as Solvents

Ethers such as diethyl ether dissolve a range of nonpolar and polar compounds. Nonpolar compounds are generally more soluble in diethyl ether than in alcohols such as ethanol because ethers do not have a hydrogen-bonding network that would have to be broken up to dissolve the solute. Because diethyl ether has a large dipole moment, polar substances readily dissolve it. Polar compounds that can serve as hydrogen-bond donors dissolve in diethyl ether because they can form hydrogen bonds to the lone-pair electrons of the ether oxygen atoms.

Ethers are aprotic. Therefore, they do not react with strong bases. We recall that protic compounds such as alcohols react with strong bases. For this reason, basic substances such as Grignard reagents can be prepared in ether solvents such as diethyl ether and tetrahydrofuran. Ethers have lone-pair electrons that stabilize electron-deficient species such as borane, BH_3, which is used in the hydroboration of alkenes (Chapter 8).

borane–THF complex

9.4 SYNTHESIS OF ETHERS

Ethers can be prepared by a method called the **Williamson synthesis.** In this reaction, a halide ion is displaced from an alkyl halide by a metal alkoxide in an S_N2 reaction. The alkoxide ion is prepared by the reaction of an alcohol with a strong base such as sodium hydride.

$$C_5H_9\text{—}\ddot{O}\text{—}H + NaH \longrightarrow C_5H_9\text{—}\ddot{O}{:}^- \, Na^+ + H_2$$

$$C_5H_9\text{—}\ddot{O}{:}^- \, Na^+ + CH_3\text{—}Br \longrightarrow C_5H_9\text{—}\ddot{O}\text{—}CH_3 + NaBr$$

The Williamson synthesis gives the best yields with primary halides because the reaction occurs by an S_N2 displacement in which a halide ion is the leaving group. The yield is lower for secondary alkyl halides because they can also react with the alkoxide ion in a competing elimination reaction. The Williamson synthesis cannot be used with tertiary alkyl halides because they cannot participate in S_N2 reactions; they undergo elimination reactions instead. Thus, to make an unsymmetrical ether that has a primary and a tertiary alkyl group, a primary alkyl halide and a tertiary alkoxide ion are the best reagents. For example, *tert*-butyl methyl ether can be prepared by the reaction of sodium *t*-butoxide with methyl iodide, but not by the reaction of sodium methoxide with 2-chloro-2-methylpropane.

$$(CH_3)_3C\text{—}\ddot{O}{:}^- + CH_3\text{—}\ddot{I}{:} \xrightarrow{S_N2} (CH_3)_3C\text{—}\ddot{O}\text{—}CH_3 + {:}\ddot{I}{:}^-$$

primary halide

$$CH_3\text{—}\ddot{O}{:}^- + H\text{—}CH_2\text{—}C(CH_3)_2\text{—}\ddot{Cl}{:} \xrightarrow{E2} H_2C{=}C(CH_3)_2 + CH_3\ddot{O}H + {:}\ddot{Cl}{:}^-$$

tertiary halide

EXAMPLE 9.2 Propose a synthesis of phenyl propyl ether.

Solution Consider the following two combinations of reagents.

$$C_6H_5\text{—Br} + CH_3CH_2CH_2\text{—}O^-$$

$$C_6H_5\text{—}O^- + CH_3CH_2CH_2\text{—Br}$$

The first combination will not give the ether product because S_N2 reactions cannot occur at the sp^2-hybridized carbon atom of bromobenzene. However, the reaction of the phenoxide ion with 1-bromopropane occurs readily because bromopropane is a primary alkyl halide. The phenoxide ion is formed by reacting phenol with sodium hydride.

9.5 REACTIONS OF ETHERS

Ethers are very stable compounds that do not react with most common reagents. They do not react with bases but do react with strong acids when the conjugate bases are good nucleophiles. For example, ethers react with HI (or with HBr), and the carbon–oxygen bond is cleaved to produce alkyl iodides (or bromides).

$$R\text{—O—}R' \xrightarrow{HX} R\text{—X} + R'\text{—OH} \xrightarrow{HX} R\text{—X} + R'\text{—X}$$

In general, the less substituted alkyl halide is formed by an S_N2 reaction. The halide ion attacks the less hindered carbon atom, and the displaced alkoxy group has the oxygen atom bonded to the more substituted carbon atom. The cleavage reaction does not occur with a halide salt: a proton from the halogen acid is required to protonate the oxygen atom and provide an alcohol as the leaving group.

$$(CH_3)_2CH\text{—}\ddot{O}\text{—}CH_2CH_2CH_3 + H\text{—}\ddot{I}: \xrightarrow{\text{fast}} (CH_3)_2CH\text{—}\overset{+}{O}(H)\text{—}CH_2CH_2CH_3$$

$$(CH_3)_2CH\text{—}\overset{+}{O}(H)\text{—}CH_2CH_2CH_3 + :\ddot{I}:^- \xrightarrow[S_N2]{\text{slow}} (CH_3)_2CH\ddot{O}H + CH_3CH_2CH_2I$$

In a subsequent reaction, the alcohol product reacts with the hydrogen halide to give a second mole of an alkyl halide. Both alkyl groups of the ether are eventually converted into alkyl halides.

EXAMPLE 9.3 Based on the mechanism of ether cleavage, write the products of the reaction of HI with phenyl propyl ether.

Solution First, the strong acid protonates the ether oxygen atom. The resulting cation is an oxonium ion.

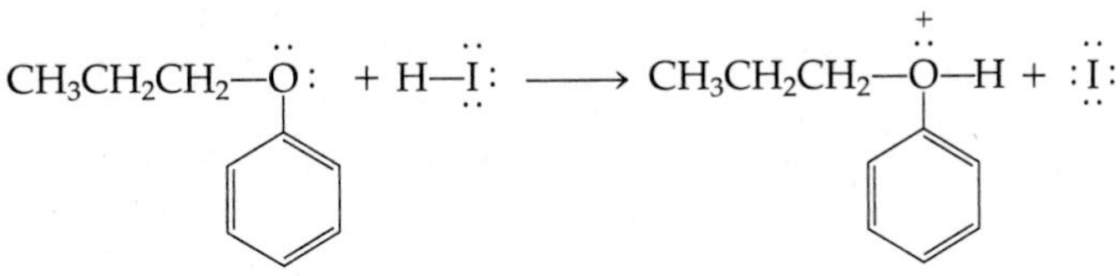

$$CH_3CH_2CH_2{-}\ddot{O}(C_6H_5){:} + H{-}\ddot{I}{:} \longrightarrow CH_3CH_2CH_2{-}\ddot{O}^{+}(C_6H_5){-}H + {:}\ddot{I}{:}^{-}$$

Subsequent nucleophilic attack by the iodide ion can occur only at the methylene carbon atom of the propyl group bearing the oxygen atom. An S_N2 reaction at the carbon atom of the benzene ring that bears the oxygen atom is not possible, so phenol will not react further with HI.

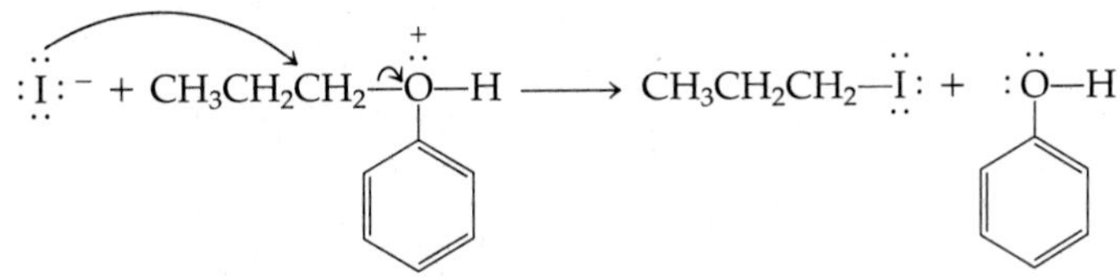

$${:}\ddot{I}{:}^{-} + CH_3CH_2CH_2{-}\ddot{O}^{+}(C_6H_5){-}H \longrightarrow CH_3CH_2CH_2{-}\ddot{I}{:} + {:}\ddot{O}(C_6H_5){-}H$$

9.6 SYNTHESIS OF EPOXIDES

The synthesis of and reactions of cyclic ethers containing four or more atoms are similar to acyclic ethers. However three-membered cyclic ethers are very much more reactive (Section 9.7) and serve as important intermediates in synthesis. Epoxides can be synthesized by oxidizing an alkene with a peroxyacid, RCO_3H. Peroxyacetic acid, CH_3CO_3H, is used in industry but *m*-chloroperoxybenzoic acid (MCPBA) is used to prepare smaller amounts of epoxides in the laboratory.

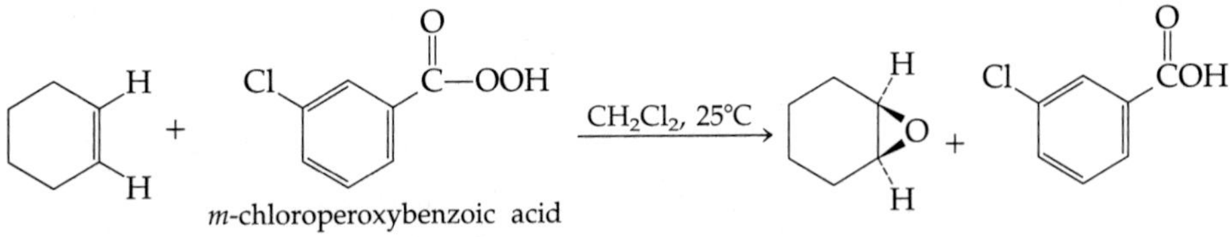

$$\text{cyclohexene} + m\text{-}ClC_6H_4C(=O){-}OOH \xrightarrow{CH_2Cl_2,\ 25^\circ C} \text{cyclohexene oxide} + m\text{-}ClC_6H_4COH(=O)$$

m-chloroperoxybenzoic acid

In the epoxidation of alkenes with *m*-chloroperoxybenzoic acid, the stereochemistry of the groups in the alkene is retained. That is, groups that are cis in the alkene are cis in the epoxide, and groups that are trans in the alkene remain trans in the epoxide.

CH3 CH3 C=C H H —MCPBA / CH_2Cl_2→ CH3 O CH3 H H

A second method of synthesizing epoxides is an intramolecular variation of the Williamson ether synthesis. First, a halohydrin is formed by reacting an alkene with an aqueous solution of a halogen. The cyclic chloronium ion formed then reacts with water as the nucleophile.

H H :Cl—Cl: → H H :Cl: —H_2O:→ H :O⁺—H H :Cl: H —$-H^+$→ H :OH :Cl: H

Treatment of the halohydrin with a base produces an alkoxide that displaces a halide from the adjacent carbon atom to form the epoxide ring.

H :O—H ⁻OH :Cl: H ⇌ H :O:⁻ :Cl: H → O H H + :Cl:⁻

9.7 REACTIONS OF EPOXIDES

The cyclic ethers tetrahydrofuran and tetrahydropyran are as unreactive as acyclic ethers and are often used as solvents. In contrast, epoxides are highly reactive because the three-membered ring has considerable bond angle strain. The products of the ring-opening reactions have normal tetrahedral bond angles and are not strained.

Acid-Catalyzed Ring Opening

Consider the reaction of water with ethylene oxide to form ethylene glycol.

O C—C H H H H —H_3O^+→ HO H H C—C H H OH

oxirane (ethylene oxide)

1,2-ethanediol (ethylene glycol)

The acid-catalyzed ring opening of epoxides occurs by an S_N2 process in which water is the nucleophile and the "leaving group" is the protonated oxygen atom of the epoxide.

$$\underset{\diagdown\ddot{\text{O}}\diagup}{CH_2{-}CH_2} + H^+ \longrightarrow \underset{\diagdown\dot{\text{O}}^+\diagup\ |\ H}{CH_2{-}CH_2}$$

$$H{-}\ddot{O}{:}(H) + CH_2{-}CH_2\ (\text{epoxide } O^+{-}H) \xrightarrow{S_N2} H{-}\overset{+}{\ddot{O}}(H){-}CH_2{-}CH_2{-}OH \longrightarrow HOCH_2CH_2OH + H^+$$

The acid-catalyzed ring opening of a cycloalkene epoxide produces a *trans*-1,2-diol because the nucleophile, water, attacks from the back of the epoxide ring, and the ring oxygen atom leaves from the opposite side.

$$\text{cyclohexene oxide} \xrightarrow{H^+} \text{protonated epoxide } (\overset{+}{O}{-}H) \xrightarrow{H_2O} \textit{trans}\text{-1,2-cyclohexanediol} + H^+$$

Ring Opening by Nucleophiles

Ethers do not generally react with nucleophiles. However, epoxides are so reactive that the ring is opened even by nucleophiles such as OH^-, SH^-, or NH_3 or the related organic species RO^-, RS^-, and RNH_2. The reaction of ethylene oxide with ammonia gives 2-aminoethanol, a compound used commercially as a corrosion inhibitor.

$$\underset{\diagdown O \diagup}{CH_2{-}CH_2} + NH_3 \longrightarrow NH_2{-}CH_2{-}CH_2{-}OH$$

Similar reactions take place in the sterilization of temperature-sensitive equipment, which is accomplished by exposing the equipment to ethylene oxide gas. The epoxide ring reacts with a variety of nucleophilic functional groups in bacterial macromolecules, and the bacteria die.

Epoxides react with Grignard reagents to produce alcohols with two more carbon atoms than the starting alkyl halide. The sequence of reactions is

$$R{-}Br + Mg \xrightarrow{\text{ether}} R{-}MgBr$$

$$R{-}MgBr + \underset{\diagdown O \diagup}{CH_2{-}CH_2} \longrightarrow R{-}CH_2{-}CH_2{-}O{-}MgBr$$

$$R{-}CH_2{-}CH_2{-}O{-}MgBr + H_2O \longrightarrow R{-}CH_2{-}CH_2{-}O{-}H + HOMgBr$$

A new carbon–carbon bond is formed, and the product alcohol has two more carbon atoms than the starting alkyl halide reactant.

Ring Opening of Biological Epoxides

In earlier chapters we saw that epoxides are produced biologically as oxidation products of alkenes and aromatic compounds. These epoxides are formed in the liver by cytochrome P-450 and then they undergo ring-opening reactions. If the epoxide reacts with a biological macromolecule, the result is potentially devastating. When epoxides are made from aromatic compounds, the products are called arene oxides. These molecules can undergo four kinds of reactions, as shown. With one exception, the reactions give products that are not harmful to the organism.

arene oxide — rearrangement → phenol (R, OH); H_2O → trans diol (R, OH, OH); GSH → (R, GS, OH); MH → (R, M, OH)

The rearrangement of an arene oxide gives a water-soluble phenol that is easily eliminated from the body. Hence, this pathway does not lead to the accumulation of toxic byproducts.

Ring opening of the arene oxide by water gives a trans diol by an S_N2 process. The diol is water-soluble and easily eliminated from the body.

We saw in Chapter 7 that glutathione contains a nucleophilic sulfhydryl group and acts as a scavenger that reacts with toxic metabolites. Glutathione (GSH) reacts with arene oxides in a ring-opening reaction. The product contains many polar functional groups, so it is water-soluble and easily eliminated.

Direction of Ring Opening

Unsymmetrical epoxides give different products under acid- and base-catalyzed conditions.

$CH_3-C(CH_3)(-O-)CH_2$ $\xrightarrow[CH_3OH]{CH_3O^-}$ $CH_3-C(OH)(CH_3)-CH_2-O-CH_3$
1-methoxy-2-methyl-2-propanol

$CH_3-C(CH_3)(-O-)CH_2$ $\xrightarrow[H^+]{CH_3OH}$ $CH_3-C(CH_3)(O-CH_3)-CH_2OH$
2-methoxy-2-methyl-1-propanol

In the case of the ring opening by a nucleophile under basic conditions, the reaction is controlled by the same features as the S_N2 displacement reac-

nucleophile

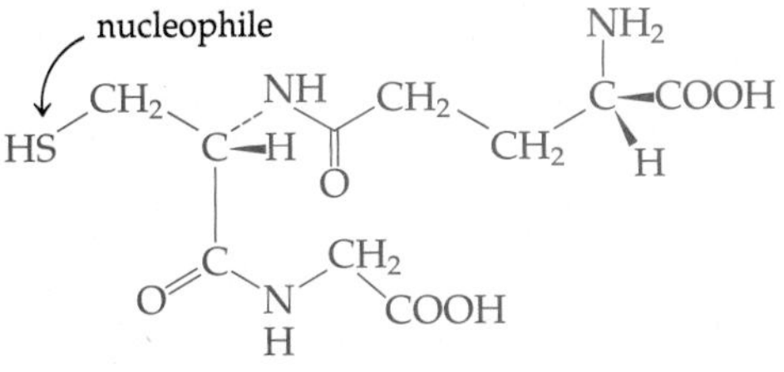

glutathione

Arene oxides react with the nucleophilic functional groups present in most macromolecules (represented by MH in the figure) including enzymes, RNA, and DNA. These reactions can cause significant alterations in biological functions. A particularly dangerous arene oxide is the epoxide of benzo[*a*]pyrene, which reacts with amino groups in DNA.

The epoxide metabolites of alkenes tend to be more stable than arene oxides. They undergo ring opening with water to give diols. One example of this type of reaction is the ring opening of the epoxide formed from the anticonvulsant drug carbamazepine.

Like arene oxides, epoxides formed from alkenes also react with glutathione to produce water-soluble compounds that are easily eliminated from the body. Epoxides derived from alkenes also undergo ring-opening reactions with macromolecules. For example, we noted earlier that aflatoxin B_1 undergoes such a reaction (Section 4.6).

It is not easy to predict whether an epoxide will react with water or glutathione and thus be nontoxic or whether it will react harmfully with macromolecules. However, it appears that relatively stable epoxides tend to undergo ring opening by water or gluathione. Also, epoxides that have sterically hindered oxirane rings—for example, benzo[*a*]pyrene—tend to react with nucleophilic groups of macromolecule.

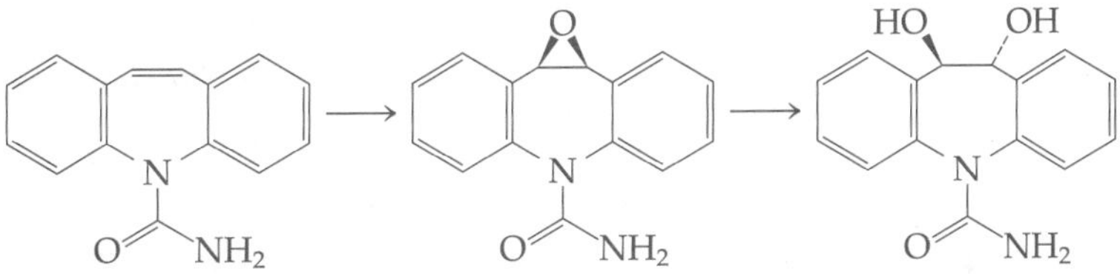

carbamazepine

tions we considered in Chapter 7. The nucleophile attacks the unhindered primary carbon atom instead of the tertiary carbon atom. The resulting alkoxide then exchanges a proton with the solvent, and the methoxide base is regenerated.

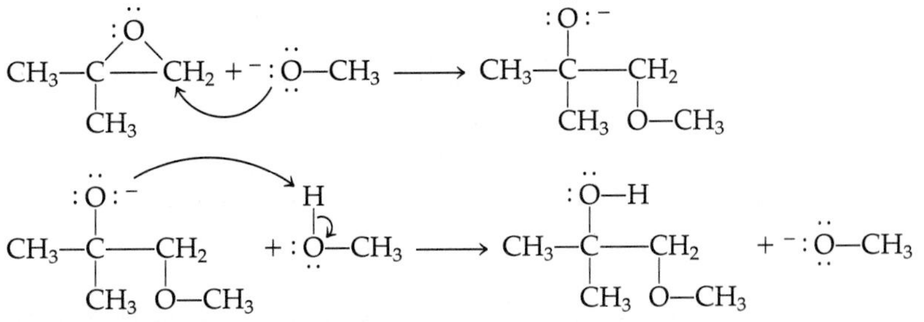

Under acidic conditions, the protonated epoxide has some positive charge on each carbon atom in the ring. Consider the following resonance forms.

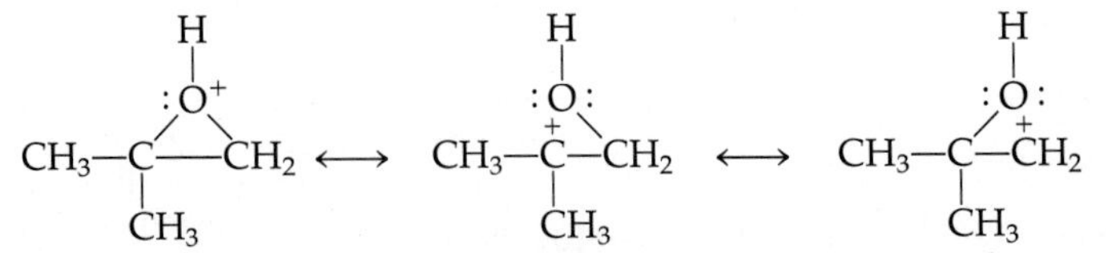

$$\underset{\text{oxonium ion}}{CH_3-\overset{\overset{H}{|}}{\underset{CH_3}{C}}(\!-\!:\overset{+}{O}H\!-\!)CH_2} \longleftrightarrow \underset{\text{tertiary carbocation}}{CH_3-\overset{+}{\underset{CH_3}{C}}-CH_2(\!-\!:\ddot{O}H)} \longleftrightarrow \underset{\text{primary carbocation}}{CH_3-\underset{CH_3}{C}-\overset{+}{C}H_2(\!-\!:\ddot{O}H)}$$

oxonium ion — tertiary carbocation — primary carbocation

The positive charge is more stable on the tertiary carbon atom than on the primary carbon atom. This tertiary carbocation resonance form is a more important contributor than the primary one. Thus, the tertiary carbocation center combines with the nucleophile to give the conjugate acid of the observed product of the acid-catalyzed ring opening. The conjugate acid then exchanges a proton with the solvent.

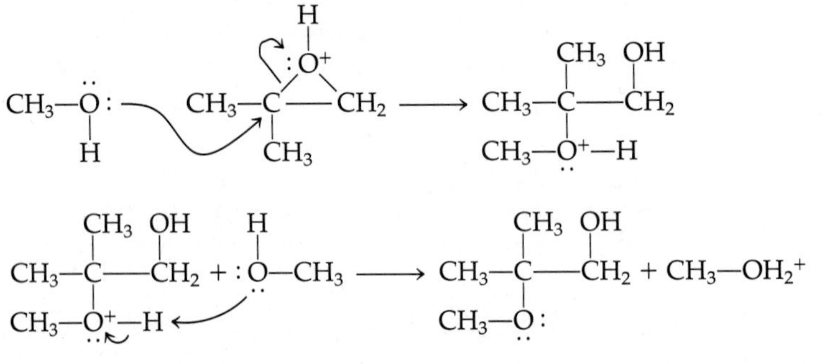

$$CH_3-\underset{H}{\ddot{O}}: + CH_3-\underset{CH_3}{C}(\!-\!:\overset{+}{O}H\!-\!)CH_2 \longrightarrow CH_3-\overset{CH_3}{\underset{CH_3-\overset{+}{O}-H}{C}}-\overset{OH}{CH_2}$$

$$CH_3-\overset{CH_3}{\underset{CH_3-\overset{+}{O}-H}{C}}-\overset{OH}{CH_2} + :\overset{H}{\ddot{O}}-CH_3 \longrightarrow CH_3-\overset{CH_3}{\underset{CH_3-\ddot{O}:}{C}}-\overset{OH}{CH_2} + CH_3-OH_2^+$$

EXAMPLE 9.4 Predict the product of the reaction of 2-methyloxirane with the Grignard reagent prepared from iodoethane.

Solution The Grignard reagent reacts as a nucleophile. Thus, the Grignard reagent of iodoethane behaves as an ethyl carbanion. Nucleophilic attack of the ethyl carbanion on methyloxirane occurs at the primary rather than the tertiary carbon atom.

$$CH_3-CH_2-MgBr + CH_2(\!-\!O\!-\!)\underset{CH_3}{CH} \longrightarrow CH_3-CH_2-CH_2-\overset{O-MgBr}{\underset{CH_3}{CH}}$$

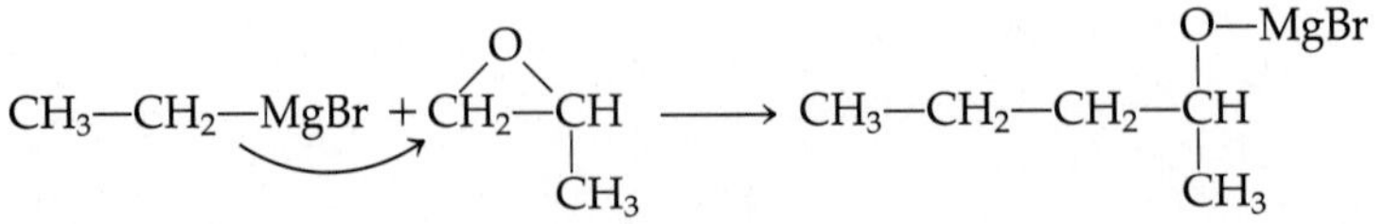

Subsequent hydrolysis of the magnesium alkoxide yields 2-pentanol.

EXAMPLE 9.5 Predict the product of the reaction of styrene oxide in an acid-catalyzed reaction with methanol.

Solution Styrene oxide is the epoxide formed by the oxidation of styrene ($C_6H_5-CH=CH_2$). Consider the following resonance forms for the protonated styrene oxide formed in acid solution.

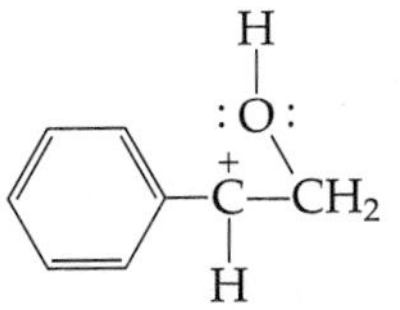

benzylic carbocation 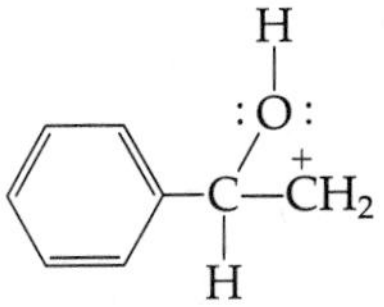   primary carbocation

The positive charge is more stable on the benzylic carbon atom than on the primary carbon atom. Thus, attack of the nucleophilic methanol at this center gives 2-methoxy-2-phenyl-1-ethanol.

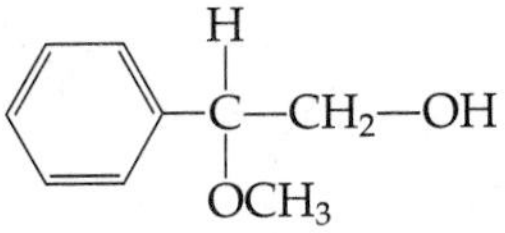

EXERCISES

Ether Isomers

9.1 Draw the three isomeric ethers with the molecular formula $C_4H_{10}O$.

9.2 Draw the four isomeric methyl ethers with the formula $C_5H_{12}O$.

9.3 Draw the isomeric saturated ethers with the molecular formula C_3H_6O.

9.4 Draw the isomeric unsaturated ethers with the molecular formula C_4H_8O.

Nomenclature of Ethers

9.5 Give the common name of each of the following compounds.

(a) 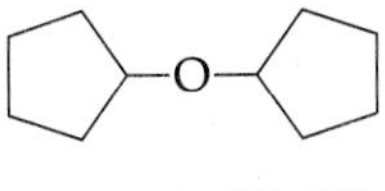 (cyclopentyl)–O–(cyclopentyl)   (b) 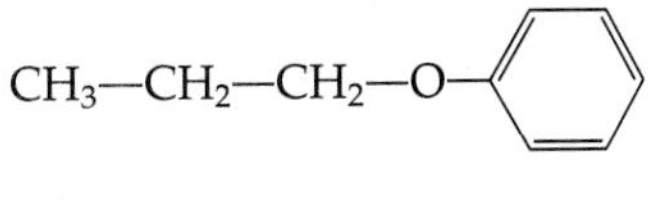 $CH_3—CH_2—CH_2—O—$(phenyl)

(c)  (cyclopentyl)$OCH_2CH_2CH_3$

9.6 Give the common name of each of the following compounds.

(a) 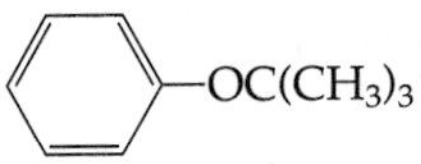 (phenyl)$—OC(CH_3)_3$ (b) 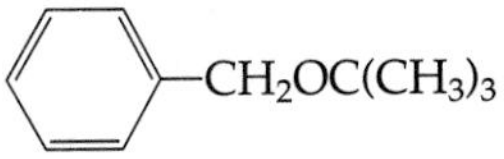 (phenyl)$—CH_2OC(CH_3)_3$

(c) 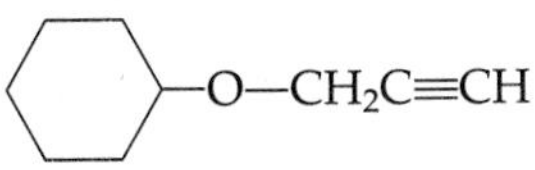 (cyclohexyl)$—O—CH_2C{\equiv}CH$

9.7 Assign the IUPAC name of each of the following compounds.

(a) 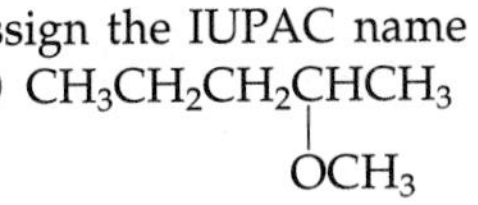 $CH_3CH_2CH_2CHCH_3$ with OCH_3 on C-2 (the CH)   (b) $CH_3CHCH_2CHCH_3$ with CH_3 and OCH_3 on the two CH carbons

(c) $CH_3CH_2CH_2CHCH_2CH_3$ with OCH_2CH_3 on the CH

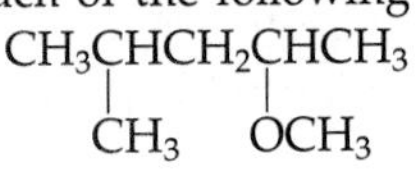

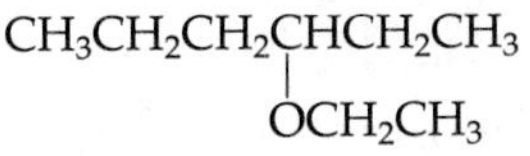

9.8 Assign the IUPAC name of each of the following compounds.

(a) $CH_3CH_2CH_2CHCH_2OCH_3$ with OCH_3 on the CH

(b) $CH_3CHCH_2CHCH_3$ with CH_3O and OCH_3 on the two CH groups

(c) $CH_3CH_2CHCH_2CHCH_3$ with CH_3CH_2O and OCH_2CH_3 on the two CH groups

9.9 The name of isoindoklon, a general anesthetic, is 1,1,1,3,3,3,-hexafluoroisopropyl methyl ether. Draw its structure.

9.10 What is the common name of the following structure for a compound that is used as an anesthetic?

$$CH_2{=}CH{-}O{-}CH{=}CH_2$$

9.11 The IUPAC name of enflurane, a general anesthetic, is 2-chloro-1,1,2-trifluoro-1-(difluoromethoxy)ethane. Draw its structure.

9.12 What is the IUPAC name of the following general anesthetic?

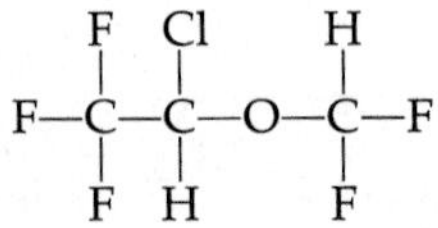

$$F{-}CF_2{-}CHCl{-}O{-}CHF{-}F$$

(structure: F–C(F)(F)–C(Cl)(H)–O–C(H)(F)–F)

Properties of Ethers

9.13 Dioxane is miscible in water. Why?

9.14 *p*-Ethylphenol is more soluble in aqueous solution than the isomeric ethoxybenzene. Explain why.

9.15 The boiling points of dipropyl ether and diisopropyl ether are 91 and 68°C, respectively. Explain why the boiling points of these isomeric ethers differ.

9.16 The boiling points of 1-ethoxypropane and 1,2-dimethoxyethane are 64 and 83 °C, respectively. Explain why.

9.17 Explain why dipropyl ether is soluble in concentrated sulfuric acid whereas heptane is insoluble.

9.18 Aluminum trichloride dissolves in tetrahydropyran and heat is liberated. Explain why.

Synthesis of Ethers

9.19 Consider the structure of the following ether and determine the best method to synthesize it.

(cyclopentane carbon bearing CH_3CH_2 and OCH_2CH_3)

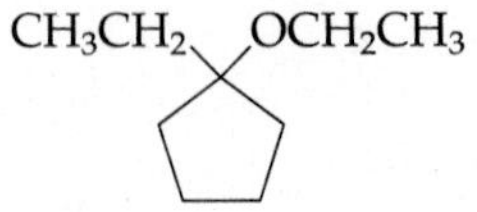

9.20 Consider the structure of the following ether and determine the best method to synthesize it.

(cyclohexyl)$-CH_2CH_2-O-$(cyclopentyl)

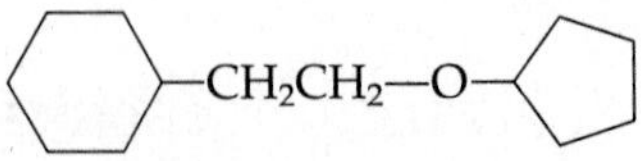

9.21 How could the local anesthetic dibucaine be prepared by using the Williamson synthesis?

O
‖ H
C—N—$CH_2CH_2N(C_2H_5)_2$

N O—$CH_2CH_2CH_2CH_3$

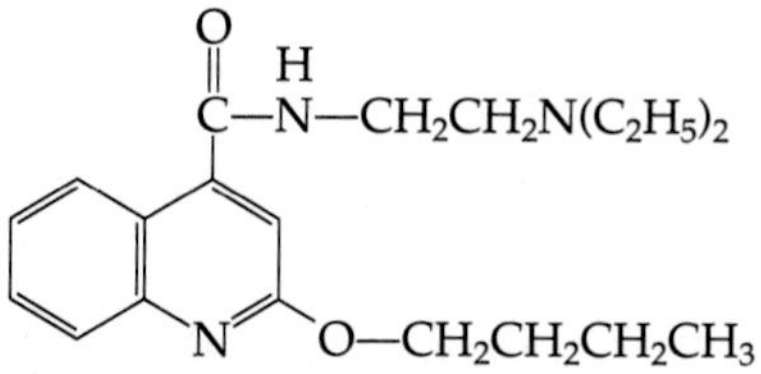

9.22 Consider synthetic methods to produce the antihistamine diphenylpyraline by using the Williamson synthesis. What difficulties might be encountered?

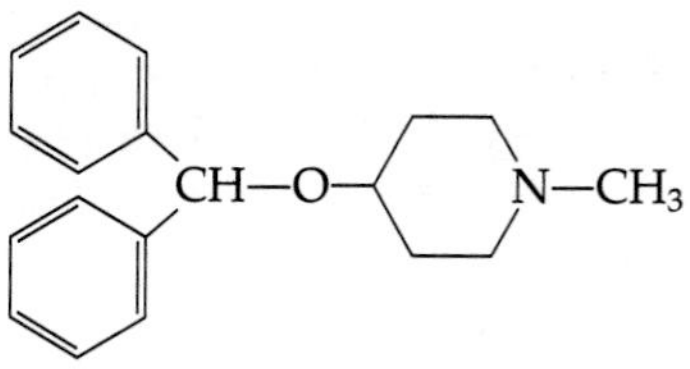

CH—O— N—CH_3

9.23 Reaction of 1-hexane with mercuric acetate in methanol as the solvent followed by reduction of the intermediate product with sodium borohydride yields 2-methoxyhexane. What is the structure of the intermediate product? How is it formed?

9.24 Reaction of 5-chloro-2-pentanol with sodium hydride yields 2-methyltetrahydrofuran. How is the product formed?

Synthetic Sequences

9.25 Write the structure of the final product of each of the following sequences of reactions.

(a) $CH_2CH{=}CH_2$ $\xrightarrow[\text{2. NaBH}_4]{\text{1. Hg(OAc)}_2}$ $\xrightarrow[CH_3I]{NaH}$

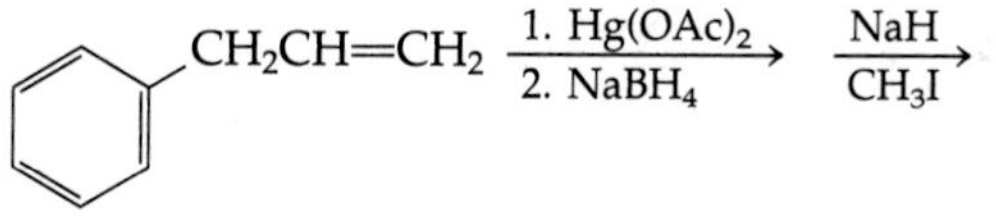

(b) —$CH{=}CH_2$ $\xrightarrow[\text{2. H}_2\text{O}_2\text{, OH}^-]{\text{1. B}_2\text{H}_6}$ $\xrightarrow[CH_3CH_2Br]{NaH}$

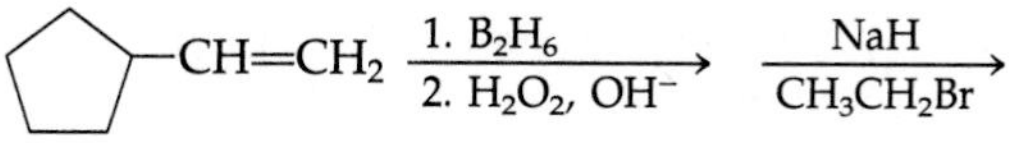

(c) —$CH_2CH{=}CH_2$ $\xrightarrow[\text{2. H}_2\text{O}_2\text{, OH}^-]{\text{1. B}_2\text{H}_6}$ $\xrightarrow[\text{ethylene oxide}]{NaH}$

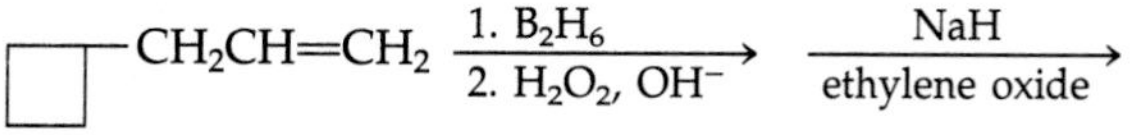

9.26 Write the structure of the final product of each of the following sequences of reactions.

(a) —$\overset{O}{\overset{\|}{C}}$—H $\xrightarrow[CH_3CH_2OH]{NaBH_4}$ $\xrightarrow[CH_3CH_2I]{K}$

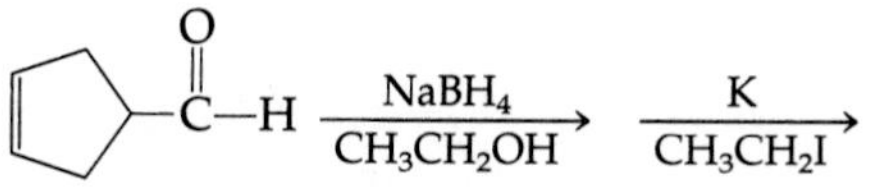

(b) 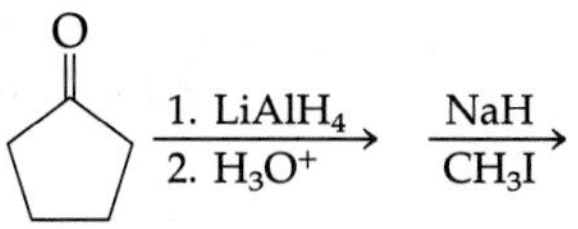 cyclopentanone $\xrightarrow[\text{2. } H_3O^+]{\text{1. } LiAlH_4}$ $\xrightarrow[CH_3I]{NaH}$

(c) 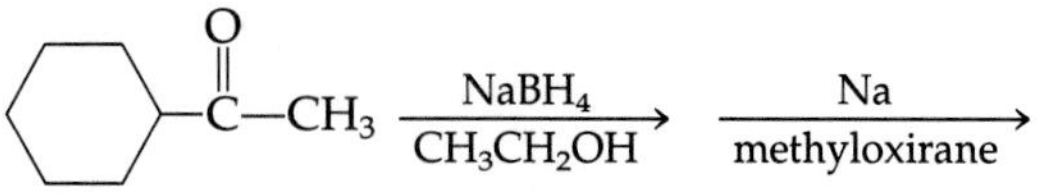 cyclohexyl–$\overset{O}{\overset{\|}{C}}$–$CH_3$ $\xrightarrow[CH_3CH_2OH]{NaBH_4}$ $\xrightarrow[\text{methyloxirane}]{Na}$

Reactions of Ethers

9.27 A compound of formula $C_5H_{12}O_2$ reacts with HI to give a mixture of iodomethane, iodoethane, and 1,2-diiodoethane. What is the structure of the compound?

9.28 A compound of formula $C_5H_{12}O_2$ reacts with HI to give a mixture of iodomethane and 1,3-diiodopropane. What is the structure of the compound?

9.29 A compound of formula $C_5H_{10}O$ reacts with HI to give only 1,5-diiodopentane. What is the structure of the compound?

9.30 A compound of formula $C_4H_8O_2$ reacts with HI to give only 1,2-diiodoethane. What is the structure of the compound?

9.31 Write the product of the reaction of the following bicyclic ether with HBr.

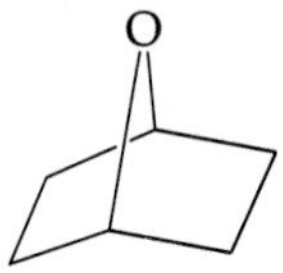

O

9.32 An ether contained in eucalyptus oil has the molecular formula $C_{10}H_{18}O$. Reaction of the compound with HCl gives the following product. What is the structure of the compound?

Cl

CH_3—C—CH_3

Cl CH_3

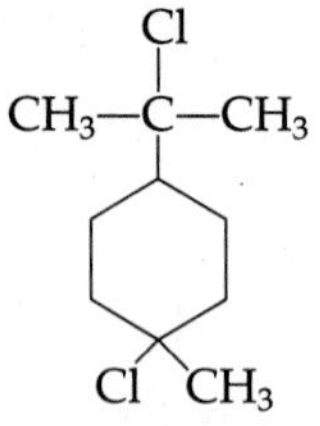

9.33 What reaction could be used to provide a visual test to distinguish between the following isomeric compounds?

O O

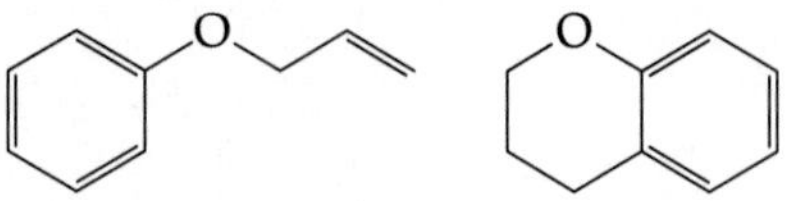

9.34 What reaction could be used to provide a visual test to distinguish between the following isomeric compounds?

cyclopentyl—CH_2—OH cyclopentyl—O—CH_3

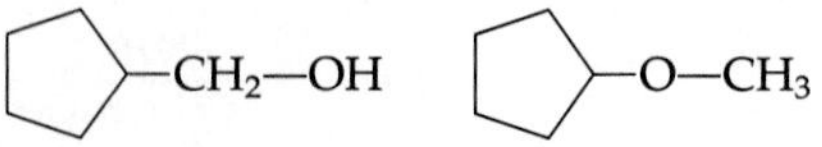

Synthesis of Epoxides

9.35 Two products can be obtained by the epoxidation of the following bicycloalkene with MCPBA. Draw their structures.

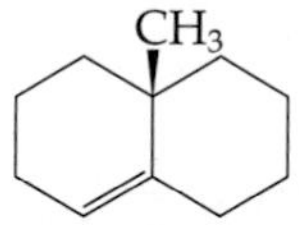

9.36 What alkene is required to synthesize the sex attractant of the gypsy moth using MCPBA?

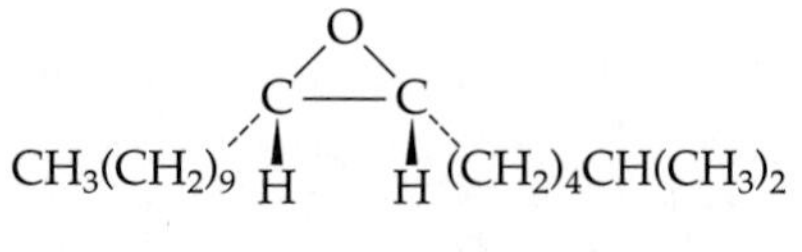

$CH_3(CH_2)_9$ H H $(CH_2)_4CH(CH_3)_2$

9.37 Draw the product of the reaction of the following compound with sodium hydroxide.

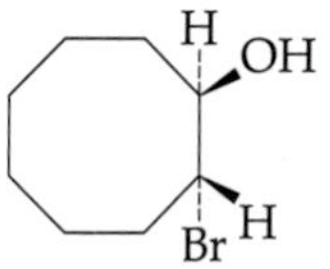

9.38 Reaction of *trans*-2-chlorocyclohexanol with sodium hydroxide yields an epoxide, but the cis isomer does not. Explain this difference.

Reactions of Epoxides

9.39 Ethyl cellosolve, $CH_3CH_2—O—CH_2CH_2—OH$, is an industrial solvent. Suggest a synthesis of this compound.

9.40 Ethyl carbitol, $CH_3CH_2—O—CH_2CH_2—O—CH_2CH_2—OH$, is an industrial solvent. Suggest a synthesis of this compound.

9.41 What is the product of the reaction of 2-methyloxirane with methanol in the presence of an acid catalyst?

9.42 Epoxide rings can be cleaved by metal hydrides. Write the product of the reaction of cyclohexene oxide and $LiAlD_4$.

9.43 A mixture of 2,2-dimethyloxirane and ethanethiol is treated with sodium hydroxide. Write the structure of the expected product.

9.44 Epoxide rings can be cleaved by phenoxides. Propose a synthesis of the muscle relaxant methocarbamol using this fact.

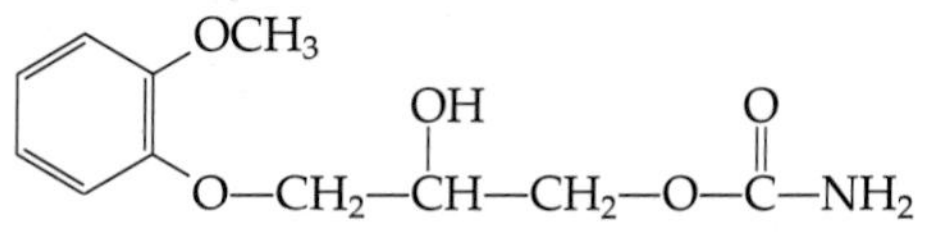

OCH_3

$O—CH_2—CH(OH)—CH_2—O—C(=O)—NH_2$

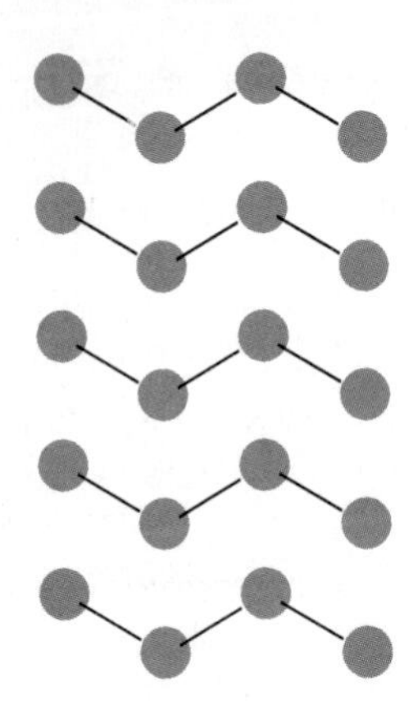

10

ALDEHYDES AND KETONES

10.1 THE CARBONYL GROUP

Formaldehyde, CH_2O, is the simplest compound with a carbonyl group. A **carbonyl group** consists of a carbon atom and an oxygen atom linked by a double bond. The oxygen atom shares two of its six valence electrons with the carbon atom. The remaining four valence electrons of the oxygen atom are present as two nonbonded electron pairs. Carbon shares two of its four valence electrons with oxygen, and the remaining two electrons form two single bonds to other atoms such as the hydrogen atoms in formaldehyde. The nonbonded pairs of electrons on the oxygen atom of the carbonyl group are not always shown in drawings of molecular structures.

$H_2C{=}\ddot{O}:$ Nonbonded electrons are shown. $H_2C{=}O$ Nonbonded electrons are not shown.

representations of formaldehyde

The carbonyl carbon atom is sp^2-hybridized. It contributes one electron to each of the three hybrid orbitals, which form three σ bonds. In the case of formaldehyde, there are two σ bonds to hydrogen atoms and one σ bond to the carbonyl oxygen atom. These bonds are at 120° to each other and are coplanar. The fourth electron of the carbonyl carbon atom is in a $2p$ orbital perpendicular to the plane of the three sp^2 hybrid orbitals. The carbonyl oxygen atom is also sp^2-hybridized. It contributes one electron of its six valence electrons to the sp^2 hybrid orbital that forms a σ bond with the carbonyl carbon atom. Four valence electrons are present as two sets of nonbonded electron pairs in the other two sp^2 hybrid orbitals at 120° to each other and to the carbon–oxygen bond (Figure 10.1). The remaining valence

FIGURE 10.1
Bonding in the Carbonyl Group

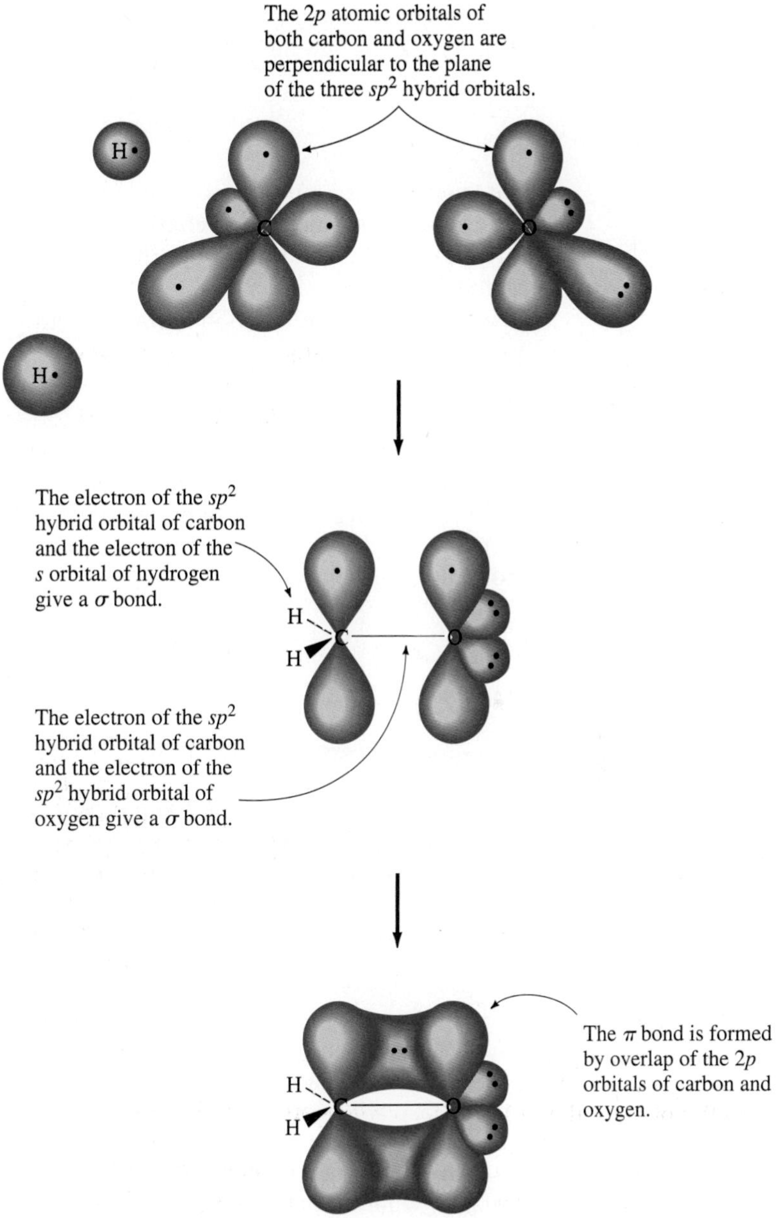

The $2p$ atomic orbitals of both carbon and oxygen are perpendicular to the plane of the three sp^2 hybrid orbitals.

H•

H•

The electron of the sp^2 hybrid orbital of carbon and the electron of the s orbital of hydrogen give a σ bond.

The electron of the sp^2 hybrid orbital of carbon and the electron of the sp^2 hybrid orbital of oxygen give a σ bond.

The π bond is formed by overlap of the $2p$ orbitals of carbon and oxygen.

electron is contained in a $2p$ orbital perpendicular to the plane of the sp^2 hybrid orbitals. The electrons in the $2p$ orbitals of the carbon and oxygen atoms overlap to form a π bond.

Oxygen is more electronegative than carbon. Hence, the electrons in the carbon–oxygen double bond are attracted to the oxygen atom, and the carbonyl bond is polar. The carbonyl group is also resonance-stabilized, as shown by the charged contributing structure (2).

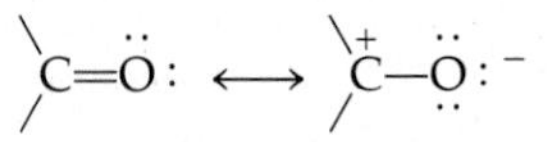

$$\mathrm{>C{=}\ddot{O}:} \longleftrightarrow \mathrm{>\overset{+}{C}{-}\ddot{\underset{..}{O}}:^{-}}$$

(1) (2)

Contributing structure (1) is more important because each atom has a Lewis octet, and there is no formal charge on either atom. However, charged structure (2) contributes to the properties of the carbonyl group as evidenced by physical properties such as dipole moment and certain chemical properties as well. The carbonyl carbon atom is electrophilic, and the carbonyl oxygen atom is nucleophilic.

Carbonyl Compounds

When a carbonyl carbon atom is bonded to at least one hydrogen atom, the resulting compound is an **aldehyde.** The aldehyde with the simplest structure is formaldehyde, in which the carbonyl carbon atom is bonded to two hydrogen atoms. Other aldehydes have structures in which the carbonyl group is bonded to a hydrogen atom and either an alkyl group (R) or an aromatic group (Ar).

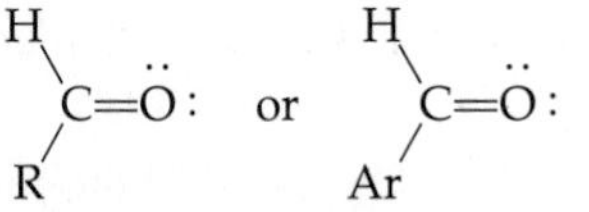

$$\mathrm{(H)(R)C{=}\ddot{O}:} \quad \text{or} \quad \mathrm{(H)(Ar)C{=}\ddot{O}:}$$

general structural formulas of aldehydes

When a carbonyl carbon atom is bonded to two other carbon atoms, the compound is a **ketone.** The bonded groups may be any combination of alkyl or aromatic groups. A ketone has 120° bond angles at the carbonyl carbon atom. However, a linear arrangement of carbon atoms is often used to draw the structure of ketones.

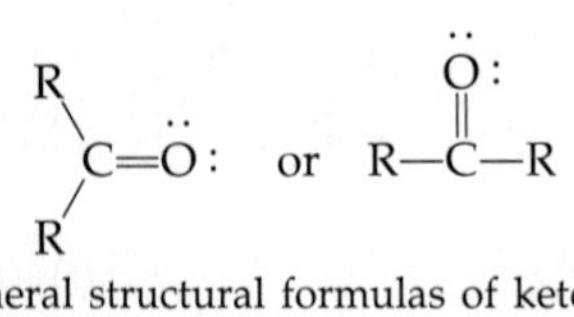

$$\mathrm{(R)(R)C{=}\ddot{O}:} \quad \text{or} \quad \mathrm{R{-}\overset{\ddot{O}:}{\overset{\|}{C}}{-}R}$$

general structural formulas of ketones

An aldehyde can be written with the condensed formula RCHO or ArCHO, where the symbol CHO indicates that both hydrogen and oxygen atoms are bonded to the carbonyl carbon atom. A ketone has the condensed formula RCOR. In this condensed formula, the symbol CO represents the carbonyl group, and the two R groups that flank the CO group are bonded to the carbonyl carbon atom.

Occurrence of Aldehydes and Ketones

The carbonyl group is the most prevalent functional group in oxygen-containing organic compounds isolated from biological sources. The presence of a carbonyl group in a molecule is sometimes indicated in common names by either of two suffixes. If the carbonyl compound is an aldehyde, the suffix *-al* is used; if the carbonyl compound is a ketone, the suffix is *-one*. For example, retinal is an aldehyde required for vision. The first part of the name indicates that this compound is present in the retina, and the suffix tells us that it is an aldehyde. Another example of a common name is α-ionone, a fragrant ketone responsible for the scent of irises. It is used in perfumes.

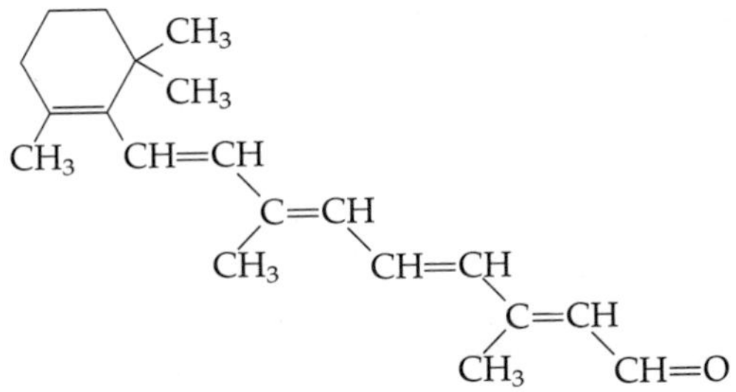

retinal

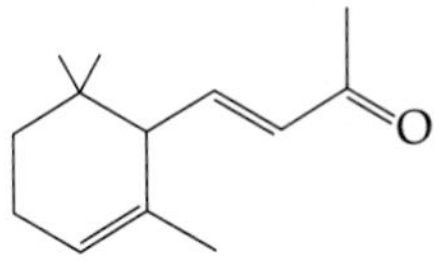

α-ionone

Carbonyl groups are also present in some steroids (Section 3.7). For example, a carbonyl group is present in the synthetic steroids norethindrone, an oral contraceptive, and methandrostenolone, an anabolic steroid.

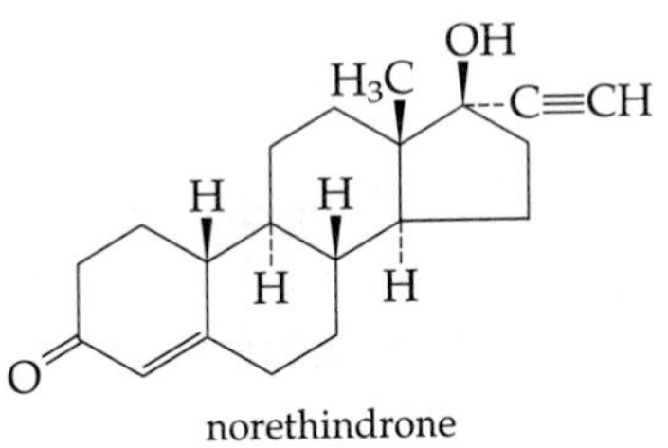

norethindrone

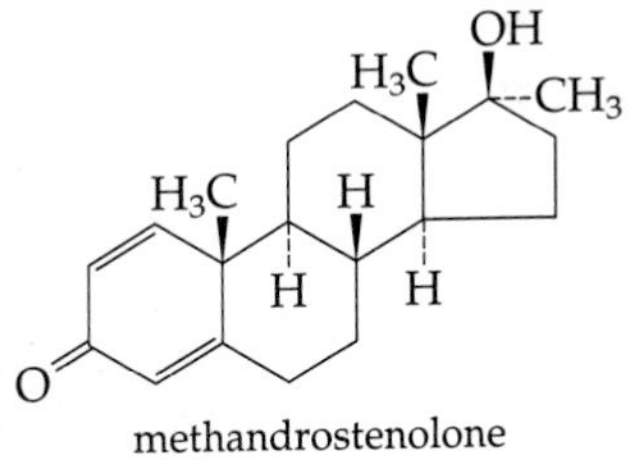

methandrostenolone

10.2 NOMENCLATURE OF ALDEHYDES AND KETONES

Aldehydes and ketones with low molecular weights are often referred to by their common names. The origins of these names will be discussed along with the related common names of acids in Chapter 12.

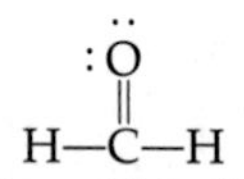

$$\underset{\text{formaldehyde}}{H-\overset{\overset{\large :\ddot{O}}{\|}}{C}-H} \qquad \underset{\text{acetaldehyde}}{CH_3-\overset{\overset{\large :\ddot{O}}{\|}}{C}-H} \qquad \underset{\text{acetone}}{CH_3-\overset{\overset{\large :\ddot{O}}{\|}}{C}-CH_3}$$

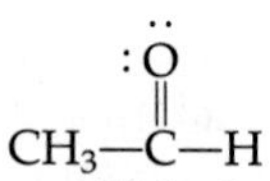

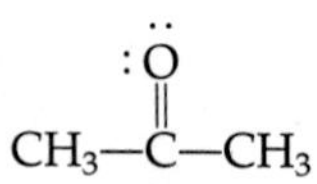

The common names of some aromatic aldehydes and ketones include

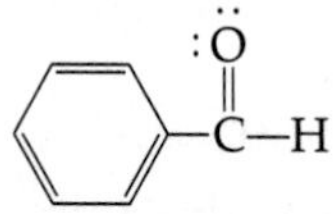

benzaldehyde 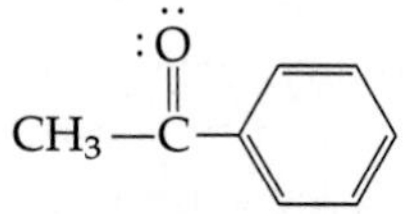   acetophenone 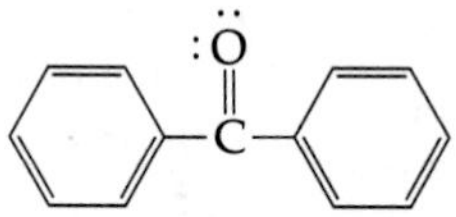   benzophenone

IUPAC Names of Aldehydes

Aldehydes are named by IUPAC rules similar to those outlined for alcohols.

1. The longest continuous carbon chain that contains the carbonyl carbon atom is the parent chain. Replace the final *-e* of the parent hydrocarbon by the ending *-al*.

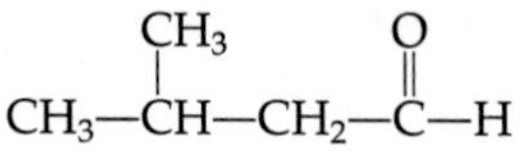

$$CH_3-\overset{\overset{CH_3}{|}}{CH}-CH_2-\overset{\overset{O}{\|}}{C}-H$$

The parent alkane is butane.

This is a substituted butanal. A methyl group is attached to the butanal chain.

2. Number the parent chain so that the carbonyl carbon atom is C—1. The number 1 is not used in the name to indicate the position of the carbonyl carbon atom because it is understood to be located at the end of the chain. Determine the name of each substituent and the number of the carbon atom to which it is attached. This information is added to the parent name as a prefix.

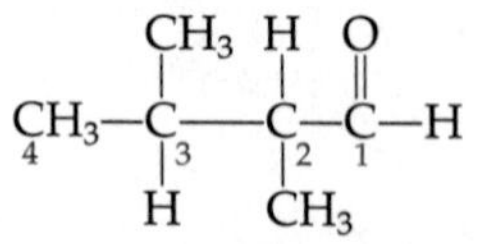

$$\underset{4}{CH_3}-\underset{\underset{H}{|}}{\overset{\overset{CH_3}{|}}{\underset{3}{C}}}-\underset{\underset{CH_3}{|}}{\overset{\overset{H}{|}}{\underset{2}{C}}}-\overset{\overset{O}{\|}}{\underset{1}{C}}-H$$

This is 2,3-dimethylbutanal, not 2,3-dimethyl-1-butanal.

3. The aldehyde functional group has a higher priority than alkyl, halogen, hydroxyl, and alkoxy groups. If any of these groups are present, their names and positions are indicated as prefixes to the name of the parent aldehyde.

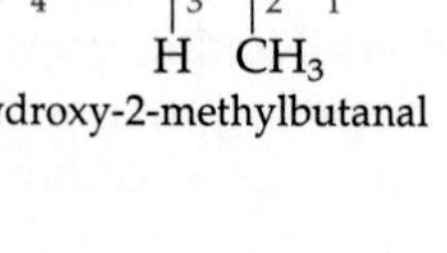

$$\underset{4}{CH_3}-\underset{\underset{H}{|}}{\overset{\overset{HO}{|}}{\underset{3}{C}}}-\underset{\underset{CH_3}{|}}{\overset{\overset{H}{|}}{\underset{2}{C}}}-\overset{\overset{O}{\|}}{\underset{1}{C}}-H$$

3-hydroxy-2-methylbutanal

4. The aldehyde functional group has a higher priority than double or triple bonds. When the parent chain contains a double or triple bond, replace the final *-e* of the name of the parent alkene or alkyne with the suffix *-al*. The position of the multiple bond is indicated by a prefix.

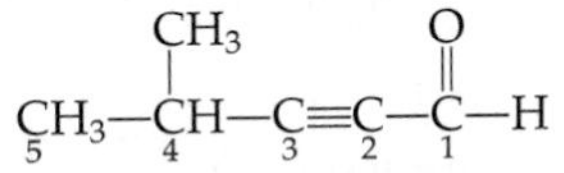

$$\overset{5}{CH_3}-\overset{4}{CH}(CH_3)-\overset{3}{C}\equiv\overset{2}{C}-\overset{1}{C}(=O)-H$$

4-methyl-2-pentynal

5. If an aldehyde or ketone contains other groups with a higher priority, such as carboxylic acids, the carbonyl group is given the prefix *oxo-*. The position of the oxo group is given by a number. The priority order is carboxylic acid > aldehyde > ketone.

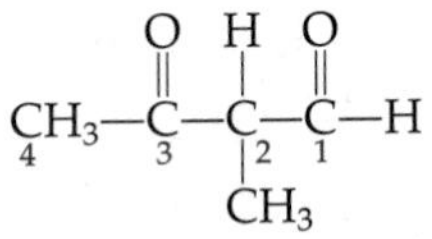

$$\overset{4}{CH_3}-\overset{3}{C}(=O)-\overset{2}{C}H(CH_3)-\overset{1}{C}(=O)-H$$

2-methyl-3-oxobutanal

6. If an aldehyde group is attached to a ring, the suffix *-carbaldehyde* is used.

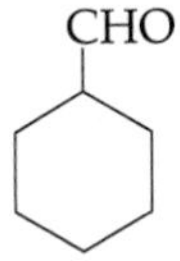

CHO — cyclohexanecarbaldehyde

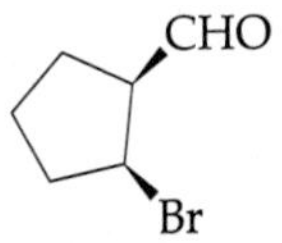

CHO, Br — *cis*-2-bromocyclopentanecarbaldehyde

EXAMPLE 10.1 Give the IUPAC name for the following compound. It is an alarm pheromone in some species of ants.

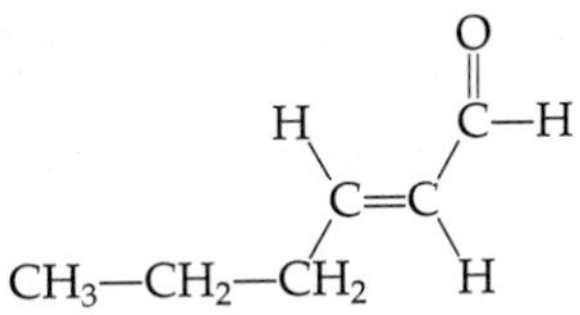

$$CH_3-CH_2-CH_2(H)C=C(H)-C(=O)-H$$

Solution The aldehyde carbon atom on the right is assigned the number 1. The double bond is therefore located at the C-2 atom, and the name is 2-hexenal. The higher priority groups bonded to the unsaturated carbon atoms—the CHO and propyl groups—are in an *E* arrangement. The name is (*E*)-2-hexenal.

IUPAC Names of Ketones

The IUPAC rules for naming ketones are similar to those used for aldehydes. However, the carbonyl group in a ketone is not on a terminal carbon atom. Therefore, the position of the carbonyl group is indicated by a number.

1. The longest continuous carbon chain that contains the ketone carbonyl carbon atom is the parent chain. Replace the final *-e* of the parent hydrocarbon with the ending *-one.*
2. Number the carbon chain so that the carbonyl carbon atom has the lower number. This number appears as a prefix to the parent name. Indicate the identity and location of each substituent as a prefix to the parent name.

$$\underset{5}{CH_3}-\underset{4}{\overset{\overset{\displaystyle CH_3}{|}}{CH}}-\underset{3}{CH_2}-\underset{2}{\overset{\overset{\displaystyle O}{\|}}{C}}-\underset{1}{CH_3}$$

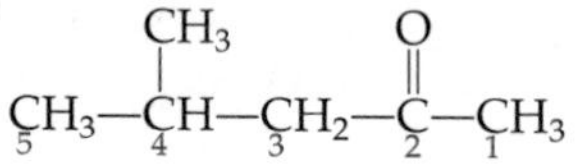

This is 4-methyl-2-pentanone, not 2-methyl-4-pentanone.

3. Cyclic ketones are named as cycloalkanones. The carbonyl carbon atom receives the number 1. The ring is then numbered in the direction that gives the lower number to the first substituent encountered.

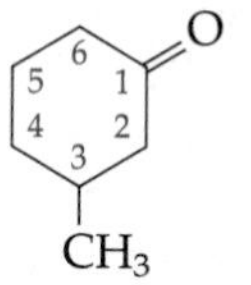

3-methylcyclohexanone 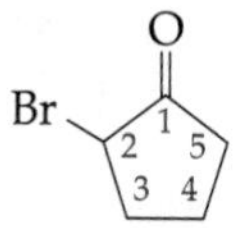   2-bromocyclopentanone

4. Halogen, hydroxyl, alkoxy groups, and multiple bonds have lower priorities than the ketone group. The name of a substituted ketone is determined by the method described for aldehydes.

EXAMPLE 10.2 The IUPAC name for capillin, which is used against skin fungi, is 1-phenyl-2,4-hexadiyn-1-one. Draw its structure.

Solution When we dissect the name, we see that it has the suffix *-1-one* and the stem name *hexa-*, indicating that the parent chain is a ketone containing six carbon atoms. We write the carbon skeleton and number the chain. Place the carbonyl oxygen atom on the C-1 atom.

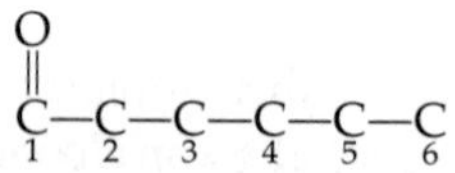

$$\underset{1}{\overset{\overset{\displaystyle O}{\|}}{C}}-\underset{2}{C}-\underset{3}{C}-\underset{4}{C}-\underset{5}{C}-\underset{6}{C}$$

The name has the prefix *1-phenyl*. Therefore, we add a phenyl group at the C-1 atom. Note that the presence of the phenyl group makes the compound a ketone. A carbonyl carbon atom at the end of a chain would otherwise be part of an aldehyde.

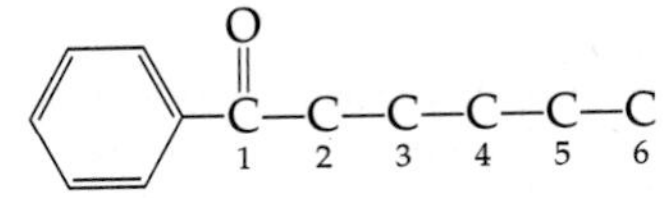

$$C_6H_5-\overset{\displaystyle O}{\overset{\|}{\underset{1}{C}}}-\underset{2}{C}-\underset{3}{C}-\underset{4}{C}-\underset{5}{C}-\underset{6}{C}$$

The *diyn* tells us that there are two triple bonds; they are located at the C-2 and C-4 atoms. Fill in the requisite hydrogen atoms.

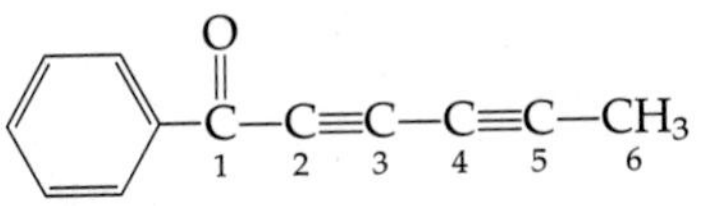

$$C_6H_5-\overset{\displaystyle O}{\overset{\|}{\underset{1}{C}}}-\underset{2}{C}\equiv\underset{3}{C}-\underset{4}{C}\equiv\underset{5}{C}-\underset{6}{CH_3}$$

10.3 PHYSICAL PROPERTIES OF ALDEHYDES AND KETONES

Because oxygen is more electronegative than carbon, the electrons in the carbonyl bond are pulled toward the oxygen atom, and the carbonyl group is polar. This property is expected based on the contributing charged structure of the resonance form described earlier. The polarity of the carbonyl group is shown by an arrow in which the arrowhead represents the negative end of the dipole. We also represent the positive and negative ends of the carbonyl bond by the symbols $\delta+$ and $\delta-$, where the lowercase Greek letter delta means "partial charge".

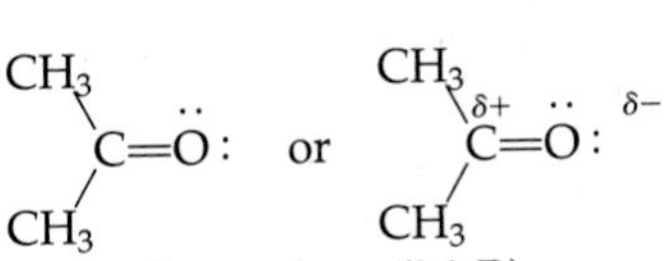

$$(CH_3)_2C{=}\ddot{O}: \quad \text{or} \quad (CH_3)_2\overset{\delta+}{C}{=}\overset{\delta-}{\ddot{O}}:$$

2-propanone (2.9 D)
(acetone)

The dipole moment for acetone, a typical ketone, is 2.9 D. The dipole moment of 2-methylpropane is 0.1 D. The high polarity of the carbonyl group reflects the contribution of the charged form of the two resonance structures. The physical properties of aldehydes and ketones reflect the polarity of the carbonyl group.

Boiling Points

Aldehydes and ketones have boiling points that are distinctly different from those of alkanes or alcohols of similar molecular weight (Table 10.1). Aldehydes and ketones have higher boiling points than the alkanes because of dipole–dipole intermolecular forces due to the carbonyl group. Alcohols have higher boiling points than aldehydes and ketones of similar molecular

TABLE 10.1 Comparative Boiling Points

Compound	Structure	Molecular weight (amu)	Boiling point (°C)
ethane	CH_3CH_3	30	−89
methanol	CH_3OH	32	64.6
methanal	HCHO	30	−21
propane	$CH_3CH_2CH_3$	44	−42
ethanol	CH_3CH_2OH	46	78.3
ethanal	CH_3CHO	44	20
butane	$CH_3CH_2CH_2CH_3$	58	− 1
1-propanol	$CH_3CH_2CH_2OH$	60	97.1
propanal	CH_3CH_2CHO	58	48.8
methylpropane	$CH_3CH(CH_3)_2$	58	−12
2-propanol	$CH_3CH(OH)CH_3$	60	82.5
propanone	CH_3COCH_3	58	56.1

weight (Figure 10.2). Thus, the dipole–dipole attractive forces of carbonyl compounds are weaker than hydrogen-bonding interactions between alcohol molecules. As the molecular weights of the carbonyl compounds increase, their dipole–dipole attractive forces become less important compared to London forces of the hydrocarbon skeleton. As a result, the physical properties of aldehydes and ketones become more like those of

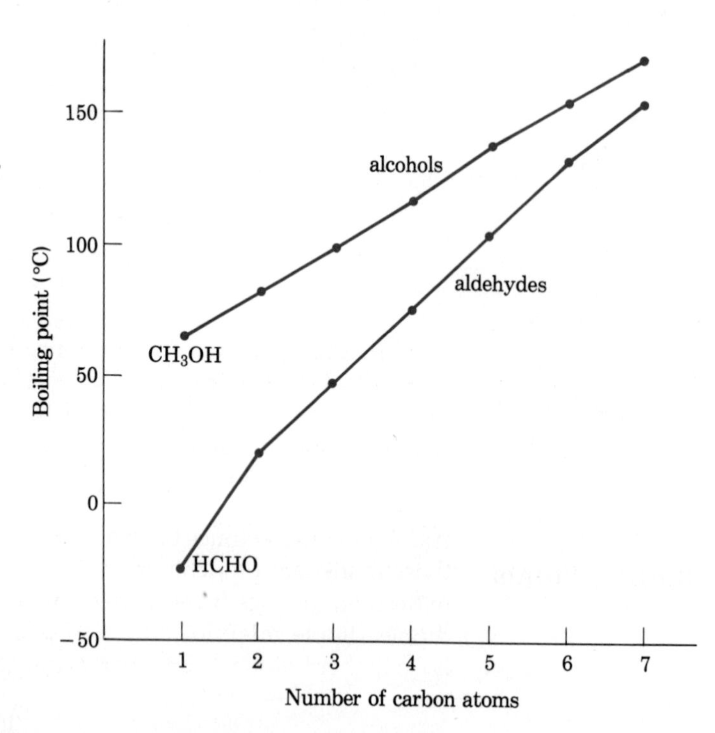

FIGURE 10.2
Boiling Points of Aldehydes and Alcohols
The boiling points of both aldehydes and alcohols increase with the number of carbon atoms. However, alcohols have higher boiling points than aldehydes.

hydrocarbons as chain length increases. The boiling point differences become smaller, although the order of boiling points is still alcohol > carbonyl compound > alkane.

Solubility in Water

Aldehydes and ketones cannot form hydrogen bonds with one another because they cannot function as hydrogen-bond donors. However, the carbonyl oxygen atom has lone-pair electrons that can serve as hydrogen-bond acceptors. Thus, carbonyl groups can form hydrogen bonds with water. Hence, the lower molecular weight compounds formaldehyde, acetaldehyde, and acetone are soluble in water in all proportions.

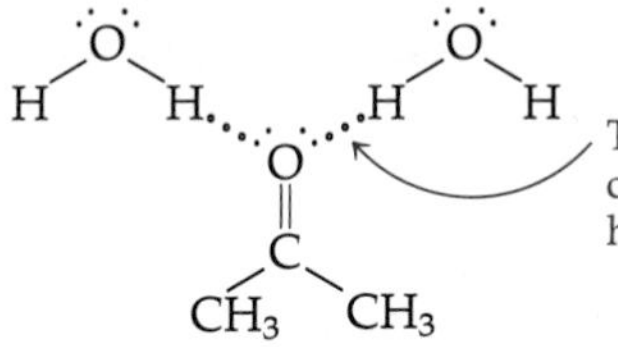

H–Ö–H···Ö=C(CH_3)(CH_3)···H–Ö–H

The lone-pair electrons of the carbonyl oxygen atom serve as hydrogen-bond acceptors.

However, the solubility of carbonyl compounds in water decreases as the chain length increases, and their solubilities become more like those of hydrocarbons.

Solvent Characteristics

Both acetone and 2-butanone (known in industry as methyl ethyl ketone or MEK) are excellent solvents for a variety of organic compounds. These polar solvents dissolve polar solutes because "like dissolves like". These solvents also readily dissolve protic solutes such as alcohols and carboxylic acids because the carbonyl group acts as a hydrogen-bond acceptor for these compounds.

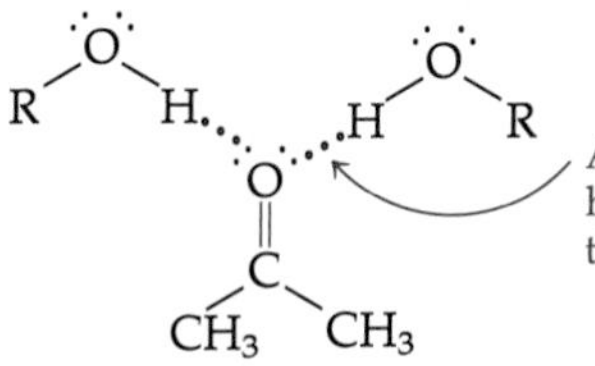

R–Ö–H···Ö=C(CH_3)(CH_3)···H–Ö–R

A hydrogen bond forms between the hydroxyl group of an alcohol and the lone-pair electrons of acetone.

10.4 REDOX REACTIONS OF CARBONYL COMPOUNDS

The carbonyl group is in an oxidation state between that of an alcohol and a carboxylic acid. Thus, a carbonyl group can be reduced to an alcohol or oxidized to a carboxylic acid.

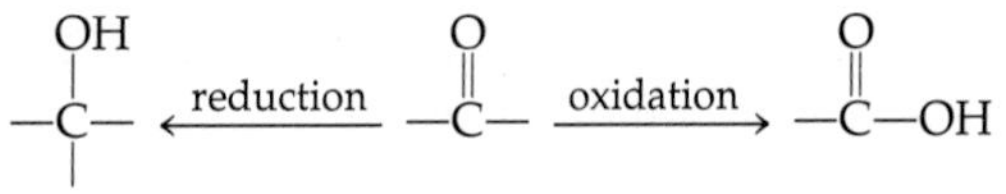

$$-\overset{OH}{\underset{|}{\overset{|}{C}}}- \xleftarrow{\text{reduction}} -\overset{O}{\overset{\|}{C}}- \xrightarrow{\text{oxidation}} -\overset{O}{\overset{\|}{C}}-OH$$

Oxidation Reactions

In Chapter 8 we saw that primary alcohols are oxidized to aldehydes, which are then easily oxidized to acids. Under the same conditions, secondary alcohols are oxidized to ketones, which are not oxidized further. This difference in reactivity distinguishes these classes of compounds.

Aldehydes react with several mild oxidizing reagents including Tollens's reagent, Benedict's solution, and Fehling's solution. Each of these converts aldehydes to carboxylic acids; none of them oxidize ketones. These reagents therefore provide a simple qualitative way of distinguishing aldehydes from ketones.

Tollens's reagent is a basic solution of a silver ammonia complex ion. When Tollens's reagent is added to a test tube that contains an aldehyde, the aldehyde is oxidized and metallic silver is deposited as a mirror on the wall of the test tube.

$$R{-}\overset{\overset{O}{\|}}{C}{-}H + 2\,Ag(NH_3)_2^{+} + 3\,OH^{-} \longrightarrow R{-}\overset{\overset{O}{\|}}{C}{-}O^{-} + 2\,Ag(s) + 4\,NH_3 + 2\,H_2O$$

$$C_5H_9{-}\underset{}{\overset{CH_3}{\overset{|}{C}H}}{-}CHO \xrightarrow{Ag(NH_3)_2^{+}} C_5H_9{-}\overset{CH_3}{\overset{|}{C}H}{-}CO_2^{-} + Ag(s)$$

$$C_5H_9{-}\overset{\overset{O}{\|}}{C}{-}CH_2{-}CH_3 \xrightarrow[\times]{Ag(NH_3)_2^{+}} \text{no oxidation product}$$

Benedict's solution contains cupric ion, Cu^{2+}, as a complex ion in a basic solution. Benedict's solution converts aldehydes to carboxylic acids. In this reaction Cu^{2+} is reduced to Cu^{+}, which forms as a brick-red precipitate, Cu_2O. Benedict's solution has the characteristic blue color of Cu^{2+}, which fades as the red precipitate of Cu_2O forms. Benedict's solution is basic, and

Benedict's Test and Diabetes

Diabetes mellitus is a disease caused by defects in insulin production or insulin receptor function. Insulin helps to regulate glucose metabolism. If insulin, which is made in the pancreas, is produced in insufficient amounts, glucose is overproduced in the liver and not metabolized fully in peripheral tissues. Hence, the concentration of glucose in the blood is elevated, and glucose is excreted in the urine. A person who suffers from diabetes mellitus can determine the timing of insulin injections required to maintain the proper blood glucose concentration by performing Benedict's test on a urine sample.

Glucose has an aldehyde group that reacts with Benedict's solution. Benedict's solution contains Cu^{2+} and is blue. If the urine sample solution turns a greenish-yellow color, the sample contained about 0.5% glucose. If the sample solution turns red (the color of Cu_2O), the concentration of glucose in the original sample was greater than 2%.

in basic solution a carboxylic acid is converted to its conjugate base, that is, a carboxylate anion. Fehling's solution, which contains Cu^{2+} as a different complex ion in a basic solution, also oxidizes aldehydes, but not ketones.

$$\underset{}{R{-}\overset{\overset{\displaystyle O}{\|}}{C}{-}H} + \underset{\text{blue solution}}{2\,Cu^{2+}} + 5\,OH^- \longrightarrow R{-}\overset{\overset{\displaystyle O}{\|}}{C}{-}O^- + \underset{\text{red precipitate}}{Cu_2O(s)} + 3\,H_2O$$

$$C_6H_5{-}CH_2{-}CHO \xrightarrow{Cu^{2+}} C_6H_5{-}CH_2{-}CO_2^- + Cu_2O(s)$$

$$C_6H_5{-}\overset{\overset{\displaystyle O}{\|}}{C}{-}CH_3 \xrightarrow[\times]{Cu^{2+}} \text{no oxidation product}$$

EXAMPLE 10.3 Can the isomeric carbonyl-containing compounds of molecular formula C_4H_8O be distinguished from each other by Tollens's reagent?

Solution There are three isomeric carbonyl-containing compounds—two aldehydes and one ketone.

$$CH_3\underset{}{\overset{\overset{\displaystyle CH_3}{|}}{C}}HCHO \qquad CH_3CH_2CH_2CHO \qquad CH_3CH_2\overset{\overset{\displaystyle O}{\|}}{C}CH_3$$

Both aldehydes react with Tollens's reagent to produce a silver mirror. Therefore, these two compounds cannot be distinguished from each other with this reagent. However, the ketone does not react with Tollens's reagent. Therefore, the compound that does not yield a silver mirror is butanone.

Reduction to Alcohols

In Chapter 4 we saw that carbon–carbon double bonds can be reduced by hydrogen gas with nickel, palladium, or platinum as catalysts. In Chapter 8 we found that aldehydes and ketones can be catalytically reduced to alcohols by hydrogen gas and Raney nickel. However, the reduction of aldehydes and ketones with hydrogen gas requires more severe conditions than are required to reduce alkenes. We also learned that both lithium aluminum hydride, $LiAlH_4$, and sodium borohydride, $NaBH_4$, reduce carbonyl groups, but neither reagent reduces carbon–carbon double or triple bonds (Section 8.9).

$$\underset{\text{butanal}}{CH_3CH_2CH_2CHO} \xrightarrow[CH_3CH_2OH]{NaBH_4} \underset{\text{1-butanol}}{CH_3CH_2CH_2CH_2OH}$$

$$\underset{\text{butanone}}{CH_3\overset{\overset{O}{\|}}{C}CH_2CH_3} \xrightarrow[2.\ H_3O^+]{1.\ LiAlH_4} \underset{\text{2-butanol}}{CH_3\overset{\overset{OH}{|}}{C}HCH_2CH_3}$$

Because neither lithium aluminum hydride nor sodium borohydride reacts with alkenes or alkynes, these reagents selectively reduce a carbonyl group in compounds with carbon–carbon multiple bonds.

$$\text{(cyclohexenyl)}-\overset{\overset{O}{\|}}{C}-H \xrightarrow[2.\ H_3O^+]{1.\ NaBH_4} \text{(cyclohexenyl)}-CH_2OH$$

Reduction to a Methylene Group

A carbonyl group can be reduced directly to a methylene group by either the Clemmensen reduction or the Wolff–Kishner reduction. The former uses a zinc amalgam (Zn/Hg) and HCl, and the latter uses hydrazine (NH_2NH_2) and base.

$$\underset{\text{cyclohexanone}}{} \xrightarrow[HCl]{Zn/Hg} \underset{\text{cyclohexane}}{}$$

$$\underset{\text{propiophenone}}{} \xrightarrow[KOH]{NH_2NH_2} \underset{\text{propylbenzene}}{}$$

We introduced the reduction of a carbonyl group to a methylene group in Chapter 5 as a method of converting the product of a Friedel–Crafts acylation to an alkyl group that could not be produced by direct Friedel–Crafts alkylation.

10.5 ADDITION REACTIONS OF CARBONYL COMPOUNDS

We recall from our discussion of alkenes in Chapter 4 that unsymmetrical reagents such as HCl add to π bonds. In these reactions, the electrophilic proton reacts with the π bond to give an intermediate carbocation. The position of the electrophilic attack on the double bond is determined by the stability of the carbocation intermediate. This carbocation subsequently reacts with a nucleophile to give the addition product.

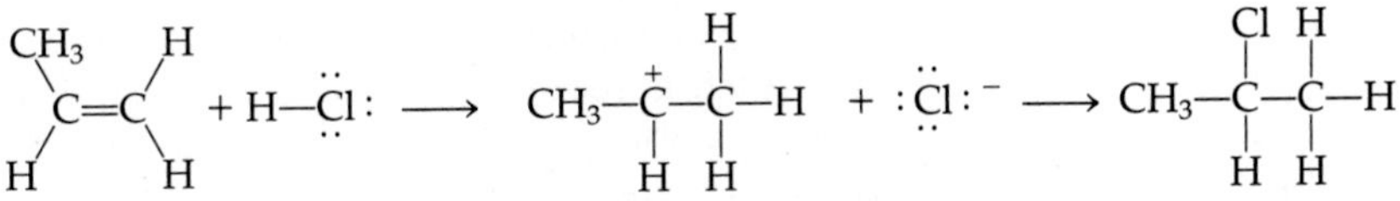

$$CH_3(H)C{=}CH_2 + H{-}\ddot{Cl}: \longrightarrow CH_3{-}\overset{+}{C}H{-}CH_3 + :\ddot{Cl}:^- \longrightarrow CH_3{-}CHCl{-}CH_3$$

isopropyl carbocation

Aldehydes and ketones also contain a π bond. Like alkenes, they too react with unsymmetrical reagents to give addition products. The carbonyl bond is polar, and an unsymmetrical reagent reacts with it so that the electrophilic part bonds to the carbonyl oxygen atom and the nucleophilic part to the carbonyl carbon atom.

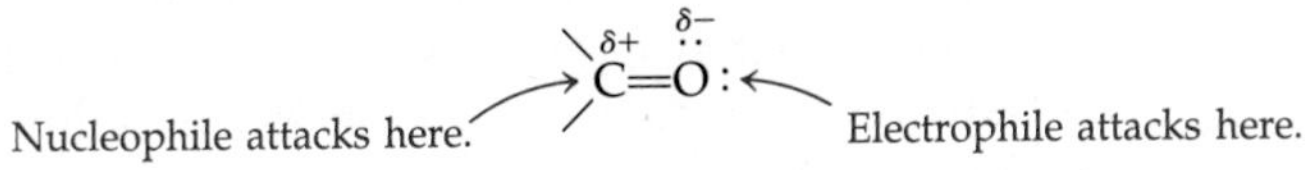

$$\overset{\delta+}{C}{=}\overset{\delta-}{\ddot{O}}:$$

Nucleophile attacks here. Electrophile attacks here.

Thus, although two isomers are structurally possible, only one compound is formed.

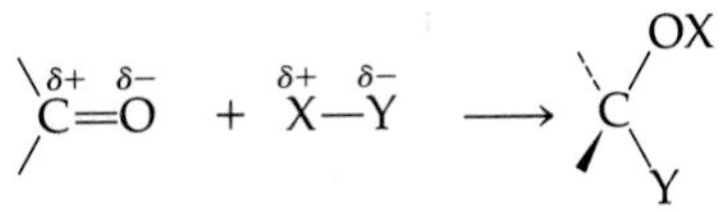

$$\overset{\delta+}{C}{=}\overset{\delta-}{O} + \overset{\delta+}{X}{-}\overset{\delta-}{Y} \longrightarrow C(OX)(Y)$$

Many reagents that add to carbonyl compounds can be represented as H—Nu. The electrophilic part of the reagent is H^+; the nucleophilic part is Nu$:^-$. (The reagent can have lone-pair electrons and be represented as H—Nu:.) The addition reaction occurs in several steps. The order of the steps depends on whether the reaction is acid- or base-catalyzed. For the acid-catalyzed reaction, the sequence is

1. A proton (an electrophile and an acid) reacts with the carbonyl oxygen atom (an electron-pair donor, or Lewis base) to produce a carbocation, which also can be represented as an oxonium ion in the alternate resonance form.

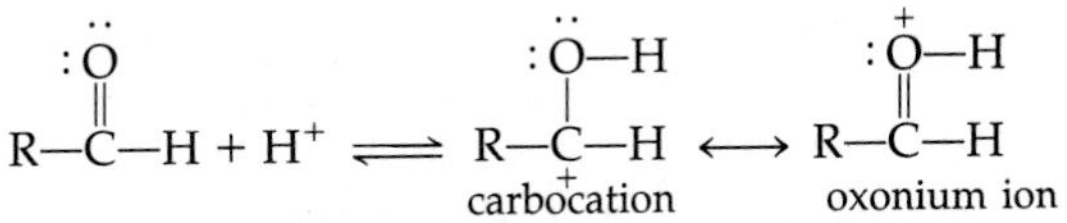

$$R{-}C(=\ddot{O}:){-}H + H^+ \rightleftharpoons R{-}\overset{+}{C}(\ddot{O}{-}H){-}H \longleftrightarrow R{-}C(=\overset{+}{O}{-}H){-}H$$

carbocation oxonium ion

2. The carbocation, which has a vacant $2p$ orbital and therefore acts as a Lewis acid, reacts with the lone-pair electrons of H—Nu:, which functions as a Lewis base.

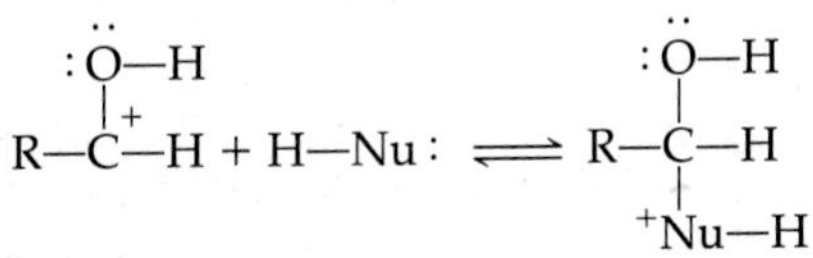

$$R{-}\overset{+}{C}(\ddot{O}{-}H){-}H + H{-}Nu: \rightleftharpoons R{-}C(\ddot{O}{-}H)({}^{+}Nu{-}H){-}H$$

3. A proton is transferred in an acid–base reaction with a hydroxylic solvent, which is represented by Sol—O—H. The transferred proton becomes available for the first step of the reaction sequence. Thus, the reaction is acid-catalyzed.

$$\text{R—C(OH)(H)—}\overset{+}{\text{N}}\text{u—H} + \text{Sol—O—H} \rightleftharpoons \text{R—C(OH)(H)—Nu} + \text{Sol—}\overset{+}{\text{O}}\text{H—H}$$

The sequence of the steps for a base-catalyzed reaction is different. The nucleophile Nu:$^-$ attacks the carbonyl carbon atom, which has a partial positive charge and is, therefore, electrophilic. Next, the alkoxide ion is protonated in an acid–base reaction with a hydroxylic solvent represented as Sol—O—H. The conjugate base of the solvent removes a proton from H—Nu to regenerate Nu:$^-$.

$$\text{R—C(=O)—H} + :\text{Nu}^- \rightleftharpoons \text{R—C(O}^-\text{)(H)—Nu}$$

$$\text{R—C(O}^-\text{)(H)—Nu} + \text{Sol—O—H} \rightleftharpoons \text{R—C(OH)(H)—Nu} + \text{Sol—O:}^-$$

$$\text{Sol—O:}^- + \text{H—Nu} \rightleftharpoons \text{Sol—O—H} + \text{Nu:}^-$$

Equilibria in Addition Reactions

The addition products of alkenes with reagents such as bromine and hydrogen bromide are very stable, and the equilibrium constants for these addition reactions are large. The equilibrium constants for the addition reactions of aldehydes and ketones are usually much smaller. Thus, to obtain high yields it is often necessary to consider Le Châtelier's principle and adjust the reaction conditions to "drive" the reaction toward the products.

Arrows to represent an equilibrium are used above in the general description of addition to a carbonyl group to emphasize the reversibility of the reaction. This notation will be used in many subsequent reactions.

Relative Reactivities of Aldehydes and Ketones

Nucleophiles react faster with aldehydes than with ketones because of both electronic and steric effects. First, we will consider electronic effects. A ketone has two alkyl groups attached to the carbonyl carbon atom. These groups donate electron density to the carbonyl carbon atom and stabilize its partial positive charge. (We recall that alkyl groups stabilize carbocations in the same way.) The carbonyl carbon atom of an aldehyde is attached to only one alkyl group. Therefore, the carbonyl carbon atom of an aldehyde has a

FIGURE 10.3 Effect of Steric Hindrance on Addition Reactions
The reaction of a nucleophile with a carbonyl carbon atom depends on the number and size of the groups bonded to that atom. Aldehydes have only one alkyl group and react faster than ketones, which have two alkyl groups.

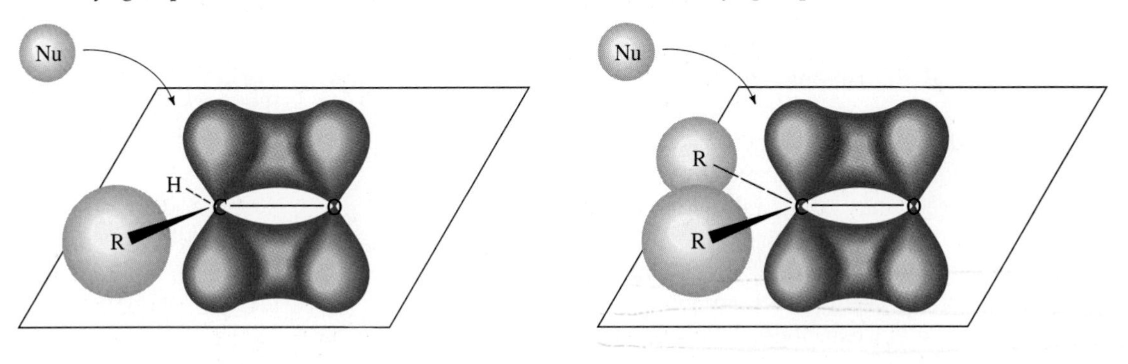

larger partial positive charge than that of a ketone. As a consequence, nucleophiles react faster with aldehydes than with ketones.

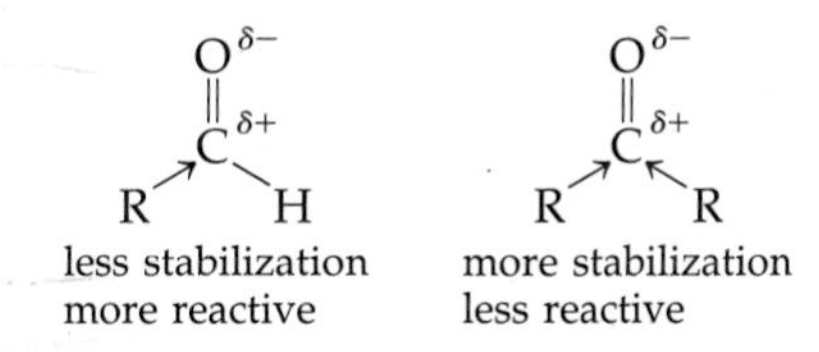

Next, let's consider the role of steric effects—the sizes of groups—in the reactivity of aldehydes and ketones. Because a ketone has two alkyl groups attached to the carbonyl carbon atom, it is sterically hindered relative to the carbonyl carbon atom of an aldehyde, which has only a hydrogen atom and one alkyl group bonded to it. Thus, a nucleophile can approach the carbonyl group of an aldehyde more readily, and the reaction is faster (Figure 10.3).

10.6 SYNTHESIS OF ALCOHOLS FROM CARBONYL COMPOUNDS

In Chapter 7 we discussed the reaction of haloalkanes with magnesium to produce **organometallic compounds** called **Grignard reagents.**

$$\text{R—X} \xrightarrow[\text{ether}]{\text{Mg}} \text{R—Mg—X}$$

Grignard reagents are very reactive and versatile reactants used to form carbon–carbon bonds in the synthesis of complex molecules from simpler molecules. Grignard reagents contain a very strongly polarized carbon–magnesium bond in which the carbon atom has a partial negative charge.

Oral Contraceptives

The female sex hormones are collectively called estrogens. Estrogens are released after menstruation. They cause growth of the lining of the uterus and the ripening of the ovum. High levels of estrogen are maintained during pregnancy, and one consequence is that ovulation is inhibited. Oral contraceptives mimic one effect of pregnancy: they inhibit ovulation.

The most potent estrogen is estradiol. However, estradiol itself is not an effective oral contraceptive because its C-17 hydroxyl group is rapidly oxidized in metabolic reactions. The oxidation product, estrone, has only one-third the activity of estradiol. Continued reaction of estrone produces estriol, which has only one-sixteenth the activity of estradiol.

estrone

estriol

estradiol

The hormonal action of estradiol is related to the C-17 hydroxyl group, which is located above the plane of the five-membered ring. If a structurally related compound that is a tertiary alcohol with the correct stereochemistry could be synthesized, then the compound could survive metabolic oxidation. Based on the reactivity of Grignard reagents, one might propose to add methyl Grignard

$$\overset{\delta-}{\mathrm{C}}-\overset{\delta+}{\mathrm{MgX}}$$

The carbon atom of the Grignard reagent resembles a carbanion. It reacts as a nucleophile and adds to the electrophilic carbon atom of a carbonyl group in an aldehyde or ketone. The magnesium ion forms a salt with the negatively charged oxygen atom. The resulting product is a magnesium alkoxide, which is hydrolyzed to obtain an alcohol.

$$\mathrm{R'{-}MgCl} + \mathrm{C{=}O} \longrightarrow \underset{\text{a magnesium alkoxide}}{\mathrm{R'{-}C{-}OMgCl}} \xrightarrow{H_2O} \mathrm{R'{-}C{-}OH} + \mathrm{ClMgOH}$$

A Grignard reagent adds to various types of carbonyl compounds to give primary, secondary, and tertiary alcohols. Primary alcohols are synthesized by reacting the Grignard reagent, R′—MgX, with formaldehyde.

reagent and produce a tertiary alcohol. However, estrone has a phenolic hydroxyl group that would react with the Grignard reagent. This difficulty has been cleverly bypassed. Instead of making a tertiary alcohol by adding an alkyl group derived from a Grignard reagent, a tertiary alcohol is made by adding acetylide anion—the conjugate base of acetylene (Section 4.5). The reaction of acetylide with cyclopentanone produces an acetylenic alcohol, as shown below. The group derived from acetylene is called ethynyl.

By analogy, sodium acetylide reacts with estrone to give ethynyl estradiol. Because the C-17 carbonyl group of estrone is planar, acetylide could potentially attack from either side, but the keto group is flanked by a methyl group that extends above the plane of the ring. The acetylide anion thus approaches the "bottom" of the ring and the product has its —OH group "up" as required for hormonal activity. This tertiary alcohol cannot be oxidized.

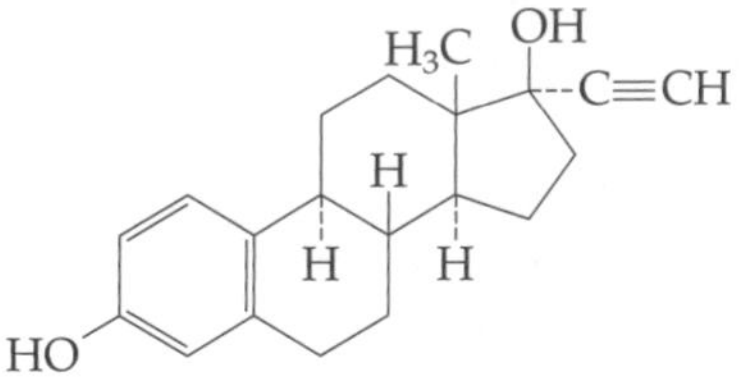

H_3C OH —C≡CH; H; H H; HO

ethinyl estradiol

Thus, oral contraceptives are manufactured with an ethynyl group at the C-17 atom to prevent oxidation and retain biological activity.

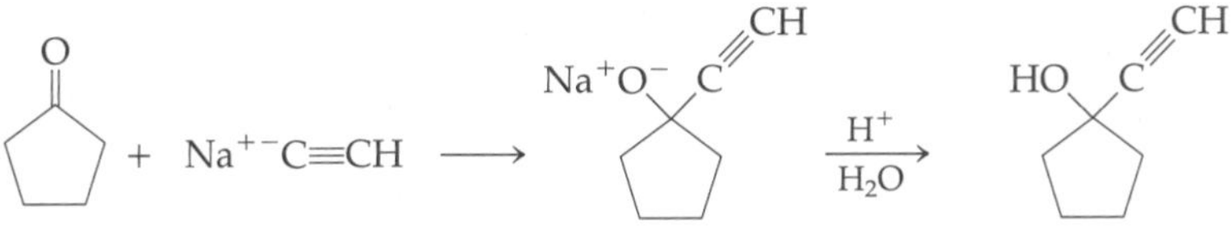

$$\text{cyclopentanone} + Na^{+-}C{\equiv}CH \longrightarrow Na^+O^-\text{—}C(C{\equiv}CH)\text{(cyclopentane)} \xrightarrow[H_2O]{H^+} HO\text{—}C(C{\equiv}CH)\text{(cyclopentane)}$$

sodium acetylide — 1-ethynylcyclopentanol

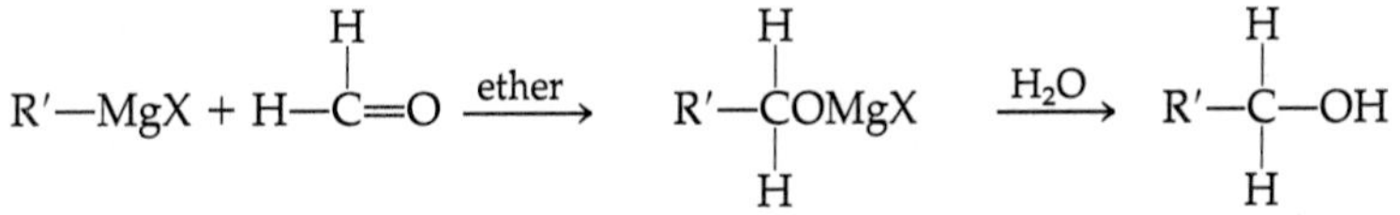

$$R'\text{—MgX} + H\text{—}\underset{}{\overset{H}{C}}{=}O \xrightarrow{\text{ether}} R'\text{—}\underset{H}{\overset{H}{C}}OMgX \xrightarrow{H_2O} R'\text{—}\underset{H}{\overset{H}{C}}\text{—OH}$$

magnesium alkoxide — primary alcohol

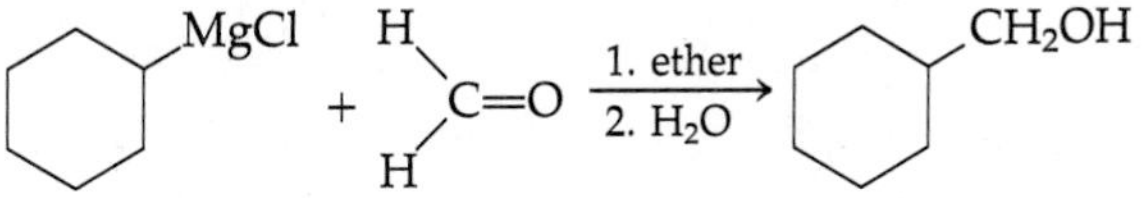

$$C_6H_{11}\text{—MgCl} + H_2C{=}O \xrightarrow[\text{2. } H_2O]{\text{1. ether}} C_6H_{11}\text{—}CH_2OH$$

Secondary alcohols are obtained by reacting the Grignard reagent, R′—MgX, with an aldehyde, R—CHO. Note that the carbon atom bearing the hydroxyl group is bonded to the alkyl groups from both the Grignard reagent and the aldehyde.

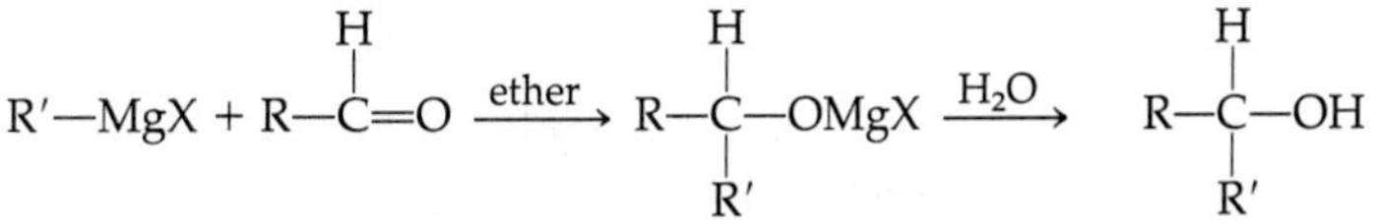

$$R'\text{—MgX} + R\text{—}\overset{H}{C}{=}O \xrightarrow{\text{ether}} R\text{—}\underset{R'}{\overset{H}{C}}\text{—OMgX} \xrightarrow{H_2O} R\text{—}\underset{R'}{\overset{H}{C}}\text{—OH}$$

secondary alcohol

$$CH_3{-}MgBr + O{=}\overset{\displaystyle H}{\overset{|}{C}}{-}C_6H_5 \xrightarrow[2.\ H_2O]{1.\ ether} CH_3{-}\overset{\displaystyle OH}{\overset{|}{C}H}{-}C_6H_5$$

Tertiary alcohols are made by reacting the Grignard reagent, R′—MgX, with a ketone. Two of the alkyl groups bonded to the carbon atom bearing the hydroxyl group were part of the ketone; one alkyl group is provided from the Grignard reagent.

$$R'{-}MgX + R{-}\overset{\displaystyle R}{\overset{|}{C}}{=}O \xrightarrow{ether} R{-}\underset{\displaystyle R'}{\underset{|}{\overset{\displaystyle R}{\overset{|}{C}}}}{-}OMgX \xrightarrow{H_2O} R{-}\underset{\displaystyle R'}{\underset{|}{\overset{\displaystyle R}{\overset{|}{C}}}}{-}OH$$

tertiary alcohol

$$CH_3\overset{\displaystyle CH_3}{\overset{|}{C}H}{-}MgI + \text{cyclopentanone} \xrightarrow[2.\ H_2O]{1.\ ether} \text{1-(1-methylethyl)cyclopentanol (HO and } \overset{\displaystyle CH_3}{\overset{|}{C}H}{-} \text{ on ring carbon)}$$

The use of the Grignard reagent has some experimental limitations. Certain functional groups cannot be present in the reactants because Grignard reagents are destroyed by reaction with acidic hydrogen atoms of water, alcohols, phenols, or carboxylic acid groups, —COOH.

$$R{-}OH + R'{-}MgBr \longrightarrow R{-}O^{-}MgBr^{+} + R'{-}H$$

Thus, the carbonyl compound cannot also have a hydroxyl group because the Grignard reagent would react first with that functional group and be destroyed.

EXAMPLE 10.4 The European bark beetle produces a pheromone that causes beetles to aggregate. Describe two ways that the compound could be synthesized in the laboratory using a Grignard reagent.

$$CH_3{-}CH_2{-}CH_2{-}\overset{\displaystyle CH_3}{\overset{|}{C}H}{-}\underset{\displaystyle OH}{\underset{|}{C}H}{-}CH_2{-}CH_3$$

Solution The compound is a secondary alcohol that can be made from an aldehyde and a Grignard reagent. The carbon atom bonded to the hydroxyl group must be chosen as the carbonyl carbon atom. One of the two alkyl groups bonded to this carbon atom must be part of the aldehyde. The other must be the alkyl group of the Grignard reagent. The two possible components that could be introduced by a Grignard reagent are an ethyl group or a 1-methylbutyl group.

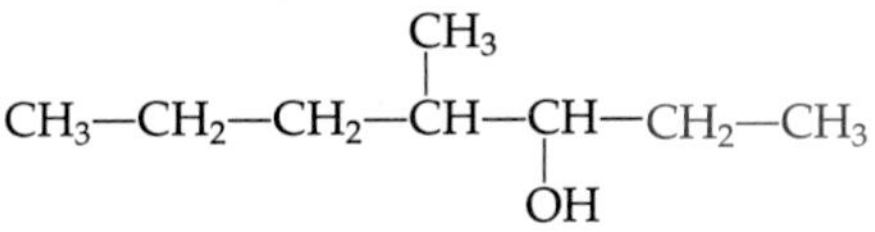

$$CH_3{-}CH_2{-}CH_2{-}\underset{}{\overset{CH_3}{\overset{|}{C}H}}{-}\underset{OH}{\underset{|}{C}H}{-}CH_2{-}CH_3 \qquad CH_3{-}CH_2{-}CH_2{-}\overset{CH_3}{\overset{|}{C}H}{-}\underset{OH}{\underset{|}{C}H}{-}CH_2{-}CH_3$$

ethyl 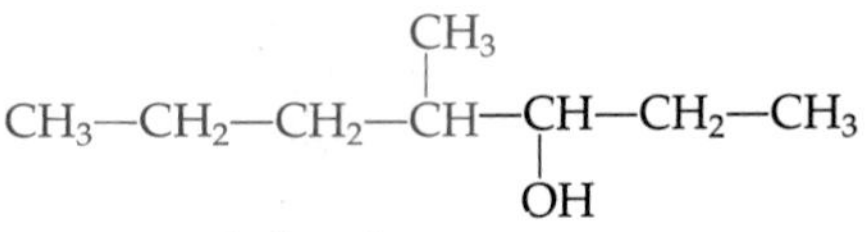 1-methylbutyl

If ethylmagnesium bromide is the Grignard reagent, the required aldehyde for the reaction is 2-methylpentanal. If the Grignard reagent is 1-methylbutylmagnesium bromide, propanal is required as the aldehyde.

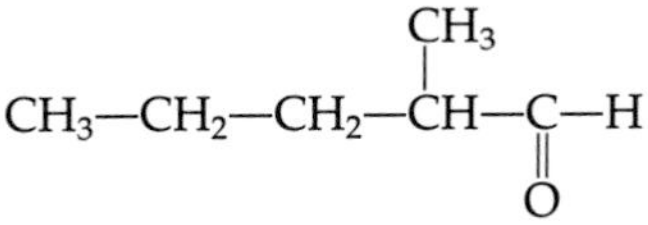

$$CH_3{-}CH_2{-}CH_2{-}\overset{CH_3}{\overset{|}{C}H}{-}\underset{O}{\underset{\|}{C}}{-}H \qquad H{-}\underset{O}{\underset{\|}{C}}{-}CH_2{-}CH_3$$

2-methylpentanal 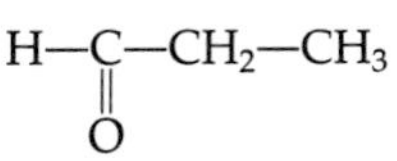 propanal

10.7 ADDITION OF OXYGEN COMPOUNDS

The nucleophile that adds to a carbonyl group can be the oxygen atom of water or an alcohol. Addition of water yields a hydrate; addition of an alcohol yields a hemiacetal or hemiketal. Each of these reactions is reversible.

Addition of Water

Water adds to aldehydes and ketones to form **hydrates.** The proton of water bonds to the oxygen atom of the carbonyl group; the hydroxide ion adds to the carbon atom.

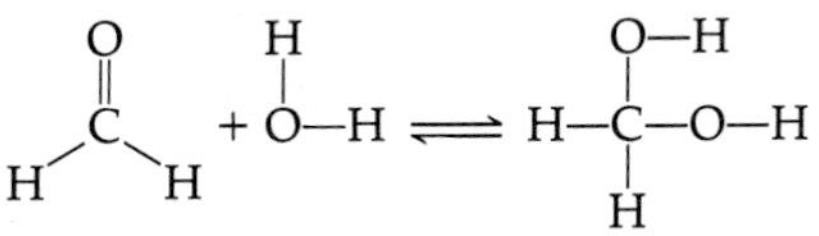

$$\underset{H\quad H}{\overset{O}{\overset{\|}{C}}} + \overset{H}{\overset{|}{O}}{-}H \rightleftharpoons H{-}\underset{H}{\underset{|}{\overset{O-H}{\overset{|}{C}}}}{-}O{-}H$$

Aldehydes with low molecular weights readily form hydrates. Formaldehyde is over 99% hydrated, whereas other aldehydes are substantially less hydrated. The hydrate of formaldehyde is called formalin. This substance is used to preserve biological specimens; it is a 37% by weight solution of formaldehyde in water.

Ketones are hydrated to a small extent, usually less than 1%. The hydrates of aldehydes and ketones usually cannot be isolated. They exist only in solution, where the large amount of water forces the equilibrium to the right.

EXAMPLE 10.5 Write the steps for the base-catalyzed hydration of CH_3CHO in aqueous solution.

Solution In the first step, the nucleophilic hydroxide ion attacks the electrophilic carbonyl carbon atom.

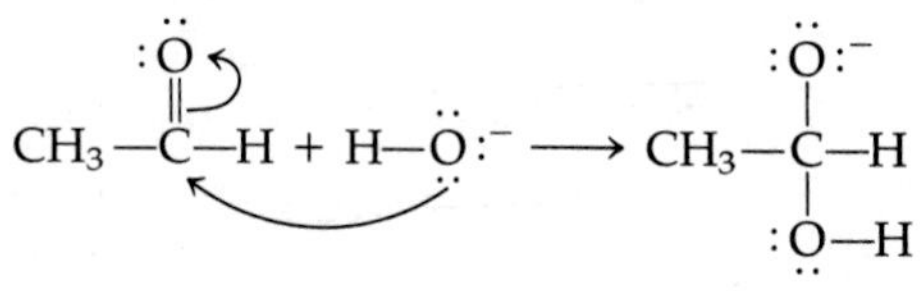

$$CH_3{-}\overset{\overset{\ddot{O}:}{\|}}{C}{-}H + H{-}\ddot{\underset{..}{O}}:^- \longrightarrow CH_3{-}\overset{\overset{:\ddot{O}:^-}{|}}{\underset{\underset{:\underset{..}{O}-H}{|}}{C}}{-}H$$

Then, a proton is transferred from water to the alkoxide ion. As a consequence, the base catalyst—the hydroxide ion—is regenerated.

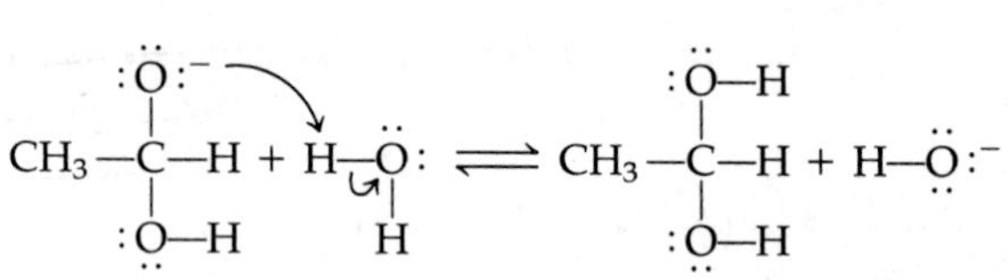

$$CH_3{-}\overset{\overset{:\ddot{O}:^-}{|}}{\underset{\underset{:\underset{..}{O}-H}{|}}{C}}{-}H + H{-}\underset{\underset{H}{|}}{\ddot{O}}: \rightleftharpoons CH_3{-}\overset{\overset{:\ddot{O}-H}{|}}{\underset{\underset{:\underset{..}{O}-H}{|}}{C}}{-}H + H{-}\ddot{\underset{..}{O}}:^-$$

Addition of Alcohols

Alcohols add to carbonyl compounds in an acid-catalyzed reaction. The reaction does not occur in basic or neutral solution. The hydrogen atom of the alcohol adds to the carbonyl oxygen atom, and the —OR′ portion (alkoxy group) adds to the carbon atom. The product is called a **hemiacetal** if the carbonyl compound is an aldehyde and a **hemiketal** if the carbonyl compound is a ketone. These molecules have both an —OH group and an —OR′ group attached to the same carbon atom. The hemiacetal has a hydrogen atom and an alkyl group attached to the original carbonyl carbon atom, whereas the hemiketal has two alkyl groups attached.

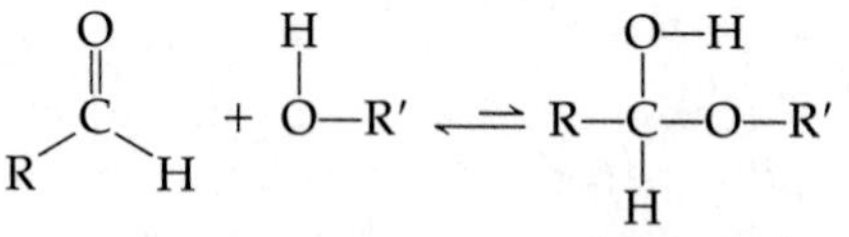

$$\underset{R\quad H}{\overset{O}{\overset{\|}{C}}} + \overset{H}{\overset{|}{O}}{-}R' \rightleftharpoons R{-}\overset{\overset{O-H}{|}}{\underset{\underset{H}{|}}{C}}{-}O{-}R'$$

hemiacetal

$$\underset{R\quad R}{\overset{O}{\overset{\|}{C}}} + \overset{H}{\overset{|}{O}}{-}R' \rightleftharpoons R{-}\overset{\overset{O-H}{|}}{\underset{\underset{R}{|}}{C}}{-}O{-}R'$$

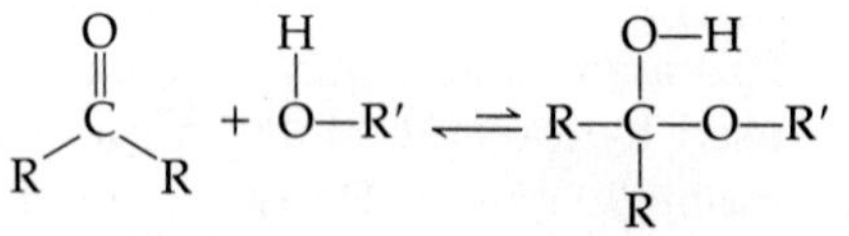

hemiketal

Hemiacetals and hemiketals are often unstable compounds. That is, the equilibrium constant for formation of either a hemiacetal or a hemiketal is less than 1, and the equilibria for the above reactions lie to the left. However, when both the carbonyl group and the alcohol are part of the same molecule, the equilibrium constant is larger. As a result, stable cyclic products form. Cyclization is favorable because the two functional groups are close to each other. Carbohydrates (Chapter 11), which contain both carbonyl and hydroxyl groups, exist to only a small extent as open-chain molecules. They exist largely as cyclic hemiacetals or hemiketals.

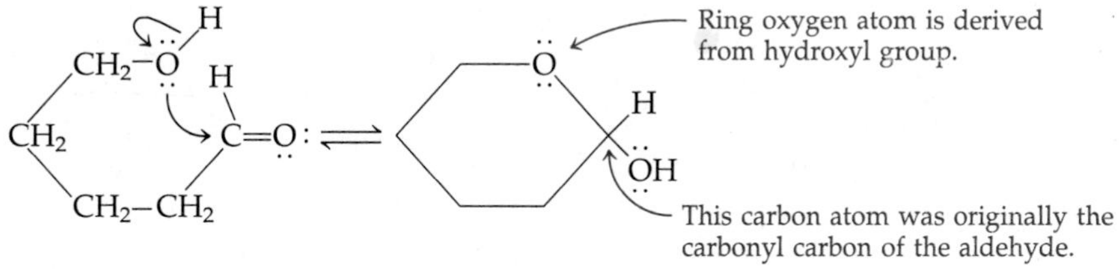

CH_2–O–H, H–C=O, CH_2, CH_2–CH_2 ⇌ (cyclic ring with O; H; OH)

Ring oxygen atom is derived from hydroxyl group.

This carbon atom was originally the carbonyl carbon of the aldehyde.

Mechanism of Addition of Alcohols

The first step in the acid-catalyzed addition of an alcohol to an aldehyde or ketone is protonation of the carbonyl oxygen atom (an electron-pair donor, or Lewis base) to produce a carbocation.

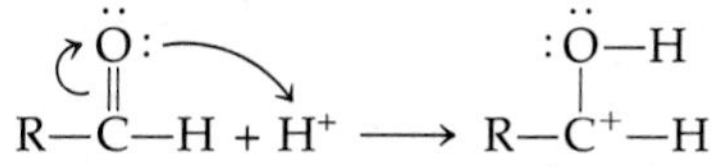

$$R{-}C({=}O){-}H + H^+ \longrightarrow R{-}C^+(OH){-}H$$

carbocation

Then the carbocation, which has a vacant 2*p* orbital and therefore acts as a Lewis acid, reacts with the lone-pair electrons of the oxygen atom of the alcohol R′—OH.

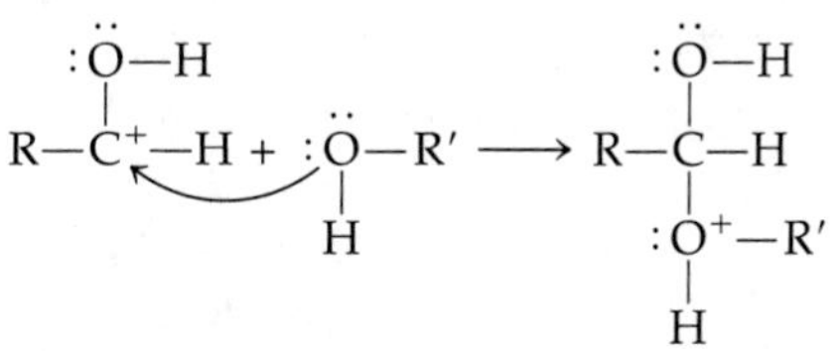

$$R{-}C^+(OH){-}H + H{-}O{-}R' \longrightarrow R{-}C(OH)(H){-}O^+(H){-}R'$$

Finally a proton is transferred in an acid–base reaction with R′—OH. The transferred proton becomes available for the first step of the reaction sequence. Thus, the reaction is acid-catalyzed.

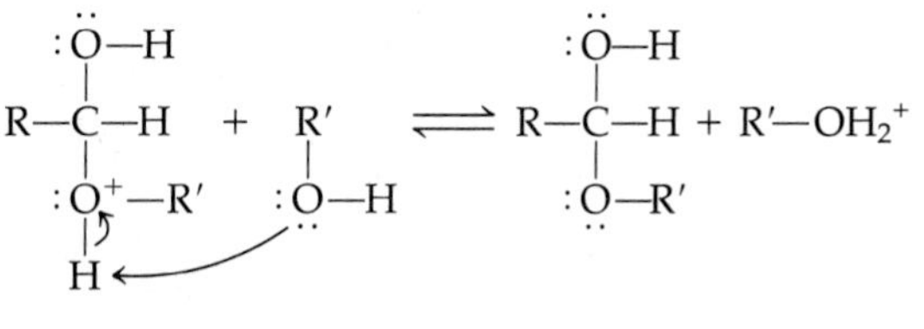

$$R{-}C(OH)(H){-}O^+(H){-}R' + R'{-}O{-}H \rightleftharpoons R{-}C(OH)(H){-}O{-}R' + R'{-}OH_2^+$$

10.8 FORMATION OF ACETALS AND KETALS

The —OH group in either a hemiacetal or a hemiketal can be replaced by substitution of —OH by another alkoxy group —OR′. This reversible reaction occurs readily in acidic solution to produce an acetal or ketal. The reaction is shown below.

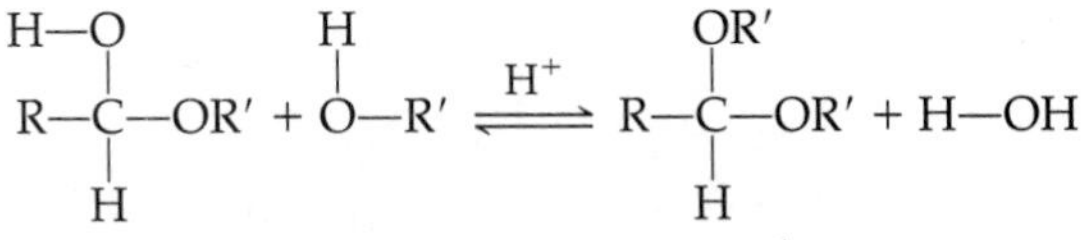

$$\text{R—C(OH)(H)—OR}' + \text{H—O—R}' \overset{H^+}{\rightleftharpoons} \text{R—C(OR}')\text{(H)—OR}' + \text{H—OH}$$

acetal

$$\text{R—C(OH)(R)—OR}' + \text{H—O—R}' \overset{H^+}{\rightleftharpoons} \text{R—C(OR}')\text{(R)—OR}' + \text{H—OH}$$

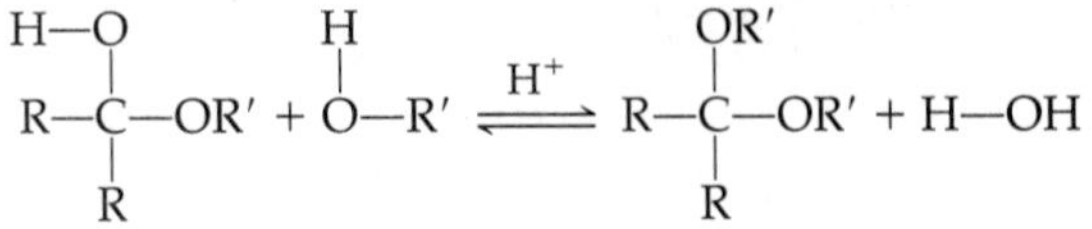

ketal

Note that both **acetals** and **ketals** have two alkoxy groups (—OR′) attached to the same carbon atom. An acetal also has a hydrogen atom and an alkyl group attached to the carbon atom, whereas the ketal has two alkyl groups attached. The formation of an acetal or a ketal requires two molar equivalents of alcohol per mole of the original carbonyl compound.

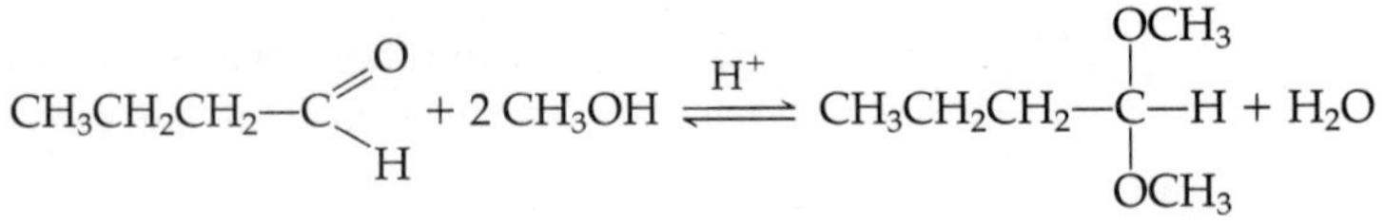

$$CH_3CH_2CH_2\text{—CHO} + 2\,CH_3OH \overset{H^+}{\rightleftharpoons} CH_3CH_2CH_2\text{—C(OCH}_3)_2\text{—H} + H_2O$$

The acetal shown in the above reaction is an acyclic compound. However, some hemiacetals and hemiketals have cyclic structures. Cyclic hemiacetals or hemiketals react with alcohols to produce cyclic acetals or ketals. The cyclic hemiacetal from 5-hydroxypentanal is shown below. Its ring oxygen atom was originally the 5-hydroxyl oxygen atom. When this cyclic hemiacetal reacts with an alcohol, R′—OH, the product is a cyclic acetal. The oxygen atom in the —OR′ group of the acetal originated in the alcohol R′—OH. We will see this reaction again when we consider carbohydrates in Chapter 11.

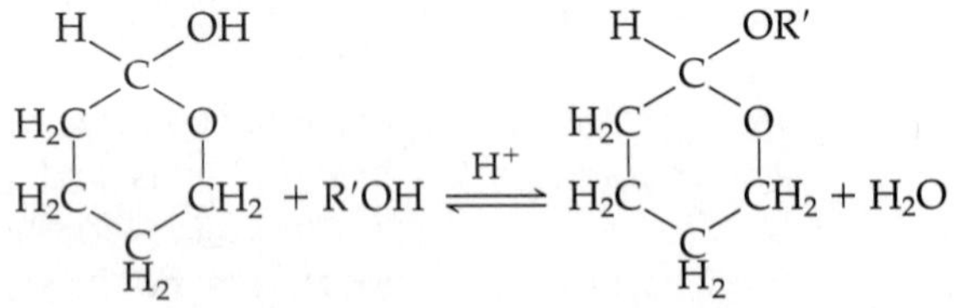

$$\text{cyclic hemiacetal (ring: C(H)(OH)—O—CH}_2\text{—CH}_2\text{—H}_2\text{C—H}_2\text{C)} + \text{R}'\text{OH} \overset{H^+}{\rightleftharpoons} \text{cyclic acetal (ring: C(H)(OR}')\text{—O—CH}_2\text{—CH}_2\text{—H}_2\text{C—H}_2\text{C)} + H_2O$$

cyclic hemiacetal   cyclic acetal

Reactivity of Acetals and Ketals

The conversion of a hemiacetal to an acetal and the conversion of a hemiketal to a ketal are reversible in acid solution. The position of the equilibria can be shifted to the right—that is, toward formation of an acetal or ketal—by removing the water formed in the reaction or by increasing the amount of the alcohol.

$$\text{hemiacetal} + \text{alcohol} \rightleftharpoons \text{acetal} + \text{water}$$

Adding alcohol "pushes" equilibrium to the right.

Removing water "pulls" equilibrium to the right.

The reverse reaction, the acid-catalyzed hydrolysis of acetals or ketals, is favored when water is added. Acetals and ketals react with water in a hydrolysis reaction to give a carbonyl compound and the alcohol. However, acetals and ketals do not react in neutral or basic solution.

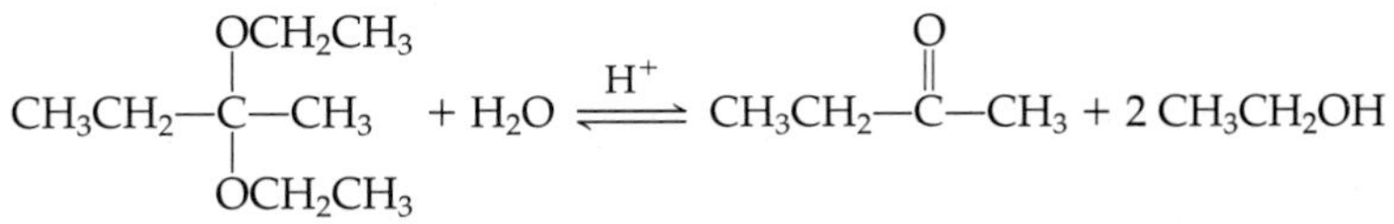

$$CH_3CH_2—C(OCH_2CH_3)_2—CH_3 + H_2O \underset{}{\overset{H^+}{\rightleftharpoons}} CH_3CH_2—C(=O)—CH_3 + 2\ CH_3CH_2OH$$

EXAMPLE 10.6 Identify the class to which each of the following compounds belongs.

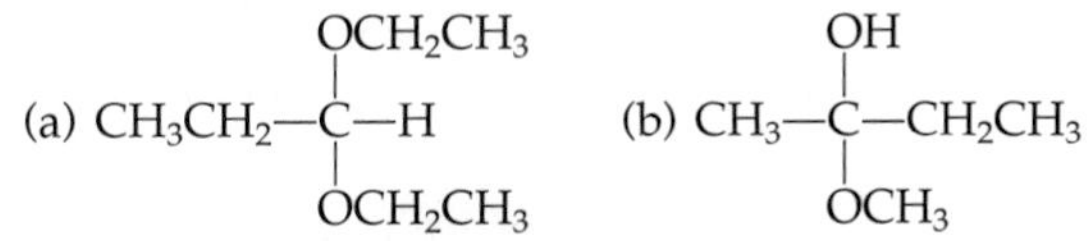

(a) $CH_3CH_2—CH(OCH_2CH_3)_2$ (central C bonded to two OCH_2CH_3 groups and H)

(b) $CH_3—C(OH)(OCH_3)—CH_2CH_3$

Solution A carbon atom in (a) is bonded to two $—OCH_2CH_3$ groups, an alkyl group, and a hydrogen atom. Compound (a) is therefore an acetal made from propanal and ethanol. In (b) a carbon atom is linked to an $—OCH_3$ group, an —OH group, and two alkyl groups. Compound (b) is a hemiketal that can exist in equilibrium with 2-butanone and methanol.

Mechanism of Acetal and Ketal Formation

The conversion of hemiacetals and hemiketals to acetals and ketals occurs in four reversible, acid-catalyzed steps. These steps are shown below for the conversion of a hemiketal to a ketal. In step 1, the acid protonates the oxygen atom of the hydroxyl group. In step 2, water leaves and a carbocation forms. In step 3, the carbocation (a Lewis acid) combines with the alcohol (a Lewis base). In step 4, the proton bonded to the oxygen atom is lost to give a ketal. Note that H^+ is a catalyst: it starts the reaction by protonating the hemiketal and is regenerated in the last step when the ketal forms.

$$\mathrm{R{-}\underset{R}{\overset{R'-\ddot{O}:}{C}}{-}\ddot{O}{-}H + H^+ \xrightleftharpoons{\text{step 1}} R{-}\underset{R}{\overset{R'-\ddot{O}:}{C}}{-}\overset{H}{\ddot{O}^+}{-}H}$$

$$\mathrm{R{-}\underset{R}{\overset{R'-\ddot{O}:}{C}}{-}\overset{H}{\ddot{O}^+}{-}H \xrightleftharpoons{\text{step 2}} R{-}\underset{R}{\overset{R'-\ddot{O}:}{C^+}} + :\overset{H}{\ddot{O}}{-}H}$$

$$\mathrm{R{-}\underset{R}{\overset{R'-\ddot{O}:}{C^+}} + :\overset{H}{\ddot{O}}{-}R' \xrightleftharpoons{\text{step 3}} R{-}\underset{R}{\overset{R'-\ddot{O}:}{C}}{-}\overset{H}{\ddot{O}^+}{-}R'}$$

$$\mathrm{R{-}\underset{R}{\overset{R'-\ddot{O}:}{C}}{-}\overset{H}{\ddot{O}^+}{-}R' \xrightleftharpoons{\text{step 4}} R{-}\underset{R}{\overset{R'-\ddot{O}:}{C}}{-}\ddot{O}{-}R' + H^+}$$

10.9 ADDITION OF NITROGEN COMPOUNDS

The carbonyl groups of aldehydes and ketones react with nucleophiles that contain nitrogen, that is, they react with ammonia, NH_3, and with amines of the general formula RNH_2 (Chapter 14). In these reactions, the carbonyl carbon atom becomes bonded to the nitrogen atom by a double bond. The reaction of a carbonyl compound with an amine occurs in two steps called an **addition–elimination reaction.** In the addition step, the nitrogen atom bonds to the carbonyl carbon atom and a hydrogen atom from the amine bonds to the carbonyl oxygen atom. This step therefore resembles the addition of an alcohol to a carbonyl compound. The initial addition product loses a molecule of water in an elimination reaction to give an **imine.**

$$\mathrm{C_6H_5{-}\overset{O}{\overset{\|}{C}}H + CH_3NH_2 \xrightarrow[\text{reaction}]{\text{addition}} C_6H_5{-}\overset{OH}{\overset{|}{C}}H{-}\overset{H}{\overset{|}{N}}CH_3 \xrightarrow[\text{reaction}]{\text{elimination}} \underset{\text{an imine}}{C_6H_5{-}CH{=}NCH_3} + H_2O}$$

The net result of reacting an aldehyde or ketone with GNH_2, where G represents any group, is the replacement of the carbonyl oxygen atom with G—N=.

$$\mathrm{GNH_2 + \underset{\substack{\text{aldehyde}\\ \text{or ketone}}}{R\overset{O}{\overset{\|}{C}}R'} \rightleftharpoons R\overset{NG}{\overset{\|}{C}}R' + H_2O}$$

Addition Reactions and Vision

We learned at our mothers' knees that "carrots are good for us". This homely injunction is true because carrots contain β-carotene, which is needed by mammals for vision. β-Carotene is a pigment that is largely responsible for the color of carrots. Persons who do not have adequate β-carotene in their diets—it can be obtained from egg yolk, liver, and various fruits and vegetables in addition to carrots—suffer from *night blindness*. Mammals have a liver enzyme system that splits β-carotene in half to give two molecules of an aldehyde named retinal. Retinal has a series of alternating single and double bonds. Geometric isomers can exist about each of the double bonds. The all-trans compound and the isomer with a cis orientation about the C-11 and C-12 atoms play an important role in vision. *cis*-11-Retinal undergoes an addition reaction with a protein in the retina called opsin to form a substance called rhodopsin. The aldehyde group of *cis*-11-retinal reacts with a specific amino group in the protein to form an imine. The shape of the imine adduct of *cis*-11-retinal allows it to "fit" into the protein.

Rhodopsin is a visual receptor in the retina that absorbs visible light. When light strikes rhodopsin, the cis double bond at C-11 is isomerized to a trans double bond, a process called photoisomerization. The resulting all-trans isomer no longer fits into the opsin; the imine spontaneously hydrolyzes, and the all-trans retinal is released from opsin. This process occurs in about one millisecond. During that time a nerve impulse is generated and travels to the brain, where it is translated into a visual image.

If *cis*-11-retinal cannot bind opsin to give rhodopsin, vision is impaired. We recall from earlier discussions that formaldehyde, which is produced by the oxidation of methyl alcohol, can cause blindness. Blindness occurs because formaldehyde competes with *cis*-11-retinal for the reactive amine group of opsin. If no rhodopsin is formed, then no "light-induced" messages will get to the brain.

β-carotene

1 2 3 4 5 6 7 15 CHO

retinal

11 CHO

cis-11-retinal

11 CHO + opsin—NH_2 $\xrightarrow{H^+}$ 11 H C N—opsin H +

cis-11-retinal

rhodopsin

Note that the formation of an imine is accompanied by the release of water as a product. Imines can be isolated, but they react with water in the reverse of the above two reactions to produce the original carbonyl compound and the nitrogen compound. Thus, one way to isolate an imine is to remove water from the solution as the imine forms.

Aldehydes and ketones react with some amines to give stable imines, which are crystalline substances. Thus, the addition–elimination reactions of certain nitrogen compounds allow us to identify aldehydes and ketones. For example, 2,4-dinitrophenylhydrazine forms a stable adduct with many aldehydes and ketones.

$$O_2N{-}C_6H_3(NO_2){-}NHNH_2 + CH_3\overset{\overset{\displaystyle O}{\|}}{C}CH_3 \longrightarrow O_2N{-}C_6H_3(NO_2){-}NHN{=}C(CH_3)_2 + H_2O$$

When this reagent reacts with an aldehyde or ketone, the product is a bright yellow-to-orange crystalline solid called a 2,4-dinitrophenylhydrazone (2,4-DNP). Its melting point is used to identify the original aldehyde or ketone. The melting point of the 2,4-DNP of an "unknown" carbonyl compound is compared to those of the 2,4-DNPs of known carbonyl compounds.

10.10 REACTIVITY OF THE α-CARBON ATOM

Up to this point, we have seen several characteristic reactions of the carbonyl group. However, the carbonyl group itself is not the only reactive site of carbonyl compounds. Many important reactions occur at the carbon atom directly attached to the carbonyl carbon atom. This carbon atom is called the **α-carbon atom.**

α-carbon atom → $-\overset{|}{\underset{|}{C}}-\overset{\overset{\displaystyle O}{\|}}{C}-\overset{\overset{\displaystyle H}{|}}{\underset{|}{C}}-$ ← α-hydrogen atom; ← α-carbon atom

Acidity of α-Hydrogen Atoms

We noted earlier that the carbonyl carbon atom has a partial positive charge and that it attracts electrons in neighboring bonds by an inductive effect. As a result, the α-carbon atom loses electron density and acquires a partial positive charge. This effect is transmitted in turn to the bonds holding hydrogen atoms to the α-carbon atom. Thus, an **α-hydrogen atom** is more acidic than a hydrogen atom in a C—H bond of a hydrocarbon. This difference in acidity is reflected in the pK_a values. The pK_a of ethane is about 50, whereas that of acetone is about 20.

$$CH_3CH_3 + H_2O \rightleftharpoons CH_3CH_2^- + H_3O^+ \qquad K_a = 10^{-50}$$

$$CH_3\overset{O}{\overset{\|}{C}}CH_3 + H_2O \rightleftharpoons CH_3\overset{O}{\overset{\|}{C}}CH_2^- + H_3O^+ \qquad K_a = 10^{-20}$$

This tremendous difference in K_a—30 powers of ten—is not just the result of the inductive effect of the carbonyl group. When a carbonyl compound loses its α-hydrogen atom in an acid–base reaction, the resulting anion is stabilized by resonance. The resonance-stabilized anion is called an **enolate anion.** One of its contributing resonance structures has a negative charge on the oxygen atom, the other has a negative charge on the carbon atom. Because the charge on the anion is delocalized, an enolate anion is more stable than a carbanion, such as $CH_3CH_2^-$, in which no such resonance stabilization is possible.

$$R\overset{-}{C}H-\overset{:\ddot{O}}{\overset{\|}{C}}R' \longleftrightarrow RCH=\overset{:\ddot{O}:^-}{\overset{|}{C}}R'$$

resonance structures of the conjugate base

Keto–Enol Equilibria

The acidity of the α-hydrogen atoms of carbonyl compounds has another consequence: both aldehydes and ketones exist as an equilibrium mixture of compounds called the **keto** and the **enol** forms.

$$\underset{\substack{\text{acetone}\\ \text{(keto form)}}}{CH_3\overset{O}{\overset{\|}{C}}CH_3} \rightleftharpoons \underset{\substack{\text{propen-2-ol}\\ \text{(enol form)}}}{CH_2=\overset{OH}{\overset{|}{C}}CH_3}$$

A hydrogen atom bonded to an α-carbon atom is removed and transferred to the carbonyl oxygen atom. The carbon–oxygen double bond becomes a single bond, and the carbon–carbon single bond becomes a double bond. The isomeric keto and enol forms are structural isomers called **tautomers.** They differ in the location of a hydrogen atom and a double bond. For most aldehydes and ketones, the keto form predominates, and less than 1% of the mixture is the isomeric enol form. Nevertheless, the enol form is often responsible for the reactivity of carbonyl compounds.

Tautomerism is important in the chemistry of carbohydrates (Chapter 11) and in the metabolism of these compounds. For example, one of the intermediates in carbohydrate metabolism is a three-carbon compound called dihydroxyacetone phosphate. It undergoes an enzyme-catalyzed isomerization reaction to form another three-carbon molecule called D-glyceraldehyde 3-phosphate. The isomerization reaction occurs by transfer of a hydrogen atom from the α-carbon atom of dihydroxyacetone phosphate to the carbonyl group. The first step of the reaction—tautomerization—

yields an **enediol intermediate.** The second step of the reaction, a second tautomerization, yields D-glyceraldehyde 3-phosphate.

$$\underset{\substack{\text{dihydroxyacetone}\\\text{phosphate}}}{\begin{array}{c}CH_2OH\\|\\C{=}O\\|\\CH_2OPO_3H_2\end{array}} \rightleftharpoons \underset{\substack{\text{enediol}\\\text{intermediate}}}{\begin{array}{c}H{-}C{-}OH\\\|\\C{-}OH\\|\\CH_2OPO_3H_2\end{array}} \rightleftharpoons \underset{\substack{\text{D-glyceraldehyde}\\\text{3-phosphate}}}{\begin{array}{c}{}^{1}CHO\\|\\H{-}{}^{2}C{-}OH\\|\\{}^{3}CH_2OPO_3H_2\end{array}}$$

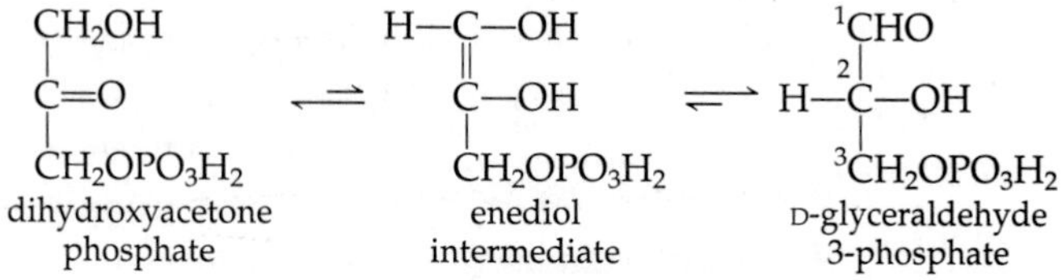

Isomerization occurs because the enediol intermediate is in equilibrium with dihydroxyacetone phosphate and D-glyceraldehyde 3-phosphate. Similar isomerization reactions occur in many enzyme-catalyzed reactions of carbohydrates.

10.11 THE ALDOL CONDENSATION

The acidity of the α-hydrogen atoms of carbonyl compounds makes it possible for two carbonyl compounds to react with one another to give a product that is both an aldehyde and an alcohol. Hence, the product is called an **aldol,** and the reaction is called an **aldol condensation.** This reaction is base-catalyzed.

In an aldol condensation, a new carbon–carbon bond is formed between the α-carbon atom of one carbonyl compound and the carbonyl carbon atom of the other one. Note that the product has just one carbon atom between the aldehyde and alcohol carbon atoms.

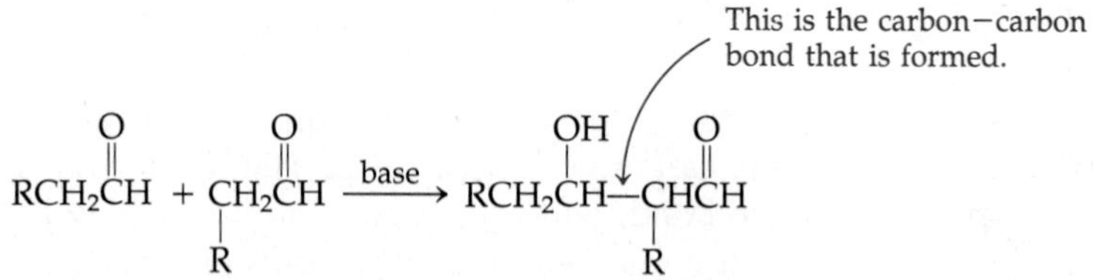

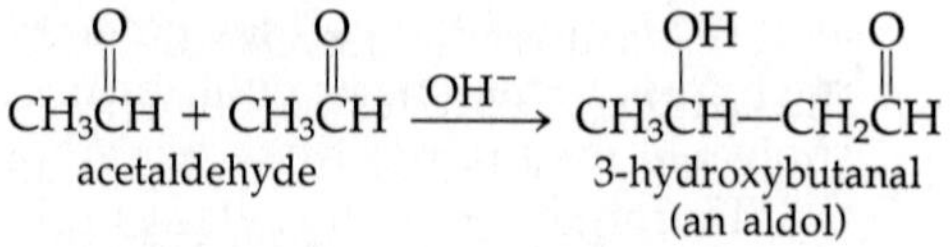

This is the carbon–carbon bond that is formed.

$$RCH_2\overset{O}{\overset{\|}{C}}H + \underset{R}{\underset{|}{C}}H_2\overset{O}{\overset{\|}{C}}H \xrightarrow{\text{base}} RCH_2\overset{OH}{\overset{|}{C}}H{-}\underset{R}{\underset{|}{C}}H\overset{O}{\overset{\|}{C}}H$$

$$\underset{\text{acetaldehyde}}{CH_3\overset{O}{\overset{\|}{C}}H + CH_3\overset{O}{\overset{\|}{C}}H} \xrightarrow{OH^-} \underset{\substack{\text{3-hydroxybutanal}\\\text{(an aldol)}}}{CH_3\overset{OH}{\overset{|}{C}}H{-}CH_2\overset{O}{\overset{\|}{C}}H}$$

The aldol condensation occurs at room temperature when an aldehyde is treated with an aqueous solution of sodium hydroxide. The reaction occurs in a three-step mechanism.

1. One aldehyde molecule reacts with base (OH^-) at its α-C—H bond to give a nucleophilic enolate anion.

2. The nucleophilic enolate anion reacts with the carbonyl carbon atom of another aldehyde molecule. The product is the alkoxide anion of an aldol.
3. The alkoxide anion extracts a proton from the solvent, water and regenerates a hydroxide ion.

The sequence of steps is shown for acetaldehyde, but the same reactions occur for any other aldehyde with α-hydrogen atoms.

$$CH_3{-}\overset{\overset{\ddot{O}:}{\|}}{C}{-}H + OH^- \rightleftharpoons :\overset{-}{C}H_2{-}\overset{\overset{\ddot{O}:}{\|}}{C}H + HOH$$

enolate anion

$$CH_3{-}\overset{\overset{\ddot{O}:}{\|}}{C}H + :\overset{-}{C}H_2{-}\overset{\overset{\ddot{O}:}{\|}}{C}H \rightleftharpoons CH_3\overset{\overset{:\ddot{O}:^-}{|}}{C}H{-}CH_2\overset{\overset{\ddot{O}:}{\|}}{C}H$$

nucleophile — an alkoxide ion

$$CH_3\overset{\overset{:\ddot{O}:^-}{|}}{C}H{-}CH_2\overset{\overset{\ddot{O}:}{\|}}{C}H + HOH \rightleftharpoons CH_3\overset{\overset{:\ddot{O}H}{|}}{C}H{-}CH_2\overset{\overset{\ddot{O}:}{\|}}{C}H + OH^-$$

aldol

Aldol products easily dehydrate to give a double bond in conjugation with the carbonyl group. Dehydration of the above aldol followed by reduction gives 1-butanol in a commercial process.

$$CH_3\overset{\overset{OH}{|}}{C}HCH_2\overset{\overset{O}{\|}}{C}H \xrightarrow[-H_2O]{H^+} CH_3CH{=}CH\overset{\overset{O}{\|}}{C}H \xrightarrow[\text{catalyst}]{H_2} CH_3CH_2CH_2CH_2OH$$

1-butanol

Mixed Aldol Condensation

In the aldol condensation reaction we described above, the same aldehyde provided the enolate and the substrate attacked by the enolate anion. But, if two different aldehydes are mixed and heated in a basic solution, the enolate anion of one can react with the carbonyl form of the other. Mixtures of products can result because any two aldehydes can react with each other. Thus, if the aldehydes are A_1 and A_2, aldol condensation can produce A_1–A_1, A_2–A_2, A_1–A_2, and A_2–A_1. In short, a dreadful mixture is produced. This unhappy outcome can be avoided if one of the aldehydes does not have any α-hydrogen atoms and if one of the aldehydes is less reactive than the other. For example, an aldol condensation occurs between benzaldehyde and acetaldehyde to give a high yield of a single aldol product. Reaction between two benzaldehyde molecules cannot occur because benzaldehyde does not have any α-hydrogen atoms. Therefore, benzaldehyde is mixed with a base, and acetaldehyde is then slowly added. The acetaldehyde is rapidly converted to an enolate anion. Because benzaldehyde is more reac-

FIGURE 10.4

Biological Aldol Reaction to Form Fructose
The enzyme aldolase catalyzes the aldol condensation of dihydroxyacetone phosphate and D-glyceraldehyde 3-phosphate.

$_1CH_2OPO_3^{2-}$ / $_2C{=}O$ / $_3CH_2OH$ dihydroxyacetone phosphate + H–$_4$C(=O) / H–$_5$C–OH / $_6CH_2OPO_3^{2-}$ D-glyceraldehyde 3-phosphate $\xrightleftharpoons{\text{aldolase}}$ $_1CH_2OPO_3^{2-}$ / $_2C{=}O$ / HO–$_3$C–H / H–$_4$C–OH / H–$_5$C–OH / $_6CH_2OPO_3^{2-}$ D-fructose 1,6-bisphosphate

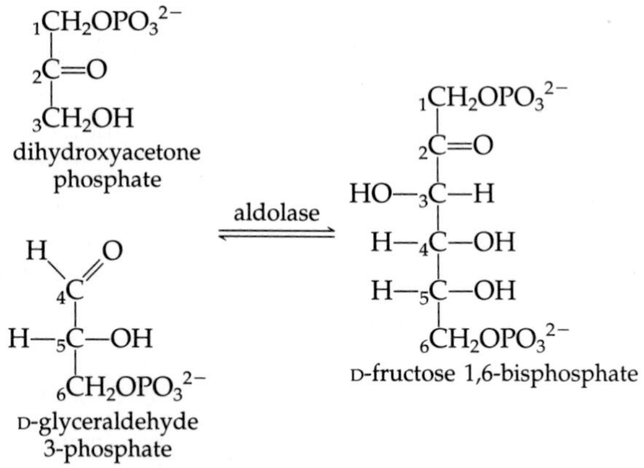

tive than acetaldehyde and the concentration of free acetaldehyde in the mixture is much smaller than that of benzaldehyde, the enolate of acetaldehyde reacts with benzaldehyde. Only one product is obtained. Because the product is derived from two different aldehydes, the reaction is called a **mixed aldol condensation.**

$$C_6H_5CHO + CH_3CHO \xrightarrow{OH^-} C_6H_5CH(OH)CH_2CHO$$

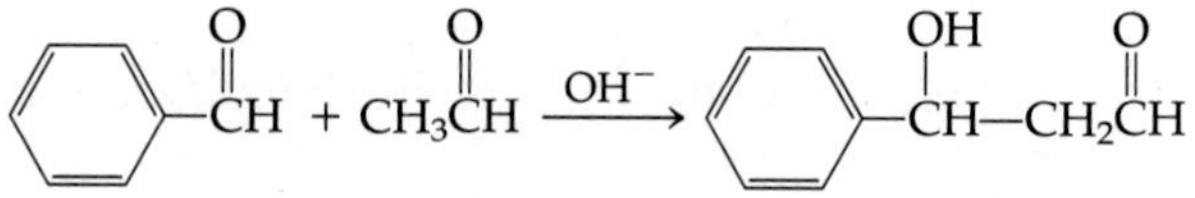

a mixed aldol

Mixed aldol condensations are important in metabolism. For example, an aldol condensation occurs in the synthesis of glucose from three-carbon precursors. In this reaction, which is catalyzed by the enzyme aldolase, the enolate anion derived from dihydroxyacetone phosphate reacts with D-glyceraldehyde to give fructose 1,6-bisphosphate (Figure 10.4). This molecule is subsequently converted to glucose.

EXAMPLE 10.7 Draw the product of the aldol condensation of propanal.

Solution First draw the structural formula of propanal with the carbonyl group on the right.

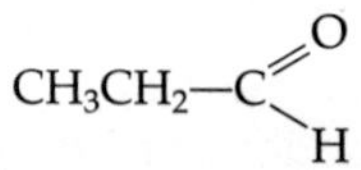

$$CH_3CH_2{-}C({=}O)H$$

Next, draw a second structural formula of propanal with the α-carbon atom near the carbonyl carbon atom of the first structure. Place the α-hydrogen atom of the second structure so that it is close to the carbonyl oxygen atom of the first molecule.

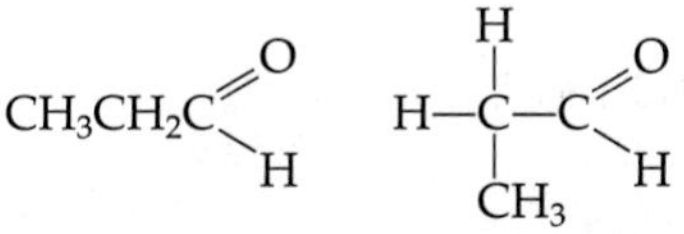

$$CH_3CH_2C(=O)H \qquad H-\underset{CH_3}{\overset{H}{C}}-C(=O)H$$

Now form the bond between the α-carbon atom of the right molecule and the carbonyl carbon atom of the left molecule.

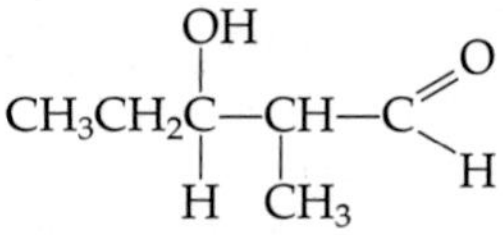

$$CH_3CH_2\underset{H}{\overset{OH}{C}}-\underset{CH_3}{CH}-C(=O)H$$

EXERCISES

Nomenclature of Aldehydes and Ketones

10.1 Draw the structural formula for each of the following compounds.
(a) 2-methylbutanal (b) 3-ethylpentanal (c) 2-bromopentanal
(d) 3,4-dimethyloctanal

10.2 Draw the structural formula for each of the following compounds.
(a) 3-bromo-2-pentanone (b) 2,4-dimethyl-3-pentanone
(c) 4-methyl-2-pentanone (d) 3,4-dimethyl-2-pentanone

10.3 Give the IUPAC name for each of the following compounds.

(a) $CH_3CH_2CH_2CHO$ (b) $CH_3\underset{CH_3}{\overset{CH_3}{C}}CH_2CHO$

(c) $CH_3\overset{CH_3}{CH}CHO$ (d) $CH_3CH_2\overset{CH_3}{CH}\underset{CH_2CH_3}{CH}CHO$

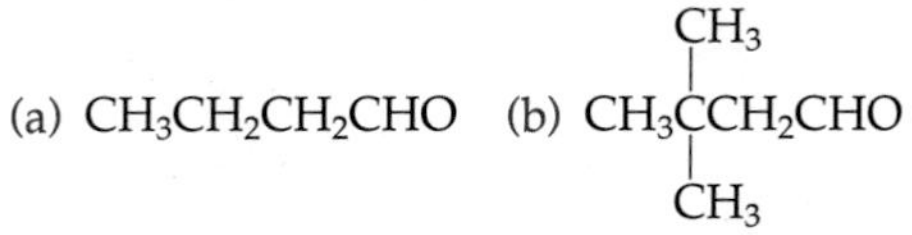

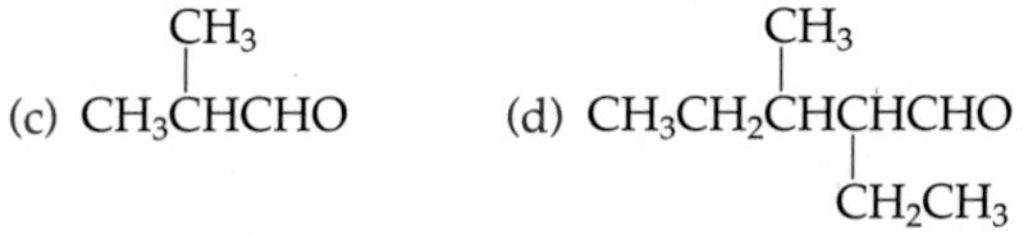

10.4 Give the IUPAC name for each of the following compounds.

(a) $CH_3CH_2\overset{O}{\overset{\|}{C}}CH_2CH_3$ (b) $CH_3\overset{O}{\overset{\|}{C}}-\underset{CH_3}{\overset{CH_3}{C}}CH_3$

(c) $CH_3\underset{CH_3}{CH}\overset{O}{\overset{\|}{C}}CH_2CH_3$ (d) $CH_3\overset{CH_3}{CH}CH_2\overset{O}{\overset{\|}{C}}CH_2CH_3$

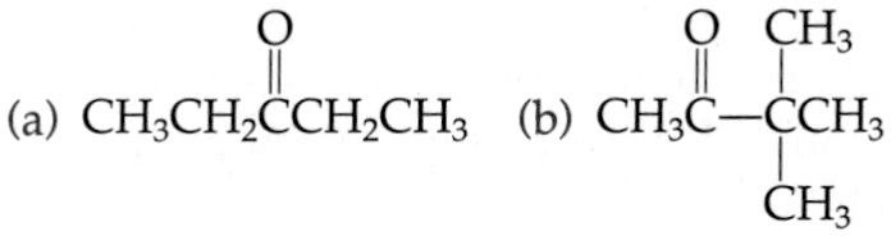

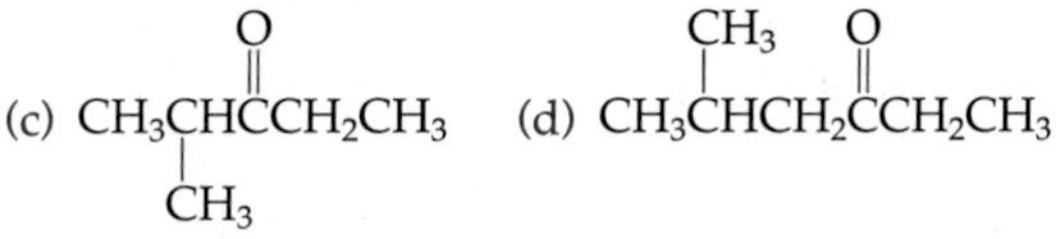

10.5 Give the IUPAC name for each of the following compounds.

(a) Cl C—H O

(b) O

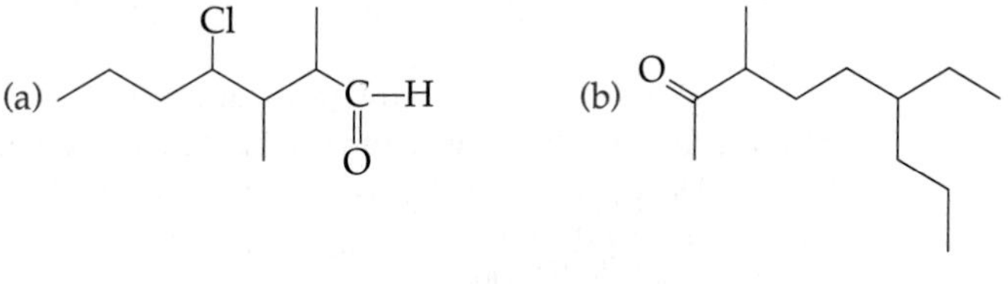

(c) C—H, O (d) O

10.6 Give the IUPAC name for each of the following compounds.

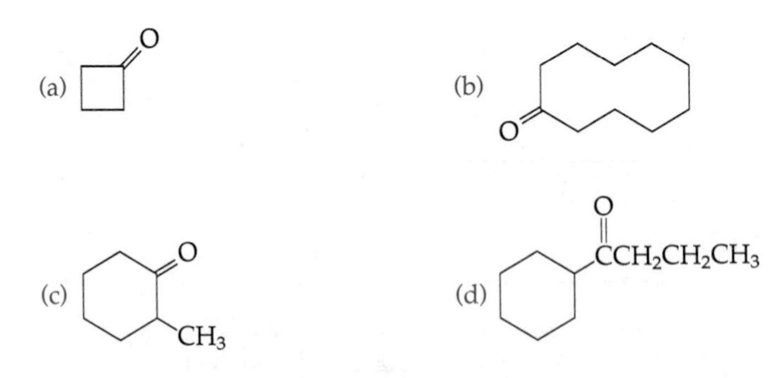

10.7 Draw the product of ozonolysis (Section 4.7) obtained from 1-methylcyclohexene. Name the compound.

10.8 Draw the products of ozonolysis of vitamin K_1. Name the carbonyl compound derived from the side chain.

O, CH_3, O, CH_3, CH_3, CH_3, CH_3, CH_3

Properties of Aldehydes and Ketones

10.9 The H—C—H bond angle of formaldehyde is 116.5°. The H—C—C bond angle of acetaldehyde is 117.2°. Explain this difference.

10.10 The C=C bond length in alkenes and the C=O bond length in aldehydes are 1.34 and 1.23 Å, respectively. Explain this difference.

10.11 The dipole moments of 1-butene and butanal are 0.34 and 2.52 D, respectively. Explain this difference.

10.12 The dipole moments of acetone and isopropyl alcohol are 2.7 and 1.7 D, respectively. Explain this difference.

10.13 The boiling points of butanal and 2-methylpropanal are 75 and 61 °C, respectively. Explain this difference.

10.14 The boiling points of 2-heptanone, 3-heptanone, and 4-heptanone are 151, 147, and 144 °C, respectively. What is responsible for this trend?

10.15 The solubilities of butanal and 1-butanol in water are 7 and 9 g/100 mL, respectively. Explain this difference.

10.16 The solubilities of butanal and 2-methylpropanal in water are 7 and 11 g/100 mL, respectively. Explain this difference.

Oxidation and Reduction

10.17 What is observed when an aldehyde reacts with Benedict's solution? What is observed when an aldehyde reacts with Tollens's reagent?

10.18 What class of compounds results from the reduction of ketones with sodium borohydride? What class of compounds results from the reduction of aldehydes with lithium aluminum hydride?

10.19 Draw the structure of the product of the following reaction.

OCH_3, HO, CHO; $\xrightarrow{Ag(NH_3)_2^+}$

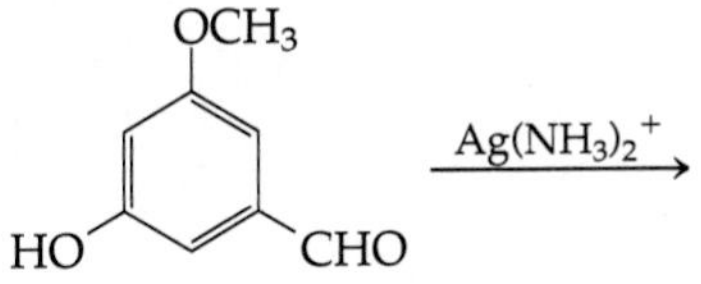

10.20 Draw the structure of the product of the following reaction.

O, CHO; $\xrightarrow{Cu^{2+}}$

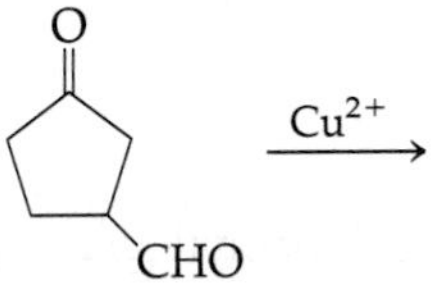

10.21 What is the product when each of the following reacts with lithium aluminum hydride?

(a) $C_6H_5C(=O)CH_3$ (b) cyclohexanone (c) $C_6H_{11}C(=O)H$

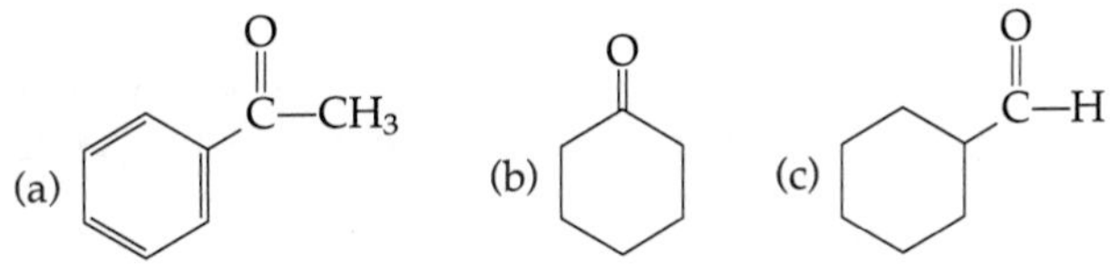

10.22 What is the product when each of the following reacts with sodium borohydride?

(a) $CH_3CH(CH_3)CHO$ (b) $CH_3C(CH_3)_2CH_2CHO$ (c) $CH_3C(=O)C(CH_3)_2CH_3$

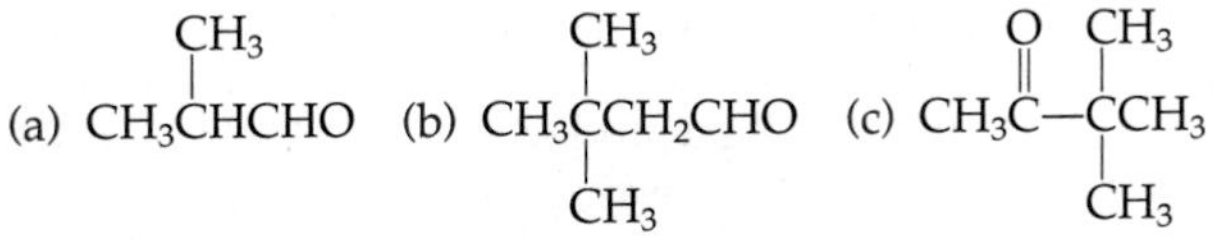

10.23 The reduction of carvone by lithium aluminum hydride yields two products. Explain why.

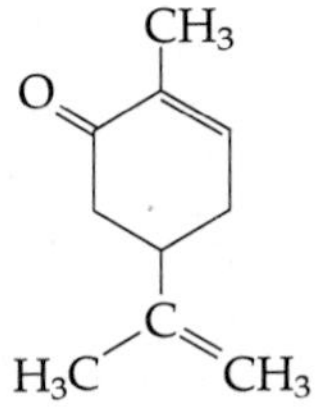

CH_3, O, C, H_3C, CH_2

10.24 The reduction of the following compound by sodium borohydride yields two products. Explain why.

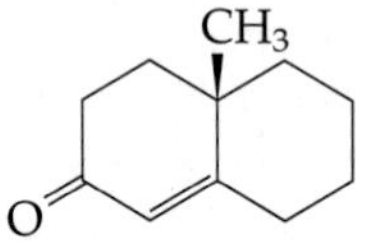

CH_3

O

10.25 Reduction by sodium borohydride of the aldol product of butanal produces a mosquito repellent. Write its structure and name it.

10.26 What is the product of the reaction of the following compound with sodium borohydride?

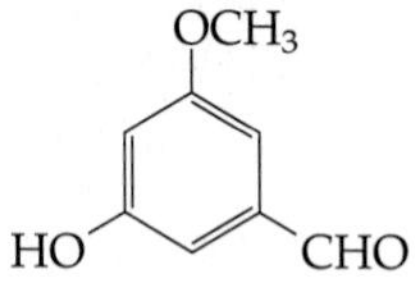

OCH_3

HO CHO

Addition Reactions

10.27 Formaldehyde has been used to disinfect rooms and surgical instruments. Why is this compound so effective compared to other carbonyl compounds?

10.28 Glutaraldehyde is used in a sterilizing solution for instruments that cannot be heated in an autoclave. Explain its action in sterilizing objects.

$$H-\overset{\overset{\displaystyle O}{\|}}{C}-CH_2CH_2CH_2-\overset{\overset{\displaystyle O}{\|}}{C}-H$$

10.29 The equilibrium constants for the hydration of ethanal and 2,2-dimethylpropanal are 1.8×10^{-2} and 4.1×10^{-3}, respectively. Explain this difference.

10.30 The equilibrium constants for the hydration of acetone and hexafluoroacetone are 2.5×10^{-5} and $2.2 \times 10^{+4}$, respectively. Explain this difference.

10.31 Hydrogen cyanide, $H-C{\equiv}N:$, reacts with aldehydes under basic conditions to give an addition product. Write the mechanism of the reaction.

10.32 Hydrogen cyanide, $H-C{\equiv}N:$, reacts with 2-propanone to give a good yield of an addition product but 2,2,4,4-tetramethyl-3-pentanone gives a poor yield in the same reaction. Why?

Grignard Reactions

10.33 What carbonyl compound and Grignard reagent are required to produce each of the following compounds?

(a) $CH_3CH_2\overset{OH}{\overset{|}{C}}HCH_2CH_3$ (b) $CH_3\overset{OH}{\overset{|}{C}}H-\underset{\underset{CH_3}{|}}{C}HCH_3$

(c) $CH_3\underset{\underset{CH_3}{|}}{C}H\overset{OH}{\overset{|}{C}}HCH_2CH_3$ (d) $CH_3\overset{CH_3}{\overset{|}{C}}HCH_2\overset{OH}{\overset{|}{C}}HCH_2CH_3$

10.34 What carbonyl compound and Grignard reagent are required to produce each of the following compounds?

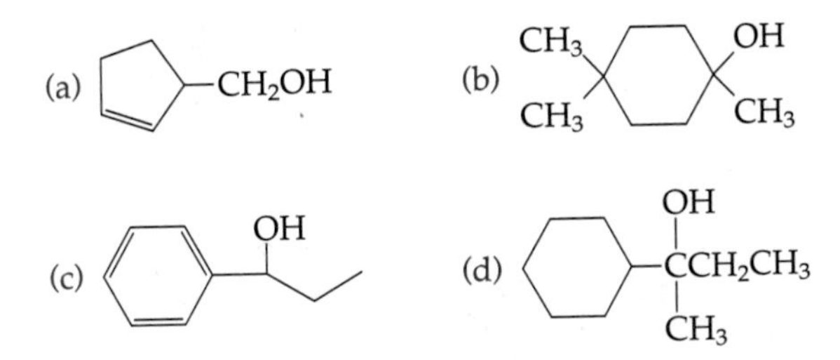

10.35 Propose a two-step sequence of reactions that could be used to convert cyclohexanone into 1-methylcyclohexene.

10.36 Propose a two-step sequence of reactions that could be used to convert 2-bromopentane into the following carboxylic acid.

$$CH_3CH_2CH_2CH(CH_3)CO_2H$$

Addition of Alcohols

10.37 Identify each of the following as a hemiacetal, hemiketal, acetal, or ketal.

(a) $CH_3CH_2CH(OCH_3)_2$ (b) $CH_3CH_2C(OCH_3)_2CH_3$

(c) $CH_3CH(OCH_2CH_3)OCH_2CH_3$ (d) $CH_3CH_2CH(OCH_3)OH$

10.38 Identify each of the following as a hemiacetal, hemiketal, acetal, or ketal.

(a) $CH_3CH_2C(OCH_3)_2CH_3$ (b) $(CH_3O)_2CHCH_2CH(OCH_3)_2$

(c) cyclohexane bearing OCH_3 and $-OH$ on the same carbon (d) cyclohexane bearing OCH_3 and $-OCH_3$ on the same carbon

10.39 Identify each of the following as a hemiacetal, hemiketal, acetal, or ketal.

(a) tetrahydropyran ring (O) with OCH_3 on the carbon adjacent to O (b) tetrahydropyran ring (O) with OH on the carbon adjacent to O

(c) tetrahydropyran ring (O) with $-OH$ and CH_3 on the carbon adjacent to O (d) tetrahydropyran ring (O) with $-OCH_3$ and CH_3 on the carbon adjacent to O

10.40 Identify each of the following as a hemiacetal, hemiketal, acetal, or ketal.

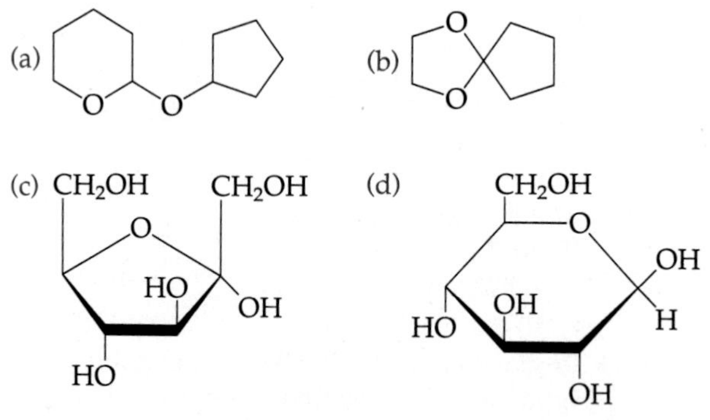

(a) (b)

(c) CH_2OH CH_2OH HO OH HO (d) CH_2OH OH OH HO H OH

10.41 Identify the functional groups in talaromycin A, a substance found in the fungus that grows in poultry litter.

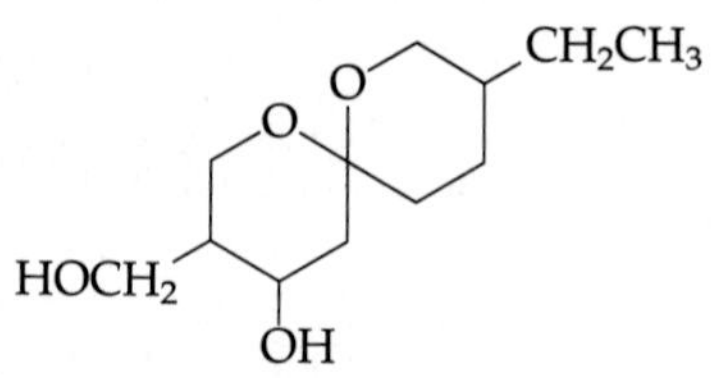

CH_2CH_3 $HOCH_2$ OH

10.42 Identify the functional groups in brevicomin, the sex attractant of the Western pine beetle.

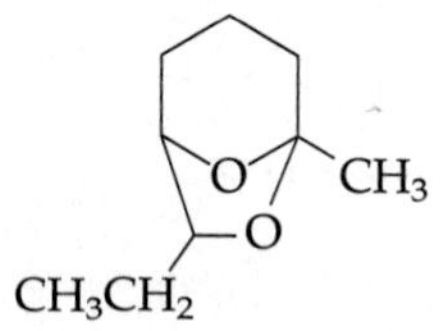

CH_3 CH_3CH_2

10.43 What carbonyl compound and alcohol are required to form the following compound?

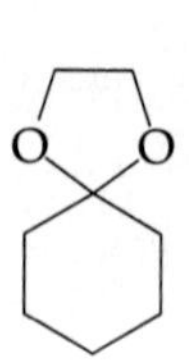

10.44 Reduction of the following compound by the Wolff–Kishner method gives $C_{11}H_{18}O_2$, but reduction by the Clemmensen method gives C_9H_{18}. Explain the difference in the products formed.

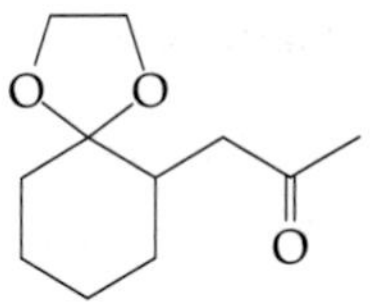

Addition of Nitrogen Compounds

10.45 Write the structure of the product of reaction of ethanal and CH_3NH_2.
10.46 Write the structure of the product of reaction of acetone and $CH_3CH_2NH_2$.
10.47 Write the structure of the product of reaction of cyclohexanone and 2,4-dinitrophenylhydrazine.
10.48 Write the structure of the product of reaction of 3-pentanone and 2,4-dinitrophenylhydrazine.
10.49 What reactants are required to produce the following isomeric imines?

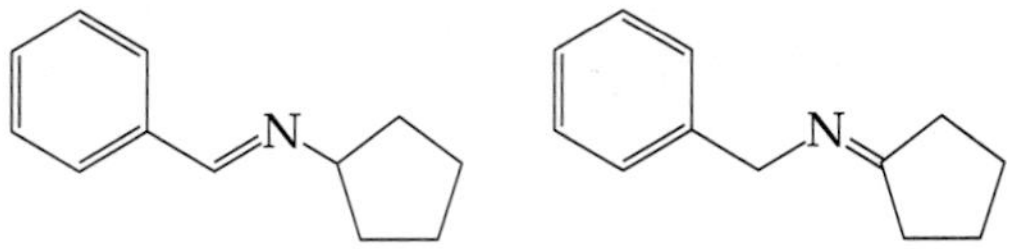

10.50 What reactants are required to produce the following isomeric imines?

N N CH_3

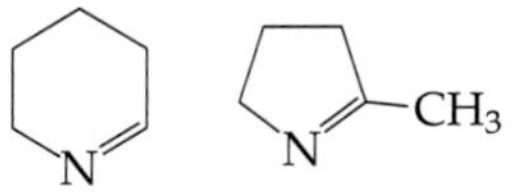

Chemistry of the α-Carbon Atom

10.51 Formaldehyde cannot exist as an enol. Why?
10.52 2,2-Dimethylpropanal cannot exist as an enol. Why?
10.53 Write the enol form of each of the following compounds.
(a) acetone (b) butanal (c) cyclohexanone (d) acetaldehyde
10.54 There are three possible enols of 2-butanone. Explain why and write the structures.
10.55 Draw the keto structure for the following enol.

CH_3 OH

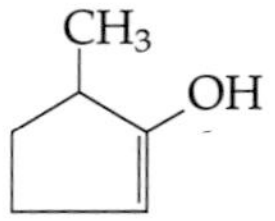

10.56 Draw the keto structure for the following enol.

OH C CH_3

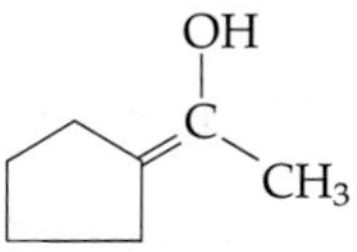

10.57 Compound I reacts with Benedict's solution, but II does not. Explain why.

$$CH_3CH_2\overset{\overset{\large O}{\|}}{C}CH_2OH \qquad CH_3\overset{\overset{\large O}{\|}}{C}-\overset{\overset{\large OH}{|}}{C}HCH_3$$

(I) (II)

10.58 Compound I reacts with Benedict's solution, but II does not. Explain why.

(I): cyclohexyl–$\overset{\overset{\large O}{\|}}{C}CH_2OH$ (II): 2-oxocyclohexyl–CH_2OH

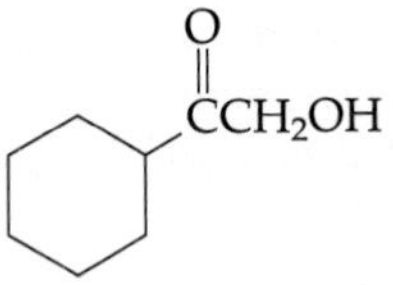

(I) 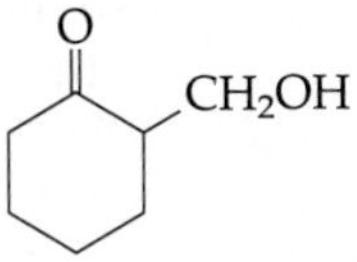 (II)

Aldol Condensation

10.59 Draw the structure of the aldol product made by reacting propanal with a base.

10.60 Draw the structure of the aldol product made by reacting phenylethanal with a base.

10.61 Acetone will form a mixed aldol with benzaldehyde. Name the product.

10.62 Acetophenone will form a mixed aldol with formaldehyde. Name the product.

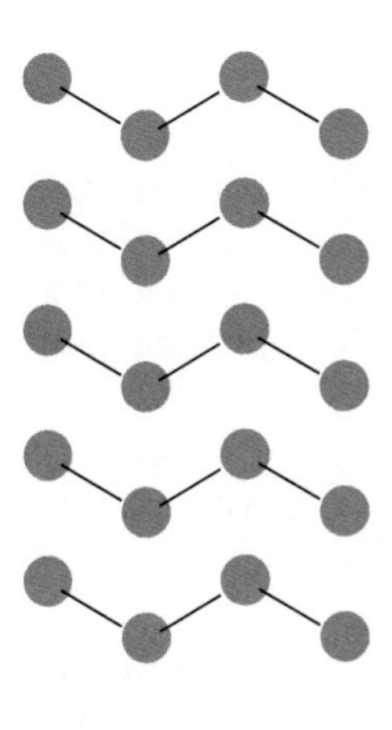

11

CARBOHYDRATES

11.1 CARBOHYDRATES AND ENERGY

If importance were measured by abundance, the carbohydrates would hold first prize, for they are far and away the most abundant molecules in the biological world (excepting only water). The generic name **carbohydrates** refers to molecules whose empirical formulas are $(CH_2O)_n$. Molecules with this general formula span a wide range of structures—from giant molecules containing thousands of sugar units to single molecules containing three carbon atoms. Carbohydrates are polyhydroxy aldehydes or ketones or compounds that can be hydrolyzed to form them. The functions of carbohydrates are as varied as their structures. They are a major source of metabolic energy and are important structural components in many cells.

Because carbohydrates are so plentiful in the biological world, they are known by a host of names. Thus, the molecules we call carbohydrates include sugars such as glucose (blood sugar), sucrose (table sugar), starch, and cellulose. Many carbohydrates with very complex structures are found in the roots, stems, and leaves of all plants. These molecules are made in reactions whose ultimate energy source is the sun.

Carbohydrates can be regarded as a form of "stored sunlight". The sun pours about 5×10^{21} kcal of radiant energy onto the Earth each year. One-third of this energy is reflected back into space by dust and clouds, and the rest eventually returns to space as heat. Less than 0.05% of the solar energy reaching the Earth is used for photosynthesis in plants. During photosynthesis, plants convert carbon dioxide and water into glucose and oxygen by a very complex series of reactions. Glucose subsequently is converted into cellulose or starch.

$$6\ CO_2 + 6\ H_2O \xrightarrow{\text{light}} 6\ O_2 + \underset{\text{glucose}}{C_6H_{12}O_6} \longrightarrow \begin{matrix}\text{cellulose}\\ \text{or}\\ \text{starch}\end{matrix} + H_2O$$

The starch stored in plants is a source of metabolic energy for animals. A closely related substance called glycogen is used by animals to store metabolic energy. Although terrestrial plants contribute significantly to our source of carbohydrates, the major photosynthetic source of carbohydrates and oxygen is the phytoplankton of the oceans. They convert about 8×10^{10} pounds of carbon dioxide into carbohydrates and oxygen each year.

Carbohydrates are a major source of metabolic energy for life forms. Solar energy is captured and changed into chemical energy stored in the carbohydrates. Carbon dioxide is converted into organic compounds, which are more reduced than carbon dioxide. These reduced compounds store chemical energy in carbon–hydrogen bonds. Animals eat the reduced carbon compounds and obtain the stored energy by oxidative metabolic reactions.

$$\begin{matrix}\text{cellulose}\\ \text{or}\\ \text{starch}\end{matrix} + H_2O \longrightarrow C_6H_{12}O_6$$

$$C_6H_{12}O_6 + 6\ O_2 \xrightarrow{\text{metabolism}} 6\ CO_2 + 6\ H_2O$$

Humans convert starch into glucose by an enzyme-catalyzed hydrolysis reaction, but we lack the enzyme necessary to hydrolyze cellulose. Some animals such as cows and the ungulates (hoofed animals) have microorganisms in their rumen that hydrolyze cellulose to produce glucose. As a consequence, the energy stored in cellulose is used by ungulates for metabolic energy and is transferred in the food chain to humans. The difference in the structures of starch and cellulose will be discussed in Section 11.10.

11.2 CLASSIFICATION OF CARBOHYDRATES

Carbohydrates can be divided into three large structural classes—monosaccharides, oligosaccharides, and polysaccharides. Carbohydrates that cannot be hydrolyzed into smaller molecules are **monosaccharides.** Examples are glucose and fructose.

Carbohydrates that are made from a "few" monosaccharides—typically 2 to 10 or so—are **oligosaccharides.** Hydrolysis of oligosaccharides may yield identical monosaccharides or two or more different monosaccharides. Oligosaccharides are called **disaccharides, trisaccharides,** and so forth, depending on the number of linked monosaccharide units. The disaccharide lactose, also called "milk sugar", contains one molecule of glucose and one of galactose. Maltose, another disaccharide, contains two glucose units.

Polysaccharides contain thousands of covalently linked monosaccharides. Those that contain only one type of monosaccharide are **homopolysaccharides.** Examples include starch and cellulose made by plants. They yield only glucose when hydrolyzed. Glycogen, found in animals, is also a homopolysaccharide containing glucose. Polysaccharides that contain more than one type of monosaccharide are called **heteropolysaccharides.**

The monosaccharides in oligo- and polysaccharides are linked by acetal or ketal bonds, which are called **glycosidic bonds** in carbohydrate chemistry. These bonds link the aldehyde or ketone site of one monosaccharide and a hydroxyl group of another monosaccharide. Hydrolysis of the glycosidic bonds yields the component monosaccharides.

Monosaccharides can be further classified by their most highly oxidized functional group. Monosaccharides are called **aldoses** if their most highly oxidized functional group is an aldehyde. They are **ketoses** if their most highly oxidized functional group is a ketone group. The suffix *-ose* indicates that a compound is a carbohydrate. The prefix *aldo-* or *keto-* indicates that the compound is an aldehyde or ketone. The number of carbon atoms in an aldose or ketose is indicated by the prefix *tri-*, *tetr-*, *pent-*, and *hex-*. Aldoses are numbered from the carbonyl carbon atom, whereas ketoses are numbered from the end of the carbon chain closer to the carbonyl carbon atom.

$$\begin{array}{c} {}^{1}CHO \\ | \\ {}^{2}CHOH \\ | \\ {}^{3}CHOH \\ | \\ {}^{4}CH_2OH \end{array} \qquad \begin{array}{c} {}^{1}CH_2OH \\ | \\ {}^{2}C{=}O \\ | \\ {}^{3}CHOH \\ | \\ {}^{4}CHOH \\ | \\ {}^{5}CHOH \\ | \\ {}^{6}CH_2OH \end{array}$$

an aldotetrose   a ketohexose

EXAMPLE 11.1 Classify the carbohydrate D-xylulose.

$$\begin{array}{c} CH_2OH \\ | \\ C{=}O \\ | \\ HO{-}C{-}H \\ | \\ H{-}C{-}OH \\ | \\ CH_2OH \end{array}$$

Solution The compound contains five carbon atoms, so it is classified as a pentose. Because the carbonyl group is a ketone, this compound is a ketopentose.

11.3 CHIRALITY OF MONOSACCHARIDES

Monosaccharides are conveniently represented by their Fischer projection formulas (Chapter 6). We recall that in a Fischer projection formula the carbon chain is represented by a vertical line. Groups attached at the top and bottom represent bonds going into the page, and horizontal lines represent bonds coming out of the page. By convention, the carbonyl carbon atom, the most oxidized carbon atom in these compounds, is placed near the "top" in the Fischer projection formula. The simplest aldose, glyceraldehyde, has three carbon atoms, one of which is a chiral center. This aldotriose can exist in two enantiomeric forms. The *R* isomer, which is naturally occurring, rotates light in a clockwise direction.

CHO
H—C—OH
CH_2OH
(*R*)-(+)-glyceraldehyde

CHO
H—|—OH
CH_2OH
Fischer projection

Monosaccharides with multiple chiral centers are represented with the carbon backbone continually pushed back behind the plane of the page. A curve of atoms in a C shape results, with the attached hydrogen atoms and hydroxyl groups pointing out from the backbone of the carbon chain.

carbonyl group at the top

CHO
H—C—OH
HO—C—H
HO—C—H
H—C—OH
CH_2OH
galactose

CHO
H—|—OH
HO—|—H
HO—|—H
H—|—OH
CH_2OH
Fischer projection

carbonyl group near the top

CH_2OH
C=O
HO—C—H
H—C—OH
H—C—OH
CH_2OH
fructose

CH_2OH
=O
HO—|—H
H—|—OH
H—|—OH
CH_2OH
Fischer projection

D and L Series of Monosaccharides

The stereochemistry of each chiral center of a monosaccharide can be assigned by the *R,S* notation. However, late in the 19th century, the German chemist Emil Fischer devised a stereochemical nomenclature that preceded the *R,S* notation. It is still in common use for carbohydrates and amino acids (Chapter 15). In the Fischer stereochemical system, the configurations of all chiral centers are based on their relation to the naturally occurring stereoisomer of glyceraldehyde. This aldotriose has the hydroxyl group on its chiral C-2 atom located on the right in the projection formula. Its chirality is symbolized as D, so the naturally occurring isomer is designated as D-glyceraldehyde. Its enantiomer, called L-glyceraldehyde, has the hydroxyl group on the left at the chiral carbon atom in the Fischer projection formula.

CHO
H—C—OH
CH_2OH
D-glyceraldehyde

CHO
HO—C—H
CH_2OH
L-glyceraldehyde

The configuration of the hydroxyl groups in a monosaccharide relative to one another determines its chemical identity. Thus, in ribose, an aldopentose, the hydroxyl groups at the three chiral carbon atoms are all on the same side in the Fischer projection formula. The name D-ribose has two components: the term D defines the configuration at the chiral carbon atom farthest from the aldehyde; the term *ribo* by itself defines the relative configuration of the three chiral centers at the C-2, C-3, and C-4 atoms. Taken together, the name D-ribose defines the absolute configuration at every chiral center in the molecule. Thus, in D-ribose, the hydroxyl groups at the C-2, C-3, and C-4 atoms are all on the right. In the enantiomer, L-ribose, the hydroxyl groups at the C-2, C-3, and C-4 atoms are all on the left.

Monosaccharides are made in cells from the "building block" D-glyceraldehyde, so nearly all naturally occurring monosaccharides have the same configuration as D-glyceraldehyde at the chiral carbon atom farthest from the carbonyl group. For example, the hydroxyl groups on the C-4 atom of ribose and on the C-5 atoms of glucose and fructose are on the right side in the following projection formulas. Each compound is a member of the D-series of monosaccharides.

CHO
H—C—OH
CH_2OH
D-glyceraldehyde

CHO
H—C—OH
H—C—OH
H—C—OH
CH_2OH
D-ribose

CHO
H—C—OH
HO—C—H
H—C—OH
H—C—OH
CH_2OH
D-glucose

CH_2OH
C=O
HO—C—H
H—C—OH
H—C—OH
CH_2OH
D-fructose

Note that the configuration of all centers of an L-monosaccharide is reversed compared to the D-monosaccharide, because they are mirror images of each other.

mirror plane

D-glucose		L-glucose
CHO		CHO
H—C—OH		HO—C—H
HO—C—H		H—C—OH
H—C—OH		HO—C—H
H—C—OH ← determines D-configuration		determines L-configuration → HO—C—H
CH_2OH		CH_2OH

The Fischer projections of the aldotetroses, aldopentoses, and aldohexoses of the D-series are shown in Figure 11.1. D-Glyceraldehyde, at the top of the "tree", is the parent aldose. When we insert a new chiral center (H—C—OH) between the carbonyl carbon atom and the chiral center below, the resulting molecules are D-aldotetroses. Because the new —CHOH group

FIGURE 11.1 Structures of the D-Aldoses

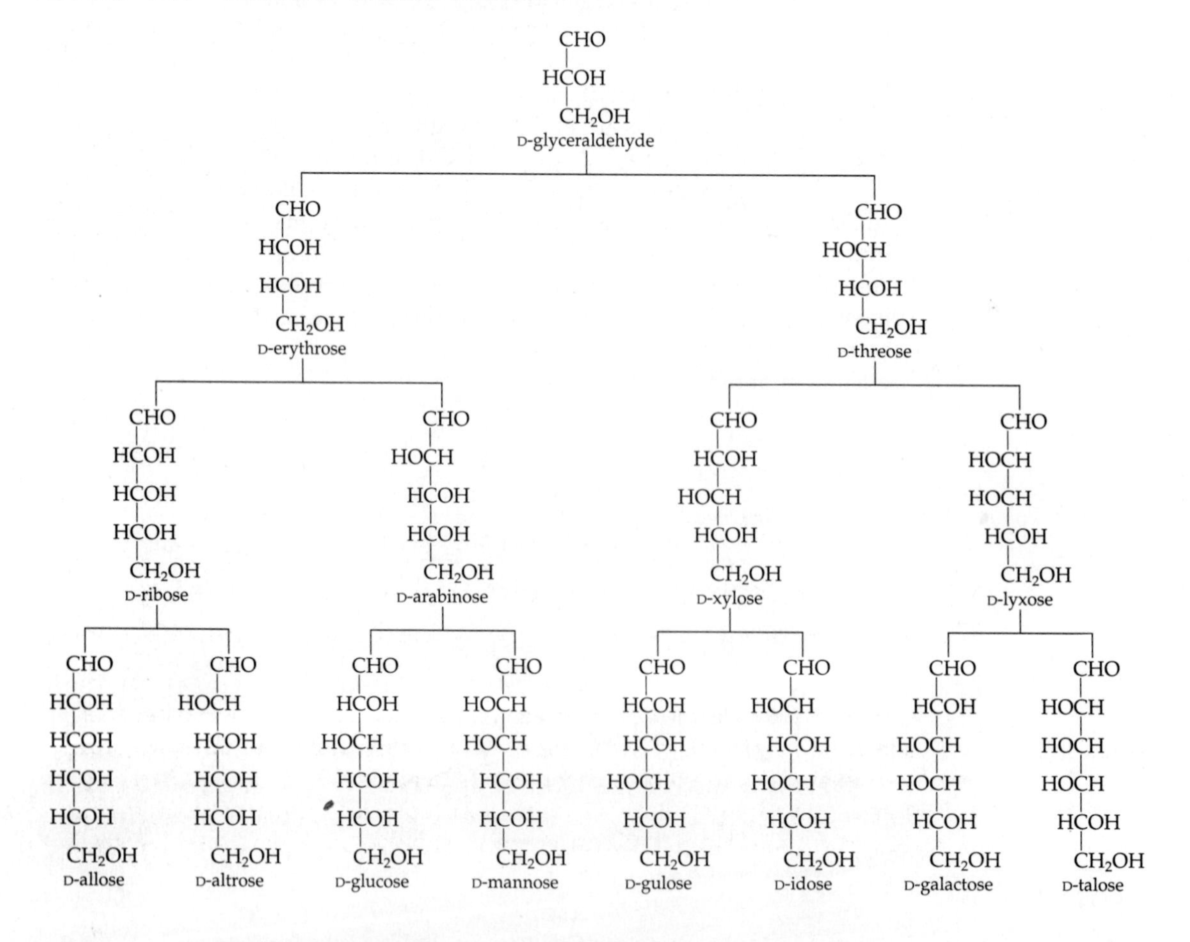

can have its —OH group on the right or left, two aldotetroses are possible: D-erythrose and D-threose. Note that aldotetroses contain two nonequivalent chiral centers, so $2^2 = 4$ stereoisomers are possible. The two L-aldotetroses are not shown in Figure 11.1. D-Erythrose and L-erythrose are enantiomers, as are D-threose and L-threose.

Inserting a new chiral center (H—C—OH), which can have either of two configurations, between the carbonyl carbon atom and the chiral center at the C-2 atom in D-erythrose leads to two D-aldopentoses: D-ribose and D-arabinose. Similarly, inserting a new chiral center (H—C—OH) between the carbonyl carbon atom and the chiral center at the C-2 atom in D-threose gives D-xylose and D-lyxose. Repeating the process one more time in each of the four D-aldopentoses gives a total of eight D-aldohexoses. D-Glucose and D-galactose are the most widely found in nature; D-mannose and D-talose occur in smaller amounts. The others are extremely rare.

Any of the members of a group of isomeric monosaccharides shown in Figure 11.1 are diastereomers of each other. They are not enantiomers because they are not mirror images.

EXAMPLE 11.2 What is the structure of L-arabinose, which is present in some antiviral drugs?

Solution L-Arabinose is the enantiomer of D-arabinose. Therefore, the L-isomer can be drawn by reflecting the planar projection formula in an imagined mirror perpendicular to the plane of the page and parallel to the carbon chain. Each hydroxyl group that is on the right in D-arabinose is on the left in L-arabinose and vice versa.

mirror plane

D-arabinose	L-arabinose
CHO	CHO
HO—C—H	H—C—OH
H—C—OH	HO—C—H
H—C—OH (determines D-configuration)	HO—C—H (determines L-configuration)
CH_2OH	CH_2OH

Epimers

We recall that diastereomers are stereoisomers, but not enantiomers. Diastereomers that contain two or more chiral carbon atoms but differ in configuration at only one chiral center are called **epimers.** Thus, the diastereomers D-glucose and D-galactose are epimers because they differ in configuration only at the C-4 atom. D-Glucose and D-mannose are epimers that differ in configuration at the C-2 atom.

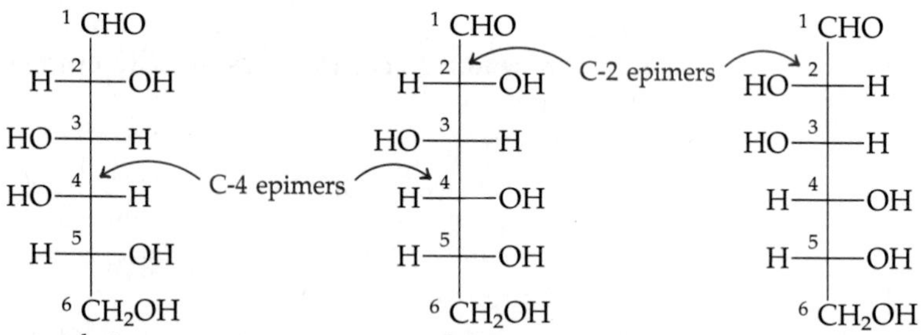

	D-galactose	D-glucose	D-mannose
C-1	CHO	CHO	CHO
C-2	H—C—OH	H—C—OH	HO—C—H
C-3	HO—C—H	HO—C—H	HO—C—H
C-4	HO—C—H	H—C—OH	H—C—OH
C-5	H—C—OH	H—C—OH	H—C—OH
C-6	CH_2OH	CH_2OH	CH_2OH

C-4 epimers (D-galactose and D-glucose); C-2 epimers (D-glucose and D-mannose)

The interconversion of epimers such as D-glucose and D-mannose at the C-2 atom illustrates the chemical reactions we described in Chapter 10. The α-hydrogen atom of an aldehyde is slightly acidic, and in the presence of a weak base it undergoes a keto–enol tautomerization reaction to produce a small amount of an isomeric enol.

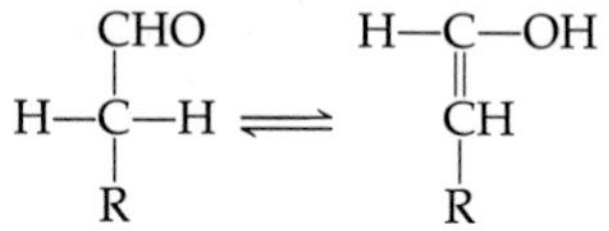

$$\mathrm{R{-}CH_2{-}CHO} \rightleftharpoons \mathrm{R{-}CH{=}CH{-}OH}$$

In an aldose the α-carbon atom is chiral and has a hydroxyl group bonded to it. Tautomerization yields an **enediol** in which the α-carbon atom is not chiral. In the reverse reaction to regenerate the aldose, a chiral center is

Galactosemia

Epimers are interconverted in cells by enzymes called **epimerases.** For example, an epimerase catalyzes the conversion of D-galactose into D-glucose in an important metabolic process. The necessary epimerase is not present in some newborn children. The lack of this enzyme is responsible for the genetic disease galactosemia.

Galactose is a component of the disaccharide lactose. Hydrolysis of lactose produces D-galactose and D-glucose. When the epimerase that converts galactose to glucose is missing, D-galactose accumulates in the body.

Galactose has a C-1 aldehyde group that is reduced to an alcohol called galactitol. Galactitol accumulates in the lens of the eye and causes cataracts. It also causes severe mental retardation. These disastrous consequences can be prevented entirely if the diet of affected newborns does not contain milk or milk products.

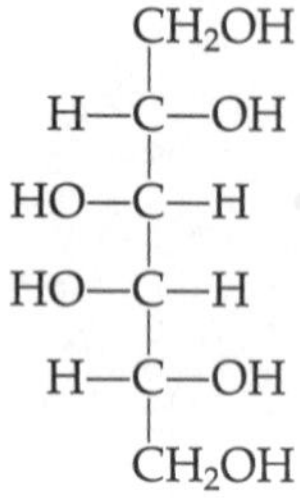

CH_2OH
H—C—OH
HO—C—H
HO—C—H
H—C—OH
CH_2OH
galactitol

Note that although galactitol has several chiral carbon atoms, the molecule as a whole has a plane of symmetry. It is therefore an optically inactive meso compound.

formed again at the α-carbon atom, which can have either of two configurations. The resultant compounds are C-2 epimers.

$$\begin{array}{ccccc} \text{CHO} & & \text{H—C—OH} & & \text{CHO} \\ | & & \| & & | \\ \text{H—C—OH} & \rightleftharpoons & \text{C—OH} & \rightleftharpoons & \text{HO—C—H} \\ | & & | & & | \\ \text{R} & & \text{R} & & \text{R} \\ \text{one epimer} & & \text{an enediol} & & \text{second epimer} \end{array}$$

Thus, D-glucose can be converted to D-mannose by way of an enediol intermediate. This reaction is catalyzed by a specific epimerase in cells.

EXAMPLE 11.3 Which aldopentose is an epimer of D-arabinose at the C-2 atom?

Solution Consider the structure of D-arabinose and examine the configuration at the C-2 atom. The hydroxyl group is on the left side. Rewrite the structure so that it has the same configuration at the C-3 and C-4 atoms, but place the hydroxyl group at the C-2 atom on the right side. This compound is D-ribose.

$$\begin{array}{cc} \text{CHO} & \text{CHO} \\ | & | \\ \text{HO—C—H} & \text{H—C—OH} \\ | & | \\ \text{H—C—OH} & \text{H—C—OH} \\ | & | \\ \text{H—C—OH} & \text{H—C—OH} \\ | & | \\ \text{CH}_2\text{OH} & \text{CH}_2\text{OH} \\ \text{D-arabinose} & \text{D-ribose} \end{array}$$

Ketoses

Our focus to this point has been the aldoses, which play an important role in many biological processes. However, several ketoses also play a pivotal role in metabolism.

The Fischer projections of the ketotetroses, ketopentoses, and ketohexoses of the D-series are shown in Figure 11.2. The "parent" ketose is the ketotriose called dihydroxyacetone. We can construct ketoses from dihydroxyacetone by consecutively inserting chiral centers (H—C—OH) between the ketone carbonyl carbon atom and the carbon atom directly below it.

The simplest ketose, dihydroxyacetone, does not contain any chiral carbon atoms. This ketose is produced in the metabolism of glucose as a phosphate ester at the C-3 hydroxyl group. Fructose, the most important ketohexose, is produced by isomerization of glucose in the glycolysis of glucose—a metabolic pathway that all cells use to degrade glucose and produce energy. The ketopentoses ribulose and xylulose are both intermediates in the pentose phosphate pathway, an important metabolic pathway that produces the ribose necessary for ribonucleic acids (Chapter 16).

FIGURE 11.2
Structures of the D-2-Ketoses

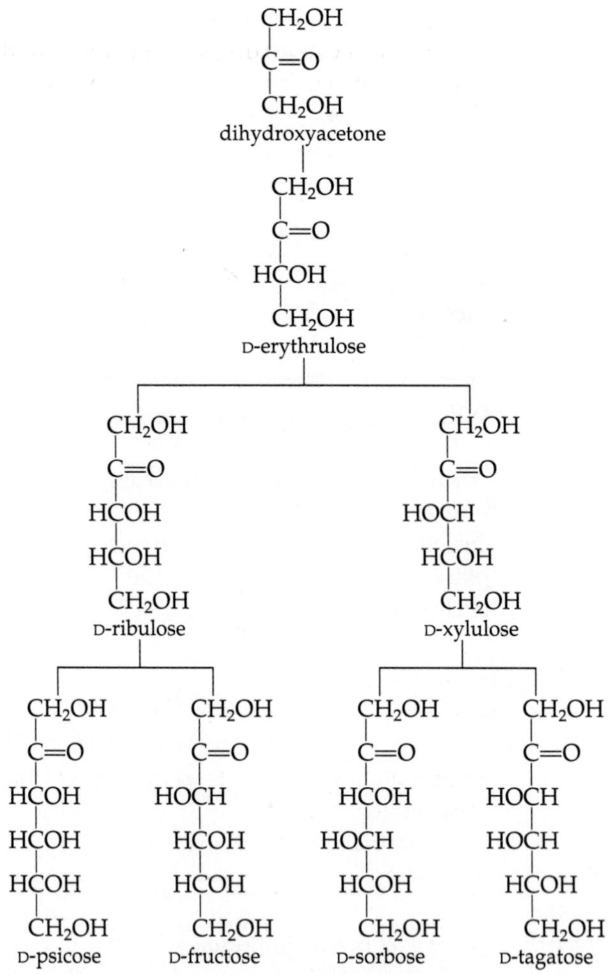

CH_2OH
C=O
CH_2OH
dihydroxyacetone

CH_2OH
C=O
HCOH
CH_2OH
D-erythrulose

D-ribulose	D-xylulose
CH_2OH	CH_2OH
C=O	C=O
HCOH	HOCH
HCOH	HCOH
CH_2OH	CH_2OH

D-psicose	D-fructose	D-sorbose	D-tagatose
CH_2OH	CH_2OH	CH_2OH	CH_2OH
C=O	C=O	C=O	C=O
HCOH	HOCH	HCOH	HOCH
HCOH	HCOH	HOCH	HOCH
HCOH	HCOH	HCOH	HCOH
CH_2OH	CH_2OH	CH_2OH	CH_2OH

11.4 HEMIACETALS AND HEMIKETALS

We recall that aldehydes and ketones react reversibly with alcohols to form hemiacetals and hemiketals, respectively (Chapter 10).

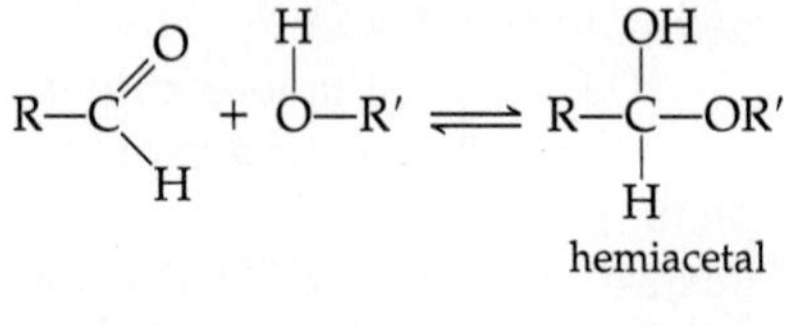

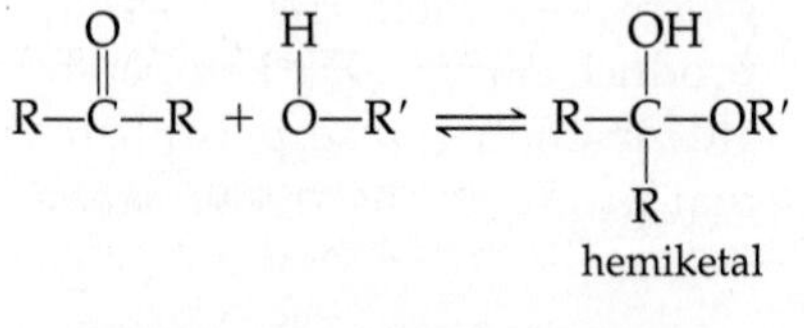

$$\mathrm{R{-}C(=O)H} + \mathrm{H{-}O{-}R'} \rightleftharpoons \mathrm{R{-}C(OH)(H){-}OR'}$$

hemiacetal

$$\mathrm{R{-}C(=O){-}R} + \mathrm{H{-}O{-}R'} \rightleftharpoons \mathrm{R{-}C(OH)(R){-}OR'}$$

hemiketal

When the hydroxyl group and the carbonyl group are part of the same molecule, a cyclic compound is formed in an intramolecular reaction. Cyclic hemiacetals containing five or six atoms form readily.

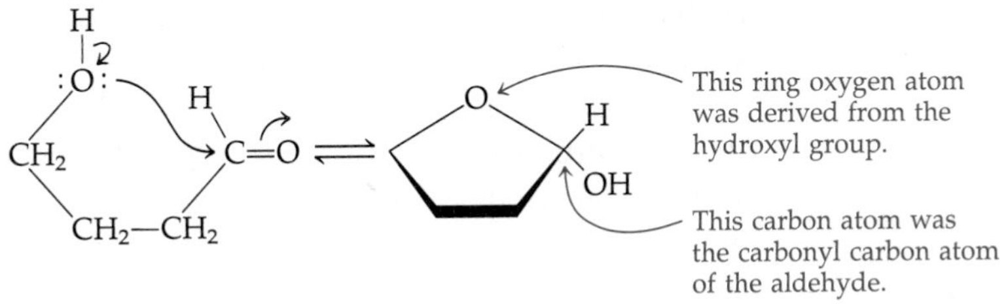

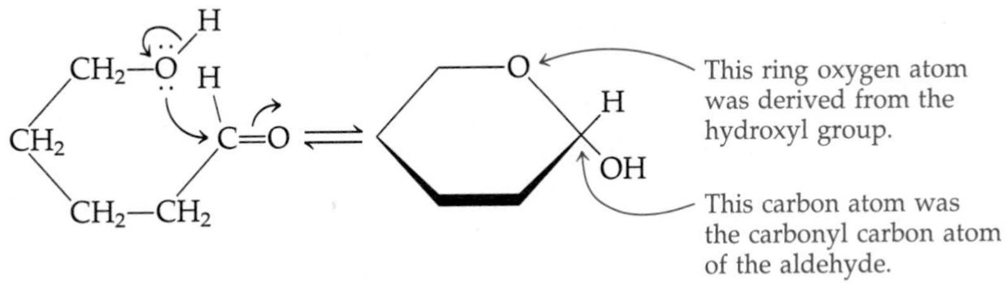

This ring oxygen atom was derived from the hydroxyl group.

This carbon atom was the carbonyl carbon atom of the aldehyde.

This ring oxygen atom was derived from the hydroxyl group.

This carbon atom was the carbonyl carbon atom of the aldehyde.

The cyclic hemiacetal or hemiketal forms of aldo- and ketohexoses and pentoses are the predominant forms of these sugars rather than the open-chain structures we have discussed to this point. Cyclic hemiacetals and hemiketals of carbohydrates that contain five-membered rings are called **furanoses;** cyclic hemiacetals and hemiketals that contain six-membered rings are called **pyranoses.** These structures are usually represented by planar structures called Haworth projection formulas.

Haworth Projection Formulas

In a Haworth projection formula, a cyclic hemiacetal or hemiketal is represented as a planar structure and viewed edge-on. Bond lines representing atoms toward the viewer are written as heavy wedges; bond lines away from the viewer are written as unaccentuated lines. The carbon atoms are arranged clockwise with the C-1 atom of the aldohexose on the right. For hemiketals the C-2 atom is placed on the right.

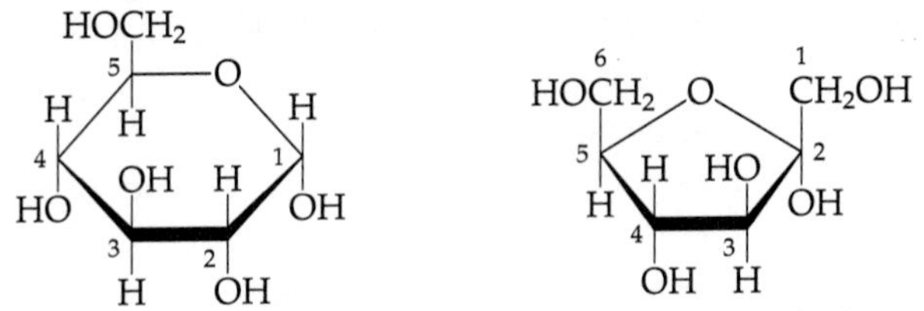

Haworth projection of a pyranose

Haworth projection of a furanose

Let's see how the Fischer projection formula of D-glucose can be converted into a hemiacetal written as a Haworth projection formula. The open-chain form of D-glucose is shaped like a "C" that is arranged vertically on the page (Figure 11.3). (We recall that the carbon chain in a Fischer projection formula is directed away from the reader.) Now, tilt this curved chain to the right so that it is horizontal. Groups on the right in the Fischer projection are then directed downward, whereas groups on the left are directed upward.

In this conformation, the C-5 —OH group is not near enough to the carbonyl carbon atom to form a ring. To bring the C-5 —OH group near the carbonyl carbon atom, rotate the structure about the bond between the C-4 and C-5 atoms. The —CH_2OH group is now above the plane of the curved carbon chain, and the C-5 hydrogen atom is below the plane. The oxygen atom of the C-5 —OH group in the plane adds to the carbonyl carbon atom, and a hydrogen atom adds to the carbonyl oxygen atom. A six-membered ring that contains five carbon atoms and one oxygen atom results. All carbohydrates with a D-configuration have the —CH_2OH group located above the ring in a Haworth projection.

When glucose forms a cyclic hemiacetal, four different groups are bonded to the C-1 atom. Thus, a new chiral center is formed at the original carbonyl carbon atom, and two configurations are possible at the C-1 atom. If the hydroxyl group of the hemiacetal is directed below the plane, the compound is α-D-glucopyranose; if it is above the plane, the compound is β-D-glucopyranose.

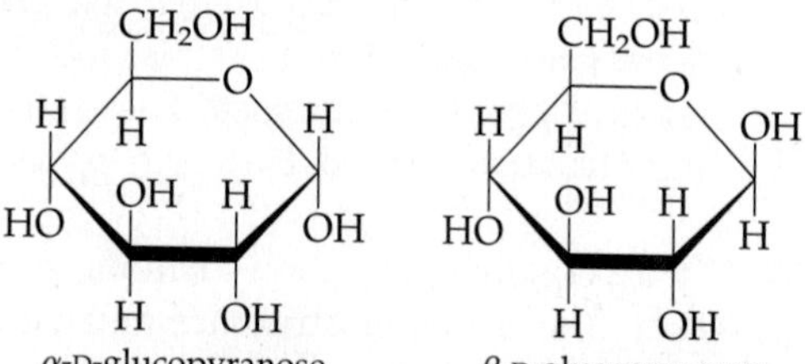

α-D-glucopyranose β-D-glucopyranose

The α and β forms of D-glucose are diastereomers that differ in configuration at one center. Hence, they are epimers. Compounds whose configurations differ only at the hemiacetal center are a special type of epimer called **anomers.** The chiral carbon atom at the hemiacetal center that forms in the cyclization reaction is called the **anomeric carbon atom.**

Now let's consider the cyclic form of the ketohexose D-fructose. D-Fructose cyclizes in aqueous solution to give a mixture that contains 20% α- and β-D-fructofuranose and 80% α- and β-D-fructopyranose. The furanose isomers are formed when the C-5 —OH group adds to the carbonyl carbon atom of the C-2 keto group. A ring of four carbon atoms and one oxygen atom results. The pyranose isomers form when the C-6 hydroxyl group adds to the C-2 carbonyl carbon atom. Again, α and β designate the configuration of the hydroxyl group at the anomeric carbon atom.

FIGURE 11.3
Conversion of Fisher Projection into Haworth Projection

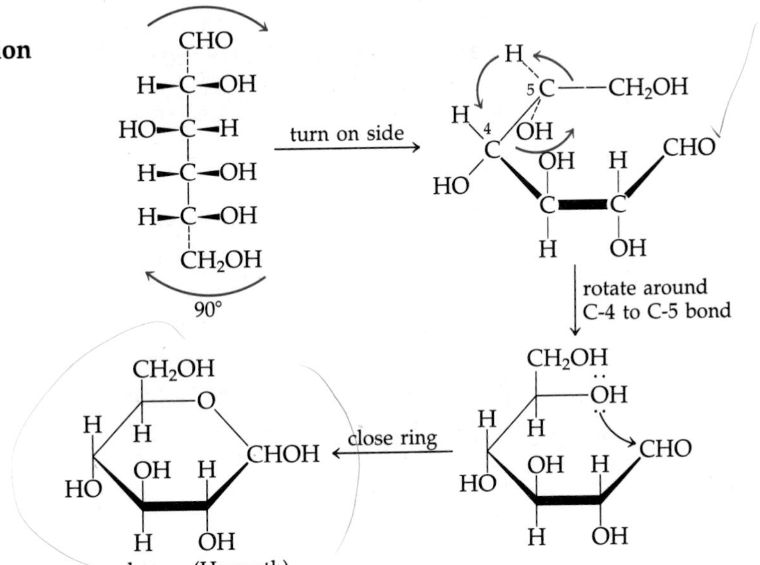

D-glucose (Haworth)

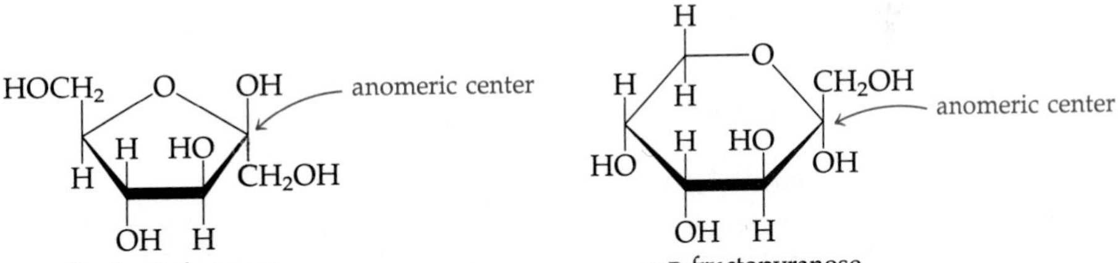

β-D-fructofuranose
(a hemiketal)

α-D-fructopyranose
(a hemiketal)

EXAMPLE 11.4 Draw the Haworth projection of the α-anomer of the pyranose form of D-galactose.

Solution First write galactose in the Fischer projection. Because the pyranose form is a six-membered ring, draw the ring of five carbon atoms and one oxygen atom.

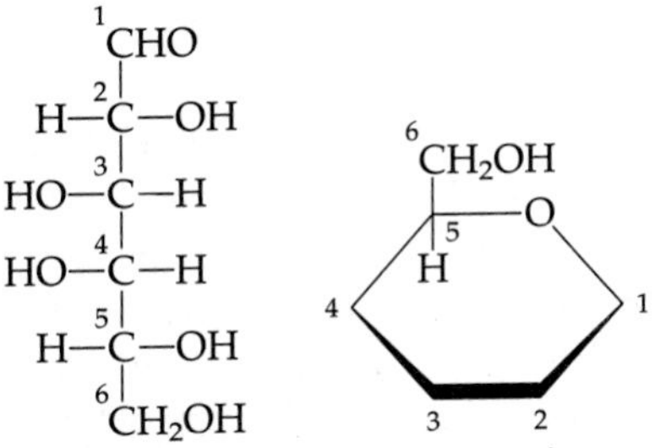

D-galactose

pyranose ring

For the D-configuration, the —CH_2OH group is above the plane of the ring. Now enter the hydroxyl groups and hydrogen atoms at C-2, C-3, and C-4. An atom or group on the right in the Fischer projection is below the ring of the Haworth projection, and an atom or group on the left is above the ring.

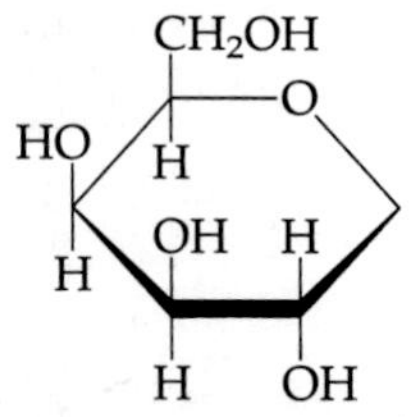

Finally, the α-anomer must have a hydroxyl group below the plane of the ring at the anomeric carbon atom, C-1.

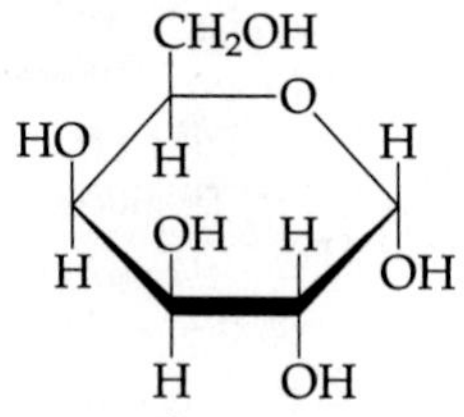

α-D-galactopyranose

Mutarotation

Now let's consider the experimental consequences of the formation of anomers of monosaccharides. When D-glucose is crystallized from methanol, α-D-glucopyranose, which melts at 146 °C, is obtained. It has $[\alpha]_D$ = +112.2. On the other hand, when D-glucose is crystallized from acetic acid,

FIGURE 11.4 Interconversion of Anomers and Mutarotation

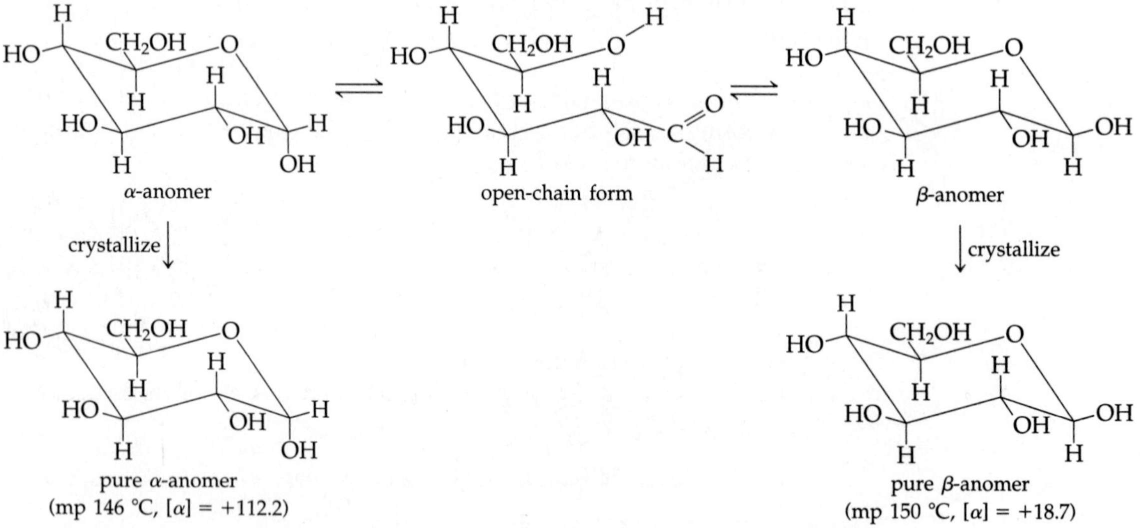

α-anomer

open-chain form

β-anomer

crystallize

crystallize

pure α-anomer
(mp 146 °C, [α] = +112.2)

pure β-anomer
(mp 150 °C, [α] = +18.7)

the β-anomer, which melts at 150 °C, is obtained. It has $[\alpha]_D = +18.7$ (Figure 11.4). We recall that diastereomers have different chemical and physical properties, so these data are not surprising.

When α-D-glucopyranose is dissolved in water, the rotation of the solution slowly changes from the initial value of +112.2 to an equilibrium value of +54. If β-D-glucopyranose is dissolved in water, the rotation of the solution slowly changes from the initial value of +18.7 to the same equilibrium value of +54. The gradual changes in rotation to an equilibrium point are known as **mutarotation.** Mutarotation results from the interconversion of the cyclic hemiacetals with the open-chain form in solution. Ring opening followed by reclosure can form either the α- or β-anomer. At equilibrium there are 36% of the α-anomer and 64% of the β-anomer for glucose; less than 0.01% of the open-chain form is present. The mutarotation of glucose in cells is catalyzed by an enzyme called mutarotase.

11.5 CONFORMATIONS OF MONOSACCHARIDES

Haworth projection formulas are easy to draw but do not give an accurate three-dimensional representation of carbohydrates. Pyranose rings contain six atoms and exist in chair conformations just like cyclohexane (Chapter 3). Any hydroxyl group (or other group) that is up in the Haworth projection is also up in the chair conformation. However, we recall that "up" and "down" do not correspond to axial and equatorial, respectively. Each carbon atom must be individually examined. On one set of alternating carbon atoms, an "up" substituent is axial; on the intervening carbon atoms, an "up" substituent is equatorial.

Haworth projection formulas are converted into chair representations by "moving" two carbon atoms. The anomeric carbon atom is lowered below the plane of the ring, and the C-4 atom is raised above the plane of the ring. The remaining four atoms—three carbon atoms and the ring oxygen atom—are unchanged. This process is shown in Figure 11.5 for both α-D-glucopyranose and β-D-glucopyranose. Both the hydrogen atoms and the hydroxyl groups can be shown. However, a more condensed form that eliminates the C—H bonds is often used.

Note the changes in the locations of the hydroxyl groups in the Haworth projection compared to those in the chair conformation. Although the hydroxyl groups were both up and down in the Haworth projection formula, all hydroxyl groups are equatorial in β-D-glucopyranose. We recall that this anomer is the more stable; that is, it is the major anomer in an equilibrium mixture of glucose. Note also that glucopyranose, the most abundant aldohexose, is the only aldohexose that has all of its hydroxyl groups in equatorial positions.

EXAMPLE 11.5 Draw the chair conformation of α-D-galactopyranose.

FIGURE 11.5 Conversion of Haworth Projections into Chair Representations

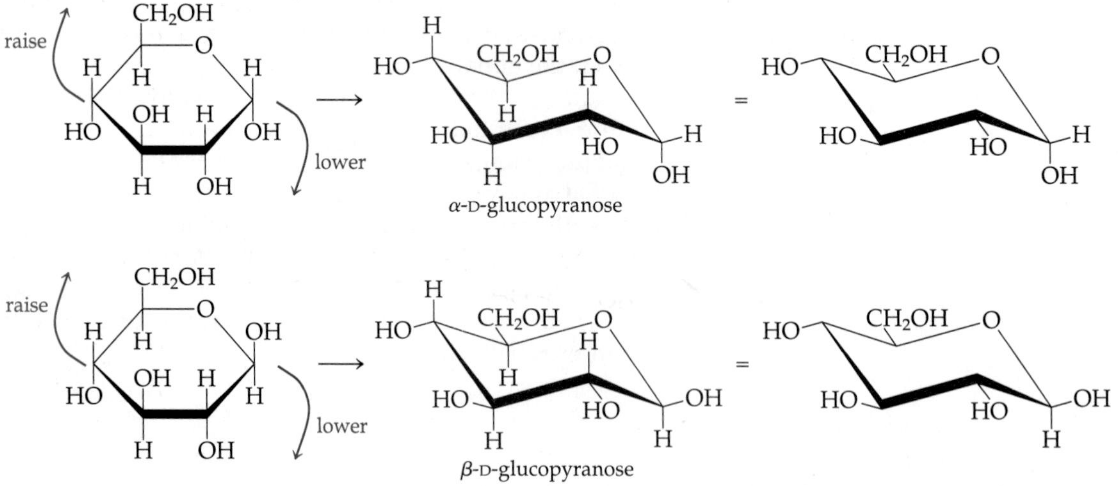

α-D-glucopyranose

β-D-glucopyranose

Solution Although the chair conformation could be derived from the Haworth projection—which can in turn be derived from the open-chain formula—there is an easier way to obtain the structure. We need only recall that the β-anomer of glucose has all of its hydroxyl groups in equatorial positions. The α-anomer of galactose must have an axial hydroxyl group at the C-1 atom. We also recall that galactose is the C-4 epimer of glucose. Therefore, the hydroxyl group at the C-4 atom must be axial.

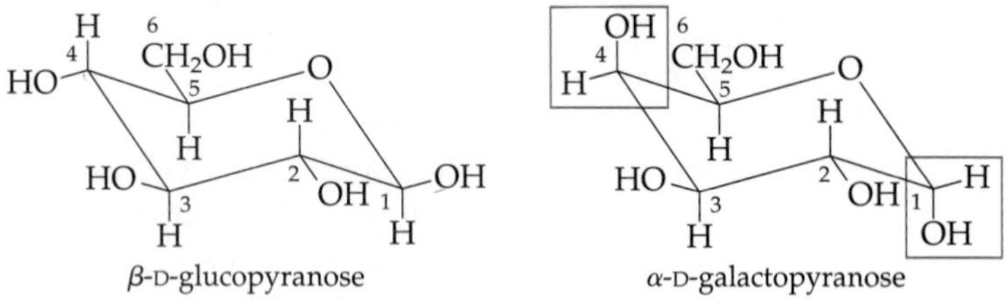

β-D-glucopyranose

α-D-galactopyranose

11.6 REDUCTION OF MONOSACCHARIDES

Although five- and six-carbon monosaccharides exist predominately as hemiacetals and hemiketals, they undergo the characteristic reactions of simple aldehydes and ketones. One such reaction is reduction. Treating an aldose or ketose with sodium borohydride reduces it to a polyalcohol called an **alditol.** The reduction reaction occurs via the aldehyde group in the small amount of the open-chain form of the aldose in equilibrium with its cyclic hemiacetal. As the aldehyde is reduced, the equilibrium shifts to produce more aldehyde until eventually all of the monosaccharide is reduced. The alditol derived from D-glucose is called D-glucitol. D-Glucitol occurs in some fruits and berries. Produced and sold commercially as a sugar substitute, it is also called sorbitol.

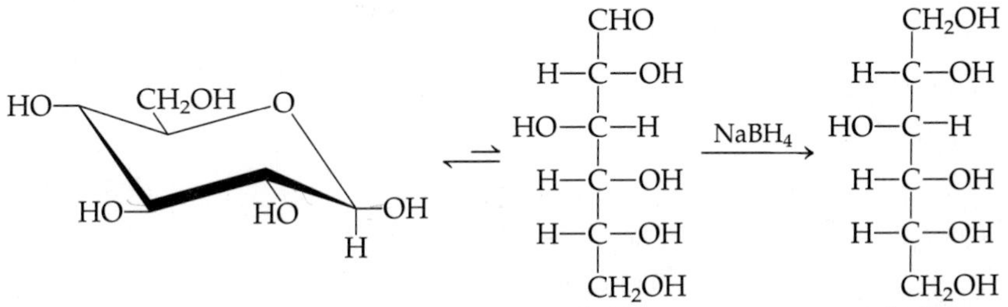

HO— CH_2OH O
HO— HO —OH
H

$$\rightleftharpoons \begin{array}{c} CHO \\ H-C-OH \\ HO-C-H \\ H-C-OH \\ H-C-OH \\ CH_2OH \end{array} \xrightarrow{NaBH_4} \begin{array}{c} CH_2OH \\ H-C-OH \\ HO-C-H \\ H-C-OH \\ H-C-OH \\ CH_2OH \end{array}$$

D-glucose

D-glucitol (an alditol)

EXAMPLE 11.6 D-Xylitol is used as a sweetener in some chewing gums that are said to have a lower probability of causing cavity formation in teeth compared to those containing glucose and fructose. Deduce the structure of D-xylitol from its name.

Solution The name resembles D-xylose, which can be reduced by sodium borohydride to an alditol. The configurations of the hydroxyl groups in D-xylitol are the same as those in D-xylose.

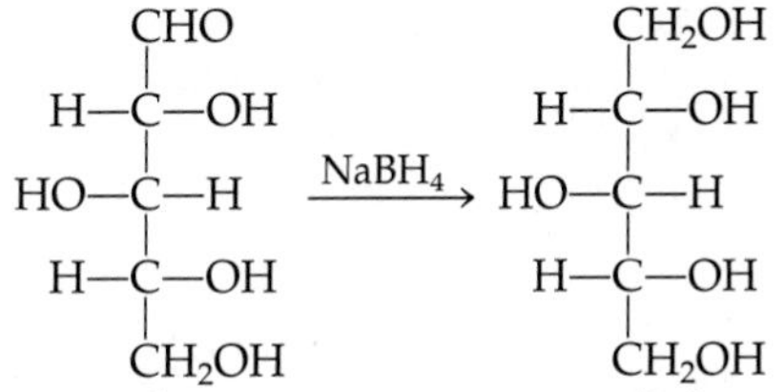

$$\begin{array}{c} CHO \\ H-C-OH \\ HO-C-H \\ H-C-OH \\ CH_2OH \end{array} \xrightarrow{NaBH_4} \begin{array}{c} CH_2OH \\ H-C-OH \\ HO-C-H \\ H-C-OH \\ CH_2OH \end{array}$$

D-xylose

D-xylitol (an alditol)

11.7 OXIDATION OF CARBOHYDRATES

In Chapter 10 we saw that aldehydes are oxidized by Tollens's reagent, Benedict's solution, and Fehling's solution. These reagents also oxidize open-chain aldoses that exist in equilibrium with the cyclic hemiacetal form. When some of the open-chain form reacts, the equilibrium shifts to form more compound for subsequent oxidation and eventually all the aldose is oxidized. Oxidation yields a product with a carboxylic acid at the original C-1 atom. This product is called an **aldonic acid.**

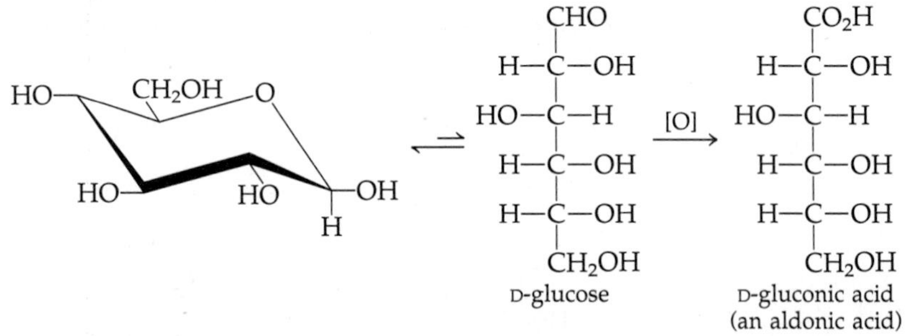

D-glucose ⇌ [O] → D-gluconic acid (an aldonic acid)

If Tollens's reagent is used as the oxidizing agent, metallic silver forms a mirror on the walls of the test tube. If Benedict's solution is used, a red precipitate of Cu_2O indicates that a reaction has occurred. We noted earlier that Benedict's solution is used to detect glucose in urine. If no glucose is present in the urine, the Benedict's solution remains blue. But with increasing glucose concentration the mixture of precipitate and solution may vary in color from green to yellow to orange to red.

Aldoses can also be oxidized to aldonic acids with aqueous bromine at approximately pH 6. This reagent is the preferred method for the laboratory synthesis of aldonic acids. It is sufficiently mild to prevent the oxidation of any of the hydroxyl groups in an aldose.

Benedict's solution also oxidizes ketoses. We certainly do not expect this, since ketones are not oxidized by Benedict's solution. However, we recall that α-hydroxy ketones tautomerize in basic solution—and Benedict's solution is basic. The tautomer of a ketone is an enediol that not only reverts to the α-hydroxy ketone, but also forms an isomeric α-hydroxy aldehyde.

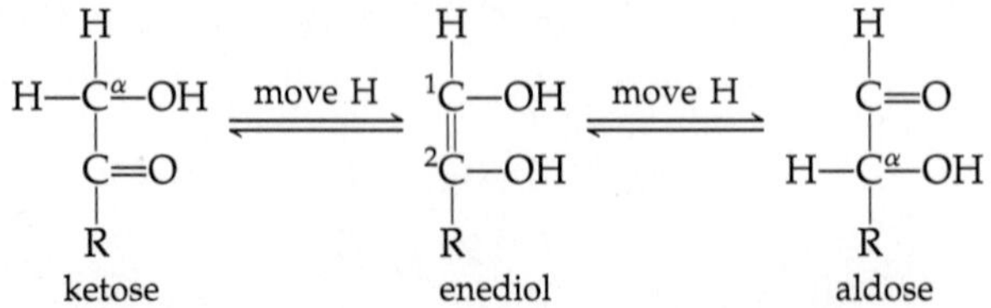

ketose ⇌ (move H) enediol ⇌ (move H) aldose

Shifting a hydrogen atom from the C-2 hydroxyl group to the C-1 atom in the enediol regenerates the original ketose. However, shifting a hydrogen atom from the C-1 hydroxyl group to the C-2 atom forms an aldose. In basic solution, then, a ketose is in equilibrium with an aldose. The aldose reacts with Benedict's solution, and more ketose is converted into aldose. The equilibrium shifts as predicted by Le Châtelier's principle, and eventually all of the original ketose is oxidized.

Carbohydrates that react with Benedict's solution are called **reducing sugars.** The term *reducing* refers to the effect of the carbohydrate on Benedict's solution. The carbohydrate is oxidized, but the Benedict's solution is reduced. Both aldoses and ketoses are reducing sugars.

EXAMPLE 11.7 Is ribulose a reducing sugar?

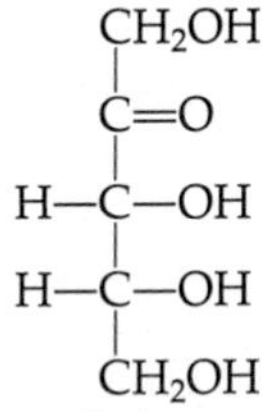

$$\begin{array}{c} CH_2OH \\ | \\ C{=}O \\ | \\ H{-}C{-}OH \\ | \\ H{-}C{-}OH \\ | \\ CH_2OH \end{array}$$

ribulose

Solution This ketose exists in equilibrium with an enediol intermediate in basic solution. The enediol is also in equilibrium with ribose. Because ribose contains an aldehyde group, ribulose will give a reaction with Benedict's solution and is classified as a reducing sugar.

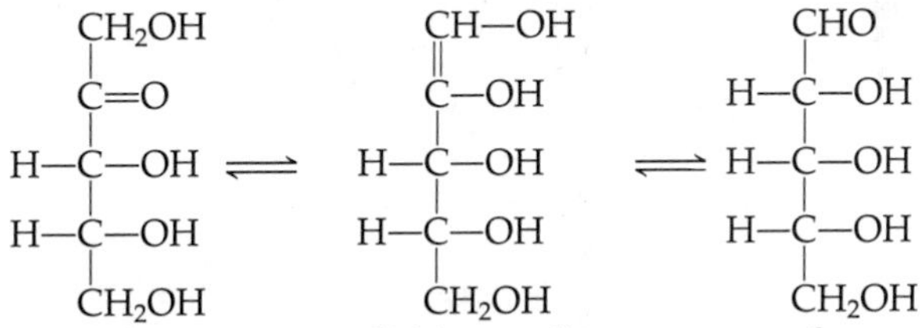

$$\begin{array}{c} CH_2OH \\ | \\ C{=}O \\ | \\ H{-}C{-}OH \\ | \\ H{-}C{-}OH \\ | \\ CH_2OH \end{array} \rightleftharpoons \begin{array}{c} CH{-}OH \\ \| \\ C{-}OH \\ | \\ H{-}C{-}OH \\ | \\ H{-}C{-}OH \\ | \\ CH_2OH \end{array} \rightleftharpoons \begin{array}{c} CHO \\ | \\ H{-}C{-}OH \\ | \\ H{-}C{-}OH \\ | \\ H{-}C{-}OH \\ | \\ CH_2OH \end{array}$$

D-ribulose enediol intermediate D-ribose

Other hydroxyl groups of aldoses may be oxidized if stronger oxidizing agents are used. For example, dilute nitric acid oxidizes both the aldehyde group and the primary alcohol of aldoses to give **aldaric acids.**

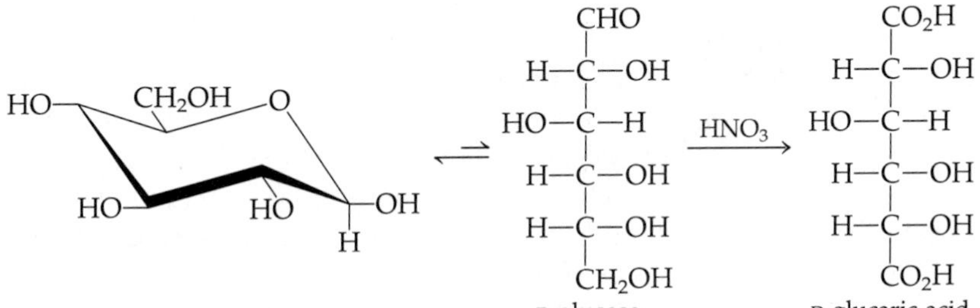

$$\text{(cyclic D-glucose: HO, } CH_2OH\text{, O, HO, HO, OH, H)} \rightleftharpoons \begin{array}{c} CHO \\ | \\ H{-}C{-}OH \\ | \\ HO{-}C{-}H \\ | \\ H{-}C{-}OH \\ | \\ H{-}C{-}OH \\ | \\ CH_2OH \end{array} \xrightarrow{HNO_3} \begin{array}{c} CO_2H \\ | \\ H{-}C{-}OH \\ | \\ HO{-}C{-}H \\ | \\ H{-}C{-}OH \\ | \\ H{-}C{-}OH \\ | \\ CO_2H \end{array}$$

D-glucose D-glucaric acid (an aldaric acid)

The terminal $—CH_2OH$ group of an aldose can be oxidized enzymatically in cells without oxidation of the aldehyde group. The product is a

uronic acid. The enzyme responsible for this reaction uses $NADP^+$ (nicotinamide adenine dinucleotide phosphate, a close structural relative of NAD^+) as the oxidizing agent. An example of this reaction is the oxidation of D-glucose to give D-glucuronic acid, a component in the polysaccharide hyaluronic acid, which is found in the vitreous humor of the eye.

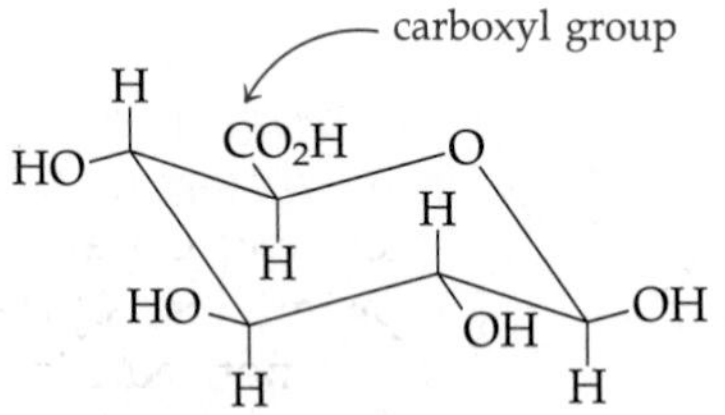

carboxyl group

D-glucuronic acid

11.8 GLYCOSIDES

In Chapter 10 we saw that hemiacetals and hemiketals react with alcohols to yield acetals and ketals, respectively. The reaction is acid-catalyzed, and the equilibrium is shifted to the right in the presence of excess alcohol, or by removal of the water formed. In this substitution reaction an —OR′ group replaces the —OH group.

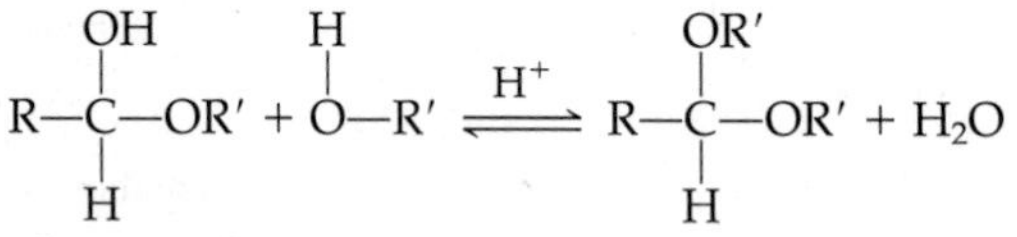

$$\underset{\text{hemiacetal}}{R-\underset{H}{\overset{OH}{C}}-OR'} + H-O-R' \xrightleftharpoons{H^+} \underset{\text{acetal}}{R-\underset{H}{\overset{OR'}{C}}-OR'} + H_2O$$

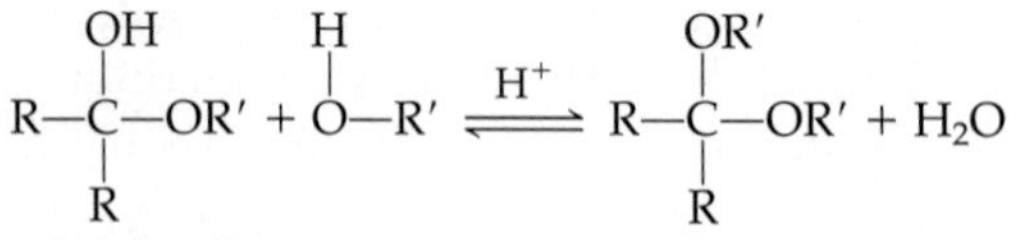

$$\underset{\text{hemiketal}}{R-\underset{R}{\overset{OH}{C}}-OR'} + H-O-R' \xrightleftharpoons{H^+} \underset{\text{ketal}}{R-\underset{R}{\overset{OR'}{C}}-OR'} + H_2O$$

The hemiacetal and hemiketal forms of monosaccharides also react with alcohols to form acetals and ketals. These acetals and ketals are called **glycosides,** and the carbon–oxygen bond formed is called a **glycosidic bond.** The group bonded to the anomeric carbon atom of a glycoside is an **aglycone.** In most aglycones, an oxygen atom from an alcohol or phenol is linked to the anomeric carbon atom. Aglycones containing a nitrogen atom are found in nucleosides, nucleotides, nucleic acids, and several coenzymes (Chapter 16).

Glycosides are named by citing the aglycone group first and then replacing the *-ose* ending of the carbohydrate with *-oside.* The configuration at the glycosidic carbon atom must be indicated.

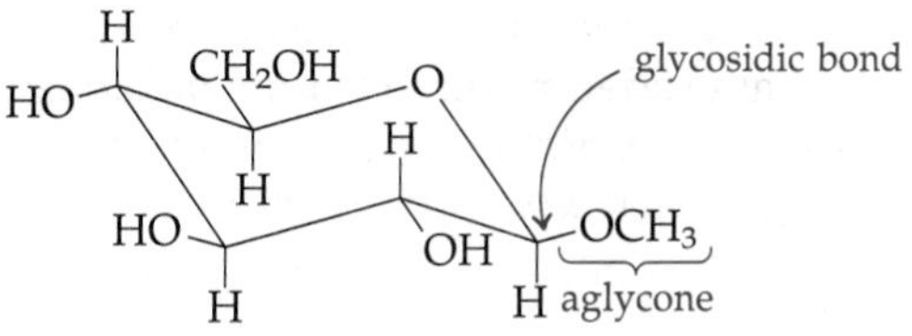

H
HO CH$_2$OH O glycosidic bond
H
H
HO OH OCH$_3$
H H aglycone
methyl β-D-glucopyranoside

Because hemiacetals or hemiketals exist in equilibrium as α or β anomers, two possible glycosides may form (Figure 11.6). Glycosides are stable compounds, and the anomers—which are diastereomers—have different physical properties. Glycosides are stable in neutral or basic solution. Therefore, they are not reducing sugars because they do not hydrolyze to form a free aldehyde group in Benedict's solution, which is basic. However, glycosides are hydrolyzed in acid solution by the reverse of the reactions shown in Figure 11.6.

EXAMPLE 11.8 Examine the following molecule to determine its component functional groups. From what compounds can the substance be formed?

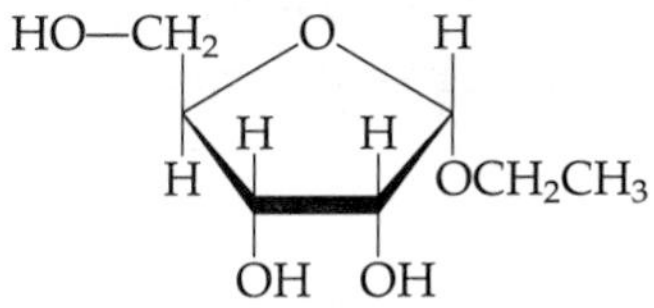

HO—CH$_2$ O H
H H H OCH$_2$CH$_3$
OH OH

FIGURE 11.6 Formation of Anomeric Glycosides

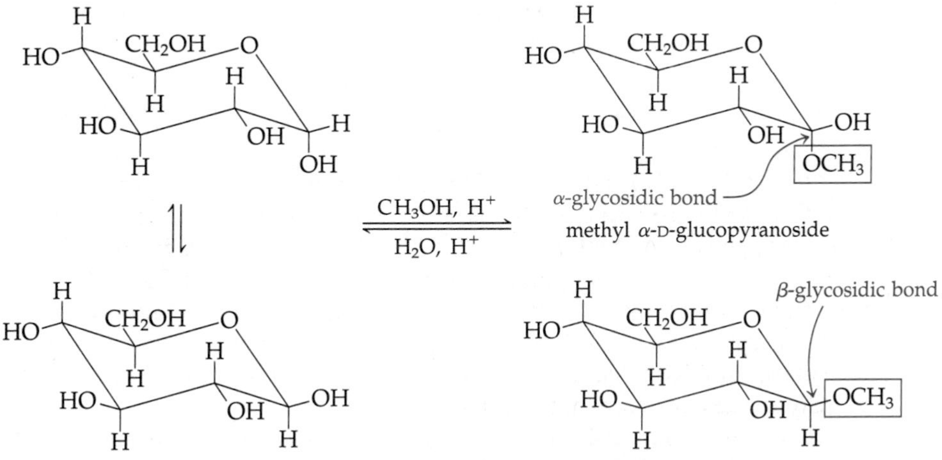

H
HO CH$_2$OH O
H
H
HO OH H
H OH

⇌

CH$_3$OH, H$^+$
H$_2$O, H$^+$

H
HO CH$_2$OH O
H
H
HO OH OH
H OCH$_3$
α-glycosidic bond
methyl α-D-glucopyranoside

H
HO CH$_2$OH O
H
H
HO OH OH
H H

H β-glycosidic bond
HO CH$_2$OH O
H
H
HO OH OCH$_3$
H H
methyl β-D-glucopyranoside

Solution The compound is a furanose form of a monosaccharide because there are four carbon atoms and one oxygen atom in a five-membered ring. The ring carbon atom on the right is an acetal because there are one hydrogen atom and two —OR groups bonded to it.

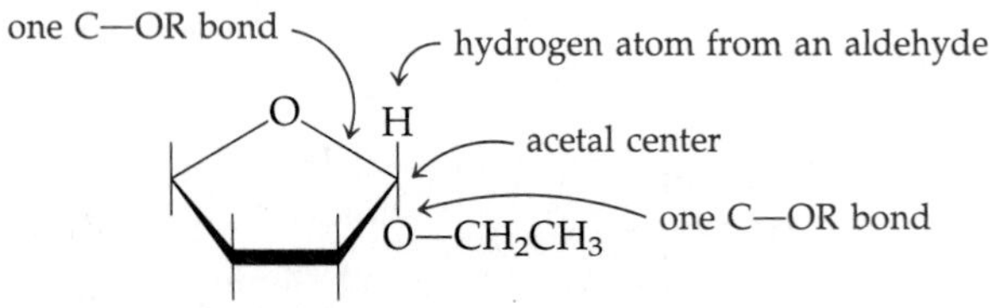

one C—OR bond
hydrogen atom from an aldehyde
O H
acetal center
O—CH_2CH_3 one C—OR bond

The acetal has the α-configuration, and the alcohol used to form the acetal is ethanol.

The carbohydrate has the D-configuration because the CH_2OH group is "up" in the Haworth projection. The other two chiral carbon atoms of the pentose have hydroxyl groups "down", which corresponds to the right in the Fischer projection. The carbohydrate component of the compound is ribose.

11.9 DISACCHARIDES

Disaccharides are glycosides formed from two monosaccharides. One monosaccharide unit is a hemiacetal or hemiketal linked through its anomeric center to the hydroxyl group of the second monosaccharide unit, which is the aglycone. Disaccharides often are linked by a glycosidic bond between the C-1 atom of the hemiacetal of an aldose to the C-4 atom of the second monosaccharide. Such bonds are designated (1,4′). The prime superscript indicates the carbon atom of the monosaccharide that provides the hydroxyl group. Maltose, cellobiose, and lactose all have (1,4′) glycosidic bonds. The configuration of the anomeric carbon atom is designated by the usual α or β.

In principle, any of the carbon atoms of a monosaccharide could provide the hydroxyl group of the aglycone. And in fact, (1,1′), (1,2′), (1,3′), (1,4′), and (1,6′) glycosidic bonds have all been found in naturally occurring disaccharides containing aldohexoses. Note that a (1,1′) glycosidic bond connects the anomeric carbon atoms of two aldoses.

Maltose

Maltose is composed of two molecules of D-glucose. The glycosidic oxygen atom of one glucose is α, and it is bonded to the C-4 atom of another glucose unit that is the aglycone. Therefore, maltose is an α-1,4′-glycoside.

Maltose is produced by the enzymatic hydrolysis of starch (a homopolysaccharide) catalyzed by the enzyme **amylase.** Maltose is further hydrolyzed by the enzyme maltase to produce two molecules of D-glucose. Maltose is formed from starch when treated with barley malt in the process of brewing beer.

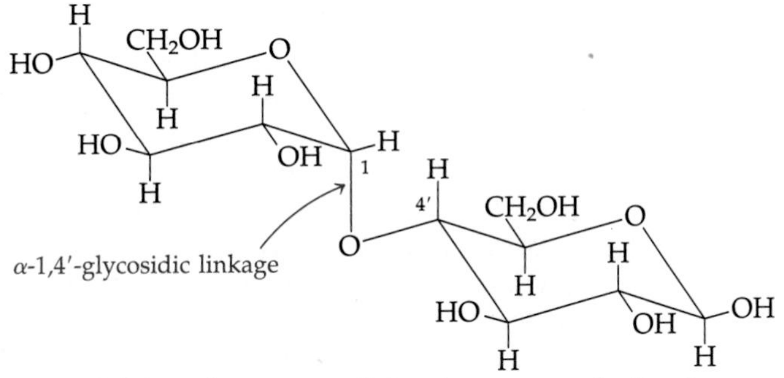

4-*O*-(α-D-glucopyranosyl)-β-D-glucopyranose (maltose)

Maltose has a more formal name: 4-*O*-(α-D-glucopyranosyl)-β-D-glucopyranose. This rather forbidding name is not quite so bad as it looks. The term in parentheses refers to the glucose unit at the left that contributes the acetal portion of the glycosidic bond. The suffix *-pyrano-* tells us that this part of the structure is a six-membered ring, and the suffix *-syl* tells us that the ring is linked to a partner by a glycosidic bond. The term α gives the configuration of the glycosidic bond. The prefix 4-*O*- refers to the position of the oxygen atom of the aglycone—the right-hand ring. The *β-D-glucopyranose* describes the aglycone. It too is a pyranose. When we look at the formal name, it is clear why some people prefer the simpler name maltose.

The right-hand glucose ring of maltose is shown as the β-anomer but the hydroxyl group can be either α or β. Because this center is a hemiacetal, both anomeric forms of maltose can exist in equilibrium in solution. This designation for the configuration of the aglycone ring should not be confused with the glycosidic bond at the acetal center, which is always α in maltose.

Because one of the monosaccharide units of maltose is a hemiacetal, maltose undergoes mutarotation. For the same reason, maltose is a reducing sugar. The free aldehyde formed by ring opening can react with Benedict's solution. If we do not want to specify the configuration of the hemiacetal center of maltose, the name 4-*O*-(α-D-glucopyranosyl)-D-glucopyranose is used.

Cellobiose

Cellobiose is a disaccharide that has two molecules of D-glucose linked by a β-1,4′-glycosidic bond. Cellobiose thus differs from maltose in the configuration of its glycosidic bond.

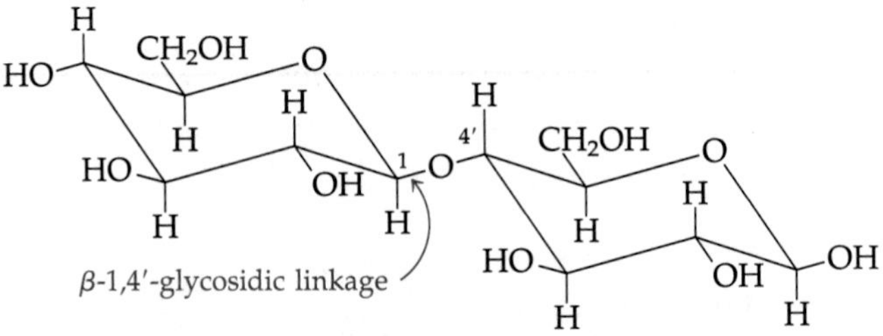

β-1,4′-glycosidic linkage

4-*O*-(β-D-glucopyranosyl)-β-D-glucopyranose (cellobiose)

As in maltose, the aglycone of cellobiose is a hemiacetal, which can be either α or β. Thus, cellobiose mutarotates and is a reducing sugar. In solution, the two forms of cellobiose exist in equilibrium. Again, do not confuse the configuration of the hemiacetal center with that of the glycosidic bond, which is always β in cellobiose.

Cellobiose is produced by hydrolysis of cellulose, a homopolysaccharide of glucose in which all units are linked by 1,4′-glycosidic bonds. Humans do not have an enzyme to hydrolyze cellobiose. Small differences in configuration at the 1,4′ linkage result in remarkable differences in the chemical reactivity of these biomolecules. Glycosidic bonds are hydrolyzed by enzymes called glycosidases. A glycosidase that hydrolyzes α-1,4′-glycosidic bonds completely ignores the molecules that have β-1,4′-glycosidic bonds (and vice versa).

Lactose

Lactose, a disaccharide found in the milk of many mammals including both humans and cows, is often called milk sugar. The IUPAC name of lactose is 4-*O*-(β-D-galactopyranosyl)-D-glucopyranose. The configuration at the C-4 atom of galactose is opposite that of glucose (glucose and galactose are C-4 epimers). The hydroxyl group at the C-4 atom of the ring on the left is equatorial in the glucose ring of maltose, but axial in the galactose ring of lactose. The ring on the left in both lactose and cellobiose is linked by a β glycosidic linkage to the C-4 atom of a D-glucopyranose ring on the right. In both lactose and cellobiose, the glycosidic bond is β-1,4′. Humans have an enzyme called β-galactosidase (also known as lactase) that hydrolyzes the β-1,4′-galactosidic linkage. However, β-galactosidase does not hydrolyze the β-1,4′-glucosidic linkage of cellobiose.

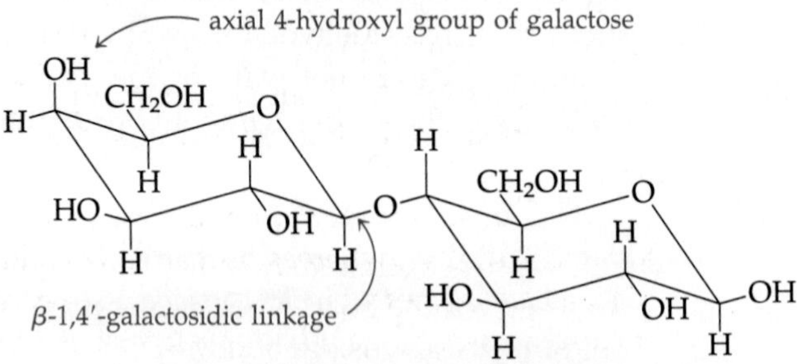

axial 4-hydroxyl group of galactose

β-1,4′-galactosidic linkage

4-*O*-(β-D-galactopyranosyl)-β-D-glucopyranose (lactose)

Lactose Intolerance

People with lactose intolerance cannot eat food that contains lactose. These individuals lack the lactase needed to hydrolyze lactose. If they ingest food that contains lactose, the high level of unhydrolyzed lactose in their intestinal fluids draws water from tissues by osmosis. The result is abdominal distention, cramping, and diarrhea. Although lactose intolerance is not life-threatening, it is quite unpleasant.

The level of the enzyme lactase in humans varies with both age and race. Most humans have sufficient lactase for the early years of life when milk is a major part of their diet. However, in adulthood the lactase level decreases, and lactose intolerance results. This trait shows remarkable genetic variations. For example, most northern Europeans have high lactase levels, as do several nomadic pastoral tribes in Africa (see the figure). The ability to digest milk as adults may be the result of an evolutionary process in societies that consumed large amounts of milk and milk products such as cheese. Those individuals with the enzyme necessary to digest milk may have had an adaptational advantage.

Some peoples, such as the Thai and the Chinese, have a high lactose intolerance. Similarly, the Ibo and Yoruba of Nigeria cannot tolerate lactose as adults. The Fula and Hausa of the Sudan differ greatly in the extent of their lactose intolerance. The Fula raise and milk a breed of cattle called fulani, whereas the Hausa, who show lactose intolerance, do not raise cattle. The Tussi, a cattle-owning class of the Rundi of east Africa, also can digest lactose.

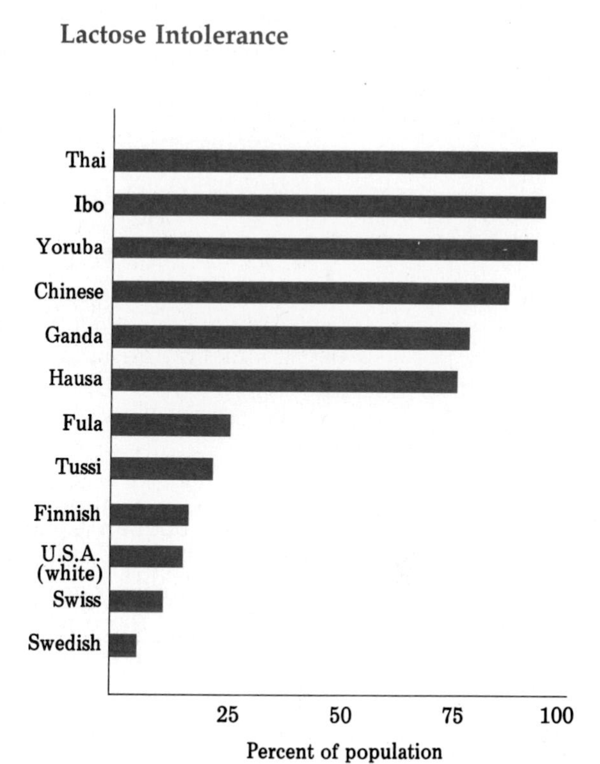

As in the cases of cellobiose and maltose, the aglycone component of lactose is a hemiacetal, which can be either α or β. Thus, lactose undergoes mutarotation and is a reducing sugar. In solution, the two forms of lactose exist in equilibrium.

The lactose content of milk varies with species; cow's milk contains about 5% lactose, whereas human milk contains about 7%. The enzyme lactase, which is present in the small intestine, catalyzes hydrolysis of lactose to form glucose and galactose. Galactose is then isomerized into glucose in a reaction catalyzed by the enzyme UDP-galactose-4-epimerase.

Sucrose

We noted at the start of this section that some disaccharides have a glycosidic linkage between both anomeric centers. Sucrose, common table sugar, is a disaccharide of α-glucose and β-fructose in which the anomeric centers are linked 1,2′.

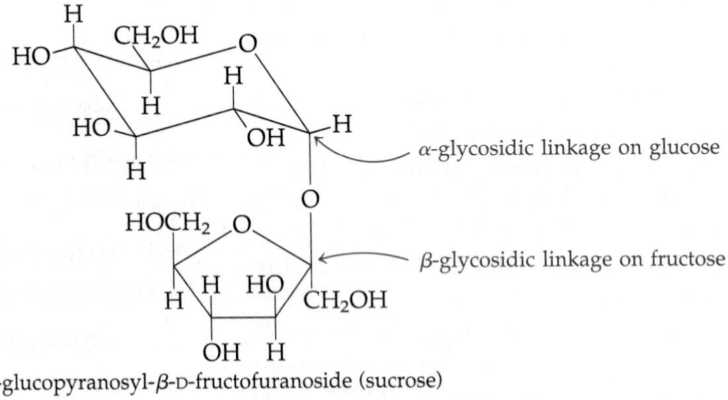

α-glycosidic linkage on glucose

β-glycosidic linkage on fructose

α-D-glucopyranosyl-β-D-fructofuranoside (sucrose)

Sucrose has both an acetal and a ketal functional group. Neither ring can exist in equilibrium with either an aldehyde or ketone. As a result, sucrose cannot mutarotate, nor is it a reducing sugar. The systematic name, α-D-glucopyranosyl-β-D-fructofuranoside, ends in the suffix *-oside,* which indicates that sucrose is not a reducing sugar.

EXAMPLE 11.9 Describe the structure of the following disaccharide.

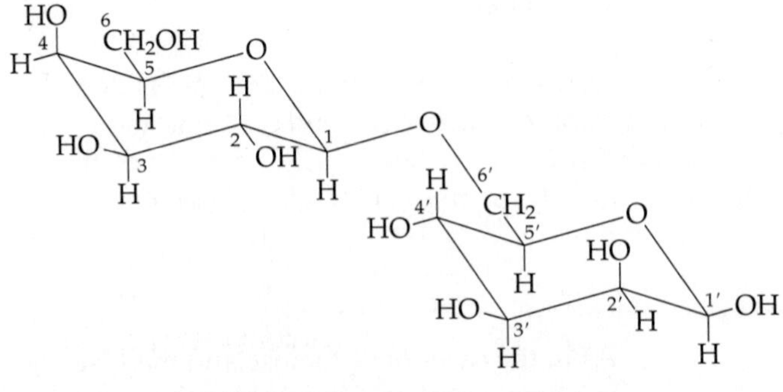

Solution The hemiacetal center located on the aglycone ring (at the right) has a hydroxyl group in the β-configuration. The glycosidic bond is from the C-1 atom of the acetal ring (on the left) to the C-6 atom of the aglycone ring (on the right). Furthermore, the oxygen bridge is formed through the β-glycosidic bond. Thus, the bridge is β-1,6′.

Next, we examine both rings to determine the identity of the monosaccharides. The ring on the left is galactose: all of its hydroxyl groups are

Sweeteners

Although all of us have different sensitivities to various taste sensations, we agree that the monosaccharides and disaccharides taste sweet. The relative degree of sweetness can be determined by comparing different concentrations of sugars or other sweeteners. For example, if a 0.1% solution of a test compound tastes as sweet as a 0.1% solution of sucrose, the two compounds have the same sweetness. If a 0.1% solution of a test compound tastes as sweet as a 10% solution of sucrose, we conclude that the test compound is 100 times as sweet as sucrose. Examples of relative sweetness values are given in the table. Exactly how we sense "sweetness" is not well-understood. For example, galactose is not very sweet, but glucose is almost as sweet as sucrose. Even though glucose and fructose are related at a number of chiral centers, fructose is considerably sweeter. The disaccharides lactose and maltose are less sweet than sucrose.

Sweeteners

Compound	Type	Relative sweetness
lactose	disaccharide	0.16
galactose	monosaccharide	0.32
glucose	monosaccharide	0.75
sucrose	disaccharide	1.00
fructose	monosaccharide	1.75
cyclamate	artificial	300
aspartame	artificial	1500
saccharin	artificial	3500

Artificial sweeteners differ considerably in structure from carbohydrates. Two common artificial sweeteners are cyclamate and saccharin. These two substances have few structural features in common and are distinctly different from sugars.

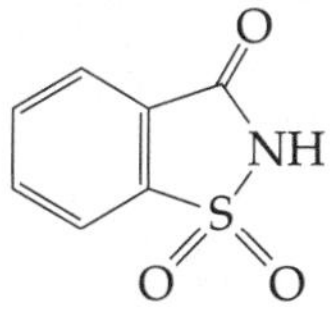

O NH S O O
saccharin

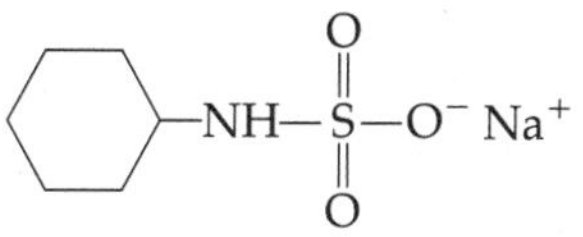

NH—S—O⁻ Na⁺ (with S=O above and below)
cyclamate

Cyclamate was banned in the United States some years ago because experiments with rats showed a link to some cancers. However, cyclamate is still used in Canada. On the other hand, saccharin is currently allowed in the United States but banned in Canada.

Aspartame is the methyl ester of a dipeptide (two amino acids). It is currently the product of choice among artificial sweeteners. Aspartame is the sweetening ingredient of NutraSweet.

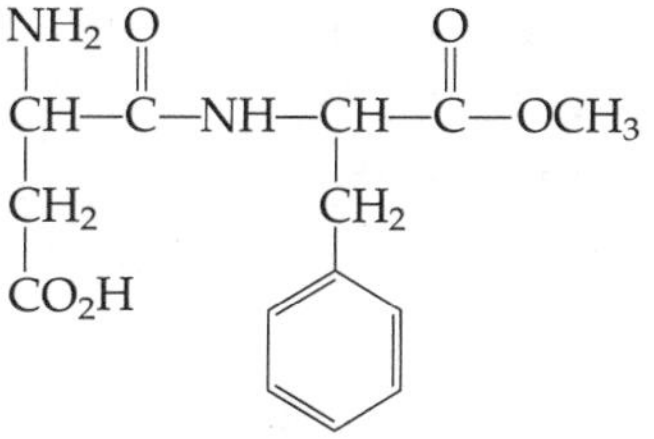

NH_2 O O
CH—C—NH—CH—C—OCH_3
CH_2 CH_2
CO_2H
aspartame

All sweeteners contain calories. However, because they are extremely sweet, much less sweetener is used compared to sucrose. Thus, the calorie content is very small. Aspartame has about the same number of calories per gram as sucrose. However, aspartame is about 1500 times sweeter than sucrose. In place of a heaping teaspoon of sucrose—about 10 grams—only 0.006 gram of aspartame is required for the same sweetness.

equatorial except the one at the C-4 atom, which is axial. The ring on the right is mannose, the C-2 epimer of glucose. Mannose has an axial hydroxyl group at the C-2 position. The compound is 6-*O*-(β-D-galactopyranosyl)-β-D-mannopyranose.

11.10 POLYSACCHARIDES

Monosaccharides can form polymers linked to each other by glycosidic bonds like those that link monosaccharides in disaccharides. These high molecular weight substances are called **polysaccharides.** Those made from one type of monosaccharide are called homopolysaccharides; those made from two or more different monosaccharides are called heteropolysaccharides. The latter class includes hyaluronic acid, found in the vitreous humor of the eye; heparin, an anticoagulant in blood; and chondroitin, a component of cartilage and tendons. Because the structures of heteropolysaccharides are more complex than those of homopolysaccharides, only homopolysaccharides will be considered in this book.

The homopolysaccharides starch and cellulose contain only glucose. About 20% of starch is **amylose,** which is soluble in cold water; the remaining 80%, called **amylopectin,** is insoluble in water. Starch is present in potatoes, rice, wheat, and other cereal grains. Because starch is a mixture of amylose and amylopectin, it has a highly variable "molecular weight" that depends on its source.

Starch and cellulose differ by one structural feature, but this difference has great biological importance. Starch, whose glucosyl units are linked α-1,4′, can be digested by most animals. Cellulose, whose glucosyl units are linked β-1,4′, can be digested by cattle and other herbivores whose digestive tracts contain microorganisms with enzymes that hydrolyze β-glycosides. Termites can also digest cellulose.

Amylose is a linear polymer with 200–2000 α-linked glucose units that serves as a major source of food for some animals. The molecular weight of amylose ranges from 40,000 to 400,000 amu. Cellulose is a β-linked polymer of glucose (Figure 11.7) that can contain 5000 to 10,000 glucose units. Certain algae produce cellulose molecules that contain more than 20,000 glucose units.

Amylopectin contains chains similar to those in amylose, but only about 25 glucose units occur per chain. Amylopectin has branches of glucose-containing chains interconnected by a glycosidic linkage between the C-6 hydroxyl group of one chain and the C-1 atom of another glucose chain (Figure 11.7). The molecular weight of amylopectin may be as high as 1 million. Because each chain has an average molecular weight of 3000, there may be as many as 300 interconnected chains.

Glycogen is synthesized by animals as a storage form of glucose. Its

FIGURE 11.7 Structures of Polysaccharides

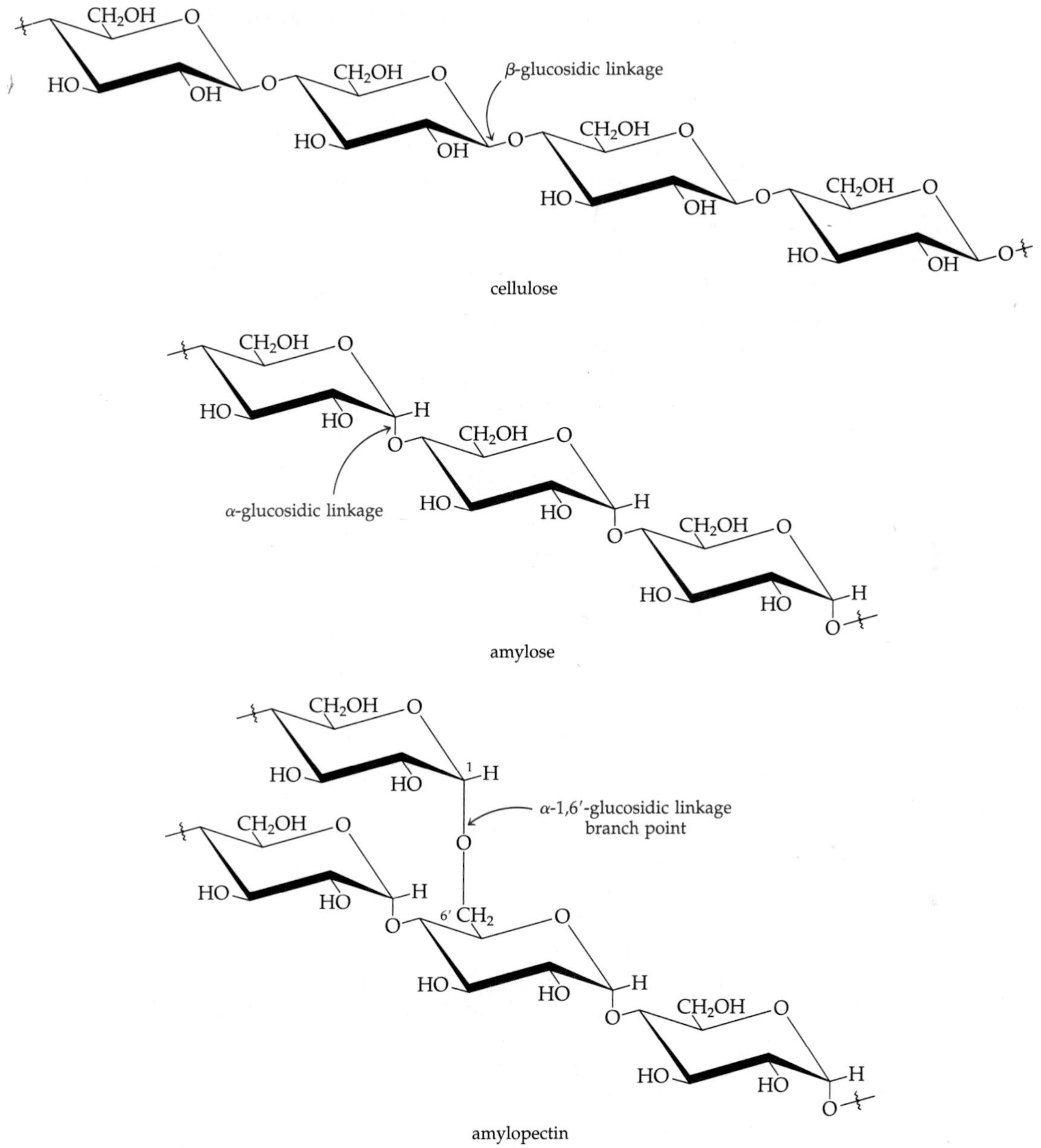

CH_2OH O HO OH O CH_2OH O β-glucosidic linkage HO OH O CH_2OH O HO OH O CH_2OH O HO OH O

cellulose

CH_2OH O HO HO H O α-glucosidic linkage CH_2OH O HO HO H O CH_2OH O HO HO H O

amylose

CH_2OH O HO HO 1 H α-1,6′-glucosidic linkage branch point O CH_2OH O HO HO H O 6′ CH_2 O HO HO H O CH_2OH O HO HO H O

amylopectin

structure is similar to that of amylopectin, but glycogen has more branches, and the branches are shorter than those of amylopectin. The average chain length in glycogen is 12 glucose units. Glycogen has a molecular weight greater than 3 million.

Glycogen is a source of metabolic energy during periods of diminished food intake. Although glycogen is found throughout the body, the largest amounts are in the liver. An average adult has enough glycogen for about 15 hours of normal activity.

EXERCISES

Classification of Monosaccharides

11.1 What is an aldose? How does it differ from a ketose?

11.2 To what carbon atom do the letters D and L refer in monosaccharides?

11.3 Classify each of the following monosaccharides.

(a)
CHO
H—OH
H—OH
H—OH
CH_2OH

(b)
CHO
HO—H
HO—H
H—OH
H—OH
CH_2OH

(c)
CHO
=O
H—OH
H—OH
H—OH
CH_2OH

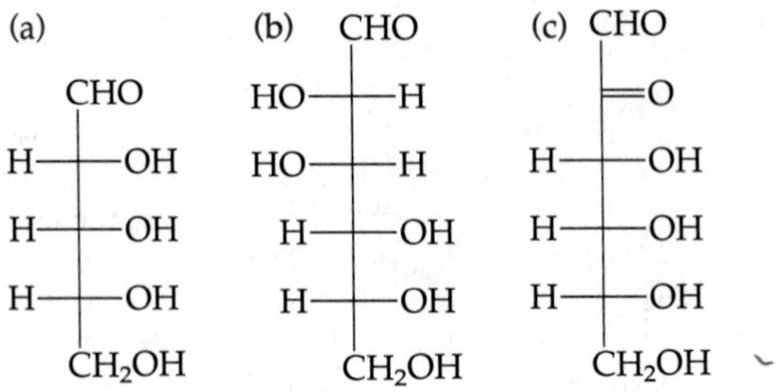

11.4 Classify each of the following monosaccharides.

(a)
CH_2OH
=O
HO—H
H—OH
CH_2OH

(b)
CHO
HO—H
H—OH
H—OH
H—OH
CH_2OH

(c)
CH_2OH
=O
H—OH
CH_2OH

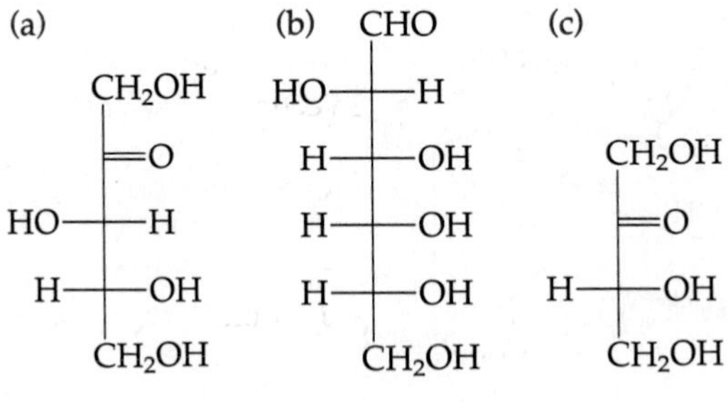

11.5 Classify each of the following monosaccharides as D or L.

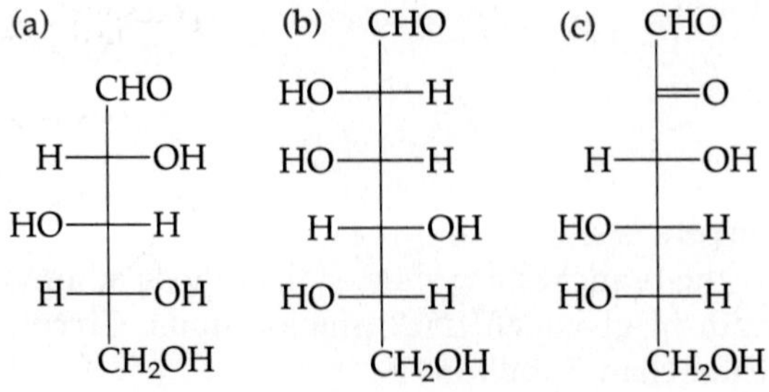

(a)
CHO
H—OH
HO—H
H—OH
CH_2OH

(b)
CHO
HO—H
HO—H
H—OH
HO—H
CH_2OH

(c)
CHO
=O
H—OH
HO—H
HO—H
CH_2OH

11.6 Classify each of the following monosaccharides as D or L.

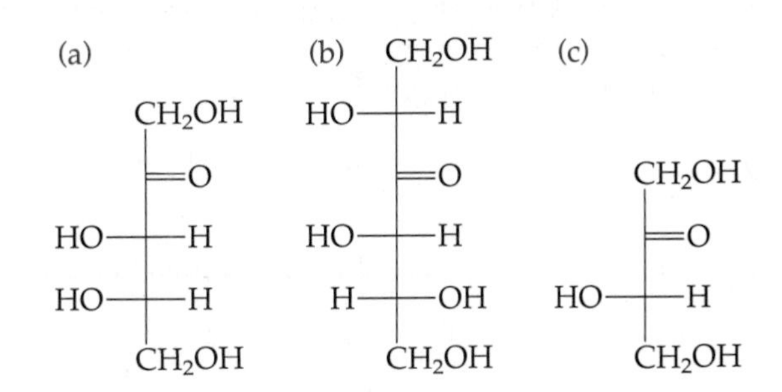

11.7 Draw the Fischer projection formulas of the D-ketopentoses with the carbonyl group at the C-2 atom.
11.8 Draw the Fischer projection formula of a ketopentose that is achiral.
11.9 Draw the Fischer projection formula of each of the following.
(a) L-xylose (b) L-erythrose (c) L-galactose
11.10 Draw the Fischer projection formula of each of the following.
(a) L-ribose (b) L-threose (c) L-mannose
11.11 Draw the Fischer projection formula of the C-2 epimer of each of the following.
(a) D-xylose (b) D-erythrose (c) D-galactose
11.12 Draw the Fischer projection formula of the C-3 epimer of each of the following.
(a) D-ribose (b) D-glucose (c) D-mannose

Haworth Projection Formulas

11.13 Draw the Haworth projection formula of the hemiacetal of 5-hydroxyhexanal.
11.14 Draw the Haworth projection formula of the hemiketal of 5-hydroxy-2-hexanone.
11.15 Draw the Haworth projection formula of the pyranose form of each of the following compounds.
(a) α-D-mannose (b) β-D-galactose (c) α-D-glucose (d) α-D-galactose
11.16 Draw the Haworth projection formula of the furanose form of each of the following compounds.
(a) α-D-fructose (b) β-D-fructose (c) α-D-ribulose (d) β-D-xylulose
11.17 Can D-erythrose exist as a pyranose? Explain.
11.18 What carbon atom contains the hydroxyl group required to form a pyranose form of D-ribose?
11.19 Identify the monosaccharide represented by each of the following structures.

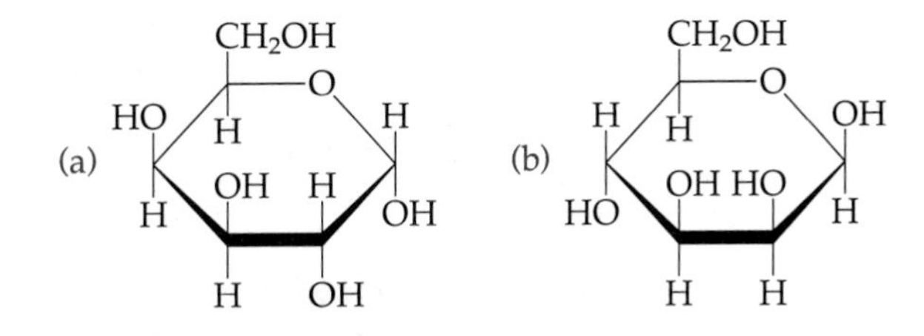

(c)

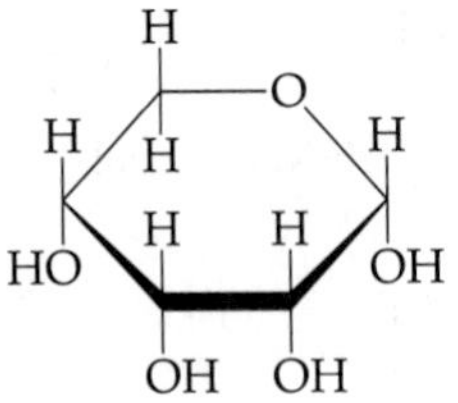

11.20 Identify the monosaccharide represented by each of the following structures.

(a) (b)

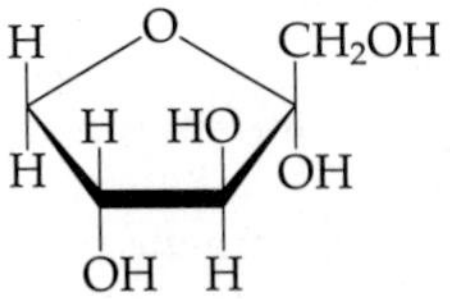

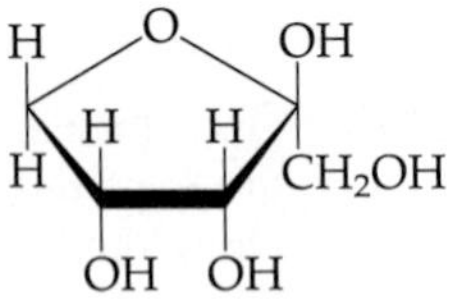

(c)

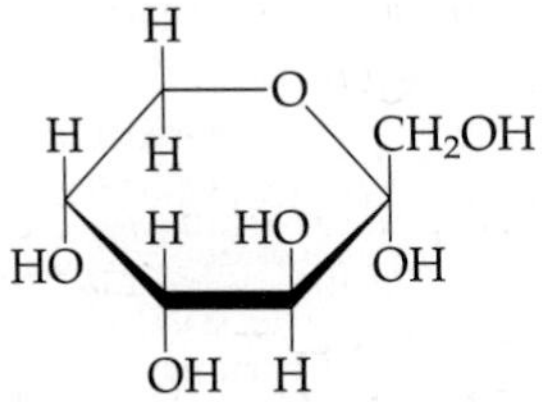

11.21 Name each compound in Exercise 11.19 and indicate the type of ring and configuration of the anomeric center.

11.22 Name each compound in Exercise 11.20 and indicate the type of ring and configuration of the anomeric center.

Conformations of Monosaccharides

11.23 Draw the chair conformations of β-galactopyranose and β-mannopyranose and compare the number of axial hydroxyl groups in each compound.

11.24 Draw the chair conformations of β-talopyranose and β-allopyranose and compare the number of axial hydroxyl groups in each compound.

Mutarotation

11.25 Can all aldopentoses and aldohexoses mutarotate? Why?

11.26 Can L-glucose mutarotate?

11.27 Which of the following compounds can mutarotate?

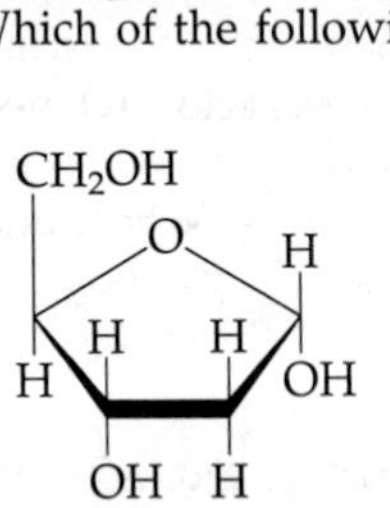

(I) 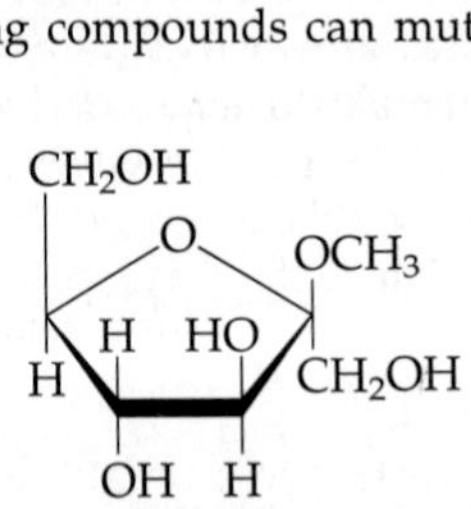 (II) 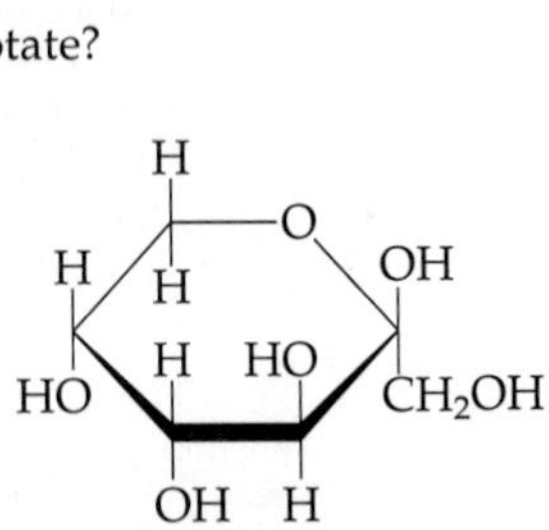 (III)

11.28 Which of the following compounds can mutarotate?

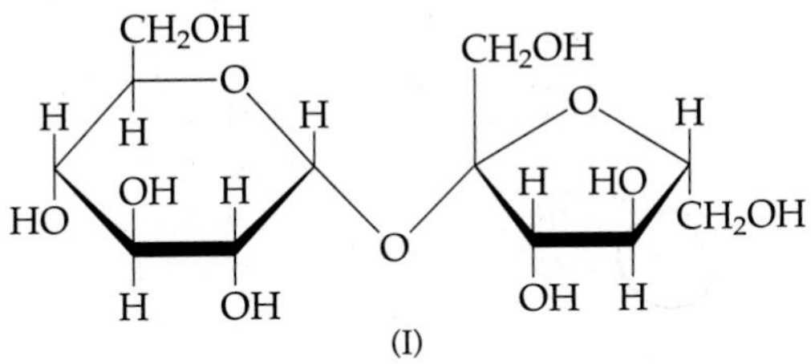

(I)

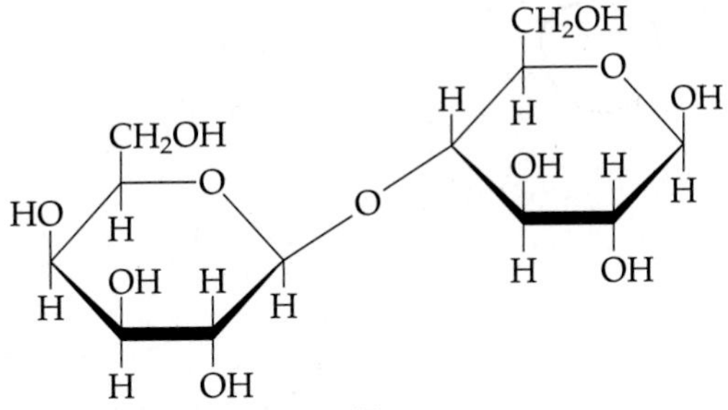

(II)

11.29 The $[\alpha]_D$ of the α- and β-anomers of D-galactose are +150.7 and +52.8, respectively. In water, mutarotation of D-galactose results in a specific rotation of +80.2. Which anomer predominates?

11.30 In solution, D-ribose forms an equilibrium mixture containing 6% α-furanose, 18% β-furanose, 20% α-pyranose, and 56% β-pyranose. Explain why the β-pyranose form predominates at equilibrium.

Reduction of Monosaccharides

11.31 Draw the Fischer projections of the alditols of D-erythrose and D-threose. One compound is optically active and the other is a meso compound. Explain why.

11.32 Which of the alditols of the D-pentoses are optically inactive? Explain why.

11.33 Reduction of D-fructose with sodium borohydride yields a mixture of two alditols. Explain why. Name the two alditols.

11.34 Reduction of D-tagatose with sodium borohydride yields a mixture of galactitol and talitol. What is the structure of D-tagatose?

Oxidation of Monosaccharides

11.35 Draw the structures for each of the following compounds.
(a) D-mannonic acid (b) D-galactonic acid (c) D-ribonic acid
(d) D-arabonic acid

11.36 Draw the structures for each of the following compounds.
(a) D-allonic acid (b) D-talonic acid (c) D-xylonic acid (d) D-lyxonic acid

11.37 Oxidation of D-erythrose and D-threose with nitric acid yields aldaric acids, one of which is optically inactive. Which one? Explain why.

11.38 Which of the D-aldopentoses will yield optically inactive aldaric acids when oxidized with nitric acid?

Isomerization of Monosaccharides

11.39 Draw the structures of the aldose and ketose that can exist in equilibrium with D-allose in basic solution.

11.40 Draw the structures of the aldose and ketose that can exist in equilibrium with D-galactose in basic solution.

11.41 Draw the structures of one aldose and one ketose that can exist in equilibrium with D-ribose in basic solution.

11.42 Draw the structures of two aldoses that can exist in equilibrium with D-xylulose in basic solution.

Glycosides

11.43 Draw the Haworth projection formulas of the two glycosides formed from the pyranose forms of glucose and ethyl alcohol.

11.44 Draw the Haworth projection formulas of the two glycosides formed from the furanose forms of fructose and ethyl alcohol.

Disaccharides

11.45 Determine the component monosaccharides of each of the following compounds and describe the type of glycosidic linkage in each.

(a)

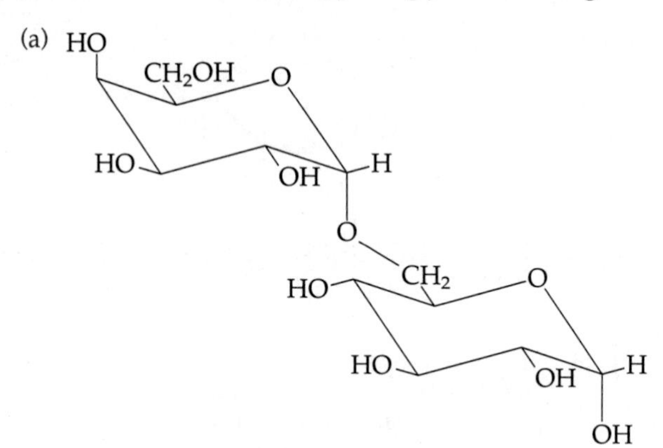

(b)

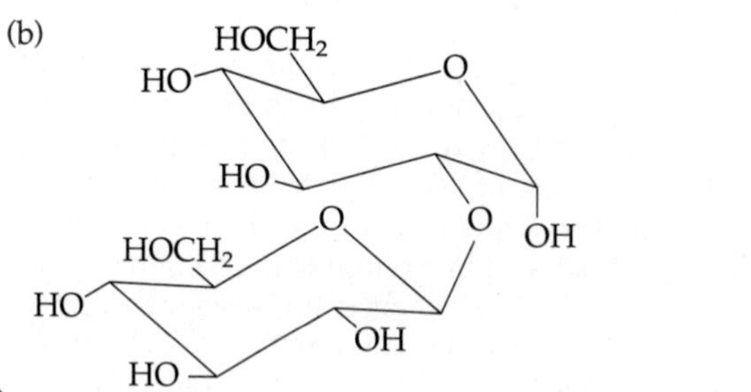

11.46 Determine the component monosaccharides of each of the following compounds and describe the type of glycosidic linkage in each.

(a)

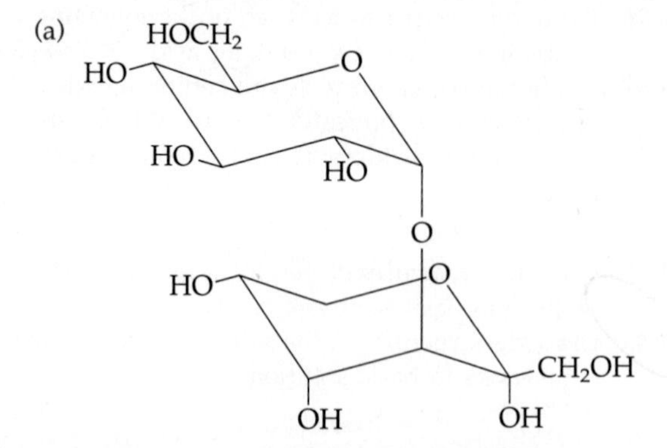

(b) 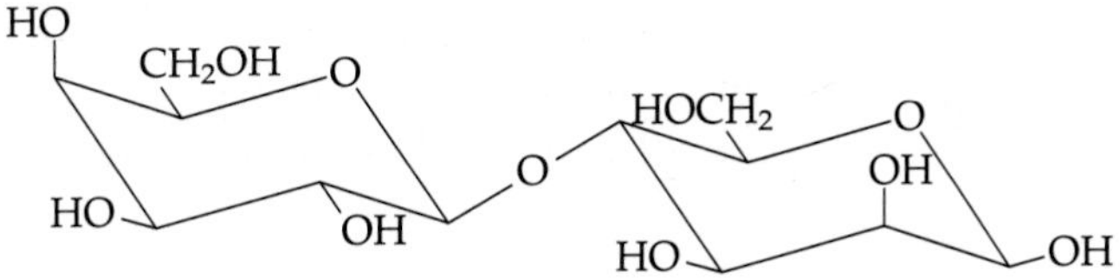 HO, CH_2OH, O, $HOCH_2$, O, OH, HO, OH, O, HO, OH

Structural Variations of Monosaccharides

11.47 The following "unusual" compounds are found in nature. What unique structural features make these compounds different from the monosaccharides studied in this chapter?

(a) 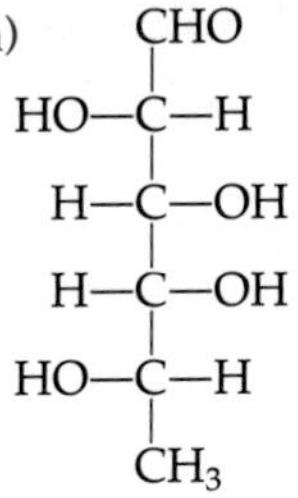
CHO
HO—C—H
H—C—OH
H—C—OH
HO—C—H
CH_3

(b) 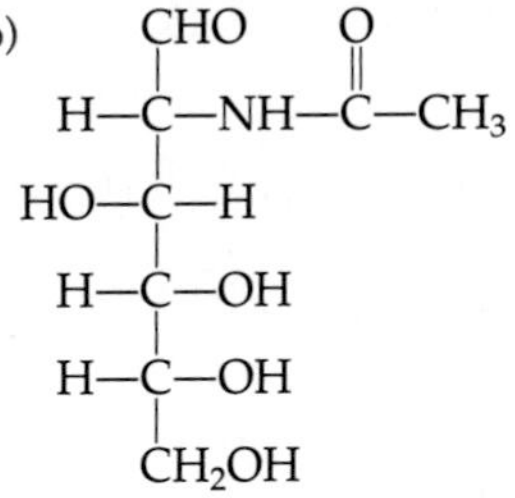
CHO O
H—C—NH—C—CH_3
HO—C—H
H—C—OH
H—C—OH
CH_2OH

11.48 The following compounds are obtained by hydrolysis of some naturally occurring polysaccharides. From what monosaccharides are they derived?

(a) 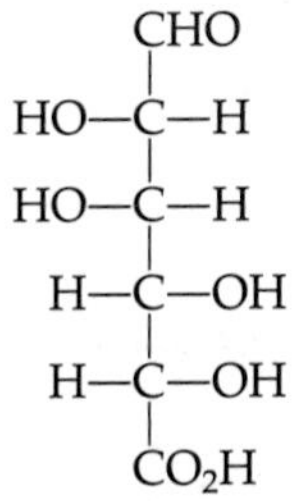
CHO
HO—C—H
HO—C—H
H—C—OH
H—C—OH
CO_2H

(b) 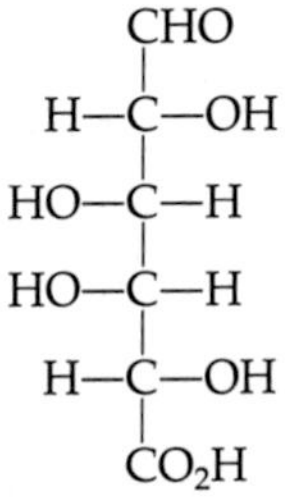
CHO
H—C—OH
HO—C—H
HO—C—H
H—C—OH
CO_2H

11.49 The term *deoxy* is used to indicate the replacement of a hydroxyl group by a hydrogen atom. The compound 3-deoxy-D-ribose is incorporated in cordycepin, an antibiotic. Draw the structure of 3-deoxy-D-ribose.

11.50 The carbohydrate daunosamine is contained in the antibiotic Adriamycin. Is daunosamine a D or L carbohydrate?

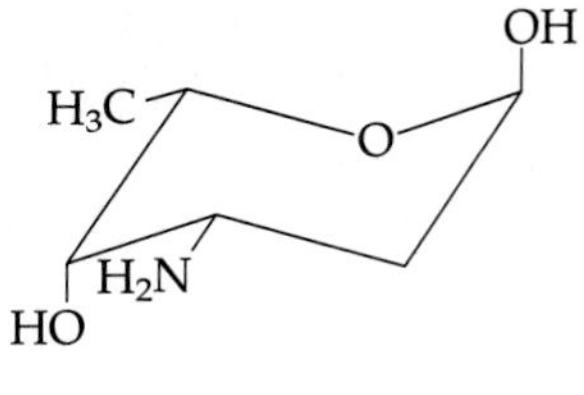

OH, H_3C, O, H_2N, HO

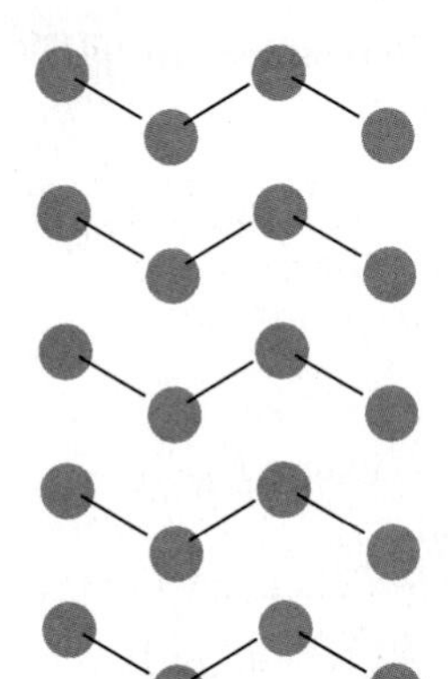

12

CARBOXYLIC ACIDS AND ESTERS

12.1 ACYL DERIVATIVES

In this chapter we will focus upon the chemistry of carboxylic acids and a family of closely related derivatives called esters. Both of these classes of compounds have an "RCO" unit called an **acyl group.** In a carboxylic acid, an —OH group is bonded to the acyl group. In an ester, an alkoxy (—OR) or phenoxy group (—OAr) is bonded to the acyl group.

$$\underset{\text{acyl group}}{R{-}\overset{\overset{\displaystyle O}{\|}}{C}{-}} \qquad \underset{\text{carboxylic acid}}{R{-}\overset{\overset{\displaystyle O}{\|}}{C}{-}O{-}H} \qquad \underset{\text{ester}}{R{-}\overset{\overset{\displaystyle O}{\|}}{C}{-}O{-}R}$$

Carboxylic acids contain a **carboxyl group.** The carboxyl group is represented in several ways.

$$-\overset{\overset{\displaystyle O}{\|}}{C}-OH \quad \text{or} \quad -COOH \quad \text{or} \quad -CO_2H$$

The carboxyl carbon atom is sp^2-hybridized. Three of its valence electrons participate in three σ bonds at 120° angles to one another (Figure 12.1). One

FIGURE 12.1
Bonding in Carboxylic Acids
The atoms bonded to the carboxyl carbon atom of a carboxylic acid all lie in a plane with the carboxyl carbon atom. The carboxyl carbon atom is sp^2-hybridized, and there is a π bond between the carbonyl oxygen atom and the carbon atom.

The π bond is formed by overlap of the $2p$ orbitals of carbon and oxygen.
An electron of the sp^2 hybrid orbital of the carbonyl carbon atom and an electron of the sp^3 orbital of the carbon atom of the alkyl group form a σ bond.
R
σ
σ
H—O
An electron of the sp^2 hybrid orbital of the carbonyl carbon atom and an electron of the sp^3 hybrid orbital of the oxygen atom form a σ bond.

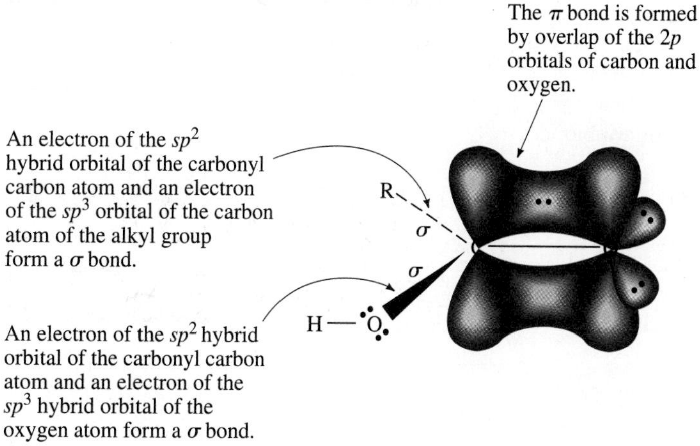

of the σ bonds is to a hydrogen atom, an alkyl group, or an aromatic group. The other two σ bonds are to oxygen atoms: one to the hydroxyl oxygen atom and the other to the carbonyl oxygen atom. The carbonyl carbon atom also has one electron in a $2p$ orbital. It forms a π bond with an electron in a $2p$ orbital of the carbonyl oxygen atom.

In this chapter, we will briefly describe other acyl derivatives. When the acyl group is linked to a substituent through a nitrogen atom, the compound is called an amide. The classification of amides is based on the number of carbon groups—including the acyl group—bonded to the nitrogen atom. Amides will be discussed later in Chapter 14 along with amines and other nitrogen-containing compounds.

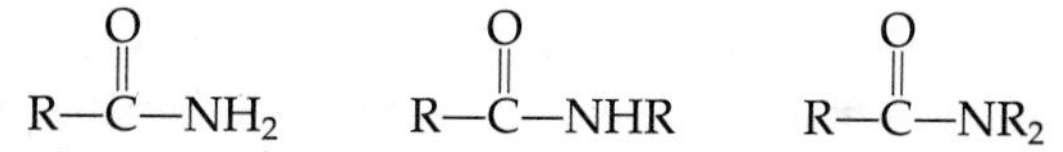

$$\underset{\text{primary amide}}{R{-}\overset{\overset{\large O}{\|}}{C}{-}NH_2} \qquad \underset{\text{secondary amide}}{R{-}\overset{\overset{\large O}{\|}}{C}{-}NHR} \qquad \underset{\text{tertiary amide}}{R{-}\overset{\overset{\large O}{\|}}{C}{-}NR_2}$$

When the acyl group is bonded to a chlorine atom, the derivative is called an **acid chloride.** These compounds are highly reactive and do not occur in nature. Acid chlorides are useful reagents for the laboratory synthesis of esters and amides. When two acyl groups are bonded to a common oxygen atom, the compound is an **acid anhydride.** These compounds are also useful reagents for laboratory synthesis of esters and amides. When a substituent is linked to an acyl group through a sulfur atom, the derivative is called a **thioester.** Thioesters are reactants in many biochemical acyl-transfer reactions.

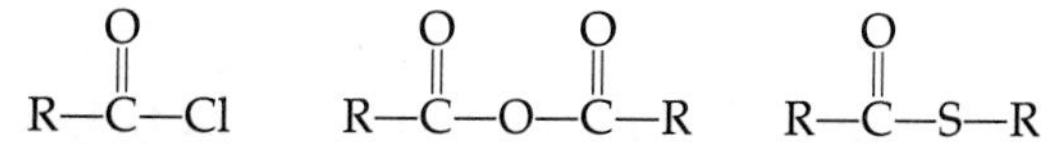

$$\underset{\text{acid chloride}}{R{-}\overset{\overset{\large O}{\|}}{C}{-}Cl} \qquad \underset{\text{acid anhydride}}{R{-}\overset{\overset{\large O}{\|}}{C}{-}O{-}\overset{\overset{\large O}{\|}}{C}{-}R} \qquad \underset{\text{thioester}}{R{-}\overset{\overset{\large O}{\|}}{C}{-}S{-}R}$$

FIGURE 12.2 Cyclic Acyl Derivatives
Coumarin is a six-membered lactone found in clover. Cantharidin, a five-membered cyclic anhydride found in a species of beetle, is a vesicant used to destroy warts on the skin. Penicillin G is a four-membered lactam used as an antibiotic.

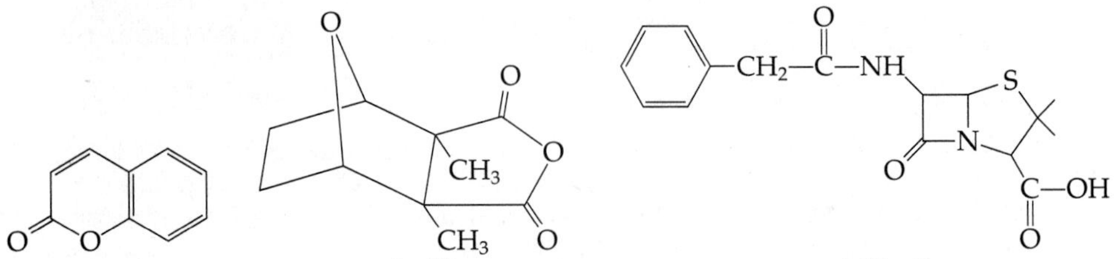

coumarin　　cantharidin　　penicillin G

Esters, amides, anhydrides, and thioesters may be part of a cyclic structure. Cyclic esters are called **lactones;** cyclic amides are called **lactams.** Cyclic acyl derivatives behave the same chemically as the acyclic compounds. Some compounds of biological interest that have acyl groups within ring structures are shown in Figure 12.2.

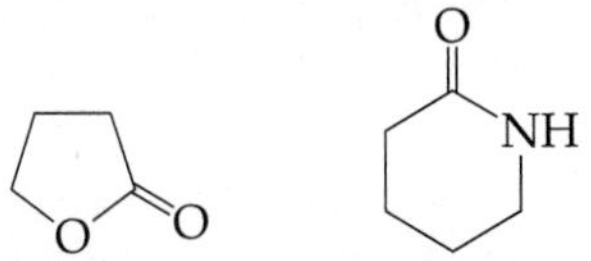

a lactone (a cyclic ester)　　a lactam (a cyclic amide)

12.2 NOMENCLATURE OF CARBOXYLIC ACID DERIVATIVES

Common Names

Carboxylic acids and their derivatives are abundant in nature and were among the first organic substances to be isolated. Because they have been known for so long, the common acids, esters, and other acyl compounds are often referred to by their common names. Some common acids are formic acid, HCO_2H, acetic acid, CH_3CO_2H, and benzoic acid, $C_6H_5CO_2H$. Both the common and the IUPAC names of a few commonly encountered carboxylic acids are given in Table 12.1.

In the common names, the positions of groups attached to the parent chain are designated alpha (α), beta (β), gamma (γ), delta (δ), and so forth. The —COOH group itself is not designated by a Greek letter.

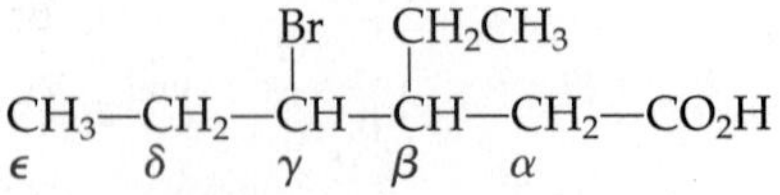

$$\underset{\epsilon}{CH_3}-\underset{\delta}{CH_2}-\underset{\gamma}{\overset{Br}{CH}}-\underset{\beta}{\overset{CH_2CH_3}{CH}}-\underset{\alpha}{CH_2}-CO_2H$$

γ-bromo-β-ethylcaproic acid

TABLE 12.1 Nomenclature of Unbranched Carboxylic Acids

Formula	Common name	IUPAC name
HCO_2H	formic acid	methanoic acid
CH_3CO_2H	acetic acid	ethanoic acid
$CH_3CH_2CO_2H$	propionic acid	propanoic acid
$CH_3(CH_2)_2CO_2H$	butyric acid	butanoic acid
$CH_3(CH_2)_3CO_2H$	valeric acid	pentanoic acid
$CH_3(CH_2)_4CO_2H$	caproic acid	hexanoic acid
$CH_3(CH_2)_6CO_2H$	caprylic acid	octanoic acid
$CH_3(CH_2)_8CO_2H$	capric acid	decanoic acid
$CH_3(CH_2)_{10}CO_2H$	lauric acid	dodecanoic acid
$CH_3(CH_2)_{12}CO_2H$	myristic acid	tetradecanoic acid
$CH_3(CH_2)_{14}CO_2H$	palmitic acid	hexadecanoic acid
$CH_3(CH_2)_{16}CO_2H$	stearic acid	octadecanoic acid

Some unbranched carboxylic acids contain a —COOH group at each end of the chain. These are *dicarboxylic acids*. Some are important metabolic intermediates.

$$HO-\overset{O}{\overset{\|}{C}}-(CH_2)_2-\overset{O}{\overset{\|}{C}}-OH$$
succinic acid

$$HO-\overset{O}{\overset{\|}{C}}-(CH_2)_3-\overset{O}{\overset{\|}{C}}-OH$$
glutaric acid

$$HO-\overset{O}{\overset{\|}{C}}-(CH_2)_4-\overset{O}{\overset{\|}{C}}-OH$$
adipic acid

IUPAC Names

The IUPAC rules to name carboxylic acids are similar to those outlined for aldehydes.

1. Select the longest continuous carbon chain that contains the carbonyl carbon atom as the parent chain. Replace the final *-e* of the parent hydrocarbon with the ending *-oic acid*.

$$CH_3-\overset{CH_3}{\overset{|}{C}H}-CH_2-\overset{O}{\overset{\|}{C}}-OH$$

The parent alkane is butane.
This is a substituted butanoic acid.
A methyl group is attached to the butanoic acid chain.

2. Number the parent chain by assigning the number 1 to the carboxylic acid carbon atom. The number 1 is not used in the name to indicate the position of the carboxylic acid carbon atom because it is understood to be located at the end of the chain. The names and locations of any substituents are added as prefixes to the parent name.

$$\underset{5}{CH_3}-\underset{\underset{H}{|}}{\overset{\overset{Br}{|}}{\underset{4}{C}}}-\underset{\underset{CH_3}{|}}{\overset{\overset{H}{|}}{\underset{3}{C}}}-\underset{2}{CH_2}-\overset{O}{\overset{\|}{\underset{1}{C}}}-OH$$
4-bromo-3-methylpentanoic acid

3. The carboxylic acid functional group has a higher priority than aldehyde, ketone, halogen, hydroxyl, and alkoxyl groups. The names and locations of these substituents are indicated as prefixes to the name of the parent carboxylic acid.

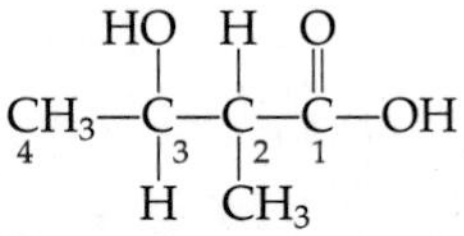

```
      HO  H   O
      |   |   ||
CH3—C—C—C—OH
 4    |3  |2  1
      H   CH3
```

3-hydroxy-2-methylbutanoic acid

4. The carboxylic acid group has a higher priority than double or triple bonds. The name of a carboxylic acid that contains a double or triple bond is obtained by replacing the final *-e* of the name of the parent alkene or alkyne name with the suffix *-oic acid*. The position of the multiple bond is indicated by a prefix.

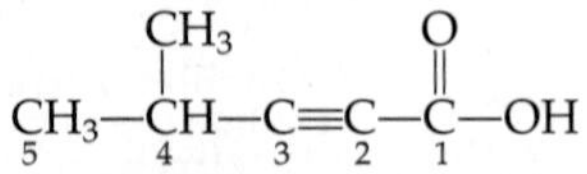

```
      CH3            O
      |              ||
CH3—CH—C≡C—C—OH
 5    4   3  2  1
```

4-methyl-2-pentynoic acid

5. If a carboxylic acid contains an aldehyde or ketone, the carbonyl group is indicated by the prefix *oxo-*. The priority order is carboxylic acid > aldehyde > ketone.

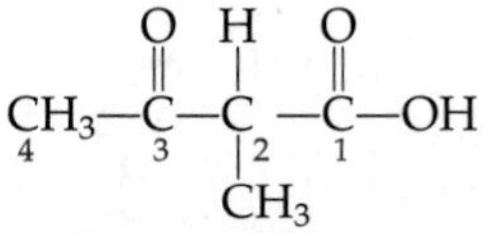

```
      O   H   O
      ||  |   ||
CH3—C—C—C—OH
 4    3   |2  1
          CH3
```

2-methyl-3-oxobutanoic acid

6. Compounds that have a $—CO_2H$ group bonded to a cycloalkane ring are named as derivatives of the cycloalkane, and the suffix *carboxylic acid* is added. The carbon atom to which the carboxylic carbon atom is bonded is assigned the number 1, but this number is not included in the name.

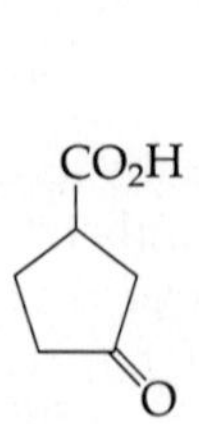

CO_2H O

3-oxocyclopentanecarboxylic acid

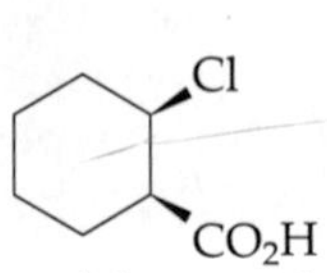

Cl CO_2H

cis-2-chlorocyclohexanecarboxylic acid

EXAMPLE 12.1 The structure of oleic acid, an unsaturated carboxylic acid present as an ester in vegetable oils, is shown. What is the IUPAC name of oleic acid?

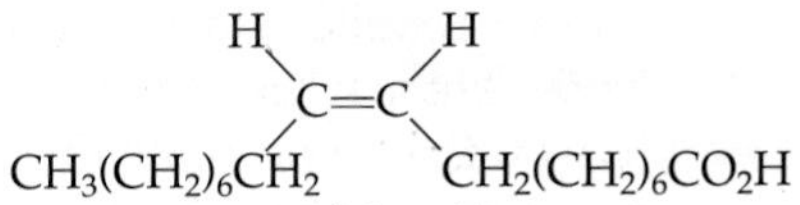

$$\underset{\text{oleic acid}}{\overset{H\qquad\qquad H}{CH_3(CH_2)_6CH_2-C=C-CH_2(CH_2)_6CO_2H}}$$

Solution First, we determine the length of the longest continuous chain that contains the —COOH group: it contains 18 carbon atoms. The double bond is located at the C-9 atom in the chain, numbering from the carboxyl group on the right. Thus, the compound is a 9-octadecenoic acid. The configuration about the double bond is Z, and thus the complete name is (Z)-9-octadecenoic acid.

Names of Carboxylates

The conjugate base of a carboxylic acid is a **carboxylate** anion. The name of the conjugate base is obtained by changing the *-oic acid* ending to *-oate.* A metal salt of the conjugate base is named by the name of the metal ion followed by the name of the carboxylate anion.

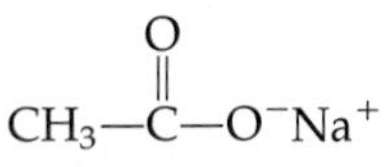

$CH_3-C(=O)-O^-Na^+$

sodium ethanoate
(sodium acetate)

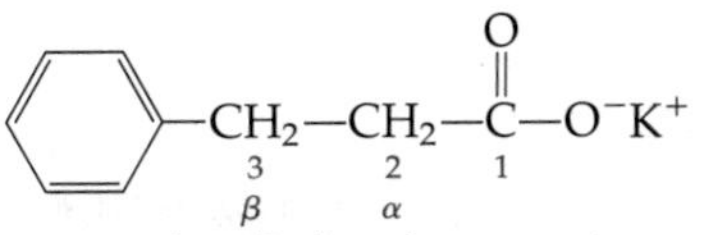

$C_6H_5-\underset{3,\ \beta}{CH_2}-\underset{2,\ \alpha}{CH_2}-\underset{1}{C}(=O)-O^-K^+$

potassium 3-phenylpropanoate
(potassium β-phenylpropionate)

Names of Esters

To name an ester, we first write the name of the alkyl group bonded to oxygen (—OR or —OAr). The acyl portion of the ester is derived from a carboxylic acid and is named as a carboxylate. Three examples are given below. The alkyl portion of each molecule is shown on the right side of the structure. The alkyl name is written first in the name of the ester regardless of how the structure is drawn.

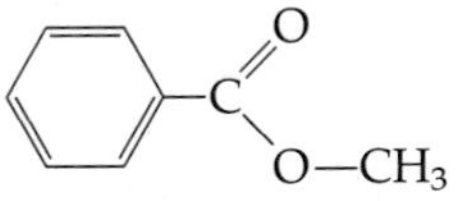

$C_6H_5-C(=O)-O-CH_3$

methyl benzoate

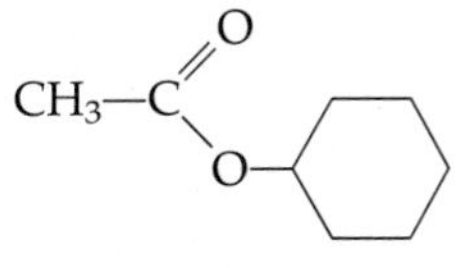

$CH_3-C(=O)-O-C_6H_{11}$

cyclohexyl ethanoate

$C_6H_{11}-C(=O)-O-CH_3$

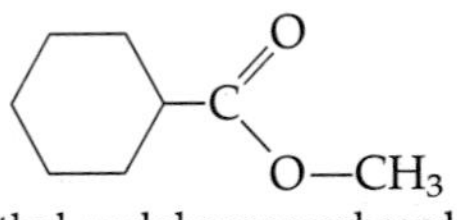

methyl cyclohexanecarboxylate

Like carboxylic acids, some commonly encountered esters are often referred to by their common names. This is especially true of esters derived from acetic acid and formic acid. Two examples are given below.

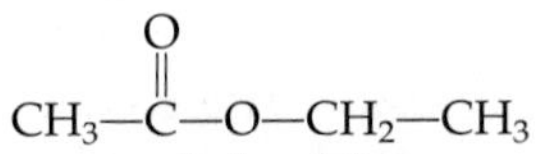

$CH_3-C(=O)-O-CH_2-CH_3$

ethyl acetate

$H-C(=O)-O-CH(CH_3)-CH_3$

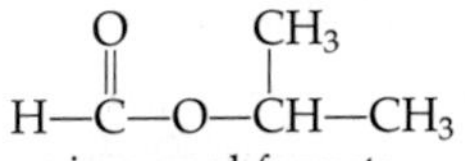

isopropyl formate

Names of Acid Halides

In an acid halide, a halogen atom is attached to an acyl group. To name halides, we change the ending *-oic acid* of carboxylic acids to the ending *-oyl halide* where the name of the halide is fluoride, chloride, bromide, or iodide. An acid halide functional group bonded to a cycloalkane ring is named as a *carbonyl halide.*

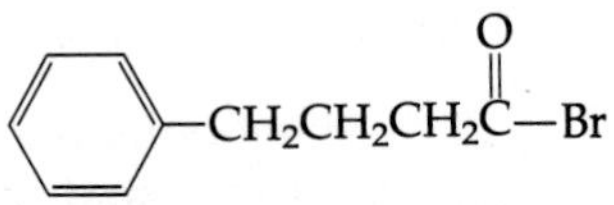

$C_6H_5-CH_2CH_2CH_2C(=O)-Br$

4-phenylbutanoyl bromide

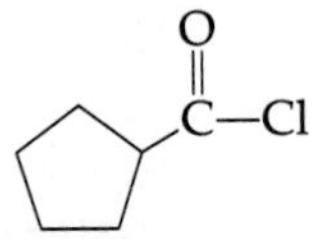

$C_5H_9-C(=O)-Cl$

cyclopentanecarbonyl chloride

Names of Acid Anhydrides

An acid anhydride consists of two acyl groups bonded through a bridging oxygen atom. Although acid anhydrides can have two different acyl groups, compounds containing identical acyl groups are more common. They are named by replacing the word suffix *-oic acid* with *-oic anhydride.*

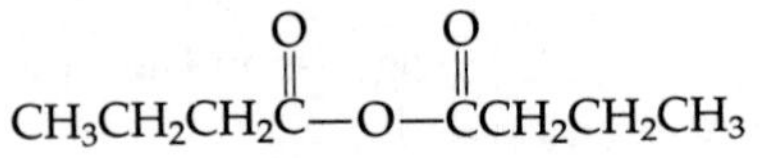

$CH_3CH_2CH_2C(=O)-O-C(=O)CH_2CH_2CH_3$

butanoic anhydride

$C_6H_5-C(=O)-O-C(=O)-C_6H_5$

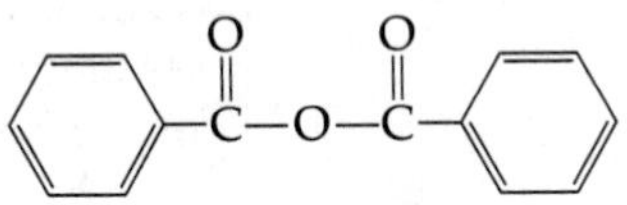

benzoic anhydride

Names of Amides

In an amide, a nitrogen atom, as in groups such as $-NH_2$, is attached to an acyl group. Amides are named by replacing the suffix for the acid (*-oic acid*) with the name *-amide.* For example, the amide derived from ethanoic acid is ethanamide. The names of two primary amides with slightly more complex structures are shown below.

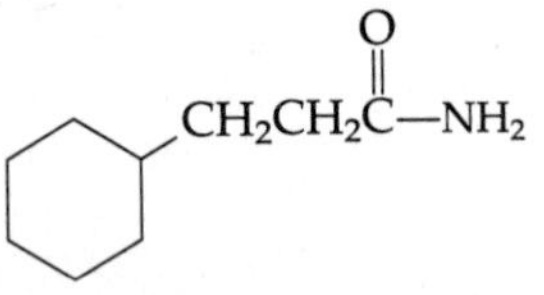

$C_6H_{11}-CH_2CH_2C(=O)-NH_2$

3-cyclohexylpropanamide

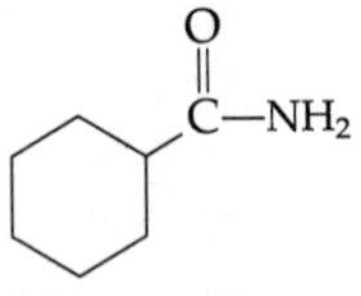

$C_6H_{11}-C(=O)-NH_2$

cyclohexanecarboxamide

In secondary and tertiary amides, the nitrogen atom is bonded to one or more alkyl or aryl groups instead of hydrogen atoms. The names of amides with groups bonded to nitrogen will be discussed along with names of amines in Chapter 14.

12.3 PHYSICAL PROPERTIES OF CARBOXYLIC ACIDS AND ESTERS

The physical properties of carboxylic acids and esters are distinctly different. We can understand these differences by considering the types of intermolecular interactions that are possible in each class of compounds. Differences in biological properties—such as odor—that depend on physiological responses are not well understood.

Carboxylic Acids: Boiling Points and Solubilities

Carboxylic acids with low molecular weights are liquids at room temperature; those with higher molecular weights are waxlike solids. The boiling points of carboxylic acids are high (Table 12.2). Carboxylic acids interact very strongly by forming hydrogen-bonded dimers. These dimers have higher boiling points than substances of comparable molecular weights that do not form dimers.

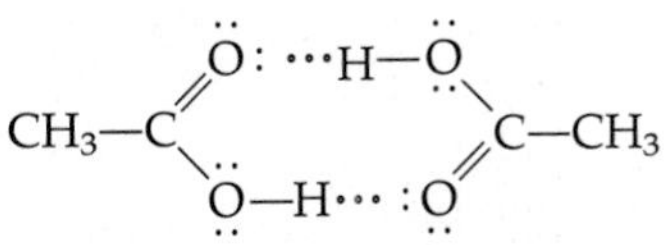

Carboxylic acids with low molecular weights are soluble in water because the carboxyl group forms several hydrogen bonds with water. A carboxylic acid serves both as a hydrogen-bond donor through its hydroxyl hydrogen atom and as a hydrogen-bond acceptor through the lone-pair electrons of both oxygen atoms. The solubility of carboxylic acids, like that of alcohols, decreases with increasing chain length because the long, nonpolar hydrocarbon chain dominates the physical properties of the acid.

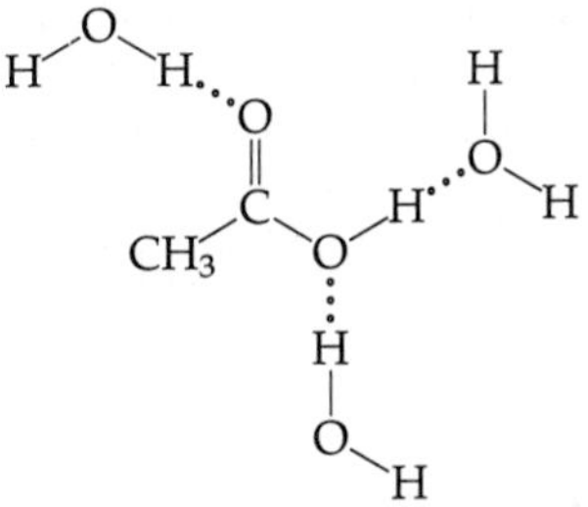

hydrogen bonds between acetic acid and water

TABLE 12.2 Physical Properties of Carboxylic Acids

IUPAC name	Melting point (°C)	Boiling point (°C)	Solubility (g/100 g H_2O, at 20 °C)
methanoic acid	8	101	miscible
ethanoic acid	17	118	miscible
propanoic acid	−21	141	miscible
butanoic acid	−5	164	miscible
pentanoic acid	−34	186	4.97
hexanoic acid	−3	205	0.96
octanoic acid	17	239	0.068
decanoic acid	32	270	0.015
dodecanoic acid	44	299	0.0055

Esters: Boiling Points and Solubilities

Esters are polar molecules, but their boiling points are lower than those of carboxylic acids and alcohols of similar molecular weight because intermolecular hydrogen bonding between ester molecules is impossible.

$$\underset{\substack{\text{1-pentanol} \\ \text{(bp 138 °C)}}}{CH_3CH_2CH_2CH_2CH_2OH} \qquad \underset{\substack{\text{butanoic acid} \\ \text{(bp 164°C)}}}{CH_3CH_2CH_2\overset{\overset{\Large O}{\|}}{C}{-}OH} \qquad \underset{\substack{\text{ethyl acetate} \\ \text{(bp 77 °C)}}}{CH_3\overset{\overset{\Large O}{\|}}{C}{-}O{-}CH_2CH_3}$$

Esters can form hydrogen bonds through their oxygen atoms to the hydrogen atoms of water molecules. Thus, esters are slightly soluble in water. However, because esters do not have a hydrogen atom to form a hydrogen bond to an oxygen atom of water, they are less soluble than carboxylic acids. The solubilities and boiling points of some esters are listed in Table 12.3.

Odors of Acids and Esters

Liquid carboxylic acids have sharp, unpleasant odors. For example, butanoic acid occurs in rancid butter and aged cheese. Caproic, caprylic, and capric acids have the smell of goats. (The Latin word for goat, *caper*, is the source of the common names of these acids.)

The odors of esters are distinctly different from those of the corresponding acids. Acids have unpleasant smells, but esters have fruity smells. In fact, the odors of many fruits are due to esters. For example, ethyl ethanoate is found in pineapples, 3-methylbutyl ethanoate in apples and bananas, 3-methylbutyl 3-methylbutanoate in apples, and octyl acetate in oranges.

The demand for processed foods in our society and the expectation that such foods will taste and smell "fresh" has created problems for the food industry. Esters have low boiling points and are driven off when foods are

TABLE 12.3 Physical Properties of Esters

Common name	Formula	Boiling point (°C)	Solubility (g/100 g H_2O, at 20 °C)
methyl formate	HCO_2CH_3	32	miscible
methyl acetate	$CH_3CO_2CH_3$	57	24.4
methyl propionate	$CH_3CH_2CO_2CH_3$	80	1.8
methyl butyrate	$CH_3(CH_2)_2CO_2CH_3$	102	0.5
methyl valerate	$CH_3(CH_2)_3CO_2CH_3$	126	0.2
methyl caproate	$CH_3(CH_2)_4CO_2CH_3$	151	0.06
methyl caprylate	$CH_3(CH_2)_6CO_2CH_3$	208	0.007
ethyl acetate	$CH_3CO_2CH_2CH_3$	77	7.4
propyl acetate	$CH_3CO_2CH_2CH_2CH_3$	102	1.9
butyl acetate	$CH_3CO_2CH_2CH_2CH_2CH_3$	125	0.9

TABLE 12.4 Esters Used as Flavoring Agents

Name of ester	Formula	Flavor
methyl butyrate	$CH_3{-}O{-}\overset{\overset{\displaystyle O}{\parallel}}{C}{-}CH_2CH_2CH_3$	apple
pentyl butyrate	$CH_3(CH_2)_4{-}O{-}\overset{\overset{\displaystyle O}{\parallel}}{C}{-}CH_2CH_2CH_3$	apricot
pentyl acetate	$CH_3(CH_2)_4{-}O{-}\overset{\overset{\displaystyle O}{\parallel}}{C}{-}CH_3$	banana
octyl acetate	$CH_3(CH_2)_7{-}O{-}\overset{\overset{\displaystyle O}{\parallel}}{C}{-}CH_3$	orange
ethyl butyrate	$CH_3CH_2{-}O{-}\overset{\overset{\displaystyle O}{\parallel}}{C}{-}CH_2CH_2CH_3$	pineapple
ethyl formate	$CH_3CH_2{-}O{-}\overset{\overset{\displaystyle O}{\parallel}}{C}{-}H$	rum

heated. Thus, to make processed food more attractive, esters are added back to the food. In some cases the esters are the same as those lost in heating. Nevertheless, government regulations require that the added esters be identified as additives on the label. Thus, some individuals claim that the product is not "natural" and should not be consumed.

The esters used in some products are not necessarily the same as those in natural fruits, but they produce the same odor or taste. The choice of esters may be dictated by their cost and availability. Some of these flavoring agents are listed in Table 12.4. Although the esters are not the same as those that occur naturally in the fruit, the product is not dangerous. The structures are similar to those of "natural" esters.

EXAMPLE 12.2 Isobutyl formate has the odor of raspberries. Draw its structural formula.

Solution The ester is given a common name. The acid portion of the ester is formic acid. The alcohol portion is isobutyl alcohol.

$$\underset{\text{formic acid}}{H{-}\overset{\overset{\displaystyle O}{\parallel}}{C}{-}OH} \qquad \underset{\text{isobutyl alcohol}}{CH_3{-}\overset{\overset{\displaystyle CH_3}{|}}{CH}{-}CH_2OH}$$

The ester can be represented in two ways depending on which component is drawn on the left of the structure. The two structures are identical.

$$H-\overset{\overset{O}{\|}}{C}-O-CH_2-\overset{\overset{CH_3}{|}}{CH}-CH_3 \quad \text{or} \quad CH_3-\overset{\overset{CH_3}{|}}{CH}-CH_2-O-\overset{\overset{O}{\|}}{C}-H$$

isobutyl formate

12.4 ACIDITY OF CARBOXYLIC ACIDS

To discuss the acidity of carboxylic acids, let's consider the general case in which an acid, H—A, donates a proton to water to give the conjugate base of the acid, A^-, and H_3O^+. In this process, the H—A bond breaks heterolytically, and the conjugate base A^- acquires a negative charge. Thus, the ionization of the acid, which is reflected in the acid dissociation constant, K_a, depends on both the strength of the H—A bond and the stability of A^- in the solvent. Although acetic acid and other carboxylic acids are weak acids, they are far more acidic than alcohols or phenols. The K_a for acetic acid is about 10^{11} times larger than the K_a for ethanol.

$$CH_3CH_2OH + H_2O \rightleftharpoons CH_3CH_2O^- + H_3O^+ \qquad K_a = 1 \times 10^{-16}$$

$$CH_3CO_2H + H_2O \rightleftharpoons CH_3CO_2^- + H_3O^+ \qquad K_a = 1.8 \times 10^{-5}$$

Resonance Stabilization of the Carboxylate Ion

Carboxylic acids are much more acidic than alcohols because the product of the ionization reaction, the carboxylate anion, is resonance-stabilized. Stabilization of the conjugate base increases the equilibrium constant. The acid dissociations of both ethanol and acetic acid produce conjugate bases that have a negative charge on oxygen. Each oxygen atom in the acetate ion bears one-half the negative charge, whereas in the ethoxide ion ($CH_3CH_2O^-$) the negative charge is concentrated on a single oxygen atom. Acetic acid is far more acidic than ethanol because the acetate ion is resonance-stabilized.

$$CH_3-C(=O)-O^- \longleftrightarrow CH_3-C(-O^-)=O$$

resonance in a carboxylate ion

Inductive Effect on Acidity

The acidity of carboxylic acids is also partly the result of an inductive effect. That is, the carbonyl group polarizes the H—O bond by attracting electrons through the σ-bonding network. The withdrawal of electron density from the H—O bond weakens it and thus increases the acidity of the ionizable hydrogen atom.

The acidity of carboxylic acids is also affected by an inductive effect caused by the alkyl or aryl group attached to the carbonyl carbon atom. An alkyl group is electron-releasing compared to hydrogen. This release of electrons to the carboxyl group stabilizes the acid and slightly destabilizes the

TABLE 12.5 Effect of Substituents on the Acidity of Carboxylic Acids

Name of acid	Formula	K_a	pK_a
acetic acid	CH_3CO_2H	1.8×10^{-5}	4.7
butyric acid	$CH_3CH_2CH_2CO_2H$	1.3×10^{-5}	4.9
isobutyric acid	$(CH_3)_2CHCO_2H$	1.6×10^{-5}	4.8
fluoroacetic acid	FCH_2CO_2H	2.5×10^{-3}	2.6
chloroacetic acid	$ClCH_2CO_2H$	1.4×10^{-3}	2.9
dichloroacetic acid	Cl_2CHCO_2H	5.0×10^{-2}	1.3
trichloroacetic acid	Cl_3CCO_2H	2.5×10^{-1}	0.6

conjugate base. Thus, acetic acid ($pK_a = 4.72$) is a weaker acid than formic acid ($pK_a = 3.75$). An aryl group is electron-withdrawing relative to an alkyl group. Thus, benzoic acid ($pK_a = 4.19$) is a stronger acid than acetic acid. The pK_a values of some carboxylic acids are given in Table 12.5.

Many commonly encountered carboxylic acids have electron-withdrawing groups bonded to their carbon skeletons. If a carboxylic acid has an electronegative group attached to its α-carbon atom, its acidity increases. For example, halogen atoms pull electrons away from the carbon skeleton and indirectly from the O—H bond. As a consequence, the proton is more easily removed, and the K_a value is larger (Table 12.5).

Electrons are pulled toward the chlorine atom.

$Cl \leftarrow CH_2 \leftarrow C(=O) \text{—} O \leftarrow H$

Electrons are pulled toward the oxygen atom owing to the net effect of the chlorine atom pulling electrons toward itself.

Electrons are pulled toward the carbon atom bonded to the chlorine atom.

EXAMPLE 12.3 Pyruvic acid is a key metabolic intermediate in oxidative processes that provide energy for the growth and maintenance of cells. Its pK_a is 2.5, indicating that it is about 100 times more acidic than propanoic acid ($pK_a = 4.7$). Explain why.

$$CH_3\text{—}C(=O)\text{—}C(=O)\text{—}OH$$

pyruvic acid

Solution Because the pK_a of pyruvic acid is smaller than that of propanoic acid, pyruvic acid is a stronger acid. Thus, the carbonyl group α to the carboxyl group pulls electrons away from the H—O bond. We recall that the carbonyl group is very polar, and that the carbonyl carbon atom has a partial positive charge. As a consequence, the carbonyl atom of the ketone group inductively pulls electron density from the carboxylic acid group, the polarity of the H—O bond increases, and the acidity increases.

$$\overset{\delta-}{O}{=}\underset{\delta+}{C}(CH_3)-C(=O)-OH$$

Salts of Carboxylic Acids

Because carboxylic acids are weak acids, carboxylate anions are relatively strong bases. Carboxylate salts react with hydronium ion to form the carboxylic acid.

$$\underset{\text{cyclobutanecarboxylate}}{C_4H_7-CO_2^-} + H_3O^+ \longrightarrow \underset{\text{cyclobutanecarboxylic acid}}{C_4H_7-CO_2H} + H_2O$$

The reactions of carboxylic acids with hydroxide ions and the reactions of carboxylate salts with hydronium ions have some practical applications in separating acids from mixtures. Because they are ionic, carboxylate salts are more soluble in water than their corresponding carboxylic acids. Carboxylic acids are often separated from other nonpolar organic compounds in the laboratory by adding a solution of sodium hydroxide to form the more soluble carboxylate salt. Consider, for example, a mixture of decanol and decanoic acid. Decanol is not soluble in water and does not react with sodium hydroxide. However, decanoic acid reacts with sodium hydroxide and thus dissolves in the basic solution.

$$\underset{\text{(insoluble in water)}}{CH_3(CH_2)_8CH_2OH} + OH^- \not\longrightarrow \text{no reaction}$$

$$\underset{\text{(insoluble in water)}}{CH_3(CH_2)_8CO_2H} + OH^- \longrightarrow \underset{\text{(soluble in water)}}{CH_3(CH_2)_8CO_2^-} + H_2O$$

Undissolved decanol is physically separated from the basic solution. Then, HCl is added to neutralize the basic solution, and insoluble decanoic acid separates from the aqueous solution.

$$\underset{\text{(soluble in water)}}{CH_3(CH_2)_8CO_2^-} + H_3O^+ \longrightarrow \underset{\text{(insoluble in water)}}{CH_3(CH_2)_8CO_2H} + H_2O$$

This procedure is used to isolate acids from mixtures in nature. It is also used to purify acids produced by chemical synthesis in the laboratory.

12.5 SYNTHESIS OF CARBOXYLIC ACIDS

We have previously described two oxidative methods that are commonly used to prepare carboxylic acids. We recall that either aldehydes or alcohols are oxidized by Jones reagent (Section 8.8) to produce carboxylic acids.

$$\text{3-cyclohexenecarbaldehyde} \xrightarrow[H_2SO_4]{CrO_3} \text{3-cyclohexenecarboxylic acid}$$

$$\underset{\text{1-decanol}}{CH_3(CH_2)_8CH_2OH} \xrightarrow[H_2SO_4]{CrO_3} \underset{\text{decanoic acid}}{CH_3(CH_2)_8CO_2H}$$

Alkylbenzenes are oxidized by potassium permanganate to give benzoic acids. We recall that the entire side chain is oxidized in this reaction. The oxidation of tetralin to produce phthalic acid is an example.

$$\text{tetralin} \xrightarrow{KMnO_4} \text{phthalic acid}$$

Carboxylic acids can also be made by two more general methods starting from haloalkanes. These methods provide structures that have one more carbon atom than the reactant haloalkane.

In Chapter 10 we saw that the Grignard reagent acts as a nucleophile, and that it reacts with the carbonyl group of aldehydes or ketones. A similar reaction occurs between a Grignard reagent and the carbon–oxygen double bond of carbon dioxide to yield the magnesium salt of a carboxylic acid. Adding aqueous acid to the salt gives the carboxylic acid.

$$R{-}MgBr + O{=}C{=}O \longrightarrow R{-}\overset{O}{\overset{\|}{C}}{-}O{-}MgBr \xrightarrow{H_3O^+} R{-}\overset{O}{\overset{\|}{C}}{-}OH$$

Starting from the haloalkane, the reaction sequence requires three steps. First, the haloalkane is converted to a Grignard reagent. Second, the ether solution of the Grignard reagent is poured over solid carbon dioxide (dry ice). Finally, the reaction mixture is acidified with aqueous acid.

$$R{-}X \xrightarrow[\text{ether}]{Mg} R{-}MgX \xrightarrow[2.\ H_3O^+]{1.\ CO_2} R{-}CO_2H$$

The reaction sequence can be presented by using a single arrow with the three steps listed.

$$p\text{-}BrC_6H_4CH_3 \xrightarrow[\text{2. } CO_2 \;\; \text{3. } H_3O^+]{\text{1. Mg, ether}} p\text{-}HO_2CC_6H_4CH_3$$

The second synthesis that adds one carbon atom to the parent chain of the reacting haloalkane is an S_N2 displacement of halogen by cyanide ion (Chapter 7). The resulting product, called a nitrile (RCN), can be hydrolyzed to produce a carboxylic acid. The chemistry of nitriles will be presented in Chapter 14.

$$R{-}X + {}^{-}{:}C{\equiv}N{:} \longrightarrow R{-}C{\equiv}N{:} + X^{-}$$

$$R{-}C{\equiv}N \xrightarrow{H_3O^+} R{-}\overset{\overset{\large O}{\|}}{C}{-}OH + NH_4^{+}$$

We recall that substitution reactions of the S_N2 type are most effective with primary haloalkanes; elimination reactions decrease the yield for secondary haloalkanes.

$$C_6H_5CH_2{-}Br \xrightarrow{CN^-} C_6H_5CH_2{-}CN \xrightarrow{H_3O^+} C_6H_5CH_2CO_2H$$

benzyl bromide → phenylacetic acid

EXAMPLE 12.4 Suggest a synthesis that accomplishes the following transformation.

$$m\text{-}HOC_6H_4CH_2Br \longrightarrow m\text{-}HOC_6H_4CH_2CO_2H$$

Solution To carry out this synthesis, one carbon atom must be added to the side chain of the aromatic compound. Which of the two methods given in this section should be selected? Can a Grignard be made from the *m*-hydroxybenzyl bromide? No, because the phenolic hydroxyl group has an acidic proton that would destroy the Grignard reagent.

Consider the substitution reaction of the starting material with cyanide ion to give the nitrile. Because benzyl bromide is a primary haloalkane, it readily reacts in an S_N2 reaction. The resulting nitrile is hydrolyzed to form the desired acid. This sequence is the method of choice.

$$HO\text{-}C_6H_4\text{-}CH_2Br \xrightarrow{CN^-} HO\text{-}C_6H_4\text{-}CH_2CN \xrightarrow{H_3O^+} HO\text{-}C_6H_4\text{-}CH_2CO_2H$$

12.6 NUCLEOPHILIC ACYL SUBSTITUTION

In Chapter 7 we saw that a nucleophile (Nu: or Nu:$^-$) can displace a leaving group (L) bonded to an alkyl carbon atom in an S_N2 mechanism.

$$Nu:^- + R\text{—}L \longrightarrow Nu\text{—}R + :L^-$$

In Chapter 10 we discussed nucleophilic addition reactions in which the nucleophile attacks the electrophilic carbonyl carbon atom of aldehydes and ketones. In each case a tetrahedral product is formed. Examples include the formation of hemiacetals with alcohols and the synthesis of alcohols using the Grignard reagent.

$$R_2C{=}\ddot{O}: \xrightarrow[2.\ H_3O^+]{1.\ Nu:^-} R_2C(\ddot{O}H)Nu$$

tetrahedral product

Acyl derivatives also react with nucleophiles. In these reactions, the nucleophile attacks the carbonyl carbon atom to generate a tetrahedral intermediate. However, the tetrahedral intermediate is unstable, and a leaving group departs to form a different acyl derivative. The overall process is called **nucleophilic acyl substitution.** The process is also called an **acyl-group transfer reaction** because it transfers an acyl group from one group (the leaving group) to another (the nucleophile).

$$RC(=\ddot{O})L \xrightarrow{Nu:^-} \left[RC(\ddot{O}:^-)(L)Nu \right] \longrightarrow RC(=\ddot{O})Nu + :L^-$$

tetrahedral intermediate

The net result is a substitution reaction whose stoichiometry resembles that of an S_N2 substitution reaction of haloalkanes. However, the resemblance is only superficial. An S_N2 reaction is a single-step process in which the nucleophile bonds to the carbon atom as the leaving group leaves. Nucleophilic

FIGURE 12.3

Comparison of the S_N2 and the Nucleophilic Acyl Substitution Mechanisms

The S_N2 mechanism is a single step process in which the nucleophile and the leaving group are both bonded to a carbon atom in the transition state. In the nucleophilic acyl substitution mechanism, a tetrahedral intermediate is formed between the first and second steps.

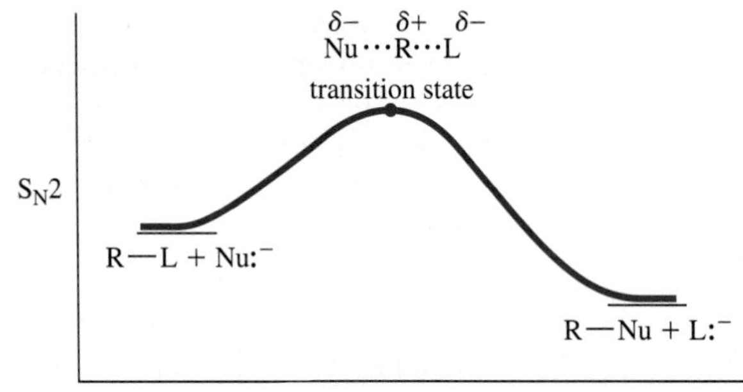

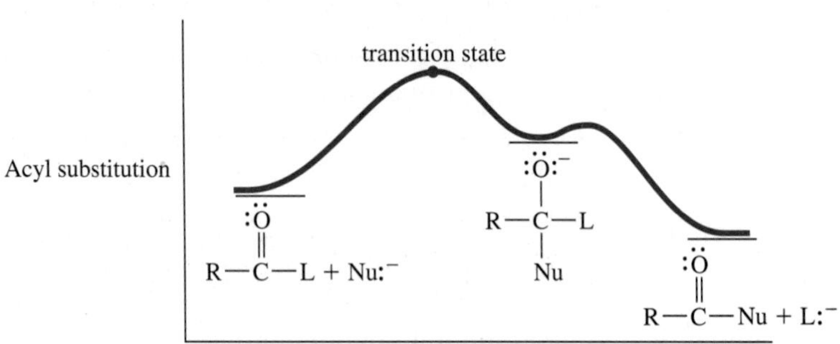

acyl substitution occurs in two steps (Figure 12.3). The rate-determining step is most commonly nucleophilic attack at the carbonyl atom to form a tetrahedral intermediate. The loss of the leaving group occurs in a second step.

Why don't acyl derivatives behave like aldehydes and ketones and form stable tetrahedral products? The answer is that the intermediate formed from an acyl derivative has a good leaving group. In the case of an acid chloride the leaving group is a weakly basic chloride ion. We recall that leaving group abilities are inversely related to base strength. An intermediate derived from a ketone does not have a good leaving group. A carbanion, the conjugate base of a hydrocarbon, is an extremely strong base and, therefore, a very poor leaving group.

$$R-\overset{\overset{\ddot{O}:}{\|}}{C}-R + Nu:^- \longrightarrow R-\overset{\overset{:\ddot{O}:^-}{|}}{\underset{\underset{Nu}{|}}{C}}-R \not\longrightarrow R-\overset{\overset{:\ddot{O}}{\|}}{C}-Nu + R^-$$

Relative Reactivity of Acyl Derivatives

The order of reactivity of acyl derivatives is acid chloride > acid anhydride > ester = acid > amide. This order of reactivity might appear to reflect the leaving-group abilities as reflected by their basicities. We know that HCl is a strong acid and NH_3 is a very weak acid. Thus, Cl^- is a weak base and NH_2^- is a strong base. Thus, Cl^- is a better leaving group than NH_2^-. However, the rate-determining step is not loss of the leaving group, but rather attack of the nucleophile.

The order of reactivities is explained by resonance stabilization of the reactant. Donation of an electron pair of the atom bonded to the acyl carbon atom decreases the partial positive charge on the carbon atom and decreases its electrophilicity. In the tetrahedral intermediate, no resonance stabilization is possible. Consider the general equation for the first step of the reaction.

$$\text{R—C(=\ddot{O}:)—\ddot{L}} \longleftrightarrow \text{R—C(—\ddot{O}:}^-\text{)=L}^+ \xrightarrow{\text{Nu:}^-} \text{R—C(—\ddot{O}:}^-\text{)(—Nu)—\ddot{L}}$$

resonance-stabilized reactant — no resonance stabilization

We recall from the discussion of substituent effects on electrophilic aromatic substitution that second-row elements such as oxygen and nitrogen effectively donate electrons by resonance. Furthermore, nitrogen is a better donor of electrons by resonance than oxygen because nitrogen is less electronegative. Thus, a nitrogen atom stabilizes amides better than an oxygen atom stabilizes esters. Finally, we recall that chlorine, a third-row element, is not effective in donating electrons by resonance. Thus, acid chlorides have little resonance stabilization by the chlorine atom. As a consequence, the highest partial positive charge at the carbonyl carbon atom of acyl derivatives is found in the acid chloride.

The order of reactivity means that a more reactive acyl derivative can be converted into a less reactive acyl derivative. The relative reactivity of acyl derivatives enables us to understand most of the chemical reactions of these compounds. Because acids and esters have similar reactivities, these two classes of compounds can be readily interconverted in equilibrium processes. These important reactions will be discussed in later sections.

$$\text{R'—C(=O)—OH} + \text{R—OH} \rightleftharpoons \text{R'—C(=O)—O—R} + H_2O$$

Acid Chlorides

Acid chlorides, the most reactive acyl derivatives, do not exist in nature. They are prepared in the laboratory and are used to synthesize less reactive acyl compounds. Acid chlorides are made by treating carboxylic acids with thionyl chloride, $SOCl_2$. This reaction takes advantage of Le Châtelier's principle. Thionyl chloride reacts with carboxylic acids to form acid chlorides because the reaction is driven to the right by the formation of HCl and SO_2, which are released as gases and escape from the reaction mixture. Phosphorus trichloride, PCl_3, can also be used to prepare acyl chlorides.

$$\text{C}_5\text{H}_9\text{—C(=O)—OH} \xrightarrow{SOCl_2} \text{C}_5\text{H}_9\text{—C(=O)—Cl}$$

cyclopentanecarboxylic acid — cyclopentanecarbonyl chloride

Acid chlorides react with most nucleophiles and are hydrolyzed by the moisture in air. Reaction of an acid chloride with an alcohol gives an ester. Amines are easily converted into amides by reaction with acid chlorides.

$$R{-}\overset{O}{\overset{\|}{C}}{-}Cl \xrightarrow{H-O-H} R{-}\overset{O}{\overset{\|}{C}}{-}O{-}H + HCl$$

$$R{-}\overset{O}{\overset{\|}{C}}{-}Cl \xrightarrow{H-O-R'} R{-}\overset{O}{\overset{\|}{C}}{-}O{-}R' + HCl$$

$$R{-}\overset{O}{\overset{\|}{C}}{-}Cl \xrightarrow{H-NH-R'} R{-}\overset{O}{\overset{\|}{C}}{-}NH{-}R' + HCl$$

Esters are easily produced from acids by first converting the acid to an acid chloride. The acid chloride reacts with an alcohol to give an excellent yield of an ester. The reaction produces HCl, and a base such as pyridine is used to react with the HCl formed to prevent side reactions.

$$\text{c-C}_3\text{H}_5{-}\overset{O}{\overset{\|}{C}}{-}OH \xrightarrow{SOCl_2} \text{c-C}_3\text{H}_5{-}\overset{O}{\overset{\|}{C}}{-}Cl$$

$$\text{c-C}_3\text{H}_5{-}\overset{O}{\overset{\|}{C}}{-}Cl + HO{-}\text{c-C}_5\text{H}_9 \xrightarrow{\text{pyridine}} \text{c-C}_3\text{H}_5{-}\overset{O}{\overset{\|}{C}}{-}O{-}\text{c-C}_5\text{H}_9$$

Acid Anhydrides

Acid anhydrides are less reactive than acid chlorides, but they are still very active acylating agents. Water hydrolyzes acid anhydrides to acids, alcohols react to give esters, and amines give amides. Note that the byproduct in each case is one molar equivalent of a carboxylic acid.

$$R{-}\overset{O}{\overset{\|}{C}}{-}O{-}\overset{O}{\overset{\|}{C}}{-}R \xrightarrow{H-O-H} R{-}\overset{O}{\overset{\|}{C}}{-}O{-}H + H{-}O{-}\overset{O}{\overset{\|}{C}}{-}R$$

$$R{-}\overset{O}{\overset{\|}{C}}{-}O{-}\overset{O}{\overset{\|}{C}}{-}R \xrightarrow{H-O-R'} R{-}\overset{O}{\overset{\|}{C}}{-}O{-}R' + H{-}O{-}\overset{O}{\overset{\|}{C}}{-}R$$

$$R{-}\overset{O}{\overset{\|}{C}}{-}O{-}\overset{O}{\overset{\|}{C}}{-}R \xrightarrow{H-NH-R'} R{-}\overset{O}{\overset{\|}{C}}{-}NH{-}R' + H{-}O{-}\overset{O}{\overset{\|}{C}}{-}R$$

The most common acid anhydride is acetic anhydride; over a million tons are produced each year. Acetic anhydride reacts with the phenol func-

Thioesters Are Nature's Active Acyl Compounds

The interconversion of acyl compounds in cells occurs by transferring acyl groups from one molecule to another. The acyl group of the donor or acceptor molecule is often in the form of a thioester. Thioesters are more reactive than esters, and the thiol group is easily replaced by an alkoxy group from an alcohol.

$$CH_3\overset{O}{\overset{\|}{C}}—SR + H—OR' \rightarrow CH_3\overset{O}{\overset{\|}{C}}—OR' + HS—R$$

The most important thioester is acetyl coenzyme A, which is formed from the thiol group of coenzyme A, a complex thiol that we will abbreviate as CoA—SH. The thiol group of CoA—SH is bonded to an acyl group in acyl-S-CoA derivatives.

When an acetyl group is linked to CoA—SH, the adduct is acetyl coenzyme A. This extremely important metabolite is produced by the degradation of long-chain carboxylic acids contained in fats. Acetyl-CoA is also produced by metabolic degradation of many amino acids and carbohydrates. Acetyl-CoA is a donor of the two-carbon acetyl group in the biosynthesis of long-chain carboxylic acids.

Acetyl coenzyme A reacts with nucleophiles in biological reactions to give new acyl compounds.

$$CH_3\overset{O}{\overset{\|}{C}}—SCoA + Nu: \rightarrow CH_3\overset{O}{\overset{\|}{C}}—Nu + CoA—SH$$

For example, acetyl coenzyme A provides the acetyl group in the biosynthesis of the neurotransmitter acetylcholine. Choline contains a hydroxyl group that is acetylated by acetyl coenzyme A to make acetylcholine.

$$CH_3\overset{O}{\overset{\|}{C}}—SCoA + \underset{\text{choline}}{(CH_3)_3\overset{+}{N}CH_2CH_2OH} \longrightarrow \underset{\text{acetylcholine}}{CH_3\overset{O}{\overset{\|}{C}}—OCH_2CH_2\overset{+}{N}(CH_3)_3} + CoA—SH$$

tional group of salicylic acid to yield an ester called acetylsalicylic acid, or aspirin. The annual production of aspirin in the United States is 12,000 tons.

$$\underset{\text{salicylic acid}}{C_6H_4(OH)(CO_2H)} + \underset{\text{acetic anhydride}}{CH_3\overset{O}{\overset{\|}{C}}—O—\overset{O}{\overset{\|}{C}}CH_3} \longrightarrow \underset{\text{acetylsalicylic acid}}{C_6H_4(O\overset{O}{\overset{\|}{C}}CH_3)(CO_2H)} + CH_3CO_2H$$

12.7 REDUCTION OF ACYL DERIVATIVES

In Chapter 10 we saw that aldehydes (or ketones) can be reduced by either sodium borohydride or lithium aluminum hydride. These reactions involve nucleophilic attack of hydride ion on the carbonyl carbon atom. A similar process occurs with acyl derivatives. In this section, the reduction of esters, acids, and acid chloride is considered. The reduction of amides is presented in Chapter 14.

Reduction of Esters

The reduction of esters requires the strong reducing agent lithium aluminum hydride. The product is a primary alcohol. The milder reagent sodium borohydride does not reduce esters.

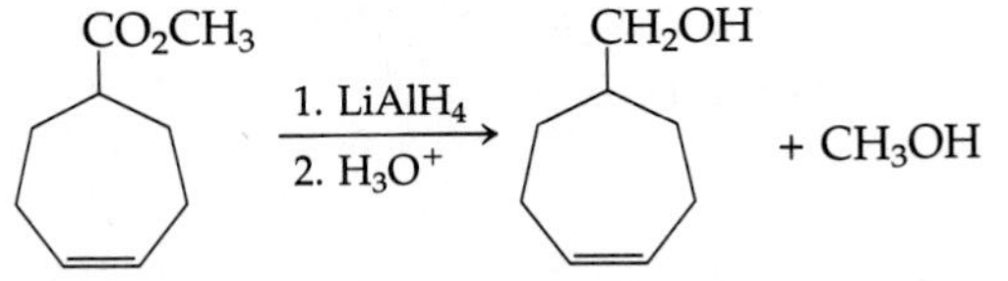

$$\text{cycloheptenyl-}CO_2CH_3 \xrightarrow[2.\ H_3O^+]{1.\ LiAlH_4} \text{cycloheptenyl-}CH_2OH + CH_3OH$$

Note that the alcohol portion of the ester is a byproduct of the reaction. The esters that are typically reduced by lithium aluminum hydride contain a low molecular weight alkyl group introduced in the conversion of an acid to an ester. The alcohol obtained by reduction of the acid portion of the ester is easily separated from the low molecular weight, water-soluble alcohol.

The mechanism of the reduction of an ester is pictured as simply a nucleophilic attack of a one molar equivalent of a "hydride" ion on the carbonyl carbon atom. However, the aluminum atom participates in the reaction and is bonded to the oxygen atom. For simplicity, the structures shown have eliminated the aluminate ion. Attack by the hydride ion produces a tetrahedral intermediate whose oxygen atom is bonded to the aluminum atom. The tetrahedral intermediate loses an alkoxide ion, and the resulting aldehyde is even more rapidly reduced by a second molar equivalent of hydride ion than the original ester. In total, two molar equivalents of hydride ion or one-half molar equivalent of lithium aluminum hydride are required for the reduction.

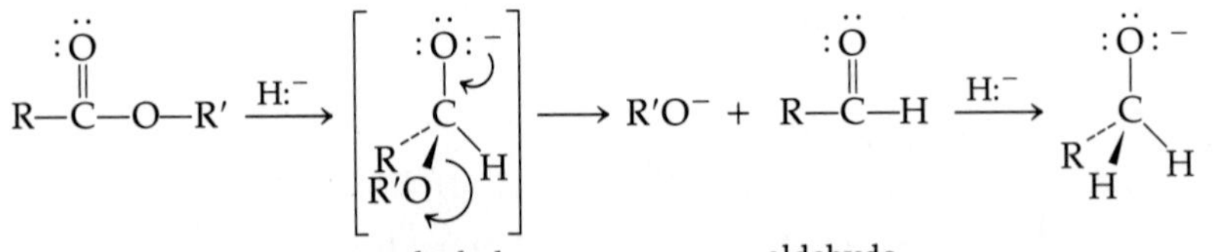

$$R{-}C({=}O){-}O{-}R' \xrightarrow{H:^-} \left[R(R'O)(H)C{-}O^- \right] \longrightarrow R'O^- + R{-}C({=}O){-}H \xrightarrow{H:^-} R(H)(H)C{-}O^-$$

tetrahedral intermediate; aldehyde intermediate

Reduction of Acids

Carboxylic acids are reduced by lithium aluminum hydride but not by the milder reagent sodium borohydride. As in the case of the reduction of esters, carboxylic acids are always completely reduced to an alcohol; no intermediate aldehyde can be isolated.

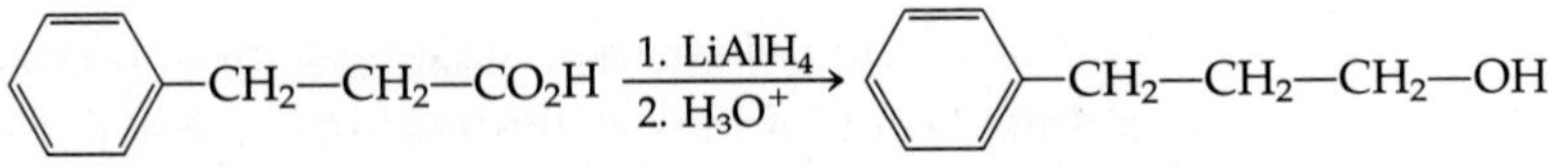

$$C_6H_5{-}CH_2{-}CH_2{-}CO_2H \xrightarrow[2.\ H_3O^+]{1.\ LiAlH_4} C_6H_5{-}CH_2{-}CH_2{-}CH_2{-}OH$$

3-phenylpropanoic acid; 3-phenyl-1-propanol

Lithium aluminum hydride is a strong base. It reacts with the acidic proton of the carboxylic acid to give hydrogen gas. One molar equivalent of hydride ion is used in the reaction.

$$R-\overset{O}{\overset{\|}{C}}-OH \xrightarrow{LiAlH_4} R-\overset{O}{\overset{\|}{C}}-O^- + H_2$$

The carboxylate salt is then reduced. This process requires another two molar equivalents of hydride ion. Therefore, reduction of a carboxylic acid requires a total of three molar equivalents of hydride ion.

Reduction of Acid Chlorides

Esters or carboxylic acids cannot be reduced to aldehydes by hydride reducing agents because aldehydes are more reactive than esters or carboxylic acids. Reagents that reduce esters or acids then rapidly reduce the product aldehyde to a primary alcohol.

Acid chlorides are more reactive than esters toward nucleophilic hydride ion. As a consequence, acid chlorides are more rapidly reduced than aldehydes. Lithium aluminum hydride is such a strong reducing agent that acid chlorides are reduced all the way to primary alcohols. However, the milder reducing agent lithium aluminum tri(*t*-butoxy) hydride reacts with acid chlorides to give aldehydes, which are not further reduced.

$$\text{(indanyl)}-CH_2-\overset{O}{\overset{\|}{C}}-Cl \xrightarrow{LiAlH_4} \text{(indanyl)}-CH_2-CH_2-OH$$

$$\text{(indanyl)}-CH_2-\overset{O}{\overset{\|}{C}}-Cl \xrightarrow{LiAlH(O\text{-}t\text{-}Bu)_3} \text{(indanyl)}-CH_2-\overset{O}{\overset{\|}{C}}-H$$

12.8 SYNTHESIS OF ESTERS

Carboxylic acids are converted to esters by a two-step method (Section 12.4). In the first step, the acid is converted to an acid chloride. In the second step, the acid chloride reacts with an alcohol to give an ester.

$$R-\overset{O}{\overset{\|}{C}}-OH \xrightarrow{SOCl_2} R-\overset{O}{\overset{\|}{C}}-Cl \xrightarrow{R'OH} R-\overset{O}{\overset{\|}{C}}-O-R'$$

Carboxylic acids are directly converted to esters in a single step called the **Fischer esterification** reaction. This reaction is an example of a condensation reaction; it entails joining two reactants into one larger product with the simultaneous formation of a second smaller product, such as water. A carboxylic acid reacts with an alcohol in a condensation reaction to give an ester and water. Thus, the ester contains a part of both the alcohol and the acid. In the general reaction given below, H—O—A represents any acid—inorganic or organic. For an organic acid, the A represents an acyl group.

Water comes from here.

part of the acid — part of the alcohol

$$\text{A—}\boxed{\text{O—H}} + \boxed{\text{H}}\text{—O—R} \longrightarrow \text{A—O—R} + H_2O$$

an acid — an alcohol — an ester

Esters can be made from many different acids. When the acid is nitric acid, the ester is called a nitrate ester. Many of these esters, such as glyceryl trinitrate (nitroglycerin), are powerful explosives. Glyceryl trinitrate is also a smooth-muscle relaxant and vasodilator. It is used to lower blood pressure and to treat angina pectoris, a heart disorder.

$$\begin{matrix} O_2N\text{—O—H} & & \text{H—O—}CH_2 & & O_2N\text{—O—}CH_2 & \\ O_2N\text{—O—H} & + & \text{H—O—CH} & \longrightarrow & O_2N\text{—O—CH} & + 3\ H_2O \\ O_2N\text{—O—H} & & \text{H—O—}CH_2 & & O_2N\text{—O—}CH_2 & \\ \text{nitric acid} & & \text{glycerol} & & \text{glyceryl trinitrate} & \end{matrix}$$

A carboxylic acid reacts with an alcohol to form an ester. The reaction is catalyzed by inorganic acids. Both the acid and its ester are present in substantial amounts at equilibrium.

Water comes from here.

part of the acid — part of the alcohol

$$\text{R—}\overset{\overset{\displaystyle O}{\|}}{\text{C}}\text{—}\boxed{\text{O—H}} + \boxed{\text{H}}\text{—O—R}' \longrightarrow \text{R—}\overset{\overset{\displaystyle O}{\|}}{\text{C}}\text{—O—R}' + H_2O$$

a carboxylic acid — an alcohol — an ester

The yield of ester can be increased by distilling the water out of the reaction mixture. The ester yield also is increased by using excess alcohol. Ethyl esters of acids are obtained by using ethanol as a solvent. Under such conditions, the high concentration of ethanol favors a high conversion of the acid to the ester. Thus, removing water or adding alcohol shifts the equilibrium to the side of the product, as predicted by Le Châtelier's principle.

$$CH_3\text{—}\overset{\overset{\displaystyle O}{\|}}{\text{C}}\text{—O—H} + CH_3CH_2OH \rightleftharpoons CH_3\text{—}\overset{\overset{\displaystyle O}{\|}}{\text{C}}\text{—O—}CH_2CH_3 + H_2O$$

ethyl acetate

Adding alcohol "pushes" the reaction to the right.

Removal of water "pulls" the reaction to the right.

Esters are common industrial products. Aspirin and oil of wintergreen, which are derived from salicylic acid, are two such examples. Salicylic acid is both a phenol and a carboxylic acid. It acts as an antipyretic (fever reducer) and an analgesic (pain reliever). However, it irritates the stomach lining, and the ester of the phenolic group—called acetylsalicylic acid, better known as aspirin—is preferred for use as a medicine. The methyl ester of the carboxylic acid is methyl salicylate (oil of wintergreen) and is used in liniments to soothe skin irritations.

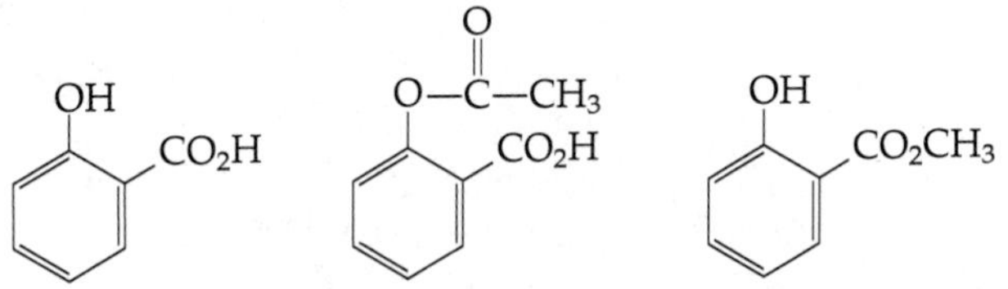

OH, CO_2H — salicylic acid
$O-\overset{O}{\overset{\|}{C}}-CH_3$, CO_2H — acetylsalicylic acid
OH, CO_2CH_3 — methyl salicylate

EXAMPLE 12.5 Clofibrate is a drug that can lower the concentration of blood triacylglycerols and cholesterol. What is its IUPAC name?

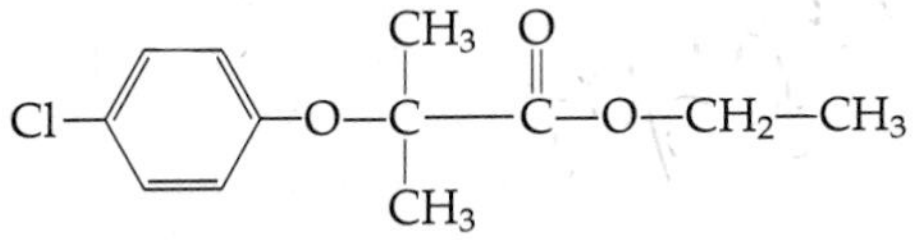

$$Cl-C_6H_4-O-C(CH_3)_2-\overset{O}{\overset{\|}{C}}-O-CH_2-CH_3$$

Solution First, identify the alcohol portion of the ester; it is located at the right of the molecule; it contains two carbon atoms. This is an ethyl ester.

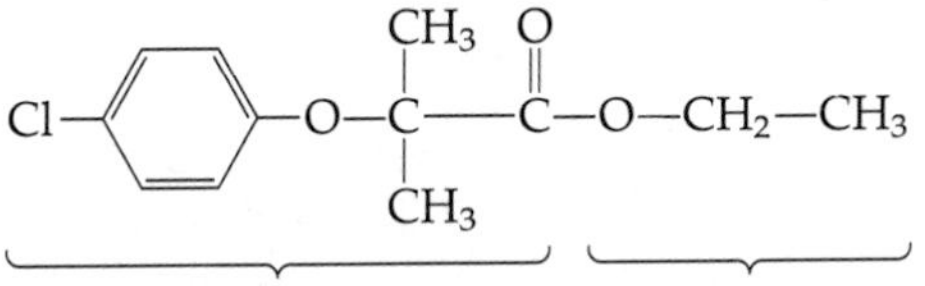

$$\underbrace{Cl-C_6H_4-O-C(CH_3)_2-\overset{O}{\overset{\|}{C}}}_{\text{acyl portion}}\underbrace{-O-CH_2-CH_3}_{\text{alcohol portion}}$$

The acyl portion is a substituted propanoic acid with a methyl group and an aryl-containing group at the C-2 atom. Imagine removing the aryl-containing group from the acid and adding a hydrogen atom to its oxygen atom. The compound is *p*-chlorophenol. The original group is *p*-chlorophenoxy.

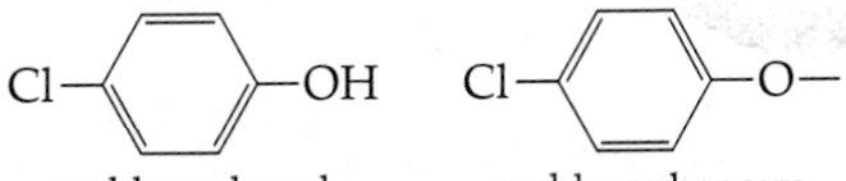

$Cl-C_6H_4-OH$ — *p*-chlorophenol
$Cl-C_6H_4-O-$ — *p*-chlorophenoxy

Polyesters

Many commercial products, called **condensation polymers,** are made by reacting monomers to give large molecules and some small molecule, such as water, as a byproduct. This polymerization process differs from addition polymerization in which the entire monomer is included in the polymer.

In condensation polymerization reactions, each monomer has two functional groups (each is said to be difunctional). An example of condensation polymerization is the reaction of terephthalic acid and ethylene glycol. One step in the reaction sequence is shown below.

Although the product looks like a simple ester, it is also both a carboxylic acid and an alcohol. Its acid end can react with ethylene glycol, and its alcohol end can react with terephthalic acid. Each reaction produces a larger molecule that continues to react to form a high molecular weight polyester. The polymer is industrially processed into a fiber called Dacron.

Dacron is used in woven and knitted fabrics in combination with cotton or wool. Dacron fabric is physiologically inert, so it can also be used in the form of a mesh to replace diseased sections of blood vessels.

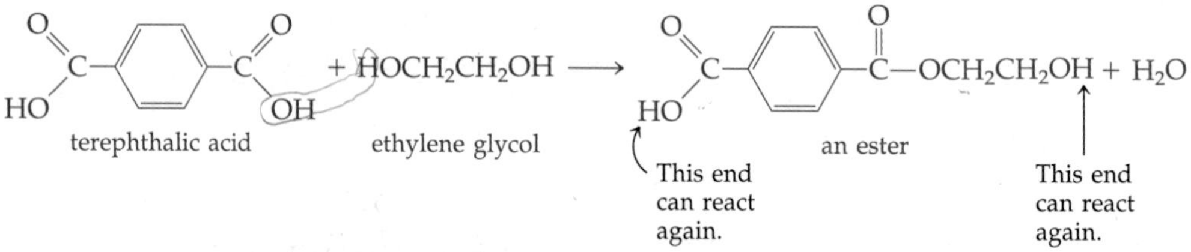

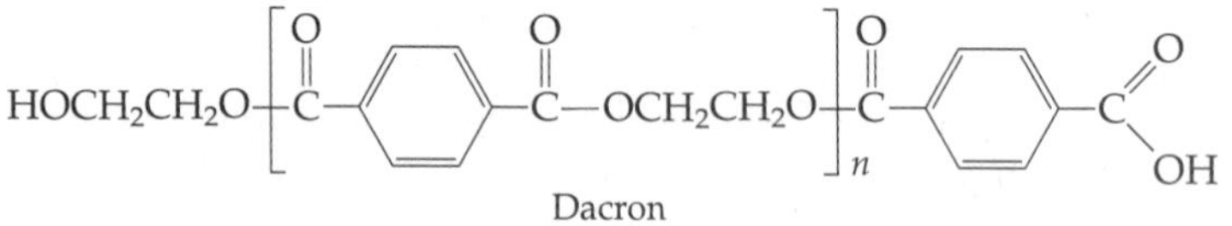

$$HOOC-C_6H_4-COOH + HOCH_2CH_2OH \longrightarrow HOOC-C_6H_4-COOCH_2CH_2OH + H_2O$$

terephthalic acid ethylene glycol an ester

This end can react again. This end can react again.

$$HOCH_2CH_2O\left[-\overset{O}{\overset{\|}{C}}-C_6H_4-\overset{O}{\overset{\|}{C}}-OCH_2CH_2O-\right]_n\overset{O}{\overset{\|}{C}}-C_6H_4-COOH$$

Dacron

The name of the acid is 2-methyl-2-(*p*-chlorophenoxy)propanoic cid. Now change the *-ic* ending of the acid to *-ate* and write the name of the alkyl group of the alcohol as a separate word in front of the modified acid name. The ester is named ethyl 2-methyl-2-(*p*-chlorophenoxy)propanoate.

Mechanism of Esterification

Does the oxygen atom linking the acyl carbon atom and the alkyl carbon atom of an ester "come from" the carboxylic acid or the alcohol? From a different perspective, does the water formed come from the hydroxyl group of the alcohol and the hydrogen of the acid or from the hydrogen of the acid and the hydroxyl group of the alcohol? These related questions were answered by carrying out the esterification reaction using ^{18}O-labeled metha-

FIGURE 12.4
Mechanism of Esterification

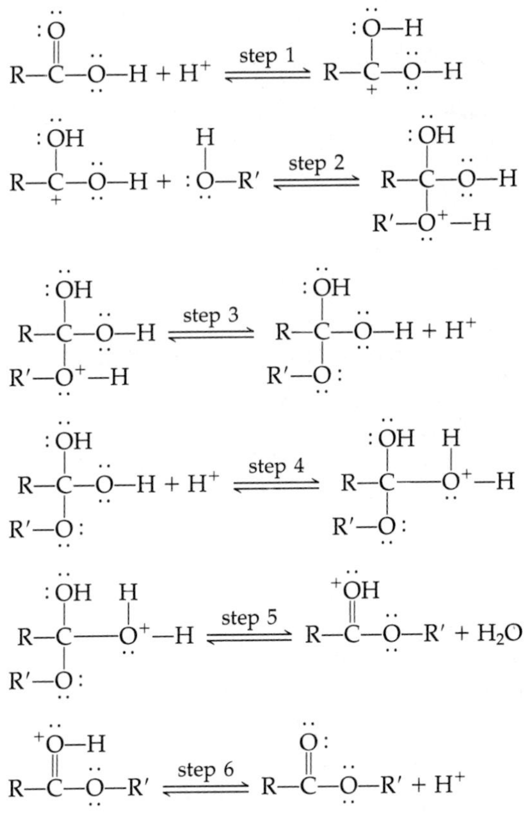

$$\mathrm{R{-}C({=}O){-}OH + H^+ \overset{\text{step 1}}{\rightleftharpoons} R{-}\overset{+}{C}(OH){-}OH}$$

$$\mathrm{R{-}\overset{+}{C}(OH){-}OH + H{-}O{-}R' \overset{\text{step 2}}{\rightleftharpoons} R{-}C(OH)(OH){-}\overset{+}{O}H{-}R'}$$

$$\mathrm{R{-}C(OH)(OH){-}\overset{+}{O}H{-}R' \overset{\text{step 3}}{\rightleftharpoons} R{-}C(OH)(OH){-}O{-}R' + H^+}$$

$$\mathrm{R{-}C(OH)(OH){-}O{-}R' + H^+ \overset{\text{step 4}}{\rightleftharpoons} R{-}C(OH)(\overset{+}{O}H_2){-}O{-}R'}$$

$$\mathrm{R{-}C(OH)(\overset{+}{O}H_2){-}O{-}R' \overset{\text{step 5}}{\rightleftharpoons} R{-}C({=}\overset{+}{O}H){-}O{-}R' + H_2O}$$

$$\mathrm{R{-}C({=}\overset{+}{O}{-}H){-}O{-}R' \overset{\text{step 6}}{\rightleftharpoons} R{-}C({=}O){-}O{-}R' + H^+}$$

nol. When methanol reacts with benzoic acid, the oxygen-18 is contained in the ester—not in the water.

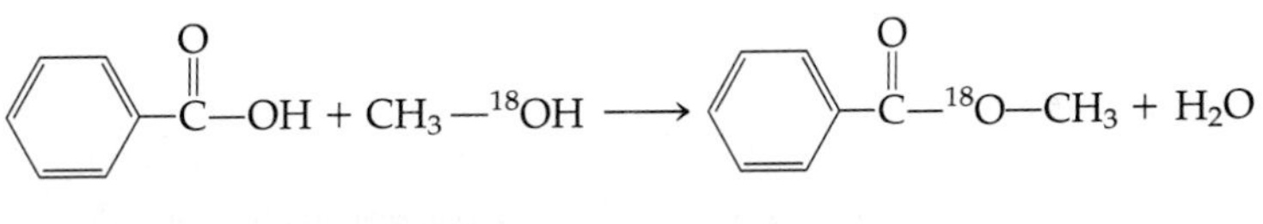

$$\mathrm{C_6H_5{-}C({=}O){-}OH + CH_3{-}{}^{18}OH \longrightarrow C_6H_5{-}C({=}O){-}{}^{18}O{-}CH_3 + H_2O}$$

Therefore, the CO—OH bond of the acid is cleaved rather than the COO—H bond. In addition, the O—H bond of the alcohol is cleaved rather than the C—O bond. The mechanism of this nucleophilic acyl substitution reaction is shown in Figure 12.4.

12.9 HYDROLYSIS OF ESTERS

Hydrolysis reactions split a reactant into two smaller products with the net incorporation of water into the two products (Section 2.5). The hydrolysis of an ester produces a carboxylic acid and an alcohol. Ester hydrolysis, then, is just the reverse of esterification.

The C—O bond is cleaved. The O—H group is bonded to the carbonyl carbon atom.

$$\mathrm{H{-}\overset{\displaystyle H}{\underset{\displaystyle H}{C}}{-}\overset{\displaystyle O}{\overset{\|}{C}}{-}O{-}\overset{\displaystyle H}{\underset{\displaystyle H}{C}}{-}H + H_2O \longrightarrow H{-}\overset{\displaystyle H}{\underset{\displaystyle H}{C}}{-}\overset{\displaystyle O}{\overset{\|}{C}}{-}O{-}H + H{-}O{-}\overset{\displaystyle H}{\underset{\displaystyle H}{C}}{-}H}$$

This O—H bond is formed.

The hydrolysis of an ester is catalyzed by strong acids. The reaction is favored by a large excess of water.

$$\mathrm{R{-}\overset{\displaystyle O}{\overset{\|}{C}}{-}O{-}R' + H_2O \overset{H^+}{\rightleftharpoons} R{-}\overset{\displaystyle O}{\overset{\|}{C}}{-}OH + R'OH}$$

Adding water "pushes" the reaction to the right.

Saponification of Esters

The hydrolysis of an ester by a strong base is called **saponification** (L. *sapon*, soap) because this reaction is used to make soaps from esters of long-chain carboxylic acids (Section 12.10).

Methyl acetate reacts with a strong base to give acetate ion and methanol. (In contrast, the acid-catalyzed hydrolysis of methyl acetate yields acetic acid and methanol.) Because the reaction mixture is basic, any carboxylic acid formed in the reaction is converted to its conjugate base by loss of a proton to the hydroxide ion.

$$\mathrm{R{-}\overset{\displaystyle O}{\overset{\|}{C}}{-}O{-}R' + OH^- \rightleftharpoons R{-}\overset{\displaystyle O}{\overset{\|}{C}}{-}O^- + R'OH}$$

$$\mathrm{CH_3{-}\overset{\displaystyle O}{\overset{\|}{C}}{-}O{-}CH_3 + OH^- \rightleftharpoons CH_3{-}\overset{\displaystyle O}{\overset{\|}{C}}{-}O^- + CH_3OH}$$

There is another important distinction between hydrolysis and saponification: hydrolysis is acid-catalyzed, whereas in the saponification reaction, hydroxide is a reagent. It is not a catalyst because equal numbers of moles of hydroxide and ester are required. Hydroxide is consumed in the reaction, and, because it is a strong base, the position of equilibrium lies overwhelmingly to the right.

EXAMPLE 12.6 The antibiotic chloramphenicol has a bitter taste. Its palatability for children is improved by using a suspension of the palmitate ester. Once orally administered, the ester is enzymatically hydrolyzed in the intestine. Given the structure of the ester, write the structure of the antibiotic.

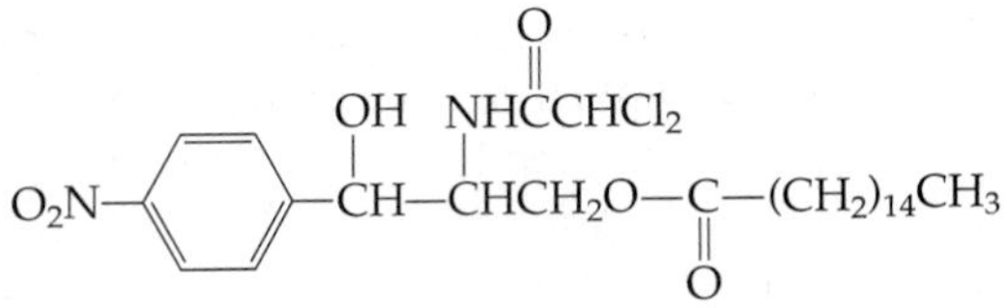

$$O_2N-C_6H_4-\underset{OH}{CH}-\underset{NHCOCHCl_2}{CH}CH_2O-\overset{O}{\overset{\|}{C}}-(CH_2)_{14}CH_3$$

Solution First, locate the ester functional group by examining the carbonyl carbon atoms. The carbonyl group at the top of the structure is bonded to a nitrogen atom. This is an amide group, a very stable acyl compound. The carbonyl carbon atom toward the right of the structure is bonded to an oxygen atom. This is the ester functional group. The carbon atom chain to the right of the carbonyl group is part of the acid. A total of 16 carbon atoms are contained in palmitic acid. Chloramphenicol is bonded in the ester through its primary hydroxyl group.

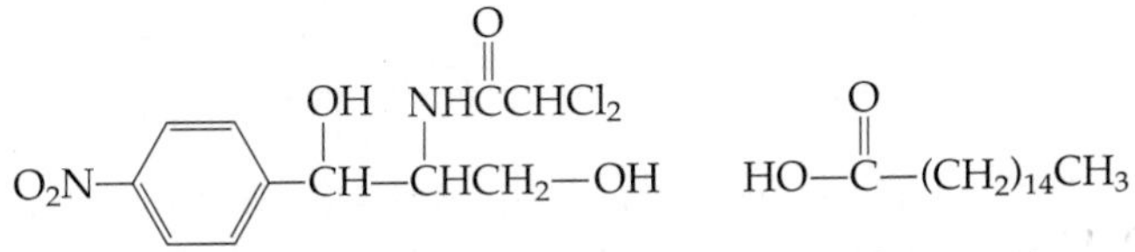

$$O_2N-C_6H_4-\underset{OH}{CH}-\underset{NHCOCHCl_2}{CH}CH_2-OH \qquad HO-\overset{O}{\overset{\|}{C}}-(CH_2)_{14}CH_3$$

chloramphenicol palmitic acid

12.10 SOAPS AND DETERGENTS

Soaps are salts of long-chain carboxylic acids called fatty acids. The best soaps are carboxylate salts made from saturated acids that have 14–18 carbon atoms. Soaps fabricated as bars are usually sodium salts, whereas the potassium salts, which are softer, are used in shaving creams.

$$CH_3(CH_2)_{16}CO_2^-Na^+$$

sodium stearate
(a soap)

Soaps were originally produced from animal fats, which are triesters of glycerol and carboxylic acids containing 12–18 carbon atoms.

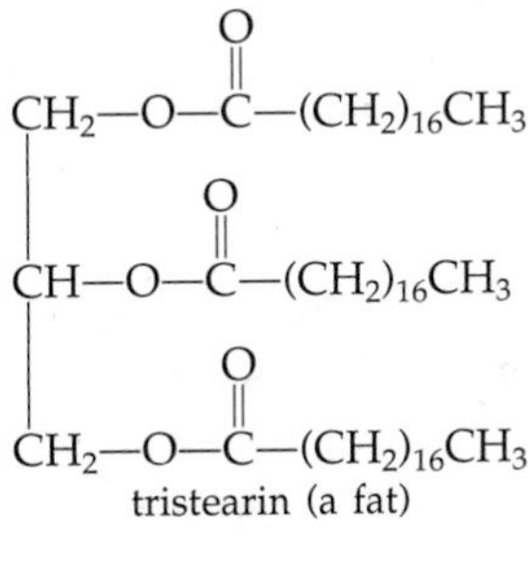

$$\begin{array}{l} CH_2-O-\overset{O}{\overset{\|}{C}}-(CH_2)_{16}CH_3 \\ | \\ CH-O-\overset{O}{\overset{\|}{C}}-(CH_2)_{16}CH_3 \\ | \\ CH_2-O-\overset{O}{\overset{\|}{C}}-(CH_2)_{16}CH_3 \end{array}$$

tristearin (a fat)

The carboxylate salts of fatty acids have long, nonpolar hydrocarbon chains. Therefore, they do not form a true solution but are dispersed as structures called micelles. A **micelle** is a spherical aggregation of molecules or ions. In a micelle of carboxylate salts, the nonpolar hydrocarbon chains point toward the interior of the sphere and the polar carboxylate "heads" lie on the surface of the sphere (Figure 12.5).

The nonpolar hydrocarbon chain of a fatty acid repels water and is said to be **hydrophobic.** In contrast, the polar "head" of the carboxylate group is attracted to water and is said to be **hydrophilic;** it forms hydrogen bonds to water. The micelle is held together by London forces between hydrocarbon chains. The tendency of nonpolar solutes to aggregate in aqueous solution is called the **hydrophobic effect.** The micelle surface, which may contain several hundred carboxylate groups, has many negative charges. As a result, individual micelles repel each other and remain suspended in water.

An example of a hydrophobic substance is grease, which does not dissolve in water because it is nonpolar. However, grease will dissolve in the hydrocarbon region of the micelle. This process accounts for the cleansing action of a soap.

Micelles also interact with the ions in hard water, which contains relatively high concentrations of Ca^{2+} and Mg^{2+} ions. They react with the carboxylate ions of soaps and form precipitates that reduce the cleansing power of the soap. For this reason detergents—salts of organic sulfate esters—are used in place of soaps.

$CH_3(CH_2)_{11}—O—SO_3^-Na^+$

sodium dodecyl sulfate

$CH_3(CH_2)_9—CH(CH_3)—C_6H_4—SO_3^-Na^+$

sodium *p*-(2-dodecyl)benzene sulfonate

FIGURE 12.5 Micelle of a Soap
The long hydrocarbon chains of carboxylates are nonpolar and associate within the micelle. The polar carboxylate heads are located on the exterior and are hydrogen-bonded to water.

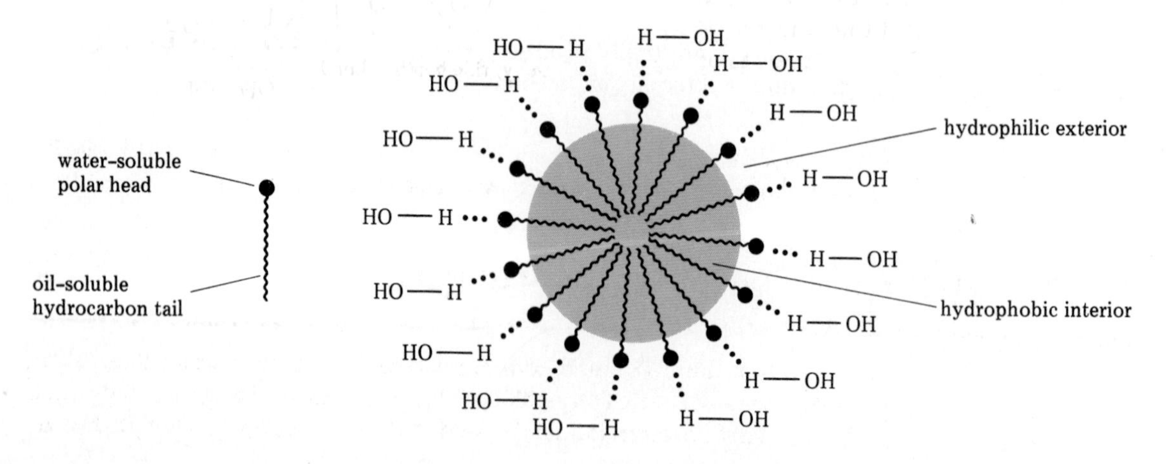

Esters and Anhydrides of Phosphoric Acid

Phosphoric acid, pyrophosphoric acid, and triphosphoric acid occur in living cells as esters.

ester bond

R—O—P(=O)(OH)—OH

alkyl phosphate

ester bond

R—O—P(=O)(OH)—O—P(=O)(OH)—OH

alkyl diphosphate

ester bond

R—O—P(=O)(OH)—O—P(=O)(OH)—O—P(=O)(OH)—OH

alkyl triphosphate

Note that these esters are also acids. At physiological pH, the protons in the —OH groups of these esters are ionized. For this reason, phosphoric acid derivatives are soluble in the aqueous environment of living systems.

The esters are also acid anhydrides of phosphoric acid. The oxygen atoms between the phosphorus atoms constitute the anhydride bonds. These compounds are analogs of acid anhydrides of carboxylic acids.

anhydride bond

R—O—P(=O)(OH)—O—P(=O)(OH)—OH

anhydride bonds

R—O—P(=O)(OH)—O—P(=O)(OH)—O—P(=O)(OH)—OH

The anhydride bonds of diphosphates and triphosphates are hydrolyzed in many biological reactions. Adenosine triphosphate (ATP), a triphosphate ester of adenosine, stores some of the energy released in the degradation of carbohydrates, fats, and amino acids. When one anhydride bond of ATP is hydrolyzed, adenosine triphosphate (ADP) is produced and 7.3 kcal of energy are released per mole of ADP produced.

Like soaps, detergents have long hydrophobic tails and hydrophilic heads. However, they do not form precipitates with Ca^{2+} and Mg^{2+} ions, so they work well even in hard water.

12.11 THE CLAISEN CONDENSATION

We recall that the hydrogen atom bonded to the α-carbon atom in an aldehyde or ketone is acidic (pK_a = 20). As a consequence, carbonyl compounds with an α-carbon atom undergo condensation reactions in which the α-

carbon atom of one molecule bonds to the carbonyl carbon atom of another molecule. The result is an aldol (Section 10.11).

Two molecules of an ester also react with each other in the presence of a base such as ethoxide ion to produce a condensation product. The reaction, which produces a β-keto ester, is called the **Claisen condensation.**

$$2\,CH_3-\overset{\overset{\Large :\ddot{O}}{\|}}{C}-\ddot{O}CH_2CH_3 \xrightarrow[HOCH_2CH_3]{NaOCH_2CH_3} CH_3-\overset{\overset{\Large :\ddot{O}}{\|}}{C}-CH_2-\overset{\overset{\Large :\ddot{O}}{\|}}{C}-\ddot{O}CH_2CH_3 + CH_3CH_2OH$$

ethyl acetate — ethyl acetoacetate

The mechanism of the Claisen condensation resembles that of the aldol condensation, but there are important differences. A full equivalent of base is required for the Claisen condensation rather than the catalytic amount required for the aldol condensation. There are several equilibria in the sequence of reactions, and some are not favorable. The entire sequence is made favorable by driving the final step to completion by an acid–base reaction that requires a molar equivalent of base.

Like aldehydes, esters have an acidic α-hydrogen atom. The acid dissociation constants of esters ($pK_a = 25$) are about 10^5 times smaller than those of aldehydes. The pK_a of ethanol is 16. Thus, the reaction of sodium ethoxide with an ester produces only a small amount of the enolate at equilibrium.

$$CH_3CH_2\ddot{O}:^- + H-CH_2-\overset{\overset{\Large :\ddot{O}}{\|}}{C}-\ddot{O}CH_2CH_3 \rightleftharpoons {}^-:CH_2-\overset{\overset{\Large :\ddot{O}}{\|}}{C}-\ddot{O}CH_2CH_3 + CH_3CH_2\ddot{O}-H$$

weaker base — weaker acid — stronger base — stronger acid

The conjugate base of the ester reacts with another molecule of ester to form a carbon–carbon bond. The addition product is the conjugate base of a hemiketal, which loses an alkoxide to give a β-keto ester.

$$CH_3-\overset{\overset{\Large :\ddot{O}}{\|}}{C}-\ddot{O}-CH_2CH_3 + {}^-:CH_2-\overset{\overset{\Large :\ddot{O}}{\|}}{C}-\ddot{O}CH_2CH_3 \rightleftharpoons CH_3-\underset{\underset{\Large :\ddot{O}CH_2CH_3}{|}}{\overset{\overset{\Large :\ddot{O}:^-}{|}}{C}}-CH_2-\overset{\overset{\Large :\ddot{O}}{\|}}{C}-\ddot{O}CH_2CH_3$$

$$CH_3-\underset{\underset{\Large :\ddot{O}CH_2CH_3}{|}}{\overset{\overset{\Large :\ddot{O}:^-}{|}}{C}}-CH_2-\overset{\overset{\Large :\ddot{O}}{\|}}{C}-\ddot{O}CH_2CH_3 \rightleftharpoons CH_3-\overset{\overset{\Large :\ddot{O}}{\|}}{C}-CH_2-\overset{\overset{\Large :\ddot{O}}{\|}}{C}-\ddot{O}CH_2CH_3 + CH_3CH_2\ddot{O}:^-$$

β-Keto esters have pK_a values around 11 because the negative charge of the conjugate base can be delocalized over both carbonyl groups. Ethanol is a weaker acid than β-keto esters, and, therefore, ethoxide ion essentially completely removes a proton from the product of the Claisen condensation. This final step drives the overall sequence of reactions to completion.

$$CH_3{-}\overset{\overset{\displaystyle :\ddot{O}}{\|}}{C}{-}\underset{H}{CH}{-}\overset{\overset{\displaystyle :\ddot{O}}{\|}}{C}{-}\ddot{O}CH_2CH_3 + {}^{-}{:}\ddot{O}CH_2CH_3 \rightleftharpoons CH_3{-}\overset{\overset{\displaystyle :\ddot{O}}{\|}}{C}{-}\bar{C}H{-}\overset{\overset{\displaystyle :\ddot{O}}{\|}}{C}{-}\ddot{O}CH_2CH_3 + H{-}\ddot{O}CH_2CH_3$$

stronger acid — stronger base — weaker base — weaker acid

Finally, dilute acid added at the end of the reaction converts the conjugate base of the β-keto ester to the product.

$$CH_3{-}\overset{\overset{\displaystyle :\ddot{O}}{\|}}{C}{-}\bar{C}H{-}\overset{\overset{\displaystyle :\ddot{O}}{\|}}{C}{-}\ddot{O}CH_2CH_3 + H{-}\underset{H}{\ddot{O}}{-}H \rightleftharpoons CH_3{-}\overset{\overset{\displaystyle :\ddot{O}}{\|}}{C}{-}\underset{H}{CH}{-}\overset{\overset{\displaystyle :\ddot{O}}{\|}}{C}{-}\ddot{O}CH_2CH_3 + {:}\underset{H}{\ddot{O}}{-}H$$

Biochemical Condensation Reactions

A condensation of acetyl coenzyme A and oxaloacetic acid occurs in the first step of the citric acid cycle, which involves a reaction between an ester and a ketone. The α-carbon atom of acetyl coenzyme A bonds to the carbonyl carbon atom of oxaloacetic acid in a reaction that resembles the aldol condensation. In this reaction, an acetyl group is transferred to oxaloacetic acid to form a compound that is subsequently hydrolyzed to citric acid.

$$\underset{\text{oxaloacetic acid}}{HO_2CCH_2\overset{\overset{\displaystyle O}{\|}}{\underset{CO_2H}{C}}} + \underset{\text{acetyl coenzyme A}}{\overset{H}{CH_2}C(=O){-}S{-}CoA} \longrightarrow HO_2CCH_2\overset{OH}{\underset{CO_2H}{C}}{-}CH_2C(=O){-}S{-}CoA$$

$$HO_2CCH_2\overset{OH}{\underset{CO_2H}{C}}{-}CH_2C(=O){-}S{-}CoA + H_2O \longrightarrow \underset{\text{citric acid}}{HO_2CCH_2\overset{OH}{\underset{CO_2H}{C}}{-}CH_2C(=O){-}OH} + CoA{-}SH$$

Most of the acetyl coenzyme A produced in metabolic reactions reacts with oxaloacetic acid to form citric acid. However, in certain illnesses, such as diabetes, the metabolism of fats predominates over the metabolism of carbohydrates. When there is not enough oxaloacetic acid to react with all of the acetyl coenzyme A produced, it reacts with itself in a Claisen condensation.

$$CH_3{-}C(=O){-}S{-}CoA + \underset{H}{CH_2}C(=O){-}S{-}CoA \longrightarrow CH_3{-}\overset{\overset{\displaystyle O}{\|}}{C}{-}CH_2C(=O){-}S{-}CoA + CoA{-}SH$$

Hydrolysis of the resulting β-keto thioester yields acetoacetic acid (3-ketobutanoic acid). Subsequent reactions produce 3-hydroxybutanoic acid and acetone, which are collectively called ketone bodies. Detection of these compounds in the urine is indicative of diabetes.

EXERCISES

Nomenclature

12.1 Give the common name for each of the following acids.
(a) $CH_3CH_2CO_2H$ (b) HCO_2H (c) $CH_3(CH_2)_4CO_2H$
(d) $CH_3(CH_2)_{10}CO_2H$ (e) $CH_3(CH_2)_{16}CO_2H$ (f) $CH_3(CH_2)_{14}CO_2H$

12.2 Draw the structure of each of the following esters.
(a) octyl acetate (b) *tert*-butyl formate (c) ethyl butyrate
(d) propyl valerate

12.3 Give the common name for each of the following esters.

(a) $CH_3CH_2{-}O{-}\overset{O}{\overset{\|}{C}}{-}H$ (b) $CH_3{-}O{-}\overset{O}{\overset{\|}{C}}{-}CH_2CH_2CH_3$

(c) $CH_3(CH_2)_7{-}O{-}\overset{O}{\overset{\|}{C}}{-}CH_3$

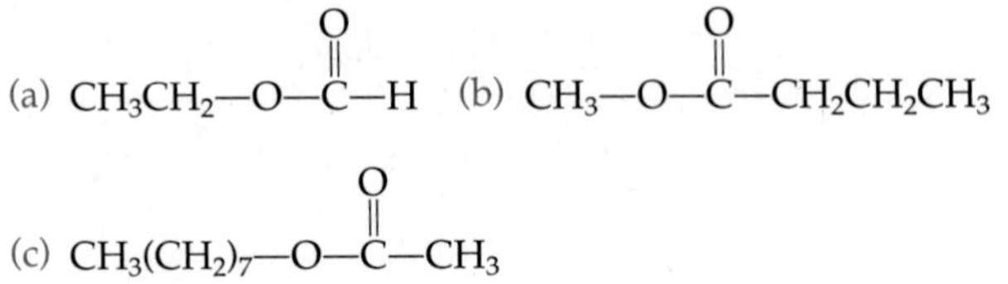

12.4 Give the common name for each of the following acids.

(a) $CH_3\underset{Cl}{\underset{|}{C}}HCH_2CO_2H$ (b) $Br\underset{CH_3}{\underset{|}{C}}HCO_2H$

(c) $CH_3\overset{CH_3}{\overset{|}{C}}H\underset{Br}{\underset{|}{C}}HCH_2CO_2H$ (d) $CH_3\overset{CH_3}{\overset{|}{\underset{CH_3}{\underset{|}{C}}}}CH_2\overset{Cl}{\overset{|}{C}}HCO_2H$

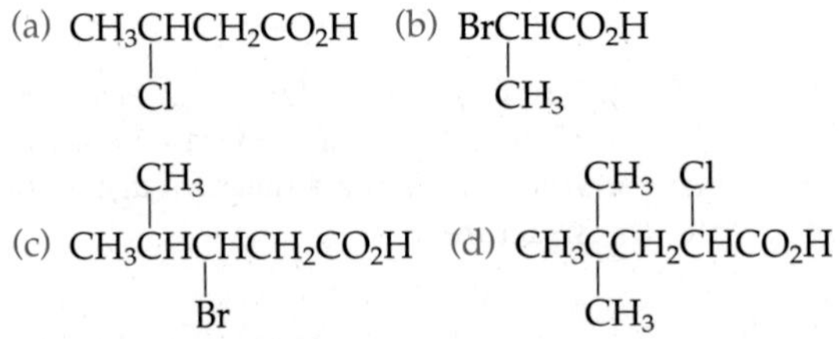

12.5 Give the IUPAC name for each of the following acids.

(a) [cyclodecene ring with COOH on the double-bond carbon] COOH
(b) [cyclopentane]—CH_2CO_2H
(c) [cyclohexane with COOH and OCH_3 at positions 1,4]
(d) CH_3—O—[benzene ring]—$\underset{CH_3}{\underset{|}{CH}}$—COOH

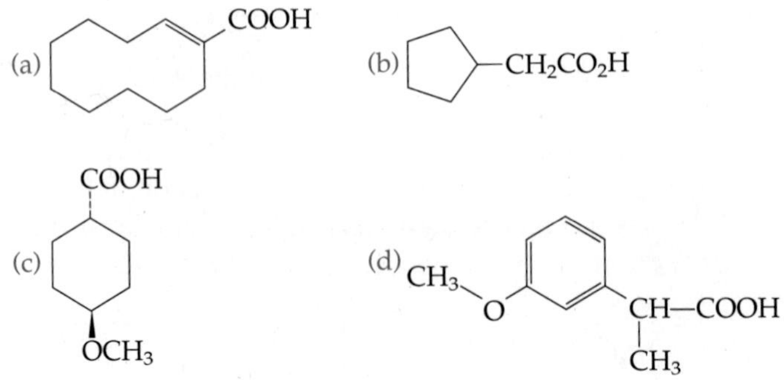

12.6 Give the IUPAC name for each of the following esters.

(a) [benzene ring]—$\overset{O}{\overset{\|}{C}}$—O—[benzene ring]
(b) [benzene ring]—$\overset{O}{\overset{\|}{C}}$—O—$C(CH_3)_3$

(c) 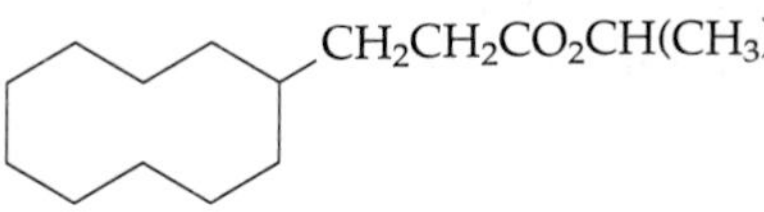 $CH_2CH_2CO_2CH(CH_3)_2$ (cyclododecyl)

(d) 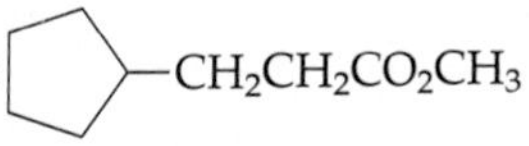 $CH_2CH_2CO_2CH_3$ (cyclopentyl)

12.7 The common name of the vasodilator cyclandelate is 3,5,5-trimethylcyclohexyl mandelate. What is the structure and name of the acid contained in the ester?

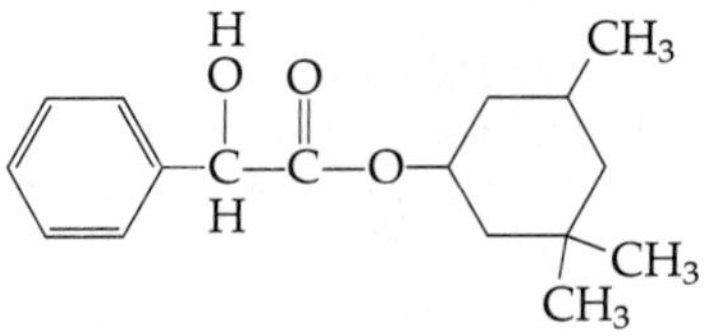

12.8 Hydrolysis of diloxanide furanoate in the body is required for it to be effective against intestinal amebiasis. What is the acid component of the drug? Based on the name of the drug, what is the name of the acid?

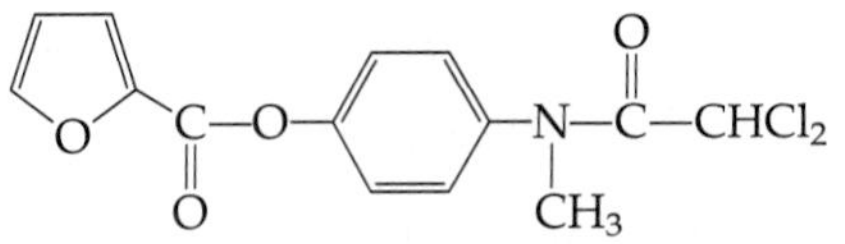

12.9 The IUPAC name of ibuprofen, the analgesic in Motrin, Advil, and Nuprin, is 2-(4-isobutylphenyl)propanoic acid. Draw the structure.

12.10 10-Undecenoic acid is the antifungal agent contained in Desenex and Cruex. Write the structure.

Cyclic Acid Derivatives

12.11 Identify the oxygen-containing functional group in the following structure of a sex pheromone of the female Japanese beetle. What is the configuration about the carbon–carbon bond?

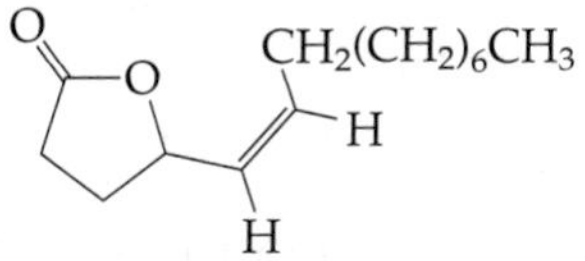

$CH_2(CH_2)_6CH_3$

12.12 Identify the nitrogen-containing functional group within the four-membered ring of cephalosporin C, an antibiotic.

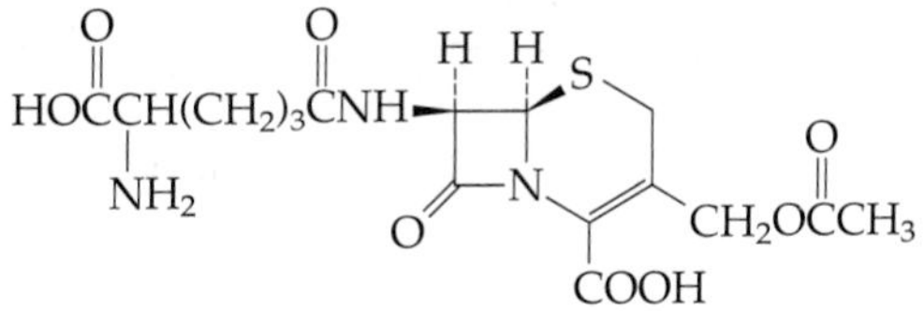

$HOCCH(CH_2)_3CNH$ NH_2 CH_2OCCH_3 COOH

12.13 The IUPAC names of lactones are derived by adding the term *lactone* at the end of the name of the parent hydroxy acid. Name each of the following compounds.

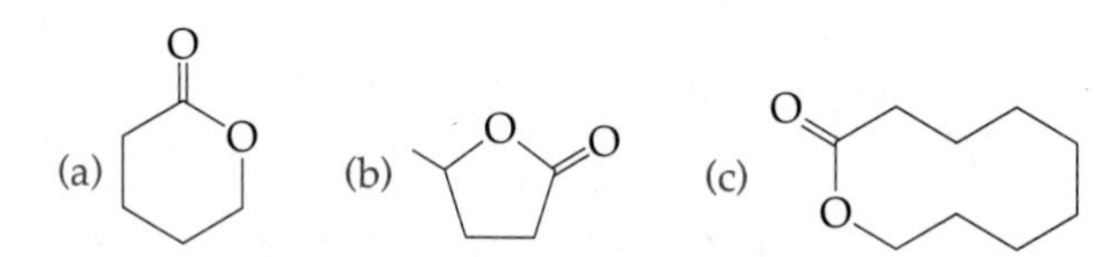

12.14 The IUPAC names of lactams are derived by adding the term *lactam* at the end of the name of the parent amino acid. Write the structure of each of the following lactams.
(a) 3-aminopropanoic acid lactam (b) 4-aminopentanoic acid lactam
(c) 5-aminopentanoic acid lactam

12.15 Which of the following compounds are lactones?

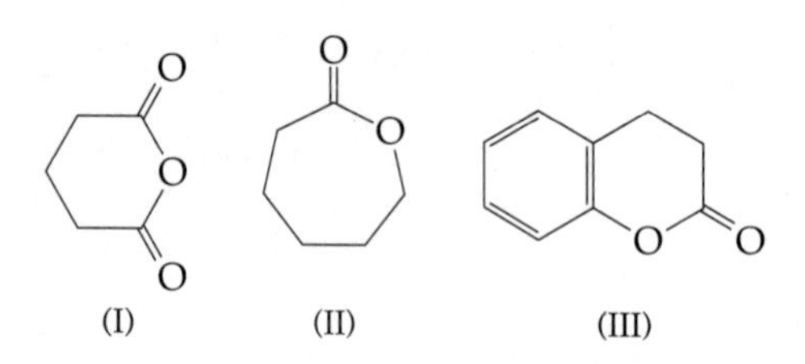

12.16 Which of the following compounds are lactams?

O N—CH_3 CH_2CH_3 (I)
O N—H CH_3 (II)
CH_3CH_2 H N O (III)

Molecular Formulas

12.17 What is the general molecular formula for a saturated carboxylic acid?
12.18 What is the general molecular formula for a saturated dicarboxylic acid?
12.19 How many isomeric acids have the molecular formula $C_4H_8O_2$?
12.20 How many isomeric esters have the molecular formula $C_4H_8O_2$?

Physical Properties

12.21 Why is 1-butanol less soluble in water than butanoic acid?
12.22 Adipic acid is much more soluble in water than hexanoic acid. Why?
12.23 The boiling point of decanoic acid is higher than that of nonanoic acid. Explain why.
12.24 The boiling points of 2,2-dimethylpropanoic acid and pentanoic acid are 164 and 186°C, respectively. Explain why.
12.25 The boiling points of methyl pentanoate and butyl ethanoate are 126 and 125°C, respectively. Explain why the values are similar.
12.26 The boiling points of methyl pentanoate and methyl 2,2-dimethylpropanoate are 126 and 102°C, respectively. Explain why these values differ.

Acidity of Carboxylic Acids

12.27 The K_a values of formic acid and acetic acid are 1.8×10^{-4} and 1.8×10^{-5}, respectively. Which compound is the stronger acid?
12.28 The pK_a values of acetic acid and benzoic acid are 4.74 and 4.19, respectively. Which acid is stronger?

12.29 The K_a of methoxyacetic acid is 2.7×10^{-4}. Explain why this value differs from the K_a of acetic acid (1.8×10^{-5}).

12.30 The K_a values of benzoic acid and *p*-nitrobenzoic acid are 6.3×10^{-5} and 3.8×10^{-4}, respectively. Explain why these values differ as they do.

12.31 The pK_a of benzoic acid is 4.2. The pK_a of probenecid is 3.4. Explain why.

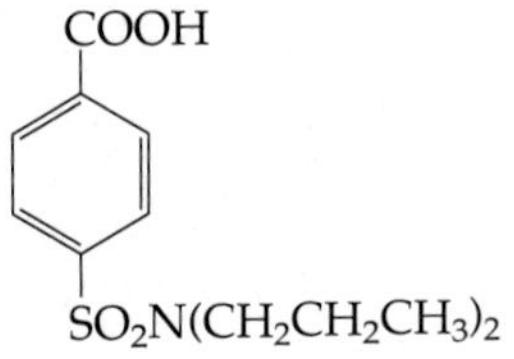

COOH

$SO_2N(CH_2CH_2CH_3)_2$

12.32 Predict the pK_a of indomethacin, an anti-inflammatory agent.

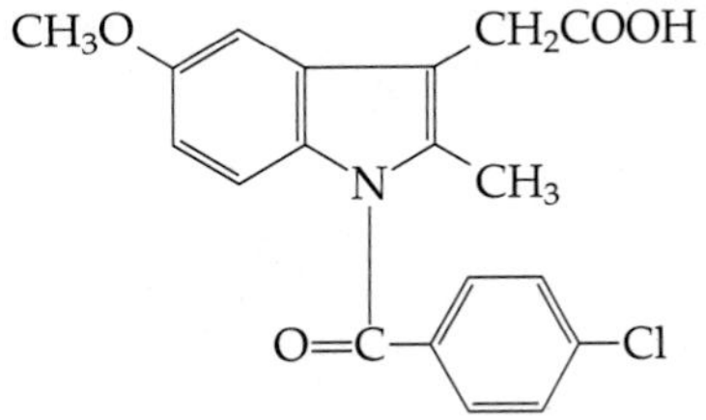

CH_3O CH_2COOH N CH_3 O=C— —Cl

Nucleophilic Acyl Substitution

12.33 Indicate whether each of the following reactions will occur.

(a) $CH_3—\overset{O}{\overset{\|}{C}}—Cl + CH_3OH \longrightarrow CH_3—\overset{O}{\overset{\|}{C}}—O—CH_3 + HCl$

(b) $CH_3—\overset{O}{\overset{\|}{C}}—NH_2 + CH_3OH \longrightarrow CH_3—\overset{O}{\overset{\|}{C}}—O—CH_3 + NH_3$

(c) $CH_3—\overset{O}{\overset{\|}{C}}—OCH_3 + CH_3NH_2 \longrightarrow CH_3—\overset{O}{\overset{\|}{C}}—NHCH_3 + CH_3OH$

12.34 Indicate whether each of the following reactions will occur.

(a) $CH_3—\overset{O}{\overset{\|}{C}}—O—\overset{O}{\overset{\|}{C}}—CH_3 + NH_3 \longrightarrow CH_3—\overset{O}{\overset{\|}{C}}—NH_2 + H—O—\overset{O}{\overset{\|}{C}}—CH_3$

(b) $CH_3—\overset{O}{\overset{\|}{C}}—O—\overset{O}{\overset{\|}{C}}—CH_3 + HCl \longrightarrow CH_3—\overset{O}{\overset{\|}{C}}—OH + Cl—\overset{O}{\overset{\|}{C}}—CH_3$

(c) $CH_3—\overset{O}{\overset{\|}{C}}—O—\overset{O}{\overset{\|}{C}}—CH_3 + CH_3OH \longrightarrow CH_3—\overset{O}{\overset{\|}{C}}—OCH_3 + H—O—\overset{O}{\overset{\|}{C}}—CH_3$

12.35 Based on the stability of the reactant, explain why thioesters are more reactive than esters in acyl substitution reactions.

12.36 Based on the stability of the reactant, explain why thioesters are less reactive than acid chlorides in acyl substitution reactions.

Reduction of Acyl Derivatives

12.37 Draw the structure of the product of each of the following reactions.

(a) cyclopentyl–C(=O)–Cl $\xrightarrow{LiAlH_4}$

(b) cyclobutyl–CH_2–C(=O)–O–CH_3 $\xrightarrow{LiAlH_4}$

(c) cyclohexenyl–C(=O)–Cl $\xrightarrow{LiAlH(O\text{-}t\text{-}Bu)_3}$

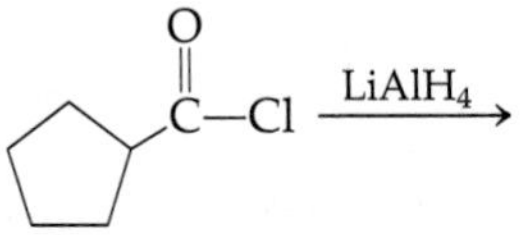

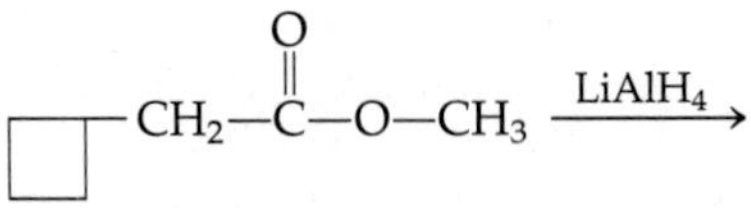

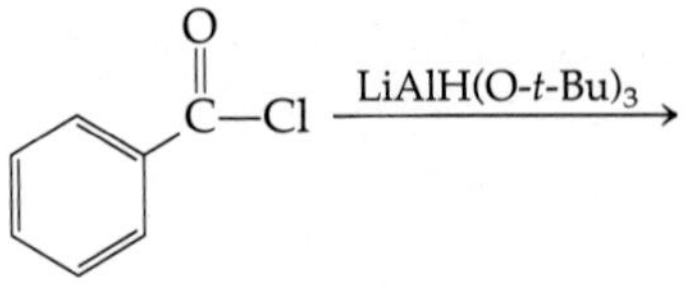

12.38 Draw the structure of the product of each of the following reactions.

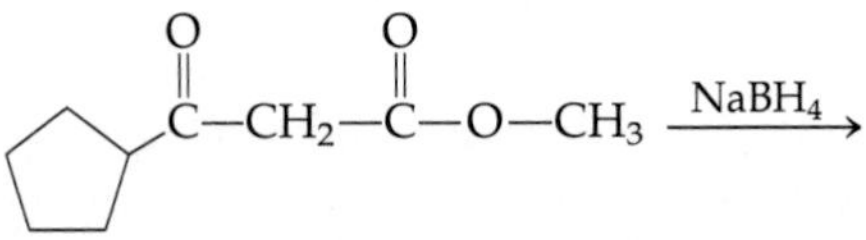

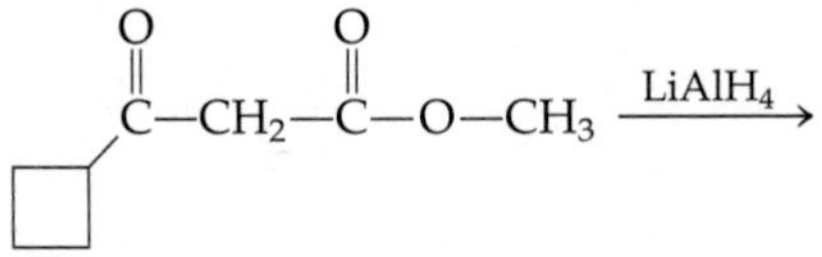

(a) cyclopentyl–C(=O)–CH_2–C(=O)–O–CH_3 $\xrightarrow{NaBH_4}$

(b) cyclobutyl–C(=O)–CH_2–C(=O)–O–CH_3 $\xrightarrow{LiAlH_4}$

(c) macrocyclic lactone (ring O, C=O) $\xrightarrow{LiAlH_4}$

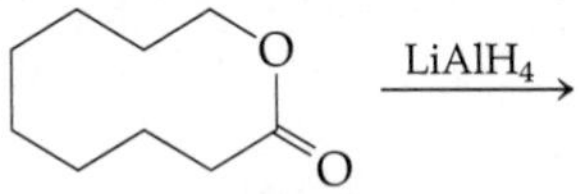

Chemical Reactions

12.39 Write the product of reaction of hexanoic acid with each of the following reagents.
(a) thionyl chloride (b) potassium hydroxide (c) methanol (d) lithium aluminum hydride

12.40 Write the product of reaction of propanoyl chloride with each of the following compounds.
(a) 1-propanol (b) 2-butanethiol (c) benzylamine ($C_6H_5CH_2NH_2$)
(d) water

12.41 What product should result from the reaction of butyrolactone with methylamine, CH_3NH_2?

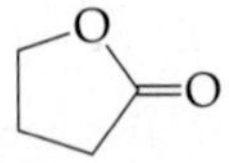

butyrolactone

12.42 What product should result from the reaction of phthalic anhydride with methanol?

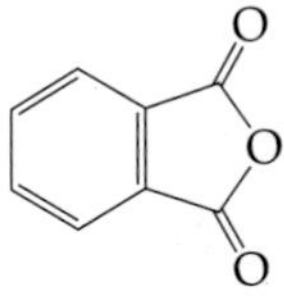

phthalic anhydride

12.43 When succinic acid is heated, a compound with the formula $C_4H_4O_3$ is formed. Suggest a structure for the compound.

12.44 When 3-(*o*-hydroxyphenyl)propanoic acid is heated in the presence of an acid catalyst, a compound with molecular formula $C_9H_8O_2$ forms. Draw its structure.

12.45 What are the products of the following reaction?

$$CH_3{-}CH_2{-}\overset{O}{\overset{\|}{C}}{-}S{-}CH_2{-}CH_3 + CH_3{-}OH \longrightarrow$$

12.46 What are the products of the following reaction?

$$CH_3{-}CH_2{-}\overset{O}{\overset{\|}{C}}{-}S{-}CH_3 + CH_3{-}CH_2{-}OH \longrightarrow$$

Multistep Synthesis

12.47 Write the structure of the final product of each of the following sequences of reactions.

(a) cyclopentyl–$CH_2CH_2OH \xrightarrow[H_2SO_4]{CrO_3} \xrightarrow{PCl_3}$

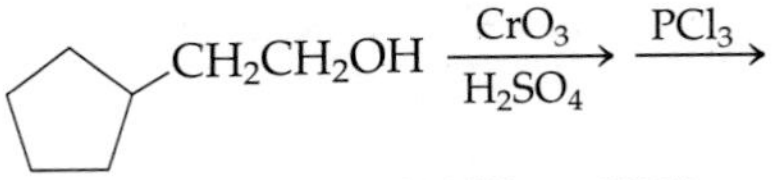

(b) phenyl–$CO_2CH_3 \xrightarrow{LiAlH_4} \xrightarrow{SOCl_2}$

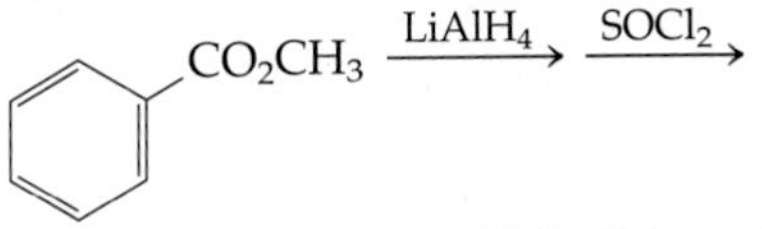

(c) phenyl–$CO_2CH_3 \xrightarrow{LiAlH(O\text{-}t\text{-}Bu)_3} \xrightarrow[H_3O^+]{CH_3OH}$

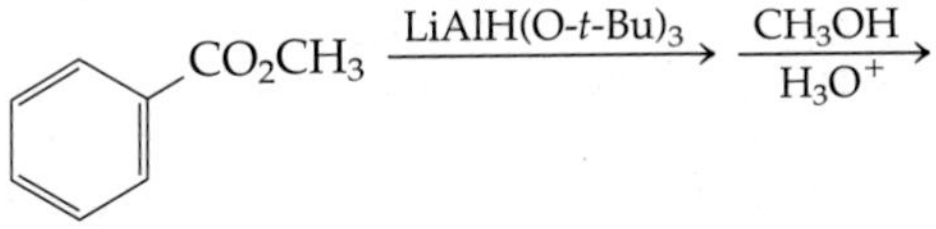

12.48 Write the structure of the final product of each of the following sequences of reactions.

(a) cyclopentyl–$CHO \xrightarrow[H_2SO_4]{CrO_3} \xrightarrow{SOCl_2}$

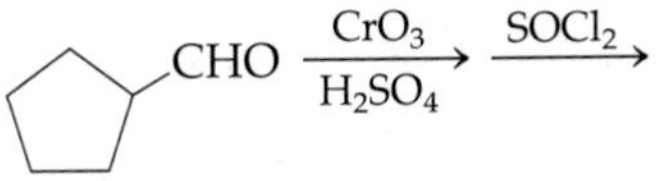

(b) phenyl–$\overset{O}{\overset{\|}{C}}{-}Cl \xrightarrow{CH_3OH} \xrightarrow{LiAlH_4}$

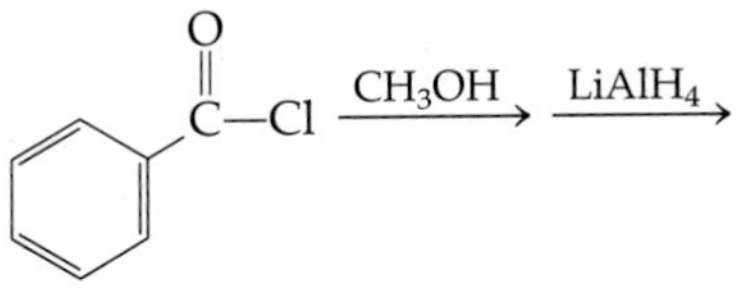

(c) Cyclooctyl–C(=O)–Cl $\xrightarrow{\text{LiAlH(O-}t\text{-Bu)}_3}$ $\xrightarrow[\text{HCl}]{\text{Zn(Hg)}}$

12.49 Outline the step(s) required to prepare cyclohexanecarboxylic acid from each of the following reactants.
(a) bromocyclohexane (b) cyclohexanol (c) cyclohexene
(d) vinylcyclohexene (e) cyclohexylmethanol

12.50 Outline the steps required to prepare hexanoic acid from each of the following reactants.
(a) 1-chlorohexane (b) 1-hexanol (c) hexanal
(d) ethyl hexanoate (e) hexanoyl chloride

12.51 Fatty acids from natural sources are long-chain unbranched carboxylic acids that contain an even number of carbon atoms. Outline steps to convert a fatty acid into a homolog containing one additional carbon atom.

12.52 Pivalic acid, $(CH_3)_3CCO_2H$, can be prepared from *t*-butyl chloride. What method should be used?

12.53 Outline the steps necessary to prepare the following compound from cyclohexanecarboxylic acid.

Cyclohexyl–C(=O)–O–CH_2–Cyclohexyl

12.54 Outline the steps necessary to prepare the following compound from benzoic acid.

Phenyl–C(=O)–O–CH_2–CH_2–Phenyl

Esters

12.55 Write the products of hydrolysis of each of the following esters.

(a) $CH_3-\overset{O}{\overset{\|}{C}}-O-CH_2CH_2CH_3$

(b) $CH_3CH_2CH_2CH_2-\overset{O}{\overset{\|}{C}}-O-CH_2CH_2CH_2CH_3$

(c) $CH_3CH_2-\overset{O}{\overset{\|}{C}}-O-CH_2CH_3$

12.56 Write the products of hydrolysis of each of the following esters.

(a) Decalin–C(=O)–OCH_2CH_3 (b) Cyclopentyl–$CO_2CH(CH_3)_2$

(c) 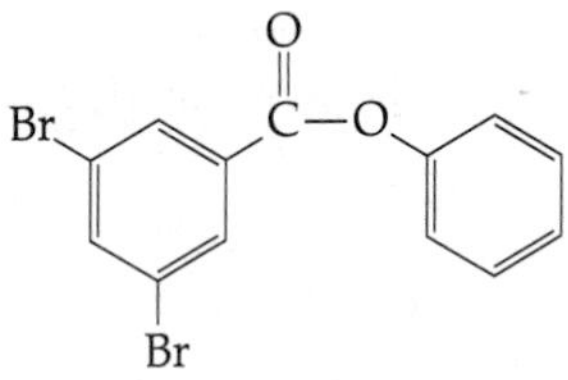 3,5-dibromophenyl ring–C(=O)–O–phenyl (Br, Br substituents)

(d) 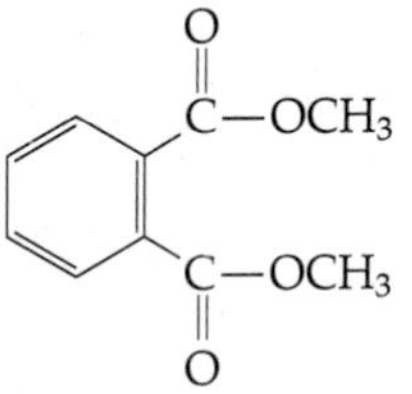 benzene ring with two adjacent –C(=O)–OCH$_3$ groups

12.57 What alcohol and acid are required to form each of the following esters?

(a) $CH_3CH_2CH_2CH_2{-}O{-}\overset{O}{\overset{\|}{C}}{-}CH_2CH_2CH_3$

(b) $CH_3{-}O{-}\overset{O}{\overset{\|}{C}}{-}CH_2CH_2CH_3$

(c) $CH_3CH_2{-}O{-}\overset{O}{\overset{\|}{C}}{-}CH_2CH_2CH_3$

12.58 What alcohol and acid are required to form each of the following esters?

(a) 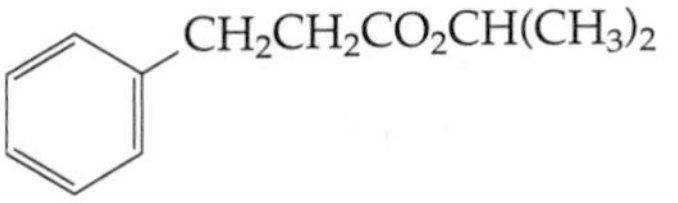 benzene ring–$CH_2CH_2CO_2CH(CH_3)_2$

(b) 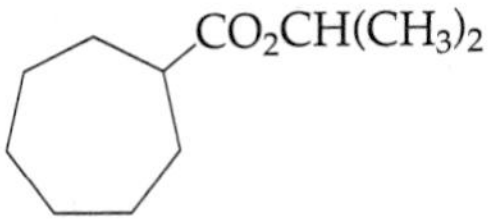 cycloheptane ring–$CO_2CH(CH_3)_2$

(c) 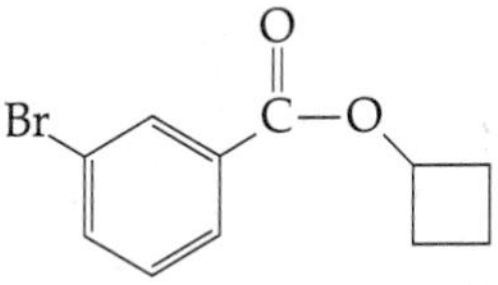 3-bromophenyl ring–C(=O)–O–cyclobutyl (Br substituent)

(d) 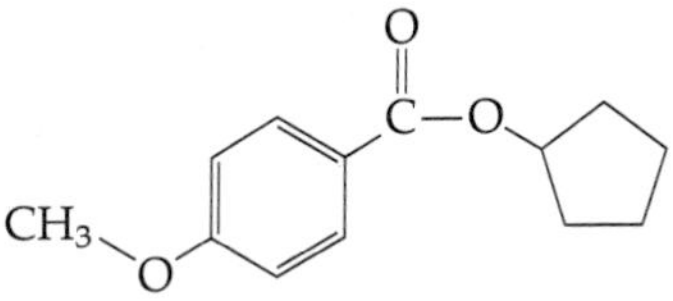 CH_3–O–phenyl ring–C(=O)–O–cyclopentyl

Polyesters

12.59 What monomers are needed to form the following polyester?

$$-O(CH_2)_4O\overset{O}{\overset{\|}{C}}(CH_2)_3\overset{O}{\overset{\|}{C}}O(CH_2)_4O\overset{O}{\overset{\|}{C}}(CH_2)_3\overset{O}{\overset{\|}{C}}-$$

12.60 What monomers are needed to form the following polyester?

$$-\overset{O}{\overset{\|}{C}}(CH_2)_4\overset{O}{\overset{\|}{C}}O(CH_2)_3O\overset{O}{\overset{\|}{C}}(CH_2)_4\overset{O}{\overset{\|}{C}}O(CH_2)_3O-$$

Phosphoric Acid Esters and Anhydrides

12.61 Draw the structure of the ester formed by reacting 1 mole of methanol with phosphoric acid. How many acidic protons are there in the product?

12.62 Draw the structure of the ester formed by reacting 1 mole of ethanol with pyrophosphoric acid. How many acidic protons are there in the product?

12.63 Determine the number of anhydride and ester bonds in the following compound. How many acidic protons are present?

$$CH_3{-}CH_2{-}O{-}\underset{OH}{\underset{|}{\overset{O}{\overset{\|}{P}}}}{-}O{-}\underset{OH}{\underset{|}{\overset{O}{\overset{\|}{P}}}}{-}O{-}CH_2{-}CH_3$$

12.64 Determine the number of anhydride and ester bonds in the following compound. How many acidic protons are present?

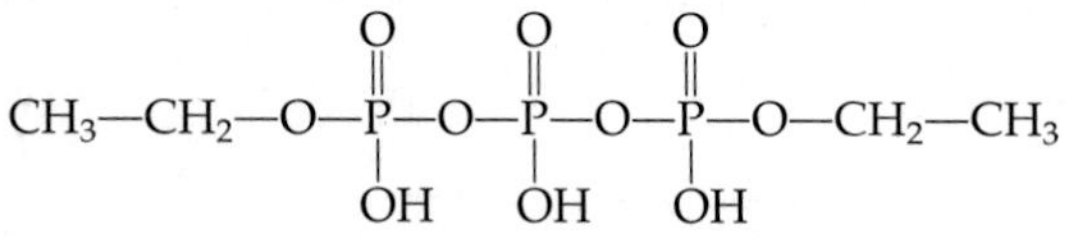

$$CH_3-CH_2-O-\overset{\overset{\displaystyle O}{\|}}{\underset{\underset{\displaystyle OH}{|}}{P}}-O-\overset{\overset{\displaystyle O}{\|}}{\underset{\underset{\displaystyle OH}{|}}{P}}-O-\overset{\overset{\displaystyle O}{\|}}{\underset{\underset{\displaystyle OH}{|}}{P}}-O-CH_2-CH_3$$

Claisen Condensation

12.65 What ester is required to form each of the following compounds by a Claisen condensation?

(a) 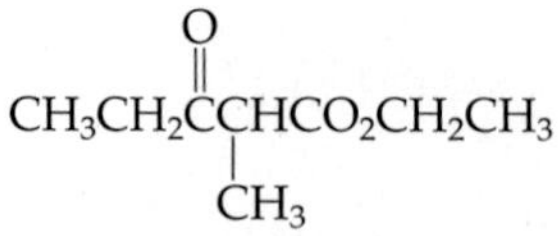 $CH_3CH_2\overset{\overset{\displaystyle O}{\|}}{C}\underset{\underset{\displaystyle CH_3}{|}}{C}HCO_2CH_2CH_3$

(b) 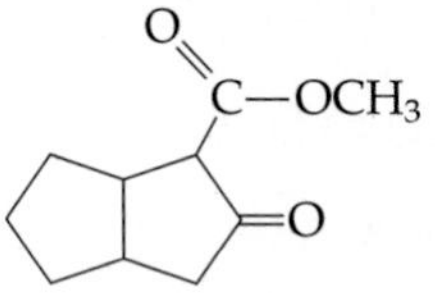 [bicyclic structure: cyclopentane fused to cyclopentanone bearing $\overset{O}{\|}C-OCH_3$ adjacent to the ring $C=O$]

12.66 Under some circumstances, Claisen condensations can take place between two different esters. Explain why the reaction of ethyl acetate and ethyl benzoate occurs readily to give a good yield of a single product. What is the structure of the product?

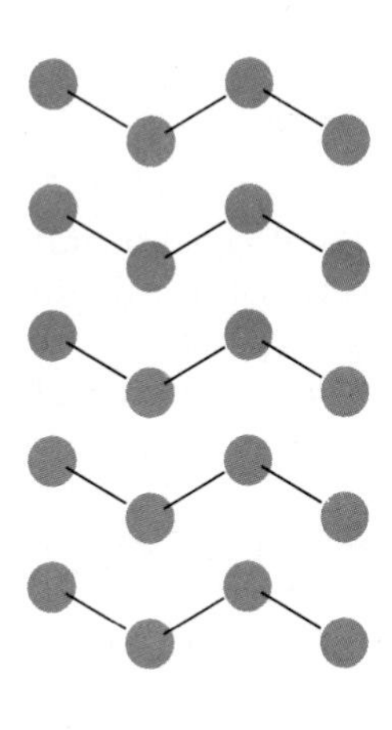

13

LIPIDS

13.1 CLASSIFICATION OF LIPIDS

The name lipid encompasses a wide range of molecular structures and an extraordinary range of biochemical functions. As fats, lipids are a major source of metabolic energy; as components of biological membranes, lipids provide a partition between a cell and its environment; and as hormones, lipids regulate a huge spectrum of cellular activities.

Lipids are relatively nonpolar compounds, and they can be separated from more polar cellular substances by their solubility in nonpolar organic solvents. In fact, lipids were historically classified as compounds of biological origin that are soluble in organic solvents. The term *lipid* is sometimes used as a synonym for fat (G. *lipos*, fat). However, fat is only one of the various types of lipids.

Lipids are divided into two groups based on their hydrolysis reactions. **Simple lipids** are not hydrolyzed in aqueous basic solution. These include terpenes (Chapter 4), which are produced in plants, and steroids (Chapter 3), which are important hormones in animals. **Complex lipids** are hydrolyzed by aqueous basic solution. The major hydrolysis products of complex lipids are long-chain carboxylic acids called fatty acids.

The fatty acids in complex lipids almost always have an even number of carbon atoms. The other components obtained in the hydrolysis of a complex lipid determine its subclass.

1. **Waxes** are esters of long-chain alcohols and fatty acids.
2. **Triacylglycerols,** also known as triglycerides, are esters of glycerol and long-chain fatty acids. Triglycerides that are relatively saturated are semisolids, whereas the more unsaturated compounds are oils.
3. **Glycerophospholipids** are composed of glycerophosphate (an ester of glycerol and phosphoric acid), long-chain fatty acids, and certain low molecular weight alcohols.
4. **Sphingophospholipids** are composed of a phosphate ester of sphingosine, long-chain fatty acids, and choline.
5. **Glycosphingolipids** are composed of sphingosine, fatty acids, and a carbohydrate that may be a monosaccharide or an oligosaccharide.

13.2 WAXES

Waxes are esters of fatty acids and long-chain alcohols, both of which contain an even number of carbon atoms. Waxes are low-melting solids that cover the surface of plant leaves and fruits. They also coat the hair and feathers of some animals to provide a water barrier. If an aquatic bird comes into contact with hydrocarbons, which can happen as a result of an oil spill, the wax dissolves, the feathers become wet, and the bird cannot maintain its buoyancy.

We often encounter waxes in our daily lives. For example, carnauba wax, which coats the leaves of palm trees, is widely used in floor polish and car wax. Beeswax is secreted by bees and is the structural material for the beehive.

Whale Oil

Whale oil isn't an oil—that is, it isn't a triacylglycerol. It is actually a mixture of waxes. As much as 4 tons of whale "oil" is contained in the head of a sperm whale. The whale uses this mixture to control its buoyancy. One of the compounds is an ester of a 16-carbon acid and a 16-carbon alcohol. Because this compound has fewer carbon atoms than carnauba wax and beeswax, it has a lower melting point.

$$CH_3(CH_2)_{14}C(=O)-O-CH_2(CH_2)_{14}CH_3$$

whale oil

Most solids have a greater density than liquids. Therefore, the volume of a substance generally decreases when it is converted from a liquid to a solid. Whales take advantage of this property of liquids and solids. Whale oil tends to freeze at the depths of the ocean where the whales feed. As a consequence the density of the whale increases, and it is able to stay submerged without expending energy. The whale controls the amount of liquid and solid oil and, therefore, its average density by passing cold seawater through chambers in its head or by increasing the circulation of warm blood in the same area.

$$CH_3(CH_2)_{24}C(=O)\text{—}O\text{—}CH_2(CH_2)_{28}CH_3$$

carnauba wax

$$CH_3(CH_2)_{12}C(=O)\text{—}O\text{—}CH_2(CH_2)_{24}CH_3$$

beeswax

EXAMPLE 13.1 Half of the dry weight of a copepod that lives in the waters off British Columbia is the compound $C_{36}H_{62}O_2$. Hydrolysis of this compound yields an unbranched acid, $C_{20}H_{30}O_2$, and an unbranched alcohol, $C_{16}H_{34}O$. Hydrogenation of the acid yields $C_{20}H_{40}O_2$. Describe the structure of the $C_{36}H_{62}O_2$ compound.

Solution Because the compound yields an acid and an alcohol when hydrolyzed, it must be an ester. The formula $C_{16}H_{34}O$ corresponds to that of a saturated alcohol ($C_nH_{2n+2}O$). The acid, $C_{20}H_{30}O_2$, must contain five double bonds because hydrogenation results in the addition of 5 moles of hydrogen gas. The location of the double bonds cannot be determined from the information given. A general representation of the wax is

$$C_{19}H_{29}\overset{\overset{\displaystyle O}{\|}}{C}\text{—}O\text{—}(CH_2)_{15}CH_3$$

13.3 FATTY ACIDS

Triacylglycerols, glycerophospholipids, sphingophospholipids, and glycosphingolipids all yield fatty acids upon hydrolysis. **Fatty acids** are long-chain carboxylic acids that contain an even number of carbon atoms. These acids range from 14 to 22 carbon atoms, but the 16- and 18-carbon acids are the most abundant. Fatty acids can be either saturated or unsaturated.

In the nomenclature of fatty acids, the long chain provides the name of the parent. Thus, a 16-carbon fatty acid is hexadecanoic acid, and an 18-carbon fatty acid is octadecanoic acid. These two acids have the common names palmitic acid and stearic acid, respectively.

The most common unsaturated fatty acids also have either 16 or 18 carbon atoms. The configuration about the double bonds is cis. The position of the double bond is indicated by the Greek letter delta (Δ) followed by a superscript to indicate the position of the double bond. Thus, the C-16 fatty acid with a cis double bond between the C-9 and C-10 atoms is called *cis*-Δ^9-hexadecenoic acid. Its common name is palmitoleic acid. The C-18 fatty acid with a cis double bond between the C-9 and C-10 atoms is called *cis*-Δ^9-octadecenoic acid. Its common name is oleic acid.

Fatty acids that contain more than one double bond are said to be **polyunsaturated.** The most common polyunsaturated fatty acids contain 18 carbon atoms and have either two or three double bonds. They are *cis-cis*-$\Delta^{9,12}$-octadecadienoic acid (linoleic acid) and *cis-cis-cis*-$\Delta^{9,12,15}$-octadecatrienoic

Prostaglandins

Prostaglandins are very potent biological compounds. They are derived from arachidonic acid—a 20-carbon polyunsaturated fatty acid with four double bonds. Prostaglandins are 20-carbon fatty acids that contain a trans-substituted, five-membered ring. They are classified according to the number and arrangement of double bonds, hydroxyl groups, and ketone groups.

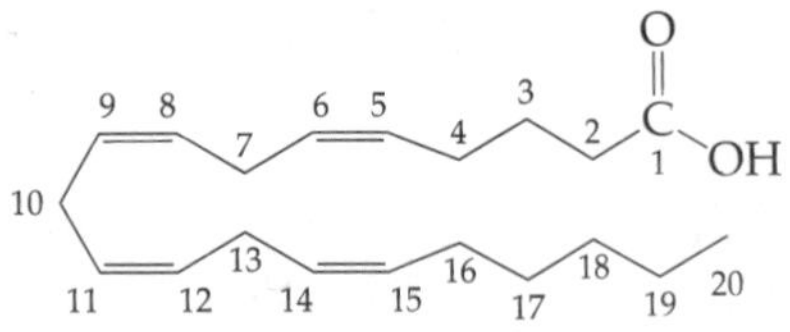

arachidonic acid

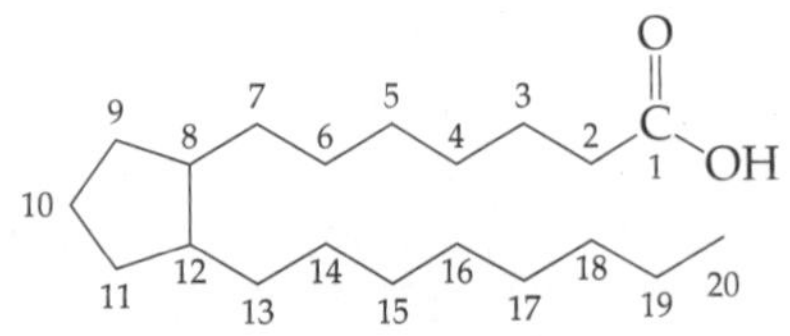

prostanoic acid

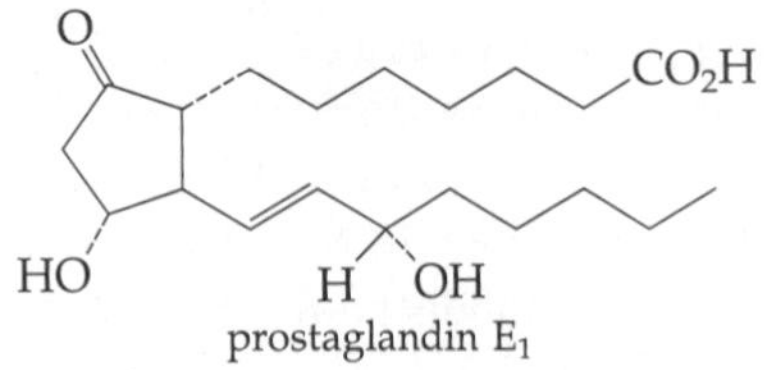

prostaglandin E_1

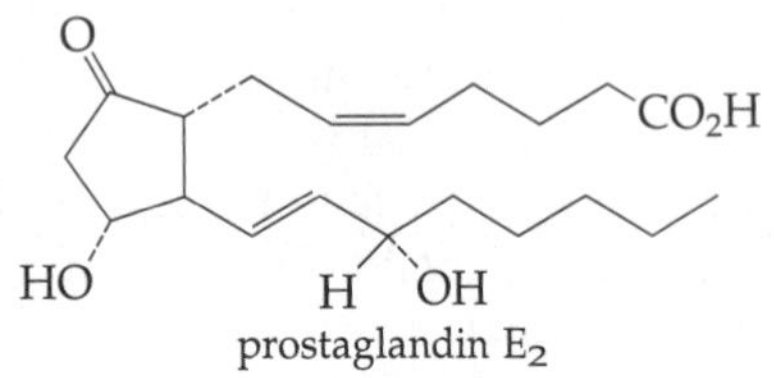

prostaglandin E_2

The prostaglandins, which were originally isolated from the prostate gland, occur in many body tissues and are physiologically active at very low concentrations. Prostaglandins alter the effects of many hormones. For example, hormone-sensitive enzymes that hydrolyze lipids respond to insulin and other hormones that also regulate the concentration of blood glucose. Prostaglandin E_1 inhibits these enzymes at concentrations as low as 10 nM (10^{-8} M).

Prostaglandins affect virtually every aspect of reproduction. They regulate menstruation and control fertility and conception. Prostaglandin E_2 stimulates smooth muscle contraction in the uterus and has been used clinically to induce labor and to abort pregnancies prematurely.

The release of prostaglandins also causes an inflammation of tissues, an effect inhibited by aspirin. Aspirin prevents the first step in prostaglandin synthesis in which the five-membered ring forms. Prostaglandin biosynthesis is also inhibited by a class of steroid hormones called corticosteroids. These steroids act in the following way. Arachidonic acid is not free in cells but esterified to a phospholipid and actually incorporated in this form in cell membranes. Arachidonate is released by the action of enzymes called phospholipases, and these enzymes are inhibited by corticosteroids. Hence, corticosteroids such as cortisone are used to reduce inflammation in tissues damaged by injury.

Much research is being done to develop synthetic prostaglandins for use as therapeutic drugs. Natural prostaglandins cannot be taken orally because they are rapidly degraded and do not survive long enough for effective action. Thus, one research goal is to develop modified prostaglandins that can be administered orally.

TABLE 13.1 Formulas and Melting Points of Common Fatty Acids

Name	Formula	Melting point (°C)
Saturated		
lauric	$CH_3(CH_2)_{10}CO_2H$	44
myristic	$CH_3(CH_2)_{12}CO_2H$	58
palmitic	$CH_3(CH_2)_{14}CO_2H$	63
stearic	$CH_3(CH_2)_{16}CO_2H$	69
arachidic	$CH_3(CH_2)_{18}CO_2H$	77
Unsaturated		
oleic	$CH_3(CH_2)_7CH{=}CH(CH_2)_7CO_2H$	13
linoleic	$CH_3(CH_2)_4CH{=}CHCH_2CH{=}CH(CH_2)_7CO_2H$	−5
linolenic	$CH_3CH_2CH{=}CHCH_2CH{=}CHCH_2CH{=}CH(CH_2)_7CO_2H$	−11
arachidonic	$CH_3(CH_2)_4CH{=}CHCH_2CH{=}CHCH_2CH{=}CHCH_2CH{=}CH(CH_2)_3CO_2H$	−50

acid (linolenic acid). These polyunsaturated fatty acids cannot be synthesized by mammals, and they are required in the diet.

The formulas and melting points of some fatty acids are given in Table 13.1. The melting points of the saturated fatty acids increase with increasing chain length as expected because London forces increase with increasing chain length. The hydrocarbon chains of saturated acids pack together tightly in the solid.

Cis unsaturated fatty acids are "bent" molecules because of the geometry about the double bonds. These "bends" hinder efficient molecular packing, and London forces are weaker than those in saturated fatty acids. As a result, unsaturated acids have lower melting points than saturated acids. The melting point of stearic acid is 69 °C. In contrast, the melting points of oleic, linoleic, and linolenic acid are 13, −5, and −11 °C, respectively. Space-filling models of oleic, linoleic, and stearic acids are shown in Figure 13.1.

EXAMPLE 13.2 The melting point of palmitoleic acid is −1 °C. Compare this melting point to that of palmitic acid and explain the difference.

$$\begin{array}{ccc} CH_3(CH_2)_4CH_2 & & (CH_2)_7CO_2H \\ & C{=}C & \\ H & & H \end{array}$$

palmitoleic acid

Solution The melting point of palmitic acid is 63 °C, which is 64 degrees higher than the melting point of palmitoleic acid. This difference is due to

FIGURE 13.1 **Space-Filling Models of Fatty Acids**

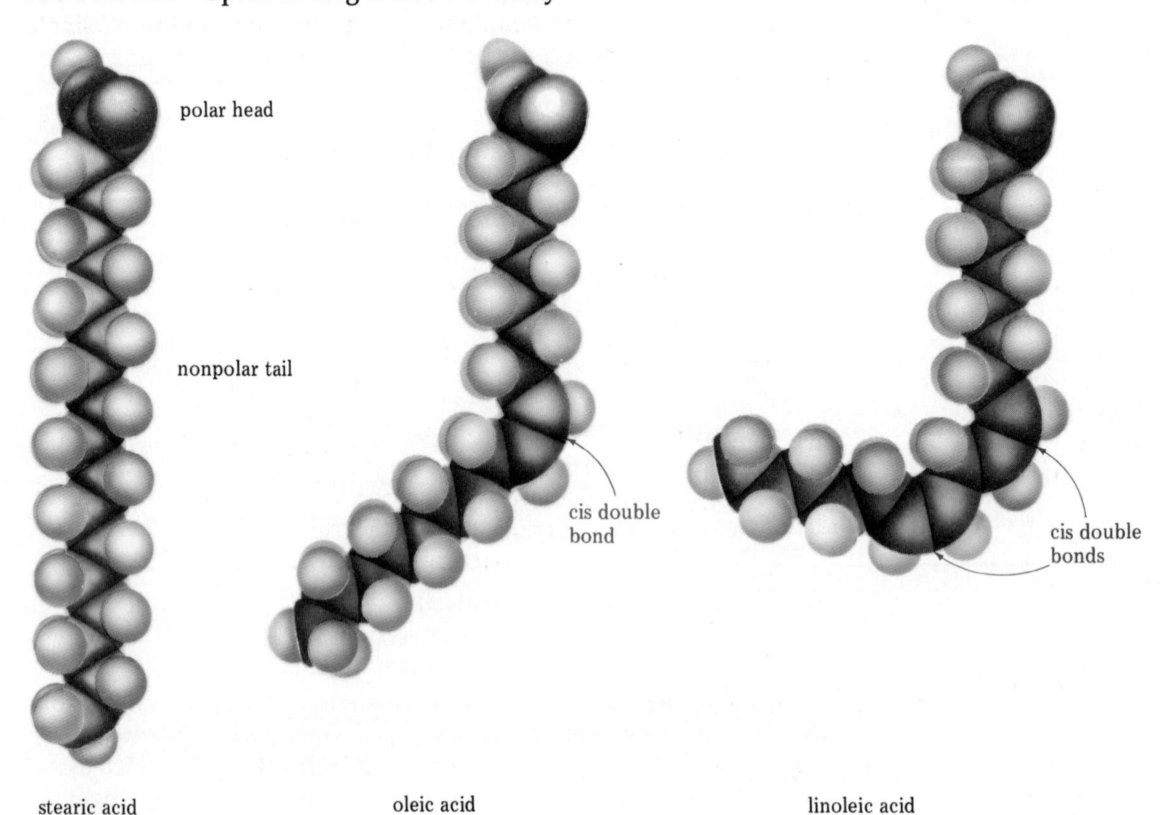

the presence of a cis double bond in palmitoleic acid, which prevents close approach of the hydrocarbon chains and results in weaker London forces.

13.4 TRIACYLGLYCEROLS

Triacylglycerols are triesters of glycerol and fatty acids. They are also known as fats and oils. Triacylglycerols are represented by the following general formula and block diagram.

$$\begin{array}{l} CH_2{-}O{-}\overset{\overset{\large O}{\|}}{C}{-}R \\ | \\ CH{-}O{-}\overset{\overset{\large O}{\|}}{C}{-}R \\ | \\ CH_2{-}O{-}\overset{\overset{\large O}{\|}}{C}{-}R \end{array}$$

glycerol	
	fatty acid unit
	fatty acid unit
	fatty acid unit

Fats and oils are mixtures of compounds. The fatty acid components of these mixtures vary in chain length and degree of unsaturation. A single molecule of a fat or oil may contain three different acid residues.

Animal fats have a high percentage of saturated acids, whereas plant oils have a high percentage of unsaturated acids. Fats are solids or semisolids and are usually obtained from animals. These fats are colloquially known as "lard". The important acids in these sources are myristic, palmitic, and stearic acids (Table 13.2). The unsaturated acids found in oils are oleic, linoleic, and linolenic acid; all contain 18 carbon atoms, but they differ in their degree of unsaturation. Examples of a fat and an oil are given below.

oleic acid

$$\begin{array}{l} CH_2-O-\overset{O}{\overset{\|}{C}}-(CH_2)_7-CH=CH-(CH_2)_7-CH_3 \\ | \\ CH-O-\overset{O}{\overset{\|}{C}}-(CH_2)_7-CH=CH-(CH_2)_7-CH_3 \\ | \\ CH_2-O-\overset{O}{\overset{\|}{C}}-(CH_2)_7-CH=CH-(CH_2)_7-CH_3 \end{array}$$

an oil

stearic acid

$$\begin{array}{l} CH_2-O-\overset{O}{\overset{\|}{C}}-(CH_2)_{16}-CH_3 \\ | \\ CH-O-\overset{O}{\overset{\|}{C}}-(CH_2)_{16}-CH_3 \\ | \\ CH_2-O-\overset{O}{\overset{\|}{C}}-(CH_2)_{16}-CH_3 \end{array}$$

a fat

Oils are typically derived from vegetable sources such as olives, peanuts, corn, and soybeans. These oils contain a high proportion of unsaturated acid residues.

TABLE 13.2 Composition of Fats and Oils

	Melting point (°C)	Saturated fatty acids (%)				Unsaturated fatty acids (%)		
		Myristic	Palmitic	Stearic	Arachidic	Oleic	Linoleic	Linolenic
Animal fats								
butter	32	11	29	9	2	27	4	—
lard	30	1	28	12	—	48	6	—
human fat	15	3	24	8	—	47	10	—
Plant oils								
corn	−20	1	10	3	—	50	34	—
cottonseed	−1	1	23	1	1	23	48	—
linseed	−24	—	6	2	1	19	24	47
olive	−6	—	7	2	—	84	5	—
peanut	3	—	8	3	2	56	26	—
soybean	−16	—	10	2	—	29	51	6

Triacylglycerols Store Energy

Mammals have several sources of metabolic energy. Two sources are blood glucose and liver glycogen. This chemical energy is readily available during strenuous exercise, but its quantity is limited.

Blood sugar can sustain metabolic activity for only a few minutes. Glycogen, a readily available storage form of glucose, would be expended in a few hours of moderate activity. Therefore, larger reserves of chemical energy are needed for survival. Unabsorbed food in the digestive tract is one reserve that can be converted into blood sugar and glycogen. However, the most important energy reserves are triacylglycerols in adipose tissue, which constitutes about 15% of body weight. The lipid reserve is sufficient to maintain life for about 40 days providing water is available. A person weighing 150 lb has chemical energy available in the indicated amounts.

blood sugar	40 kcal
glycogen	600 kcal
triacylglycerols	80,000 kcal

Triacylglycerols are more reduced than carbohydrates. Therefore, the metabolic oxidation of fatty acids releases more energy than that of carbohydrates. Oxidation of fatty acids yields about 9.3 kcal/g, whereas that of carbohydrates yields about 3.7 kcal/g.

Triacylglycerols are a more concentrated store of metabolic energy than carbohydrates. Triacylglycerols, which are nonpolar, are stored in a nearly anhydrous form in adipose tissue. Glycogen, which contains many hydroxyl groups, is very polar and binds about 2 g of water per gram of compound. Thus, 1 g of anhydrous fat stores about six times as much energy as a gram of hydrated glycogen.

Migratory birds efficiently store chemical energy as triacylglycerols. These birds sometimes travel for several days over water. If the energy required for flight was provided from glycogen, the bird would have to carry six times as much weight for fuel.

Animals accumulate fat (adipose tissue) when their intake of food exceeds their demand for energy. Adipose tissue surrounds vital organs such as the kidneys and forms a protective cushion. A subcutaneous layer of fat helps insulate the animal against heat loss. Although plants do not generally store fats and oils for energy requirements, some (such as peanuts and olives) produce triacylglycerols in abundance.

The degree of unsaturation in plant oils depends on their growing conditions. For example, linseed oil obtained from flaxseed grown in warm climates may be twice as unsaturated as oil obtained from seed grown in cold climates. Similarly, the composition of lard from hogs depends on their diet; the fat of corn-fed hogs is more saturated than that of peanut-fed hogs.

The relationship between consumption of saturated fats and arterial disease has been the object of extensive medical research. Unsaturated fats do not tend to accumulate in arterial deposits. Safflower oil, because of its high content of unsaturated fatty acids, is now a popular product.

EXAMPLE 13.3 Soybean oil is 51% linoleic acid. Draw a structure for one of the possible components of soybean oil.

Solution Although all oils are mixtures of triacylglycerols, the large percentage of linoleic acid means that there must be a large amount of triacylglycerols containing only linoleic acid.

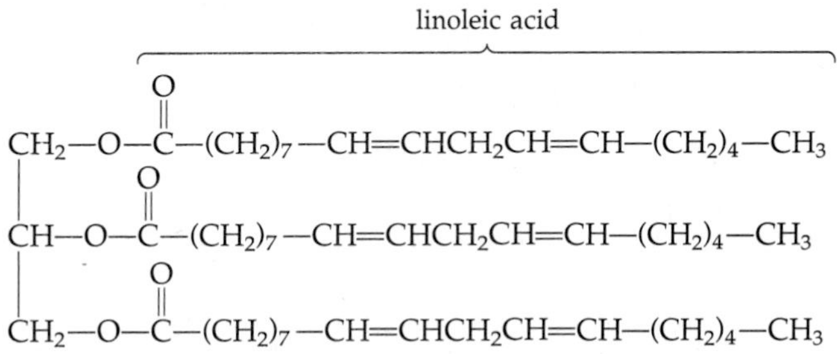

linoleic acid

$$
\begin{array}{l}
CH_2-O-\overset{O}{\overset{\|}{C}}-(CH_2)_7-CH{=}CHCH_2CH{=}CH-(CH_2)_4-CH_3 \\
\,| \\
CH-O-\overset{O}{\overset{\|}{C}}-(CH_2)_7-CH{=}CHCH_2CH{=}CH-(CH_2)_4-CH_3 \\
\,| \\
CH_2-O-\overset{O}{\overset{\|}{C}}-(CH_2)_7-CH{=}CHCH_2CH{=}CH-(CH_2)_4-CH_3
\end{array}
$$

13.5 GLYCEROPHOSPHOLIPIDS

All cells are surrounded by a thin membrane. This cell membrane, also called the plasma membrane, contains a high proportion of lipids called glycerophospholipids. The "backbone" of glycerophospholipids is L-3-phosphoglycerate. The C-1 and C-2 hydroxyl groups of L-3 phosphoglycerate are esterified with fatty acids to give a molecule called a phosphatidic acid. At physiological pH, phosphatidic acid exists in an ionized form called phosphatidate.

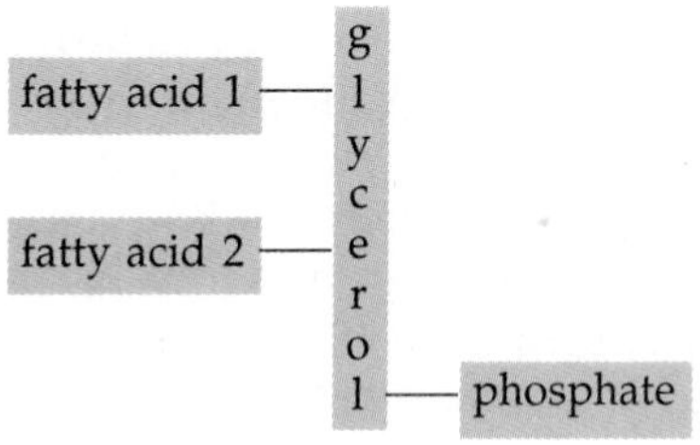

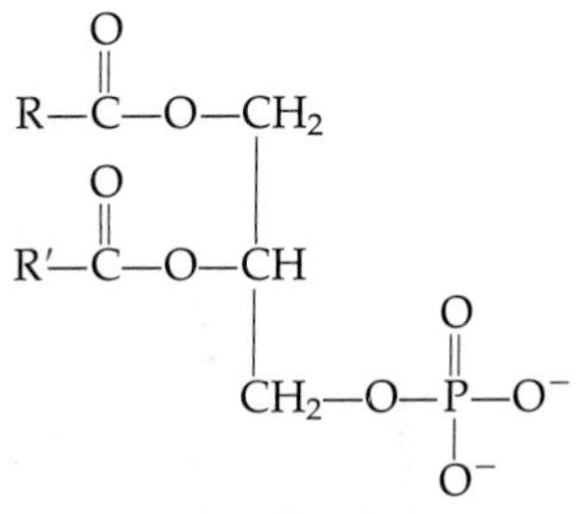

fatty acid 1 — glycerol
fatty acid 2 —
phosphate

$$
\begin{array}{l}
R-\overset{O}{\overset{\|}{C}}-O-CH_2 \\
R'-\overset{O}{\overset{\|}{C}}-O-CH \\
\qquad\qquad\;\; CH_2-O-\overset{O}{\overset{\|}{P}}-O^- \\
\qquad\qquad\qquad\qquad\;\; O^-
\end{array}
$$

phosphatidate

A phosphatidate group is linked to various alcohols—including choline, ethanolamine, inositol, or serine—through its phosphoryl group.

$HO-CH_2-CH_2-NH_2$
ethanolamine

$HO-CH_2-CH_2-N^+(CH_3)_3$
choline

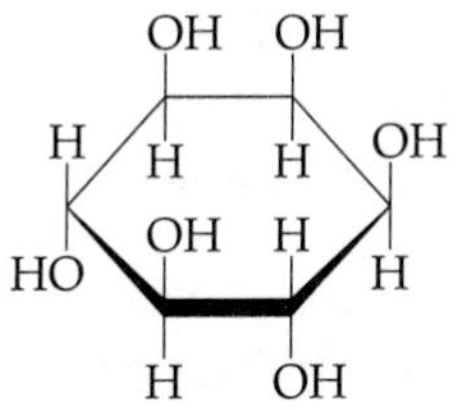

inositol

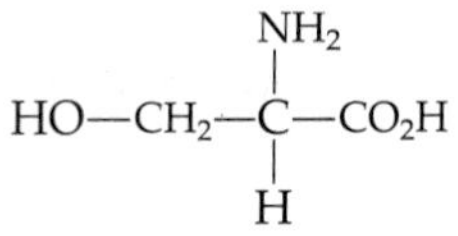

$HO-CH_2-\overset{NH_2}{\underset{H}{C}}-CO_2H$
serine

FIGURE 13.2 **Structure of Phosphoglycerides**

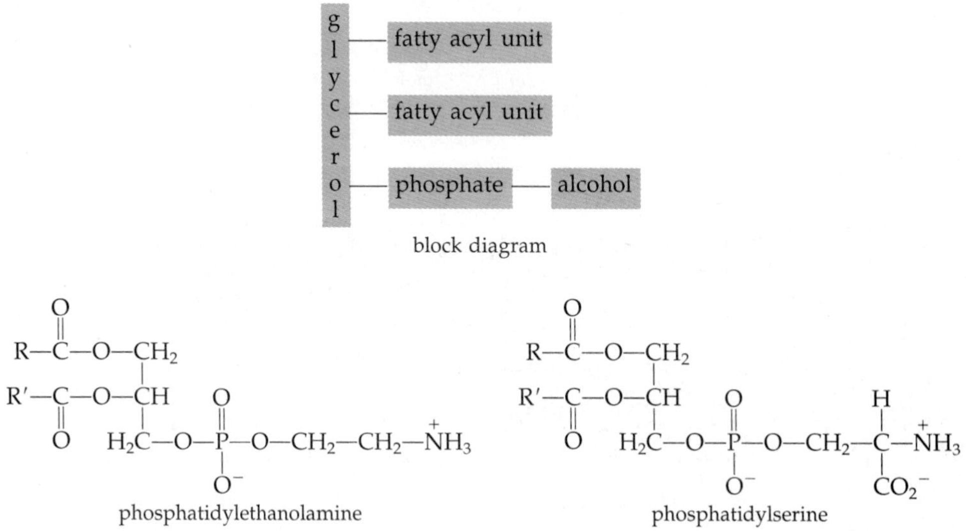

glycerol — fatty acyl unit
glycerol — fatty acyl unit
glycerol — phosphate — alcohol

block diagram

$$
\begin{array}{l}
R{-}\overset{O}{\overset{\|}{C}}{-}O{-}CH_2 \\
R'{-}\underset{\underset{O}{\|}}{C}{-}O{-}CH \\
\qquad\quad H_2C{-}O{-}\overset{\overset{O}{\|}}{\underset{\underset{O^-}{|}}{P}}{-}O{-}CH_2{-}CH_2{-}\overset{+}{N}H_3
\end{array}
$$

phosphatidylethanolamine

$$
\begin{array}{l}
R{-}\overset{O}{\overset{\|}{C}}{-}O{-}CH_2 \\
R'{-}\underset{\underset{O}{\|}}{C}{-}O{-}CH \\
\qquad\quad H_2C{-}O{-}\overset{\overset{O}{\|}}{\underset{\underset{O^-}{|}}{P}}{-}O{-}CH_2{-}\overset{\overset{H}{|}}{\underset{\underset{CO_2^-}{|}}{C}}{-}\overset{+}{N}H_3
\end{array}
$$

phosphatidylserine

A block diagram of a phosphatide and general formulas for two of the types of phosphatides are given in Figure 13.2. The phosphatides exist as anions at physiological pH (approx. 7).

When the phosphoryl group of phosphatidate is linked to the hydroxyl group of ethanolamine in a phosphate ester, the resulting molecule is a phosphatidylethanolamine, also called cephalin. Its amine nitrogen atom is protonated at pH 7.0. Phosphatidylethanolamine is a major component of cell membranes in the heart and liver and in brain tissue.

When the phosphoryl group of phosphatidate is linked to the hydroxyl group of choline in a phosphate ester, the resulting molecule is a phosphatidylcholine, also known as lecithin. Lecithins are particularly abundant in eggs and are present in most cell membranes. Phosphatidylcholine has a positive charge at the nitrogen atom, a quaternary ammonium ion. As a result, both phosphatidylcholine and phosphatidylethanolamine are dipolar, but have no net charge.

Phosphatidate is linked to the amino acid serine in phosphatidylserine. This molecule has three charged sites at pH 7.0. The phosphate oxygen atom bears a negative charge because the proton is dissociated at pH 7.0. Also, the carboxyl group of serine exists as the carboxylate ion, and the amine group of serine is protonated. In total, phosphatidylserine bears a net negative charge.

A molecular model of a phosphatidylcholine containing one unsaturated and one saturated acid is given in Figure 13.3. All phospholipids have polar sites, which are represented by a circle in the simplified model. The two nonpolar hydrocarbon chains of the fatty acids are represented as wavy lines or "tails" attached to the polar "head". These structural features con-

FIGURE 13.3 **Representation of a Phosphoglyceride**

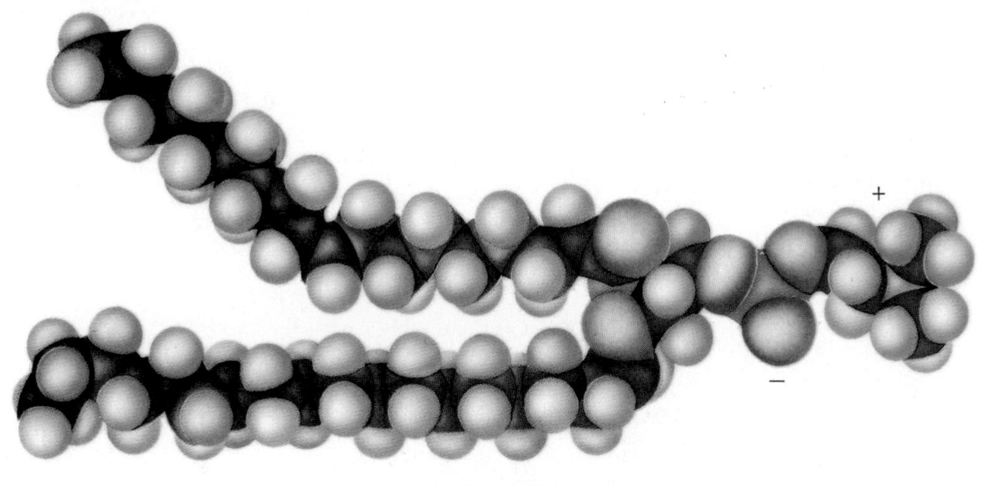

space-filling molecular model

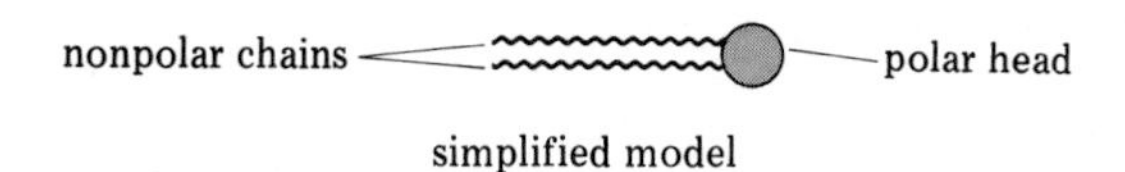

simplified model

stitute a "head-and-tail" model that is used in representing the structure of cell membranes (Section 13.8).

13.6 SPHINGOPHOSPHOLIPIDS

Sphingophospholipids are constructed upon a backbone molecule called sphingosine (Figure 13.4). Sphingosine contains an amino group that is called a **ceramide** when converted to an amide with a fatty acid. Sphingosine is both a primary and a secondary alcohol. When the primary alcohol is linked to phosphorylcholine, the product is called sphingomyelin. Sphingomyelins are present in most cell membranes and are the major component of the myelin sheath surrounding nerve cells.

Compare the structures of a sphingophospholipid (Figure 13.4) and a phosphoglyceride (Figure 13.2). Although the components are different, the overall structures are similar. Both compounds have a polar head and two nonpolar tails. However, there are significant chemical differences. Sphingophospholipids have a single amide group and are stable to hydrolysis; glycerophospholipids have two carboxylic esters that are easily hydrolyzed.

Sphingomyelins have fatty acid residues that are 20–26 carbon atoms long. These long chains have strong London forces and form a very stable coating for nerve fibers. In individuals with some genetic diseases, the carbon chains are shorter, which results in defects in the myelin sheath. Gauch-

FIGURE 13.4
Sphingosine and Sphingophospholipids
A ceramide contains an amide group formed from the amine of sphingosine and the carboxyl group of an acid. A sphingophospholipid is a phosphate ester of the ceramide. The phosphate group also forms an ester with the alcohol choline.

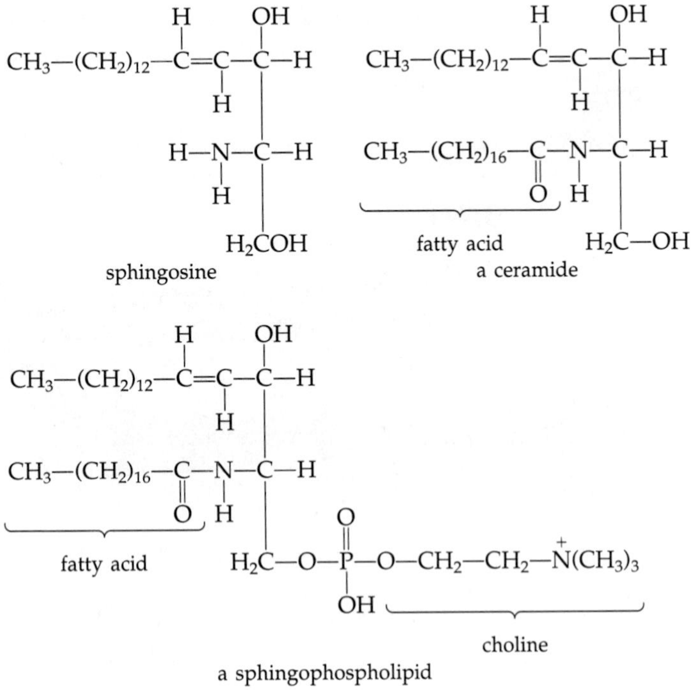

$CH_3—(CH_2)_{12}—CH=CH—CH(OH)—CH(NH_2)—CH_2OH$

sphingosine

$CH_3—(CH_2)_{12}—CH=CH—CH(OH)—CH(NH—C(=O)—(CH_2)_{16}—CH_3)—H_2C—OH$

fatty acid

a ceramide

$CH_3—(CH_2)_{12}—CH=CH—CH(OH)—CH(NH—C(=O)—(CH_2)_{16}—CH_3)—H_2C—O—P(=O)(OH)—O—CH_2—CH_2—\overset{+}{N}(CH_3)_3$

fatty acid

choline

a sphingophospholipid

er's disease, Niemann–Pick disease, multiple sclerosis, and leukodystrophy are all the result of unstable myelin membranes.

13.7 GLYCOSPHINGOLIPIDS

Glycosphingolipids are similar to sphingophospholipids; they contain sphingosine and a fatty acid residue bonded as an amide in a ceramide. However, glycosphingolipids contain no phosphate. Instead, a carbohydrate is bonded via a glycosidic linkage to the primary alcohol oxygen atom of the ceramide.

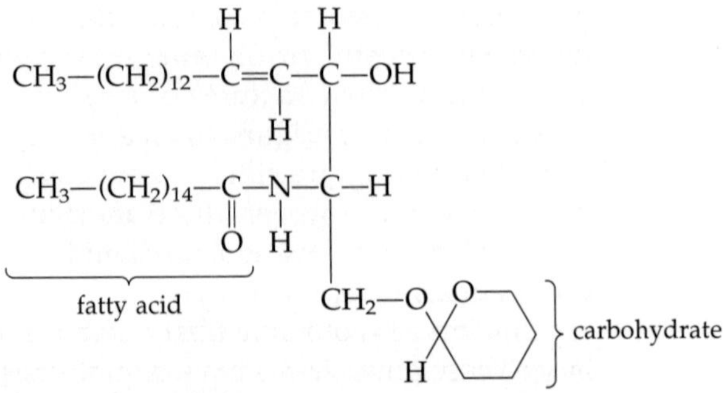

$CH_3—(CH_2)_{12}—CH=CH—CH(OH)—CH(NH—C(=O)—(CH_2)_{14}—CH_3)—CH_2—O—$ carbohydrate

fatty acid

carbohydrate

Cerebrosides and gangliosides are glycosphingolipids. Cerebrosides contain only glucose or galactose, whereas gangliosides contain an oligosaccharide. Both are found in the myelin sheath. Cerebrosides are in the white matter of the central nervous system; gangliosides occur in the gray matter of the brain.

Gangliosides are synthesized or degraded by the sequential addition or removal of monosaccharide units. The degradation of a ganglioside occurs inside subcellular organelles called lysosomes, which contain the necessary enzymes for the reaction. Gangliosides accumulate abnormally in the brain because of a deficiency of the enzymes necessary for their degradation in a genetic disease called Tay-Sachs disease. The consequences of this enzyme deficiency are retarded development, blindness, and death at an early age. Tay-Sachs disease occurs in American Jews at an incidence 100 times higher than that for other Americans.

EXAMPLE 13.4 Identify the components of the following compound and classify it.

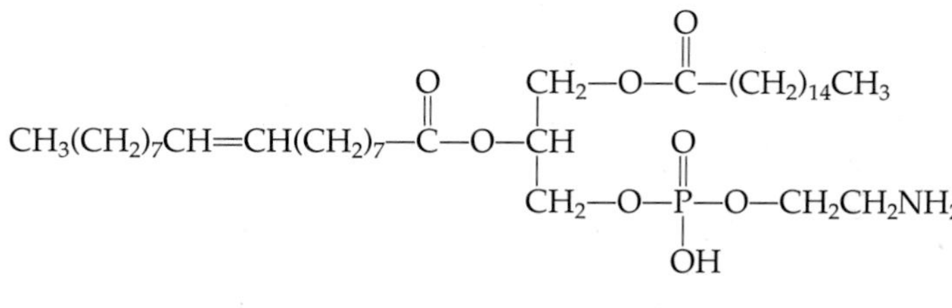

$$\begin{array}{l} \phantom{CH_3(CH_2)_7CH{=}CH(CH_2)_7{-}C{-}O{-}}CH_2{-}O{-}\overset{O}{\overset{\|}{C}}{-}(CH_2)_{14}CH_3 \\ CH_3(CH_2)_7CH{=}CH(CH_2)_7{-}\overset{O}{\overset{\|}{C}}{-}O{-}CH \\ \phantom{CH_3(CH_2)_7CH{=}CH(CH_2)_7{-}C{-}O{-}}CH_2{-}O{-}\underset{OH}{\underset{|}{\overset{O}{\overset{\|}{P}}}}{-}O{-}CH_2CH_2NH_2 \end{array}$$

Solution The carboxylic acid residue at the top right of the structure is palmitic acid, and the carboxylic acid residue at the left is oleic acid. A phosphate ester of ethanolamine appears at the bottom right. The center portion of the structure is derived from glycerol. This structure is a phosphatidylethanolamine or cephalin.

13.8 BIOLOGICAL MEMBRANES

Biological membranes separate cells from their environment, and separate organelles from each other within the cytoplasm of the cell. However, membranes are more than just sacks to contain the cells and organelles. They contain proteins that regulate the flow of molecules into and out of the cell.

Composition

Cell membranes are about 75 μm thick and consist of phospholipids—glycerophospholipids and sphingophospholipids—and proteins. The fraction of each component is related to the function of the membrane. The myelin covering of nerve fibers is about 80% lipid by weight. This covering is very nonpolar and serves a protective function. Most other cell membranes must allow certain ions and polar molecules to pass in and out of the cells, and their lipid content is about 50% by weight. The transport function of these membranes is mediated by proteins. Inner mitochondrial membranes are only about 20% lipid by weight. Mitochondria play an important role in energy conversions within cells, and many molecules must cross the mitochondrial membrane. The proteins that are the major part of the inner mitochondrial membrane control the transport of molecules into and out of the mitochondria.

Structure

Each lipid in a membrane has a polar or ionic "head" and two nonpolar "tails" (Figure 13.5). The phospholipids form bilayers in aqueous solutions so that each side of the membrane has polar heads toward the exterior and nonpolar tails toward the interior. The polar heads are hydrophilic and are exposed to the fluid surrounding and contained in cells. The hydrophobic tails are kept away from water molecules and intermingle within the bilayer. The properties of the membrane are determined by the type of fatty acid and the kind of polar group in the phospholipids. Cell membranes differ on the inner and outer surfaces. For example, phosphatidylcholine and sphingomyelin tend to be located on the outer surface of a membrane, whereas phosphatidylethanolamine and phosphatidylserine are usually on the inner surface.

Proteins in Membranes

Protein molecules are present in the bilayer in two different ways. Proteins extending from the edge of the bilayer into the interior of the membrane are **integral membrane proteins.** These proteins may be embedded in only one side of the membrane or they may extend through to the other side. Proteins

FIGURE 13.5 **Model of a Biological Membrane with Integral Proteins**
The lipids keep their hydrophilic tails in the interior while the hydrophilic heads are in contact with the surrounding aqueous solution. The proteins may be embedded in only one side of the membrane or extend across both sides of the membrane.

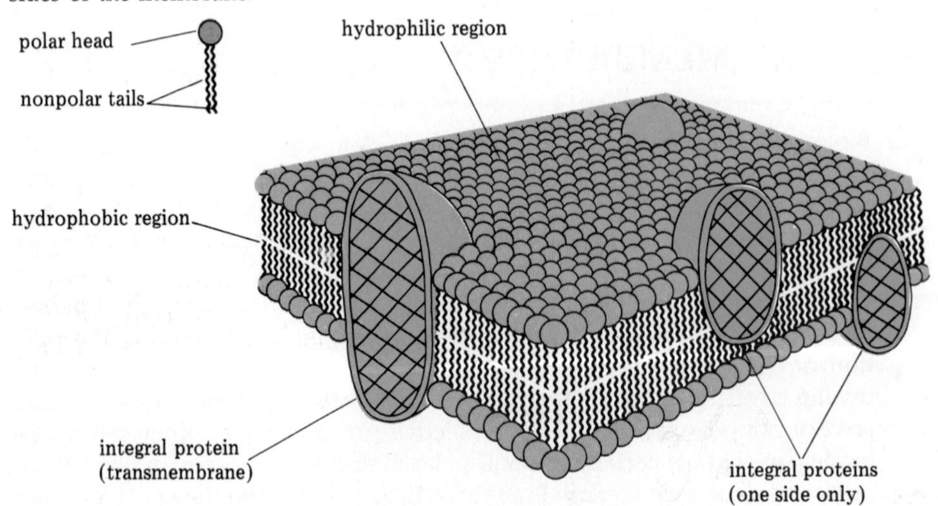

that extend across the membrane are called **transmembrane proteins.** Integral membrane proteins interact with the interior of the membrane by hydrophobic forces between certain nonpolar portions of a protein and the tails of the lipid.

Proteins that are associated with only one surface of the bilayer are called **peripheral proteins.** Peripheral proteins may be bound to the membrane by electrostatic forces or hydrogen bonds.

Carbohydrates in Membranes

Membranes contain carbohydrates combined with lipids in glycolipids and combined with proteins in glycoproteins. In mammals, these glycolipids and glycoproteins are always located on the outer surface of the cell membrane. The carbohydrate portion is hydrophilic and remains directed toward the water in the external environment. The protein is anchored in the membrane. The arrangement provides some intercellular recognition. It not only allows the grouping of cells to form tissue but also aids the immune system in the recognition of foreign cells.

Membrane Fluidity

Membranes are stabilized by London forces. These forces depend on both the length and the geometry of the hydrocarbon chain. The longer the chain, the stronger the London forces are and the more rigid the membrane is. The degree of saturation affects the flexibility of the membrane. Unsaturated fatty acids have bends in the chain, and they do not pack together efficiently in the bilayer. As a result, membranes with a high degree of unsaturation are more flexible.

The lipids and proteins in membranes can move laterally (on one side) in the membrane. A phospholipid can move about 10^{-4} cm/s, but the lateral mobility of proteins varies considerably. Some proteins are almost as mobile as lipids, whereas others are essentially immobile because they are anchored to a cellular structure called the cytoskeleton.

Lipid movement from one side of a membrane to the other is very slow. Such transverse diffusion or "flip-flop" of the polar head from one side to the other occurs at a rate only 10^{-9} as fast as lateral movement. Proteins, which are much more polar than lipids, do not undergo transverse diffusion.

Transport Across Membranes

Molecules and ions pass through a biological membrane to provide food for the cell and to release waste products from the cell. Smaller hydrophobic molecules such as O_2 and polar but uncharged molecules such as H_2O, urea, and ethanol can diffuse across the membrane relatively rapidly. Charged ions such as Na^+ and K^+ and large polar molecules diffuse very slowly. Polar molecules such as glucose diffuse so slowly that a cell could not obtain sufficient energy to maintain its processes if it relied upon the unaided diffusion of glucose across the cell membrane. There are two other ways to move molecules across membranes—facilitated diffusion and active transport. The processes differ in energy requirements.

Facilitated diffusion occurs without consuming cellular energy and in a direction from high to low concentration. Materials move across the membrane faster than in simple diffusion because a "carrier" facilitates the process. These carriers are transmembrane proteins with molecular weights in the range of 9000–40,000. There are a variety of carriers, each specific for certain molecules. The carrier protein meets a specific molecule or ion at one surface of the membrane and forms a complex. Formation of the complex causes a conformational change in the protein that allows the molecule that it is carrying to slip through a "channel" to the other side of the membrane. Once the molecule is released, the protein returns to its original conformation. Glucose enters cells in this manner.

The transport of anions in human erythrocytes occurs in the same way. In one anion transport system, a protein exchanges the bicarbonate ion inside the cell for the chloride ion outside the cell. Carbon dioxide produced by cellular metabolism dissolves in aqueous media as the bicarbonate ion. Its concentration within the cell is higher than outside the cell, and it is transported spontaneously to the region of lower concentration. To maintain charge balance, the chloride ion flows into the cell.

Active transport is similar to facilitated diffusion; a specific interaction takes place between a transmembrane protein and the molecule to be transported. However, active transport occurs against the "natural" flow expected from concentration differences, and material moves from a region of low concentration to one of high concentration. The energy needed for active transport is provided by hydrolysis of ATP.

One of the most important active transport processes in animal cells is the sodium ion–potassium ion transport system known as the "sodium pump". Sodium ions are pumped out of the cell to maintain a concentration of 0.1 M within the cell, whereas the extracellular sodium ion concentration is 0.14 M. At the same time, potassium ions are pumped into the cell from an extracellular concentration of 0.005 M to provide an intracellular concentration of about 0.15 M. The energy required for this process is provided by hydrolysis of ATP. The entire process is carefully controlled because any imbalance in the total concentration of ions would cause a change in the osmotic pressure. The cell would swell if its osmotic pressure increased or shrink if its osmotic pressure decreased.

EXERCISES

Waxes

13.1 What are the structures and molecular formulas of the products of hydrolysis of carnauba wax?

13.2 Describe how whale oil could be converted into a soap.

13.3 State whether each of the following structures could be a naturally occurring wax.

(a) $CH_3(CH_2)_{30}CO_2CH_2CH_2CH_3$ (b) $CH_3(CH_2)_{28}CO_2CH_2(CH_2)_{19}CH_3$
(c) $CH_3(CH_2)_{27}CO_2CH_2(CH_2)_{18}CH_3$

13.4 The wax of a particular copepod is unsaturated. This species lives in cold water and uses the wax as a source of metabolic energy. Explain the benefit of the unsaturation in the acid portion of this ester.

Fatty Acids

13.5 What do stearic acid and oleic acid have in common? How do they differ?

13.6 Write the structures of four naturally occurring carboxylic acids that have 18 carbon atoms.

13.7 Cod liver oil is a triglyceride containing palmitoleic acid. Suggest a structure for the acid.

13.8 Predict the melting point of $CH_3(CH_2)_{20}CO_2H$.

13.9 Stearolic acid is named 9-octadecynoic acid by the IUPAC method. The molecular formula is $C_{18}H_{32}O_2$. Write its structure.

13.10 A compound called hypogeic acid is prepared in the laboratory and is named 7-hexadecenoic acid. Its melting point is 33 °C. What is the geometry at the double bond?

13.11 Why does linoleic acid have a lower melting point than oleic acid?

13.12 The melting point of elaidic acid (*trans*-9-octadecenoic acid) is 45 °C. Compare this value to the melting points of stearic acid and oleic acid and explain the differences.

Triglycerides

13.13 Write a balanced equation for the hydrolysis of a fat molecule using a base.

13.14 A sample of one oil is hydrolyzed to produce 50% oleic acid and 35% linoleic acid. A second oil produces 25% oleic and 50% linoleic acid. Which oil is more unsaturated?

13.15 Identify the following compound as a fat or an oil.

$$\begin{array}{l} CH_2-O-\overset{\displaystyle O}{\overset{\|}{C}}-(CH_2)_7CH=CHCH_2CH=CH(CH_2)_4CH_3 \\ | \\ CH-O-\overset{\displaystyle O}{\overset{\|}{C}}-(CH_2)_7CH=CH(CH_2)_7CH_3 \\ | \\ CH_2-O-\overset{\displaystyle O}{\overset{\|}{C}}-(CH_2)_7CH=CHCH_2CH=CHCH_2CH=CHCH_2CH_3 \end{array}$$

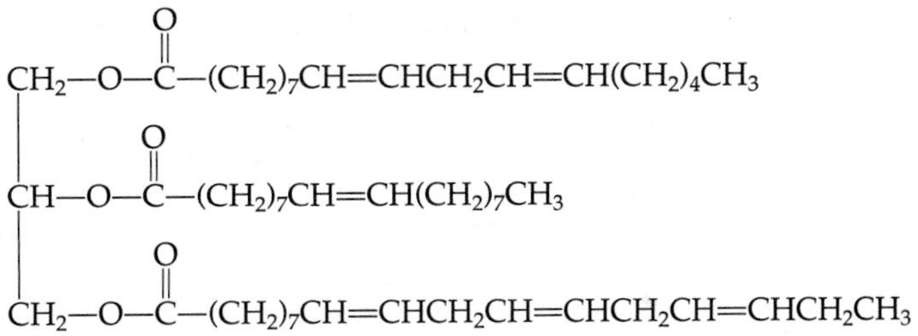

13.16 Identify the following compound as a fat or an oil.

$$\begin{array}{l} CH_2-O-\overset{\displaystyle O}{\overset{\|}{C}}-(CH_2)_{14}CH_3 \\ | \\ CH-O-\overset{\displaystyle O}{\overset{\|}{C}}-(CH_2)_{10}CH_3 \\ | \\ CH_2-O-\overset{\displaystyle O}{\overset{\|}{C}}-(CH_2)_7CH=CH(CH_2)_7CH_3 \end{array}$$

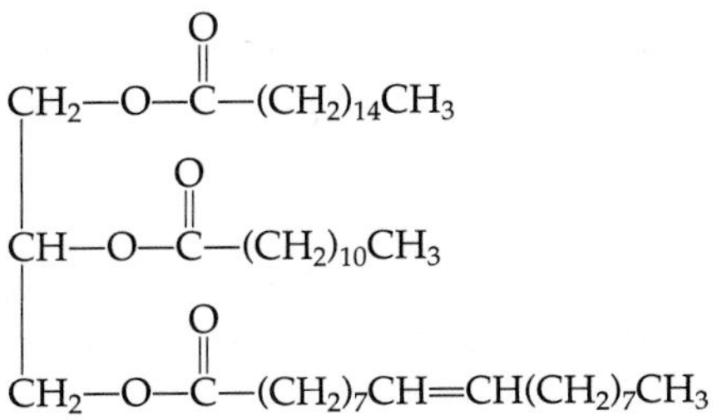

13.17 Identify the fatty acids in the compound of Exercise 13.15.

13.18 Identify the fatty acids in the compound of Exercise 13.16.

13.19 Draw the structure of a triacylglycerol containing palmitic acid as an ester at the secondary carbon atom and stearic acid as esters at the two primary carbon atoms of glycerol. Can this compound exist in an optically active form?

13.20 State whether each of the following is likely to be found as an ester in a naturally occurring compound.
(a) $CH_3(CH_2)_{15}CO_2H$ (b) $CH_3(CH_2)_{20}CO_2H$ (c) $(CH_3)_2CH(CH_2)_{14}CO_2H$

13.21 Can the following compound be optically active?

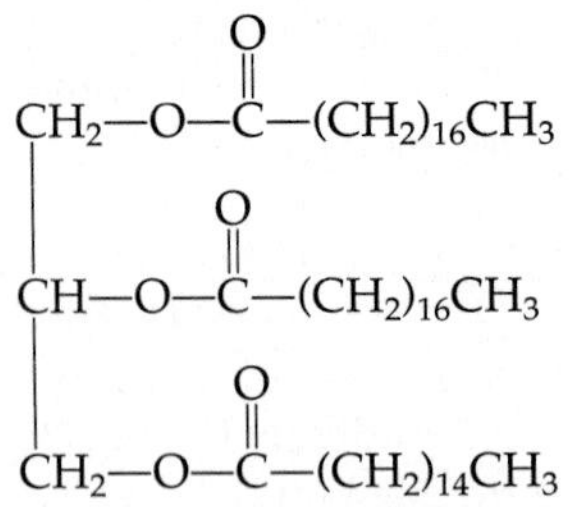

$$\begin{array}{l} CH_2{-}O{-}\overset{\overset{O}{\|}}{C}{-}(CH_2)_{16}CH_3 \\ | \\ CH{-}O{-}\overset{\overset{O}{\|}}{C}{-}(CH_2)_{16}CH_3 \\ | \\ CH_2{-}O{-}\overset{\overset{O}{\|}}{C}{-}(CH_2)_{14}CH_3 \end{array}$$

13.22 Hydrolysis of an optically active triacylglycerol gives 1 mole each of glycerol and oleic acid and 2 moles of stearic acid. Write a structure for the triglyceride.

Glycerophospholipids

13.23 What products result from hydrolysis of a glycerophospholipid?

13.24 What alcohols are found in glycerophospholipids?

13.25 Identify the components of the following glycerophospholipid.

$$\begin{array}{l} CH_3(CH_2)_{16}{-}\overset{\overset{O}{\|}}{C}{-}O{-}CH_2 \\ CH_3(CH_2)_7\underset{H}{C}{=}\underset{H}{C}(CH_2)_7{-}\underset{\underset{O}{\|}}{C}{-}O{-}CH \\ \qquad\qquad\qquad\qquad\qquad\quad CH_2{-}O{-}\overset{\overset{O}{\|}}{\underset{\underset{^-O}{|}}{P}}{-}OCH_2CH_2\overset{+}{N}(CH_3)_3 \end{array}$$

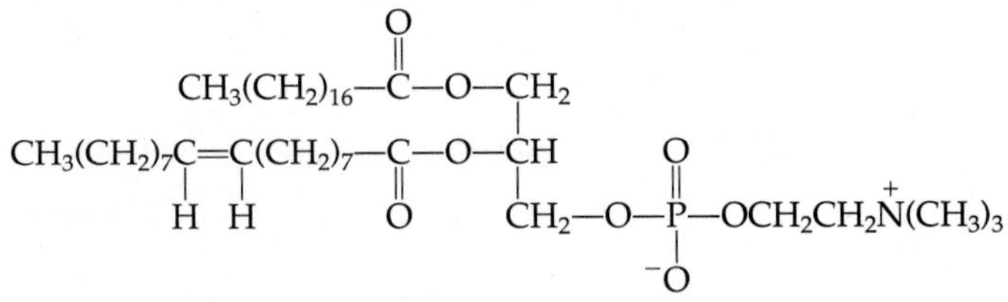

13.26 What are the hydrolysis products of the following glycerophospholipid?

$$\begin{array}{l} CH_3(CH_2)_{14}{-}\overset{\overset{O}{\|}}{C}{-}O{-}CH_2 \\ CH_3(CH_2)_7\underset{H}{C}{=}\underset{H}{C}(CH_2)_7{-}\underset{\underset{O}{\|}}{C}{-}O{-}CH \\ \qquad\qquad\qquad\qquad\qquad\quad CH_2{-}O{-}\overset{\overset{O}{\|}}{\underset{\underset{^-O}{|}}{P}}{-}OCH_2\overset{\overset{\overset{+}{N}H_3}{|}}{\underset{\underset{H}{|}}{C}}{-}CO_2^- \end{array}$$

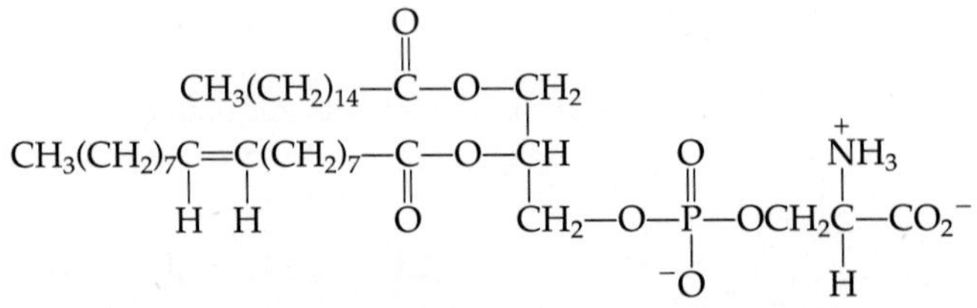

13.27 What charges exist on the polar head of each of the types of glycerophospholipids at physiological pH?

13.28 How many ionizable hydrogen atoms are there in a phosphatidic acid?

Sphingophospholipids

13.29 How do sphingophospholipids differ from glycerophospholipids?

13.30 Why are sphingophospholipids not hydrolyzed as readily as glycerophospholipids?

13.31 Sphingophospholipids are said to have two nonpolar tails. One is a fatty acid residue. What is the structure of the second chain?

13.32 What are sphingomyelins? What structural feature allows them to serve a uniquely important biological function?

Glycosphingolipids

13.33 How are glycosphingolipids similar to sphingophospholipids? In what ways do the two types of compounds differ?

13.34 What is the difference between a cerebroside and a ganglioside? Where are these compounds found?

13.35 What type of bond joins the sugar unit to the sphingosine part of a glycosphingolipid?

13.36 Based on the structure of glycosphingolipids, predict whether these molecules are more stable in acidic or basic solution.

Biological Membranes

13.37 How does the structure of the fatty acid affect the rigidity of a cell membrane?

13.38 How are proteins incorporated in a cell membrane?

13.39 What kind of forces hold a cell membrane together?

13.40 What relationship exists between the protein content and the permeability of a cell membrane?

13.41 Peripheral proteins can be removed from a membrane by washing it with a detergent solution. Explain why.

13.42 Where is the sugar portion of a glycolipid located in a cell membrane?

Diffusion and Transport

13.43 Describe two ways in which materials cross cell membranes.

13.44 Why is an active transport system necessary to maintain the potassium ion content of a cell?

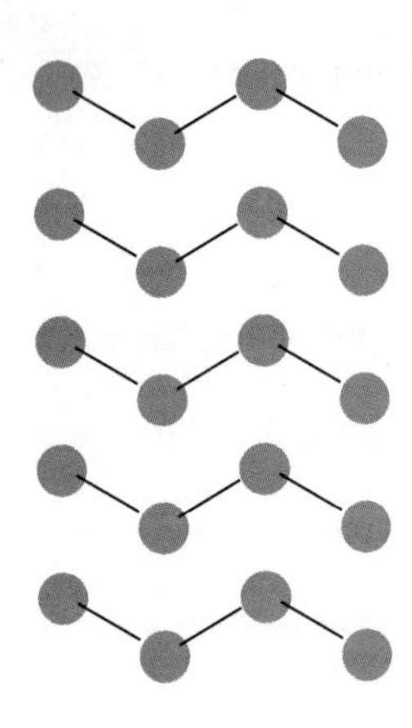

14

AMINES AND AMIDES

14.1 ORGANIC NITROGEN COMPOUNDS

Organic compounds that contain nitrogen are everywhere in the world around us. Some nitrogen-containing compounds are important industrial products. Among these are polymers such as nylon, many dyes, explosives, and pharmaceutical agents. Other nitrogen-containing compounds are essential for life. Nitrogen is present in many vitamins and hormones. Nitrogen is essential in amino acids and proteins (Chapter 15) and in nucleotides and polynucleotides (Chapter 16).

A nitrogen atom has five valence electrons and forms a total of three covalent bonds to carbon or hydrogen. A nitrogen atom in a functional group can have single, double, or triple bonds. In this chapter we will consider amines and amides, and, to a lesser extent, imines and nitriles.

$R{-}\ddot{N}H_2$	$R{-}C({=}\ddot{O}{:}){-}\ddot{N}H_2$	$R{-}C(R){=}N(R){:}$	$R{-}C{\equiv}N{:}$
an amine	an amide	an imine	a nitrile

In the simplest amine, methylamine (CH_3NH_2), one hydrogen atom of ammonia has been replaced by a methyl group. Methylamine has the shape shown in Figure 14.1. In methylamine and other amines, the nitrogen atom

FIGURE 14.1 Structure of Methylamine

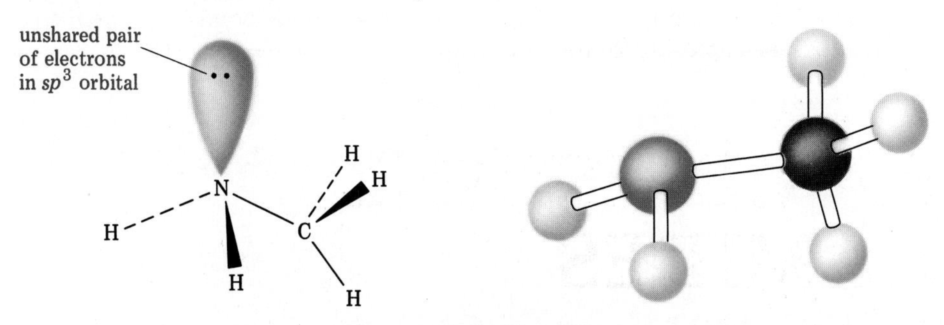

perspective structural formula **ball and stick model**

has five valence electrons in four sp^3 hybrid orbitals. These orbitals are directed to the corners of a tetrahedron. Three of these orbitals are half-filled; the fourth contains a pair of nonbonded electrons that plays an important role in the chemical properties of amines.

Many amines are physiologically active. They affect the brain, spinal cord, and nervous system. These compounds include the neurotransmitters epinephrine, serotonin, and dopamine. Epinephrine, commonly called adrenaline, stimulates the conversion of stored glycogen into glucose. Serotonin is a hormone that causes sleep, and serotonin deficiency is responsible for some forms of mental depression. When the dopamine concentration is low, the result is Parkinson's disease.

HO, HO, OH, H, $CHCH_2NCH_3$ — epinephrine

HO, $CH_2CH_2NH_2$, N, H — serotonin

HO, HO, $CH_2CH_2NH_2$ — dopamine

Heterocyclic compounds that contain multiple nitrogen atoms are required for the transmission of genetic information. DNA (deoxyribonucleic acid) and RNA (ribonucleic acid) contain substituted pyrimidine and purine rings. This chemistry will be presented in Chapter 16.

N, N — pyrimidine

N, N, N, N, H — purine

Proteins, one of the most important and versatile classes of biological compounds, are made from nitrogen-containing molecules called α-amino acids. The amine functional group of one α-amino acid reacts with the carboxyl group of another α-amino acid to form an amide bond. This chemistry will be presented in Chapter 15.

amide bonds

$$-NH\underset{\substack{|\\R}}{C}H\overset{\substack{O\\||}}{C}-NH\underset{\substack{|\\R'}}{C}H\overset{\substack{O\\||}}{C}-NH\underset{\substack{|\\R''}}{C}H\overset{\substack{O\\||}}{C}-$$

14.2 STRUCTURE AND CLASSIFICATION OF AMINES AND AMIDES

Just as we can regard alcohols and ethers as organic derivatives of water, we can regard amines as organic derivatives of ammonia. However, amines are not classified like alcohols. The classification of alcohols is based on the number of groups attached to the carbon atom bearing the hydroxyl group. Amines are classified by the number of alkyl (or aryl) groups attached to the nitrogen atom.

$$H-\underset{\substack{|\\H}}{\overset{\substack{H\\|}}{N}}: \qquad R-\underset{\substack{|\\H}}{\overset{\substack{H\\|}}{N}}: \qquad R-\underset{\substack{|\\H}}{\overset{\substack{R\\|}}{N}}: \qquad R-\underset{\substack{|\\R}}{\overset{\substack{R\\|}}{N}}:$$

ammonia — primary (1°) amine — secondary (2°) amine — tertiary (3°) amine

For example, *tert*-butylamine has a *tert*-butyl group attached to an $-NH_2$ group. However, the amine is primary because only one alkyl group is bonded to the nitrogen atom. In contrast *tert*-butyl alcohol is a tertiary alcohol because the carbon atom bonded to the —OH group is bonded to three alkyl groups. Trimethyamine is a tertiary amine because the nitrogen atom is bonded to three alkyl groups.

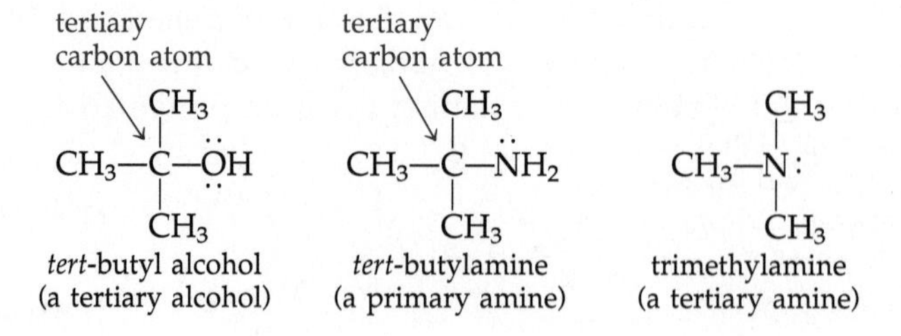

tert-butyl alcohol (a tertiary alcohol) — *tert*-butylamine (a primary amine) — trimethylamine (a tertiary amine)

Amines in which a nitrogen atom is part of a ring are common in nature. Compounds that have one or more atoms other than carbon in the ring are heterocyclic compounds (Section 3.1). For example, pyrrolidine and piperidine are five- and six-membered nitrogen-containing heterocyclic compounds.

pyrrolidine (a secondary amine)

piperidine (a secondary amine)

Amides have an amino group or a substituted amino group bonded to a carbonyl carbon atom. The other two bonds of the nitrogen atom may be to hydrogen atoms, alkyl groups, or aryl groups. Amides are classified based on the number of carbon groups bonded to the nitrogen atom.

primary amide — secondary amide — tertiary amide

The structures of amides resemble those of other carbonyl compounds: the three atoms bonded to carbon are in the same plane (Figure 14.2). The nitrogen atom of an amide has an unshared pair of electrons that are delocalized with the π electrons of the carbonyl group. Thus, an amide such as formamide has two resonance structures.

The bond between carbon and nitrogen is therefore intermediate between a single bond and a double bond. The partial double bond character of the

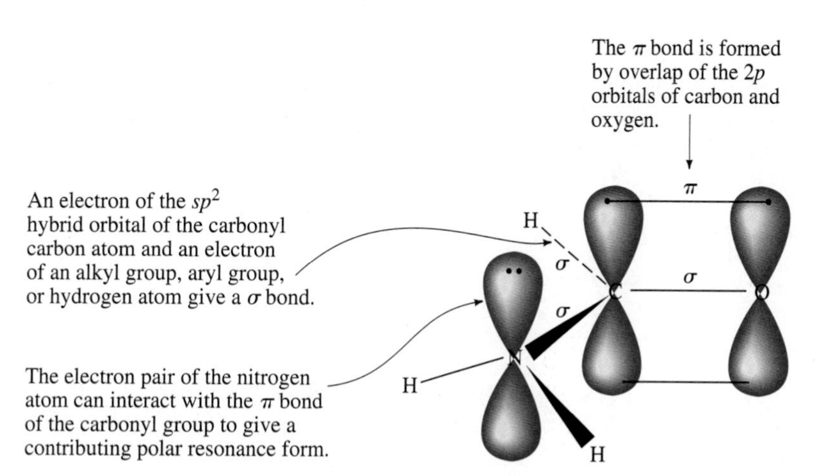

The π bond is formed by overlap of the $2p$ orbitals of carbon and oxygen.
An electron of the sp^2 hybrid orbital of the carbonyl carbon atom and an electron of an alkyl group, aryl group, or hydrogen atom give a σ bond.
The electron pair of the nitrogen atom can interact with the π bond of the carbonyl group to give a contributing polar resonance form.
π
σ
σ
σ
H
H
H

FIGURE 14.2
Bonding in Formamide
The geometry of the atoms bonded to the carbonyl carbon atom of an amide is planar. The interaction of the unshared pair of electrons of the nitrogen atom and the π-bonding electrons of the carbonyl group causes restriction of rotation about the carbon–nitrogen bond.

C—N bond of amides makes rotation about the carbon–nitrogen bond somewhat restricted. In Chapter 15 we will study the chemistry of proteins, which contain amide bonds. These compounds have some properties that reflect the restricted rotation about the carbon–nitrogen bond.

EXAMPLE 14.1 Classify Demerol, a synthetic narcotic analgesic, as a primary, secondary, or tertiary amine.

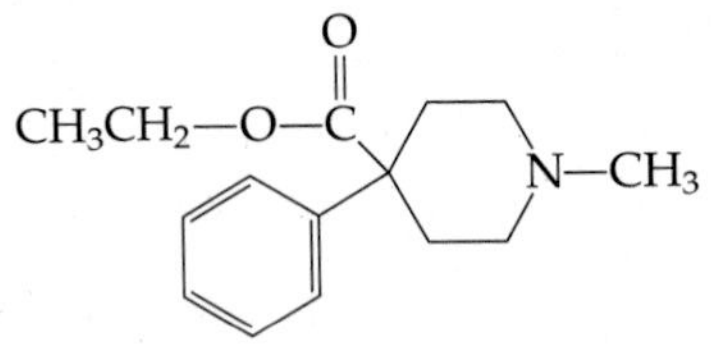

CH_3CH_2—O—C(=O)— ... N—CH_3

Demerol

Designer Drugs

Addictive drugs cannot be sold to an individual if they are on the U.S. Drug Enforcement Administration (DEA) list of controlled substances. However, they may be administered by doctors for medical treatment or used in surgical procedures in the United States. Fentanyl, whose structure is shown below, is an addictive drug that is about 150 times more powerful than morphine. Note that the two nitrogen atoms are contained in amine and amide functional groups.

Other compounds structurally related to fentanyl are also potent narcotics. Thus, some criminals with a knowledge of chemistry have produced compounds similar to fentanyl and sold them to drug addicts. These compounds have come to be known as "designer drugs" because they have been designed to be similar to legal drugs but made to be sold on the street. Many of these substances are far more potent than the drugs they replace. For example, heroin addicts who inadvertently use fentanyl or its derivatives often die from overdoses. The 3-methylfentanyl compound is about 3000 times more potent than morphine! Nevertheless, the continued synthesis of compounds similar to fentanyl is legal. Unfortunately the sale of compounds to individuals cannot legally be banned until it is established that they are being used as drugs. It takes time for a compound to be detected and for legislation to specifically ban it. Many compounds are designed one after the other to stay one step ahead of the law.

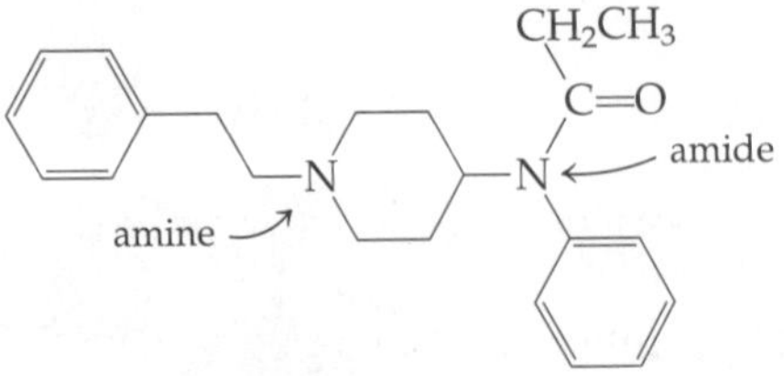

CH_2CH_3, C=O, N, N, amide, amine

fentanyl

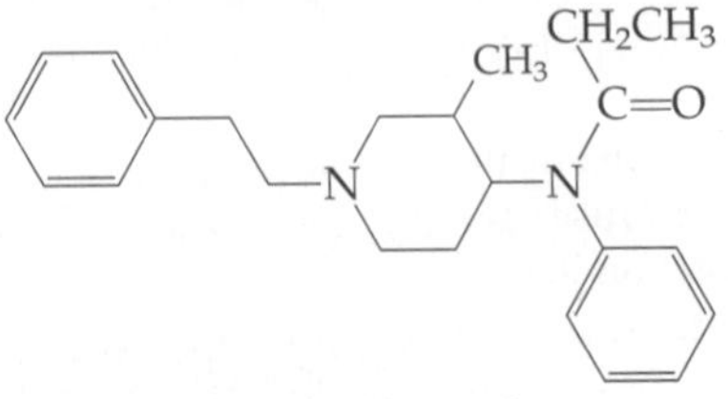

CH_3, CH_2CH_3, C=O, N, N

3-methylfentanyl

Solution There are three carbon atoms bonded to the nitrogen atom. Two of the carbon atoms are in the heterocyclic ring. The third carbon atom bonded to the nitrogen atom is a methyl group. Demerol is a tertiary amine.

EXAMPLE 14.2 Identify the functional groups in penicillin V that contain nitrogen.

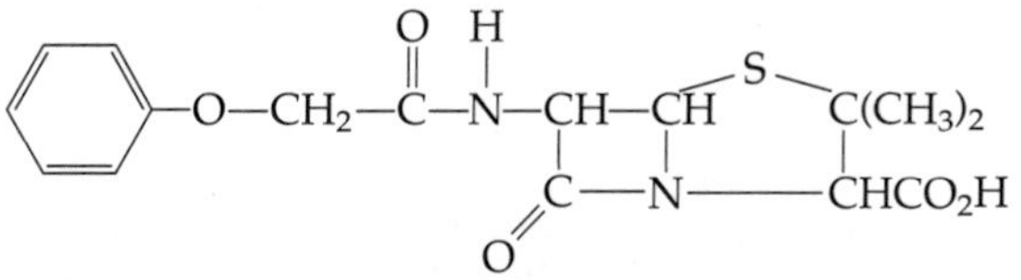

$$C_6H_5{-}O{-}CH_2{-}\overset{O}{\overset{\|}{C}}{-}\overset{H}{\overset{|}{N}}{-}CH{-}CH{-}S{-}C(CH_3)_2 \quad (\text{ring: } C(=O){-}N{-}CHCO_2H)$$

penicillin V

Solution There are two nitrogen atoms in this compound, both of which are bonded to carbonyl carbon atoms. Both nitrogen atoms are part of amide functional groups. The nitrogen atom bonded to the four-membered ring also has one bonded hydrogen atom and one alkyl group; this amide is secondary. The nitrogen atom that is part of the four-membered ring—a lactam (Section 12.1)—has two other carbon atoms bonded to nitrogen; this is a tertiary amide.

14.3 NOMENCLATURE OF AMINES AND AMIDES

Although amines and amides have quite different physical and chemical properties, the IUPAC method of naming them is similar. Amides are produced from amines. Thus, the amine part of amides is named by the same method used for amines. Like all organic compounds, many amines are known by their common names as well as by their IUPAC names.

Amines

The common name of a primary amine is obtained by naming the alkyl group bonded to the amino group ($-NH_2$) and adding the suffix *-amine.* The entire name is written as one word. The common name for a secondary or tertiary amine is obtained by listing the alkyl groups alphabetically. When two or more identical alkyl groups are present, the prefixes *di-* and *tri-* are used.

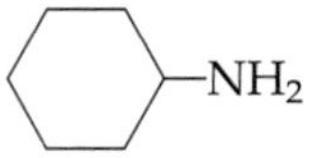

$C_6H_{11}{-}NH_2$	$CH_3{-}CH_2{-}NH{-}CH_3$	$CH_3{-}CH_2{-}NH{-}CH_2{-}CH_3$
cyclohexylamine	ethylmethylamine	diethylamine

To give a common name to a more complex primary amine, the amino group is treated as a substituent. The nitrogen-containing substituent in complex secondary and tertiary amines is named as an *N*-alkylamino (—NHR) or *N,N*-dialkylamino (—NRR′) group. The capital *N* indicates that the alkyl group is bonded to the nitrogen atom and not to the parent chain. The largest or most complicated group is used as the parent molecule.

$NH_2—CH_2—CH_2—CH_2—CO_2H$
γ-aminobutyric acid

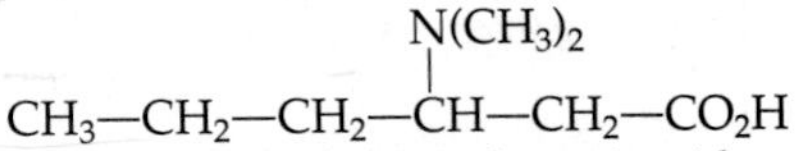

$CH_3—CH_2—CH_2—CH(N(CH_3)_2)—CH_2—CO_2H$
β-(*N,N*-dimethylamino)caproic acid

Amines are given IUPAC names by rules similar to those used to name alcohols. The longest continuous chain to which the amino group is attached is the parent alkane. The *-e* ending of the alkane is changed to *-amine.* Substituents on the carbon chain are designated by number. The prefix *N*- is used for each substituent on the nitrogen atom.

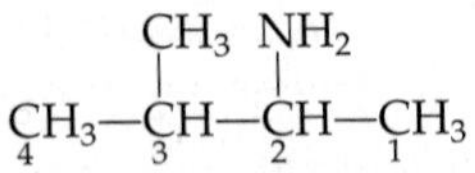

$\overset{4}{C}H_3—\overset{3}{C}H(CH_3)—\overset{2}{C}H(NH_2)—\overset{1}{C}H_3$
3-methyl-2-butanamine

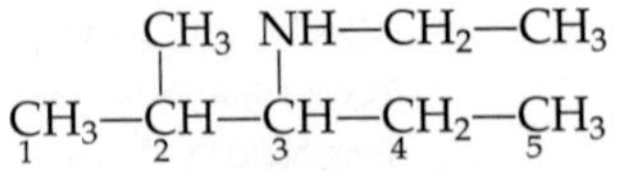

$\overset{1}{C}H_3—\overset{2}{C}H(CH_3)—\overset{3}{C}H(NH—CH_2—CH_3)—\overset{4}{C}H_2—\overset{5}{C}H_3$
N-ethyl-2-methyl-3-pentanamine

Amines in which the nitrogen atom is part of an aromatic ring are called **heterocyclic aromatic amines.** In these compounds, the positions of substituents are established by using the numbering system indicated below. Note that a nitrogen atom is assigned the number 1 and that the direction of numbering is selected to provide the lower possible number if more than one nitrogen atom is present in the ring.

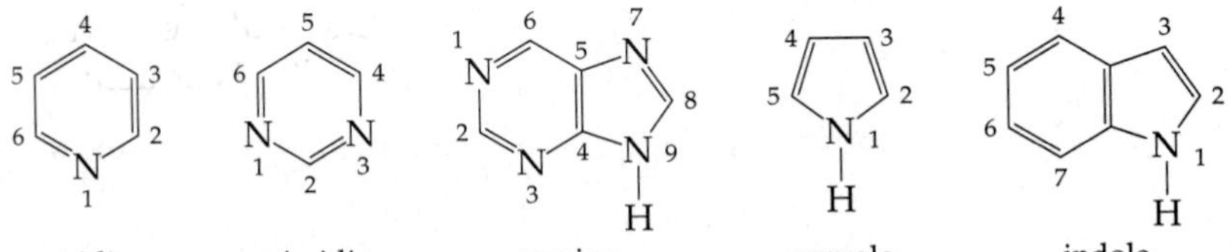

pyridine   pyrimidine   purine   pyrrole   indole

EXAMPLE 14.3 Write the IUPAC name for baclofen, a muscle relaxant.

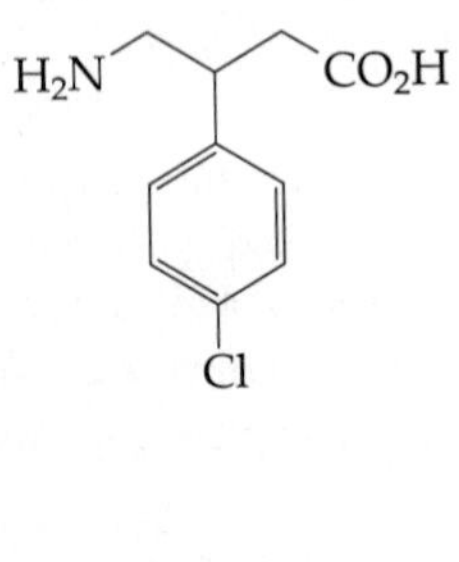

H_2N CO_2H Cl

Solution The parent chain of Baclofen is butanoic acid. It contains an amino group and an aryl group. The amino group is located at the C-4 atom. The aryl group located at the C-3 atom is named *p*-chlorophenyl. Placing the groups in alphabetical order, the name is 4-amino-3-(*p*-chlorophenyl)butanoic acid. The name of the aryl group is written within parentheses to clearly identify it.

Amides

The common names of amides are formed by dropping the suffix *-ic* of the related acid and adding the suffix *-amide.* When there is a substituent on the nitrogen atom, the prefix *N*- followed by the name of the group bonded to nitrogen is attached to the name. Substituents on the acyl group are designated by Greek letters α, β, γ, and so on, as in the common names of carboxylic acids.

In the IUPAC system, the longest chain that contains the amide functional group is the parent. The final *-e* of the alkane is replaced by *-amide.* The substituents on nitrogen are indicated by the same method as in the common system. However, numbers are used for substituents on the parent chain.

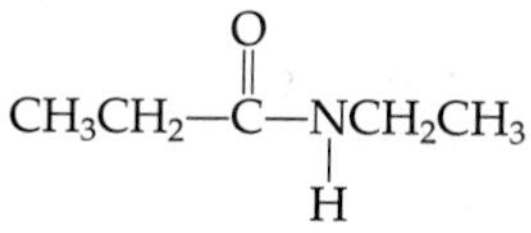

$CH_3CH_2-C(=O)-N(H)CH_2CH_3$

N-ethylpropanamide
(*N*-ethylpropionamide)

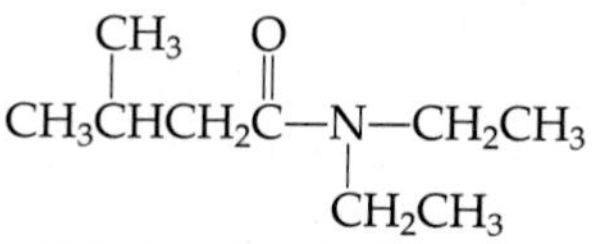

$CH_3CH(CH_3)CH_2C(=O)-N(CH_2CH_3)-CH_2CH_3$

N,N-diethyl-3-methylbutanamide
(*N,N*-diethyl-β-methylbutyramide)

EXAMPLE 14.4 The IUPAC name for lidocaine, a local anesthetic, is 2-(diethylamino)-*N*-(2,6-dimethylphenyl)ethanamide. Write its structure.

Solution The parent is the two-carbon amide, ethanamide. The group contained within parentheses and preceded by an *N* is 2,6-dimethylphenyl. Place this group on the nitrogen atom of ethanamide.

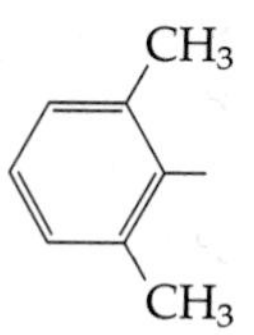

2,6-dimethylphenyl

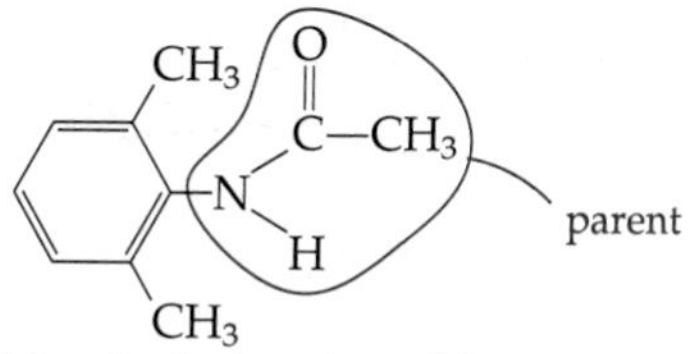

(*N*-aryl-substituted amide) — parent

The compound is also an amine. The diethylamino group is bonded to the C-2 atom.

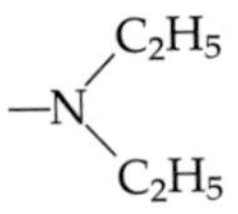

$-N(C_2H_5)_2$

diethylamino

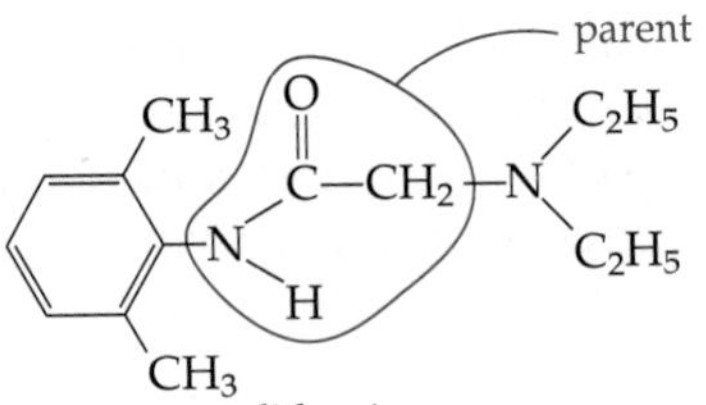

$$2,6-(CH_3)_2C_6H_3-NH-\overset{O}{\overset{\|}{C}}-CH_2-N(C_2H_5)_2$$

parent

lidocaine

14.4 PHYSICAL PROPERTIES OF AMINES AND AMIDES

Amines with low molecular weights are gases at room temperature, but amines with higher molecular weights are liquids or solids (Table 14.1). Amines have boiling points that are higher than those of alkanes of similar molecular weight, but lower than those of alcohols.

	$CH_3-CH_2-CH_3$	$CH_3-CH_2-NH_2$	CH_3-CH_2-OH
bp	−42 °C	17 °C	78 °C

With the exception of formamide, primary amides are solids at room temperature. Substituted amides have lower melting points. All amides have high boiling points.

	$CH_3-\overset{O}{\overset{\|}{C}}-NH_2$	$CH_3-\overset{O}{\overset{\|}{C}}-\underset{}{N}(CH_3)-H$	$CH_3-\overset{O}{\overset{\|}{C}}-N(CH_3)-CH_3$
bp	221 °C	204 °C	165 °C
mp	82 °C	28 °C	−20 °C

TABLE 14.1 Boiling Points of Some Representative Amines

Name	Structure	Boiling point (°C)
ammonia	NH_3	−33
methylamine	CH_3NH_2	−6
ethylamine	$C_2H_5NH_2$	17
propylamine	$C_3H_7NH_2$	49
butylamine	$C_4H_9NH_2$	77
tert-butylamine	$t\text{-}C_4H_9NH_2$	44
dimethylamine	$(CH_3)_2NH$	7
trimethylamine	$(CH_3)_3N$	3
aniline	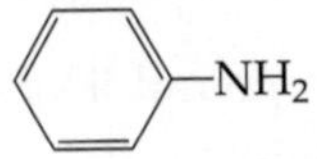 $C_6H_5-NH_2$	184

Hydrogen Bonding in Nitrogen Compounds

Amines have higher boiling points than hydrocarbons of comparable molecular weight because the C—N bond is more polar than a C—C bond. Also, primary and secondary amines can form intermolecular hydrogen bonds because they can serve as both hydrogen-bond donors and acceptors.

$$\mathrm{R{-}\overset{\displaystyle H}{\underset{\displaystyle R}{N}}{:}\cdots H{-}\underset{\displaystyle R}{\ddot{N}}{-}R}$$

a hydrogen bond

Tertiary amines have no hydrogen atoms bonded to the nitrogen atom and cannot serve as hydrogen-bond donors. Thus, these amines cannot form intermolecular hydrogen bonds. As a consequence, they have lower boiling points than primary and secondary amines of comparable molecular weight.

Amines have lower boiling points than alcohols because nitrogen is less electronegative than oxygen. As a result, the N—H bond is less polar than the O—H bond, and the N—H· · ·N hydrogen bond in amines is weaker than the O—H· · ·O hydrogen bond in alcohols.

$$\mathrm{R{-}\underset{\displaystyle R}{\ddot{N}}{-}H\cdots{:}\overset{\displaystyle H}{\underset{\displaystyle R}{N}}{-}R} \qquad \mathrm{R{-}\ddot{\underset{..}{O}}{-}H\cdots{:}\underset{\displaystyle H}{\ddot{O}}{-}R}$$

less polar bond

weaker hydrogen bond

more polar bond

stronger hydrogen bond

Amides form strong intermolecular hydrogen bonds between the amide hydrogen atom of one molecule and the carbonyl oxygen atom of a second molecule (C=O· · ·H—N). This intermolecular interaction is responsible for the high melting and boiling points of primary amides. Substitution of the hydrogen atoms on the nitrogen atom by alkyl or aryl groups reduces the number of possible intermolecular hydrogen bonds and lowers the melting and boiling points. Tertiary amides cannot form intermolecular hydrogen bonds.

$$\mathrm{CH_3{-}\overset{\displaystyle :\ddot{O}}{\overset{\|}{C}}{-}\underset{\displaystyle H}{\ddot{N}}{-}H} \qquad \mathrm{CH_3{-}\overset{\displaystyle :\ddot{O}}{\overset{\|}{C}}{-}\underset{\displaystyle H}{\ddot{N}}{-}CH_3} \qquad \mathrm{CH_3{-}\overset{\displaystyle :\ddot{O}}{\overset{\|}{C}}{-}\underset{\displaystyle CH_3}{\ddot{N}}{-}CH_3}$$

hydrogen bond　　hydrogen bond　　no hydrogen bond possible

$$\mathrm{CH_3{-}\overset{\displaystyle :\ddot{O}}{\overset{\|}{C}}{-}\underset{\displaystyle H}{\ddot{N}}{-}H} \qquad \mathrm{CH_3{-}\overset{\displaystyle :\ddot{O}}{\overset{\|}{C}}{-}\underset{\displaystyle H}{\ddot{N}}{-}CH_3} \qquad \mathrm{CH_3{-}\overset{\displaystyle :\ddot{O}}{\overset{\|}{C}}{-}\underset{\displaystyle CH_3}{\ddot{N}}{-}CH_3}$$

Solubility in Water

Primary and secondary amines function as both donors and acceptors of hydrogen bonds, and they readily form hydrogen bonds with water. Amines with five or fewer carbon atoms are miscible with water. Even tertiary amines are soluble in water because the nonbonded electron pair of the nitrogen atom is a hydrogen-bond acceptor of a hydrogen atom of water.

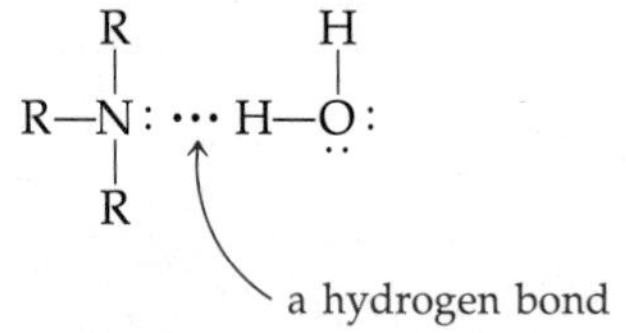

$$R_3N: \cdots H{-}\ddot{O}{-}H$$

a hydrogen bond

The solubilities of toluene and aniline illustrate the effect of hydrogen bonding on the solubilities of amines. Aniline forms hydrogen bonds with water; toluene does not.

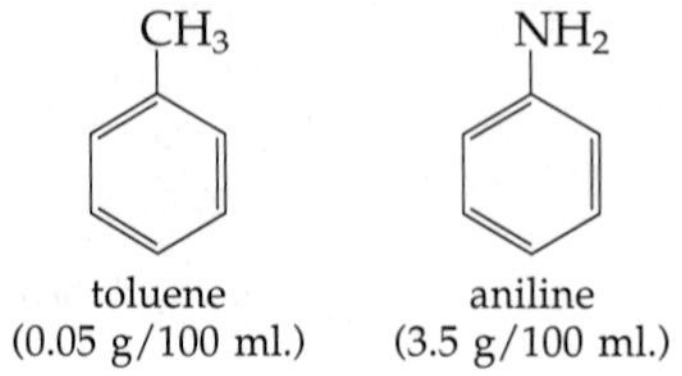

CH_3 NH_2

toluene (0.05 g/100 ml.) aniline (3.5 g/100 ml.)

As we have seen for other types of compounds, the solubility of amines decreases with increasing molecular weight because the functional group is a less significant part of the structure.

Amides having low molecular weights are soluble in water because hydrogen bonds form between the amide group and water. Even low molecular weight tertiary amides are water-soluble because the carbonyl oxygen atom can form hydrogen bonds to the hydrogen atoms of water.

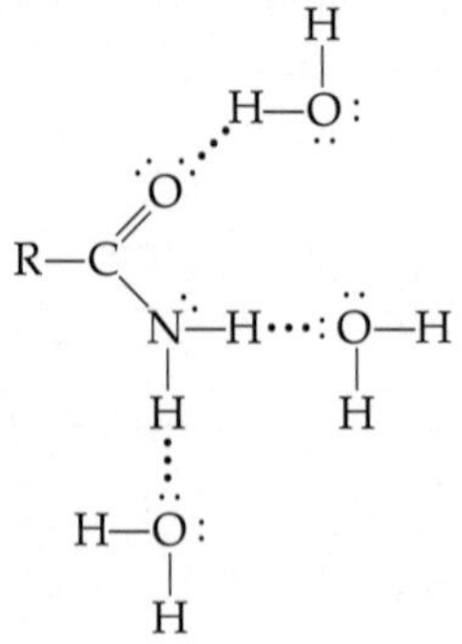

$$R{-}C({=}O\cdots H{-}OH){-}N(H\cdots OH_2){-}H\cdots :OH_2$$

Odors of Amines

Amines with low molecular weights have sharp penetrating odors similar to ammonia. Amines with higher molecular weights smell like decaying fish. Two compounds responsible for the odor of decaying animal tissue are appropriately named putrescine and cadaverine.

$NH_2CH_2CH_2CH_2CH_2NH_2$
putrescine

$NH_2CH_2CH_2CH_2CH_2CH_2NH_2$
cadaverine

14.5 BASICITY OF NITROGEN COMPOUNDS

The nitrogen atom in both amines and amides has an unshared pair of electrons. However, there is a substantial difference in the chemistry of these two classes of compounds. In this section we consider the ability of these nitrogen compounds to donate an electron pair to a proton—that is, the basicity of the compounds. In the following section we will examine the donation of the electron pair to an electrophile other than a proton.

Amines

Amines are fairly weak bases. They accept a proton from water to form an ammonium ion and a hydroxide ion. The equilibrium constant for the reaction of a base with water is the **base ionization constant,** symbolized by K_b. The K_b values of alkyl-substituted amines are smaller than 10^{-3}. The reaction of methylamine with water is typical.

$$CH_3NH_2 + H_2O \rightleftharpoons \underset{\text{methylammonium ion}}{CH_3NH_3^+} + OH^-$$

$$K_b = \frac{[OH^-][CH_3NH_3^+]}{[CH_3NH_2]} = 4.4 \times 10^{-4}$$

The base ionization constants for several amines are given in Table 14.2. We recall that alkyl groups are electron-donating toward carbocations (Section 4.9), and that they are electron-donating in electrophilic aromatic substitution reactions (Section 5.7). Similarly, alkyl-substituted amines are slightly stronger bases than ammonia because the inductive donation of electrons to

TABLE 14.2 Basicity of Amines and Acidity of Ammonium Salts

	K_b	K_a	pK_b	pK_a
ammonia	1.8×10^{-5}	5.5×10^{-10}	4.74	9.26
methylamine	4.6×10^{-4}	2.2×10^{-11}	3.34	10.66
ethylamine	4.8×10^{-4}	2.1×10^{-11}	3.20	10.80
dimethylamine	4.7×10^{-4}	2.1×10^{-11}	3.27	10.73
diethylamine	3.1×10^{-4}	3.2×10^{-11}	3.51	10.49
triethylamine	1.0×10^{-3}	1.0×10^{-11}	2.99	11.01
cyclohexylamine	4.6×10^{-4}	2.2×10^{-11}	3.34	10.66
aniline	4.3×10^{-10}	2.3×10^{-5}	9.37	4.63

the nitrogen atom by alkyl groups makes the unshared pair of electrons more available to a proton.

Aryl-substituted amines are much weaker bases than ammonia and alkyl-substituted amines. Their K_b values are less than 10^{-9} (Table 14.2). For example, the K_b value of aniline is 10^{-6} smaller than the K_b value for cyclohexylamine.

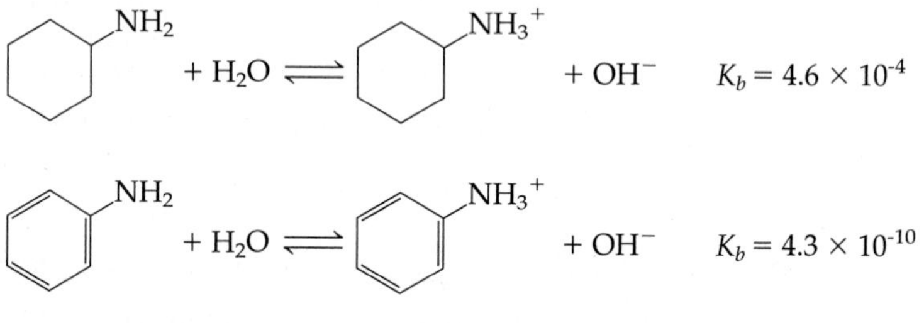

$$C_6H_{11}NH_2 + H_2O \rightleftharpoons C_6H_{11}NH_3^+ + OH^- \quad K_b = 4.6 \times 10^{-4}$$

$$C_6H_5NH_2 + H_2O \rightleftharpoons C_6H_5NH_3^+ + OH^- \quad K_b = 4.3 \times 10^{-10}$$

Aryl-substituted amines are weaker bases than ammonia because the unshared pair of electrons of the nitrogen atom is resonance-delocalized over the π-orbital system of the benzene ring. As a result, the unshared electron pair of nitrogen is less available for bonding with a proton.

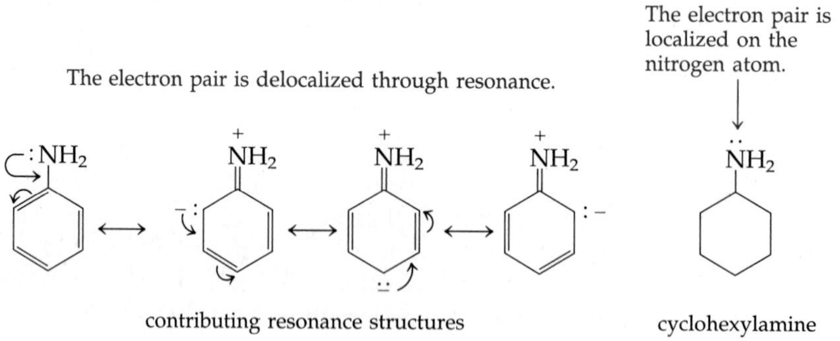

The electron pair is delocalized through resonance.

The electron pair is localized on the nitrogen atom.

NH_2 $\longleftrightarrow$ $\overset{+}{N}H_2$ $\longleftrightarrow$ $\overset{+}{N}H_2$ $\longleftrightarrow$ $\overset{+}{N}H_2$ $\quad$ NH_2

contributing resonance structures

cyclohexylamine

Amides

In contrast to amines, amides are not basic. This difference in basicity is due to the carbonyl group, which draws electrons away from the nitrogen atom. Thus, the unshared electron pair of the amide nitrogen atom is delocalized and not available to react with a proton. An amide is polar, planar, and resonance-stabilized.

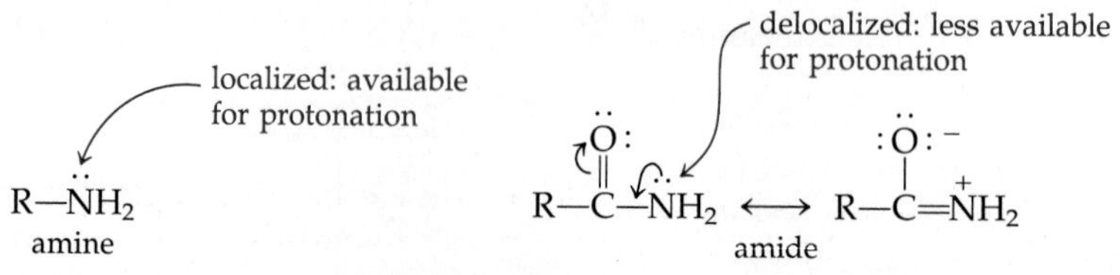

localized: available for protonation

$R{-}NH_2$

amine

delocalized: less available for protonation

$R{-}C(=O){-}NH_2 \longleftrightarrow R{-}C(O^-){=}\overset{+}{N}H_2$

amide

Acidity of Ammonium Ions

When an amine is protonated, the conjugate acid product is a positively charged, substituted ammonium ion. The ionization constant of the conjugate acid of methylamine, the methylammonium ion, is derived from the following reaction.

$$\underset{\text{methylammonium ion}}{CH_3NH_3^+} + H_2O \rightleftharpoons CH_3NH_2 + H_3O^+$$

$$K_a = \frac{[CH_3NH_2][H_3O^+]}{[CH_3NH_3^+]} = 2.2 \times 10^{-11}$$

The conjugate acid has a K_a value that is related to the K_b of the base. If a conjugate acid is relatively weak, it has a small K_a value. The base from which it is derived has a relatively large K_b, and the base is relatively strong.

If this ammonium salt is a weaker acid (small K_a), then this amine is a stronger base (large K_b).

$$R{-}NH_3^+ + H_2O \rightleftharpoons R{-}NH_2 + H_3O^+$$

The K_a value of the methylammonium ion and the K_b value of methylamine illustrate the relationship between a base and its conjugate acid. The K_a value for the methylammonium ion is relatively small; the K_b value for methylamine is relatively large. The values of K_a and K_b for a conjugate acid–base pair are related as follows:

$$(K_a)\ (K_b) = K_w = 1 \times 10^{-14}$$

pK_b and pK_a

The basicity of an amine is usually listed as a pK_b, the negative logarithm of K_b. For an amine with $K_b = 10^{-4}$, the pK_b is 4. The pK_b values of strong bases are small. Thus, as pK_b increases, base strength decreases.

It is also common practice to indicate the relative base strength of amines in terms of the pK_a of their conjugate acids. (Recall that pK_a = −log K_a.) The values of pK_a and pK_b for a conjugate acid–base pair are related as follows.

$$pK_a + pK_b = 14$$

Thus, a strong amine base has a small pK_b, but its conjugate ammonium ion has a large pK_a.

EXAMPLE 14.5 The pK_b values for diethylamine and triethylamine are 3.51 and 2.99, respectively. Which compound is the stronger base?

Solution The stronger base has the larger K_b value and hence the smaller pK_b value. Thus, triethylamine is the stronger base.

14.6 SOLUBILITY OF AMMONIUM SALTS

When an amine is added to a solution of a strong acid such as hydrochloric acid, the amine nitrogen atom is protonated to produce an ammonium salt.

$$RNH_2 + HCl \longrightarrow RNH_3^+ + Cl^-$$

Because the nitrogen atom of an ammonium salt has a positive charge, ammonium salts are more soluble than amines. This property is used by drug companies to manufacture compounds that are soluble in body fluids. Drugs containing an amino group are often prepared as ammonium salts to improve their solubility.

Amines can be separated from other substances by converting them to ammonium salts. Consider, for example, the separation of 1-chlorooctane from 1-octanamine. Both compounds are insoluble in water. Adding HCl to a solution that contains both compounds converts the 1-octanamine into its ammonium salt, whereas 1-chlorooctane is not affected.

$$\underset{\text{(insoluble in water)}}{CH_3(CH_2)_6CH_2NH_2} + HCl \longrightarrow \underset{\text{(soluble in water)}}{CH_3(CH_2)_6CH_2NH_3^+} + Cl^-$$

$$\underset{\text{(insoluble in water)}}{CH_3(CH_2)_6CH_2Cl} + HCl \not\longrightarrow \text{no reaction}$$

The 1-chlorooctane is physically separated from the aqueous acid solution. Then, the acid solution is neutralized with sodium hydroxide to form the free amine. The amine can then be physically separated from the aqueous solution.

$$\underset{\text{(soluble in water)}}{CH_3(CH_2)_6CH_2NH_3^+} + OH^- \longrightarrow \underset{\text{(insoluble in water)}}{CH_3(CH_2)_6CH_2NH_2} + H_2O$$

14.7 NUCLEOPHILIC REACTIONS OF AMINES

We described some reactions of amines in earlier chapters. These reactions occur because the nitrogen atom of an amine is nucleophilic. We will review each type of reaction in this section.

Reaction with Carbonyl Compounds

In Chapter 10, we described the addition–elimination reaction of amines with carbonyl compounds. An amine adds to the electrophilic carbonyl carbon atom to give a tetrahedral intermediate. This product is unstable and loses water to form an imine. In general, imines are less stable than carbonyl compounds. Thus, the reaction is favorable only if water is removed from the reaction mixture. Imines are not stable; they rapidly hydrolyze in aqueous solution to give carbonyl compounds.

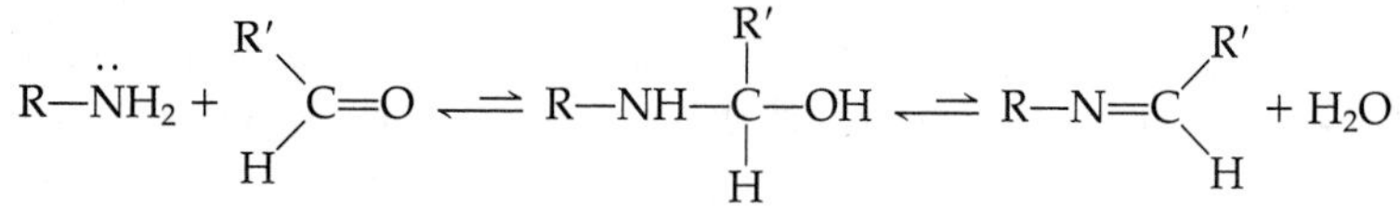

$$R{-}\ddot{N}H_2 + R'(H)C{=}O \rightleftharpoons R{-}NH{-}C(R')(H){-}OH \rightleftharpoons R{-}N{=}C(R')(H) + H_2O$$

aldehyde or ketone — a carbinol amine — an imine

Reaction with Acyl Derivatives

In Chapter 12, we noted that an amide can be made by treating an amine with an acid halide. We recall that acid halides are very reactive acyl derivatives of acids; amides are very stable.

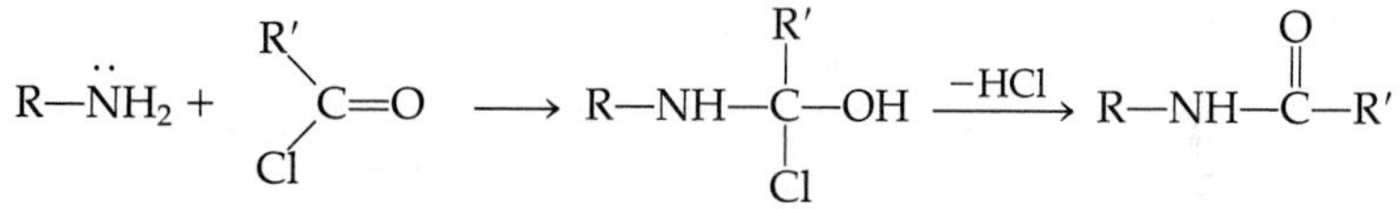

$$R{-}\ddot{N}H_2 + R'(Cl)C{=}O \longrightarrow R{-}NH{-}C(R')(Cl){-}OH \xrightarrow{-HCl} R{-}NH{-}\overset{O}{\overset{\|}{C}}{-}R'$$

acid chloride — reactive tetrahedral intermediate — an amide

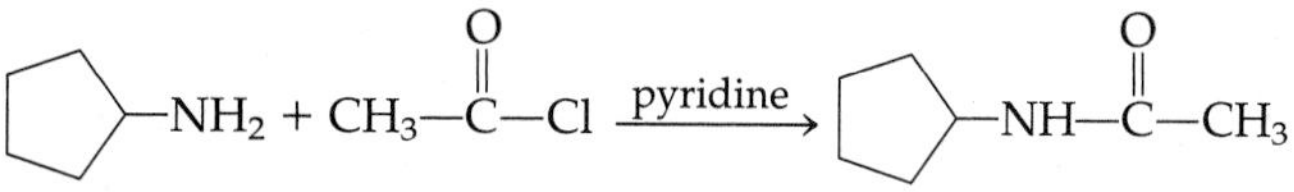

$$C_5H_9{-}NH_2 + CH_3{-}\overset{O}{\overset{\|}{C}}{-}Cl \xrightarrow{\text{pyridine}} C_5H_9{-}NH{-}\overset{O}{\overset{\|}{C}}{-}CH_3$$

Only ammonia and primary or secondary amines can form amides. Note that pyridine, which cannot form an amide (it can be regarded as a tertiary amine), is used as a base to react with the HCl formed in the reaction.

EXAMPLE 14.6 Flecainide, an antiarrhythmic drug, is an amide. Draw the structures of the compounds that could be used to produce the drug. What possible complications might occur with this combination of reactants?

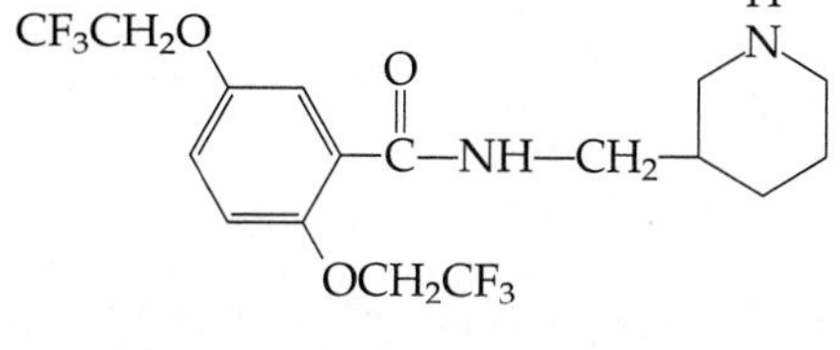

[Structure: benzene ring bearing CF_3CH_2O and OCH_2CF_3 substituents, with $-\overset{O}{\overset{\|}{C}}{-}NH{-}CH_2{-}$ attached to a piperidine ring (N–H)]

Solution Mentally separate the amide into two components by breaking the bond between the nitrogen atom and the carbonyl carbon atom. Place a

Quaternary Ammonium Salts Are Invert Soaps

Quaternary ammonium salts are ammonium salts that have four alkyl or aryl groups bonded to a nitrogen atom. Some quaternary ammonium salts containing a long carbon chain are **invert soaps.**

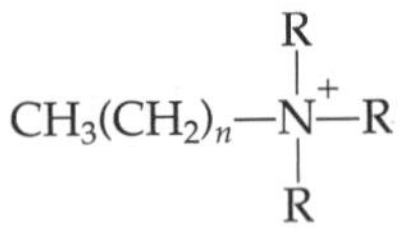

$$CH_3(CH_2)_n{-}\overset{R}{\underset{R}{\overset{|}{\underset{|}{N^+}}}}{-}R$$

an invert soap

Invert soaps differ from soaps and detergents because the polar end of the ion is positive rather than negative. As in soaps, the long hydrocarbon tail associates with nonpolar substances, and the polar head dissolves in water. Thus invert soaps act by the same cleansing mechanism described in Chapter 12 for soaps and detergents.

Invert soaps are widely used in hospitals. They are active against bacteria, fungi, and protozoa, but they are not effective against spore-forming microorganisms. One type of invert soap is the family of benzalkonium chlorides. The alkyl groups of these compounds contain from 8 to 16 carbon atoms. These compounds are effective at concentrations of 1:750 to 1:20,000. The more complex benzethonium chloride is also an effective antiseptic.

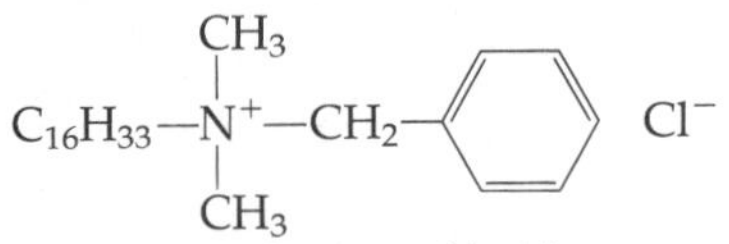

$$C_{16}H_{33}{-}\overset{CH_3}{\underset{CH_3}{N^+}}{-}CH_2{-}C_6H_5 \quad Cl^-$$

benzalkonium chloride

$$CH_3{-}\overset{CH_3}{\underset{CH_3}{C}}{-}CH_2{-}\overset{CH_3}{\underset{CH_3}{C}}{-}C_6H_4{-}OCH_2CH_2OCH_2CH_2{-}\overset{CH_3}{\underset{CH_3}{N^+}}{-}CH_2{-}C_6H_5 \quad Cl^-$$

benzethonium chloride

hydrogen atom on the nitrogen atom. Place a chlorine atom on the carbonyl carbon atom.

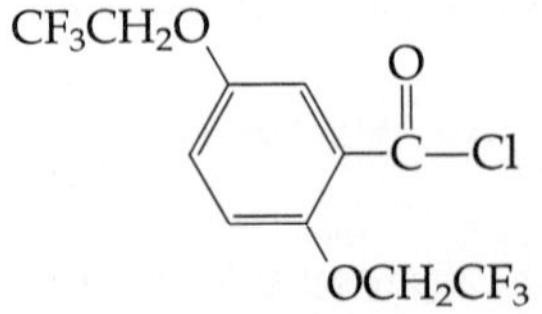

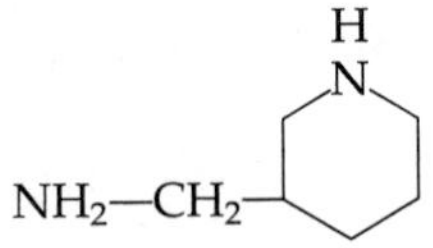

$$CF_3CH_2O{-}C_6H_3(OCH_2CF_3){-}\overset{O}{\overset{\|}{C}}{-}Cl \qquad NH_2{-}CH_2{-}C_5H_9NH$$

Note that the "amine" is actually a diamine: one is a primary amine, the other a secondary amine. Thus, the diamine could react at either nitrogen atom and form two isomeric amides. The primary amine is more reactive because a secondary amine is more sterically hindered.

Reaction of Amines with Alkyl Halides

We described nucleophilic substitution reactions of alkyl halides in Chapter 7. Primary and secondary alkyl halides react with nucleophiles by an S_N2 mechanism. Amines are nucleophiles that can displace a halide ion from a primary or secondary alkyl halide to form an ammonium halide salt after neutralization.

$$R{-}\ddot{N}H_2 + R'{-}X \longrightarrow R{-}\overset{+}{N}H_2{-}R' \xrightarrow{\text{base}} R{-}\ddot{N}H{-}R'$$

$$C_6H_5\ddot{N}H_2 + C_5H_9CH_2Cl \longrightarrow C_6H_5\ddot{N}H{-}CH_2{-}C_5H_9$$

The initial product of the nucleophilic substitution reaction is a secondary ammonium ion. It can lose a proton in an equilibrium reaction with the reactant primary amine.

$$R{-}\ddot{N}H_2 + R{-}\overset{+}{N}H_2{-}R' \rightleftharpoons R{-}NH_3^+ + R{-}\ddot{N}H{-}R'$$

The secondary amine then can continue to react with the alkyl halide to give a tertiary amine and eventually a quaternary ammonium salt.

$$R{-}\ddot{N}H_2 \xrightarrow{R'X} R\ddot{N}HR' \xrightarrow{R'X} R\ddot{N}R'_2 \xrightarrow{R'X} R\overset{+}{N}R'_3$$

14.8 REACTION OF AMINES WITH NITROUS ACID

Nitrous acid (HNO_2) is an unstable compound produced by the reaction of a nitrite salt and a strong acid such as sulfuric acid.

$$H_3O^+ + NO_2^- \longrightarrow H_2O + HNO_2$$

Under the reaction conditions, nitrous acid is a source of the nitrosonium ion. This species is an electrophile that reacts with amines. The products of reaction with primary, secondary, and tertiary amines differ and thus enable us to classify an unknown amine.

$$H{-}\ddot{\underset{..}{O}}{-}\ddot{N}{=}\ddot{\underset{..}{O}}{:} + H_2SO_4 \longrightarrow H{-}\underset{..}{\overset{+}{O}}H{-}\ddot{N}{=}\ddot{\underset{..}{O}}{:} + HSO_4^-$$

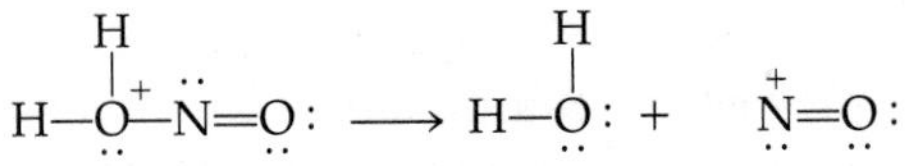

$$H-\overset{\overset{H}{|}}{O}^{+}-N{=}O: \longrightarrow H-\overset{\overset{H}{|}}{O}: + \overset{+}{N}{=}O:$$

nitrosonium ion

Primary amines react with HNO_2 to produce unstable diazonium salts, which decompose to produce nitrogen gas and a mixture of organic products. The evolution of nitrogen gas is a visual confirmation that the amine is primary.

$$R-NH_2 + HNO_2 + H_3O^+ \longrightarrow R-N{\equiv}N^+ + 3\ H_2O$$
$$R-N{\equiv}N^+ \longrightarrow N_2 + R^+$$

The mechanism of the reaction of the nitrosonium ion with a primary amine is given in Figure 14.3. The organic products of the reaction are derived from

FIGURE 14.3
Mechanism of the Reaction of Amines with Nitrous Acid

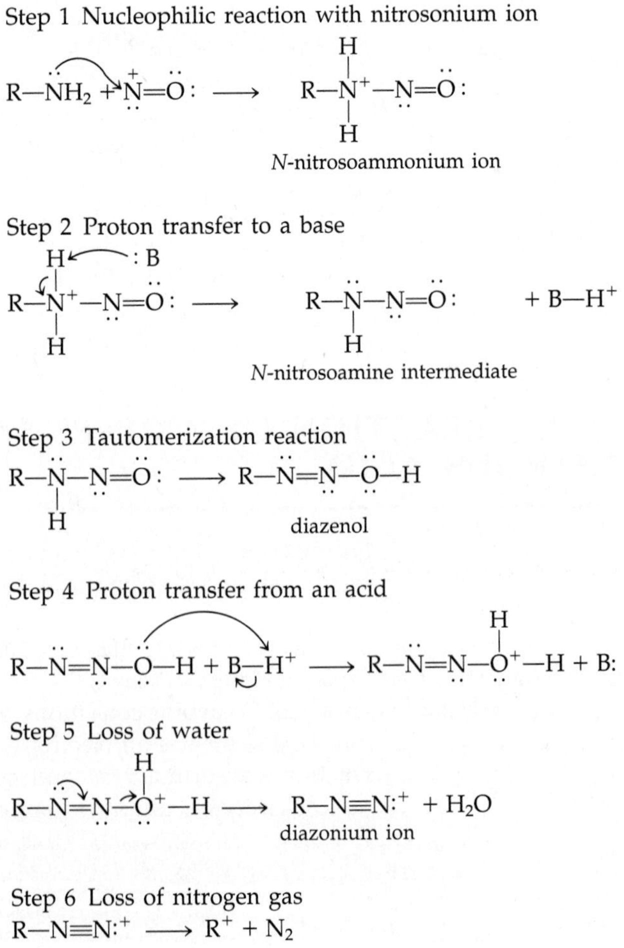

Step 1 Nucleophilic reaction with nitrosonium ion

$$R-NH_2 + \overset{+}{N}{=}O: \longrightarrow R-\underset{\underset{H}{|}}{\overset{\overset{H}{|}}{N^+}}-N{=}O:$$

N-nitrosoammonium ion

Step 2 Proton transfer to a base

$$R-\underset{\underset{H}{|}}{\overset{\overset{H \leftarrow :B}{|}}{N^+}}-N{=}O: \longrightarrow R-\underset{\underset{H}{|}}{N}-N{=}O: \quad + B-H^+$$

N-nitrosoamine intermediate

Step 3 Tautomerization reaction

$$R-\underset{\underset{H}{|}}{N}-N{=}O: \longrightarrow R-N{=}N-O-H$$

diazenol

Step 4 Proton transfer from an acid

$$R-N{=}N-O-H + B-H^+ \longrightarrow R-N{=}N-\overset{\overset{H}{|}}{O^+}-H + B:$$

Step 5 Loss of water

$$R-N{=}N-\overset{\overset{H}{|}}{O^+}-H \longrightarrow R-N{\equiv}N:^+ + H_2O$$

diazonium ion

Step 6 Loss of nitrogen gas

$$R-N{\equiv}N:^+ \longrightarrow R^+ + N_2$$

the carbocation intermediate; they include substitution and elimination products.

Secondary amines react with the nitrosonium ion from HNO_2 according to the first two steps listed in Figure 14.3. Tautomerization cannot occur because there is no N—H bond in the intermediate. The *N*-nitrosoamine formed separates from the reaction mixture as a yellow oil that floats on the acid solution.

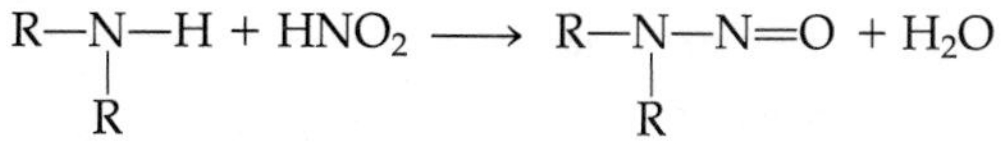

$$\underset{\displaystyle\mathrm{R}}{\mathrm{R{-}\overset{}{N}{-}H}} + HNO_2 \longrightarrow \underset{\displaystyle\mathrm{R}}{\mathrm{R{-}N{-}N{=}O}} + H_2O$$

a *N*-nitrosamine

The first step shown in Figure 14.3 occurs with tertiary amines. The second step cannot occur because there is no acidic N—H bond in a tertiary nitrosoammonium ion. Thus, the ion exists in equilibrium with the amine. Because both species are soluble, there is no visible reaction, and the amine appears to simply dissolve in the acid solution.

In summary, primary amines react with HNO_2 to liberate nitrogen gas; secondary amines form insoluble *N*-nitroso compounds; tertiary amines give no visible reaction. As a result, the three classes of amines are easily distinguished from each other.

EXAMPLE 14.7 How can the following isomeric compounds be distinguished?

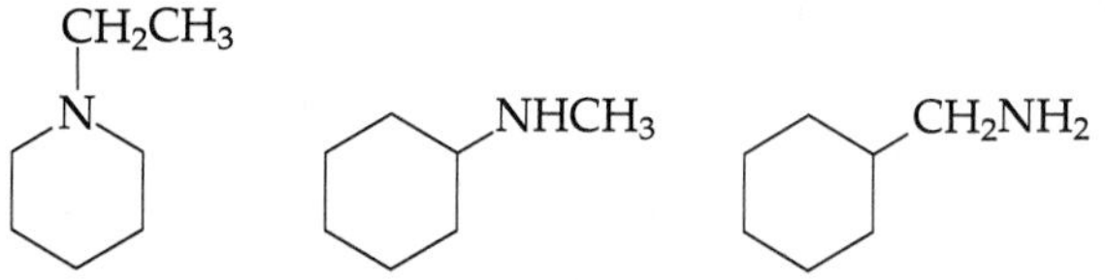

(I) N-ethylpiperidine (CH_2CH_3 on N of a six-membered ring)

(II) cyclohexyl–$NHCH_3$

(III) cyclohexyl–CH_2NH_2

Nitrites in Food

Studies of the effects of nitrosoamines on animals indicate that they are carcinogenic. It has been suggested that nitrites added to bacon, hot dogs, and sandwich meats to retard spoilage may cause stomach cancer. Nitrite ions in the stomach react with gastric acid (HCl) to form nitrous acid, which can react with amines present in foods to produce nitrosoamines.

There is considerable controversy about the use of nitrites in foods and their possible effect on the human body. Processed meats must have preservatives to have a reasonable shelf life. However, how much exposure to nitrites can we tolerate? The amount of nitrites used in laboratory tests with rats is many powers of 10 higher than that used in commercial food products. Thus, the amount of nitrites ingested by most individuals is small, and the benefit is large relative to the risk. At this point, it does not appear that nitrites in foods pose a serious risk of cancer.

Solution Compound III is a primary amine that releases nitrogen gas when treated with HNO_2. Compound II is a secondary amine that forms a yellow oil with the same reagent. Compound I is a tertiary amine and will simply dissolve in the HCl solution used to produce HNO_2.

14.9 SYNTHESIS OF AMINES

Many of the general methods to synthesize amines have already been discussed in preceding sections and chapters. Except for the displacement reaction of an alkyl halide by ammonia or an amine, the remaining methods involve compounds that already have a nitrogen atom contained in a functional group that is then transformed into an amine functional group.

Alkylation of Amines by Alkyl Halides

In Section 14.7 we saw that a nucleophilic substitution reaction of ammonia with an alkyl halide yields a mixture of products resulting from multiple alkylation. The chances for multiple alkylation can be diminished somewhat by selecting the proper reaction conditions. For example, if the reaction of an alkyl halide with ammonia is carried out with excess ammonia, an alkyl halide can be converted to a primary amine. When the concentration of ammonia is greater than the concentration of the primary amine product, the probability is reduced that the primary amine will continue to react with the alkyl halide.

$$C_6H_{11}\text{—}CH_2\text{—}Br \xrightarrow[\text{excess}]{NH_3} C_6H_{11}\text{—}CH_2\text{—}NH_2$$

By analogy, we expect that secondary amines could be prepared by reaction of an alkyl halide with an excess of a primary amine. In general, this reaction is not used because of the waste involved with the use of excess amines, which are more expensive than the ammonia used to prepare primary amines.

Reduction of Imines

We recall that the carbonyl group of either aldehydes or ketones is reduced to an alcohol by either catalytic hydrogenation or metal hydrides. Imines are the nitrogen analogs of carbonyl compounds. They are reduced in the same way.

$$C_6H_5\text{—}CH{=}N\text{—}C_6H_{11} \xrightarrow[Ni]{H_2} C_6H_5\text{—}CH_2\text{—}NH\text{—}C_6H_{11}$$

$$CH_3(CH_2)_4\text{—}CH{=}N\text{—}CH_2CH_3 \xrightarrow{NaBH_4} CH_3(CH_2)_4\text{—}CH_2\text{—}NH\text{—}CH_2CH_3$$

Imines do not have to be prepared and isolated for subsequent reduction. A mixture of a carbonyl compound and ammonia or the appropriate amine reacts in the presence of hydrogen gas and a metal catalyst. The imine initially formed is reduced to an amine. The overall process is called **reductive amination.**

$$C_6H_5-CHO + NH_3 \xrightarrow[Ni]{H_2} C_6H_5-CH_2-NH_2$$

Reductive amination can also be accomplished by a modified borohydride—sodium cyanoborohydride. This reagent reduces the intermediate imine functional group but not the carbonyl group of the reactant.

$$C_5H_9-CH_2-NH_2 + CH_3-\overset{O}{\overset{\|}{C}}-CH_2CH_3 \xrightarrow[CH_3OH]{NaBH_3CN} C_5H_9-CH_2-NH-\overset{CH_3}{\overset{|}{C}}HCH_2CH_3$$

Reduction of Amides

Reduction of amides is one of the most frequently used methods of preparing amines. The method is very versatile because primary, secondary, and tertiary amines are easily prepared from the corresponding classes of amides. Amides are prepared by acylation of amines using activated acyl derivatives such as acid chlorides or acid anhydrides (Section 12.6).

$$\underset{\text{primary amide}}{R-\overset{O}{\overset{\|}{C}}-NH_2} \xrightarrow[2.\ H_3O^+]{1.\ LiAlH_4} \underset{\text{primary amine}}{R-CH_2-NH_2}$$

$$\underset{\text{secondary amide}}{R-\overset{O}{\overset{\|}{C}}-NH-R'} \xrightarrow[2.\ H_3O^+]{1.\ LiAlH_4} \underset{\text{secondary amine}}{R-CH_2-NH-R'}$$

$$\underset{\text{tertiary amide}}{R-\overset{O}{\overset{\|}{C}}-\underset{R''}{\underset{|}{N}}-R'} \xrightarrow[2.\ H_3O^+]{1.\ LiAlH_4} \underset{\text{tertiary amine}}{R-CH_2-\underset{R''}{\underset{|}{N}}-R'}$$

Reduction of Nitriles

Nitriles can be prepared from primary alkyl halides by a direct S_N2 displacement reaction using sodium cyanide as the nucleophile (Section 7.6). Then, the nitrile is reduced to a primary amine with lithium aluminum hydride.

$$R-X \xrightarrow{CN^-} R-C\equiv N \xrightarrow[2.\ H_3O^+]{1.\ LiAlH_4} R-CH_2-NH_2$$

$$CH_3\text{—}C(CH_3)_2\text{—}CH_2\text{—}CH_2\text{—}Br \xrightarrow[\text{2. } LiAlH_4 \;\; \text{3. } H_3O^+]{\text{1. } CN^-} CH_3\text{—}C(CH_3)_2\text{—}CH_2\text{—}CH_2\text{—}CH_2\text{—}NH_2$$

3,3-dimethyl-1-bromobutane → 4,4-dimethyl-1-pentanamine

Reduction of Nitro Compounds

There is no synthetic procedure to introduce an amino group onto an aromatic ring in one step. However, it is possible to substitute an amino group onto an aromatic ring in two steps. First the ring is nitrated. Then, the nitro group is reduced to an amino group. We described this process in Chapter 5.

$$C_6H_5CH(CH_3)_2 \xrightarrow[H_2SO_4]{HNO_3} p\text{-}O_2N\text{—}C_6H_4\text{—}CH(CH_3)_2 \xrightarrow{Sn/HCl} p\text{-}H_2N\text{—}C_6H_4\text{—}CH(CH_3)_2$$

14.10 HYDROLYSIS OF AMIDES

Hydrolysis of an amide breaks the carbon–nitrogen bond and produces an acid and either ammonia or an amine. This reaction resembles the hydrolysis of esters, which we discussed in Chapter 12. There are, however, important differences. The hydrolysis of esters occurs relatively easily, whereas amides are very resistant to hydrolysis. Amides are hydrolyzed only by heating for hours with a strong acid or strong base. When amide hydrolysis is carried out in basic solution, the salt of the carboxylic acid forms; a mole of base is required per mole of amide. When amide hydrolysis is carried out under acidic conditions, the ammonium salt of the amine is formed, and a mole of acid is required per mole of amide.

$$C_6H_5\text{—}\overset{O}{\overset{\|}{C}}\text{—}NHCH_3 + OH^- \longrightarrow C_6H_5\text{—}\overset{O}{\overset{\|}{C}}\text{—}O^- + CH_3NH_2$$

$$C_6H_5\text{—}\overset{O}{\overset{\|}{C}}\text{—}NHCH_3 + HCl + H_2O \longrightarrow C_6H_5\text{—}\overset{O}{\overset{\|}{C}}\text{—}OH + CH_3NH_3^+\ Cl^-$$

The great stability of amides toward hydrolysis has an important biological consequence because amino acids in proteins are linked by amide bonds. Because amides are stable, proteins do not readily hydrolyze at physiological pH and at body temperature in the absence of a specific enzyme catalyst. However, in the presence of specific enzymes, the hydrolysis of amides is rapid. These reactions will be discussed in Chapter 15.

EXAMPLE 14.8 What are the products of the hydrolysis of phenacetin by a base? Phenacetin was formerly used in APC analgesic tablets consisting of aspirin, phenacetin, and caffeine.

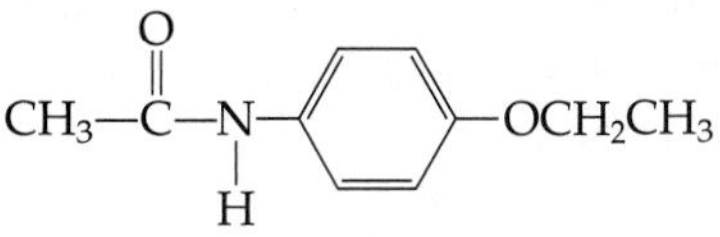

$$CH_3-\overset{\overset{\large O}{\|}}{C}-\underset{\underset{\large H}{|}}{N}-C_6H_4-OCH_2CH_3$$

phenacetin

Solution The functional group on the right side of the benzene ring is an ether, which does not react with base (Section 7.5). The functional group on the left is an amide.

Hydrolysis of an amide breaks the bond between the nitrogen atom and the carbonyl group. The acid fragment is acetic acid. The amine fragment is a substituted aniline containing an ether substituent.

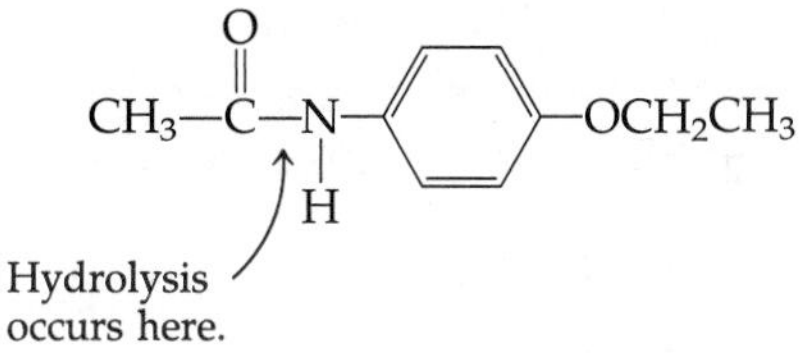

$$CH_3-\overset{\overset{\large O}{\|}}{C}-\underset{\underset{\large H}{|}}{N}-C_6H_4-OCH_2CH_3$$

Hydrolysis occurs here.

Because a base is used in the hydrolysis, the acid product is present in the reaction mixture as the acetate ion. The amine is *p*-ethoxyaniline.

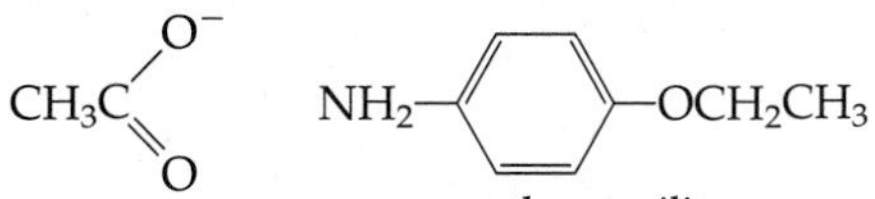

$$CH_3C(=O)O^- \qquad NH_2-C_6H_4-OCH_2CH_3$$

p-ethoxyaniline

14.11 SYNTHESIS OF AMIDES

Carboxylic acids react to form an amide and water when heated with ammonia, a primary amine, or a secondary amine. Tertiary amines do not form amides because they have no hydrogen atom bonded to the nitrogen atom.

$$R-\overset{\overset{\large O}{\|}}{C}-OH + NH_3 \longrightarrow R-\overset{\overset{\large O}{\|}}{C}-NH_2 + H_2O$$

$$R-\overset{O}{\overset{\|}{C}}-OH + R'NH_2 \longrightarrow R-\overset{O}{\overset{\|}{C}}-NHR' + H_2O$$

$$R-\overset{O}{\overset{\|}{C}}-OH + R'R''NH \longrightarrow R-\overset{O}{\overset{\|}{C}}-\underset{R''}{\underset{|}{N}}-R' + H_2O$$

The high temperature of this direct reaction often affects other functional groups in the molecule. An amide can be synthesized under milder conditions by the reaction of an acyl chloride with ammonia, a primary amine, or a secondary amine. This reaction readily occurs at lower temperatures. We described this reaction previously with a different focus—the reactions of amines (Section 14.7).

$$C_8H_{15}-\overset{O}{\overset{\|}{C}}-Cl + NH_2-\underset{CH_3}{\underset{|}{\overset{CH_3}{\overset{|}{C}}}}-CH_3 \longrightarrow C_8H_{15}-\overset{O}{\overset{\|}{C}}-NH-\underset{CH_3}{\underset{|}{\overset{CH_3}{\overset{|}{C}}}}-CH_3$$

Proteins are polyamides of α-amino acids. The amino group of one amino acid reacts with the acid group of another amino acid in a process that requires many enzyme-catalyzed reactions. The resultant compound has both an amine and a carboxyl functional group. Continued reaction with the same or different amino acids forms proteins.

$$NH_2-\underset{R}{\underset{|}{\overset{H}{\overset{|}{C}}}}-\overset{O}{\overset{\|}{C}}-OH + NH_2-\underset{R'}{\underset{|}{\overset{H}{\overset{|}{C}}}}-\overset{O}{\overset{\|}{C}}-OH \xrightarrow{\text{enzyme}} NH_2-\underset{R}{\underset{|}{\overset{H}{\overset{|}{C}}}}-\overset{O}{\overset{\|}{C}}-\underset{H}{\underset{|}{N}}-\underset{R'}{\underset{|}{\overset{H}{\overset{|}{C}}}}-\overset{O}{\overset{\|}{C}}-OH$$

14.12 POLYAMIDES

Many commercial products are high molecular weight polyamides. Perhaps the most famous of these is nylon. Synthetic polyamides such as nylon are produced from diamines and dicarboxylic acids by condensation polymerization. One type of nylon is made from adipic acid and hexamethylenediamine (1,6-diaminohexane), which condense to form an amide.

$$HO-\overset{O}{\overset{\|}{C}}(CH_2)_4\overset{O}{\overset{\|}{C}}-OH + H_2N(CH_2)_6NH_2 \longrightarrow HO-\overset{O}{\overset{\|}{C}}(CH_2)_4\overset{O}{\overset{\|}{C}}-\underset{H}{\underset{|}{N}}(CH_2)_6NH_2 + H_2O$$

adipic acid — hexamethylenediamine — an amide

This end can react with an amine. (the HO–C=O end)

This end can react with an acid. (the NH_2 end)

The product of the first condensation reaction is an amide that also contains a free amino group and a free carboxylic acid group. The amine end of this molecule reacts with another molecule of adipic acid to produce another amide linkage. The carboxylic acid end of the molecule reacts with another molecule of hexamethylenediamine. This sequence of reactions occurs again and again to produce a polyamide.

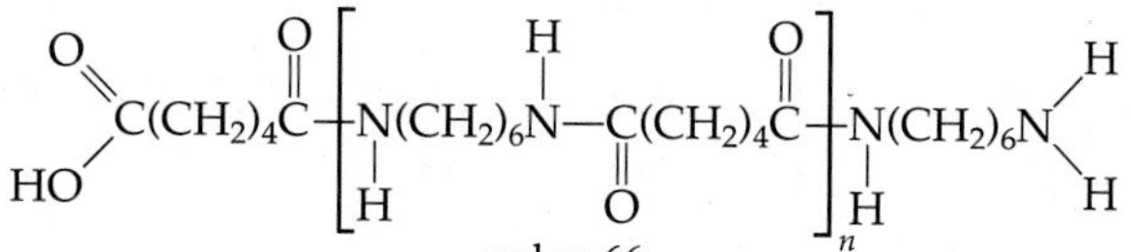

$$HO-\overset{O}{\overset{\|}{C}}(CH_2)_4\overset{O}{\overset{\|}{C}}\left[\underset{H}{\underset{|}{N}}(CH_2)_6\overset{H}{\overset{|}{N}}-\underset{O}{\underset{\|}{C}}(CH_2)_4\overset{O}{\overset{\|}{C}}\right]_n\underset{H}{\underset{|}{N}}(CH_2)_6NH_2$$

nylon 66

The nylon formed from adipic acid and hexamethylenediamine is called nylon 66. The "66" refers to the six-carbon diacid and six-carbon diamine reactants.

Polyamides containing aromatic rings (aramides) have many special properties. The presence of aromatic rings in the polymer produces a stiff and tough fiber. One commercially important aramide is Kevlar, a polyamide made from *p*-phenylenediamine and terephthaloyl chloride. It is used in place of steel in bullet-resistant vests. These vests are so light and flexible that they can be worn inconspicuously under normal clothing.

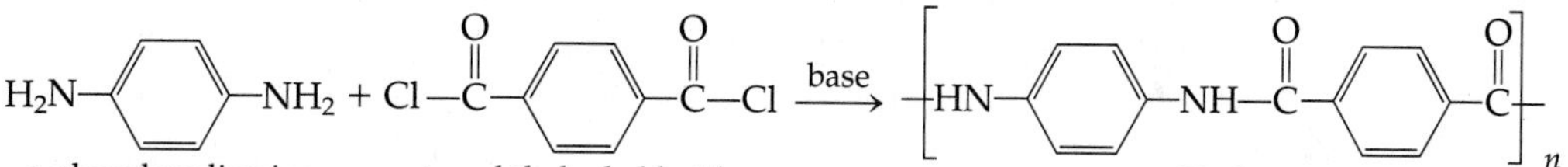

$$H_2N-C_6H_4-NH_2 + Cl-\overset{O}{\overset{\|}{C}}-C_6H_4-\overset{O}{\overset{\|}{C}}-Cl \xrightarrow{\text{base}} \left[HN-C_6H_4-NH-\overset{O}{\overset{\|}{C}}-C_6H_4-\overset{O}{\overset{\|}{C}}\right]_n$$

p-phenylenediamine — terephthaloyl chloride — Kevlar

An aramide called Nomex has a structure that resembles that of Kevlar. The monomers in Nomex are meta rather than para isomers. Nomex is used in flame-resistant clothing for fire fighters and racing-car drivers; it is so strong that it can be used in flame-resistant building materials.

EXERCISES

Classification of Amines and Amides

14.1 Classify each of the following amines.

(a) $CH_3-\overset{H}{\overset{|}{N}}-CH_2CH_3$ (b) $CH_3CH_2-\overset{CH_2CH_3}{\overset{|}{N}}-CH_2CH_3$

(c) $CH_3CH_2-\overset{H}{\overset{|}{N}}-\overset{CH_3}{\overset{|}{C}}HCH_3$ (d) $CH_3\overset{CH_3}{\overset{|}{C}}HCH_2-NH_2$

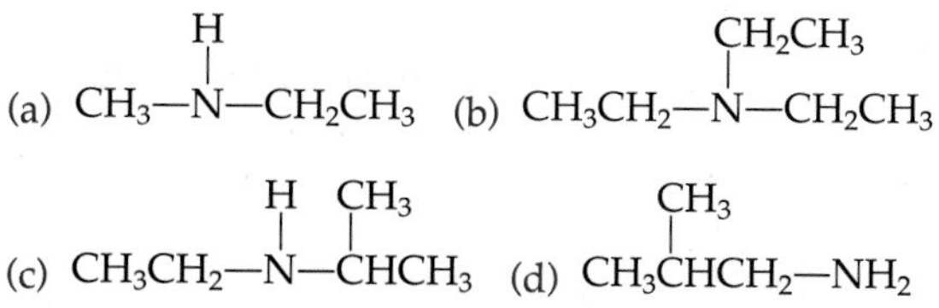

14.2 Classify each of the following amines.

(a) cyclohexane ring containing N, with N—CH_3 (b) $C_6H_5-\overset{H}{\overset{|}{N}}-C_6H_5$

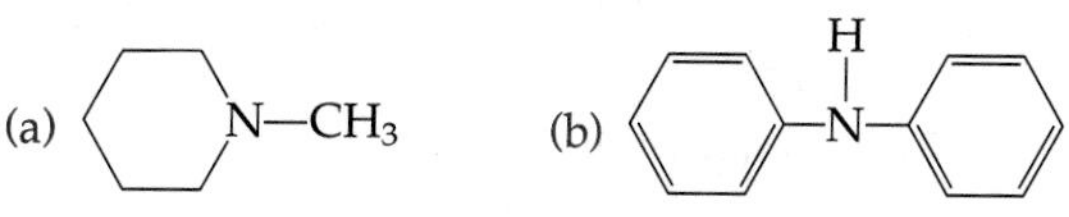

(c) 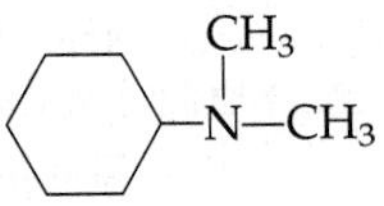 $C_6H_{11}\text{—}N(CH_3)\text{—}CH_3$ (d) 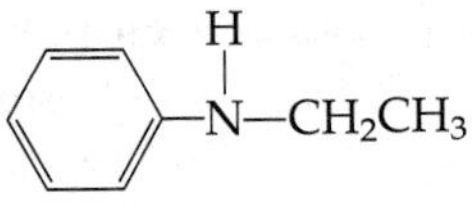 $C_6H_5\text{—}NH\text{—}CH_2CH_3$

14.3 Classify each of the following amides.

(a) 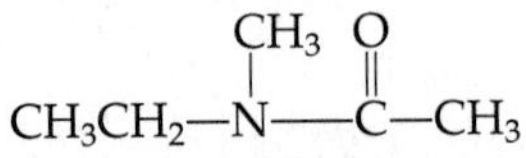 $CH_3CH_2\text{—}N(CH_3)\text{—}C(=O)\text{—}CH_3$ (b) $CH_3CH_2CH_2\text{—}C(=O)\text{—}NH\text{—}CH_3$

(c) $NH_2\text{—}C(=O)\text{—}CH_2CH(CH_3)CH_2CH_3$

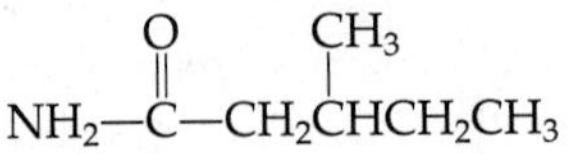

14.4 Classify each of the following amides.

(a) 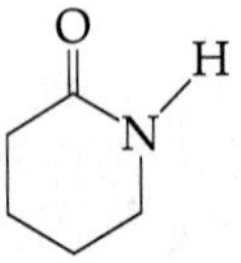 (b)

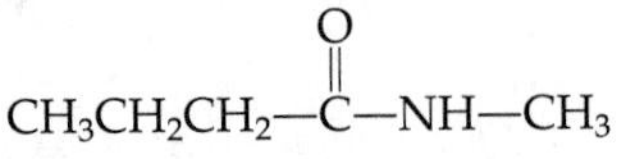

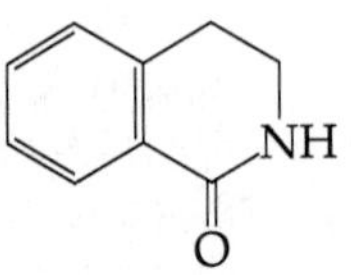

(c) 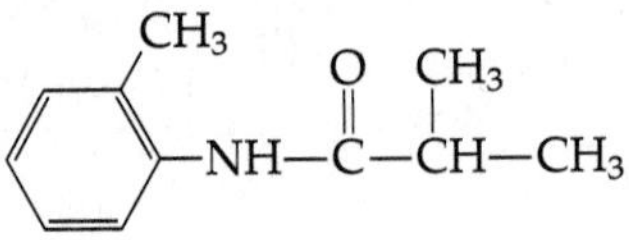 CH_3-substituted benzene ring $\text{—}NH\text{—}C(=O)\text{—}CH(CH_3)\text{—}CH_3$ (d) 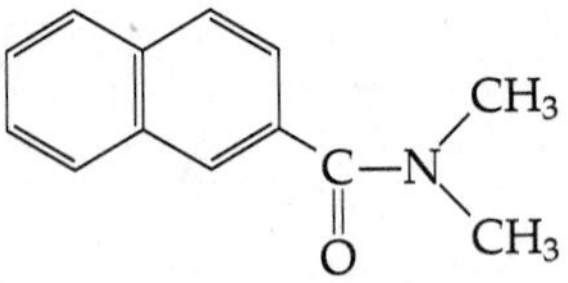 naphthalene ring $\text{—}C(=O)\text{—}N(CH_3)\text{—}CH_3$

14.5 Classify the nitrogen-containing functional group in each of the following structures.

(a) acetaminophen, the analgesic in Tylenol

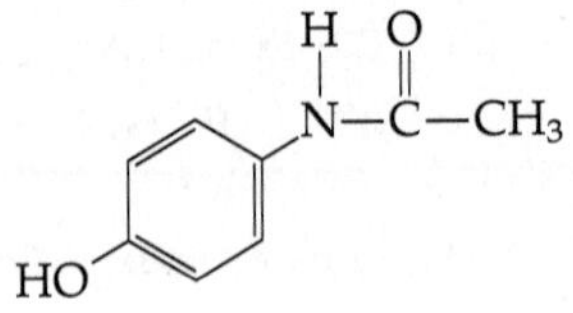

$HO\text{—}C_6H_4\text{—}NH\text{—}C(=O)\text{—}CH_3$

(b) methadone, a heroin substitute used in treating addicts

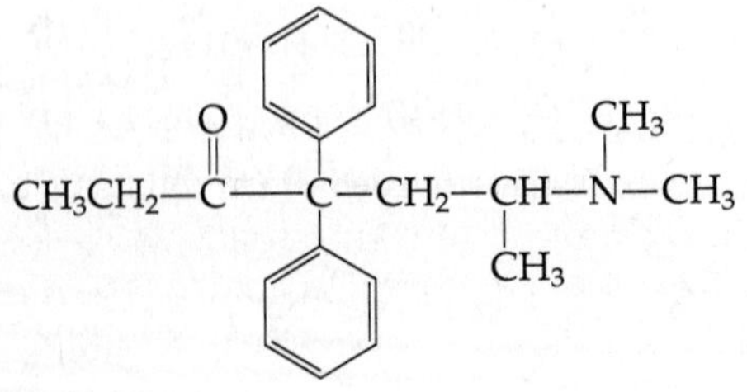

$CH_3CH_2\text{—}C(=O)\text{—}C(C_6H_5)_2\text{—}CH_2\text{—}CH(CH_3)\text{—}N(CH_3)\text{—}CH_3$

(c) coniine, the hemlock poison drunk by Socrates

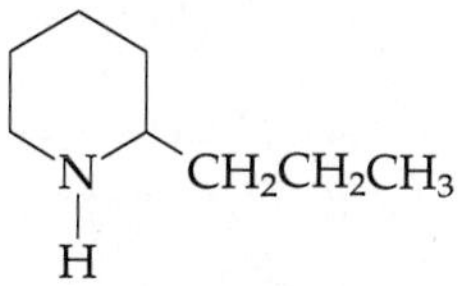

N–H ring with $CH_2CH_2CH_3$

14.6 Classify the nitrogen-containing functional group in each of the following structures.

(a) pantothenic acid, vitamin B_5

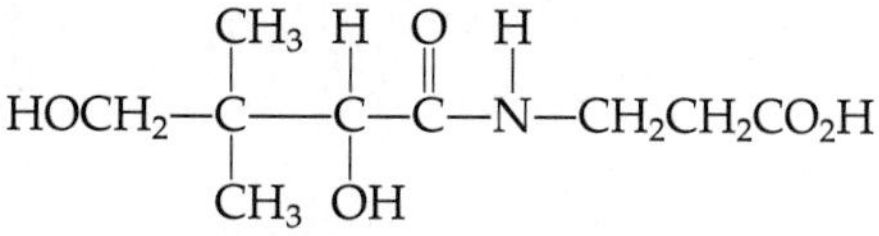

$$HOCH_2-C(CH_3)_2-CH(OH)-C(=O)-NH-CH_2CH_2CO_2H$$

(b) DEET, a insect repellent

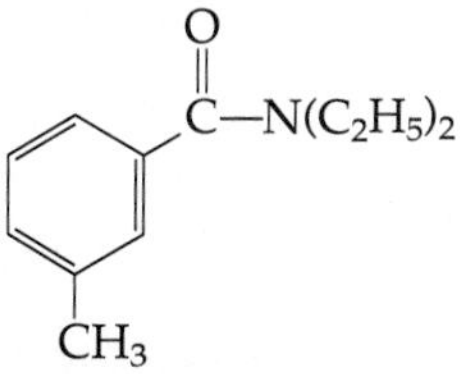

$C(=O)-N(C_2H_5)_2$ on benzene ring with CH_3

(c) phencyclidine, a hallucinogen

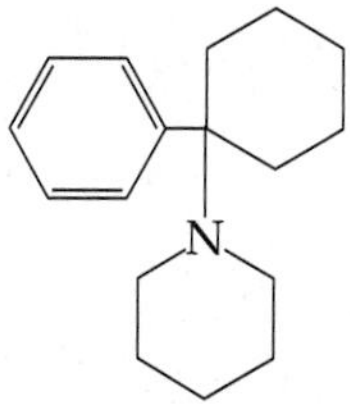

N

14.7 Classify the nitrogen-containing functional groups in encainide, an antiarrhythmic drug.

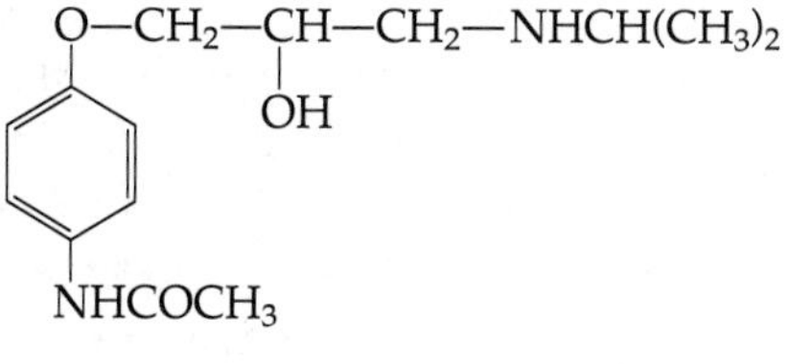

$$O-CH_2-CH(OH)-CH_2-NHCH(CH_3)_2$$

NHCOCH$_3$

14.8 Classify the nitrogen-containing functional groups in practolol, an antihypertensive drug.

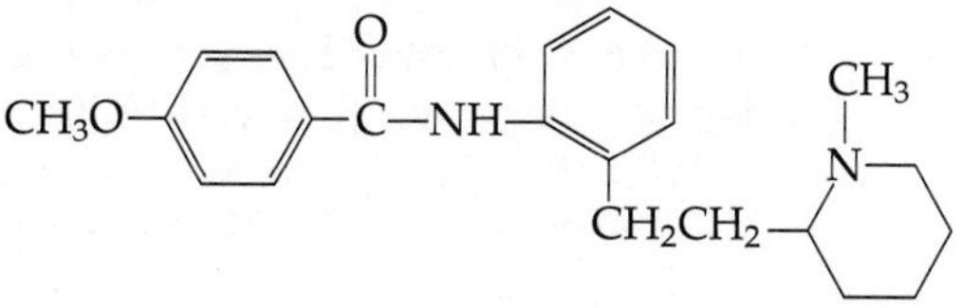

$CH_3O-C_6H_4-C(=O)-NH-C_6H_4-CH_2CH_2-$ (N-CH_3 piperidine)

Nomenclature

14.9 Give the IUPAC name for each of the following compounds.

(a) $CH_3CH_2\overset{NH_2}{\overset{|}{C}}HCH_2CH_2CH_3$ (b) $CH_3CH_2CH_2CH_2{-}\overset{CH_3}{\overset{|}{N}}{-}CH_3$

(c) $CH_3\overset{CH_3}{\overset{|}{C}}HCH_2\overset{NH_2}{\overset{|}{C}}HCH_3$

14.10 Give the IUPAC name for each of the following compounds.

(a) cyclohexane ring$-CH_2CH_2NH_2$ (b) cyclohexene ring$-NH_2$

(c) cycloheptane ring$-N(CH_3)_2$ (d) cyclohexane ring$-\underset{NH_2}{\underset{|}{C}}HCH_3$

14.11 Give the IUPAC name of each of the following compounds.

(a) $CH_3CH_2\overset{O}{\overset{\|}{C}}{-}NH_2$ (b) $CH_3CH_2\overset{O}{\overset{\|}{C}}{-}NHCH_2CH_3$

(c) $CH_3CH_2\underset{CH_3}{\underset{|}{C}}H\overset{O}{\overset{\|}{C}}{-}NH_2$ (d) $CH_3CH_2CH_2\overset{O}{\overset{\|}{C}}{-}N(CH_3)_2$

14.12 Give the IUPAC name for each of the following compounds.

(a) cyclobutane ring$-\overset{H}{\overset{|}{N}}{-}\overset{O}{\overset{\|}{C}}{-}CH_3$ (b) $Cl-$benzene ring$-\overset{O}{\overset{\|}{C}}{-}NH_2$

(c) cyclopentane ring$-\overset{CH_3}{\overset{|}{N}}{-}\overset{O}{\overset{\|}{C}}{-}\underset{CH_3}{\underset{|}{C}}H{-}Cl$ (d) cyclononane ring$-\overset{O}{\overset{\|}{C}}{-}N(CH_3)_2$

14.13 An antidepressant drug is named *trans*-2-phenylcyclopropylamine. Draw its structure.

14.14 Tranexamic acid is a drug that aids blood clotting. Its IUPAC name is *trans*-4-(aminomethyl)cyclohexanecarboxylic acid. Draw its structure.

14.15 Draw the structure of each of the following compounds.
(a) 2-ethylpyrrole (b) 3-bromopyridine
(c) 2,5-dimethylpyrimidine (d) 2,6,8-trimethylpurine

14.16 Name each of the following compounds.

(a) 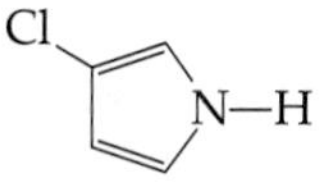 (b) 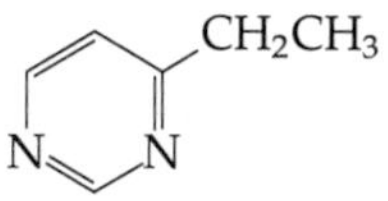 (c) 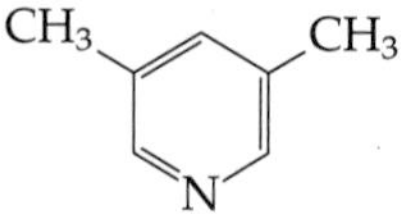 (d)

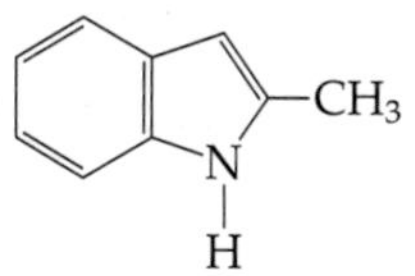

Isomers and Molecular Formulas of Amines

14.17 What is the general molecular formula for a saturated amine?
14.18 What is the general molecular formula for a saturated cyclic amine?
14.19 How many isomers are possible with the molecular formula C_2H_7N?
14.20 How many isomers are possible with the molecular formula C_3H_9N?
14.21 How many isomers are possible for primary amines with the molecular formula $C_4H_{11}N$?
14.22 How many isomers are possible for tertiary amines with the molecular formula $C_5H_{13}N$?

Properties of Amines

14.23 The boiling points of the isomeric compounds propylamine and trimethylamine are 49 and 3.5 °C, respectively. Explain this large difference.
14.24 The boiling point of 1,2-diaminoethane is 116 °C. Explain why this compound boils at a much higher temperature than propylamine (49 °C).
14.25 Based on the physical properties of amines, explain why lemon juice, which contains citric acid, is put on fish dishes.
14.26 Ephedrine, which is used in some cold and allergy medications, melts at 79 °C and has an unpleasant odor. It is sold in the form of the hydrochloride salt, which melts at 217 °C and has no odor. Explain the difference in the melting points and odor.

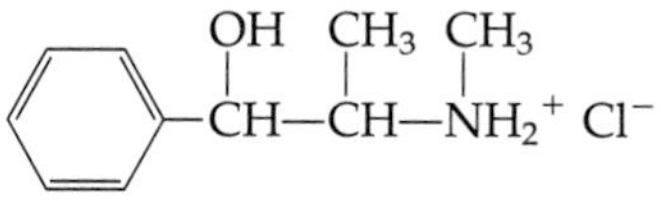

Basicity of Amines

14.27 The pK_b values for cyclohexylamine and triethylamine are 3.34 and 2.99, respectively. Which compound is the stronger base?
14.28 The K_b values for dimethylamine and diethylamine are 4.7×10^{-4} and 3.1×10^{-4}, respectively. Which compound is the stronger base?
14.29 Estimate the K_b of each of the following.

(a) 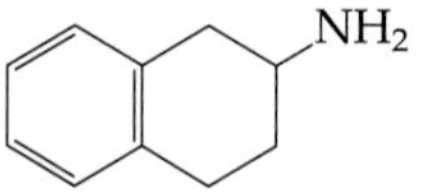 (b)

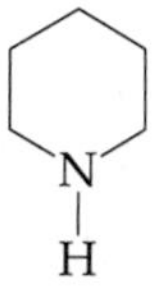

(c) 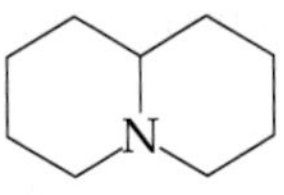 (d) 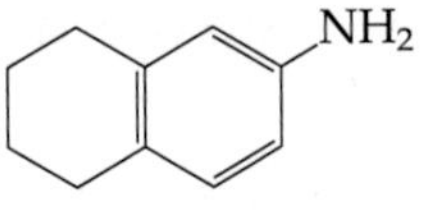 NH_2

14.30 Estimate the K_b of each of the following.

(a) 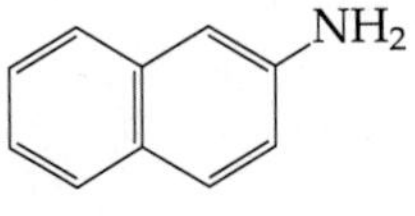 NH_2 (b) 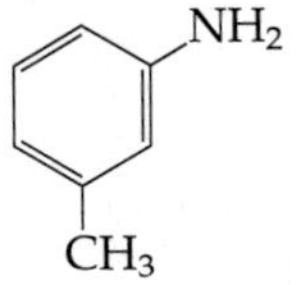 NH_2 CH_3

(c) 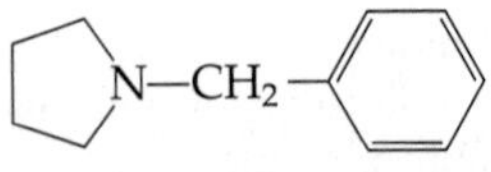 $N-CH_2-$ (d) 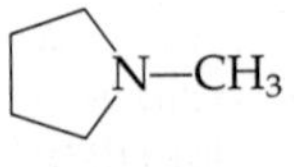 $N-CH_3$

14.31 Explain why the pK_b of aniline (9.4) is different than the pK_b of *p*-nitroaniline (13.0).

14.32 Explain the difference in the pK_a of the conjugate acids of the following bases.

$N{\equiv}C-CH_2-CH_2-NH_2$ (7.8) $\quad N{\equiv}C-CH_2-NH_2$ (5.3)

14.33 Explain the difference in the pK_a of the conjugate acids of the following bases.

$CH_3(CH_2)_3NH_2$ (10.6) $\quad CH_3O(CH_2)_3NH_2$ (9.9)

14.34 The pK_b of aniline is 9.4. Would the pK_b of p-methylaniline be greater or smaller than 9.4?

14.35 Physostigmine is used in 0.1–1.0% solutions to decrease the intraocular pressure in treatment of glaucoma. Rank the three nitrogen atoms in the molecule in order of increasing basicity.

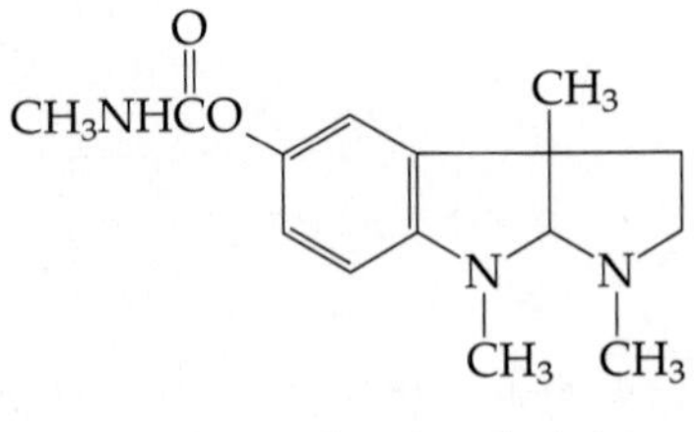

O CH_3NHCO CH_3 N N CH_3 CH_3

14.36 Nubucaine is a local anesthetic that is administered as the salt formed by reaction with hydrochloric acid. Explain why the salt is used. Which nitrogen atom is protonated?

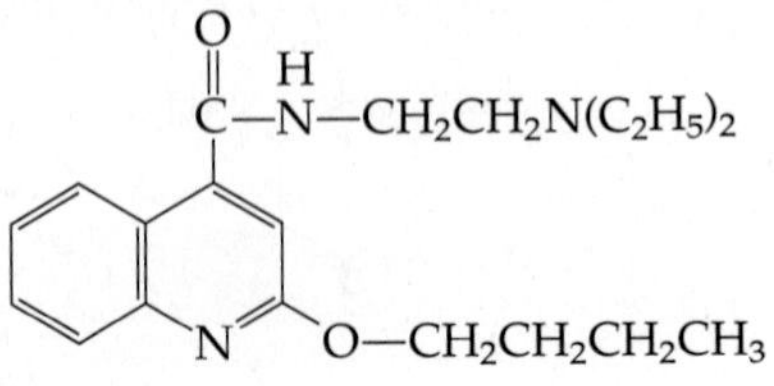

O H $C-N-CH_2CH_2N(C_2H_5)_2$ N $O-CH_2CH_2CH_2CH_3$

14.37 Which of the following compounds is the stronger base?

N   N—H

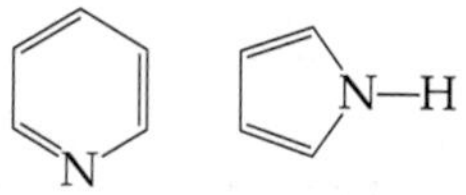

14.38 Which of the following compounds is the stronger base?

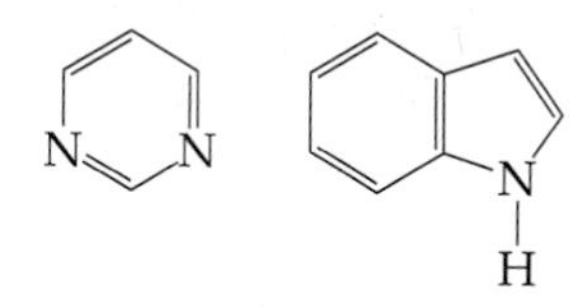

Reactions of Amines

14.39 A compound $C_4H_{11}N$ reacts with nitrous acid to yield a yellow oil. What structures are possible for $C_4H_{11}N$?

14.40 A compound $C_4H_{11}N$ reacts with nitrous acid to yield nitrogen gas. What structures are possible for $C_4H_{11}N$?

14.41 Draw the structure of the compound formed when benzylmethylamine reacts with each of the following reagents.
(a) excess methyl iodide (b) acetyl chloride (c) hydrogen iodide

14.42 Draw the structure of the compound formed when piperidine reacts with each of the following reagents.
(a) allyl bromide (b) benzoyl chloride (c) acetic anhydride

Synthesis of Amines

14.43 Write the structure of the product of each of the following reactions.

(a) $CH_3—\overset{\overset{O}{\|}}{C}—CH_2—CH_3 + CH_3—CH_2—NH_2 \xrightarrow{NaBH_3CN}$

(b) 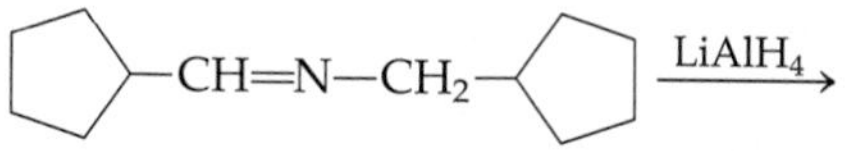 $C_5H_9—CH{=}N—CH_2—C_5H_9 \xrightarrow{LiAlH_4}$ (cyclopentyl rings)

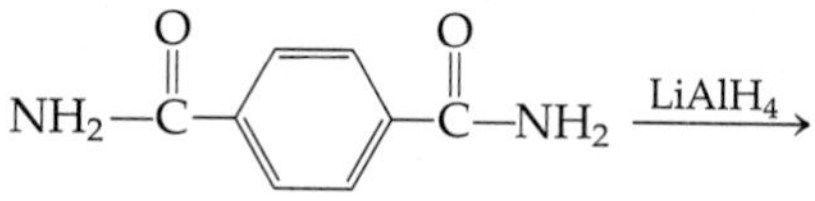

(c) $NH_2—\overset{\overset{O}{\|}}{C}—C_6H_4—\overset{\overset{O}{\|}}{C}—NH_2 \xrightarrow{LiAlH_4}$

14.44 Write the structure of the product of each of the following reactions.

(a) $CH_3—\overset{\overset{NCH_3}{\|}}{C}—CH_2—CH_2—CH_3 \xrightarrow{LiAlH_4}$

(b) 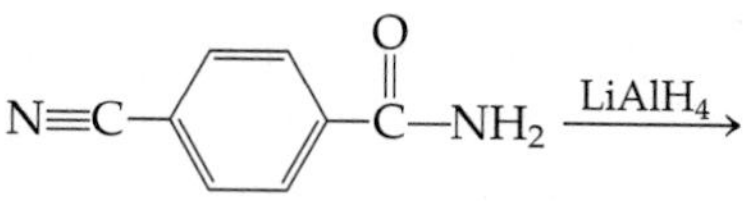 $N{\equiv}C—C_6H_4—\overset{\overset{O}{\|}}{C}—NH_2 \xrightarrow{LiAlH_4}$

(c) $CH_3—\overset{\overset{O}{\|}}{C}—CH_2—CH_2—CH_3 + CH_3—NH_2 \xrightarrow{Ni/H_2}$

14.45 Write the structure of the final product of each of the following sequences of reactions.

(a) $CH_3(CH_2)_4CO_2H \xrightarrow{SOCl_2} \xrightarrow{NH_3} \xrightarrow{LiAlH_4}$

(b) cyclopropyl$-CH_2-CH_2-Br \xrightarrow{CN^-} \xrightarrow{LiAlH_4}$

(c) $CH_3(CH_2)_3CO_2CH_3 \xrightarrow{LiAlH_4} \xrightarrow{PBr_3} \xrightarrow{NH_3\ (excess)}$

14.46 Write the structure of the final product of each of the following sequences of reactions.

(a) $CH_3-CH(OH)-CH_2-CH_2-CH_3 \xrightarrow{PCC} \xrightarrow[CH_3-NH_2]{Ni/H_2}$

(b) cyclopentyl$-CH_2-CO_2H \xrightarrow{SOCl_2} \xrightarrow{NH_3} \xrightarrow{LiAlH_4}$

(c) $HO-CH_2(CH_2)_4CH_2-OH \xrightarrow[(excess)]{HBr} \xrightarrow[(excess)]{CN^-} \xrightarrow[(excess)]{LiAlH_4}$

14.47 Outline the steps required to convert benzoic acid into *N*-ethylbenzylamine.

14.48 Outline the steps required to convert benzyl chloride into 2-phenylethanamine.

14.49 Outline the steps required to convert 1,4-dibromobutane into 1,6-diaminohexane.

14.50 Propose a synthesis of the following compound, which involves a nucleophilic substitution reaction, starting from methylamine.

$$CH_3-NH-CH_2-CH_2-OH$$

14.51 Propose a synthesis of hexafluorenium bromide, a neuromuscular-blocking agent, using 1,6-dibromohexane.

$$\text{(fluorenyl)}-N^+(CH_3)_2-CH_2CH_2CH_2CH_2CH_2CH_2-{}^+N(CH_3)_2-\text{(fluorenyl)}\quad 2Br^-$$

14.52 Propose a synthesis of benzocaine, a local anesthetic, starting from toluene.

$$H_2N-C_6H_4-C(=O)-O-CH_2-CH_3$$

Reactions of Amides

14.53 Write the products of each of the following reactions.

(a) 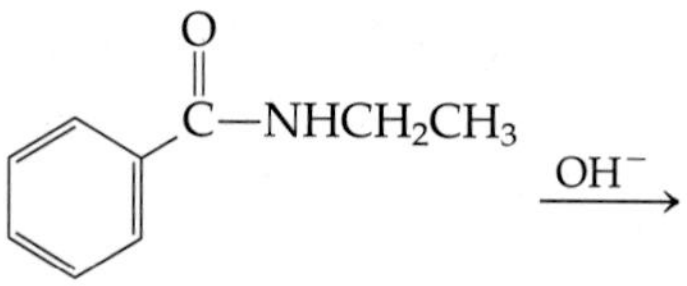 C_6H_5–$\overset{O}{\overset{\|}{C}}$–$NHCH_2CH_3$ $\xrightarrow{OH^-}$

(b) $CH_3CH_2\overset{O}{\overset{\|}{C}}$–$N(CH_3)_2$ $\xrightarrow{HCl}$

(c) 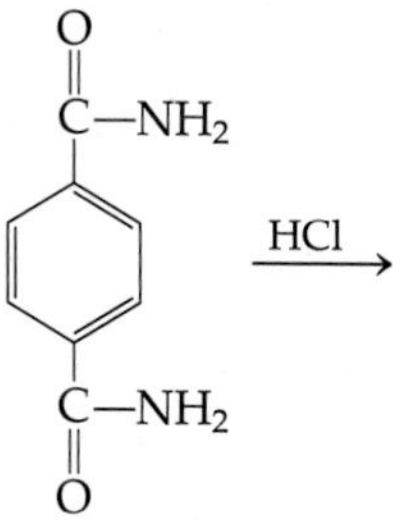 H_2N–$\overset{O}{\overset{\|}{C}}$–$C_6H_4$–$\overset{O}{\overset{\|}{C}}$–$NH_2$ (para) $\xrightarrow{HCl}$

14.54 Write the products of each of the following reactions.

(a) $CH_3CH_2\overset{O}{\overset{\|}{C}}$–$NHCH_3 + H_3O^+ \longrightarrow$

(b) 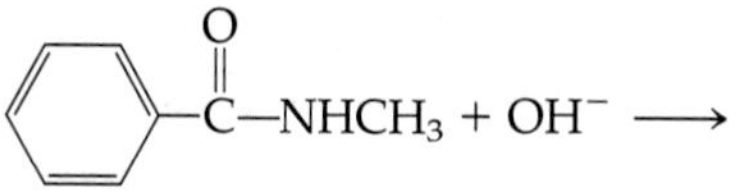 C_6H_5–$\overset{O}{\overset{\|}{C}}$–$NHCH_3 + OH^- \longrightarrow$

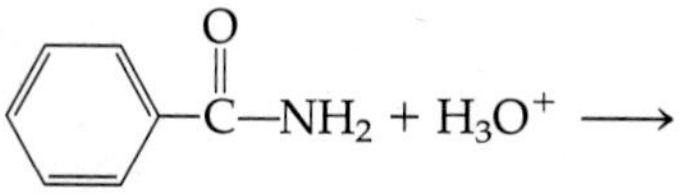

(c) C_6H_5–$\overset{O}{\overset{\|}{C}}$–$NH_2 + H_3O^+ \longrightarrow$

14.55 Write the product of reduction of each compound in 14.11 by lithium aluminum hydride.

14.56 Write the product of reduction of the following lactams by lithium aluminum hydride.

(a) 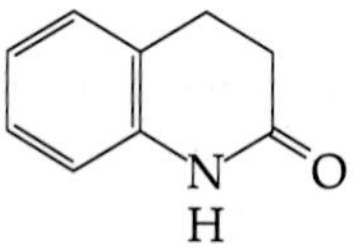 [3,4-dihydro-2(1H)-quinolinone: benzene ring fused to ring containing N–H and C=O] (b) [N-methyl-2-piperidinone: six-membered ring with C=O and N–CH_3] (c) [N-methylisoindolinone: benzene ring fused to five-membered ring with C=O and N–CH_3]

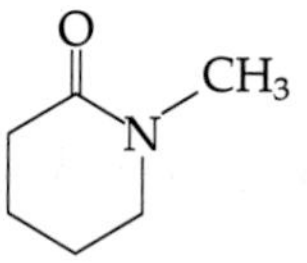

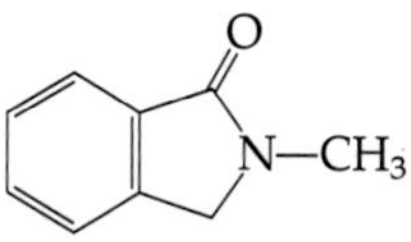

Synthesis of Amides

14.57 Write the structure of the product of each of the following reactions.

(a) $CH_3(CH_2)_3\overset{O}{\overset{\|}{C}}$–$Cl + CH_3NH_2 \longrightarrow$

(b) 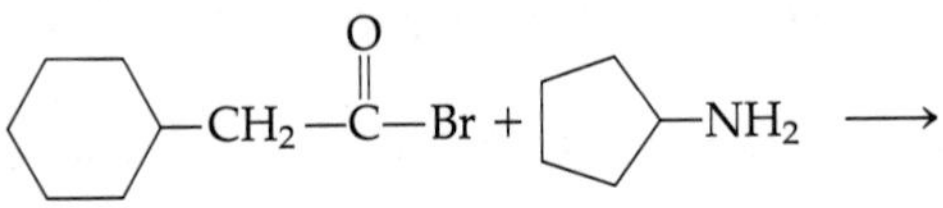 cyclohexyl—CH_2—C(=O)—Br + cyclopentyl—NH_2 ⟶

(c) cyclobutyl—CH_2—C(=O)—O—CH_3 + NH_3 ⟶

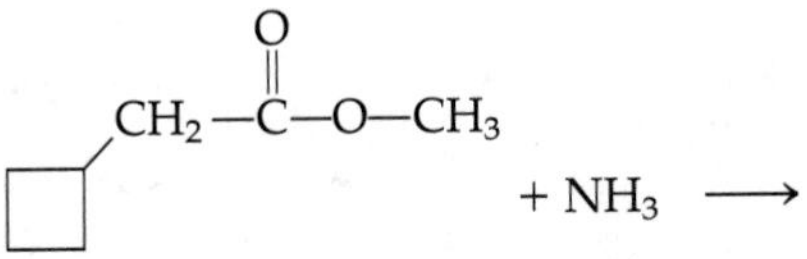

14.58 Write the structure of the product of each of the following reactions.

(a) $CH_3(CH_2)_3$C(=O)—O—CH_2CH_3 + $CH_3CH_2NH_2$ ⟶

(b) 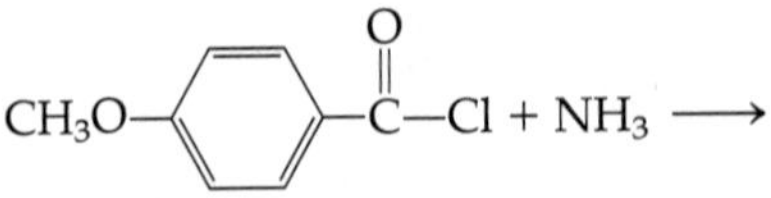 CH_3O—C_6H_4—C(=O)—Cl + NH_3 ⟶

(c) cyclopentyl—CH_2—C(=O)—S—CH_3 + CH_3NH_2 ⟶

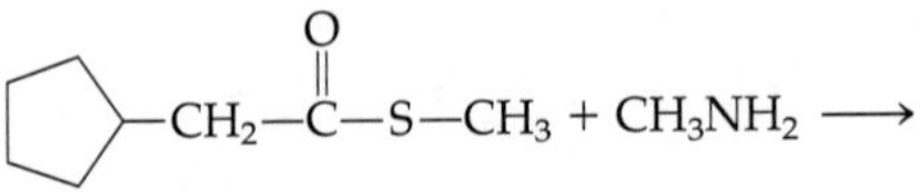

14.59 Select two reactants that could be used to prepare crotamiton, which is used to treat scabies.

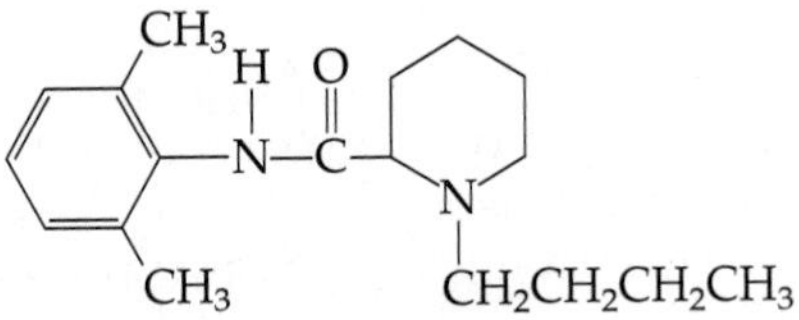

(2,6-dimethylphenyl)—N(H)—C(=O)—(piperidin-2-yl, N—$CH_2CH_2CH_2CH_3$)

14.60 Select two reactants that could be used to prepare bupivacaine, a local anesthetic.

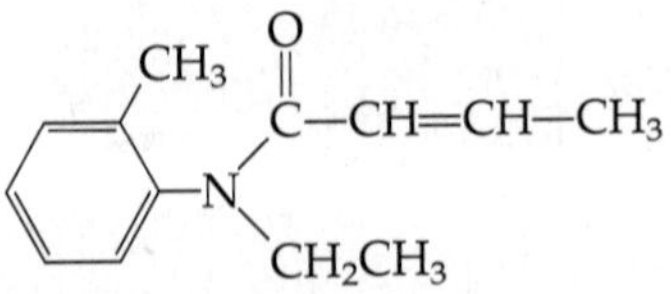

(2-methylphenyl)—N(CH_2CH_3)—C(=O)—CH=CH—CH_3

Polyamides

14.61 Nylon is resistant to dilute acids or bases, but polyesters are damaged by acids or bases. Explain this difference.

14.62 Draw a representation of the condensation polymer formed by adipic acid and 1,3-diaminopropane.

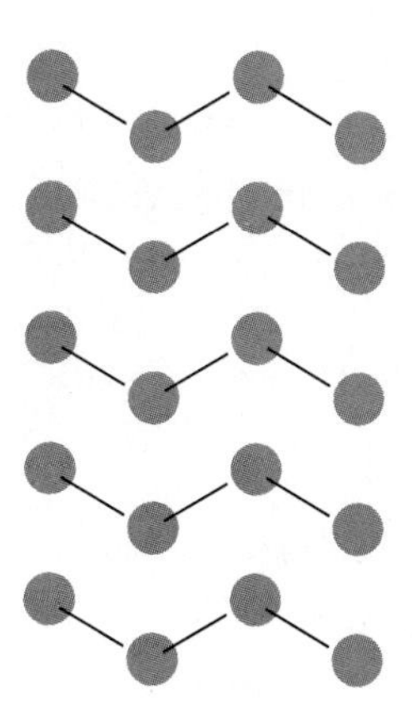

15

AMINO ACIDS AND PROTEINS

15.1 PROTEINS ARE BIOPOLYMERS

In Chapter 11 we learned how monosaccharides form polymers by linking one unit to another by acetal bonds. The resulting polysaccharides are used for energy storage in animals and maintaining cellular structure in plants. This chapter considers proteins—a second class of natural polymers contained in all cells.

The name *protein* is derived from the Greek word *proteios,* meaning preeminent or holding first place. Proteins have an extraordinary range of functions. Proteins called enzymes catalyze nearly all of the chemical reactions of the cell. Proteins are required for the transport of most substances across cell membranes. They are the major structural substances of skin, blood, muscle, hair, and other tissues of the body. Proteins in the immune system, called antibodies, resist the effects of foreign substances that enter the body.

Proteins and related structures called polypeptides are polymers of α-amino acids. Proteins are polymers of 50 or more α-amino acids; some proteins contain more than 8000 amino acid units. Polypeptides are smaller molecules that contain fewer than about 50 amino acids. Some important hormonal polypeptides with physiological functions such as pain relief and control of blood pressure contain as few as nine amino acid units.

Amino acids contain both amino and carboxylic acid functional groups. In proteins and polypeptides they are linked by amide bonds, which are also called **peptide bonds.**

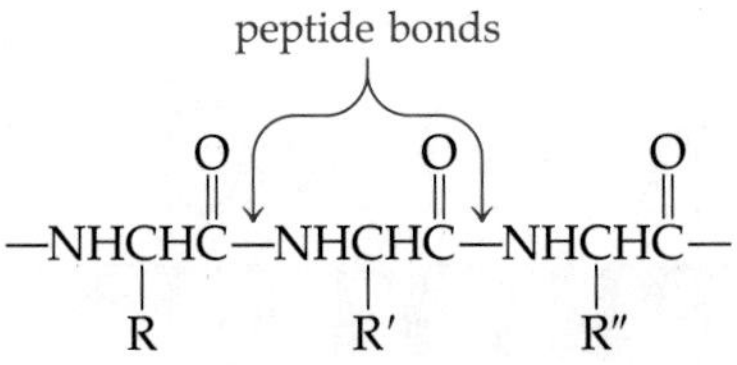

peptide bonds

$$-\mathrm{NH}\underset{\underset{\mathrm{R}}{|}}{\mathrm{CH}}\overset{\overset{\mathrm{O}}{\|}}{\mathrm{C}}-\mathrm{NH}\underset{\underset{\mathrm{R'}}{|}}{\mathrm{CH}}\overset{\overset{\mathrm{O}}{\|}}{\mathrm{C}}-\mathrm{NH}\underset{\underset{\mathrm{R''}}{|}}{\mathrm{CH}}\overset{\overset{\mathrm{O}}{\|}}{\mathrm{C}}-$$

This chapter begins by describing the structure and properties of the 20 amino acids isolated from proteins. Then, we will consider the structure and properties of polypeptides and proteins. We will also describe both the general method of synthesizing polypeptides and proteins as well as analytical methods to determine their structure.

15.2 AMINO ACIDS

Amino acids contain both an amino group and a carboxylic acid group. About 250 have been found in natural sources; however, only about 20 of them occur in large amounts in protein. The amino acids of proteins in all cells are α-amino acids. They have an amino group bonded to the α-carbon atom of a carboxylic acid. The Fischer projection formula for an α-amino acid is

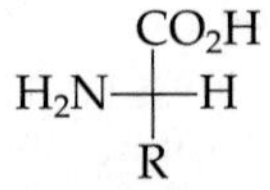

$$\begin{array}{c} CO_2H \\ H_2N-\!\!\!+\!\!\!-H \\ R \end{array}$$

an α-amino acid

In this structure, the R group is called the side chain. There are 20 different R groups in amino acids isolated from proteins. Of these 20 α-amino acids, 19 are chiral; they have the L configuration. The one that isn't chiral is glycine: its R group is hydrogen. Amino acids are either synthesized by cells when they are needed for protein synthesis or obtained from nutritional sources.

Classification of Amino Acids

The amino acids in proteins are primary amines except for proline, which is a secondary amine (Table 15.1). Three-letter abbreviations of the amino acids are used as a shorthand to describe protein structure. (An alternate one-letter shorthand method exists, but will not be used in this text.)

The amino acids are classified by their side-chain R groups as neutral, basic, and acidic. **Neutral amino acids** contain one amino group and one carboxyl group and, as we shall see below, have no net charge at physiological pH. The neutral amino acids are further divided according to the polarity

TABLE 15.1 The 20 Amino Acids Commonly Found in Proteins

Nonpolar R Groups

Glycine (Gly) $\mathrm{H{-}CH(NH_2){-}COOH}$

Alanine (Ala) $\mathrm{CH_3{-}CH(NH_2){-}COOH}$

Valine (Val) $\mathrm{CH_3{-}CH(CH_3){-}CH(NH_2){-}COOH}$

Leucine (Leu) $\mathrm{CH_3{-}CH(CH_3){-}CH_2{-}CH(NH_2){-}COOH}$

Isoleucine (Ile) $\mathrm{CH_3{-}CH_2{-}CH(CH_3){-}CH(NH_2){-}COOH}$

Proline (Pro)

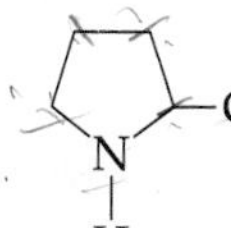

Phenylalanine (Phe)

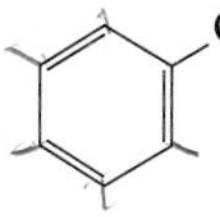

$\mathrm{CH_2{-}CH(NH_2){-}COOH}$

Methionine (Met) $\mathrm{CH_3{-}S{-}CH_2CH_2{-}CH(NH_2){-}COOH}$

Polar But Neutral R Groups

Serine (Ser) $\mathrm{HO{-}CH_2{-}CH(NH_2){-}COOH}$

Threonine (Thr) $\mathrm{CH_3{-}CH(OH){-}CH(NH_2){-}COOH}$

Cysteine (Cys) $\mathrm{HS{-}CH_2{-}CH(NH_2){-}COOH}$

Tyrosine (Tyr)

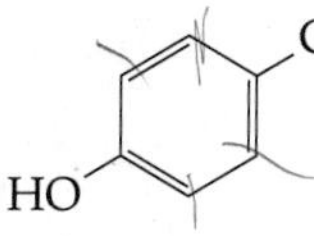

$\mathrm{CH_2{-}CH(NH_2){-}COOH}$

Asparagine (Asn) $\mathrm{NH_2{-}C({=}O){-}CH_2{-}CH(NH_2){-}COOH}$

Glutamine (Gln) $\mathrm{NH_2{-}C({=}O){-}CH_2CH_2{-}CH(NH_2){-}COOH}$

Tryptophan (Trp)

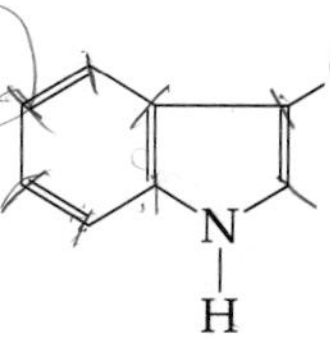

$\mathrm{CH_2{-}CH(NH_2){-}COOH}$

Acidic R Groups

Glutamic acid (Glu) $\mathrm{HO{-}C({=}O){-}CH_2CH_2{-}CH(NH_2){-}COOH}$

Aspartic acid (Asp) $\mathrm{HO{-}C({=}O){-}CH_2{-}CH(NH_2){-}COOH}$

Basic R Groups

Lysine (Lys) $\mathrm{NH_2{-}CH_2CH_2CH_2CH_2{-}CH(NH_2){-}COOH}$

Arginine (Arg) $\mathrm{NH_2{-}C({=}NH){-}NH{-}CH_2CH_2CH_2{-}CH(NH_2){-}COOH}$

Histidine (His)

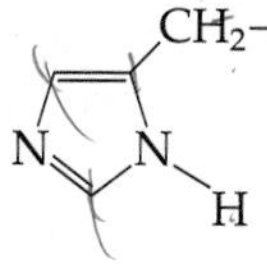

$\mathrm{CH_2{-}CH(NH_2){-}COOH}$

of the R group. Three of the neutral amino acids—serine, threonine, and tyrosine—have hydroxyl groups. Phenylalanine, tyrosine, and tryptophan contain aromatic rings. Two of the neutral amino acids, cysteine and methionine, contain a sulfur atom. The remaining neutral amino acids have hydrocarbon side chains.

Three **basic amino acids**—lysine, arginine, and histidine—have basic side chains; that is, they ionize at pH 7 to give a conjugate acid and hydroxide ion. Two **acidic amino acids**—aspartic acid and glutamic acid—have carboxylic acid side chains. The side chains of these amino acids exist predominantly as their conjugate bases ($—CO_2^-$) at pH 7. The acidic amino acids also have close relatives that exist as neutral amides, asparagine and glutamine.

Amino acids can also be classified by the tendency of their side chains to interact favorably or unfavorably with water. Those amino acids with polar side chains are said to be **hydrophilic,** that is, water-loving. Those whose side chains are nonpolar are said to be **hydrophobic.** Hydrophobic amino acids have alkyl or aromatic groups that do not hydrogen bond to water.

15.3 ACID–BASE PROPERTIES OF AMINO ACIDS

Until now, we have represented the structures of α-amino acids as uncharged molecules. However, the properties of amino acids resemble those of salts rather than uncharged molecules. Consider ethylamine, acetic acid, and glycine. Ethylamine is a gas and acetic acid is a liquid at room temperature. In contrast, glycine is a solid.

	$CH_3CH_2NH_2$	CH_3CO_2H	$NH_2CH_2CO_2H$
mp	−84 °C	16 °C	232 °C

Amino acids have low solubilities in organic solvents, but are moderately soluble in water, unlike most organic compounds.

Ionic Forms of Amino Acids

When an amino acid dissolves in water at pH 7, its carboxyl group ionizes to give a conjugate base, and the α-amino group reacts to give a conjugate acid. Thus the entire molecule exists as a dipolar ion sometimes called a **zwitterion**. The dipolar ion is amphoteric; that is, it has the properties of both an acid and a base.

$$
\begin{array}{c}
\quad\quad\quad CO_2^- \\
\quad\quad\quad | \\
{}^+NH_3—C—H \\
\quad\quad\quad | \\
\quad\quad\quad R
\end{array}
$$

structure of a dipolar ion (zwitterion)

When an amino acid dissolves in basic solution, the carboxyl group exists as a carboxylate anion and the α-amino group exists in its uncharged basic form. This species is the conjugate base of the original amino acid. It has a negative charge because the carboxylate ion has a −1 charge and the amino group is uncharged.

When an amino acid dissolves in acidic solution, the carboxyl group does not ionize, but the α-amino group is protonated. This species is the conjugate acid of the original amino acid. Because the carboxylate group and the α-amino group are both protonated in acidic solution, the conjugate acid has a net charge of +1.

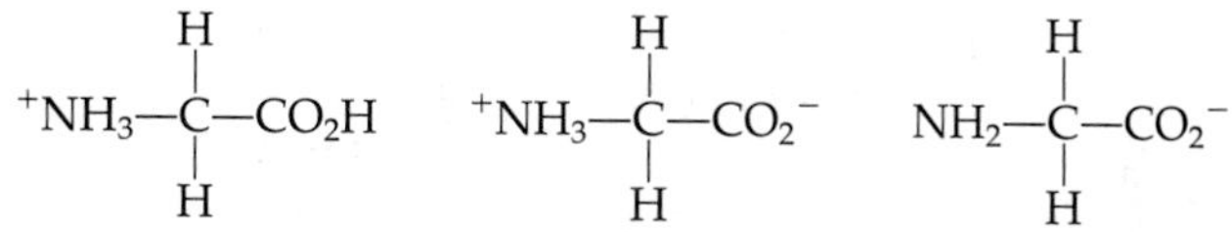

$$^{+}NH_3—\underset{H}{\overset{H}{C}}—CO_2H \qquad ^{+}NH_3—\underset{H}{\overset{H}{C}}—CO_2^{-} \qquad NH_2—\underset{H}{\overset{H}{C}}—CO_2^{-}$$

conjugate acid — zwitterion — conjugate base

EXAMPLE 15.1 Write the structure of the dipolar ion and the conjugate base of alanine.

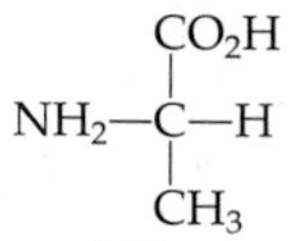

$$NH_2—\underset{CH_3}{\overset{CO_2H}{C}}—H$$

alanine

Solution The dipolar ion is written by removing a proton from the carboxyl group and transferring it to the nitrogen atom. The conjugate base is written by removing a proton from the ammonium ion site of the dipolar ion. The nitrogen atom becomes neutral, whereas the carboxylate ion retains a negative charge.

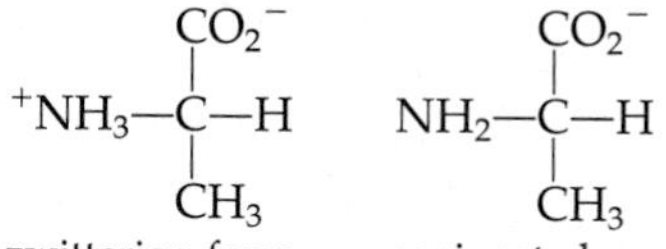

$$^{+}NH_3—\underset{CH_3}{\overset{CO_2^{-}}{C}}—H \qquad NH_2—\underset{CH_3}{\overset{CO_2^{-}}{C}}—H$$

zwitterion form — conjugate base

pK_a Values of α-Amino Acids

The pK_a values of the carboxyl and α-amino groups of amino acids depend on the structure of the amino acid. The pK_a values of the carboxyl groups of amino acids range from 1.8 for histidine to 2.6 for phenylalanine. The pK_a values of the α-ammonium groups range from 8.8 for asparagine to 10.8 for cysteine (Table 15.2).

TABLE 15.2 pK_a Values of Some Amino Acids at 25 °C

Amino acid	pK_a (COOH)	pK_a (NH_3^+)	pK_a (side chain)
glycine	2.4	9.8	
alanine	2.4	9.9	
serine	2.2	9.2	
phenylalanine	2.6	9.2	
cysteine	1.9	10.8	
asparagine	2.0	8.8	
aspartic acid	2.0	10.0	3.9
glutamic acid	2.2	9.7	4.3
histidine	1.8	9.2	6.0
lysine	2.2	9.0	10.5
arginine	1.8	9.0	12.5

$$^{+}NH_3-CH_2-CO_2H + H_2O \rightleftharpoons {}^{+}NH_3-CH_2-CO_2^- + H_3O^+ \qquad K_a = 5 \times 10^{-3},\ pK_a = 2.4$$

$$^{+}NH_3-CH_2-CO_2^- + H_2O \rightleftharpoons NH_2-CH_2-CO_2^- + H_3O^+ \qquad K_a = 1.6 \times 10^{-10},\ pK_a = 9.8$$

When an amino acid dissolves in an aqueous solution, several species exist to some extent. When the pH of the solution equals the pK_a of an ionizing group, the concentrations of the acid form and its conjugate base are equal. For example, the pK_a of the $-CO_2H$ of glycine is 2.4 and the pK_a of the $-NH_3^+$ of glycine is 9.8. Thus, when the pH of a dilute solution is 2.4, the concentrations of the species with the carboxylate ion and the carboxylic acid group are equal, while the α-amino group exists almost entirely as an $-NH_3^+$ group. Thus, at pH 2.4, the concentrations of the dipolar ion and the conjugate acid form of glycine are equal. When the pH of the glycine solution is increased to 9.8, the concentrations of the conjugate base and the dipolar ion are equal. At pH values between 2.4 and 9.8, the dipolar ion is the major ionic form of the amino acid in solution.

EXAMPLE 15.2 In what ionic form does serine exist in 0.1 M HCl?

Solution A 0.1 M solution of HCl has a pH of 1.0. The pK_a values of serine are 2.2 and 9.2 (Table 15.2). Thus, at pH = 1, serine will exist as the conjugate acid.

$$
\begin{array}{c}
\quad CO_2H \\
\quad | \\
{}^{+}NH_3\text{—}C\text{—}H \\
\quad | \\
\quad CH_2OH
\end{array}
$$

15.4 ISOIONIC POINT

The pH at which the concentration of the zwitterion is at a maximum is the **isoionic point,** abbreviated pH_i. When the pH equals the pH_i, an amino acid has no *net* charge. When the pH is greater than the pH_i, the net charge of the predominant ionic form of the amino acid is negative. When the pH is less than the pH_i, the net charge of the predominant ionic form of the amino acid is positive. The isoionic points of some amino acids are given in Table 15.3. The isoionic point of an amino acid equals one half the sum of the pK_a values of the carboxyl group and the α-amino group if it does not have an ionizing side chain. For example, the pK_a of the carboxyl group of alanine is 2.4 and the pK_a value of the α-ammonium ion is 9.9. The isoionic point of alanine is 6.1.

The isoionic points of acidic and basic amino acids are calculated as follows. The isoionic point of an acidic amino acid is one-half the sum of the pK_a of the α-CO_2H and the pK_a of the side chain —CO_2H. Similarly the isoionic point of a basic amino acid is one-half the sum of the pK_a of the α-NH_3^+ and the pK_a of the ionizing side chain. In general, acidic amino acids have pK_i values less than 7 and basic amino acids have pK_i values greater than 7.

Isoionic Points of Proteins

Because proteins are made of amino acids, a protein has an isoionic point that depends upon its amino acid composition. The isoionic points of a few proteins are listed in Table 15.3. At its isoionic point, a protein has no net charge, and its solubility is at a minimum. As a consequence, a protein tends to precipitate from solution at its isoionic pH. For example, casein, a protein in milk, has a net negative charge at pH = 6.3. Casein has many glutamic

TABLE 15.3 Isoionic Points of Amino Acids and Proteins

Amino acid	Isoionic point	Protein	Isoionic point
aspartic acid	2.8	pepsin (enzyme)	1.1
glutamic acid	3.2	casein (milk protein)	4.6
serine	5.7	egg albumin	4.7
valine	6.0	urease (enzyme)	5.0
alanine	6.1	insulin (hormone)	5.3
histidine	7.6	hemoglobin	6.8
lysine	9.7	ribonuclease (enzyme)	9.5
arginine	10.8	chymotrypsin (enzyme)	9.5

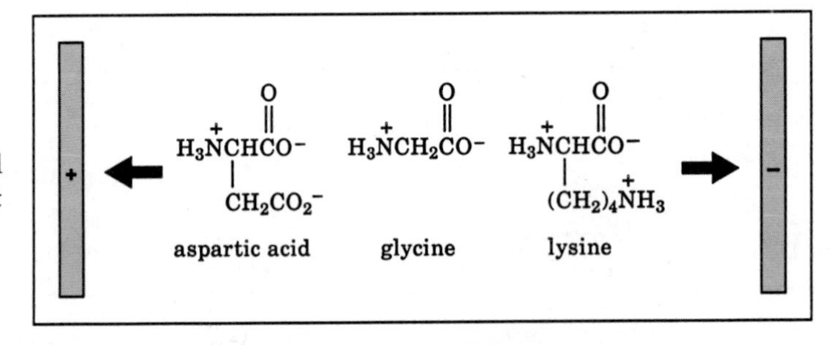

FIGURE 15.1
Electrophoresis of Amino Acids
At pH 6.0 glycine exists largely as the zwitterion and does not migrate toward either electrode. Aspartic acid is largely deprotonated and has a net negative charge at this pH; it migrates toward the positive electrode. Lysine is largely protonated and has a net positive charge at this pH; it migrates toward the negative electrode.

acid and aspartic acid residues. If milk is made more acidic, the glutamate and aspartate side chains of casein are protonated, and casein precipitates. Casein, which is used in making cheese, is obtained by adding an acid to milk or by adding bacteria that produce lactic acid.

Separation of Amino Acids and Proteins

Mixtures of amino acids can be separated and identified by a technique called **electrophoresis** (Figure 15.1). In this technique, a paper strip saturated with a buffer solution at a selected pH bridges two vessels containing the buffer. A sample of the amino acid mixture is placed at the center of the paper as a "spot," and an electric potential is applied between the two vessels. If the buffer pH equals the isoionic point of an amino acid, the dipolar ion predominates, and it does not migrate. An amino acid with a negative charge at that pH migrates toward the positive electrode, whereas an amino acid with a positive charge at that pH migrates toward the negative electrode. After a while, the original "spot" of the amino acid sample separates into two or more spots, each corresponding to an amino acid that was present in the original mixture.

Proteins can also be separated by electrophoresis. Electrophoretic separation of proteins is an important tool in clinical laboratories. Because proteins have different charges and molecular weights, they move at different rates in the electrophoresis apparatus. Electrophoresis is commonly used to analyze blood serum. For example, the identification of certain enzymes in the blood is used as a diagnostic tool for myocardial infarction.

15.5 PEPTIDES

When the α-amino group of one amino acid is linked to the carboxyl group of a second amino acid by an amide bond, the product is called a **peptide.** If the peptide contains two amino acid units, it is a **dipeptide;** a peptide that contains three amino acids is called a **tripeptide.** In general, a prefix, *di-*, *tri-*, etc., indicates the number of amino acids in a peptide. But a peptide that contains, say, 14 amino acids is more likely called a 14-peptide than a tetra-

Cholesterol and Lipoproteins

Lipoproteins are complex particles composed of several types of proteins and lipids, including triacylglycerols, cholesterol, and phospholipids. Lipoproteins transport lipids in human plasma and regulate the cholesterol level in the blood. Lipoproteins account for about 0.5–1.0% of blood serum. Lipoproteins are divided into three classes according to their densities. The three classes are: high density lipoproteins (HDL), low density lipoproteins (LDL), and very low density lipoproteins (VLDL). The density of a lipoprotein complex depends upon the lipid to protein ratio (see figure). Because proteins are more dense than lipids, increasing the protein component of a lipoprotein complex increases its density.

The densities of VLDLs range from 1.006 to 1.018 g/mL. VLDLs consist of about 95% lipid and 5% protein. LDLs have densities of 1.019–1.063 g/mL; they are about 75% lipid and 25% protein. HDLs have densities of 1.063–1.21 g/mL; they are about 50% lipid and 50% protein.

Plasma lipoproteins share a common structure. They have a hydrophobic core of triacylglycerols and cholesterol esters surrounded by a shell of phospholipids, cholesterol, and proteins.

The VLDLs are the principal carriers of triacyglycerols, whereas LDLs carry 80% of the blood serum cholesterol. HDLs carry the remaining cholesterol. The functions of LDLs and HDLs in cholesterol transport are quite different. Low density lipoproteins carry cholesterol to cells, where it is incorporated in cell membranes or used for the synthesis of other molecules. HLDs carry excess cholesterol away from cells to the liver for processing and excretion from the body. Individuals with high HDL levels have an efficient means of removing unneeded cholesterol from blood serum. This is an important function because the accumulation of cholesterol ester deposits in arteries results in atherosclerosis, that is, hardening of the arteries. If the HDL level is too low, excess cholesterol is deposited on the walls of the arteries. One consequence of this condition is coronary heart disease, which often occurs in men between 35–50 years old. So, high concentrations of HDLs tend to diminish the risk of heart disease.

The average concentration of HDLs is 45 mg/100 mL for men and 55 mg/100 mL for women. This difference in HDL concentration may partly explain why proportionately fewer women have heart attacks than men. The concentration of HDLs appears to increase if a person exercises. For example, the HDL concentration in male long-distance runners may be as high as 75 mg HDL/100 mL.

Composition of Lipoproteins
The density of lipoproteins increases as the percent of proteins increases: VLDL < LDL < HDL.

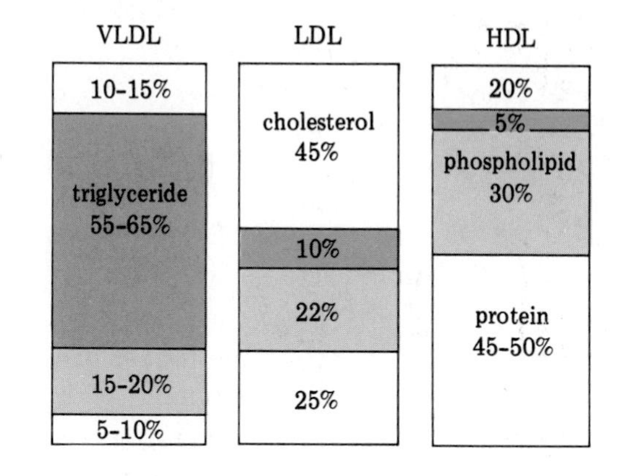

decapeptide. Peptides that contain only a "few" amino acids are called **oligopeptides.**

A peptide has two ends: the end with a free α-amino group is called the **N-terminal amino acid residue;** the end with the free carboxyl group is called the **C-terminal amino acid residue.** Peptides are named from the N-terminal amino acid to the C-terminal amino acid. Two examples of this nomenclature for isomeric dipeptides containing glycine and alanine are shown below.

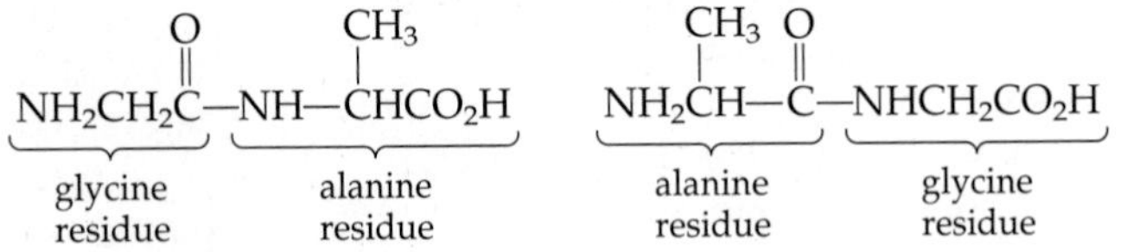

$$\underbrace{NH_2CH_2\overset{O}{\overset{\|}{C}}}_{\text{glycine residue}}\text{—}\underbrace{NH\text{—}\overset{CH_3}{\overset{|}{C}}HCO_2H}_{\text{alanine residue}}$$

glycylalanine (Gly-Ala)

$$\underbrace{NH_2\overset{CH_3}{\overset{|}{C}}H\text{—}\overset{O}{\overset{\|}{C}}}_{\text{alanine residue}}\text{—}\underbrace{NHCH_2CO_2H}_{\text{glycine residue}}$$

alanylglycine (Ala-Gly)

The number of isomeric peptides containing one molecule each of n different amino acids is equal to $n!$, where

$$n! = 1 \times 2 \times 3 \times \cdots \times (n - 1) \times n$$

Thus, there are six possible isomers of a tripeptide with three different amino acids. The isomeric tripeptides with the amino acids glycine, alanine, and valine are Gly-Ala-Val, Gly-Val-Ala, Val-Gly-Ala, Val-Ala-Gly, Ala-Gly-Val, and Ala-Val-Gly. For peptides containing one molecule each of 20 different amino acids there are 2,432,902,008,176,640,000 isomers! Polypeptides and proteins may actually contain two or more molecules of the same amino acid, and the above formula does not apply. However, the number of isomers is still astronomically large.

EXAMPLE 15.3 Identify the terminal amino acids of tuftsin, a tetrapeptide that stimulates phagocytosis and promotes the destruction of tumor cells. Write the amino acid sequence using three-letter abbreviations for the amino acids. Also write the complete name without abbreviations.

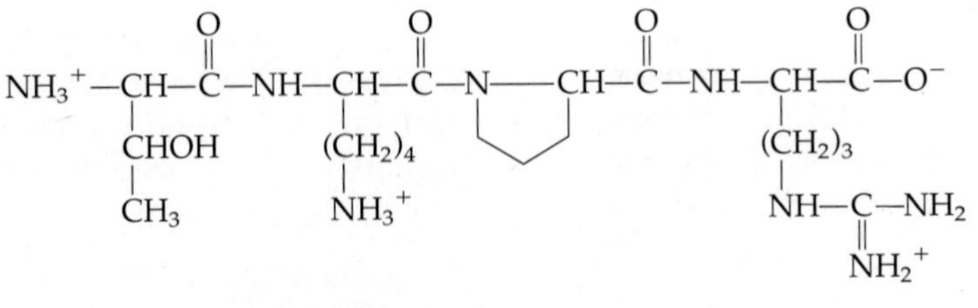

$$NH_3^+\text{—}\underset{\underset{CH_3}{|}}{\underset{CHOH}{|}}{CH}\text{—}\overset{O}{\overset{\|}{C}}\text{—}NH\text{—}\underset{\underset{NH_3^+}{|}}{\underset{(CH_2)_4}{|}}{CH}\text{—}\overset{O}{\overset{\|}{C}}\text{—}N\text{—}CH\text{(ring)}\text{—}\overset{O}{\overset{\|}{C}}\text{—}NH\text{—}\underset{\underset{NH\text{—}\underset{\underset{NH_2^+}{\|}}{C}\text{—}NH_2}{|}}{\underset{(CH_2)_3}{|}}{CH}\text{—}\overset{O}{\overset{\|}{C}}\text{—}O^-$$

Solution The residue on the left is the N-terminal amino acid threonine. The residue on the right is the C-terminal amino acid arginine. The internal amino acids are the basic amino acid lysine and the secondary amino acid proline. The abbreviated name is Thr-Lys-Pro-Arg. The complete name is threonyllysylprolylarginine.

Biological Functions of Peptides

Cells contain many relatively small peptides that have diverse functions. Some are hormones with physiological functions such as pain relief and control of blood pressure. These oligopeptides are produced and released in small amounts. They are rapidly metabolized, but their physiological action is necessary for only a short time. For example, the 14-peptide somatostatin, which inhibits the release of other hormones such as insulin, glucagon, and secretin, has a biological half-life of less than 4 minutes. Examples of some of these hormonal peptides are given in Table 15.4.

Enkephalins are peptides that bind specific receptor sites in the brain to reduce pain. The enkephalin receptor sites have a high affinity for opiates, including morphine, heroin, and other structurally similar substances. Hence, enkephalin receptors are commonly called *opiate receptors.* Opiates mimic the enkephalins that are normally present in the body to mitigate pain.

Peptides are produced in many tissues. For example, angiotensin II is made in the kidneys. It causes constriction of the blood vessels and thus increases blood pressure. Angiotensin II is the most potent vasoconstrictor

TABLE 15.4 Peptide Hormones

Hormone	Amino acid residues	Molecular weight	Function
tuftsin	4	501	stimulates phagocytosis
met-enkephalin	5	645	analgesic activity
angiotensin II	8	1031	affects blood pressure
oxytocin	8	986	affects uterine contractions
vasopressin	8	1029	an antidiuretic
bradykinin	9	1069	produces pain
somatostatin	14	1876	inhibits release of other hormones
gastrin	17	2110	promotes pepsin secretion
secretin	27	2876	stimulates pancreatic secretions
glucagon	29	3374	stimulates glucose production from glycogen
calcitonin	32	3415	decreases calcium level in blood
relaxin	48	5500	relaxation of pubic joints
insulin	51	5700	affects blood sugar level

known, and the production of excess angiotensin II is responsible for some forms of hypertension.

Oxytocin and vasopressin are important peptides. Oxytocin is formed by the pituitary gland. It causes the contraction of smooth muscle, such as that of the uterus. It is used to induce delivery or to increase the effectiveness of uterine contractions. Vasopressin is also produced in the pituitary gland. It is one of the hormones that regulate the excretion of water by the kidneys, and it affects blood pressure. The structures of oxytocin and vasopressin differ by only two amino acids. They are both cyclic peptides that result from a disulfide bond (—S—S—) between what would otherwise be the N-terminal amino acid cysteine and another cysteine five amino acid residues away. The C-terminal amino acid exists as an amide in both compounds.

```
S__________________S
|                  |
Cys-Tyr-Ile-Gln-Asn-Cys-Pro-Leu-Gly-NH2
```

Cys-Tyr-**Ile**-Gln-Asn-Cys-Pro-**Leu**-Gly-NH_2

oxytocin

```
S__________________S
|                  |
Cys-Tyr-Phe-Gln-Asn-Cys-Pro-Arg-Gly-NH2
```

Cys-Tyr-**Phe**-Gln-Asn-Cys-Pro-**Arg**-Gly-NH_2

vasopressin

The structural difference between oxytocin and vasopressin may seem small at first glance. But, in fact, the difference is enormous. When we compare oxytocin and vasopressin, we see that residue 3 in oxytocin is isoleucine and that residue 3 in vasopressin is phenylalanine. This is a relatively small change: both residues are nonpolar and about the same size. However, residue 8 in oxytocin is leucine—a nonpolar amino acid with a *sec*-butyl side chain—whereas residue 8 in vasopressin is arginine, an amino acid with a strongly basic side chain and a positive charge at pH 7. Because of this difference in charge, the receptor for oxytocin has a weak affinity for vasopressin, and the receptor for vasopressin has a very low affinity for oxytocin. These peptides bind different receptors and have different functions.

15.6 SYNTHESIS OF PEPTIDES

The synthesis of peptides and polypeptides is an important aspect of research in biochemistry and is a lucrative part of the biotechnology industry. A highly specialized set of reagents is required for this process.

We cannot simply react two amino acids under conditions that lead to the formation of amide bonds if we wish to synthesize a dipeptide. Two amino acids, such as alanine and glycine, yield a mixture of dipeptides. Each amino acid has two reactive ends. Therefore, each amino acid could randomly form bonds with its own kind to form Gly-Gly and Ala-Ala or to each other to give Gly-Ala and Ala-Gly. Also, the amino acids in the reaction mixture can continue to react in an uncontrolled manner with the dipeptide products to yield oligopeptides.

The synthesis of a dipeptide having a specific sequence requires modification of both amino acids. One amino acid is protected at its carboxyl group—by a reagent we will call P_C—leaving the amino group available for peptide bond formation. The second amino acid is protected at the amino group—by a reagent we will call P_N—leaving the carboxyl group available for peptide bond formation. Only one condensation reaction then is possible.

$$NH_2-\overset{R_1}{\overset{|}{C}H}-\overset{O}{\overset{\|}{C}}-OH \xrightarrow{P_C} NH_2-\overset{R_1}{\overset{|}{C}H}-\overset{O}{\overset{\|}{C}}-P_C$$

$$NH_2-\overset{R_2}{\overset{|}{C}H}-\overset{O}{\overset{\|}{C}}-OH \xrightarrow{P_N} P_N-NH-\overset{R_2}{\overset{|}{C}H}-\overset{O}{\overset{\|}{C}}-OH$$

$$P_N-NH\overset{R_2}{\overset{|}{C}H}\overset{O}{\overset{\|}{C}}-OH + NH_2-\overset{R_1}{\overset{|}{C}H}\overset{O}{\overset{\|}{C}}-P_C \longrightarrow P_N-NH\overset{R_2}{\overset{|}{C}H}\overset{O}{\overset{\|}{C}}-NH-\overset{R_1}{\overset{|}{C}H}\overset{O}{\overset{\|}{C}}-P_C$$

This method of peptide synthesis has several requirements.

1. The carboxyl group of one amino acid must be protected.
2. The amino group of the other amino acid must be protected.
3. A reagent must be chosen to form the amide bond.
4. Conditions must be chosen that selectively free one group so that the sequence can be repeated.

Protection of the Carboxyl Group

The carboxyl group is protected by converting it to a methyl or ethyl ester. Because esters are more reactive toward nucleophiles such as hydroxide ion than are amides, the C-terminus is easily "unprotected" at the end of the synthesis.

$$\underset{NH_2}{\underset{|}{RCH}}CO_2H + CH_3CH_2OH \xrightarrow{HCl} \underset{NH_2}{\underset{|}{RCH}}CO_2CH_2CH_3 + H_2O$$

Protection of the Amino Group

Several protecting groups have been developed to protect the amino terminus of an amino acid; the *t*-butoxycarbonyl (Boc) derivative is typical. Reaction of an amino acid with di-*tert*-butyl dicarbonate gives a Boc-amino acid.

$$\underset{\text{di-}tert\text{-butyl dicarbonate}}{(CH_3)_3CO\overset{O}{\overset{\|}{C}}-O-\overset{O}{\overset{\|}{C}}OC(CH_3)_3} + NH_2\underset{R}{\underset{|}{C}H}CO_2H \longrightarrow \underset{\text{a Boc-amino acid}}{(CH_3)_3CO-\overset{O}{\overset{\|}{C}}-NH\underset{R}{\underset{|}{C}H}CO_2H}$$

Note that the carbonyl group of the Boc derivative is bonded to both an oxygen atom and a nitrogen atom. This functional group is a carbamate; it is more easily hydrolyzed than amides and even esters. The Boc group can be removed with trifluoroacetic acid. Both the amide groups and the ester of a protected carboxyl group are unaffected by the reaction conditions. The byproducts of the reaction, CO_2 and 2-methylpropene, are gases.

$$(CH_3)_3CO-\overset{O}{\overset{\|}{C}}-NH\underset{R}{\underset{|}{C}}H-CO_2CH_3 \xrightarrow{CF_3CO_2H} (CH_3)_2C{=}CH_2 + CO_2 + NH_2\underset{R}{\underset{|}{C}}H-CO_2CH_3$$

Condensation of the Amine and Carboxyl Groups

The protecting groups of both the amino and the carboxyl groups are sensitive to acid and base. Therefore, the condensation reaction to produce a dipeptide bond must be carried out under neutral conditions. A special reagent—dicyclohexylcarbodiimide (DCCI)—causes condensation of two amino acids by removing the elements of water. The reaction has a very high yield, and no other functional groups on the amino acids are affected.

$$(CH_3)_3CO-\overset{O}{\overset{\|}{C}}-NH\underset{R}{\underset{|}{C}}H-\overset{O}{\overset{\|}{C}}-OH + NH_2-\underset{R'}{\underset{|}{C}}H-CO_2CH_3 \xrightarrow{DCCI}$$

$$(CH_3)_3CO-\overset{O}{\overset{\|}{C}}-NH\underset{R}{\underset{|}{C}}H\overset{O}{\overset{\|}{C}}NH\underset{R'}{\underset{|}{C}}H-CO_2CH_3$$

Formation of a Polypeptide

The dipeptide that is protected at both the carboxyl and amino groups is unprotected by hydrolysis of the Boc group at the N-terminal amino acid. The ester linkage of the carboxyl group is unaffected.

$$(CH_3)_3CO-\overset{O}{\overset{\|}{C}}NH\underset{R}{\underset{|}{C}}H\overset{O}{\overset{\|}{C}}-NH-\underset{R'}{\underset{|}{C}}H-CO_2CH_3 \xrightarrow{CF_3CO_2H} NH_2\underset{R}{\underset{|}{C}}H\overset{O}{\overset{\|}{C}}NH\underset{R'}{\underset{|}{C}}H-CO_2CH_3$$

This dipeptide can only react at the free amino group. Reaction with another Boc-amino acid and DCCI yields a tripeptide. Ultimately after the proper number of reaction sequences, the final polypeptide is liberated by hydrolysis of the methyl ester with base.

15.7 DETERMINATION OF AMINO ACID COMPOSITION IN PROTEINS

The first item of business when analyzing the structure of a protein is determination of its amino acid composition. The components of a polypeptide or protein are called **amino acid residues.** The amino acid composition of a polypeptide is determined in two steps. First, the protein is hydrolyzed. Then the hydrolysis products are separated by chromatography and identified. It is relatively easy to determine the amino acid composition of a peptide or protein because well-established techniques using automated instruments are available for that purpose.

A protein or peptide is hydrolyzed by heating it for 24 hours in 6 N HCl at 100 °C. Complete hydrolysis produces the constituent amino acids of the peptide or protein. For example, hydrolysis of the pentapeptide leucine enkephalin gives two molar equivalents of glycine and one each of leucine, tyrosine, and phenylalanine. The composition is represented as Gly_2,Leu,Tyr,Phe.

$$\text{leucine enkephalin} \longrightarrow 2\ \text{Gly} + \text{Leu} + \text{Tyr} + \text{Phe}$$

15.8 DETERMINATION OF AMINO ACID SEQUENCES IN PROTEINS

The linear sequence of amino acid residues in a peptide or protein is called the **primary structure.** This structure is "primary" for two reasons. First, the sequence of amino acid residues in a protein is specified by its gene. Second, all higher ordered structures of the protein automatically result from the primary structure.

The amino acid sequence is determined by a combination of methods that include

1. partial hydrolysis and fragment overlap analysis
2. selective enzymatic hydrolysis
3. end-group analysis

Partial Hydrolysis

When a peptide is heated with HCl for short time intervals, hydrolysis reactions yield oligopeptides having random sizes. For example, heating leucine enkephalin in HCl might yield the tripeptides Phe-Leu-Gly and Gly-Phe-Leu and the dipeptide Ala-Gly. The amino acid sequence can be obtained by aligning the common partial sequences.

```
Ala-Gly
    Gly-Phe-Leu
        Phe-Leu-Gly
-------------------
Ala-Gly-Phe-Leu-Gly
```

Enzymatic Hydrolysis

A protein or peptide is specifically cleaved by enzymes called **proteases.** These enzymes are used in the laboratory to determine the structure of peptides and proteins. Two common proteolytic enzymes are chymotrypsin and trypsin. Chymotrypsin hydrolyzes peptide bonds on the C-terminal side of the aromatic amino acids phenylalanine, tyrosine, and tryptophan. Trypsin hydrolyzes peptide bonds on the C-terminal side of the basic amino acids lysine and arginine.

$$\text{Lys-Glu-Tyr-Leu} \xrightarrow{\text{chymotrypsin}} \text{Lys-Glu-Tyr} + \text{Leu}$$

$$\text{Lys-Glu-Tyr-Leu} \xrightarrow{\text{trypsin}} \text{Lys} + \text{Glu-Tyr-Leu}$$

Let us see how we can use enzymatic hydrolysis to determine the structure of leucine enkephalin. We know from the total hydrolysis that the compound contains phenylalanine, an aromatic amino acid. The structure of leucine enkephalin can be partially established by the hydrolysis with chymotrypsin. The pentapeptide cleaves to give a tripeptide with a C-terminal phenylalanine residue and a dipeptide with an N-terminal leucine residue.

$$\text{leucine enkephalin} \xrightarrow{\text{chymotrypsin}} \text{Ala-Gly-Phe} + \text{Leu-Gly}$$

Phenylalanine must have been bonded to leucine, the N-terminal amino acid of the other fragment. Based on the structures of the two fragments, the overall structure must be Ala-Gly-Phe-Leu-Gly.

EXAMPLE 15.4 There are several enkephalins. Predict the products of the chymotrypsin-catalyzed hydrolysis of the following enkephalin.

Tyr-Gly-Gly-Phe-Leu

Solution Chymotrypsin catalyzes hydrolysis of peptide bonds on the C-terminal side of aromatic amino acids in peptides and proteins. The enkephalin contains both phenylalanine and tyrosine, so chymotrypsin cleaves the peptide in two places. Consider each step. Tyrosine is the N-terminal amino acid, and hydrolysis at its carboxyl end results in free tyrosine.

$$\text{Tyr-Gly-Gly-Phe-Leu} \longrightarrow \text{Tyr} + \text{Gly-Gly-Phe-Leu}$$

The phenylalanine in the tetrapeptide is bonded to the C-terminal amino acid, leucine. Hydrolysis at the carboxyl group of phenylalanine frees leucine. A tripeptide results.

$$\text{Gly-Gly-Phe-Leu} \longrightarrow \text{Gly-Gly-Phe} + \text{Leu}$$

The products of the reaction are tyrosine, leucine, and the tripeptide Gly-Gly-Phe.

End-Group Analysis

Partial hydrolysis of a polypeptide chain—either by HCl or by specific enzymes—produces peptides of varying lengths. The sequence of these peptides must be determined in order to determine the overall sequence. The sequence of the peptide fragments produced by partial hydrolysis can often be deduced by end group analysis. Consider a tripeptide that may have six isomeric structures. If the identity of the N-terminal amino acid is known, then only two possible isomeric arrangements are possible for the other two amino acid residues. Subsequent identification of the C-terminal amino acid provides the complete structure.

Enzymatic End-Group Analysis

Some enzymes hydrolyze peptide bonds at either the N- or the C-terminal amino acid, nibbling their way down the chain until they have digested the entire molecule. For example, **carboxypeptidases** sequentially remove peptides from the C-terminal end of a polypeptide chain. In contrast, **aminopeptidases** sequentially hydrolyze peptides from the N-terminal amino acid. By identifying the amino acids produced by a carboxypeptidase or an aminopeptidase at various time intervals, the sequence of amino acids can be determined. For example, hydrolysis of the pentapeptide leucine enkephalin catalyzed by carboxypeptidase first liberates glycine. The tetrapeptide remaining in solution and in contact with the enzyme then yields leucine, followed by phenylalanine, and so on.

$$\text{leucine enkephalin} \xrightarrow{\text{carboxypeptidase}} \text{tetrapeptide} + \text{Gly}$$

$$\text{tetrapeptide} \xrightarrow{\text{carboxypeptidase}} \text{tripeptide} + \text{Leu}$$

When leucine enkephalin is hydrolyzed by an aminopeptidase, the first amino acid released is alanine. The remaining tetrapeptide then yields glycine, followed by phenylalanine, and so on.

$$\text{leucine enkephalin} \xrightarrow{\text{aminopeptidase}} \text{Ala} + \text{tetrapeptide}$$

$$\text{tetrapeptide} \xrightarrow{\text{aminopeptidase}} \text{Gly} + \text{tripeptide}$$

Carboxypeptidases and aminopeptidases can only be used to determine a few residues in a polypeptide chain. Because these enzymes hydrolyze peptide bonds continuously and at different rates, the reaction mixture rapidly becomes difficult to analyze.

Chemical End-Group Analysis

The identity of the N-terminal amino acid of a polypeptide can be determined by a method invented by Pehr Edman that is called the **Edman degradation.** In the Edman degradation, the polypeptide is treated with phenyl isothiocyanate—the Edman reagent—which reacts with the N-terminal amino acid to give an *N*-phenylthiourea derivative. This derivative forms by addition of the terminal N—H bond across the C=N of the phenyl isothiocyanate. After the adduct has formed, anhydrous trifluoroacetic acid is added to the reaction mixture. This reagent cleaves the polypeptide at the

FIGURE 15.2
Use of the Edman Reagent in End-Group Analysis

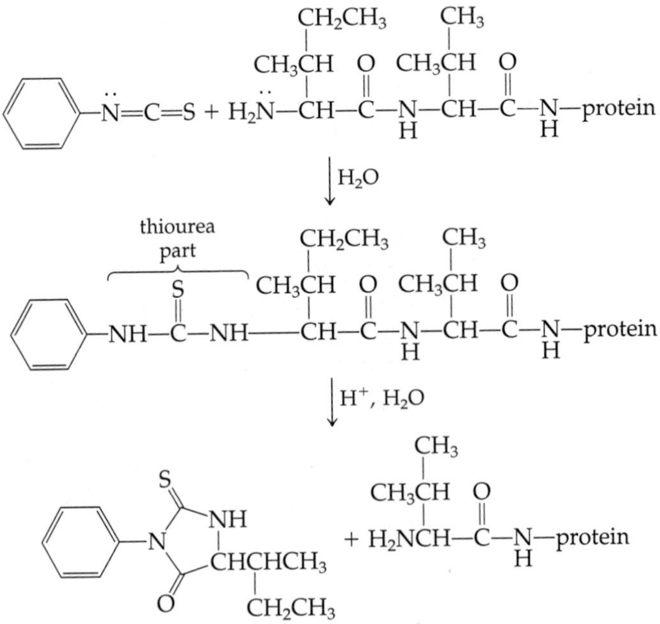

N-terminal residue. Under these conditions, the peptide bonds in the protein do not break (Figure 15.2). A complex cyclization reaction occurs to give a substituted phenylthiohydantoin. This ring contains the carbonyl carbon atom, the α-carbon atom, and the amino-nitrogen atom. The R group of the amino acid is attached to the ring. Comparison with the phenylthiohydantoin of known amino acids establishes the identity of the amino acid. This entire process can be carried out automatically by an instrument called an automatic sequenator.

Because the Edman degradation does not cleave the peptide bonds in the protein, it can be repeated to sequentially identify the amino acids from the N-terminal amino acid end of the molecule. The yield of the Edman degradation approaches 100% and sequences of 30 residues of a polypeptide can be determined from 5 picomole (5×10^{-12} mole) samples. This means that the sequence of a peptide with 30 amino acid residues, with a molecular weight of about 3000, can be determined from a 15-nanogram sample!

EXAMPLE 15.5 β-Endorphin, a 31-peptide, has analgesic effects and promotes the release of growth hormone and prolactin. Treating β-endorphin with phenyl isothiocyanate followed by hydrolysis with anhydrous trifluoroacetic acid yields the following phenylthiohydantoin. What does this information reveal about the structure of the peptide?

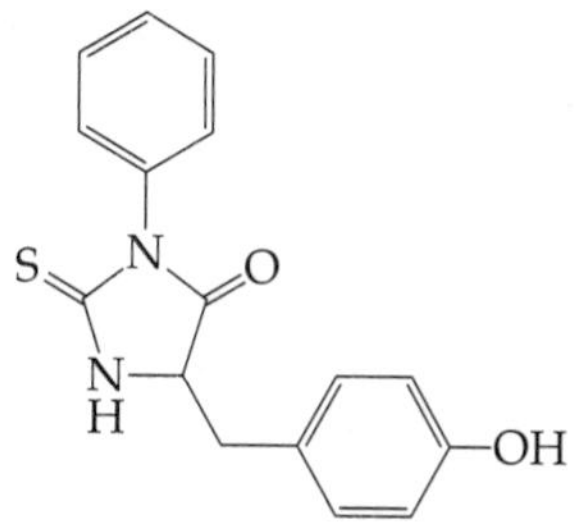

Solution The procedure described above is an Edman degradation, which removes the N-terminal amino acid from the peptide. Be careful not to confuse the two aromatic rings in the phenylthiohydantoin. The aromatic ring bonded to the nitrogen atom between the C═O and the C═S is the phenyl group of the phenyl isothiocyanate. The group bonded to the ring between the nitrogen atom and the C═O is the R group of the amino acid. The N-terminal amino acid of β-endorphin is tyrosine.

15.9 BONDING IN PROTEINS

In this section we will consider the major types of bonds that occur in proteins. These include peptide bonds, disulfide bonds, hydrogen bonds, ionic bonds, and hydrophobic interactions.

The Peptide Bond

The peptide bond is the strongest and most important bond in a protein. A peptide bond is a resonance hybrid of two contributing structures. These resonance hybrids are possible because the unshared electron pair of the amide nitrogen atom is delocalized. Hence, the carbon–nitrogen bond has partial double bond character. The peptide bond is planar, and there is restricted rotation about the carbon–nitrogen bond (Chapter 13). As a result, peptides exist almost exclusively in trans conformations about the carbon–nitrogen bond.

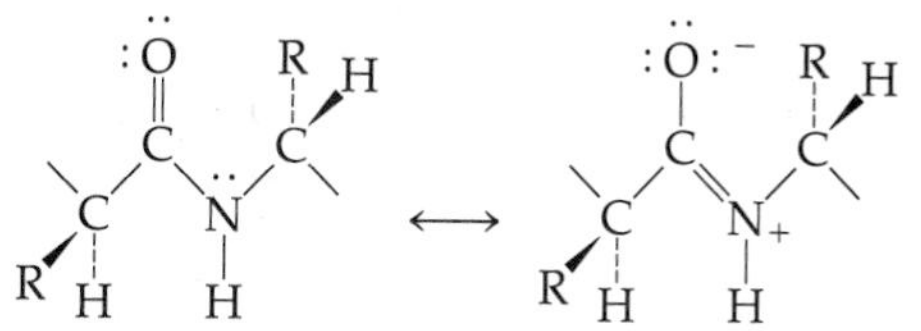

Although free rotation does not occur around the peptide bond, the bond between the α-carbon atom and the carbonyl carbon atom is a rotationally free single bond. Similarly, the single bond between the nitrogen atom and the α-carbon atom of the next amino acid is also rotationally free. There is, in addition, free rotation about the bonds between the α-carbon atoms and the R groups. Thus, a protein chain consists of rigid peptide units connected to one another by freely rotating single bonds.

The Disulfide Bond

Many proteins—especially relatively small ones containing fewer than 100 amino acid residues—have a high cysteine content. Each of these cysteine residues have a sulfhydryl group (—SH) that can form a disulfide bond. A **disulfide bond** is a covalent bond between two sulfur atoms. A disulfide

Complete Proteins in the Diet

Humans cannot synthesize all of the amino acids required to make cellular proteins. The amino acids that cannot be synthesized are said to be **essential.** The essential amino acids must be included in dietary protein.

Dietary proteins are rated in terms of biological value on a percentage scale. A so-called "complete protein" is a mixture of proteins derived from the same biological source—soybeans or milk, for example. A complete protein has a high biological value because it supplies all of the amino acids in the amounts required for growth. Protein sources and their biological value are listed in the table. Note that hen's eggs, cow's milk, and fish provide proteins of high biological value. Plant proteins vary more in biological value than animal proteins. Some plant proteins have a very low amount of one or more essential amino acids. However, not all plant proteins are deficient in the same amino acids. Gliadin, a wheat protein, is low in lysine; zein, a corn protein, is low in both lysine and tryptophan. Societies that eat large amounts of corn or wheat products must have other sources of lysine.

Vegetarians must carefully choose their food so that all of the essential amino acids are available. For example, beans are high in lysine, whereas wheat is low in lysine. Similarly, wheat is high in cysteine and methionine, whereas beans are low in these two amino acids. Eating both beans and wheat increases the percentage of usable proteins in a vegetarian's diet.

Biological Value of Dietary Protein

Food	Biological value (%)
whole hen's egg	94
whole cow's milk	84
fish	83
beef	73
soybeans	73
white potato	67
whole grain wheat	65
whole grain corn	59
dry beans	58

The diets of many people in some areas of the world fall below the minimum daily requirement of protein owing to economic and, in some cases, social and religious customs. In poor countries, costly animal protein is replaced by cereal grains and other incomplete protein sources. If a variety of plant proteins are not available, a number of diseases in young children result. One of these is Kwashiorkor, a protein-deficiency disease that develops in young children after weaning, when their diet is changed to starches. The disease is characterized by bloated bellies and patchy skin. After a certain point, death is inevitable. Some forms of mental retardation also result from incomplete nutrition.

bond results from the oxidation of the —SH (sulfhydryl) groups of two cysteine molecules.

$$\underset{\text{cysteine}}{\begin{array}{c} CO_2H \\ | \\ NH_2CH \\ | \\ CH_2{-}SH \end{array}} + \underset{\text{cysteine}}{\begin{array}{c} CO_2H \\ | \\ NH_2CH \\ | \\ HS{-}CH_2 \end{array}} \xrightarrow{[O]} \underset{\text{cystine}}{\begin{array}{cc} CO_2H & CO_2H \\ | & | \\ NH_2CH & NH_2CH \\ | & | \\ CH_2{-}S{-} & S{-}CH_2 \end{array}} + H_2O$$

Disulfide bonds form after a protein has folded into its biologically active conformation. Once they have formed, the protein conformation is much less flexible. Intrachain disulfide bonds occur in small peptides such as oxytocin and vasopressin, as we saw in Section 15.5. Disulfide bonds can also link a cysteine residue in one polypeptide chain with a cysteine residue in another polypeptide chain as in insulin (Section 15.9).

Hydrogen Bonds

Proteins contain many functional groups that can form hydrogen bonds. Although hydrogen bonds are much weaker than peptide and disulfide bonds, they help to stabilize the folded conformation of proteins. Intramolecular hydrogen bonding between the amide hydrogen atom of one peptide unit and the carbonyl oxygen atom of another peptide unit is very common. Hydrogen bonds also form between various amino acid side chains.

Ionic Bonds

At physiological pH, some of the R groups attached to the polypeptide chain are charged. Ionic attractive forces between the carboxylate groups and the ammonium groups pull portions of chains together. The intrachain ionic bond is called a **salt bridge.** Ionic bonds occur between acidic and basic amino acids.

Hydrophobic Interactions

Proteins contain many nonpolar side chains. These side chains are repelled by water and tend to associate with one another on the "inside" of a folded protein molecule, out of contact with water. This **hydrophobic effect** is similar to those in the micelle of a soap (Chapter 12) or the bilayer of lipids in membranes (Chapter 13). Hydrophobic interactions among nonpolar side chains in proteins are weak but abundant and are primarily responsible for maintaining the folded conformation of a protein.

15.10 PROTEIN STRUCTURE

The biological activity of a protein depends on the three-dimensional shape or conformation of the molecule. The overall shape and structure of a protein constitutes its **native state** or **native conformation.**

Protein structure is divided into four levels: primary, secondary, tertiary, and quaternary. These divisions are somewhat arbitrary because it is the total structure of the protein that controls its function. Nevertheless, it is useful to consider the levels of structure one by one.

Primary Structure

The linear sequence of amino acids in a protein and the location of disulfide bonds is called its **primary structure.** For example, insulin consists of two peptide chains, called the A chain and the B chain, linked by two disulfide bonds. The A chain has 21 amino acids, and the B chain has 30 amino acids (Figure 15.3). There is also an intrachain disulfide bond within the A chain. Insulins from different animals have slightly different amino acid sequences as noted in Figure 15.3. Because the sequence within the cyclic portion of the shorter chain does not affect the physiological function of the insulin, diabetic individuals who become allergic to one type of insulin can often use insulin from another animal source.

Secondary Structure

Many proteins contain regularly repeating conformations of the polypeptide backbone. These conformations are stabilized by hydrogen bonds between residues that are relatively close to one another in the sequence. These regularly repeating conformations constitute the **secondary structure** of the protein (Figure 15.4). Many proteins consist of chains coiled into a spiral known as an α helix. Like a screw, such a helix may be either right- or left-handed. The right-handed (or α) helix is more stable than the left-handed helix, which is virtually nonexistent in proteins. The spiral is held together by hydrogen bonds between the proton of the N—H group of one amino acid and the oxygen atom of the C=O group of another amino acid in the next turn of the helix.

FIGURE 15.3
Primary Structure of Insulin
The insulin structure varies according to species but the majority of the amino acids are identical.

Animal	Positions 8	9	10
Sheep	Ala	Gly	Val
Cow	Ala	Ser	Val
Pig	Thr	Ser	Ile
Horse	Thr	Gly	Ile

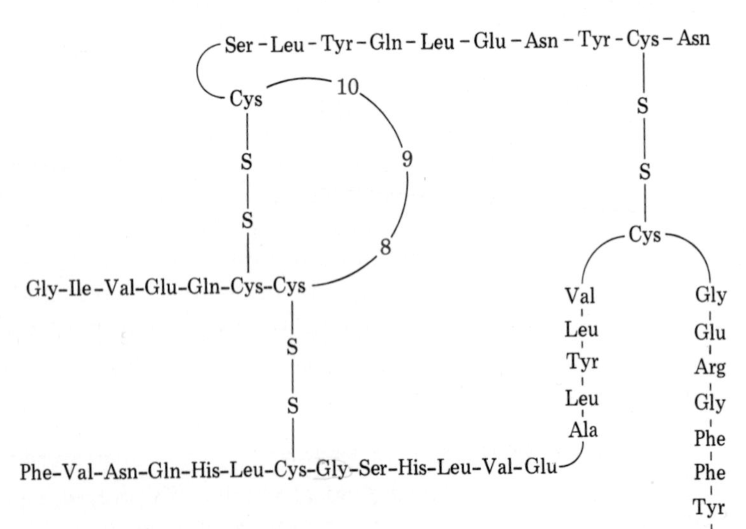

FIGURE 15.4 Hydrogen Bonding in Proteins

(a) The intramolecular hydrogen bonds between coils of the helix are shown only on the "front". This structure is found in a large variety of proteins.
(b) The intermolecular hydrogen bonds between the chains of proteins cause a regular pleated or partially folded structure. This type of structure is found in the proteins of silk.

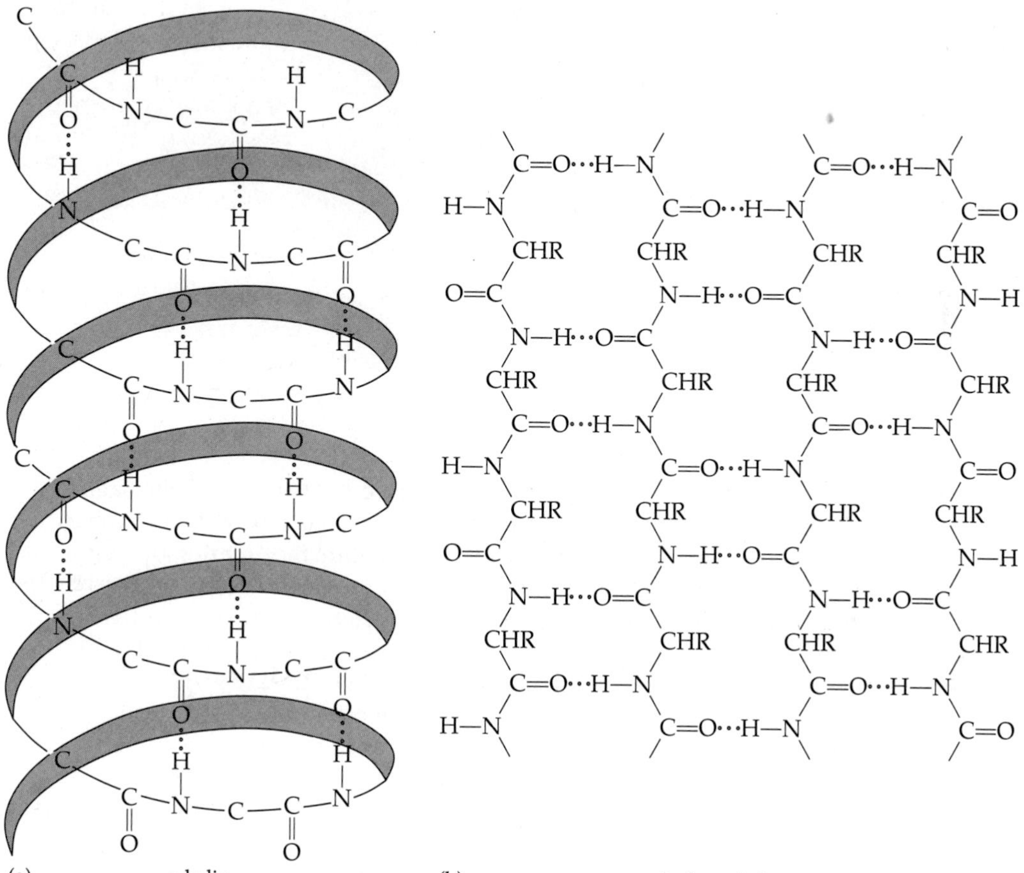

(a) α helix

(b) β-pleated sheet

Some proteins have interchain hydrogen bonds (Figure 15.4). Proteins with interchain hydrogen bonding include fibrin (the blood-clotting protein), myosin (a protein of muscle), and keratin (the protein of hair).

Tertiary Structure

The folded three-dimensional conformation of a protein constitutes its **tertiary structure.** The tertiary structure of a protein results from noncovalent interactions among the side chains of amino acid residues that are far apart in the polypeptide chain. Two amino acid residues that are separated by many intervening amino acids in the primary structure may actually be close together in the folded structure. The proximity of amino acids in the tertiary structure is responsible for the activity of many enzymes. The three-

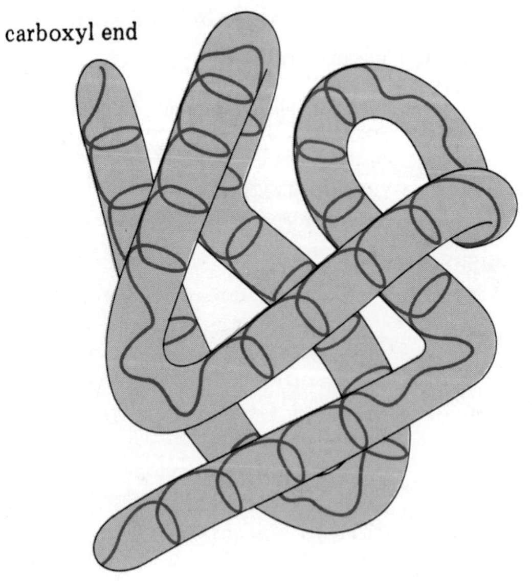

FIGURE 15.5
Tertiary Structure of a Protein
The helix is shown as a coil within the volume occupied by the protein.

dimensional folded shape of proteins (Figure 15.5) is entirely determined by the primary structure. When a protein assumes its native tertiary structure, hydrophobic residues tend to associate within the interior of the folded structure. Polar or charged (hydrophilic) groups tend to be located at the surface near water molecules. Thus, the conformation of a protein results from a balance among the different kinds of noncovalent bonds.

Because the tertiary structure of a protein depends upon its primary structure, changes in the primary structure often disrupt the tertiary structure and destroy biological activity. Changes in the primary structure result from changes in the gene coding for the protein. Some genetic diseases are caused by a mutation that changes a single amino acid residue in a protein that contains hundreds of amino acids.

Quaternary Structure

Some proteins exist as assemblies of two or more polypeptide chains that interact only by noncovalent forces. These proteins, called oligomers, are said to have a **quaternary structure.** Thus, the quaternary structure of a protein is the organization or the association of several protein chains or subunits into a closely packed arrangement. Each subunit has its own primary, secondary, and tertiary structure.

The subunits in a quaternary structure must be specifically arranged in order for the entire protein to function properly. Any alteration in the structure of the subunits or the way in which the subunits are associated results in marked changes in biological activity. A list of some proteins that have quaternary structure is given in Table 15.5.

One example of a protein with a quaternary structure is hemoglobin. Hemoglobin consists of two pairs of different proteins, designated the α and

TABLE 15.5 **Proteins with Quaternary Structure**

Protein	Molecular weight	Number of subunits	Biological function
alcohol dehydrogenase	80,000	4	enzyme of alcohol fermentation
aldolase	150,000	4	enzyme for glycolysis
fumarase	194,000	4	enzyme in the citric acid cycle
hemoglobin	65,000	4	oxygen transport in blood
insulin	11,500	2	hormone regulating metabolism of glucose

the β chains. Each is linked covalently to a molecule of heme. Heme is the site in hemoglobin at which O_2 binds. The two identical α chains and the two identical β chains are arranged tetrahedrally in a three-dimensional structure (Figure 15.6). These units are held together by hydrophobic interactions, hydrogen bonding, and salt bridges between oppositely charged amino acid side chains.

Hemoglobin Structure and Function

There are 141 and 146 amino acids in the α and β chains of hemoglobin, respectively. In some people the sixth amino acid from the N-terminal end of the β chain is valine rather than glutamic acid. This difference of a single amino acid out of 146 in the chain causes changes in the shape of the red blood cells. The cells tend to take the shape of a sickle, and as a result their passage through the blood vessels is restricted. The associated circulatory problems are known as sickle cell anemia.

The four protein subunits of hemoglobin do not behave independently. When one heme molecule binds O_2, the conformation of the surrounding protein chains is slightly altered. When a change in conformation at one site of an oligomeric protein is caused by a change in a spatially separated site of the oligomer, the change is called an **allosteric effect,** and the protein is called an **allosteric protein.** Hemoglobin is an allosteric protein.

When one heme group in hemoglobin binds oxygen, it becomes easier for successive oxygen molecules to bind at the remaining three sites. Thus, once oxygenation occurs at one heme, there is cooperation at all other sites in hemoglobin, so that hemoglobin can carry its full "load" of four oxygen molecules.

Differences in hemoglobin may not affect its oxygen-carrying capacity structure. The β chains of the gorilla and human hemoglobins are identical except for position 104; in gorillas, lysine replaces another basic amino acid, arginine, of humans. The pig β chain differs from human hemoglobin at 17 sites, and that of the horse at 26 sites. In spite of the variability of composition of hemoglobin in different animals, nine positions contain the same amino acids in all hemoglobin molecules. These positions are important to the oxygen-binding function of hemoglobin.

FIGURE 15.6
Quaternary Structure of Hemoglobin
(a) The iron atoms of the heme are shown as spheres within the folds of the four protein chains. Heme is shown by the plane around the iron atom.
(b) The structure of heme.

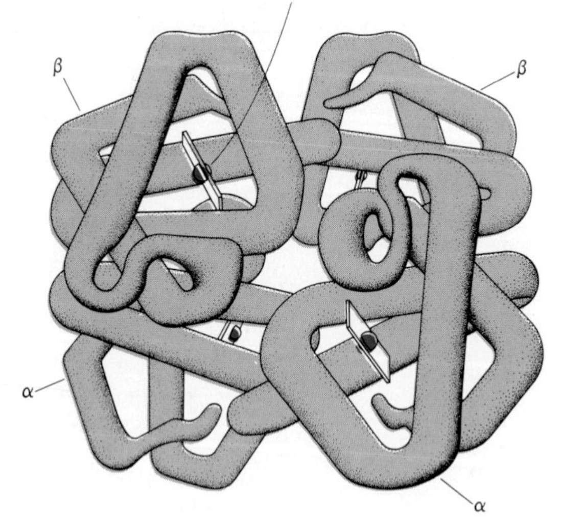

(a)

(b)

COO^- — CH_2 — CH_2 COO^- — CH_2 — CH_2

H_3C—C   C—N   HC   Fe   N—C   CH   C—CH_3 C=N   N—C   C—CH_3

H_2C=CH CH_3 H   HC=CH_2

Heme

EXERCISES

Amino Acids

15.1 A number of D-amino acids are found in bacteria. D-Glutamic acid is found in bacterial cell walls. Draw a projection formula for the amino acid.

15.2 Gramicidin S is a cyclic peptide antibiotic that contains D-phenylalanine. Draw a projection formula for the amino acid.

15.3 The following compound is an unusual amino acid found in collagen. From what amino acid could this compound be derived?

$$NH_2{-}CH_2{-}\underset{\displaystyle OH}{\underset{|}{CH}}{-}CH_2{-}CH_2{-}\underset{\displaystyle NH_2}{\underset{|}{CH}}{-}CO_2H$$

15.4 The following compound is an unusual amino acid that is a neurotransmitter. Classify this amino acid, and determine its IUPAC name.

$$NH_2{-}CH_2{-}CH_2{-}CH_2{-}CO_2H$$

15.5 The following antibacterial agent is contained in garlic. From what amino acid might it be derived?

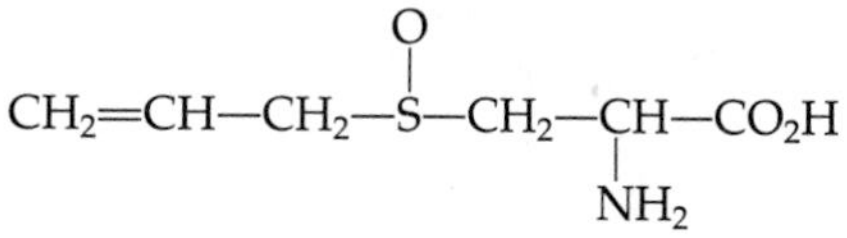

$$CH_2{=}CH{-}CH_2{-}\overset{\overset{\displaystyle O}{|}}{S}{-}CH_2{-}\underset{\underset{\displaystyle NH_2}{|}}{CH}{-}CO_2H$$

15.6 Consider the formula for penicillin G. Identify two amino acids that are required to form it.

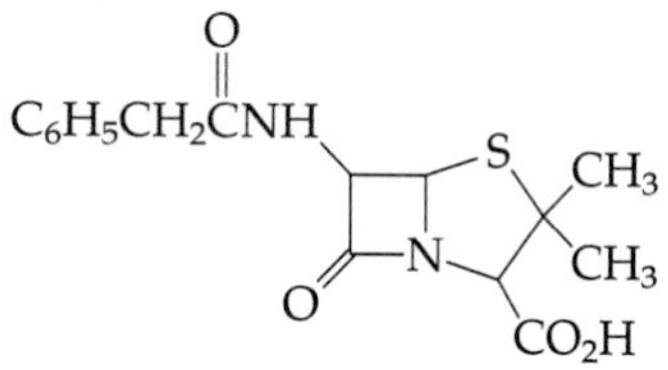

$$C_6H_5CH_2\overset{\overset{\displaystyle O}{\|}}{C}NH$$

S, CH_3, N, CH_3, O, CO_2H

Acid–Base Properties

15.7 Draw the structures of alanine and glutamic acid at pH = 1 and pH = 12.
15.8 Write the structures for the zwitterions of serine and valine.
15.9 How could you distinguish between a solution of asparagine and a solution of aspartic acid?
15.10 Would you expect an aqueous solution of lysine to be neutral, acidic, or basic? Explain.

Isoionic Points

15.11 Estimate the isoionic points of the following tripeptides.
(a) Ala-Val-Gly (b) Ser-Val-Asp (c) Lys-Ala-Val
15.12 Estimate the isoionic points of the following tripeptides.
(a) Glu-Val-Ala (b) Arg-Val-Gly (c) His-Ala-Val
15.13 Examine the structures of oxytocin and vasopressin in Section 15.5. Which should have the higher isoionic point?
15.14 Examine the structure of the enkephalin given in Section 15.8 and estimate its isoionic point.
15.15 The isoionic point of chymotrypsin is 9.5. What does this value indicate about the composition of chymotrypsin?
15.16 The isoionic point of pepsin is 1.1. What does this value indicate about the composition of pepsin?

Peptides

15.17 Write the complete formula and the condensed formula for alanylserine.
15.18 How does glycylserine differ from serylglycine?
15.19 Which amino acids can form peptides containing carboxyl groups or carboxylate groups at internal positions in the peptide chain?
15.20 Which amino acids can form peptides containing amino groups or ammonium groups at internal positions in the peptide chain?

15.21 Identify the amino acids contained in the following tripeptide. Name the compound.

$$NH_2{-}CH_2{-}\overset{\overset{\displaystyle O}{\|}}{C}{-}NH{-}\underset{\underset{\displaystyle CH_2SH}{|}}{CH}{-}\overset{\overset{\displaystyle O}{\|}}{C}{-}NH{-}\underset{\underset{\displaystyle CH(CH_3)_2}{|}}{CH}CO_2H$$

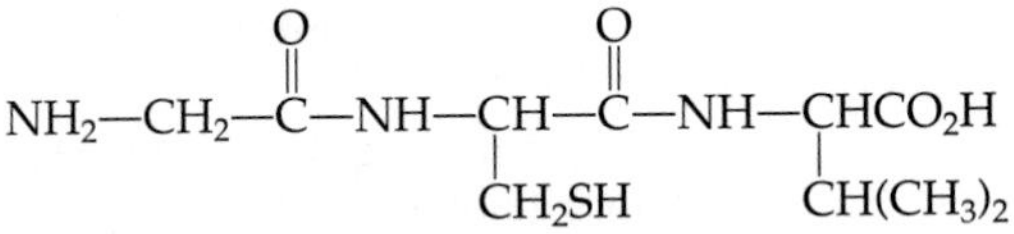

15.22 Identify the amino acids contained in the following tripeptide. Name the compound.

$$NH_2{-}\underset{\underset{\displaystyle CH_2OH}{|}}{CH}{-}\overset{\overset{\displaystyle O}{\|}}{C}{-}NH{-}CH_2{-}\overset{\overset{\displaystyle O}{\|}}{C}{-}NH{-}\underset{\underset{\displaystyle CH_3}{|}}{CH}CO_2H$$

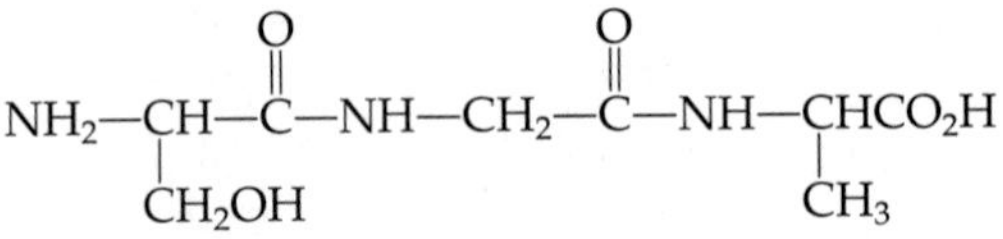

15.23 Thyrotropin-releasing hormone (TRH) causes the pituitary gland to release thyrotropin, which then stimulates the thyroid gland. Examine its structure and comment on two unusual structural features.

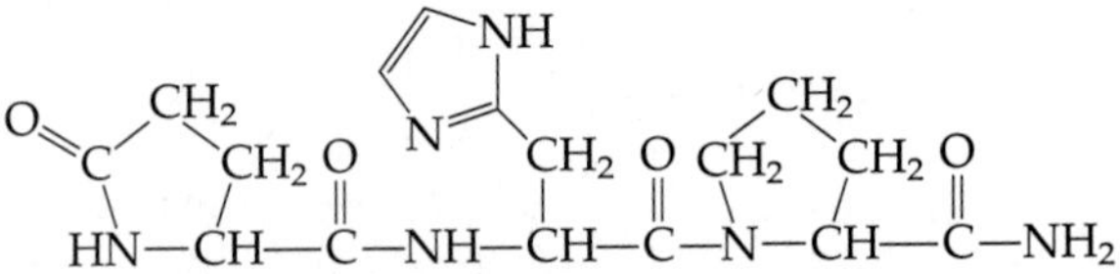

O=C, CH₂, CH₂ O, NH, N, CH₂ O CH₂, CH₂, CH₂ O

HN—CH—C—NH—CH—C—N—CH—C—NH₂

15.24 The tripeptide glutathione, which is important in detoxifying metabolites, has an unusual structural feature. Identify it.

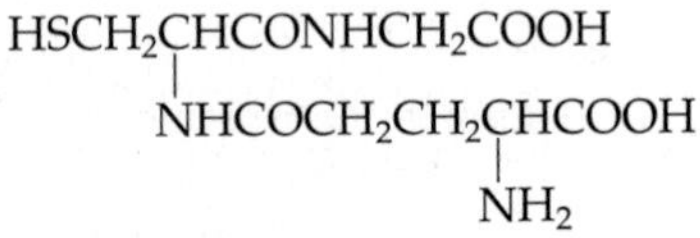

$$HSCH_2\underset{\underset{\displaystyle NHCOCH_2CH_2\underset{\underset{\displaystyle NH_2}{|}}{C}HCOOH}{|}}{C}HCONHCH_2COOH$$

15.25 How many isomeric compounds with the composition Gly_2,Ala_2 are there?

15.26 How many isomeric compounds with the composition Gly_2,Ala,Leu are there?

Hydrolysis and Structure Determination

15.27 Assuming that only dipeptides are formed by partial hydrolysis, what is the minimum number that must be identified to establish the structure of a pentapeptide?

15.28 Assuming that only tripeptides are formed by partial hydrolysis, what is the minimum number that must be identified to establish the structure of an octapeptide?

15.29 The tetrapeptide tuftsin is hydrolyzed to produce Pro-Arg and Thr-Lys. Does this information establish the structure of tuftsin?

15.30 Assume that the octapeptide angiotensin II is hydrolyzed to produce Pro-Phe, Val-Tyr-Ile, Asp-Arg-Val, and Ile-His-Pro. What is its structure?

Enzymatic Hydrolysis

15.31 Which of the following tripeptides will be cleaved by trypsin? If cleavage occurs, name the products.
(a) Arg-Gly-Tyr (b) Glu-Asp-Gly (c) Phe-Trp-Ser (d) Ser-Phe-Asp

15.32 Which of the following tripeptides will be cleaved by trypsin? If cleavage occurs, name the products.
(a) Asp-Lys-Ser (b) Lys-Tyr-Cys (c) Asp-Gly-Lys (d) Arg-Glu-Ser

15.33 Indicate which of the tripeptides in Exercise 15.31 will be cleaved by chymotrypsin and name the products.

15.34 Indicate which of the tripeptides in Exercise 15.32 will be cleaved by chymotrypsin and name the products.

15.35 The tetrapeptide tuftsin is hydrolyzed by trypsin to produce Pro-Arg and Thr-Lys. Does this information establish the structure of tuftsin?

15.36 The pentapeptide met-enkephalin is hydrolyzed by chymotrypsin to give Met, Tyr, and Gly-Gly-Phe. Does this information establish the structure of met-enkephalin?

15.37 The nonapeptide known as the sleep peptide is hydrolyzed by chymotrypsin to produce Ala-Ser-Gly-Glu and Ala-Arg-Gly-Tyr and Trp. What two structures are possible for the sleep peptide?

15.38 The sleep peptide is hydrolyzed by trypsin to produce Gly-Tyr-Ala-Ser-Gly-Glu and Trp-Ala-Arg. What is the structure of the sleep peptide?

End-Group Analysis

15.39 Hydrolysis of tuftsin with an aminopeptidase yields Thr. Using the information in Exercise 15.29, what is the structure of tuftsin?

15.40 Hydrolysis of met-enkephalin with a carboxypeptidase yields Met. Using the information in Exercise 15.36, what is the structure of met-enkephalin?

15.41 A structure determination of insulin using the Edman method yields two phenylthiohydantoin products. Why?

15.42 Cholecystokinin, a 33-peptide, plays a role in reducing the desire for food, and its production is stimulated by food intake. Its N-terminal amino acid is lysine. Draw the structure of the phenylthiohydantoin product.

15.43 Reaction of angiotensin II with the Edman reagent yields the following product. What information has been established?

S
NH
N
$CH—CH_2—CO_2H$
O

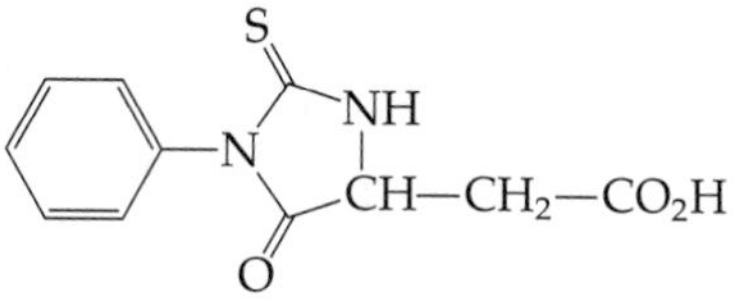

15.44 Corticotropin is released when the blood level of corticosteroids is diminished. Reaction of corticotropin with the Edman reagent yields the following product. What information has been established?

S
NH
N
$CH—CH_2—OH$
O

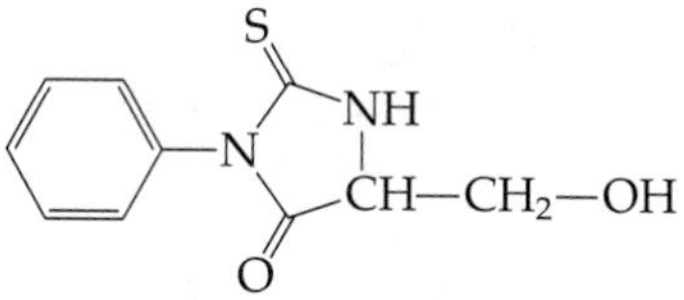

Proteins

15.45 Which amino acids can form salt bridges in proteins?

15.46 Which amino acids have R groups that form hydrogen bonds?

15.47 Which of the following amino acids are likely to exist in the interior of a protein dissolved in an aqueous solution?
(a) glycine (b) phenylalanine
(c) glutamic acid (d) arginine

15.48 Which of the following amino acids are likely to exist in the interior of a protein dissolved in an aqueous solution?
(a) proline (b) cysteine
(c) glutamine (d) aspartic acid

15.49 If a protein is embedded in a lipid bilayer, which of the amino acids listed in Exercise 15.47 will be in contact with the interior of the bilayer?

15.50 If a protein is embedded in a lipid bilayer, which of the amino acids listed in Exercise 15.48 will be in contact with the interior of the bilayer?

15.51 Noting that proline is a secondary amine, explain how proline can disrupt the α helix of a protein.

15.52 Examine the structures of valine and glutamic acid and suggest a reason why human hemoglobin is affected by the substitution of valine for glutamic acid at position 6 in the β chain.

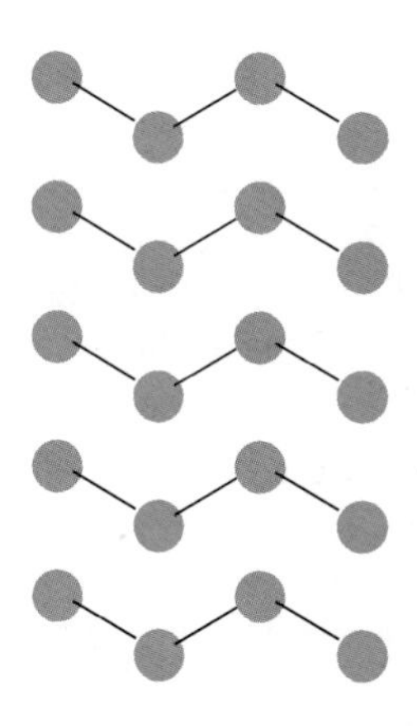

16

NUCLEIC ACIDS

16.1 NUCLEIC ACIDS AND GENETICS

One essential process for the continuance of life is the transmission of hereditary information from one generation to the next. Genetic information in higher organisms is contained in **chromosomes** in the nuclei of cells. These chromosomes consist of genes that contain the information that determines the characteristics of the living organism. Thus, carrot seeds produce carrots and not cabbages. Mice make more mice, not elephants. The number of chromosomes and the complexity of their molecular structures are unique for each species. Each cell of a species contains an even number of chromosomes.

Chromosomes are complex structures made of proteins and nucleic acid molecules called **deoxyribonucleic acid, DNA.** Each chromosome contains tens of thousands of genes. Each gene contains the instructions for the synthesis of a specific protein with a specific function. The location of genes on the various chromosomes of a species can be "mapped." Genes responsible for some genetic diseases such as cystic fibrosis have been identified, and technology is being developed to artificially alter certain genes to change genetic directions.

As will be described in this chapter, DNA undergoes self-replication and contains the information required for the synthesis of three types of ribonucleic acids (RNA) known as ribosomal RNA, messenger RNA, and transfer RNA. RNA molecules participate in the synthesis of all proteins.

DNA and RNA are polymers. In this chapter we will first examine the structures of the nucleic acid monomer components known as nucleotides. Then, we will discuss the structures of the nucleic acids and their central role in life processes.

16.2 COMPONENTS OF NUCLEIC ACIDS

DNA and RNA are polymers of nucleotides. Nucleotides consist of three simpler units: a sugar, a nitrogen base, and a phosphate group.

Sugars in Nucleic Acids

The sugar in RNA is the β-anomer of D-ribose, which accounts for the name ribonucleic acid. The sugar in DNA is the β-anomer of D-2-deoxyribose; hence, the name deoxyribonucleic acid. The absence of a hydroxyl group at the C-2 atom in D-2-deoxyribose is the major structural difference between DNA and RNA.

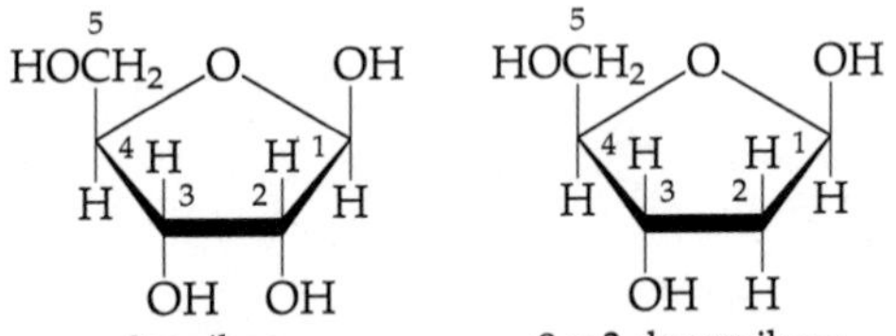

β-D-ribose

β-D-2-deoxyribose

Nitrogen Bases in Nucleic Acids

The sugar ring in DNA and RNA is attached to a nitrogen-containing purine or pyrimidine base. Two purine bases—adenine (A) and guanine (G)—occur in both DNA and RNA. The capital letters are shorthand notations used to represent the names of bases in nucleic acids.

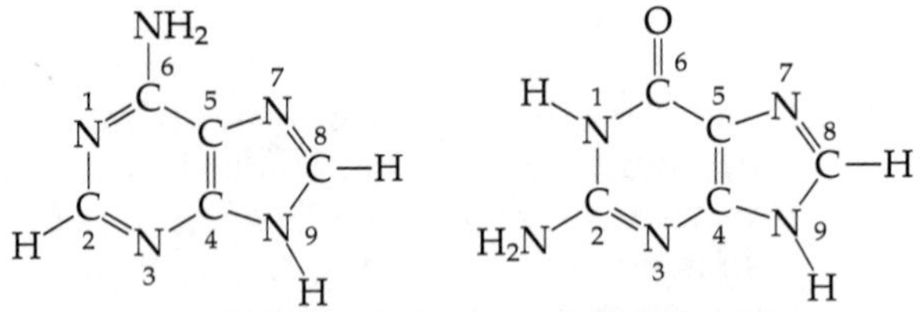

adenine (A)

guanine (G)

Three pyrimidine bases—cytosine (C), thymine (T), and uracil (U)—occur in nucleic acids. Cytosine appears in both RNA and DNA. Thymine is found exclusively in DNA; uracil is found only in RNA. Note that thymine and uracil differ only by a methyl group at the C-5 atom in the pyrimidine ring.

In summary, DNA and RNA each contain four bases: two purines and two pyrimidines. In DNA the bases are A, G, C, and T; the bases for RNA are A, G, C, and U.

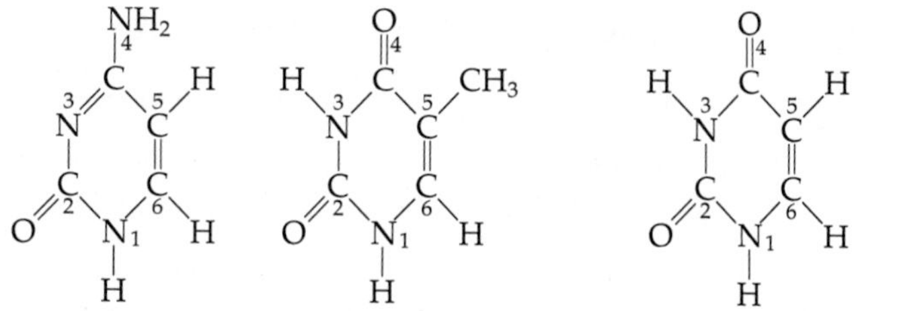

cytosine (C)

thymine (T)
(found in DNA)

uracil (U)
(replaces thymine in RNA)

Nucleosides

A **nucleoside** is a glycoside that forms when the hemiacetal center of the sugar and an —NH of a purine or pyrimidine base join with the elimination of a molecule of water. The result is a C—N glycosidic bond at the C-1 atom of the sugar. This glycosidic bond has the β configuration in both DNA and RNA.

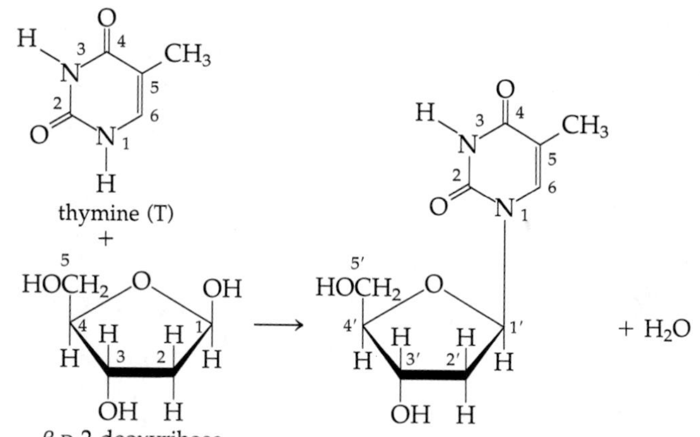

thymine (T)
+
β-D-2-deoxyribose

$\longrightarrow$ + H_2O

A nucleoside contains a sugar and a base. Each structural component is numbered; primed numbers are used for the atoms of the carbohydrate; unprimed numbers refer to ring atoms of the base. The nitrogen base is always attached to the C-1′ atom of the carbohydrate. The carbohydrate has a primary hydroxyl group located at the C-5′ atom. Ribonucleosides have secondary hydroxyl groups at the C-2′ and C-3′ atoms, whereas deoxyribonucleosides have a secondary hydroxyl group only at the C-3′ atom.

Ribonucleosides are named by modifying the name of the base. Nucleosides that contain a purine have the ending *-osine;* nucleosides that contain a pyrimidine end in *-idine.* For example, the nucleoside made with guanine and ribose is called guanosine; the nucleoside made with uracil and ribose is called uridine. The same convention is used for deoxyribonucleosides along with the prefix *deoxy-*. The names of the nucleosides are listed in Table 16.1.

TABLE 16.1 Names of Nucleosides and Nucleotides

Base	Nucleoside	Nucleotide
DNA		
adenine (A)	deoxyadenosine	deoxyadenylic acid
guanine (G)	deoxyguanosine	deoxyguanylic acid
thymine (T)	deoxythymidine	deoxythymidylic acid
cytosine (C)	deoxycytidine	deoxycytidylic acid
RNA		
adenine (A)	adenosine	adenylic acid
guanine (G)	guanosine	guanylic acid
uracil (U)	uridine	uridylic acid
cytosine (C)	cytidine	cytidylic acid

Nucleotides

A **nucleotide** is an ester of a nucleoside and phosphoric acid. The phosphate ester is formed by using the hydroxyl group at either the C-3′ or C-5′ atom. The products are 3′- and 5′-phosphate esters, respectively. These numbers are included in the names of nucleotides.

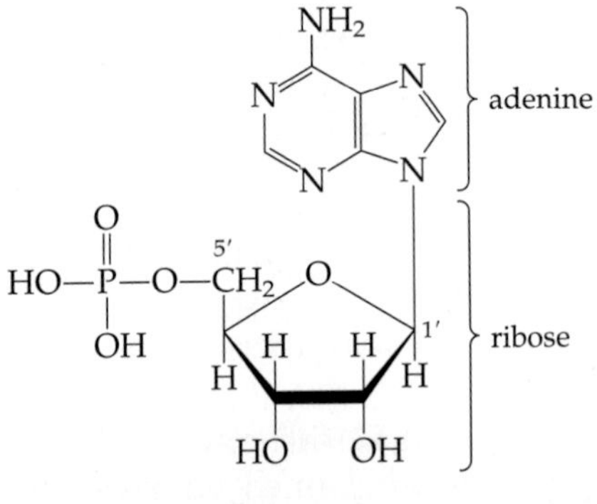

NH_2 adenine ribose HO—P—O—CH_2 O OH H H H H HO OH 5′ 1′

The two protons of the monophosphate ester are ionized at physiological pH, and the ester exists as an ion with a −2 charge in solution. The $—PO_3^{2-}$ group attached to the C-3′ or C-5′ oxygen atom of the sugar is called a **phosphoryl group.**

A nucleotide can have more than one name. For example, the 3′- and 5′-phosphate esters of a nucleoside can be named as nucleoside monophosphates (NMPs or dNMPs, where "d" means deoxy). Thus, the 5′-phosphate ester of adenosine is called adenosine-5-monophosphate, abbreviated 5′-AMP. Because the phosphate ester contains an acidic phosphoryl group, nucleotides can also be named as acids. To name a nucleotide as an acid, we replace the ending *-osine* in the name of the nucleoside with the ending *-ylic acid.* We add a number, 3′ or 5′, to indicate the position of the phosphoryl group. Thus 5′-AMP is also 5′-adenylic acid. The names of the nucleotides are listed in Table 16.1.

We noted earlier that the phosphoryl group is ionized at pH 7. To name the anion of a nucleotidylic acid, we change the ending *-ic acid* to the ending *-ate*. Thus, the anion derived from adenylic acid is adenylate.

Let's consider one more aspect of nucleotide nomenclature. We would like to name the residue of a nucleotide when it is attached to another molecule. We recall that the name of $—CH_3$ is methyl, and that we originally obtained the name methyl by replacing the alkane ending *-ane* with the ending *-yl*. We use a similar rule to name a nucleotidylic acid moiety attached to another molecule through a bridging phosphoryl group oxygen atom. We drop the *-ic acid* part of the name of the nucleotidylic acid and add the suffix *-yl* to obtain the name of a nucleotidylyl group. For example, an adenylic acid group attached to another molecule through a bond to its ester oxygen atom is called an adenylyl group. Note that this name has a "double

Nucleosides in Medicine

Several chemically modified nucleosides are used to treat diseases, including certain kinds of cancer and acquired immune deficiency syndrome (AIDS). Chemically modified nucleosides have profound physiological effects because they interfere with cell division. During cell division, DNA replication occurs. Nucleosides that interfere with DNA replication are used to treat cancer. Cancer cells divide rapidly and require a large supply of DNA. Thymidine, one of the components of DNA, is synthesized in the cell by methylation of uridine. An analog of uracil called fluorouracil is a chemotherapeutic drug. It inhibits the enzyme that catalyzes the conversion of thymidine into uridine. Fluorouracil administered as part of chemotherapy inhibits DNA replication, and therefore the growth rate of the tumor is decreased.

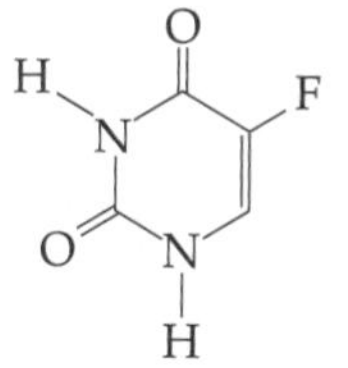

fluorouracil

However, fluorouracil also interferes with DNA synthesis in normal cells and causes side effects that must be monitored. During chemotherapy the weakened body is more susceptible to bacterial infections.

Azidothymidine (AZT) is a nucleoside used to treat AIDS. It is not a cure for this disease but delays the onset of clinical symptoms of AIDS. AZT has been approved for limited use in the United States, but it causes severe anemia. When AZT is prescribed, blood transfusions may be needed, and close medical supervision is required.

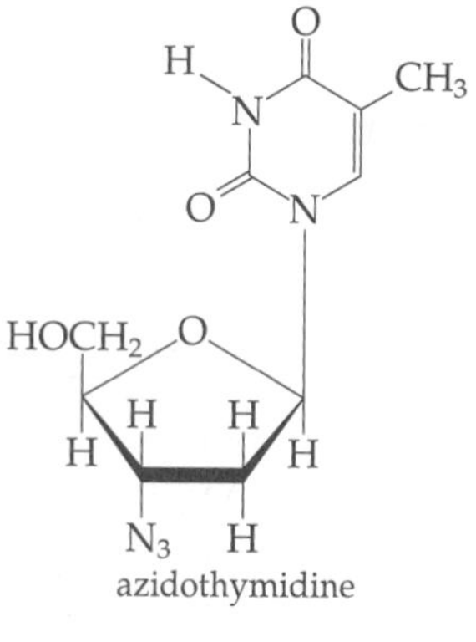

azidothymidine

yl." We will use the names of nucleotidylyl groups in Section 16.3, when we consider polynucleotides.

EXAMPLE 16.1 Classify the following structure, identify its components, and name it.

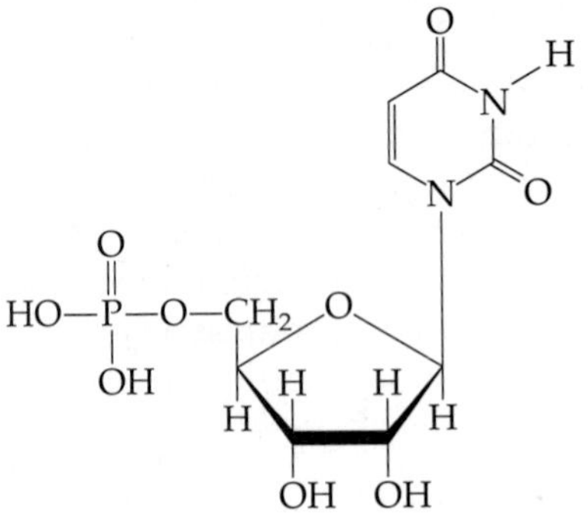

Solution The compound contains a sugar, a base, and a phosphoryl group; therefore, it is a nucleotide. The sugar is ribose, so the compound is a ribonucleotide. The base is uracil, which can be found only in ribonucleotides. The name of the compound is uridine 5′-monophosphate (5′-UMP) or 5′-uridylic acid.

16.3 POLYNUCLEOTIDES

Throughout this text we have focused upon the composition and structure of molecules. Although nucleic acids are extremely complex, we can analyze them by applying the same systematic method of identifying their important functional groups and structural properties.

Base Composition

Just as a protein is characterized by its amino acid composition, a polynucleotide is characterized by its nucleotide composition. Furthermore, because nucleotides differ only in their bases, a polynucleotide is characterized by its base composition. The amounts of A, T, G, and C in the DNA of animals, plants, and microorganisms reveal an interesting pattern (Table 16.2). The amount of C is equal to the amount of G; the amount of A is equal to the amount of T. The amount of each of the pairs differs from one life form to another. However, the sum of the percentages of A and G, the two purine bases, equals the sum of the percentages of C and T, the two pyrimidine bases. The relationships are

$$\%\,C = \%\,G \qquad \%\,A = \%\,T \qquad \%\,A + \%\,G = \%\,C + \%\,T$$

TABLE 16.2 Composition of Various DNAs

Species	Mole %			
	A	T	G	C
Escherichia coli	25	25	25	25
brewer's yeast	32	32	18	18
wheat	27	27	23	23
bovine	28	28	22	22
human	30	30	20	20

We will return to these relationships in DNA later in this section. There are no simple relationships in the percentages of the four bases in RNA.

Primary Structure

A polynucleotide is a linear chain of nucleotidylyl residues linked through phosphodiester bonds. The nucleotide sequence defines the primary structure of DNA and RNA molecules (Figure 16.1). Phosphodiester bonds are formed between the 3′-hydroxyl group of one nucleotide and the 5′-phosphate ester of another nucleotide. The link between the two nucleotides is therefore a 3′-5′ phosphodiester bond. When a phosphodiester bond forms, water is removed. The nucleotides that remain are therefore nucleotide residues. These residues are called nucleotidylyl groups; we discussed their names earlier.

DNA and RNA each have a backbone that consists of sugar residues linked by 3′–5′ phosphodiester bonds. DNA and RNA differ in the sugar (deoxyribose in DNA and ribose in RNA) and in their base compositions. We recall that thymine is found only in DNA and uracil is present only in RNA. Note that each phosphodiester has one acidic hydrogen atom, hence polynucleotides are called nucleic acids. At physiological pH, the phosphate groups of nucleic acids are ionized. The negatively charged phosphate groups are bound to Mg^{2+} ions in cells.

The sequence of nucleotides in a polynucleotide is written by convention in the 5′ → 3′ direction. Each nucleotide in the sequence is abbreviated as a single letter. If either the 5′- or 3′-end of the molecule is attached to a phosphoryl group, this is indicated by a lowercase p. An example of a sequence in an RNA molecule is written as follows.

$$5' \xrightarrow[\text{RNA}]{5'\text{pA—C—G—}\cdots\text{—U}} 3'$$

The same system is used for DNA, but a lowercase d is placed at the left of the first base to indicate the presence of deoxyribose in the backbone. The presence of thymine rather than uracil also indicates that the nucleic acid is deoxyribonucleic acid. An example of a nucleotide sequence in DNA is

$$5' \xrightarrow[\text{DNA}]{5'\text{dT—C—G—}\cdots\text{—A}} 3'$$

FIGURE 16.1 Structure of a Polynucleotide The backbone of this deoxyribonucleotide consists of deoxyribose and phosphate units. The bases are A, T, G, and C. A ribonucleotide has a similar structure; ribose replaces deoxyribose and U replaces T.

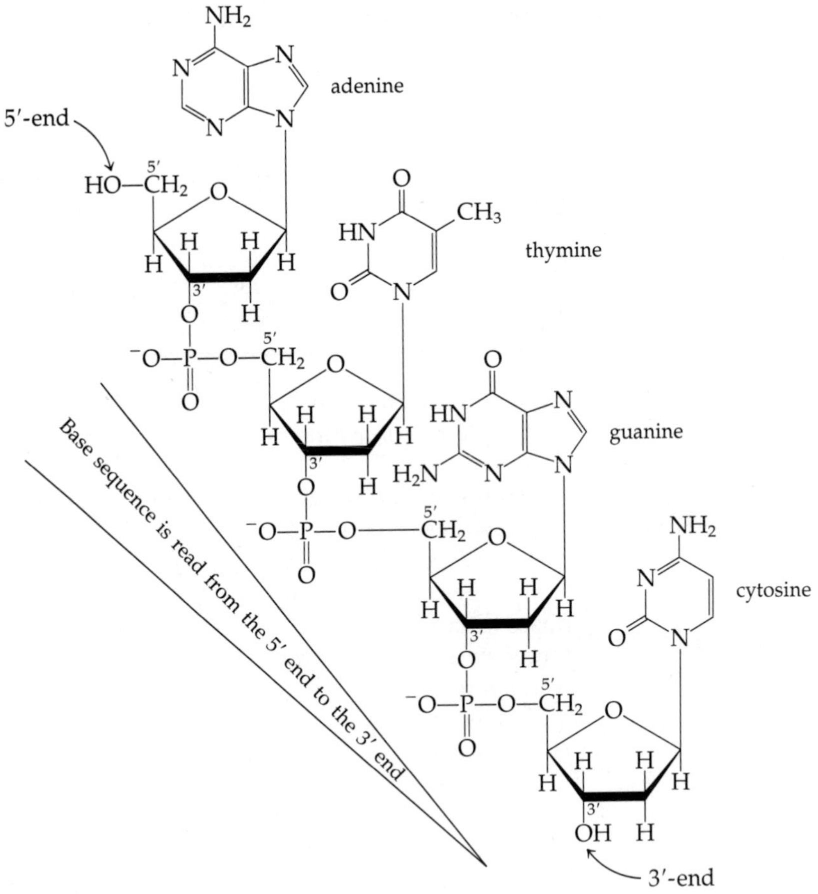

5′-end
adenine
thymine
guanine
cytosine
3′-end
Base sequence is read from the 5′ end to the 3′ end

Secondary Structure of DNA

The base composition of DNA provides an important clue to its structure. The fact that %C = %G and %A = %T suggested to J. D. Watson and F. H. C. Crick that DNA is a double-stranded molecule in which G in one strand is always paired with C in the other, and that A in one strand is always paired with T in the other. The pairing of A and T and also of C and G in DNA is the result of hydrogen bonding. Cytosine forms three hydrogen bonds to guanine, and adenine forms two hydrogen bonds to thymine (Figure 16.2). The interaction of bases through hydrogen bonds is called **complementary base-pairing.** Thus, the Watson–Crick model for the structure of DNA explained the experimentally observed relationship between bases.

Watson and Crick proposed that DNA consists of two polynucleotide chains coiled around each other in a right-handed double helix (Figure 16.2). In the double helix, the polydeoxyribonucleotide chains run in opposite

FIGURE 16.2 **Hydrogen Bonding in the Double Helix**

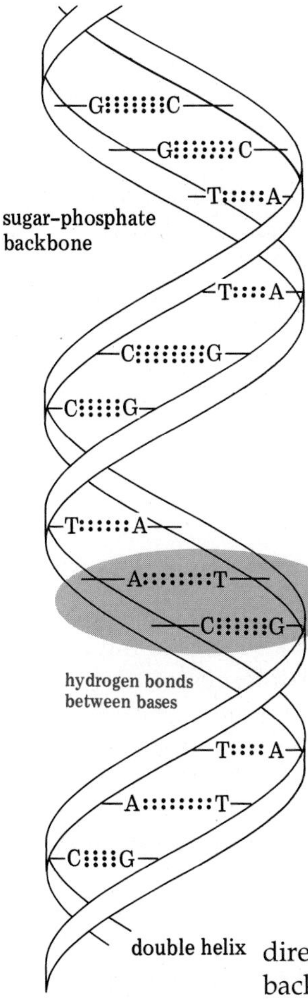

sugar-phosphate backbone

hydrogen bonds between bases

double helix

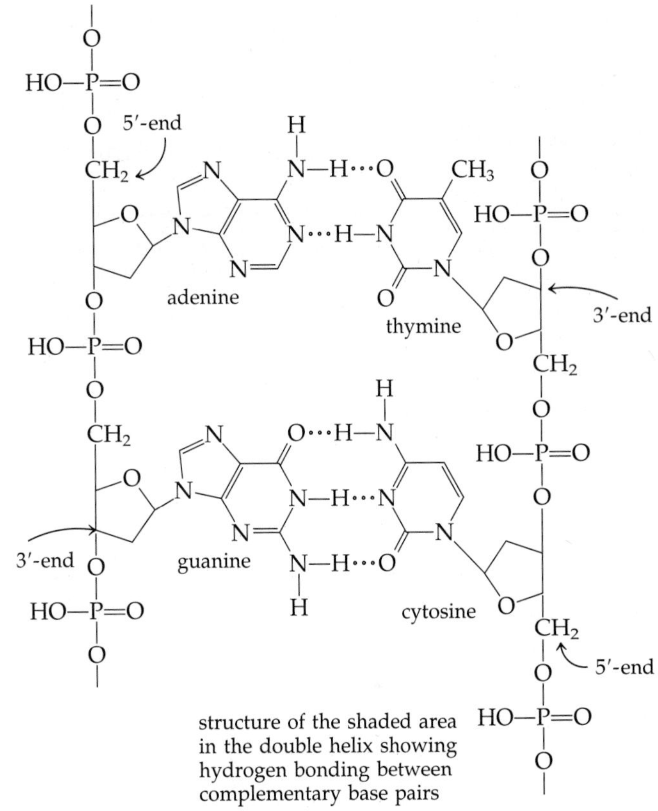

structure of the shaded area in the double helix showing hydrogen bonding between complementary base pairs

directions. Thus, they are said to be **antiparallel.** The sugar–phosphate backbone of the double helix winds outside of the double helix like the banisters of a spiral staircase. The bases attached to the sugar–phosphate backbone of each chain extend inward toward the other chain at right angles to the long axis of the chain. We can think of the bases as the "steps" of the staircase. Each base in one chain is hydrogen-bonded to a complementary base in the other chain. A shorthand representation of complementary base-pairing in the antiparallel strands is

5′ ⟶ 3′
dT···—G—C—A
A···—C—G—dT
3′ ⟵ 5′

Complementary base-pairing of A:T and G:C is necessary for steric reasons: two pyrimidine molecules would be too small to bridge the two strands, and two purine molecules across from each other would be too large for the interior of DNA. Hence A-A, G-G, and A-G pairs do not occur. Only a small pyrimidine paired with a large purine fits.

When the DNA double helix forms, the bases are stacked upon one another. These stacked bases interact by London forces. These forces are weak individually, but they are abundant and collectively provide the most important force in the stabilization of the double helical DNA.

Structure of RNA

We have seen that the structure of DNA results from complementary base-pairing between A and T and between G and C. In RNA, however, there is no relationship between the bases U and A or between C and G. RNA is a single-stranded molecule. This single strand can fold back on itself in complicated ways to give intramolecular hydrogen bonds between pairs of bases in different parts of the strand. The difference in the structures and functions of the three types of RNA will be discussed in Section 16.6.

16.4 THE FLOW OF GENETIC INFORMATION

Early experiments in molecular biology led Francis Crick to propose a set of generalizations that he called the **central dogma** of biological information flow. These generalizations state that biological information flows from DNA, the carrier of genetic information, to RNA, to protein. These generalizations are based upon the following experimental observations.

1. DNA is the repository of genetic information in all organisms.
2. DNA replication occurs when cells are ready to divide. **Replication** is a self-copying process in which a complementary copy of each strand of parental DNA is made.
3. DNA is a template for synthesis of a molecule of **messenger RNA, mRNA.** The transmission of information from DNA to mRNA is called **transcription.**
4. mRNA molecules provide the information required to synthesize proteins by a process called translation. In **translation,** the linear sequence of nucleotidylyl residues in mRNA is expressed as a linear sequence of amino acids residues in a protein chain.
5. During protein synthesis, amino acids attached to **transfer RNA, tRNA,** are brought to ribosomes, the cellular sites of protein synthesis.

16.5 DNA REPLICATION

The structure of DNA provides a clue that unlocks the secret of the mechanism of DNA replication, the process by which biological information is passed from one generation to the next. Each strand of DNA serves as a

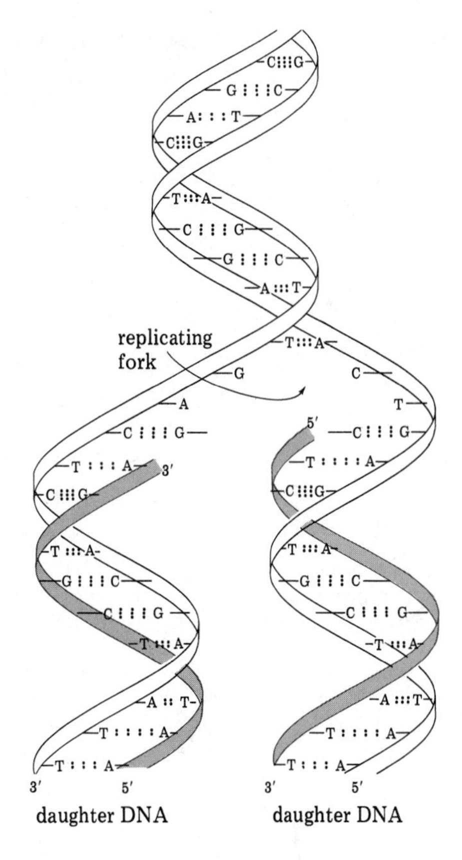

FIGURE 16.3
Replication of DNA

template upon which a new complementary strand of DNA is synthesized. This new complementary strand is identical to the original complementary strand.

DNA replication is **semiconservative;** that is, each new DNA molecule contains one strand of the parent DNA molecule and one newly synthesized strand. In DNA replication, the double strand must unwind so that the new complementary chain can be formed. The enzyme DNA polymerase catalyzes DNA replication. Many other proteins are required for this process. Although the entire process is complex, a replicating fork is a simple model (Figure 16.3). The replicating fork occurs as DNA unravels and creates a place for DNA synthesis.

EXAMPLE 16.2 Part of the nucleotide sequence in one chain of DNA is given below. Write a representation of the complementary DNA chain.

$$5' \xrightarrow{\text{dA—T—C—G}} 3'$$

Solution The complementary base pairs in DNA are A with T and G with C. For each A in one chain there is a T in the other. For each C in one chain there is a G in the other. The complementary chains are

dA—T—C—G
T—A—G—dC

16.6 DNA TRANSCRIPTION

The process by which DNA is copied onto an RNA molecule is called **transcription.** Transcription produces three kinds of RNA molecules called messenger RNA (mRNA), ribosomal RNA (rRNA), and transfer RNA (tRNA). Transcription differs from DNA replication because

1. RNA polymerase rather than DNA polymerase is the catalyst.
2. Uracil rather than thymine is incorporated in RNA.
3. The RNA molecules formed are single strands rather than double strands.

A shorthand representation of transcription is given below.

5′ —DNA strand→ 3′
dA—T—C—G
U—A—G—C
3′ ←RNA strand— 5′

Ribosomal RNA

Ribosomal RNA is present in cells in combination with about 50 different proteins in complex structures called **ribosomes.** A ribosome has a mass of about 3 million. Ribosomes are the sites of protein synthesis and serve as miniature factories that manufacture proteins when they are provided with the necessary materials and directions for assembly.

Messenger RNA

Messenger RNA carries the information, contained in a gene, necessary to synthesize a protein. Each amino acid in a protein is specified by a three-letter base sequence called a **codon.** The collection of all the codons is called the **genetic code.** The genetic code is discussed in Section 16.7.

Transfer RNA

Transfer RNA provides the "link" between nucleic acids and proteins in protein synthesis. tRNA delivers individual amino acids to the ribosomal site of protein synthesis, where they bind to mRNA and transfer an amino acid to a growing protein chain. A specific tRNA carries one type of amino acid. However, as will be shown in Section 16.7, several different tRNA molecules can transport the same amino acid. Each tRNA has an amino acid attachment site and a template (mRNA) recognition site. Every tRNA molecule has the sequence C—C—A at its 3′-end. The amino acid is attached through its carbonyl carbon atom to the 3′-oxygen atom of the 3′-adenylate residue. The base sequence of the template recognition site will be discussed in Section 16.7.

In 1965 R. Holley of Cornell University determined the entire sequence of a tRNA for alanine (abbreviated $tRNA^{Ala}$). He suggested that the single-stranded tRNA molecule could form intrachain hydrogen bonds between some complementary base pairs. By construction of the maximum number of hydrogen bonds, he postulated a "cloverleaf" model for $tRNA^{Ala}$ (Figure 16.4). The base-pairings have been confirmed experimentally, and the shape of the three-dimensional molecule has been established by X-ray crystallography.

The structure of a tRNA molecule has several base-paired loops. One loop of tRNA has a three-nucleotide sequence called an **anticodon.** An anticodon is the complement of a **codon** located in mRNA. The interaction of the codon of mRNA and the anticodon of tRNA is described in Section 16.8.

FIGURE 16.4 **Structure of tRNA**
(a) A two-dimensional simplified "cloverleaf" structure; the squares represent nucleotides. Base-pairing occurs between a number of bases.
(b) The actual three-dimensional structure. The three loops are indicated in color; the anticodon loop is at the bottom of the structure. The 3′-terminus of every tRNA contains the sequence CCA.

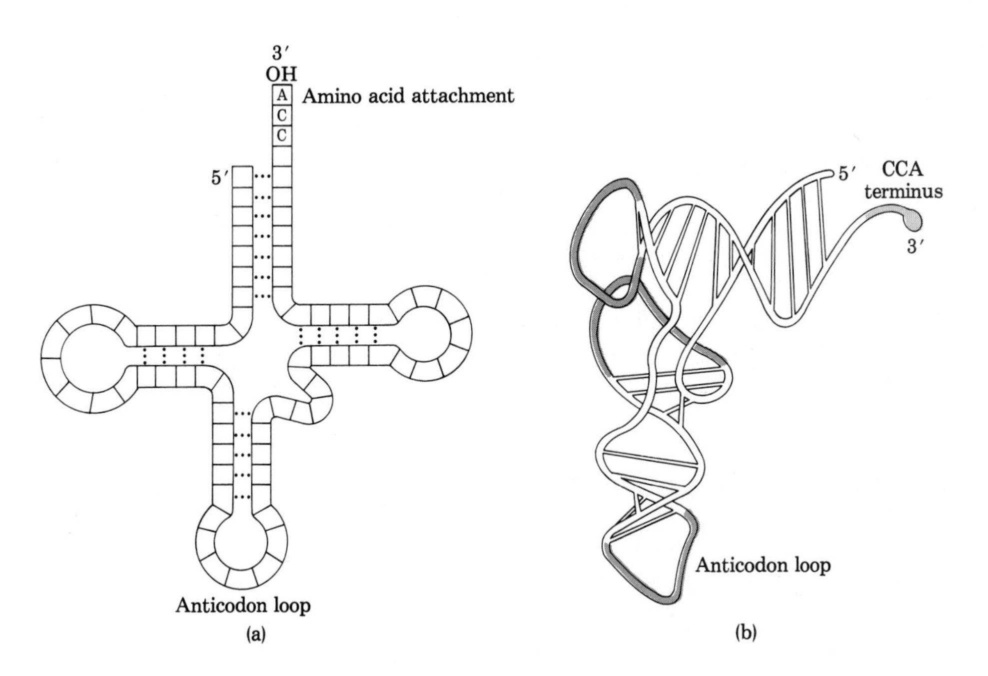

EXAMPLE 16.3 What portion of an mRNA chain will be produced from the following portion of a DNA strand?

$$5' \xrightarrow{\text{dG—C—A—T}} 3'$$

Solution Base-pairing between DNA and the bases in mRNA places C in RNA opposite G in DNA and G in RNA opposite C in DNA. The base T in DNA pairs with A in RNA, but A in DNA pairs with U in RNA.

$$3' \xleftarrow[\text{C—G—U—A}]{} 5'$$

Exons and Introns

The model of a gene as an uninterrupted series of codons that is transcribed to give an mRNA that corresponds to the linear sequence of amino acids in proteins applies only to lower organisms such as bacteria. In higher organisms, genes are not continuous. Sequences of bases that code for an amino acid sequence in a protein are interrupted by base sequences that do not code for anything. The sequence of bases (codons) that are the "genetic code" of DNA are called **exons,** and sequences that lie between the expressed sequences are called **intervening sequences** or **introns** (Figure 16.5).

Because DNA contains both exons and introns, the pre-mRNA initially produced by transcription contains introns. For example, the pre-mRNA produced by transcription of the gene coding for the production of globin has three exons separated by two introns. The noncoding introns are cut out of the pre-mRNA by enzymes, and the exons are spliced together to produce the mRNA that is translated during protein synthesis.

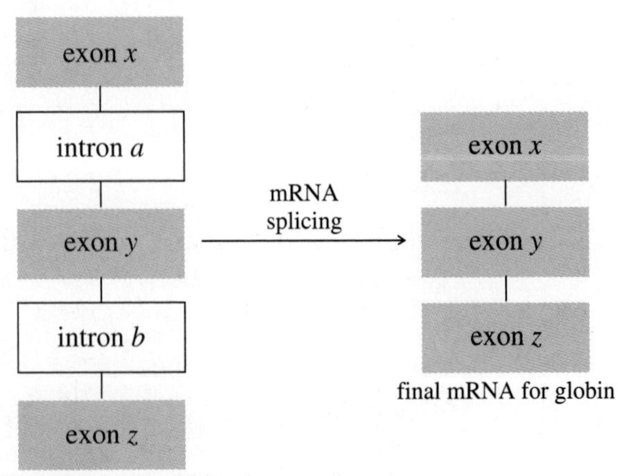

FIGURE 16.5
RNA Processing
The reaction that modifies a pre-mRNA into a final mRNA is called RNA processing. Introns are removed to produce the functional mRNA that passes information from DNA to protein.

16.7 THE GENETIC CODE

DNA is a storehouse of information. It is a passive molecule whose biological function depends entirely upon the action of the enzymes that extract information from it. The relationship of DNA to the proteins that act upon it provides us with a "chicken and egg problem". We need eggs to get chickens, and chickens to get more eggs: we need DNA to obtain the information to make proteins, but we need proteins to synthesize DNA.

How is all of the genetic information coded with only four different bases? George Gamow, of the University of Colorado, suggested that the four bases are a four-letter alphabet whose characters combine to form code words or **codons.** To provide enough codons to uniquely identify each of the 20 different amino acids used in protein synthesis, it is necessary to use three-letter words. With three-letter words, $4 \times 4 \times 4 = 64$ codons are possible (Table 16.3).

After a gene has been copied and the pre-mRNA molecule has been spliced, the final mRNA consists of a linear sequence of three-letter codons. These codons are recognized by tRNA molecules, which contain a complementary base sequence called an **anticodon.**

In 1962 the relationship between a codon in mRNA and a specific amino acid was determined. Ribosomes from the bacterium *Escherichia coli* were

TABLE 16.3 Codons and Amino Acids

First Base		Second Base: U		C		A		G		Third Base
U		UUU	Phe	UCU	Ser	UAU	Tyr	UGU	Cys	U
		UUC	Phe	UCC	Ser	UAC	Tyr	UGC	Cys	C
		UUA	Leu	UCA	Ser	UAA		UGA		A
		UUG	Leu	UCG	Ser	UAG		UGG	Trp	G
C		CUU	Leu	CCU	Pro	CAU	His	CGU	Arg	U
		CUC	Leu	CCC	Pro	CAC	His	CGC	Arg	C
		CUA	Leu	CCA	Pro	CAA	Gln	CGA	Arg	A
		CUG	Leu	CCG	Pro	CAG	Gln	CGG	Arg	G
A		AUU	Ile	ACU	Thr	AAU	Asn	AGU	Ser	U
		AUC	Ile	ACC	Thr	AAC	Asn	AGC	Ser	C
		AUA	Ile	ACA	Thr	AAA	Lys	AGA	Arg	A
		AUG	Met	ACG	Thr	AAG	Lys	AGG	Arg	G
G		GUU	Val	GCU	Ala	GAU	Asp	GGU	Gly	U
		GUC	Val	GCC	Ala	GAC	Asp	GGC	Gly	C
		GUA	Val	GCA	Ala	GAA	Glu	GGA	Gly	A
		GUG	Val	GCG	Ala	GAG	Glu	GGG	Gly	G

bound to synthetic polyuridylic acid molecules, which served as synthetic mRNA molecules. The ribosomes associated only with the tRNA of phenylalanine and formed polyphenylalanine. Therefore, the base sequence U-U-U specifies phenylalanine.

The relationship between each codon and the amino acid incorporated into a protein has been established (Table 16.3). With 64 possible codons, there is more than one codon for most amino acids. A group of codons referred to as **synonyms** can specify the same amino acid. For example, there are four codons for threonine: ACU, ACC, ACA, and ACG. Not all codons specify amino acids. The codons UAA, UAG, and UGA are signals that lead to termination of protein synthesis. This feature of the genetic code will be discussed in the next section.

EXAMPLE 16.4 How are the codons for alanine related?

Solution Table 16.3 shows the codons GCU, GCC, GCA, and GCG for alanine. These four codons are related by identical letters in the first two positions. Only the third letter differs.

16.8 PROTEIN SYNTHESIS

The synthesis of a protein from information contained in mRNA requires the following substances:

1. Amino acids
2. ATP to activate the amino acids
3. Transfer RNA
4. Messenger RNA
5. Ribosomal RNA
6. Guanosine triphosphate (GTP)

Activation of Amino Acids

To prepare an amino acid for protein synthesis, ATP reacts with the carbonyl carbon atom to yield an activated product in which adenosine monophosphate (AMP) is bonded to the amino acid. The product is called an aminoacyl adenylate (Figure 16.6). AMP is linked to the carbonyl carbon atom of an amino acid via a phosphoanhydride bond to its 5′-phosphoryl group. The reaction is catalyzed by a group of enzymes known as aminoacyl synthetases. Each amino acid is activated by a specific aminoacyl synthetase.

Esterification of tRNA

Aminoacyl adenylates react with a molecule of tRNA. Although each tRNA has a unique structure, they all terminate in the sequence CCA. The carbonyl carbon atom of the amino acid forms an ester with a hydroxyl group of the terminal 3′-adenylate residue.

FIGURE 16.6 Formation of an Activated Amino Acid and Activated tRNA

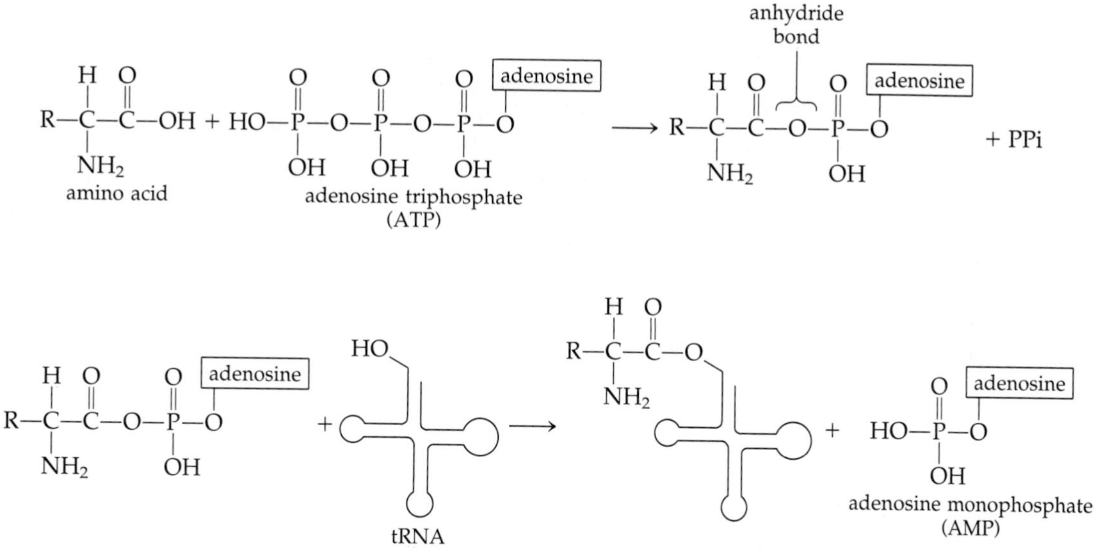

amino acid + adenosine triphosphate (ATP) ⟶ (anhydride bond) + PPi

+ tRNA ⟶ + adenosine monophosphate (AMP)

A tRNA molecule contains a three-base sequence called an anticodon that associates with the codon of mRNA. The anticodon, in one of the sections of the transfer RNA, is responsible for delivering the proper aminoacyl RNA molecule to a codon in mRNA. We have seen that complementary base-pairing is responsible for the double helical structure of DNA, for DNA replication, and for transcription. Complementary base-pairing also is required for matching codons and anticodons.

Site of Protein Synthesis

The mRNA binds a ribosome at the 5′-end of the mRNA. The ribosome then translates the message in mRNA in the 5′ ⟶ 3′ direction. A single mRNA molecule is often bound to several ribosomes, each of which translates it to make a protein. Thus, several ribosomes simultaneously translate the same mRNA molecule. Each ribosome is at a different position along the mRNA chain, and several proteins are in various stages of growth (Figure 16.7). The protein chain is synthesized from the N-terminal toward the C-terminal amino acid. The peptide bond formed is always from the carboxyl group of the last peptide in the growing peptide chain to the amino group of the

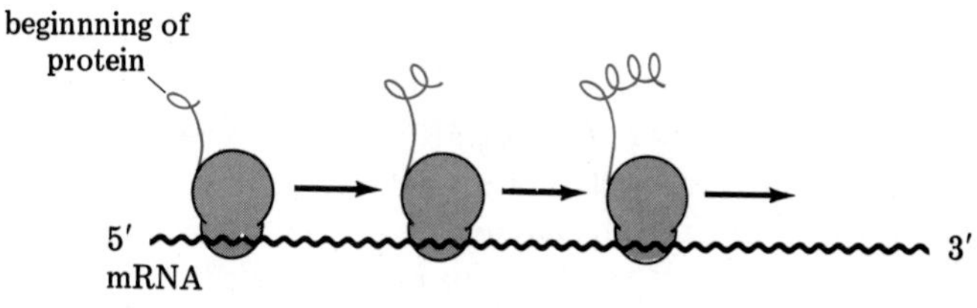

FIGURE 16.7
Ribosomes and mRNA
The ribosomes move from the 5′- to the 3′-end of the mRNA and function independently as they each synthesize a protein chain.

beginnning of protein
5′ mRNA 3′

aminoacyl adenylate. Guanosine triphosphate is required to provide energy for certain steps in the synthesis process.

EXAMPLE 16.5 What anticodon must exist in the template recognition site of an aminoacyl tRNA to bind at the site having the codon sequence AGU in mRNA?

Solution The anticodon must consist of complementary base pairs (U for A, C for G, A for U) in order to bind aminoacyl tRNA to the codon sequence. The base sequence must be U-C-A.

EXAMPLE 16.6 What peptide is synthesized from the following codon sequence in mRNA?

-G-C-U-G-A-A-U-G-G

Solution Reading the letters as sequences of three-letter words, we have the codons GCU, GAA, and UGG. These codons specify alanine, glutamic acid, and tryptophan, respectively. The peptide will be Ala-Glu-Trp.

Initiation and Termination

In the preceding discussion two important features of protein synthesis—initiation and termination—were not considered. **Initiation** refers to the start of the growth of the protein chain. **Termination** refers to the conclusion of the synthesis of the protein chain and its release from the ribosome.

Initiation of protein synthesis is a complex process. However, for many mRNA molecules, the first amino acid in the chain is a modified methionine called formylmethionine. Its codon is AUG.

Termination of the protein synthesis is indicated by the termination codons UAA, UGA, or UAG. No tRNA molecule has an anticodon that pairs

Antibiotics and Protein Synthesis

Many antibiotics—including streptomycin, erythromycin, puromycin, chloramphenicol, and tetracycline—inhibit protein synthesis. These antibiotics exploit the differences in molecular details of protein synthesis in bacteria and eukaryotic cells. These differences enable antibiotics to inhibit protein synthesis in bacteria without adversely affecting protein synthesis in eukaryotic cells.

Streptomycin interferes with the binding of formylmethionine-tRNA and prevents the initiation of protein synthesis. Erythromycin inhibits protein synthesis by interfering with steps in moving the growing protein chain between sites within ribosomal RNA. Puromycin causes premature chain termination, which results in the release of an incomplete protein. Chloramphenicol binds with sites in ribosomal RNA and prevents tRNA from binding to it. The tetracyclines also prevent binding of tRNA to mRNA.

with these trinucleotide sequences. The termination codons bind specific proteins called release factors. Binding of these factors activates the enzyme needed to hydrolyze the peptide from the ribosome.

16.9 GENE MUTATIONS AND GENETIC DISEASE

DNA is subject to a host of environmental assaults that can cause alterations in its nucleotide sequence. The changes are called **mutations.** Mutations are of three types.

1. Missense mutations occur when one nucleotide substitutes for another, altering the nucleotide sequence within a codon.
2. Frameshift mutations occur by insertion or deletion of nucleotides. They alter the "reading frame" of the nucleotide sequence.
3. Nonsense mutations change a codon that specifies an amino acid to a termination codon that prematurely terminates protein synthesis.

We can examine the nature of mutations by considering what happens to a sentence when letters are substituted, added, or deleted. We will use three-letter words in our sentence because codons contain three nucleotides. Our sentence is

THE BIG DOG SAW THE CAT

Substituting one letter for another in the message changes the meaning of that word, but does not alter the other words. For example, consider the effect of replacing "G" with "H" in "DOG".

THE BIG DOH SAW THE CAT

The meaning of the message is similar, but the sense is different. If you wanted to send information about a dog, the sentence with the word DOH is seriously flawed.

Consider adding a letter to the sentence. If the words in the sentence, like the codons in a gene, are still read three letters at a time, a major part of the sentence changes. Consider the addition of "E" after "G" of "DOG".

THE BIG DOG ESA WTH ECA T

Deletion of a letter such as the "G" of "DOG" also results in a serious loss of information.

THE BIG DOS AWT HEC AT

In the above two examples, not only is the sense of one word changed but the meaning of the sentence is obliterated after the changed word.

Substituting, adding, and deleting nucleotides change codons the same way changing letters alters sentences. If DNA is mutated, then the transcribed mRNA molecule perpetuates the error. As a consequence, protein synthesis is flawed. The effect of defective proteins on biological function depends on which base or bases are changed.

First, let's consider the consequences of substituting one nucleotide for another—a missense mutation. The consequences of changing the nucleotide corresponding to the third letter in a three-letter word sequence may not be serious because many amino acids have several codons that differ only in the third letter. For example, valine is coded by GUA, GUG, GUU, and GUC. Replacing the base corresponding to the third letter still places the appropriate amino acid in the protein. Some base mutations, however, can lead to serious biological damage. For example, if the wrong amino acid is placed at the active site of an enzyme, the enzyme will lose its catalytic activity. Consider the codon UCG for serine. Mutation to give UUG results in substitution of leucine for serine. If serine is required at the active site, the enzyme with the substituted leucine will be inactive.

Mutations in amino acids that are not required at enzyme-active sites may change the conformation of a protein and alter its biological function. For example, in normal adult hemoglobin, glutamic acid is the sixth amino acid in the β chain. The codons for glutamic acid are GAA and GAG. The codons for valine are GUA, GUG, GUU, and GUC. Substitution of U for A, the second letter in the glutamic acid codon, gives a codon for valine. The replacement of glutamic acid by valine changes the tertiary structure of hemoglobin because glutamic acid is an acidic amino acid, whereas valine is a neutral amino acid.

A mutation caused by deletion or addition—a frame shift error—has serious consequences. As we noted earlier, such a change in a sentence alters all words after the insertion or deletion. Beyond the point of change in a

A Gene Defect May Cause Osteoarthritis

Osteoarthritis is a disease in which the cartilage between bones disintegrates, and pain, stiffness, and loss of mobility results. This disease disables millions of people. Until recently it was thought that this disease was an inevitable consequence of aging. However, in 1990, researchers at the Jefferson Medical College in Philadelphia found a genetic mutation responsible for the production of the weak cartilage associated with osteoarthritis.

Cartilage contains a protein called collagen II, which contains more than 1000 amino acids. Collagen II strengthens cartilage. In some people who suffer from osteoarthritis, the gene coding for collagen II has a mutation in which a cysteine residue is replaced by an arginine residue. This mutation has serious consequences because cysteine is the amino acid that forms disulfide bridges. Furthermore, arginine is a basic amino acid, whereas cysteine is a neutral amino acid. This one substitution significantly affects the structural integrity of collagen II and causes it to disintegrate.

polynucleotide, the instructions lead to the formation of a completely unsuitable sequence of amino acids.

A nonsense mutation occurs when the codon for an amino acid is changed into a termination codon. Thus, protein synthesis terminates prematurely. One of the codons for serine is UCA. If G is substituted for C, the termination codon UGA results. When this codon is reached, the synthesis of the protein stops. Such a truncated protein is nonfunctional.

A frameshift can also generate a termination codon. Thus, frameshift mutations not only result in the improper incorporation of many amino acids in a polypeptide, but can also cause premature termination of protein synthesis.

EXAMPLE 16.7 What are the consequences of a genetic mutation that results in replacement of CGC in mRNA by CCC?

Solution The change is in the second letter and will result in the specification of a different amino acid in protein synthesis. The codon CGC specifies the basic amino acid arginine, whereas the codon CCC specifies proline, a cyclic amino acid. Because these amino acids differ substantially in structure and acid–base properties, the tertiary structure and function of the protein may be seriously altered.

16.10 RECOMBINANT DNA AND GENETIC ENGINEERING

Now that we understand the fundamentals of nucleic acid chemistry, we might ask if chemists can manipulate DNA and RNA in the same way in the laboratory. The answer is yes. The laboratory manipulation of DNA provides the foundation of the biotechnology industry. DNA can be manipulated in the laboratory to make new genes and to introduce genes from one organism to another organism. The construction of DNA to make altered genes is called **genetic engineering.**

Modification of Bacterial DNA

Bacteria have a single chromosome that contains more than 10^9 base pairs. This chromosomal DNA codes for most bacterial proteins. However, bacteria such as the intestinal bacterium *Escherichia coli (E. coli)* also contain small, circular DNA molecules called **plasmids.** Plasmids code only a few genes. They are replicated independently of the chromosomal DNA.

Plasmids can be removed from *E. coli* and "snipped" open by specific enzymes called **restriction endonucleases.** An opened plasmid has two "sticky ends" with nucleotides that can base-pair with other polynucleotides (Figure 16.8). The sticky ends of plasmid DNA can base-pair with human genes snipped from human chromosomes by the same restriction endonuclease. The human DNA is inserted into the opened bacterial plas-

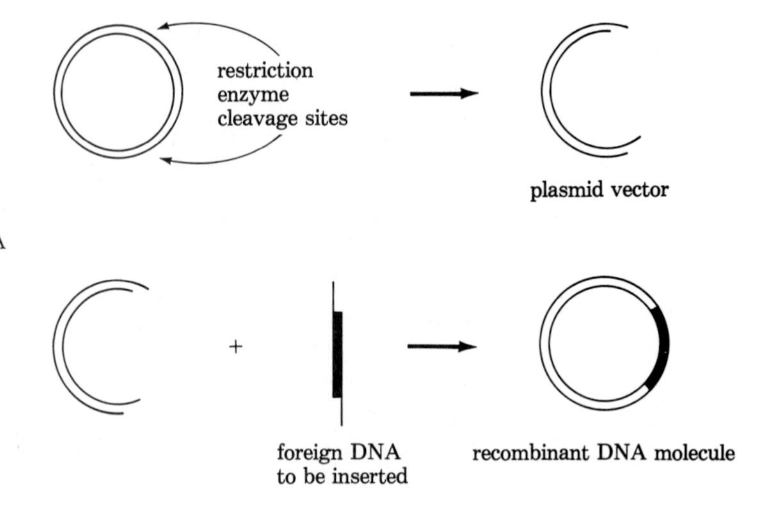

FIGURE 16.8
Recombinant DNA
In the first step a plasmid from a prokaryotic cell is cleaved with a restriction endonuclease. The plasmid vector is then joined to a section of "foreign" DNA obtained by use of a restriction enzyme on DNA of a eukaryotic cell.

mid by an enzyme called DNA ligase. The modified plasmid is called **recombinant DNA.**

The modified plasmid is reinserted in *E. coli.* Then, the bacterium multiplies and reproduces the new plasmid each time it divides. In this way, the original copy of the gene is greatly amplified. When the genes control the protein synthesis machinery of *E. coli,* the bacterium produces proteins that it normally does not make. For example, the gene for human insulin can be placed into a bacterial plasmid, and the bacterium then proceeds to make human insulin. The insulin can be "harvested" from the *E. coli* colony, which functions as a miniature chemical factory.

The human insulin produced by *E. coli* can be used by people who are diabetic. Diabetics can use insulins obtained from animal sources that differ only slightly from human insulin. However, allergic reactions sometimes occur. Because *E. coli* can make human insulin, diabetics do not have an allergic reaction to it.

16.11 VIRUSES

Viruses cause many diseases in plants and animals, including smallpox, measles, mumps, and influenza in humans. A virus is not an organism. It consists of a strand of genetic material, either DNA or RNA, enclosed in a protein coat. Sometimes this coat is surrounded by a membrane. Viruses do not have the necessary nucleic acids to replicate themselves, nor do they possess enzymes necessary to support life.

Viruses cannot independently reproduce themselves. They must attack cells and take over the chemical machinery of DNA replication and protein synthesis. The protein coat of a virus contains an enzyme that breaks down the cell membrane of the host cell. This coat remains outside the cell while the DNA or RNA is injected into the target cell. Each virus has a unique membrane-dissolving enzyme that allows it to attack only selected types of cells.

Once inside the cell, the viral nucleic acids can become a "silent gene" and remain for long periods of time before becoming active. Alternatively, they may immediately take over the operation of the cell, which then stops making its own DNA, RNA, and protein. The cell starts to replicate the nucleic acids of the virus as well as viral proteins. Many copies of the virus are made by a single cell, which is eventually destroyed because it no longer produces the materials necessary for its own survival. The new viruses are released and attack other cells. If the process is not stopped, the host organism gets sick and may die.

Mechanism of Viral Action

When a virus enters a cell, one of the first events is replication of the viral genome. If the virus has an RNA genome, the host cell has no enzymes that are able to copy the RNA genome of the virus. Viral RNA replication can occur in either of two ways. One process for replicating the viral RNA employs an enzyme called RNA replicase that makes duplicate copies of the viral RNA. The gene for RNA replicase is contained in the viral genome. Thus, viral RNA acts as a messenger RNA that is translated to produce RNA replicase. Thus, contrary to the so-called central dogma, some RNA molecules can reproduce themselves. RNA replicase is used by polio virus, rabies virus, and many other viruses to replicate genetic material.

The second mechanism for replicating the RNA genome of a virus occurs in viruses whose RNA can direct the synthesis of an enzyme called reverse transcriptase. This enzyme makes a DNA copy of the RNA genome. This single-stranded DNA serves as a template for the synthesis of double-helical DNA by the DNA replication machinery of the cell. As a consequence, RNA is responsible for the synthesis of DNA, which then can form more RNA. Viruses that operate by this mechanism are called **retroviruses.** Note that the entire process is a violation of the originally stated central dogma.

Antiviral Drugs

Antibiotics act by killing bacterial infections. These drugs are not effective against diseases induced by viruses. (A virus is not "living", so there is no way to kill it.) There are some virucidal agents, but their effectiveness is limited. An antiviral drug must stop the reproduction of viral nucleic acids inside host cells but not prevent the normal replication of the cell's DNA. Vidarabine, sold as Vira-A, is an antiviral agent. Its structure is similar to adenosine but its sugar is arabinose rather than ribose. (Arabinose and ribose have the opposite configuration at the C-2 atom.) Vira-A is used only to fight life-threatening diseases because it is also toxic if dosages are not carefully controlled. It may also cause chromosomal damage.

Many viral infections can be prevented by immunization with a specific vaccine. The vaccine contains an inactive virus that stimulates the body's

The HIV Retrovirus

AIDS is caused by the retrovirus called human immunodeficiency virus—that is, HIV. This deadly virus has an RNA genome and a gene that codes for a reverse transcriptase. The genome is surrounded by a protein coat that is contained within a lipid bilayer.

The mechanism of reproduction of the HIV virus resembles that of other retroviruses. However, the site attacked is quite unusual. The HIV virus enters the T4 lymphocyte cell, an important cell in the human immune system. HIV reverse transcriptase translates its RNA genome into DNA in the T4 cell, which can then no longer function in the immune system. Instead, the cell now produces RNA of the retrovirus and releases it to attack other T4 cells. Eventually the AIDS victim succumbs to infectious diseases, such as *Pneumocystis carinii* pneumonia, to some rare cancers, or to the ravaging effects of the HIV virus itself. HIV eventually gets to the brain and causes dementia, and, in combination with the low blood counts due to both HIV and AZT, the body ceases to function properly.

AZT appears to act against the AIDS virus by taking the place of thymidine in the DNA copy of the RNA genome.

natural immune system to produce antibodies against the virus. These antibodies circulate in the body and deactivate invading active viruses before they can enter host cells.

Smallpox has been eradicated by a worldwide program of vaccination. Similarly, the incidence of measles and polio has been greatly reduced in the United States and many other countries.

EXERCISES

DNA and RNA

16.1 How are all DNA molecules structurally alike?

16.2 How do DNA molecules differ?

16.3 What group bridges the sugars in DNA and RNA?

16.4 DNA and RNA are acidic materials. Explain why.

16.5 Indicate whether each of the following is a purine or pyrimidine base.
(a) uracil (b) cytosine (c) adenine (d) thymine

16.6 Which bases are found in DNA? Which bases are found in RNA?

Nucleosides and Nucleotides

16.7 Write the structural formula for adenosine monophosphate.

16.8 Write a structural formula for a nucleoside containing thymine and deoxyribose.

16.9 What is the configuration of the *N*-glycosidic bond of nucleotides?

16.10 What hydroxyl group of ribonucleosides is phosphorylated to give ribonucleotides?

16.11 How many free hydroxyl groups exist in a polynucleotide of deoxyribose?

16.12 How many free hydroxyl groups exist in a polynucleotide of ribose?

16.13 How many chiral centers are there in a ribonucleoside?

16.14 How many chiral centers are there in a deoxyribonucleoside?

16.15 Identify the components of the following substance.

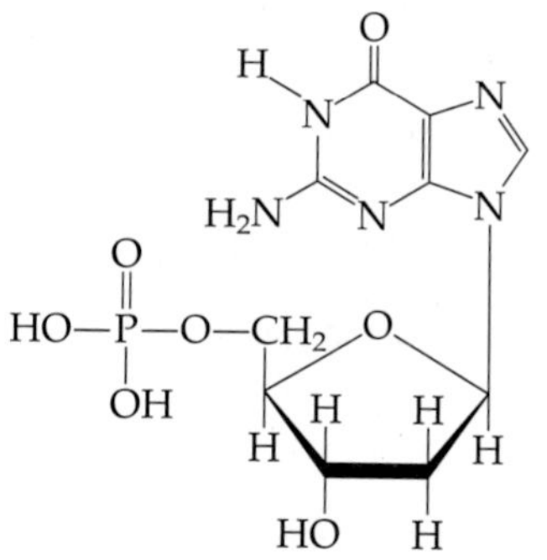

16.16 Identify the components of the following substance.

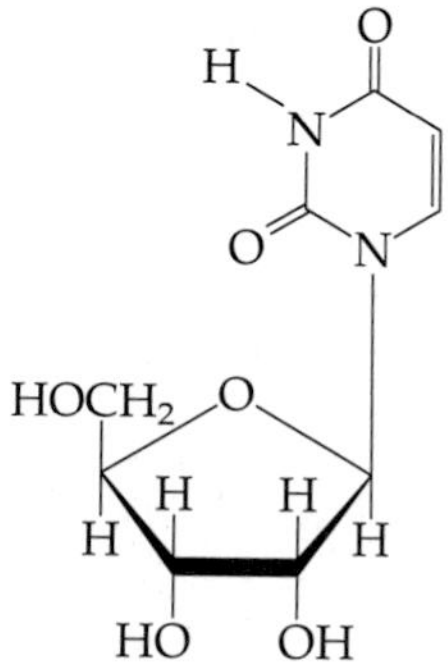

The Double Helix

16.17 Does each of the following base pairs occur in the DNA molecule? If not, explain why.
(a) G-A (b) G-T (c) C-U

16.18 Does each of the following base pairs occur in the DNA molecule? If not, explain why.
(a) A-A (b) C-T (c) A-C

16.19 What forms the backbone structure of each strand of the DNA double helix?

16.20 Where are the base pairs located in the double helix?

Replication

16.21 Describe the replication process.

16.22 What is the replicating fork?

16.23 What is the complementary strand for a DNA strand with the sequence dG-C-A-T-C-A-G?

16.24 What is the complementary strand for a DNA strand with the sequence dA-T-C-A-G-T-A?

Transcription

16.25 What mRNA would be formed from the base sequence in Exercise 16.23?

16.26 What mRNA would be formed from the base sequence in Exercise 16.24?

16.27 What is the complementary base in mRNA for A in DNA?

16.28 What base in mRNA is the complement of T in DNA?

16.29 A portion of DNA has the sequence C-C-C-T-G-T-A-C-A-C-C-T. What base sequence will form in mRNA? What peptide would be formed?

16.30 A portion of DNA has the sequence G-G-G-T-G-C-A-G-A-C-C-A. What base sequence will form in mRNA? What peptide would be formed?

Translation

16.31 How is a codon involved in the translation process?

16.32 How is an anticodon involved in the translation process?

16.33 What amino acid does each of the following codons in mRNA place in a protein chain?
(a) GUU (b) CCC (c) UUU (d) ACG

16.34 What amino acid does each of the following codons in mRNA place in a protein chain?
(a) UGU (b) GCA (c) AGG (d) CUA

16.35 What anticodons in tRNA will base-pair with the codons of Exercise 16.33?

16.36 What anticodons in tRNA will base-pair with the codons of Exercise 16.34?

16.37 What base sequence in mRNA is necessary to form Val-Asp-Ala-Gly?

16.38 What base sequence in mRNA is necessary to form Ser-Glu-Pro-Phe?

Codons and Mutations

16.39 What are the genetic consequences of a frame shift?

16.40 Why are substitution mutations generally less serious than deletion or addition mutations?

16.41 Describe two types of substitution mutations that will cause seriously defective proteins.

16.42 What are the differences between the codons for acidic amino acids and those for basic amino acids?

Recombinant DNA

16.43 What are restriction endonucleases?

16.44 Where is a plasmid found? What is it made of?

16.45 What are the sources of the DNA used in producing recombinant DNA?

16.46 Why might it be more difficult to use recombinant DNA techniques in human cells?

Viruses and Bacteria

16.47 How can we chemically fight bacterial infections?

16.48 How are viral infections different from bacterial infections?

16.49 Plants are susceptible to some viral infections. Why don't these viruses affect humans?

16.50 What enzyme must a virus have in order to produce RNA from viral RNA? What enzyme must a virus have in order to produce DNA from viral RNA?

16.51 Vidarabine is employed in the treatment of viral encephalitis. What is the carbohydrate component? What nitrogen base does the compound contain?

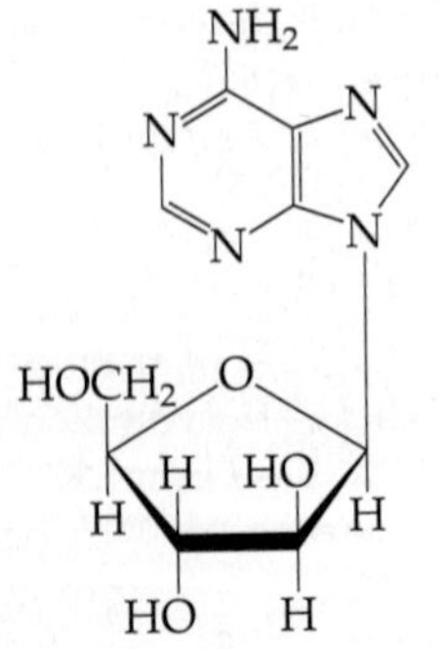

16.52 Trifluridine is employed in the treatment of herpes simplex viruses. What is the carbohydrate component? What nitrogen base does the base in the compound most closely resemble? How does it differ?

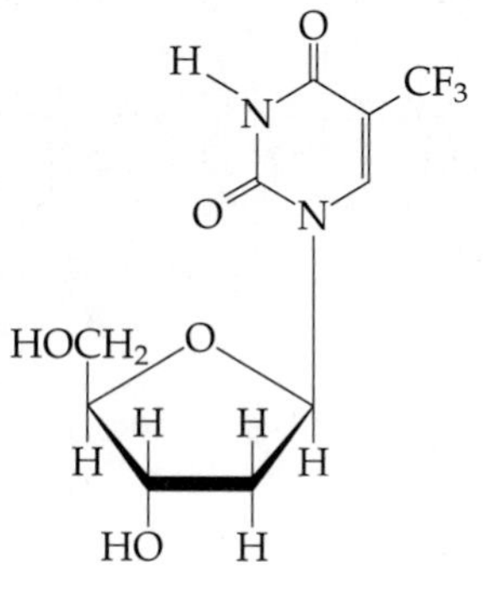

H N O CF3 O N HOCH2 O H H H H HO H

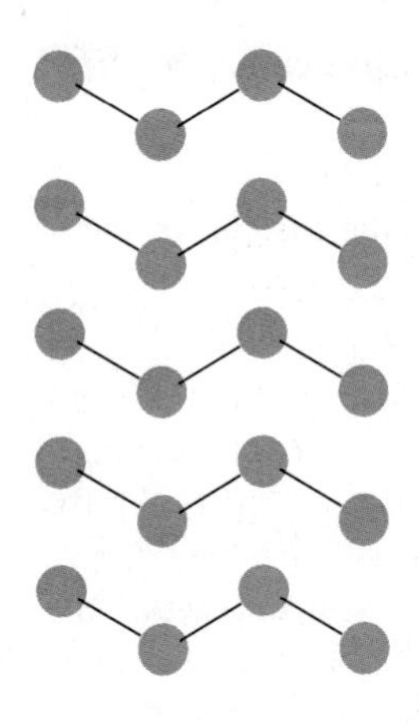

17

STRUCTURE DETERMINATION AND SPECTROSCOPY

17.1 STRUCTURE DETERMINATION

How does a chemist know the molecular structure of a compound obtained from either a natural source or as the product of a chemical reaction? The question is of paramount importance. The structure of a natural product is often very complex and may be obtained initially in only small quantities. We must first learn the structure of a compound to understand the details of its biological function or to synthesize it in the laboratory. The determination of the structure of a compound synthesized in the laboratory from known reactants is a somewhat less demanding, but still challenging, process.

At one time, organic chemists determined the structure of an organic compound by chemical reactions that related the unknown compound to other known compounds. The reactions used were oxidation, reduction, addition, substitution, or elimination reactions. To determine the structures of complex molecules, it was often necessary to use reactions that systematically degraded the large molecule into smaller molecules. Then the chemist could "reason backwards" to postulate what the structure of the original compound must have been to yield the observed products. Thus, structure determination by chemical reactions is a time-consuming process that requires experimental skills and deductive reasoning. Structural determination by chemical reactions has one severe limitation: part of the sample of the unknown compound is destroyed in each chemical reaction.

The structure determination process is now carried out by nondestructive spectroscopic instruments, and the sample can be recovered unchanged after its spectrum is determined. Spectroscopic information can be obtained by using very small amounts of the compound. Spectroscopic methods provide information about the kind of atoms present in the compound and how they are connected. These experimental methods require very little time. Although the number of sophisticated instruments increases each year, we will examine only three: (1) ultraviolet, (2) infrared, and (3) nuclear magnetic resonance spectroscopy. Each provides a different kind of information.

Ultraviolet spectroscopy provides information about the π system in a compound. Thus, the method is used to distinguish between conjugated and nonconjugated compounds. Infrared spectroscopy is used to identify the functional groups in a compound. The method can, for example, distinguish between an aldehyde and a ketone. Nuclear magnetic resonance is used to determine the carbon–hydrogen framework. This method is the most powerful of the three instrumental methods discussed in this chapter.

17.2 SPECTROSCOPY

Spectroscopy is a study of the interaction of electromagnetic radiation with molecules. Electromagnetic radiation encompasses X-rays, ultraviolet, visible, infrared, microwaves, and radio waves. Electromagnetic radiation is described as a wave that travels at the speed of light (3×10^8 m/s). Waves are characterized by a wavelength (λ, Greek lambda) and a frequency (ν, Greek nu). The **wavelength** is the length of one wave cycle, such as from trough to trough (Figure 17.1). The wavelength is expressed in the metric unit convenient for each type of electromagnetic radiation. The **frequency** is the number of waves that move past a given point in a unit of time. Frequency is usually expressed in hertz (Hz). Wavelength and frequency are inversely proportional and are related by $\lambda = c/\nu$, where c is the speed of light. Thus, as the wavelength of the electromagnetic radiation increases, the corresponding frequency decreases.

The energy (E) associated with electromagnetic radiation is quantized. The relationship is given by

$$E = h\nu = \frac{hc}{\lambda} = hc\frac{1}{\lambda}$$

where h is the proportionality constant known as Planck's constant. Thus, the energy of electromagnetic radiation is directly proportional to its frequency but inversely proportional to its wavelength. The energy of electromagnetic radiation is also directly proportional to the quantity $1/\lambda$, known as the **wavenumber.** The frequency of ultraviolet radiation is higher than

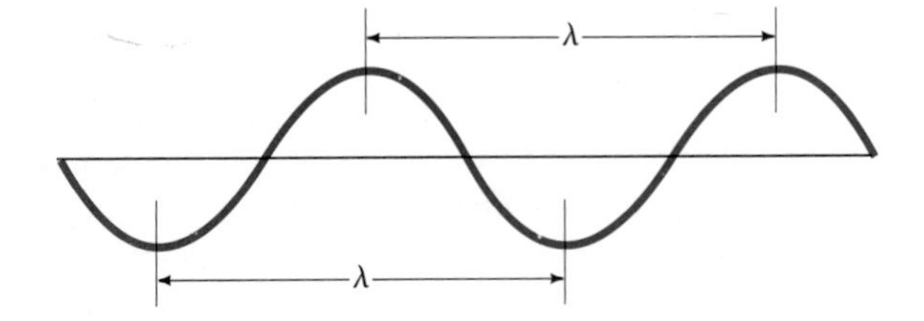

FIGURE 17.1
Electromagnetic Radiation
The wavelength of electromagnetic radiation is the distance between any two peaks or troughs of the wave.

that of infrared radiation. Alternatively expressed, the wavelength of ultraviolet radiation is shorter than the wavelength of infrared radiation. Because ultraviolet radiation has a high frequency, it has higher energy than infrared radiation (Figure 17.2).

Molecules can absorb only certain discrete amounts of energy. To change the energy content of a molecule from E_1 to E_2, the energy difference $E_2 - E_1$ is provided by a characteristic electromagnetic radiation with a specific frequency (and wavelength). The energy absorbed by the molecule can change its electronic or vibrational energy. For example, ultraviolet radiation causes changes in the electron distribution in π bonds; infrared radiation causes bonds to stretch and bond angles to bend.

In the various types of spectroscopy, radiation is passed from a source through a sample that may or may not absorb certain wavelengths of the radiation. The wavelength is systematically changed, and a detector determines which wavelengths of light are absorbed (Figure 17.3). At a wave-

FIGURE 17.2 The Electromagnetic Spectrum
The regions of the spectrum used in organic chemistry are characterized by a frequency and a wavelength. Usually the wavelength or the reciprocal of the wavelength, the wavenumber, is used to identify absorptions of organic molecules. The relationship of the visible spectrum to other spectral regions is shown with an expansion of that region.

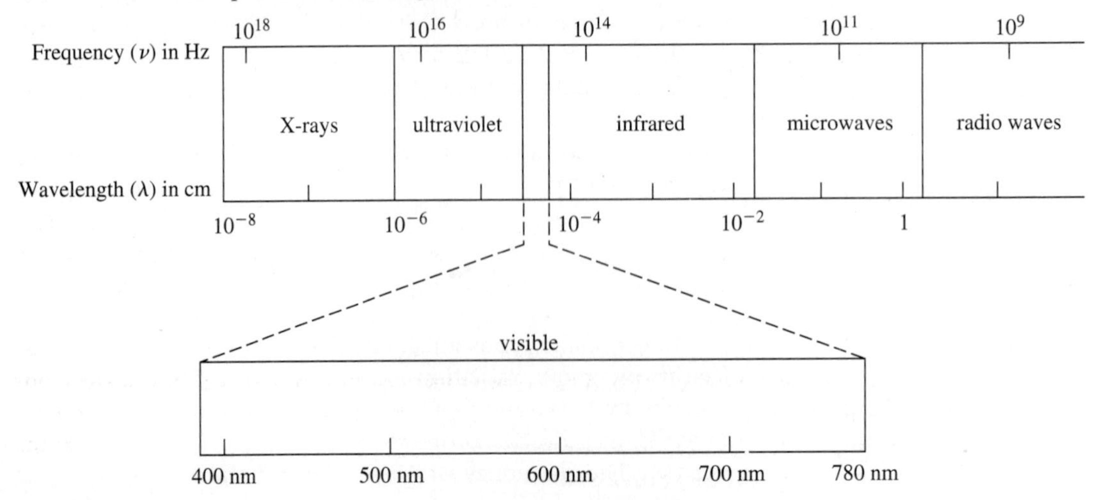

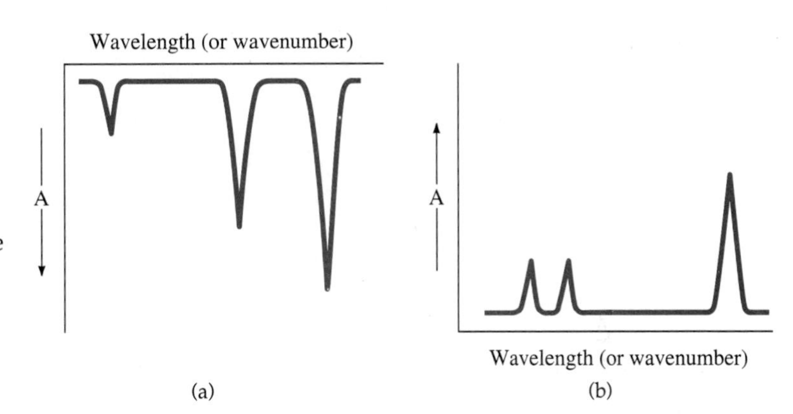

FIGURE 17.3
Features of a Spectrum
The portion of the spectrum where no absorption occurs is the base line. This horizontal line may be located at the top or bottom of the graph. Absorption then is recorded as a peak or "dip" from the base line. In an infrared spectrum (a) the base line is at top of the spectrum. In an ultraviolet spectrum (b) or an NMR spectrum, the base line is at the bottom of the spectrum.

length that corresponds to the energy $E_2 - E_1$ necessary for a molecular change, the radiation emitted by the source is absorbed by the molecule. The amount of light absorbed by the molecule (absorbance) is plotted as a function of wavelength. At most wavelengths the amount of radiation detected by the detector is equal to that emitted by the source—that is, the molecule does not absorb radiation. At such wavelengths a plot of absorbance on the vertical axis versus wavelength yields a horizontal line (Figure 17.3) When the molecule absorbs radiation of a specific wavelength, the amount of radiation that arrives at the detector is less than that emitted by the source. This difference is recorded as an absorbance.

17.3 ULTRAVIOLET SPECTROSCOPY

The ultraviolet region of the electromagnetic spectrum spans wavelengths from 200 to 400 nm (1 nm = 10^{-9} m). In the ultraviolet region of the electromagnetic spectrum a molecule with conjugated double bonds absorbs energy. Sigma bonds as well as isolated carbon–carbon double bonds require electromagnetic radiation of higher frequency to absorb energy.

Ultraviolet (UV) spectra are simple in appearance. A UV spectrum is a plot of the absorbance of light on the vertical axis and the wavelength of light in nanometers (nm) on the horizontal axis (Figure 17.4). The wavelength corresponding to the UV "peak" is called the λ_{max}. The intensity of the absorption depends on the structure of the compound and the concentration of the sample in the solution. Concentrations in the 10^{-3} to 10^{-5} M range are typically used to obtain a spectrum.

The energy absorbed by conjugated systems moves bonding π electrons into higher energy levels. The specific wavelength of ultraviolet light required to "excite" the electrons in a conjugated molecule depends on the structure of the compound.

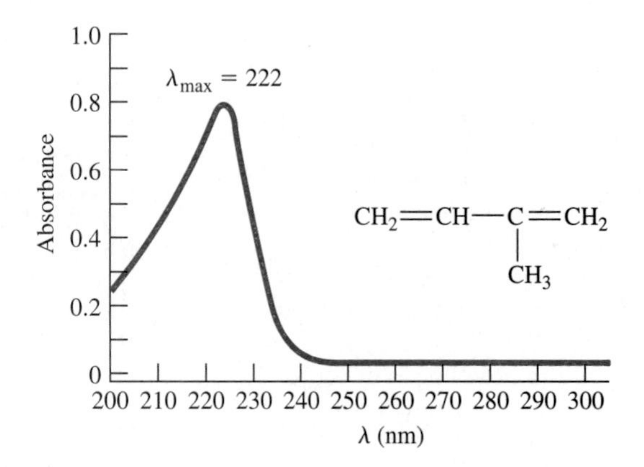

FIGURE 17.4
Ultraviolet Spectrum of Isoprene
The ultraviolet spectrum of isoprene dissolved in methanol is representative of the spectra of conjugated dienes. The position of maximum absorption occurs at 222 nm.

Ultraviolet spectroscopy provides information about the extent of the conjugation: as the number of double bonds increases, the wavelength of light absorbed also increases. For 1,3-butadiene, the λ_{max} = 217 nm. With an increased number of conjugated double bonds, compounds absorb at longer wavelengths, that is, a lower energy is required for electronic excitation. For example 1,3,5-hexatriene and 1,3,5,7-octatetraene have λ_{max} at 268 and 304 nm, respectively.

Although the extent of the conjugation is the primary feature in affecting the λ_{max}, the degree of substitution causes changes that are useful in structure determination. For example, the presence of an alkyl group such as methyl appended to one of the carbon atoms of the conjugated system causes approximately a 5 nm shift to longer wavelengths. Thus, 2-methyl-1,3-butadiene (isoprene) absorbs at 222 nm, compared to 217 nm for 1,3-butadiene.

EXAMPLE 17.1 Predict the λ_{max} of 2,4-dimethyl-1,3-pentadiene.

Solution First draw the structure of the compound and determine the structural features that can affect the position of the ultraviolet absorption.

$(H)_2C{=}C(CH_3){-}CH{=}C(CH_3)_2$

The compound contains two conjugated double bonds. Thus, the compound should absorb near 217 nm, as for butadiene. However, two branching methyl groups and the terminal C-5 methyl group are bonded to the unsaturated carbon atoms of the butadiene-type system. Thus, the compound should absorb at 217 + 3(5) = 232 nm.

Some naturally occurring compounds with extensively conjugated double bonds absorb at such long wavelengths that the λ_{max} occurs in the visible region (400–800 nm) of the spectrum. β-Carotene, which is contained in carrots, absorbs at 455 nm—the blue-green region of the spectrum. Because blue-green light is absorbed, the color that is reflected to our eyes is yellow-orange; that is, we see the complement of the absorbed light. Thus, the color of a compound provides qualitative information about its λ_{max} (Table 17.1). A compound is colored only if absorption occurs in some portion of the visible spectrum. Compounds that absorb only in the ultraviolet region are colorless because no "visible" light is absorbed.

Other kinds of conjugated molecules besides polyenes have ultraviolet absorptions. For example, benzene absorbs at 254 nm, and substituents on the aromatic ring affect the position of the absorption. By studying the effect of substituents on aromatic rings, as well as other classes of conjugated compounds, chemists have compiled information relating structural effects on ultraviolet spectra that make it possible to establish the structures of "unknown" compounds.

EXAMPLE 17.2 Naphthalene and azulene are isomeric compounds that have extensively conjugated π systems. Naphthalene is a colorless compound, but azulene is blue. Deduce information about the absorption of electromagnetic radiation by these two compounds.

naphthalene azulene

TABLE 17.1 Absorbed Light and Reflected Color

Absorbed wavelength (nm)	Reflected color
400 (violet)	yellow-green
450 (blue)	orange
510 (green)	purple
590 (orange)	blue
640 (red)	blue-green
730 (purple)	green

Solution The π electrons of naphthalene would be expected to absorb energy. However, because the compound is colorless, its λ_{max} must be in the ultraviolet region. Azulene is blue, which indicates that a portion of the visible spectrum is absorbed by its π system. The complement of blue is orange. Thus, azulene must have a λ_{max} near 590 nm.

17.4 INFRARED SPECTROSCOPY

Bonded atoms in a molecule do not remain at fixed positions with respect to each other. Molecules vibrate at various frequencies that depend on molecular structure. Similarly, the angle between two atoms bonded to a common central atom expands and contracts by a small amount at a frequency that depends on molecular structure. These vibrational and bending frequencies correspond to the frequencies of light in the infrared region in the electromagnetic spectrum.

For every type of bond or bond angle there is a specific frequency (wavelength or wavenumber) at which the molecule absorbs infrared radiation. However, the number of different absorptions is large for even the simplest of organic molecules. The infrared spectrum of 4-bromo-1-butene is shown in Figure 17.5. The wavelength, given on the bottom of the graph, is plotted against percent transmittance of light by the sample. An absorption corresponds to a "peak" pointed toward the bottom of the graph. Because the wavelength of absorbed light is inversely proportional to its energy, absorptions that occur at high wavelength (toward the right of the graph) represent molecular vibrations that require low energy. The plot also gives

FIGURE 17.5 Infrared Spectrum of 4-Bromo-1-butene
Wavenumbers are given on the top horizontal axis. Wavelengths in microns (micrometers) are given on the bottom horizontal axis. The absorption peaks are directed toward the bottom of the graph.

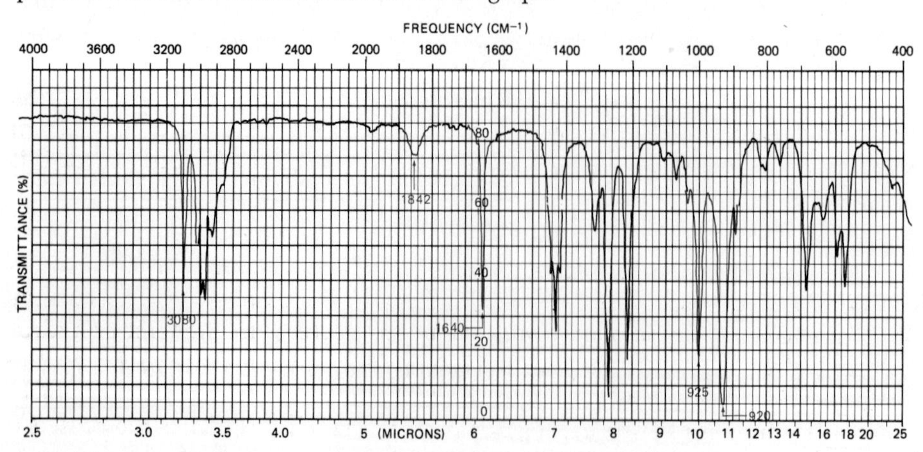

the corresponding value of the wavenumber, $1/\lambda$, for the absorption. The energy of absorbed light is directly proportional to the wavenumber. Thus, absorptions that occur at higher wavenumbers (toward the left of the graph) represent molecular vibrations that require higher energy.

Establishing the Identity of an Unknown Compound

The infrared spectrum of an organic molecule is complex and a peak-by-peak analysis is very difficult. However, the total spectrum is characteristic of the compound and can be used to clearly establish the identity of an "unknown" compound. If the spectrum of the "unknown" has all of the same absorption peaks—both wavelength and intensity—as a compound of known structure, then the two samples are identical. If the "unknown" has one or more peaks that differ from the spectrum of a known, then the two compounds are not identical or some impurity in the "unknown" sample causes the extra absorptions. On the other hand, if the "unknown" lacks even one absorption peak that is present in the known structure, the unknown has a different structure than the known. For example, if the infrared spectrum of a compound of molecular formula C_4H_7Br lacks an absorption at 1640 cm^{-1}, the compound is not 4-bromo-1-butene.

Characteristic Group Vibrations

Although the infrared spectrum of an organic compound is complex, distinctive bands appear in spectra of compounds with common functional groups. Distinctive absorptions corresponding to the vibration of specific bonds or functional groups are called **group vibrations.** Thus, we can use the presence or absence of these absorptions to characterize a compound. For example, the absorption at 1640 cm^{-1} in 4-bromo-1-butene is due to the stretching of a carbon–carbon double bond (Figure 17.5). Although the exact position of carbon–carbon double-bond absorptions varies slightly for various alkenes, they are all in the 1630–1670 cm^{-1} region. Bromocyclobutane, an isomeric compound, does not have a carbon–carbon double bond and does not have an absorption in the 1640 cm^{-1} region. Isomeric compounds such as 2-bromo-1-butene or *trans*-1-bromo-2-butene will have an absorption in the 1640 cm^{-1} region. However, the spectra will differ in some other areas, which tells us that the compounds are not identical to the known 4-bromo-1-butene.

In the following subsections, we will consider a few characteristic group vibrations. The small differences in group vibrations that result from minor but important differences in structure will not be discussed in this text.

Identifying Hydrocarbons

When we first started to study organic chemistry we learned that hydrocarbons are classified as saturated and unsaturated based on the absence or presence of multiple bonds. Multiple bonds decrease the number of hydrogen atoms in a molecular formula below the number given in C_nH_{2n+2}, the molecular formula for a saturated acylic hydrocarbon. However, the molecular formulas of hydrocarbons do not unambiguously indicate the presence or absence of a multiple bond. Both 1-octene and cyclooctane have the same molecular formula, C_8H_{16}.

$CH_3(CH_2)_5—CH{=}CH_2$

1-octene

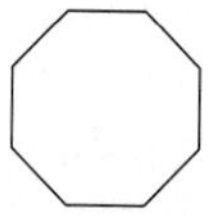

cyclooctane

The structural features present in 1-octene and absent in cyclooctane are a carbon–carbon double bond and sp^2-hybridized C—H bonds. If these features give rise to characteristic group absorptions, then 1-octene will have them and cyclooctane will not.

Now also consider the difference between 1-octyne and an isomeric bicyclic hydrocarbon.

$CH_3(CH_2)_5—C{\equiv}C—H$

1-octyne

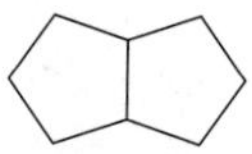

a bicyclic hydrocarbon

The structural features present in 1-octyne and absent in the bicyclic hydrocarbon are a carbon–carbon triple bond and an *sp*-hybridized C—H bond. If these features give rise to characteristic group absorptions, then 1-octyne will have them and a bicyclic hydrocarbon will not.

The energy of the infrared radiation absorbed by a C—H bond depends on the hybridization of the carbon atom (Table 17.2). Carbon-hydrogen bonds become stronger in the order of $sp^3 < sp^2 < sp$, because the increased *s* character of the carbon atom keeps the bonding electrons closer to the carbon atom. Thus, the energy required to stretch the bond also increases.

TABLE 17.2 Characteristic Infrared Group Frequencies

Class	Group	Wavenumber (cm^{-1})
alkane	C—H	2850–3000
alkene	C—H	3080–3140
	C=C	1630–1670
alkyne	C—H	3300–3320
	C≡C	2100–2140
alcohol	O—H	3400–3600
	C—O	1050–1200
ether	C—O	1070–1150
aldehyde	C=O	1725
ketone	C=O	1700–1780

The sp^3-hybridized C—H bonds in saturated hydrocarbons like octane (Figure 17.6) absorb infrared radiation in the 2850–3000 cm^{-1} region. The sp^2-hybridized C—H bonds in 1-octene (an alkene) absorb energy at 3080 cm^{-1}. This peak is well-separated from the absorptions associated with the sp^3-hybridized C—H bonds in this molecule (Figure 17.6). Note that the isomeric cyclooctane does not have the 3080 cm^{-1} absorption. The *sp*-hybridized C—H bond in 1-octyne (an alkyne) absorbs infrared radiation at 3320 cm^{-1} (Figure 17.6). An isomeric bicyclic hydrocarbon with no multiple carbon–carbon bonds and hence no sp^2- or *sp*-hybridized carbon atoms, will have absorptions only in the 2850–3000 cm^{-1} region.

Hydrocarbons can also be classified based on absorptions due to the carbon–carbon bonds. Carbon–carbon bond strength increases in the order single < double < triple. Thus, the wavenumber positions (cm^{-1}) of the absorptions corresponding to stretching these bonds increase in the same order. Saturated hydrocarbons—both alkanes and cycloalkanes—all contain many carbon–carbon single bonds that absorb in the 800–1000 cm^{-1} region, but the intensity is very low. Carbon–carbon single bonds present in unsaturated compounds also absorb in the same region. Many other molecular bond stretching vibrations and bond angle bending modes occur in the same region and are much more intense. Thus, this region has limited diagnostic value. Moreover, we already know that most organic compounds have carbon–carbon single bonds.

Unsaturated hydrocarbons are identified by the absorption for the carbon–carbon double bond, which occurs in the 1630–1670 cm^{-1} region. The intensity of the absorption decreases with increased substitution. Terminal alkenes have the most intense absorptions. The double bond in 1-octene absorbs at 1640 cm^{-1} (Figure 17.6). The isomeric cyclooctane does not have any absorption in this region.

The absorption for a carbon–carbon triple bond occurs in the 2100–2140 cm^{-1} region. Terminal alkynes have the most intense absorption; internal (disubstituted) alkynes have lower intensity absorptions. The triple bond in 1-octyne absorbs at 2120 cm^{-1} (Figure 17.6). Any isomeric bicyclic hydrocarbon would not have any absorption in this region.

EXAMPLE 17.3 Explain how you could distinguish between the following two compounds by infrared spectroscopy.

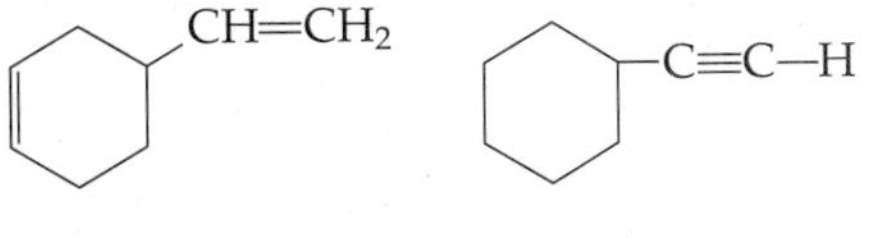

$CH{=}CH_2$ $\quad\quad$ $-C{\equiv}C{-}H$

Solution The compounds are isomeric. The compound on the left is a diene, which has two differently substituted double bonds. Thus, this compound may have two absorptions in the 1630–1670 cm^{-1} region correspond-

FIGURE 17.6 Infrared Spectra of Hydrocarbons

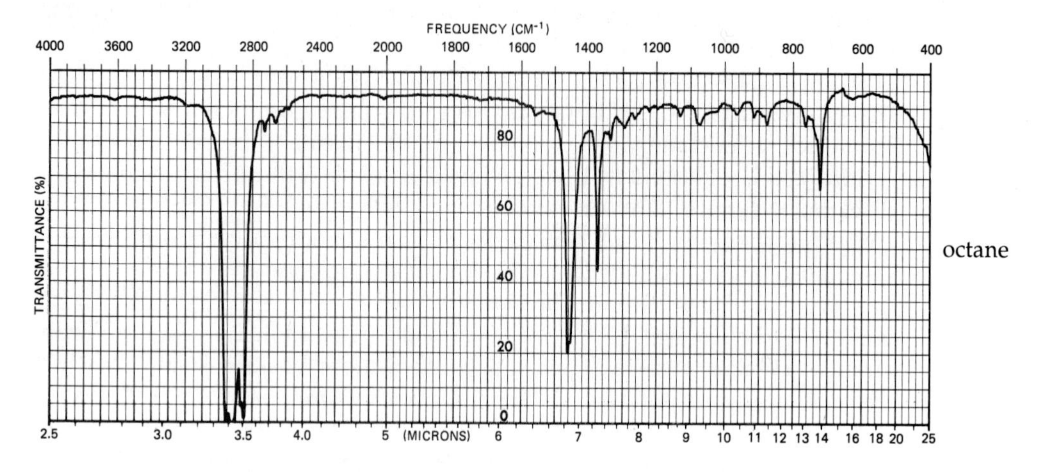

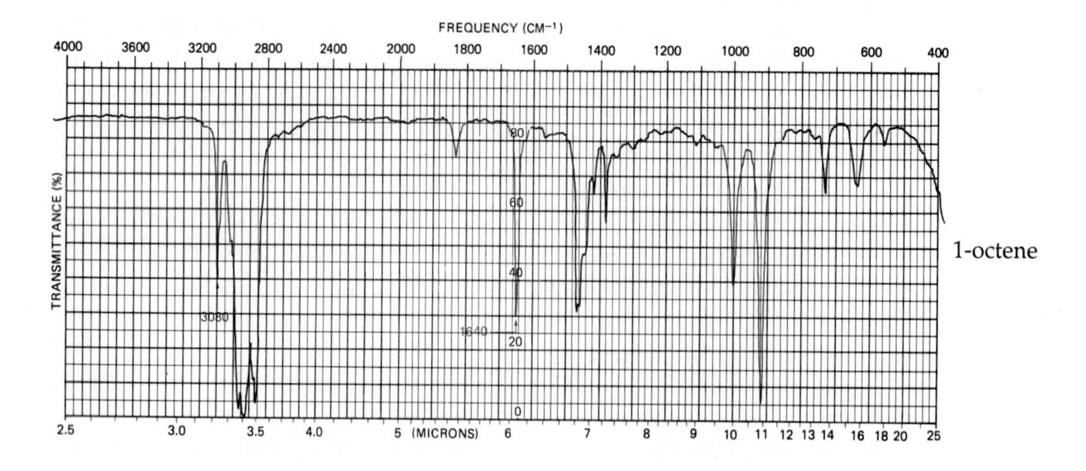

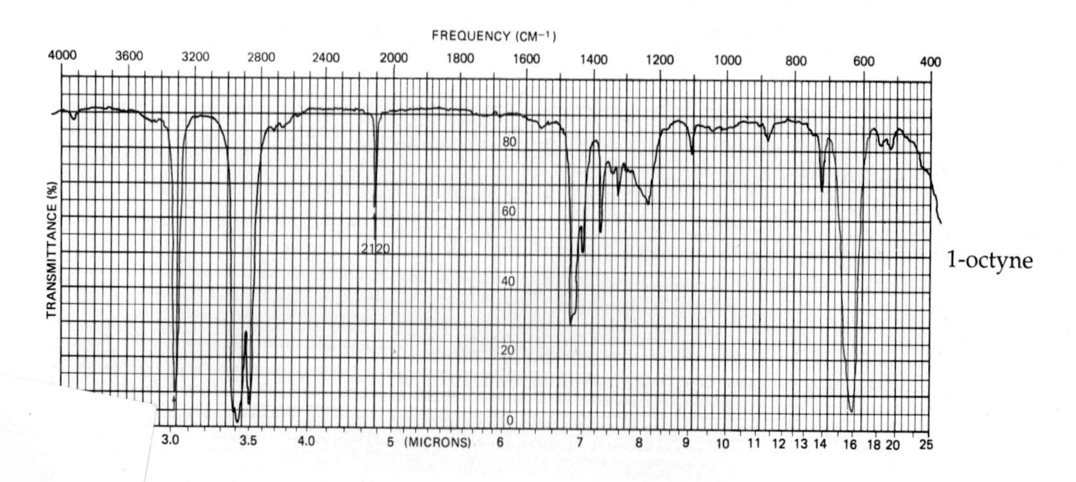

ing to the two carbon–carbon double bond stretching vibrations. There should also be C—H stretching absorptions in the 3080–3140 cm^{-1} region for the hydrogen atoms bonded to sp^2-hybridized carbon atoms.

The compound on the right is an alkyne. Thus, this compound has an absorption in the 2100–2140 cm^{-1} region that corresponds to the carbon–carbon triple bond stretching vibration. There is also a C—H stretching absorption in the 3300–3320 cm^{-1} region for the hydrogen atom bonded to the *sp*-hybridized carbon atom.

Identifying Oxygen-Containing Compounds

Many functional groups contain oxygen. These functional groups have the characteristic infrared absorptions given in Table 17.2. The characteristic group frequencies of aldehydes and ketones are from 1700–1780 cm^{-1}. The carbon–oxygen double bond of carbonyl compounds requires more energy to stretch than does the carbon–oxygen single bond of ethers and alcohols. Thus, aldehydes and ketones absorb infrared radiation at higher wavenumber positions (1700–1780 cm^{-1}) than alcohols and ethers (1050–1200 cm^{-1})

The absorption for a carbonyl group is extremely intense and is easily detected because it is in a region of the infrared spectrum that is devoid of conflicting absorptions. Note that carbon–carbon double bond stretching vibrations are at lower wavenumber positions than carbonyl group vibrations. A typical spectrum of a ketone is shown for 2-heptanone in Figure 17.7. The carbonyl stretching vibration occurs at 1712 cm^{-1}.

The position of the carbonyl group absorption of acyl derivatives depends on the inductive and resonance effects of atoms bonded to the carbonyl carbon atom. We recall that a carbonyl group is represented by two contributing resonance forms.

$$-\overset{\overset{:\ddot{\text{O}}}{\|}}{\text{C}}- \longleftrightarrow -\overset{\overset{:\ddot{\text{O}}:^-}{|\,+}}{\text{C}}-$$

FIGURE 17.7 Infrared Spectrum of 2-Heptanone

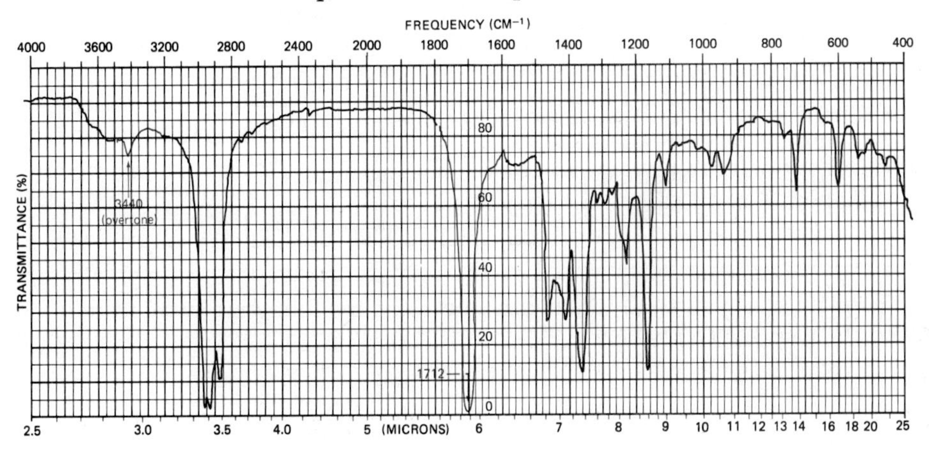

Because less energy is required to stretch a single bond than a double bond, any structural feature that stabilizes the contributing polar resonance form with a carbon–carbon single bond will cause the infrared absorption to occur at a lower wavenumber position. Thus, any group that donates electrons by resonance causes a shift in the absorption to lower wavenumbers. We recall that the nitrogen atom of amides is very effective in donation of electrons to the carbonyl carbon atom.

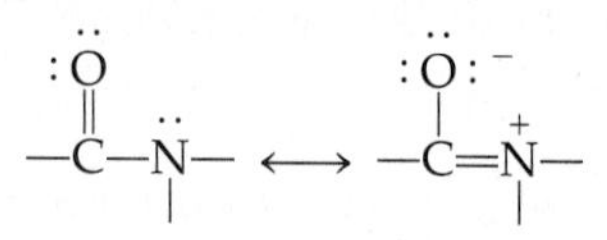

$$\overset{:\ddot{O}}{\overset{\|}{—C}}—\underset{|}{\ddot{N}}— \longleftrightarrow \overset{:\ddot{O}:^-}{\overset{|}{—C}}=\underset{|}{\overset{+}{N}}—$$

Thus, the double bond character of the carbonyl group decreases. As a result, an amide carbonyl group absorbs in the 1650–1690 cm^{-1} region—that is, at a lower wavenumber than for ketones.

EXAMPLE 17.4 The carbonyl group of an acid chloride absorbs at 1800 cm^{-1}. Explain why this value is at a higher wavenumber than for an aldehyde or ketone.

Solution We recall that a chlorine atom is not effective in donating electrons by resonance but does inductively withdraw electrons. The chlorine atom of an acid chloride destabilizes the contributing polar resonance form. As a consequence the carbonyl group of an acid chloride has more double bond character and absorbs at a higher wavenumber than the carbonyl group of an aldehyde or ketone.

The carbon–oxygen single bond stretching vibration of alcohols and ethers appears in a region complicated by many other absorptions. However, the absorption of a carbon–oxygen single bond is more intense than the absorption of carbon–carbon single bonds. The presence of a hydroxyl group is better established by the oxygen–hydrogen stretching vibration that occurs as an intense broad peak in the 3400 cm^{-1} region. This absorption is illustrated in the spectrum of 2-methyl-2-propanol (Figure 17.8).

Ethers can be identified by a process of elimination. If a compound contains an oxygen atom and the infrared spectrum lacks absorptions characteristic of a carbonyl group or a hydroxyl group, we can conclude that the compound is an ether.

EXAMPLE 17.5 Explain how you could distinguish between the following two compounds by infrared spectroscopy.

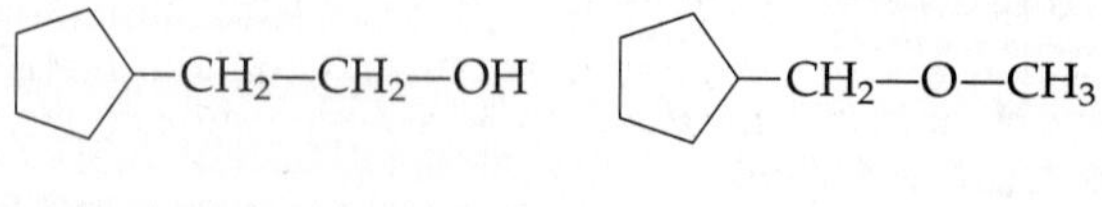

(cyclopentyl)—CH_2—CH_2—OH (cyclopentyl)—CH_2—O—CH_3

FIGURE 17.8 **Infrared Spectrum of 2-Methyl-2-propanol**

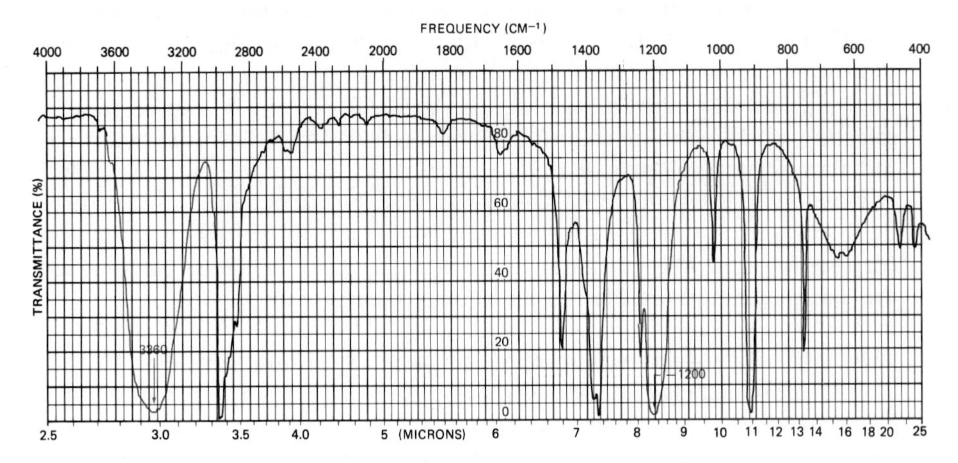

Solution These compounds are isomers. The compound on the left is an alcohol; the compound on the right is an ether. Both compounds have absorptions in the 1050–1200 cm^{-1} region due to the carbon–oxygen stretching vibration. However, the alcohol also has a broad absorption in the 3400 cm^{-1} region due to the oxygen–hydrogen stretching vibration. There is no corresponding absorption in that region for the ether.

17.5 NUCLEAR MAGNETIC RESONANCE SPECTROSCOPY

The ^{1}H nucleus spins about its axis in either of two directions described as clockwise and counterclockwise. Because the nucleus is charged, a magnetic moment results from the spinning nucleus. Thus, the hydrogen nucleus is a tiny magnet with two possible orientations in the presence of an external magnetic field. They may be aligned with the external magnetic field or against it, with the former being of lower energy. If a spinning hydrogen nucleus with its magnetic moment aligned with the external field is irradiated with electromagnetic radiation in the radio-frequency range, the absorbed energy causes the nucleus to "flip" and spin in the opposite direction (Figure 17.9). Thus, absorption of energy results in a higher energy state for the hydrogen nucleus. The process is called **nuclear magnetic resonance** (NMR).

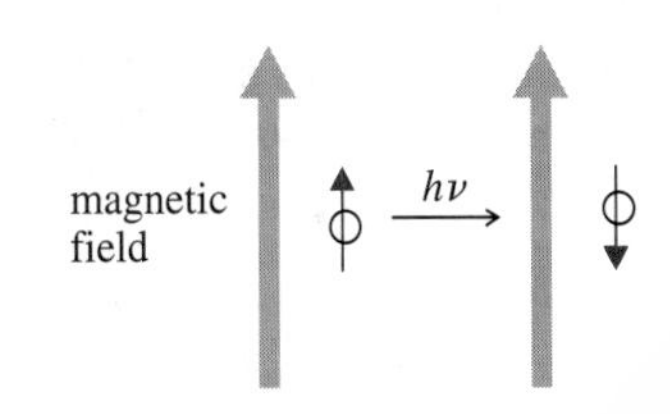

FIGURE 17.9
Absorption of Electromagnetic Radiation by a Nucleus
When the magnetic moment of a spinning nucleus is aligned with the magnetic field of an NMR spectrometer, a low energy state results. Absorption of a specific frequency causes a change in the spin of the nucleus and results in a magnetic moment opposed to the magnetic field of the instrument.

NMR spectroscopy uses electromagnetic radiation in the radio-frequency range. The energy associated with this radiation is very small. The energy required depends on the strength of the external magnetic field. Increasing the strength of the external magnetic field increases the energy difference between the two spin states. An NMR experiment can be done by selecting a magnetic field strength and then varying the radio-frequency radiation to find the proper frequency to make the hydrogen atom "flip". Alternately, a constant radio frequency may be used while the magnetic field strength is varied to cause a difference in the energy of the nuclei corresponding to the energy of the electromagnetic radiation. In practice, this latter technique is the easiest to do experimentally.

Chemical Shift

The magnetic field strength required to "flip" the spin of various hydrogen atoms within a molecule differs. If all hydrogen atoms absorbed the same electromagnetic radiation in an NMR experiment at the same magnetic field strength, then only a single absorption would be observed. As a consequence we would only know that the molecule contained hydrogen atoms.

The hydrogen nuclei in organic molecules are surrounded by electrons that also have spins. The electrons thus set up small local magnetic fields that are opposed to the applied external magnetic field. The local fields affect the magnetic environment of the hydrogen nuclei. When a local field opposes the external magnetic field, we say that the nucleus is **shielded.** The effective field felt by the nucleus is the applied magnetic field minus the local magnetic field generated by the electrons.

$$H_{effective} = H_{applied} - H_{local}$$

The local magnetic fields due to the spins of the electrons differ throughout the molecule because the bonding characteristics differ. Thus, the degree of shielding of each hydrogen nucleus is unique, and distinct resonances are obtained for each structurally nonequivalent hydrogen atom in a molecule. At a constant radio frequency, the external magnetic field required is larger for the more shielded nucleus.

The strength of the local magnetic fields for various hydrogen atoms are about 10^{-6} times that of the applied magnetic field. Thus, the magnetic fields required to flip various structurally different hydrogen nuclei in a molecule differ in the order of parts per million (ppm). Rather than using absolute values of the field strength, a relative scale is used. An NMR chart is labeled on the horizontal axis with a **delta scale** in which one delta unit (δ) is 1 ppm of the magnetic field used. The resonance for the hydrogen atoms of tetramethylsilane, $(CH_3)_4Si$, is used as a reference for the magnetic resonance spectra of hydrogen compounds. This resonance is defined as 0 δ. By convention an absorption that occurs at lower field than tetramethylsilane (TMS) appears to the left of the TMS absorption and is assigned a positive δ value.

By examining the NMR spectrum you can tell at a glance how many

FIGURE 17.10 NMR Spectrum of 1,2,2-Trichloropropane
The three equivalent hydrogen atoms bonded to the C-3 atom absorb at 2.2 δ. The two equivalent hydrogen atoms bonded to the C-1 atom absorb at 4.0 δ.

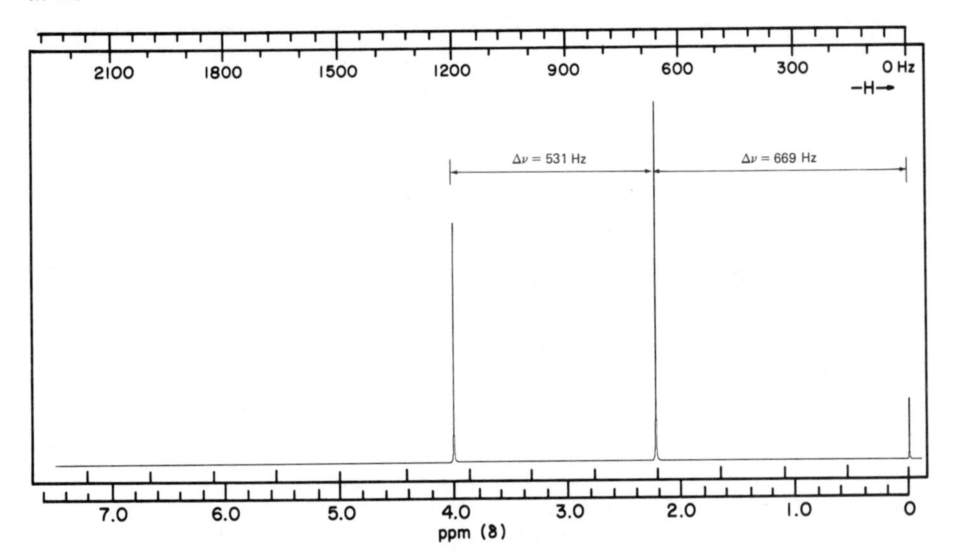

sets of structurally nonequivalent hydrogen atoms are contained in a molecule. Consider the NMR spectrum of 1,2,2-trichloropropane shown in Figure 17.10. The spectrum consists of two peaks pointed to the top of the graph. The resonances occur at 2.2 and 4.0 δ. There are two different sets of hydrogen atoms in 1,2,2-trichloropropane. Each set of hydrogen atoms gives rise to one peak.

one set of hydrogen atoms → Cl—CH₂—CCl₂—CH₃ ← a second set of hydrogen atoms

(structure: Cl—C(H)(H)—C(Cl)(Cl)—C(H)(H)—H)

For most organic molecules, the various hydrogen resonances appear between 0 and 10 δ. This range is conveniently divided into regions that reflect certain structural characteristics. Hydrogen atoms bonded to sp^2 carbon atoms absorb at lower fields than hydrogen atoms bonded to saturated sp^3 carbon atoms. Hydrogen atoms bonded to saturated carbon atoms without any directly bonded substituents absorb in the 0.5–1.7 δ range. Hydrogen atoms in alkenes absorb in the 5–6 δ region. The hydrogen atoms bonded to an aromatic ring absorb in the general region of 7–8 δ.

The exact position of absorption for hydrogen atoms bonded to either sp^3 or sp^2 carbon atoms also depends on the number and type of substituents also bonded to the atom. Hydrogen atoms bonded to carbon atoms with bound electronegative atoms such as oxygen, nitrogen, or halogens absorb at lower fields. Examples of typical chemical shifts are listed in Table

TABLE 17.3 Chemical Shifts of Hydrogen Atoms

Partial structural formula	Chemical shift (ppm)	Partial structural formula	Chemical shift (ppm)
$—CH_3$	0.7–1.3	Br—C—H	2.5–4.0
$—CH_2—$	1.2–1.4	I—C—H	2.0–4.0
—C—H	1.4–1.7	—O—C—H	3.3–4.0
$C{=}C—CH_3$	1.6–1.9	C=C—H	5.0–6.5
$—C(=O)—CH_3$	2.1–2.4	Ar—H	6.5–8.0
—C≡C—H	2.5–2.7	—C(=O)—H	9.7–10.0
Cl—C—H	3.0–4.0		

Relative Peak Areas

The set of hydrogen atoms bonded to the C-1 atom of 1,2,2-trichloropropane has an absorption at 4.0 δ and the set of hydrogen atoms bonded to the C-3 atom has its absorption at 2.2 δ. This assignment is made based on the generalization that hydrogen atoms bonded to a carbon atom that is also bonded to an electronegative atom absorb at a lower field. However, there is another method to confirm this assignment—"proton counting". The area of each resonance is proportional to the relative number of hydrogen atoms of each kind.

The relative area of a resonance is obtained from an electronic integrator used after the spectrum has been recorded. The area is equal to the vertical displacement of a "stair step" superimposed on the resonance peak. These vertical distances have a ratio equal to the ratio of the numbers of hydrogen atoms. Thus, the ratio of the integrated intensities of the two resonances of 1,2,2-trichloropropane is 3:2 (Figure 17.11).

5 Predict the chemical shift of the resonances of 1,2-dichloro-2-methylpropane. What are the relative intensities of the absorptions?

FIGURE 17.11 Integrated Intensities of an NMR Spectrum
The area of each resonance is proportional to the number of hydrogen atoms. The vertical distances of the "stair steps" shown measure those areas. The ratio of 75 to 50 nm is 3:2, which corresponds to the numbers of hydrogen atoms bonded to the C-3 and C-1 atoms, respectively, in 1,2,2-trichloropropane.

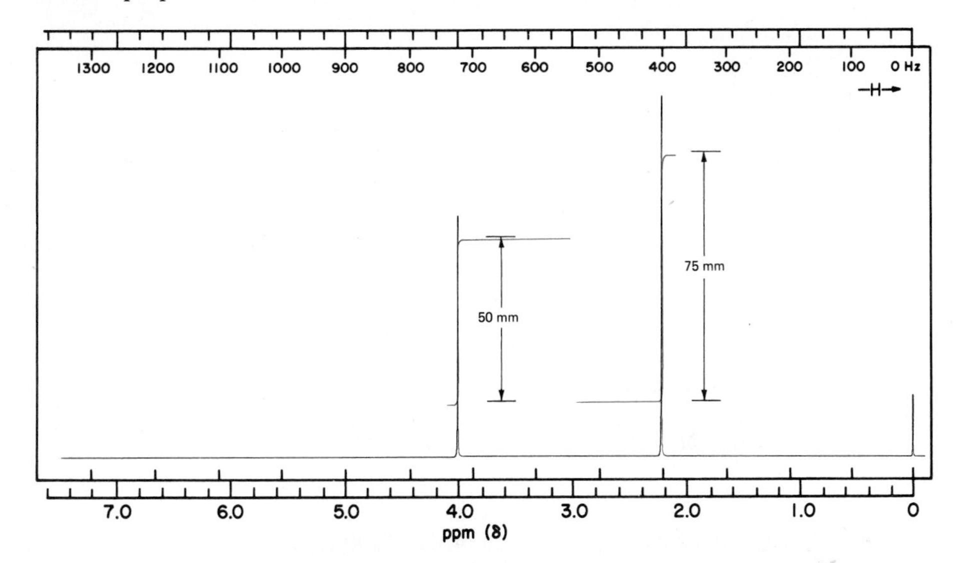

Solution Write the structure of the compound to determine the number of sets of nonequivalent hydrogen atoms.

$$\begin{array}{c} CH_3 \\ | \\ CH_3\text{—}C\text{—}CH_2\text{—}Cl \\ | \\ Cl \end{array}$$

One set of hydrogen atoms is located at the C-1 atom, which has one chlorine atom bonded to it. These hydrogen atoms should have a resonance near 4 δ. The six hydrogen atoms of the two equivalent methyl groups should have a resonance near 1 δ. The ratio of the two sets of hydrogen atoms is 6:2. The ratio of the relative areas at 1 and 4 δ should be 3:1.

Spin–Spin Splitting

In 1,2,2-trichloropropane, each of the two sets of hydrogen atoms is responsible for a single peak. Now consider the spectrum of 1,1,2-tribromo-3,3-dimethylbutane (Figure 17.12).

$$\begin{array}{c} \quad CH_3\ \ H\ \ H \\ \quad |\qquad |\quad | \\ CH_3\text{—}C\text{——}C\text{—}C\text{—}Br \\ \quad |\qquad |\quad | \\ \quad CH_3\ \ Br\ Br \end{array}$$

1,1,2-tribromo-3,3-dimethylbutane

FIGURE 17.12 NMR Spectrum of 1,1,2-Tribromo-3,3-dimethylbutane
The inserts show the doublets for the resonances at 4.4 and 6.4 δ. The total integrated area of each doublet is proportional to one hydrogen atom.

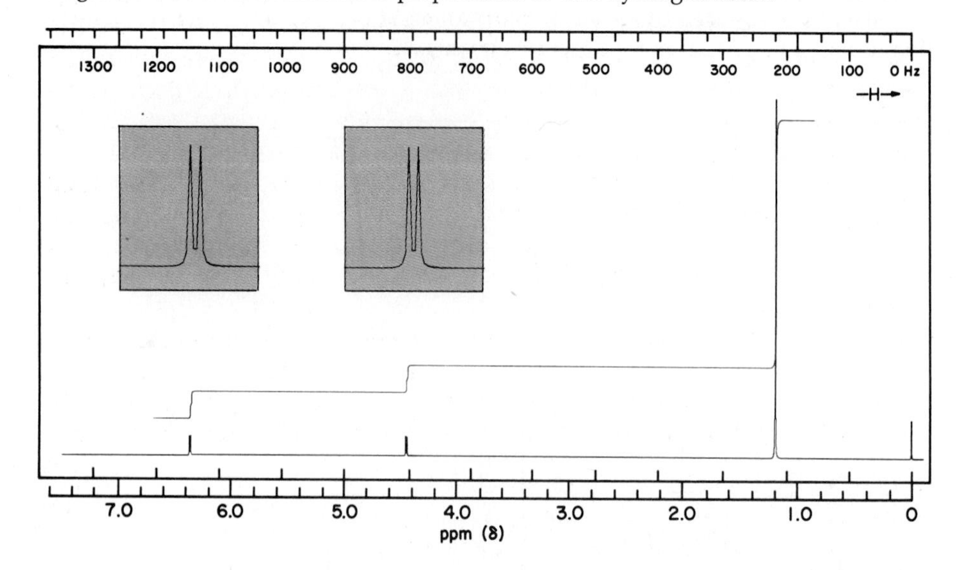

The nine equivalent hydrogen atoms of the three equivalent methyl groups give rise to the intense peak at 1.2 δ. The single hydrogen atoms bonded to the C-1 and C-2 atoms are nonequivalent. The resonance of the hydrogen atom at the C-1 atom is located at 6.4 δ as a consequence of the two electronegative bromine atoms bonded to that carbon atom. The resonance of the hydrogen atom at the C-2 atom is located at 4.4 δ because only one bromine atom is bonded to that carbon atom. Thus, the intensities of the absorptions and the chemical shifts of the hydrogen atoms are as expected based on molecular structure.

Both the 4.4 and 6.4 δ absorptions of 1,1,2-tribromo-3,3-dimethylbutane are "split" as shown in the inserts containing expanded representations of the resonances. Each area contains two peaks called **doublets.** The phenomenon of multiple peaks is common in NMR spectroscopy. Other common multiplets include **triplets** and **quartets,** meaning that resonances are split into three and four peaks, respectively.

Multiple absorptions for a set of equivalent hydrogen atoms is known as **spin–spin splitting.** It results from the interaction of the nuclear spin(s) of one or more nearby "neighboring" hydrogen atom(s) with the set of equivalent hydrogen atoms. The small magnetic field of nearby hydrogen atoms affects the magnetic field felt by other hydrogen atoms. Consider the hydrogen atom on the C-1 atom of 1,1,2-tribromo-3,3-dimethylbutane. It has a neighboring hydrogen atom on the C-2 atom that can be spinning in either of two possible directions. In those molecules where the hydrogen atom at the C-2 atom is spinning clockwise, the hydrogen atom at the C-1 atom experiences a small magnetic field that differs from that felt in molecules where the hydrogen atom at the C-2 atom is spinning counterclockwise. Thus, two slightly different external magnetic fields are needed for the C-1 hydrogen atoms in various molecules to absorb electromagnetic radiation. A doublet results. The same explanation accounts for the doublet for the hydrogen atom at the C-2 atom. In this case, the hydrogen atom on the C-1 atom is the neighboring atom and its spin affects the magnetic field experienced by the hydrogen atom at the C-2 atom. Thus, in general, sets of hydrogen atoms on neighboring carbon atoms **couple** with each other. If hydrogen atom A couples and causes splitting of the resonance for hydrogen atom B, the resonance for hydrogen atom B is also split by hydrogen atom A. The resonance for the nine methyl atoms of 1,1,2-tribromo-3,3-dimethylbutane is not "split" because the neighboring quaternary carbon atom has no hydrogen atoms.

A set of one or more hydrogen atoms that has n equivalent neighboring hydrogen atoms has $n+1$ peaks in the NMR spectrum. The appearance of several sets of multiplets resulting from $n = 1$ through $n = 5$ are shown in Figure 17.13 for some common structures. The areas of the component peaks

FIGURE 17.13 Common NMR Multiplets
The resonance of a common single hydrogen atom is shown. The number of equivalent neighboring hydrogen atoms is responsible for the multiplicity of the resonance.

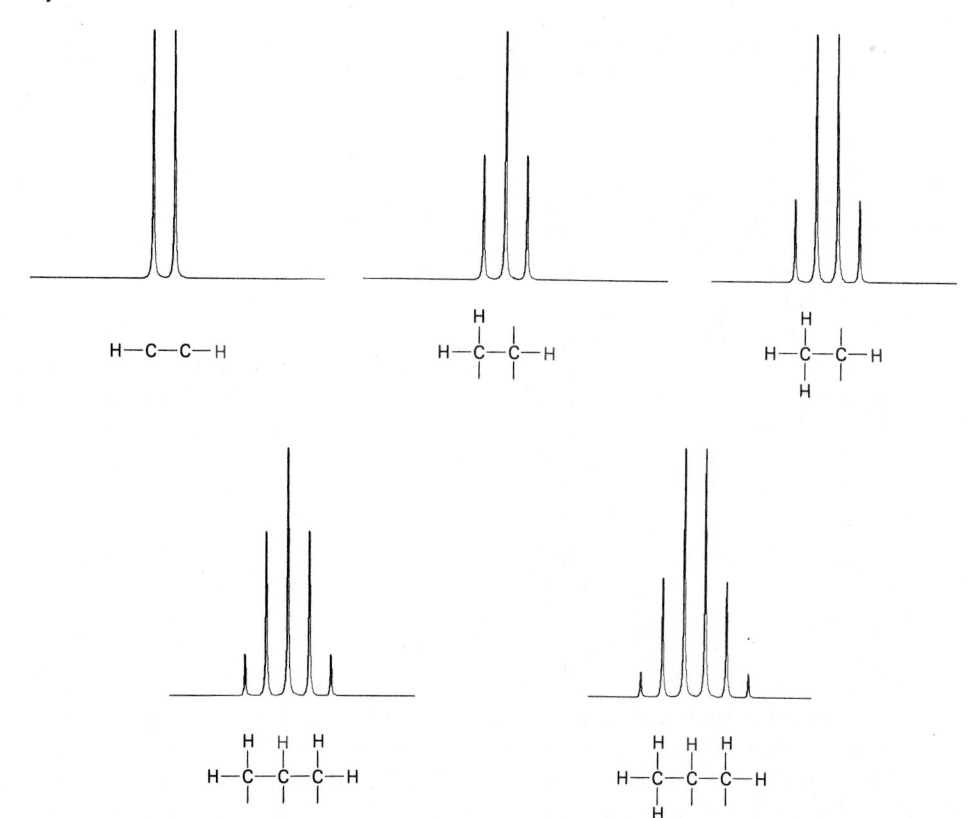

of a doublet are equal; the areas of the component peaks of other multiplets are not.

To understand the relative peak areas of multiplets resulting from more than one neighboring hydrogen atom, let's consider the spectrum of 1,1,2-trichloroethane (Figure 17.14).

```
     Cl H
     |  |
 Cl—C—C—Cl
     |  |
     H  H
```

1,1,2-trichloroethane

The doublet near 4 δ corresponds to the two hydrogen atoms bonded to the C-2 atom. These hydrogen atoms have one neighboring hydrogen atom which spins in either of two directions we can designate as α or β. As a consequence the C-2 hydrogen atoms in various molecules experience two different magnetic fields, resulting in a doublet absorption. Now let's consider the triplet resonance for the C-1 hydrogen atom. The spins of the neighboring two hydrogen atoms at the C-2 atom can be $\alpha\alpha$, $\alpha\beta$, $\beta\alpha$, and $\beta\beta$. Because the magnetic fields generated by either $\alpha\beta$ and $\beta\alpha$ sets of spins are equivalent, the hydrogen atom at the C-1 atom experiences three different magnetic fields in the ratio of 1:2:1. This ratio is the same as the ratio of the observed triplet.

Extension of the possible combinations of spins of n neighboring equivalent hydrogen atom accounts for the area ratios of multiplets in Table 17.4.

FIGURE 17.14 NMR Spectrum of 1,1,2-Trichloroethane
The inserts show the doublet for the 4.0 δ resonance of the hydrogen atoms bonded to the C-2 atom and the triplet at 5.7 δ of the hydrogen atom bonded to the C-1 atom. Note that the integrated intensities of the 4.0 and 5.7 δ resonances are in the ratio 2:1.

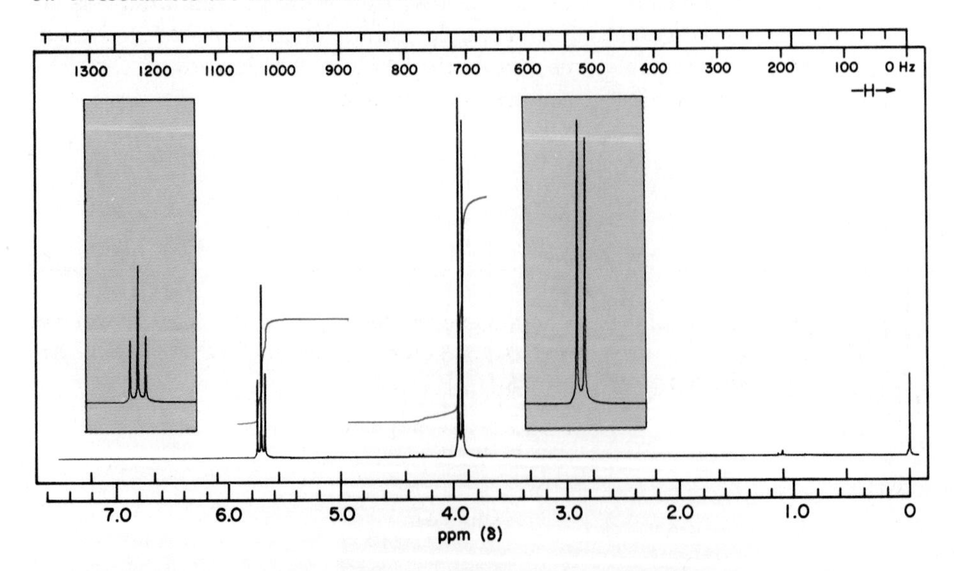

TABLE 17.4 Number of Peaks of Multiplets and Area Ratios

Number of equivalent adjacent hydrogens	Total number of peaks	Area ratios
0	1	1
1	2	1:1
2	3	1:2:1
3	4	1:3:3:1
4	5	1:4:6:4:1
5	6	1:5:10:10:5:1
6	7	1:6:15:20:15:6:1

EXAMPLE 17.7 Describe the NMR spectrum of 2-chloropropane.

Solution First draw the structure of 2-chloropropane to determine the number of sets of nonequivalent hydrogen atoms and the number of neighboring hydrogen atoms that can couple with each set.

$$CH_3\text{—}\underset{\displaystyle H}{\overset{\displaystyle Cl}{\overset{|}{\underset{|}{C}}}}\text{—}CH_3$$

The six hydrogen atoms located on the equivalent C-1 and C-3 atoms are equivalent. They have a resonance in the 1 δ region. The single hydrogen atom at the C-2 atom has a resonance near 4 δ because a chlorine atom is bonded to that carbon atom. The relative areas of the 1 and 4 δ peaks are 6:1.

The 4 δ resonance is a septet because the C-2 hydrogen atom has six neighboring hydrogen atoms that couple with it. Each of the two equivalent methyl groups have only one neighboring hydrogen atom—the C-2 hydrogen atom. Thus, the 1 δ resonance is a doublet.

EXERCISES

Ultraviolet Spectroscopy

17.1 The λ_{max} of naphthalene, anthracene and tetracene are 314, 380, and 480 nm, respectively. Suggest a reason for this order of wavelengths of absorption. Are any of the compounds colored?

tetracene

17.2 How many conjugated double bonds are contained in lycopene? Compare the conjugation in this compound to that of β-carotene. Based on this information, predict the color of lycopene.

lycopene

β-carotene

17.3 How might 2,4-hexadiyne be distinguished from 1,4-hexadiyne by ultraviolet spectroscopy?

17.4 One of the following unsaturated ketones has λ_{max} = 225 nm and the other has λ_{max} = 252 nm. Assign these values to the proper structure.

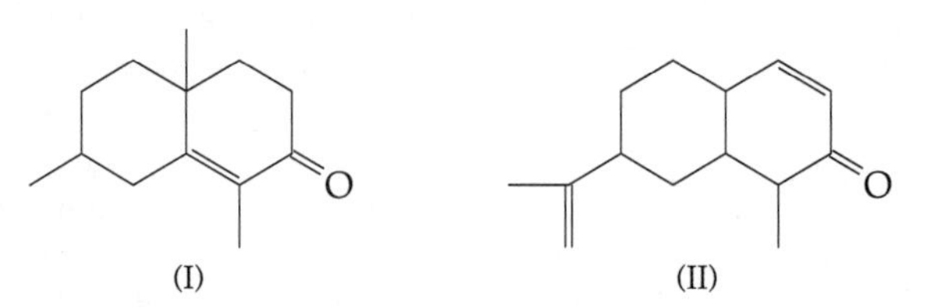

17.5 Rank the following polyenes in order of longest wavelength of absorption in the ultraviolet spectrum.

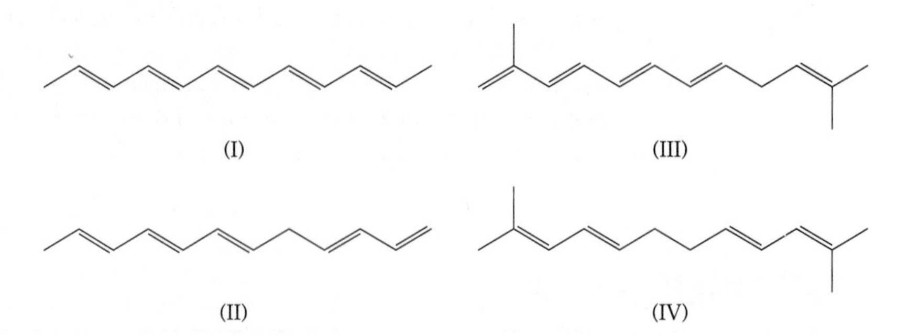

17.6 The λ_{max} of 2,4,6-octatriyne and 2,4,6,8-decatetrayne are 207 and 234 nm, respectively. Explain why these values differ.

Infrared Spectroscopy

17.7 How could infrared spectroscopy be used to distinguish between propanone and 3-propen-1-ol?

17.8 How could infrared spectroscopy be used to distinguish between 1-pentyne and 2-pentyne?

17.9 The carbonyl stretching vibration of ketones is at longer wavelength than the carbonyl stretching vibration of aldehydes. Suggest a reason for this observation.

17.10 The carbonyl stretching vibrations of esters and amides occur near 1735 and 1670 cm^{-1}, respectively. Suggest a reason for this difference.

17.11 An infrared spectrum of a compound with molecular formula $C_4H_8O_2$ has an intense broad band between 3500 and 3000 cm^{-1} and an intense peak at 1710 cm^{-1}. Which of the following compounds best fits this data?

$CH_3CH_2CO_2CH_3$ (I) $CH_3CO_2CH_2CH_3$ (II) $CH_3CH_2CH_2CO_2H$ (III)

17.12 Dehydration of 1-methylcyclohexanol gives two isomeric alkenes. The minor product has a more intense C=C stretching vibration than the major product. Assign the structures of these compounds.

Nuclear Magnetic Resonance Spectroscopy

17.13 The NMR spectrum of a compound with molecular formula $C_3H_6Cl_2$ has only a singlet at 2.8 δ. What is the structure of the compound?

17.14 The NMR spectrum of a compound with molecular formula C_4H_9Br has only a singlet at 1.8 δ. What is the structure of the compound?

17.15 The NMR spectrum of an ether with molecular formula $C_5H_{12}O$ has singlets at 1.1 and 3.1 δ whose intensities are 3:1. What is the structure of the compound?

17.16 The NMR of an ether with molecular formula $C_6H_{14}O$ consists of an intense doublet at 1.0 δ and a septet at 3.6 δ. What is the structure of the ether?

17.17 The NMR spectrum of 1-chloropropane consists of a triplet at 1.0 δ, a sextet at 1.8 δ, and a triplet at 3.5 δ. Assign the hydrogen atoms to the proper resonances.

17.18 The NMR spectrum of iodoethane has resonances at 1.9 and 3.1 δ. What are the intensities of the two resonances? What are the multiplicities?

17.19 Describe the NMR spectrum expected for each of the following compounds. (a) CH_3CHCl_2 (b) $(CH_3)_3CCH_2Cl$ (c) $(CH_3)_3CCH_2CH_2Cl$

17.20 Describe the differences in the NMR spectra of the following two esters.

$CH_3CH_2CO_2CH_3$ (I) $CH_3CO_2CH_2CH_3$ (II)

INDEX

B

D